Advances in Agricultural Engineering Technologies and Application—Volume I

Advances in Agricultural Engineering Technologies and Application—Volume I

Editors

Muhammad Sultan
Redmond R. Shamshiri
Md Shamim Ahamed
Muhammad Farooq

MDPI

Basel • Beijing • Wuhan • Barcelona • Belgrade • Novi Sad • Cluj • Manchester

Editors

Muhammad Sultan
Bahauddin Zakariya University
Multan, Pakistan

Redmond R. Shamshiri
Leibniz-Institut für Agrartechnik und Bioökonomie
Potsdam, Germany

Md Shamim Ahamed
University of California Davis
Davis, USA

Muhammad Farooq
University of Engineering and Technology
Lahore, Pakistan

Editorial Office

MDPI
St. Alban-Anlage 66
4052 Basel, Switzerland

This is a reprint of articles from the Special Issue published online in the open access journal *Agriculture* (ISSN 2077-0472) (available at: https://www.mdpi.com/journal/agriculture/special_issues/aengineeringtechno_agriculture).

For citation purposes, cite each article independently as indicated on the article page online and as indicated below:

Lastname, A.A.; Lastname, B.B. Article Title. *Journal Name* **Year**, *Volume Number*, Page Range.

Volume I
ISBN 978-3-0365-9374-6 (Hbk)
ISBN 978-3-0365-9375-3 (PDF)
doi.org/10.3390/books978-3-0365-9375-3

Set
ISBN 978-3-0365-9372-2 (Hbk)
ISBN 978-3-0365-9373-9 (PDF)

Contents

About the Editors

Muhammad Sultan

Dr. Muhammad Sultan is an Associate Professor of Agricultural Engineering at Bahauddin Zakariya University (Pakistan). He holds a B.Sc. and an M.Sc. in Agricultural Engineering from the University of Agriculture Faisalabad (Pakistan) and a Ph.D. in Energy and Environmental Engineering from Kyushu University (Japan). He also conducted postdoctoral research at Kyushu University (Japan) and Simon Fraser University (Canada). Dr. Sultan has been recognized as one of the World's Top 2% Scientists (2023) by Stanford University/Elsevier. He also ranked No. 1 as a Scientist in Agricultural Engineering (Pakistan) according to the AD Scientific Index. He has published more than 300 articles in international journals, conferences, books, and book chapters. He has been a reviewer for more than 100 renowned journals and holds editor roles for 15 SCI journals. His research focuses on developing energy-efficient temperature and humidity control systems for agricultural applications including greenhouse, fruit/vegetable storage, livestock, and poultry applications. His research involves adsorption heat pumps, desiccant air-conditioning, evaporative cooling, Maisotsenko cycle, adsorption desalination, energy recovery ventilator, atmospheric water harvesting, and wastewater treatment.

Redmond R. Shamshiri

Dr. Redmond R. Shamshiri is a senior scientist at the Leibniz-Institut für Agrartechnik und Bioökonomie working on agro-mechatronics projects. He holds a Ph.D. in Agricultural Automation with a focus on control systems and dynamics. His research fields include distributed collision avoidance for autonomous navigation of agricultural mobile robots, wireless sensors for digital farming applications, and optimization of controlled environment crop production systems. He is currently involved in the evaluation of 5G technology to reduce yield losses from potato fields.

Md Shamim Ahamed

Dr. Shamim Ahamed currently serves as an Assistant Professor in the Department of Biological and Agricultural Engineering at the University of California, Davis (UC Davis). He completed his bachelor's degree in Agricultural Engineering and master's degree in Farm Power and Machinery from Bangladesh Agricultural University (BAU) and his Ph.D. in Environmental Engineering from the University of Saskatchewan in Canada. Before joining UC Davis, he worked as a postdoctoral fellow at the City University of Hong Kong and Concordia University in Montreal. Dr. Ahamed is leading the Controlled Environment Engineering Lab at UC Davis and has been working in energy systems optimization and precision controls for controlled environment agriculture (CEA) systems (greenhouses, indoor vertical farming, and livestock barns) in 2013. His research mainly focuses on thermal environment modeling, energy-efficient controls, and HVAC systems designed for next-generation CEA. His research aims to integrate renewable energy into the operations of heating, ventilation, air conditioning, and dehumidification systems to achieve a net-zero energy status for CEA facilities. His research also spans automated fault detection and diagnosis of automatic and precision control of HVAC systems and other systems like nutrient management, CO_2 enrichment, and lighting.

Muhammad Farooq

Dr. Muhammad Farooq is an experienced professional with more than 15 years of blended experience with research, teaching, training, industry, and the management of projects around the world and has visited more than 25 countries for different activities related to his profession. Currently, he is working as an Associate Professor and Director Postgraduate Research and leading an international multidisciplinary research group focused on energy environment and sustainability research (EESR) at the department of Mechanical Engineering, University of Engineering and Technology Lahore Pakistan, with a team of 30+ researchers and international collaborators, while supervising a number of M.Sc. and PhD students. He is the author of 100+ leading international journal articles; has a cumulative impact factor of 500+; and advocates for worldwide net-zero and carbon-natural activities to achieve sustainable environments via application of the SDGs. His work has been cited 2000+ times according to Google Scholar. He is among the top 2% scientists according to the list issued by Stanford University, USA, and Elsevier/Scopus in October 2023. As an editor, he has handled more than 400 research articles of different journals and peer-reviewed international conferences. As a volunteer and recognized reviewer of leading international journals, he has reviewed more than 300 research papers and received recognition awards from top-notch journals of Elsevier, Springer, SAGE Publishing, MDPI, and the Taylor & Francis group.

Preface

Modernization in the agriculture sector is important to ensure food security and poverty alleviation, which are the primary themes of UN-SDGs. There are many challenges in developing advanced agricultural techniques, tools, and systems, by which sustainable agriculture and food security can be satisfied.

Worldwide, agricultural mechanization and modernization can be attained with advancements in agricultural engineering technologies and their associated applications. The concept is directly linked to the technological advancements in agricultural automation and robotics; precision agriculture; high-efficiency irrigation systems; farm energy systems; handling, storage, and processing of agricultural products; livestock and poultry sheds; farm water/wastewater management; biomass, biogas, and biochar; remote sensing and geographical studies; societal aspects in agriculture; and the associated bioenvironment.

Such advances in agricultural engineering technologies and applications are the need of the 21st century, particularly from the viewpoint of the agricultural food–energy–water security nexus. Therefore, this book provides a dedicated collection of original research and review studies in the abovementioned research areas.

Muhammad Sultan, Redmond R. Shamshiri, Md Shamim Ahamed, and Muhammad Farooq
Editors

agriculture

Article

Conservation Tillage Technology: A Study on the Duration from Awareness to Adoption and Its Influencing Factors—Based on the Survey of the Yellow River Basin in China

Tianzhi Gao [1,†], Yaqin Ren [2,†], Qian Lu [1,*] and Hui Feng [1]

1 College of Economics and Management, Northwest Agriculture and Forestry University, Xianyang 712100, China; gtz22@nwafu.edu.cn (T.G.); fh22@nwafu.edu.cn (H.F.)
2 School of Sociology and Population Studies, Renmin University of China, Beijing 100872, China; renyaqin@ruc.edu.cn
* Correspondence: xnlq123@126.com
† These authors contributed equally to this work.

Abstract: Studying the dynamic duration of technology adoption helps farmers weigh and select different attributes and stages of conservation tillage techniques. In this study, non-parametric K-M survival analysis and discrete duration models were employed to estimate the time taken by farmers in the Yellow River Basin region to transition from awareness to the adoption of conservation tillage techniques between 2002 and 2020. The results indicate (1) The duration from awareness to adoption of conservation tillage technology is relatively short. (2) The likelihood of farmers postponing adoption decisions is highest in the initial 10 years and gradually decreases over time, suggesting negative time dependency. (3) Controlling for proportional hazards assumptions, it was found that factors such as education level and social learning positively influence the duration from awareness to adoption of conservation tillage techniques. Extreme weather variations and household labor migration delay the adoption time for farmers. In the process of promoting and implementing conservation tillage techniques, it is essential to consider the issue of intertemporal technology choice, stimulate farmers' intrinsic demand, shorten the time interval from awareness to adoption, and ultimately improve technology adoption rates.

Keywords: conservation tillage technology; adoption of time persistence; nonparametric K-M survival analysis method; discrete duration model

Citation: Gao, T.; Ren, Y.; Lu, Q.; Feng, H. Conservation Tillage Technology: A Study on the Duration from Awareness to Adoption and Its Influencing Factors—Based on the Survey of the Yellow River Basin in China. *Agriculture* **2023**, *13*, 1207. https://doi.org/10.3390/agriculture13061207

Academic Editors: Redmond R. Shamshiri, Muhammad Sultan, Md Shamim Ahamed and Muhammad Farooq

Received: 17 March 2023
Revised: 4 June 2023
Accepted: 4 June 2023
Published: 7 June 2023

1. Introduction

The degradation of cultivated land quality and the decline in basic soil fertility are the concentrated manifestations of the ecological environmental problems of farmland in China. At present, 26% of cultivated land soil organic matter content is less than 1% in China, more than 40% of the cultivated land has been degraded, and 21.6% of the cultivated land has been seriously acidified. The annual loss of effective components such as nitrogen, phosphorus, and potassium due to wind erosion and desertification is as high as 55.9 million tons. The power of cultivated land's contribution to food production is only about 50%, which is 20–30% points lower than that of developed countries [1]. Practice has shown that conservation tillage techniques, such as an environmentally friendly soil cultivation approach integrating minimum tillage, straw return to the field, and weed and pest control measures, offer a range of benefits. These techniques have the functions of reducing soil erosion, protecting the ecological environment of farmland, saving labor costs, reducing greenhouse gas emissions, and helping to achieve agricultural transformation. Moreover, it is of great significance in ensuring arable land quality, ecology, and food security and promoting the sustainable development of modern agriculture [2,3].

In response to the increasingly severe farmland ecological environment and climate change, it can be seen from the No. 1 Central Document in the past five years that the government has put forward high standards and strict requirements for rural land management, especially cultivated land protection, and has vigorously adopted various subsidies and preferential policies to promote and encourage farmers to adopt it actively. However, conservation tillage technology has not been widely adopted by farmers [4,5], with weak demand and low adoption rate, which leads to difficulties in the promotion of this technology [6].

To solve the challenges such as the low adoption rate of conservation tillage technology, firstly, we explored the influencing factors of technology adoption. There are several factors that will influence the adoption of technology, such as characteristics of farmers (gender, age, education level, risk preference, ecological cognition, social capital, etc.); family endowment (income level, cropping scale, cropping system, part-time situation, risk and uncertainty, etc.); environmental factors (soil type, extension system, technical training, rainfall and pest shocks, government compensation, etc.) [6–11]. Secondly, we focused on the attributes of technology adoption. Some scholars have noticed that there was a strong heterogeneity in farmers' choice of technologies with different attributes, and there may be substitution or a complementarity relationship between the adoption of sub-technologies, and that technology adoption decisions aim to achieve maximum utility with a set of technology attributes, which is much more complex than the single technology adoption decision problem [12,13]. Finally, we understood that farmers' technology adoption was a gradual process from cognitive, trial adoption to sustained adoption, and the factors influencing technology adoption at different stages also differed significantly [14]. The first two stages have been studied by most scholars while the analysis of the duration of technology adoption is currently a relatively new field. However, the ecological function and the economic and social benefits of conservation tillage technology start to emerge only after a few years of adoption by farmers, and the benefits will occur in the future, which is a typical inter-term agricultural technology [15,16]. From awareness to adoption decision, farmers need to weigh and choose the costs and benefits of technology adoption at different time points, and the duration is uncertain. The long duration means that the technology adoption process has to pay frequent pesticide costs, learning costs, and transaction costs to consolidate and maintain [17,18], and farmers are not motivated to participate in demand, resulting in "short-sighted thinking", which adversely affects conservation tillage technology adoption decisions [19]. Therefore, studying the adoption duration will help to understand the dynamic spreading factors of conservation tillage technology more clearly, and provide a reasonable explanation for the phenomenon of "low adoption rate of conservation tillage technologies with economic and ecological benefits", so as to guide policy interventions to promote the adoption of conservation tillage technologies and maximize incentives for farmers to shorten the adoption interval, and to reduce production costs so as to improve agricultural efficiency.

The existing literature on duration research has focused on Chinese import and export enterprises, manufacturing listed companies, and other perspectives, and the research content has mostly focused on import and export trade products, innovation sustainability of listed companies, bilateral political relations, and VAT reform [20–24]. Fewer studies have combined the duration of adoption into the field of agricultural technology, and the influencing factors were also different from other macro perspectives. Second, the time of data used in previous research was relatively short. In this study, we used the data that had a longer time to study the problem of duration, so as to reflect the dynamic adoption process of conservation tillage technology more accurately. Third, the Cox PH model was used for most of the continuous time, resulting in a biased conclusion when the events were interrelated. The discrete duration model was used in this study to solve problems such as unobservable heterogeneity and data censoring [25], so as to reach more reliable and scientific conclusions.

2. Theoretical Framework and Research Design

Given that the farmer is currently in the farming season s_0, and the technology choice set contains two mutually exclusive conservation tillage options, denoted by $L_j\epsilon(CA, CT)$, where *CA* is the conservation tillage measure, and *CT* is the traditional ridge tillage measure. Farmers gain economic and ecological benefits through one or both of the two land tillage measures, expressed by $\pi_j(s_t); s_t\epsilon\{(s_0 + t)t = 0, 1, 2, \ldots, T\}$, where 0 indicates the start time of the new technology; i.e., the time when the farmer have the awareness of the new technology. While t and T represent the adoption date and the end date of the research, respectively. In retrospective data, the adoption time t is discontinuous and is usually recorded in units of years.

The expected net present value (NPV) of the discounted benefit flow of the i-th farmer is represented by $V_i[\pi_{CA}(s_t), \pi_{CT}(s_t)]$, indicating the farmer's investment propensity for any conservation tillage technology. If the benefit of conservation tillage technology is greater than traditional ridge farming, i.e., $V^* = V_{CA}[\pi_{CA}(s_t) - \pi_{CT}(s_t)] > 0$, then the farmers have the motivation to adopt conservation tillage technology.

In this study, we assumed that investments were uncertain and irreversible, implying that sunk costs are important. Therefore, farmers would optimize the investment and returns (V^*) related to the decision making of conservation tillage technology regardless of whether or when to adopt it. They would consider two options: if the technology is worthwhile, then use conservation tillage technology in the current season (s_1); if one wants to obtain a variety of information, then use it in the future after verifying the prospects in many ways, which was expressed as $\{s_t\epsilon[(s_1 + t)t = 2, 3, \ldots, T]\}$.

Since conservation tillage provides multiple benefits, including ecological benefits that may be hard to value (such as carbon sequestration, increased soil organic matter content, and increased water infiltration capacity), its complete benefit function was expressed as $\{V^* = (.)\}$. The adoption decision that occurs during the research time ($t = 1, 2, 3, \ldots, T$) was used as a proxy indicator. If $V^* > 0$, it was considered that the technology had positive benefits [26]. Therefore, in this study, farmers' adoption decision was shown as follows:

$$\underset{s_t|t=1,2,3\ldots T}{Adopter} = \begin{cases} 1 & if \quad V^*_{t-T} = V_{CA}(.) - V_{CT}(.) > 0 \\ 0 & if \quad V^*_{t-T} = V_{CA}(.) - V_{CT}(.) \leq 0' \end{cases} \tag{1}$$

Different from the previous binary adoption decision, we used a risk function and the discrete duration model to analyze the dynamic adoption time and measure the adoption probability of farmers who do not adopt conservation tillage in the cultivation season (s_t) but will adopt the technology in the following season (s_{t+1}).

2.1. Discrete Duration Model in Technology Adoption

As a parameter method of survival analysis, the discrete duration model can not only simulate the occurrence of continuous time events, but can also solve the problem of data censoring. In previous studies, the right-censored data could be dropped directly, which led to sample selection bias and the inefficiency of the estimation results. The discrete-time model used in this study can deal with this problem [25].

We set the time from awareness to adoption of the conservation tillage technology as the discrete duration, with the year as the time interval unit, and expressed by the random variable $T \geq 0$. In the context of technology adoption, the equation was listed as follows:

$$\Pr(T = t|T \geq t) = h(t) = \frac{f(t)}{S(t)}' \tag{2}$$

$f(t)$ is a probability density function that represents the frequency distribution of the time t to achieve technology adoption. $S \in \{t = 1, 2, \ldots, T\}$, indicating the period during which farmer i adopts the conservation tillage technology or is censored at the end of the study period T. $S(t) = \Pr(T \geq t) = 1 - h(t)$, $S(t)$ is the survival function, indicating the probability that technology adoption will not be achieved before T, and $h(t)$ is the risk function that assesses whether and when the technology can be adopted [27].

The adoption probability of the uncensored individuals in the research period ($t \leq T$) is as follows:

$$L_i = \Pr(T_i > t) = S_i(t) = h_{it}\prod_{s=1}^{t}(1 - h_{is}), \tag{3}$$

$$L_i = \Pr(T_i = t) = f_i(t) = h_{it}\prod_{s=1}^{t-1}(1 - h_{is}), \tag{4}$$

c_i is an indicator representing whether the censoring occurred, if c_i = 0, then no censoring occurred, if c_i = 1, then the individual is right-censored. The likelihood function can be expressed as follows, where n is the sample size:

$$L_i = [\Pr(T_i = t_i)]^{1-Ci}[\Pr(T_i > t_i)]^{Ci} = \prod_{i=1}^{n}\left[h_{it}\prod_{s=1}^{t-1}(1 - h_{is})\right]^{1-Ci}\left[\prod_{s=1}^{t}(1h_{is})\right]^{Ci}, \tag{5}$$

Equation (6) operates by replacing c_i by y_i; t_i is the adoption status of the farmer. Finally, for each individual in the sample, we established a binary dependent variable. If the adoption occurs in any period $t \leq T$, then $y_i \mid c_i = 0 = 1$. If the adoption does not occur until time T, then $y_i \mid c_i = 1 = 0$. Thus, the full sample likelihood function was expressed as follows:

$$L_i = \prod_{i=1}^{n}\prod_{s=1}^{t_i}[h_{is}]^{y_{it}}[(1 - h_{is})]^{1-y_{it}} = \prod_{i=1}^{n}[f(t)]^{y}[s(t)]^{1-y_i}, \tag{6}$$

2.2. Model Selection

In this study, the baseline risk and covariates were incorporated into the risk function based on proportional risk (PH), and the formula was as follows:

$$\mathrm{h}(t|x_i, \beta) = h_0(t) \cdot \exp(x_i{}'\beta), \tag{7}$$

β is a vector of unknown parameters, $x_i{}'\beta$ is called the logarithmic relative risk, $\exp(x_i{}'\beta)$ is called the relative risk, $\exp(x_i{}'\beta) > 0$. $h_0(t)$ is called the baseline hazard, which means that all farmers in the sample have a common and constant baseline hazard function, which depends on time t, but does not depend on $x_i{}'\beta$. If all explanatory variables are 0, the risk function is equal to the baseline risk, the baseline risk $h_0(t)$ is the same for each individual in the population, and the risk function of the individual is based on $\exp(x_i{}'\beta)$ and it is proportional to $h_0(t)$, so it is called proportional hazard [28].

In order to use a discrete duration model, a functional form (such as binary logit and probit models) needs to be chosen to estimate Equation (7), which can be used to investigate the factors affecting the adoption time [25]. Referring to the studies of other scholars on discrete duration models, we selected the logarithmic complementary model (Clog-log) to parameterize the formula mentioned above [20–22], as follows:

$$\begin{cases} \mathrm{Cloglog}[h_i(t, X|e_i)] \equiv \ln\{-\ln(1 - h_{it})\} = h_0(t)\exp[X_i(t)\beta + e_i) \\ \ln\{-\ln(1 - h_{it})\} = \alpha D(t) + \beta_{it}X_{it} + e_i \end{cases}, \tag{8}$$

where X is a vector of covariates; D is a time variable representing the duration-dependent effect (baseline risk); α and β are estimable parameters; and $e_i \sim \mathrm{N}(0, \sigma^2)$ is a random error term controlling for unobserved heterogeneity to reduce the error in estimation [29].

3. Data Sources and Descriptive Statistics of Variables

3.1. Data Sources

The data came from the survey of our research team. We conducted surveys in Shaanxi, Gansu, Ningxia, and Shanxi provinces in the Yellow River Basin in August 2021 and November 2016, respectively. The climate and agricultural conditions vary significantly across different regions of the Loess Plateau. However, the surveyed areas, including Yulin and Xi'an in Shaanxi Province, Qingyang in Gansu Province, Guyuan in Ningxia Province, and Yuncheng in Shanxi Province, are located in the northern region of the Loess Plateau.

These areas experience dry and less rainy weather throughout the year, with the rainy season being concentrated. The annual rainfall is only around 400 mm to 500 mm, and there is limited time for cultivation. The cropping system for grain crops follows a one-crop-per-year pattern, either with spring maize and winter wheat or with winter wheat followed by summer maize. Part of the main body of the four provinces is located in the central part of the Loess Plateau in the hilly and ravine area, and they are at the edge of the East Asian monsoon and located in the transition zone from the temperate continental monsoon climate to the temperate semi-arid climate, with the geological characteristics of the Loess Plateau. These areas are national key ecological environment construction areas in the upper and middle reaches of the Yellow River. However, the natural environment in this area is fragile, and extreme weather changes such as heavy rainfall, floods, low temperature and freezing damage, etc., have caused unstable agricultural production and changed the agricultural production conditions. At the same time, the population there is relatively concentrated and the farming income accounts for a relatively high proportion of the total income of farmers there. Meanwhile, anthropic factors such as land abuse and over-exploitation in these areas have caused serious soil erosion. Therefore, it is of high research value to take the four provinces of Shaanxi, Gansu, Ningxia, and Shanxi in the Loess Plateau region as the research site to obtain data.

The survey adopted a combination of a typical survey, stratified sampling, and simple random sampling. First of all, we selected several areas with better conservation tillage management effects as typical survey sites, they are Xi'an, Yulin, Shaanxi, Qingyang, Gansu, Guyuan, Ningxia, and Yuncheng, Shanxi. Second, a combination of stratified and random sampling methods was used to select 2–4 counties from each city randomly, and then a stratified approach was adopted to select 4–8 villages in each town, and about 20 farming households were randomly selected for each village. A one-to-one survey was conducted on the selected farmers, and the questionnaire involved information about the characteristics of the head of household, family situation, awareness of conservation tillage technology, the time of adoption, etc. A total of 1900 questionnaires were distributed in the two surveys, and 1870 valid samples were obtained after excluding samples with missing key information and inconsistent logic. The effective rate of the questionnaire was 98.42%.

Given the development of conservation tillage technology in China, we treated 2002 (since 2002, the specific funds for conservation farming were arranged by the government) as the initial time of our research, as it can help to deal with the problem of left-censored data. The left-censored data refers to data where the technology had been adopted before the initial time or had not been adopted till the end time of the research [24]. Figure 1 showed the trends in awareness and adoption of conservation tillage during the study period (2002–2020). It showed that at the beginning of the study, few farmers in the sample were aware of the existence of conservation tillage techniques, and then the popularity of these techniques began to rise from 2014 to 2016 (as shown in Figure 1).

3.2. *Variable Selection*

3.2.1. Dependent Variables

We took the duration between awareness and adoption of conservation tillage techniques as the dependent variable. Specifically, the duration referred to the number of years from when farmers began to realize that there was a specific conservation tillage technology to when any one of the three conservation tillage techniques (low-tillage and no-tillage, straw mulching, and weed and pest control) was adopted. According to our study period (2002–2020), the data can be divided into three types based on how long the technology has been adopted since awareness. First, conservation tillage techniques were adopted during the study period. Second, there was still no adoption of conservation tillage techniques till the end of the survey (2020). Third, conservation tillage techniques were adopted before the start date of the survey (2002) or remained unconscious until 2020. Among the survey samples, only 11, 8, and 9 respondents indicated that they had awareness of these three conservation tillage techniques (minimum tillage, straw return to the field, and weed and pest control) before 2002, respectively. It can also be seen from Figure 1 that the time for awareness of the conservation tillage technology was always longer than the time for adopting it. It also showed that a period of time was needed before a technology was adopted. Faced with a new technology, farmers would go through three stages: awareness, trial adoption, and continuous adoption. During this process, they would adjust the risk through risk mânagement, social learning, and investment so as to reduce the uncertainty [14].

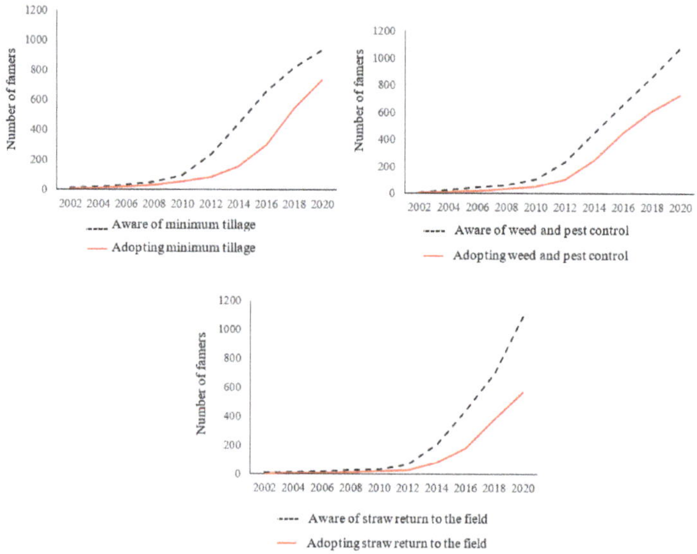

Figure 1. Farmers' awareness and adoption time trend of conservation tillage.

3.2.2. Independent Variables

Three types of conservation tillage techniques (minimum tillage, weed and pest control, and straw return to the field) were the independent variables in our study. Among all the respondents, there were 1312 farmers who adopted the technology of minimum tillage, accounting for 70.16%; 671 farmers adopted the technology of weed and pest control, accounting for 35.88%; and 1075 farmers adopted the technology of straw return to the field, accounting for 57.49%.

The adoption rate of weed and pest control was relatively low, and both physical and chemical methods are required when using it, and it is mainly based on prevention, which makes it difficult for the farmers to adopt. The adoption rates of both minimum tillage and straw return to the field technologies were more than half, indicating that the coverage rate of conservation tillage techniques in the survey area was relatively high, and farmers had a strong awareness of the importance of conservation tillage technology. However, there was still an average of 45.49% of farmers who were unwilling to adopt conservation tillage measures. Thus, the quality of cultivated land should be improved and the effectiveness of land governance should be addressed.

3.2.3. Control Variables

In order to avoid other possible factors from interfering with the results, we selected the control variables from seven aspects. They were the individual characteristics of the head of a household (gender, age, and education level), family characteristics (dependency ratio, cultivated land area, and soil fertility), social network conditions (whether to participate in cooperatives), geographical distance characteristics (distance from the nearest agricultural material sales point), government incentives (the government provides legal and regulatory policy assistance and the degree of influence of agricultural subsidy policies), extreme weather characteristics (whether the farmland suffered heavy rainfall and floods in the past three years, whether the farmland suffered drought disasters), and area characteristics (whether located in Shaanxi, Shanxi, Gansu, or Ningxia).

3.3. Variable Descriptive Statistics

It can be seen from Table 1 that the average duration from awareness to adoption of minimum tillage technology was 2 years, and the average duration of straw return to the field, weeds, and pest control was no more than 1 year. It indicated that farmers were aware of the importance of conservation tillage and may adopt the technologies timely to maximize the expected returns. Most household heads were male, mostly middle-aged (51–54 years old), with a low level of education (below junior high school). The proportion of economically inactive household members (people over 65 and children aged 1–15) to the total number of household members was used to represent the household dependency ratio, because the amount of household labor will directly affect the rate of conservation tillage technology adoption, such as straw mulching, which can be labor-intensive and increase the demand for labor when crop residues were spread in the field before planting, or when weed and pest control were necessary to increase manpower to cope with the intensity of weed removal and pest control. The dependency ratio in Table 1 did not exceed 32%, indicating that the quantity and quality of the labor force in the surveyed area were relatively high. Land size and fertility can also affect the adoption of conservation tillage technologies. The larger the land area, the more fertile the soil, and the more likely farmers are to adopt conservation technologies. The mean value of land area owned by the farmers in sample data was in the range of 0.47–0.80 hectares, with moderate soil fertility. Policy incentives had a relatively large impact on farmers. Government-provided agricultural subsidies and publicity laws and regulations can motivate farmers to adopt conservation tillage technologies. In addition, both social networks and social learning had an impact on the adoption of technology. The data showed that 43% of farmers participated in cooperatives, indicating that farmers were less involved in industrial organizations. In terms of social learning, more than 80% of farmers exchanged experiences with people around them, and received guidance and training from professionals. In addition, more than 80% of farmers obtained knowledge and experience of conservation tillage technology through internet channels such as mobile phones and computers, thereby increasing their confidence in adopting the technology independently and improving efficiency when implementing the technology. However, farmers seldom obtained information through traditional media such as newspapers and radio (less than 20%).

Table 1. Descriptive statistics of variables [1].

Variables	Meaning	Minimum Tillage		Weed and Pest Control		Straw Return to the Field	
		Adopter [2] (1312)	Non-Adopter (558)	Adopter [2] (671)	Non-Adopter (1199)	Adopter [2] (1075)	Non-Adopter (795)
		Mean	Mean	Mean	Mean	Mean	Mean
Adoption duration	Duration time from awareness to adoption (years)	2.448 (0.162)		0.654 (0.113)		0.473 (0.060)	
Age	Head of household's age (years)	52.332 (0.282)	53.513 (0.473)	51.845 (0.368)	53.153 (0.318)	52.260 (0.303)	53.258 (0.399)
Gender	Head of household's gender (1 = male, 0 = female)	0.875 (0.009)	0.928 (0.011)	0.784 (0.016)	0.951 (0.006)	0.843 (0.011)	0.956 (0.007)
Education	Head of household's education level (years)	6.752 (0.095)	5.946 (0.169)	7.753 (0.116)	5.817 (0.109)	7.021 (0.099)	5.821 (0.141)
Dependency ratio	Proportion of economically inactive members to the total number of households	0.282 (0.007)	0.313 (0.013)	0.279 (0.009)	0.297 (0.008)	0.285 (0.008)	0.299 (0.010)
Cultivated area	Household cultivated land area (hectare)	0.727 (0.286)	0.549 (0.311)	0.973 (0.468)	0.506 (0.195)	0.801 (0.333)	0.502 (0.242)

Table 1. *Cont.*

Variables	Meaning	Minimum Tillage		Weed and Pest Control		Straw Return to the Field	
		Adopter [2] (1312)	Non-Adopter (558)	Adopter [2] (671)	Non-Adopter (1199)	Adopter [2] (1075)	Non-Adopter (795)
		Mean	Mean	Mean	Mean	Mean	Mean
Soil fertility	Farmers' evaluation of soil fertility (1–5: very poor–very good)	3.530 (0.027)	3.102 (0.043)	3.444 (0.038)	3.379 (0.030)	3.445 (0.030)	3.345 (0.037)
Government incentives	The impact of regulations, policies, and agricultural subsidies provided by the government on you (1–5: very small–very large)	3.171 (0.040)	3.695 (0.049)	2.636 (0.062)	3.715 (0.031)	3.006 (0.046)	3.764 (0.039)
Social network	Whether family members participate in cooperatives (1 = yes, 0 = no)	0.350 (0.013)	0.571 (0.021)	0.235 (0.016)	0.517 (0.014)	0.334 (0.014)	0.527 (0.018)
Learn from peers	Whether to learn related technologies from peers around (1 = yes, 0 = no)	0.864 (0.009)	0.935 (0.010)	0.785 (0.016)	0.942 (0.007)	0.849 (0.011)	0.935 (0.009)
Extension staff guide learning	Whether to learn related technologies through the guidance of extension personnel (1 = yes, 0 = no)	0.642 (0.013)	0.808 (0.017)	0.434 (0.019)	0.836 (0.011)	0.580 (0.015)	0.843 (0.013)
Learning through traditional channels	Do you learn related technologies through traditional channels such as radio and television (1 = yes, 0 = no)	0.171 (0.010)	0.120 (0.014)	0.212 (0.016)	0.124 (0.010)	0.193 (0.012)	0.106 (0.011)
Digital learning	Whether to learn related technologies through digital network channels such as mobile phones and computers (1 = yes, 0 = no)	0.852 (0.010)	0.724 (0.019)	0.957 (0.008)	0.733 (0.013)	0.891 (0.010)	0.709 (0.016)
Distance from agricultural materials sale station	The distance between your home and the nearest agricultural material sales point (miles)	3.484 (0.088)	4.153 (0.180)	3.460 (0.154)	3.808 (0.095)	3.510 (0.111)	3.918 (0.121)
Area	1 = in Shaanxi, 2 = in Gansu, 3 = in Ningxia, 4 = in Shanxi	2.175 (0.030)	2.287 (0.053)	2.110 (0.037)	2.264 (0.036)	2.021 (0.026)	2.462 (0.050)
Heavy rainfall and flood disasters	Whether the farmland suffered heavy rainfall and flood disasters in the past three years (1 = yes, 0 = no)	0.768 (0.012)	0.324 (0.020)	0.598 (0.019)	0.656 (0.014)	0.626 (0.015)	0.648 (0.017)
Extreme drought disaster	Whether the farmland suffered from drought in the past three years (1 = yes, 0 = no)	0.543 (0.014)	0.622 (0.021)	0.469 (0.019)	0.621 (0.014)	0.502 (0.015)	0.653 (0.017)

[1] Note: Standard errors are in parentheses. [2] The number of adopters in Table 1 is different from the number of adopters in Figure 1, because the number of adopters in Figure 1 is the "number of adopters for the first time", while the number in Table 1 refers to the "number of conservation tillage techniques adopted last year".

Information is disseminated through multiple channels. Whether it is through passive social learning or self-directed online learning, it will help farmers become aware of the importance of conservation tillage technology, which will help increase the adoption rate of the technology, and guide farmers to use it at a suitable time. In addition, distance from the point of sale of agricultural materials and extreme weather can also affect technology adoption. The closer the point of purchase, the lower the transportation costs and the more likely the technology will be adopted. It can be seen from the data that farmers believed that heavy rainfall, floods, and extreme droughts had obviously intensified in the past three years, which did have a certain impact on agricultural production and daily life. We created a structural model for all the influencing factors, as shown in Figure 2.

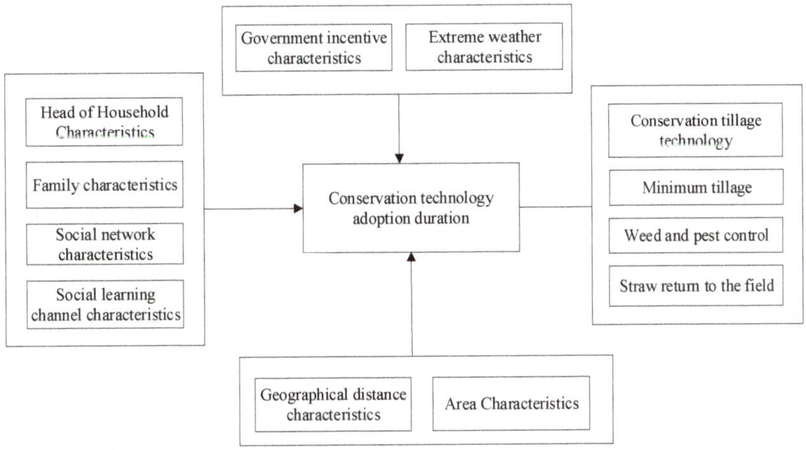

Figure 2. Structural model for all the influencing factors.

4. Empirical Analysis and Results

4.1. Non-Parametric Estimation of K-M

Kaplan–Meier's non-parametric graphical method was used to construct the survival function of conservation tillage technology in the Yellow River Basin of China and estimated the distribution of the adoption duration. Figure 3 measured the probability of survival (non-adoption) after time t. This non-parametric method can help to analyze the speed of adoption of conservation tillage technology during the study period. As shown in Figure 3, the probability of survival was decreasing and, therefore, the rate of adoption of conservation tillage technologies increased steadily over time. The nonparametric Kaplan–Meier method does not make any assumptions about the form of the survival function, and it cannot estimate the effect of covariates [26,29]. Therefore, semi-parametric and parametric methods were used in the following part to verify the stability and reliability of the conclusions.

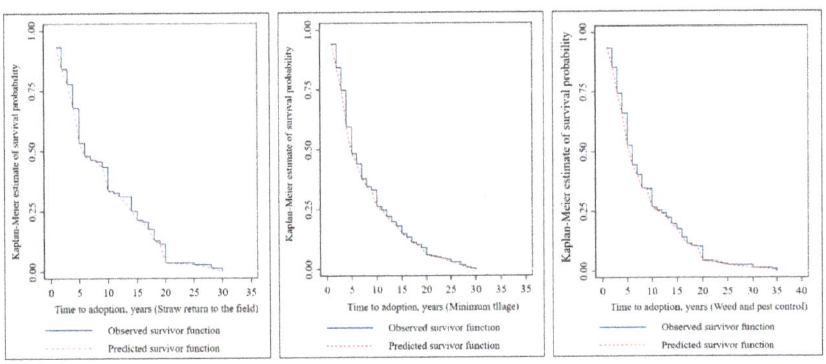

Figure 3. K-M survival estimation of the duration of conservation tillage technology adoption in the Yellow River Basin.

4.2. Clog-log Parameter Estimation and Cox PH Semi-Parametric Estimation

4.2.1. Model Diagnostic Tests

Tables 2–4 showed the model estimation results for each of the three conservation tillage techniques. In the baseline PH model, the hazard rate was regressed using the duration of adoption as the only covariate, assuming that the effects of other covariates were not important [26]. The Continuous-time Cox PH model belongs to the semi-parametric estimation method, which does not make any restrictive functional assumptions about the distribution pattern of the baseline risk. It allows for the controlling of multiple risk factors and is suitable for analyzing the effects of multiple risk factors on observed individuals at different times. So, it was suitable to use the Cox model in our study when we assumed that the adoption times were recorded at small intervals [29]. The discrete duration model (Clog-log) is a parameter estimation method that can not only solve the problem of time nodes, but can also be extended. It can not only be used to explain the unobservable heterogeneity, but can also avoid the assumption of proportional hazards and can deal with the problem of right censoring [25]. It can be seen in Tables 2–4 that the estimated results of the Clog-log model were similar to the Cox model, and based on AIC, BIC, and the loglikelihood, the Clog-log model was more suitable for our research than the Cox PH. Therefore, the discussion in this section mainly focused on the Clog-log model.

Table 2. Cox PH model and Clog-log model of minimum tillage technology adoption duration.

Variables	Minimum Tillage					
	Baseline PH		COX PH		Clog-log	
	HR	Z-Score	HR	Z-Score	HR	Z-Score
Age			1.009 (0.006)	1.60	0.996 * (0.003)	−1.23
Gender			1.153 (0.503)	0.33	0.765 ** (0.089)	−2.30
Education			1.000 (0.021)	−0.01	1.018 * (0.010)	1.74
Dependency ratio			0.970 (0.263)	−0.11	0.734 ** (0.097)	−2.34
Cultivated area			0.999 (0.020)	−0.06	1.014 *** (0.005)	3.06
Soil fertility			0.863 (0.049)	−2.57	1.281 *** (0.046)	6.87
Government incentives			0.833 (0.057)	−2.68	0.933 (0.033)	−1.93
social network			0.668 (0.050)	−5.39	0.478 (0.051)	−6.92
Learn from peers			0.902 (0.065)	−1.43	1.200 ** (0.105)	2.09
Extension staff guide learning			0.638 (0.156)	−1.84	1.145 (0.109)	1.42
Learning through traditional channels			1.710 * (0.551)	1.67	1.067 (0.105)	0.66
Digital learning			0.731 (0.059)	−3.85	1.385 ** (0.224)	2.02
Distance from agricultural materials sale station			1.012 (0.016)	0.74	0.986 (0.010)	−1.38
Area			1.030 (0.048)	0.62	0.967 (0.029)	−1.09
Heavy rainfall and flood disasters			3.196 *** (0.247)	15.02	0.240 *** (0.044)	−7.79
Extreme drought disaster			1.166 ** (0.087)	2.05	0.818 ** (0.073)	−2.24
Long-term dependence (continuous dependence)						
D1 ($1 \leq t \leq 10$)	0.092 *** (0.007)	−30.52			0.922 ** (0.032)	−2.33
D2 ($11 \leq t \leq 20$)	0.022 *** (0.003)	−25.63			1.080 (0.035)	2.37
D3 ($21 \leq t \leq 30$)	0.013 *** (0.004)	−15.60			1.003 (0.081)	0.03
D4 ($31 \leq t \leq 40$)						
Constant					0.655 (0.196)	−1.41
LLR	−1139		−954		−733	
AIC	2283		1929		1501	
BIC	2294		1970		1592	
N	1870		479		1594	

Note: Standard deviations in parentheses. * Significant at the 10% level; ** significant at the 5% level; *** significant at the 1% level.

Table 3. Cox PH model and Clog-log model of weed and pest control technology adoption duration.

Variables	Weed and Pest Control					
	Baseline PH		COX PH		Clog-log	
	HR	Z-Score	HR	Z-Score	HR	Z-Score
Age			1.004 (0.003)	1.59	1.009 (0.004)	2.01
Gender			0.927 (0.292)	−0.24	0.484 *** (0.058)	−6.05
Education			0.986 ** (0.006)	−2.47	1.063 *** (0.012)	5.32
Dependency ratio			0.778 * (0.102)	−1.92	1.086 (0.191)	0.47
Cultivated area			0.980 * (0.011)	−1.84	1.038 *** (0.004)	8.82
Soil fertility			0.948 (0.034)	−1.46	1.131 *** (0.051)	2.74
Government incentives			1.005 (0.058)	0.08	1.012 (0.049)	0.26
social network			0.782 (0.084)	−2.29	1.395 ** (0.210)	2.21
Learn from peers			1.086 (0.056)	1.62	1.098 (0.103)	1.00
Extension staff guide learning			1.102 (0.130)	0.82	1.420 ** (0.195)	2.55
Learning through traditional channels			1.190 (0.156)	1.33	1.393 *** (0.162)	2.85
Digital learning			0.932 (0.141)	−0.46	0.380 (0.036)	−10.19
Distance from agricultural materials sale station			0.988 (0.017)	−0.69	0.978 (0.014)	−1.57
Area			0.942 ** (0.023)	−2.46	0.990 (0.041)	−0.25
Heavy rainfall and flood disasters			0.884 (0.083)	−1.32	0.917 (0.107)	−0.74
Extreme drought disaster			1.476 *** (0.146)	3.93	0.978 (0.111)	−0.20
Long-term dependence (continuous dependence)						
D1 ($1 \leq t \leq 10$)	0.179 *** (0.010)	−30.01			0.765 *** (0.060)	−3.44
D2 ($11 \leq t \leq 20$)	0.055 *** (0.006)	−27.92			1.124 (0.117)	1.12
D3 ($21 \leq t \leq 30$)	0.035 *** (0.010)	−11.58			0.739 *** (0.070)	−3.21
D4 ($31 \leq t \leq 40$)	0.025 *** (0.010)	−9.07				
Constant					0.286 *** (0.107)	−3.35
LLR	−1199		−1727		−771	
AIC	2403		3476		1576.	
BIC	2414		3515		1668	
N	1870		390		1594	

Note: Standard deviations in parentheses. * Significant at the 10% level; ** significant at the 5% level; *** significant at the 1% level.

Table 4. Cox PH model and Clog-log model of straw return to the field technology adoption duration.

Variables	Straw Return to the Field					
	Baseline PH		COX PH		Clog-log	
	HR	Z-Score	HR	Z-Score	HR	Z-Score
Age			0.995 (0.009)	−0.59	1.001 (0.003)	0.34
Gender			2.730 (3.101)	0.88	0.595 *** (0.065)	−4.75
Education			1.038 (0.025)	1.53	1.019 * (0.010)	1.89
Dependency ratio			0.840 (0.713)	−0.21	0.966 (0.131)	−0.25

Table 4. *Cont.*

Variables	Straw Return to the Field					
	Baseline PH		COX PH		Clog-log	
	HR	Z-Score	HR	Z-Score	HR	Z-Score
Cultivated area			0.986 (0.023)	−0.60	1.030 *** (0.005)	6.46
Soil fertility			0.968 (0.023)	−0.33	1.092 ** (0.039)	2.44
Government incentives			0.790 (0.086)	−2.18	1.070 * (0.040)	1.82
social network			0.819 (0.136)	−1.20	0.720 (0.056)	−4.22
Learn from peers			1.241 (0.205)	1.31	0.858 (0.063	−2.08
Extension staff guide learning			0.998 (0.097)	−0.02	3.745 ** (2.252)	2.2.
Learning through traditional channels			0.728 (0.335)	−0.69	1.466 *** (0.141)	3.96
Digital learning			0.581 (0.046)	−6.80	3.421 * (2.476	1.70
Distance from agricultural materials sale station			1.073 (0.051)	1.47	0.986 (0.011)	−1.33
Area			0.907 (0.113)	−0.78	0.841 *** (0.025)	−5.72
Heavy rainfall and flood disasters			0.999 (0.075)	−0.01	1.195 (0.321)	0.66
Extreme drought disaster			1.069 (0.082)	0.86	0.508 *** (0.108)	−3.18
Long-term dependence (continuous dependence)						
D1 ($1 \leq t \leq 10$)	0.098 *** (0.010)	−27.70			0.972 ** (0.012)	−2.26
D2 ($11 \leq t \leq 20$)	0.056 *** (0.009)	−17.99			0.725 (0.193)	−1.21
D3 ($21 \leq t \leq 30$)	0.035 *** (0.016)	−7.47				
D4 ($31 \leq t \leq 40$)						
Constant					1.404 (0.419)	1.14
LLR	−1246		−189		−923	
AIC	2496		398		1881	
BIC	2507		422		1972	
N	1870		79		1594	

Note: Standard deviations in parentheses. * Significant at the 10% level; ** significant at the 5% level; *** significant at the 1% level.

4.2.2. Duration Dependence Test

The duration dependence test is a new and effective method used by most scholars to test the rational bubbles of the stock market in economics [30–33]. Chen Qiang [27] has proposed that in unemployment duration studies, the longer the duration of unemployment, the lower the probability of finding a job; that is, the risk rate decreases with time. If each individual is exactly the same, it means the duration dependence is negative. The persistence test in the duration analysis can not only improve the accuracy and consistency of parameter estimation but also deal with the problem of "unobservable heterogeneity" in the sample, and it is more robust to autoregressive phenomena [25,31].

Drawing on related research, we assumed that the effect of other covariates on the rate of adoption is zero. We used a baseline PH model for a duration dependence test to examine the effect of time, then modeled the baseline PH as a step function with some significant time points and divided it into four time periods (D1: $1 \leq t \leq 10$, D2: $11 \leq t \leq 20$, D3: $21 \leq t \leq 30$, D4: $31 \leq t \leq 40$). We then examined whether the conclusions change after significant and important changes in the technology adoption environment around these time points. If the effects of other covariates were added, a flexible full-parameter Clog-log model (columns 6, 12, and 18 of D1-D4 in Table 2) was generated to examine the sensitivity of the test results to the choice of function. The application of this model can help to reveal the pattern of duration persistence of conservation tillage technology adoption.

As shown in the results, the estimated coefficients of the baseline PH model were gradually decreasing, and the probability of delaying the adoption of minimum tillage technology, weed and pest control technology, and straw return to the field technology was the highest in the initial D1 stage (0.092–0.179), with the coefficients gradually becoming smaller in stage D3 (0.013–0.035). The coefficients of weed and pest control technologies reached the smallest (0.025) in the D1–D4 stages, and even the coefficients of minimum tillage and straw return to the field were zero in stage D4. In conclusion, the probability of delaying the adoption of technologies was the highest at the beginning; i.e., there was a negative time dependence. As a result, the farmers became more willing to adopt conservation tillage techniques over time. The surge in adoption in the later years could be due to various factors, such as increased risk perception, more knowledge, enhanced social networks, policy incentives, etc.

4.3. Estimation Results and Analysis

To make the comparison of results easier, all coefficients were reported as hazard ratios (HR). If the value is one, it means the covariate has no effect on the HR, while a value greater (less) than one indicates a positive (negative) effect on the possibility of adopting conservation tillage techniques [25].

As shown in Tables 2–4, the coefficient of gender was significantly negative (less than one), indicating that in order to obtain higher economic income and more opportunities for development, males tended to join in labor transfer and females became the main decision makers in family agricultural production [34]. This would affect the process from awareness to adoption of agricultural technology. Education level had a significant positive impact on the duration of conservation tillage technology adoption (greater than one), which was consistent with most research, such as Wollnim et al. (2010) [4] and Li Wei et al. (2017) [6], who studied the influencing factors of conservation tillage techniques on Honduran hillsides and the Chinese Loess Plateau. These findings indicate that the higher the level of education was, the shorter the duration could be. The area of cultivated land and the degree of soil fertility had a positive effect on the duration of conservation tillage technology adoption, indicating that the resource endowment of cultivated land determined farmers' production input preferences in cultivated land conservation behavior, and the larger and more fertile the cultivated land, the quicker the conservation tillage technology could be adopted, which is consistent with the results of Xu et al. (2018) [16], who conducted a study on the impact of land management practices on the adoption of straw mulching as a conservation tillage measure.

As for the technology of minimum tillage, the younger farmers were more likely to learn and innovate the technology. The higher dependency ratio would delay farmers' adoption of minimum tillage technology. It is because a higher dependency ratio implies a higher proportion of children and elderly in the household, resulting in a lack of labor for pre-production and post-production crop operations such as local deep pine or rototill, direct seeding, fertilization, application, mulching, suppression, and harvesting. As for the technology of straw return to the field, policy incentives have had a positive effect on the adoption time of the technology. This is because the promulgation of regulations, such as the straw burning ban and the implementation of the straw return subsidy policy deepened farmers' awareness and stimulated motivation for its adoption. When it came to the weed and pest control technology, farmers participating in cooperatives could broaden their social network, promote precise connections between technology adopters and the market, and shorten the time interval from awareness to technology adoption. The findings above are consistent with the conclusions of Beyene et al. (2015) [29], Zheng et al. (2018) [13], and Xu et al. (2018) [16], who investigated the importance of different attributes of conservation tillage techniques in relation to the duration of adoption.

Social learning was also an important factor influencing whether a technology was adopted and the rate of adoption. The results showed that social learning positively and significantly influenced the duration of adoption of conservation tillage technologies by farmers (greater than one), demonstrating that farmers tended to delay adoption decisions when they were uncertain or poorly informed about the profitability of a new technology. The importance of information is that it can dispel possible misconceptions that farmers may have about the technology and help to build their expertise in the technology. Farmers may be unfamiliar with technology-related information at the beginning, but as more information about the attributes of the technology becomes available, farmers are better able to objectively evaluate the benefits of the technology, so as to accelerate the time from awareness to adoption. This finding is consistent with the results of Khataza et al. (2018) [25], who studied the impact of information channels in social learning on the duration of adoption of conservation agriculture in the Malawi region.

In addition, extreme weather changes such as heavy rainfall, flooding, and extreme drought disaster negatively affected the adoption of conservation tillage technologies. This is because it is an adaptive process for farmers from perceiving the extreme weather changes to the timely adoption of farm management strategies. Compared to normal years, more tasks and management measures should be implemented to resist the agricultural risks caused by the disasters. Therefore, the occurrence of extreme weather changes prolongs the time from awareness to adoption of conservation tillage techniques by farmers. This finding is consistent with the results of Xiao et al. (2012) [35], who studied the willingness of farmers to adopt conservation agriculture practices.

5. Discussion

While duration studies are essential for assessing the effectiveness and impact of conservation tillage technology adoption, they also have certain limitations. Firstly, duration studies require tracking and evaluating conservation tillage practices, which often necessitate long time spans and significant resource investments, limiting the scale and feasibility of the research. Secondly, conservation tillage techniques are influenced by regional factors such as geography, climate, and soil conditions. Duration studies are typically conducted in specific regions, making it challenging to generalize the conclusions to other areas and limiting our understanding of the universality of conservation tillage technology across different geographical environments. Lastly, duration studies often focus on specific conservation tillage practices over a relatively long period. However, technological advancements and innovations may occur during the study period, with new techniques potentially exhibiting improved effectiveness or higher feasibility. Duration studies may not fully account for these factors of technological progress and innovation.

To gain a more comprehensive understanding and evaluation of conservation tillage technology, future research should not only involve long-term monitoring and assessment to acquire data spanning a longer time period and better understand the long-term effects of the technology but also incorporate additional indicators and research methods to comprehensively evaluate conservation tillage techniques. This will ultimately provide scientific evidence for sustainable agricultural development and offer technical and economic references for policymakers.

6. Conclusions and Policy Implications

This study utilizes survey data from 1870 households in the Loess Plateau region of China to investigate the duration of adoption of conservation tillage techniques. The results showed that the overall duration from awareness to adoption of conservation tillage technology was generally short, with an average duration of only 1.192 years, with the duration from awareness to adoption of low-tillage and no-tillage technology being 2.448 years. The duration from awareness to adoption of straw return to the field and weeds and pest control were all within 1 year. Our study analyzed the factors affecting the duration from awareness to adoption of conservation tillage technologies among smallholder farmers in the Yellow River Basin by using the Clog-log model. The results showed that the key factors in deciding to adopt conservation tillage technology were farmers' education level, social learning to access information, high-quality soil fertility, arable land scale, and policy incentives. The reduction in labor involved in agricultural production due to household labor shifts and the uncertainty of risks and benefits due to extreme weather changes may delay the adoption of these technologies. The time-dependent test found that the probability of delaying the decision time to adopt conservation tillage was highest in the initial D1 stage (0.092–0.179), but the coefficient gradually became smaller (0.013–0.035) in the later D3 stage, and even came to zero in the D4 stage. In other words, there was a negative time dependence. Through the study of the duration of adoption of conservation tillage techniques, farmers can gain a better understanding of the improvements in agricultural productivity and ecological conservation benefits associated with this technology. Based on this understanding, farmers can evaluate and select different technological attributes and stages over time, adjust their agricultural management strategies, enhance agricultural production capacity, improve farmers' income levels, enhance the health of the ecosystem, and ultimately promote sustainable development in rural areas.

This study has practical guiding significance for farmers as well as government departments. On the one hand, the article examined the factors that affect the duration of technology adoption in detail, so that the farmers can take the conclusion as a reference before choosing an advanced agricultural technology. In this way, they can weigh and select the costs and benefits of technology adoption at different times, and then make timely adoption decisions to minimize the cost of technology adoption. On the other hand, this study pointed out the impact of social learning and policy incentives on the duration of technology adoption. It was suggested that the government should consider broadening different social learning channels while promoting conservation tillage technologies. Specifically, more offline guidance and training by experts can be conducted, in addition to combining online platforms such as WeChat and Tencent meetings for learning to enhance the social learning atmosphere and promote the broad participation of potential farmers. It is also suggested that the government should provide a certain amount of ecological compensation in terms of agricultural subsidies, so that technology adoption can improve the externalities of the ecological environment. This would also stimulate farmers' enthusiasm for technology adoption, so as to shorten the time interval from awareness to adoption, and accelerate the transformation and increase the technology adoption rate.

Author Contributions: Conceptualization, T.G. and Y.R.; methodology, T.G. and Y.R.; validation, T.G. and Y.R.; investigation, H.F.; writing—original draft preparation, T.G. and Y.R.; writing—review and editing, H.F.; visualization, T.G. and Y.R.; supervision, Q.L.; funding acquisition, Q.L., T.G. and Y.R. contributed equally to this work, both of them are regarded as the first author. All authors have read and agreed to the published version of the manuscript.

Funding: This research was funded by the National Natural Science Foundation of China (grant number:71673223), and the National Natural Science Foundation of China (grant number: 71973105).

Institutional Review Board Statement: Not applicable.

Informed Consent Statement: Not applicable.

Data Availability Statement: Please contact the corresponding author based on reasonable grounds if necessary.

Conflicts of Interest: The authors declare no conflict of interest.

References

1. Shen, R.; Wang, C.; Sun, B. Soil science and technology in the implementation of the strategy of "storing food in the land and storing food in technology". *J. Chin. Acad. Sci.* **2018**, *33*, 135–144.
2. FAO. *Conservation Agriculture: Conserving Resources above-and below-the Ground*; FAO: Rome, Italy, 2012.
3. Kassie, M.; Marenya, P.; Tessema, Y.; Jaleta, M.; Zeng, D.; Erenstein, O.; Rahut, D. Measuring farm and market level economic impacts of improved maize production technologies in Ethiopia: Evidence from Panel Data. *J. Agric. Econ.* **2018**, *69*, 76–95. [CrossRef]

4. Wollni, M.; Lee, D.R.; Thies, J.E. Conservation agriculture, organic marketing and collective action in the Honduran hillsides. *Agric. Econ.* **2010**, *41*, 373–384. [CrossRef]
5. Abdulai, A.N. Impact of conservation agriculture technology on household welfare in Zambia. *Agric. Econ.* **2016**, *47*, 729–741. [CrossRef]
6. Wei, L.; Caixia, X.; Shunbo, Y.; Ruixiang, Z. Farmers' Conservation Tillage Technology Adoption and Its Influencing Factors: Based on the Analysis of 476 Farmers in the Loess Plateau. *Chin. Rural Econ.* **2017**, *385*, 44–57+94–95.
7. Cherylr, R.D. Analyzing technology adoption using micro studies: Limitation, challenges and opportunities for improvement. *Agric. Econ.* **2006**, *34*, 207–219.
8. Mccord, P.F.; Cox, M.; Schmitt-Harsh, M. Crop diversification as a smallholder livelihood strategy within semiarid agricultural systems near Mount Kenya. *Land Use Policy* **2015**, *42*, 738–750. [CrossRef]
9. Teklwold, H.; Kassie, M.; Shiferaw, B. Cropping system diversification, conservation tillage and modern seed adoption in Ethiopia: Impacts on household income, agrochemical use and demand for labor. *Ecol. Econ.* **2013**, *93*, 85–93. [CrossRef]
10. Cai, R.; Cai, S. Empirical Analysis of the Adoption of Conservation Tillage Technology and Its Impact on Crop Yield—Based on the Survey Data of Rice Planters in Anhui Province. *Resour. Sci.* **2012**, *34*, 1705–1711.
11. Hu, W. Social capital, cultivated land value cognition and farmers' willingness to pay for cultivated land protection—An empirical study based on a moderated mediation effect model. China Population. *Resour. Environ.* **2019**, *29*, 12.
12. Wen, C.; Wang, B.; Wu, J. Analysis of Influencing Factors of Farmers Adopting Different Attributes of "Two-Type Agriculture" Technology—Based on a Survey of Liaoning Province Farmers' Questionnaire. *Agric. Mod. Res.* **2016**, *37*, 701–708.
13. Zheng, X.; Wang, F.; Ying, R. Peasants' Endowment Constraints, Technical Attributes and Agricultural Technology Selection Bias—An Analysis Framework of Farmers' Technology Adoption Based on Incomplete Factor Market Conditions. *China Rural Econ.* **2018**, *399*, 105–122.
14. Lambrecht, I.; Vanlauwe, B.; Merckx, R.; Maertens, M. Understanding the process of agricultural technology adoption: Mineral fertilizer in Eastern DR Congo. *World Dev.* **2014**, *59*, 132–146. [CrossRef]
15. Omotilewa, O.J.; Ricker-Gilbert, J.; Ainembabazi, J.H. Subsidies for Agricultural Technology Adoption: Evidence from a Randomized Experiment with Improved Grain Storage Bags in Uganda. *Am. J. Agric. Econ.* **2019**, *101*, 753–772. [CrossRef]
16. Xu, Z.; Zhang, J.; Lv, K. Operation scale, land tenure and inter-temporal agricultural tech-nology adoption—Taking straw returning directly to the field as an example. *Chin. Rural Econ.* **2018**, *399*, 61–74.
17. Zhao, L.; Cai, S. Analysis of pesticide cost savings and grain production increase effect of IPM technology adopted by late rice farmers. *Chin. Rural Econ.* **2013**, *341*, 78–87.
18. Chen, R.; Han, K.; Wei, W. The Influence of Hedonic Value and Learning Cost on New Product Adoption. *Sci. Sci. Manag. Sci. Technol.* **2012**, *33*, 121–128.
19. Liang, J.; Gao, K.; Gao, Q. Transaction Costs, Production Costs and Outsourcing of Agricultural Production-Based on the Adjustment Perspective of Farmland Endowment Effect. *Resour. Sci.* **2021**, *43*, 1589–1604.
20. Hu, B.; Su, Z. Research on Innovation Sustainability of Chinese Manufacturing Listed Companies—Based on Discrete Duration Model. *J. Harbin Univ. Commer.* **2014**, *139*, 77–87.
21. Du, Y.; Wang, L. Research on the Duration of China's Export Trade and Its Influencing Factors—Based on the Cloglog Model. *Sci. Res. Manag.* **2015**, *36*, 7.
22. Chen, Y.; Li, Y.; Zhou, S. Export Duration of Chinese Enterprises and Its Determinants. *Econ. Res.* **2012**, *47*, 48–61.
23. Sun, C.; Li, M.; Chen, J. Can the Improvement of Bilateral Political Relations Prolong the Duration of Export Products of Enterprises? *Int. Econ. Trade Explor.* **2022**, *38*, 4–24.
24. Li, D.; Chen, J.; Sun, C. Value-added tax reform and export duration of Chinese manufacturing enterprises. *Int. Trade Issues* **2022**, *474*, 73–89.
25. Khataza, R.R.; Doole, G.J.; Kragt, M.E.; Hailu, A. Information acquisition, learning and the adoption of conservation agriculture in Malawi: A discrete-time duration analysis. *Technol. Forecast. Soc. Chang.* **2018**, *132*, 299–307. [CrossRef]
26. Tiller, K.J.; Feleke, S.T.; Starnes, J.H. A discrete-time hazard analysis of the exit of burley tobacco growers in Tennessee, North Carolina, and Virginia. *Agric. Econ.* **2010**, *41*, 397–408.
27. Conley, T.G.; Udry, C.R. Learning about a new technology: Pineapple in Ghana. *Am. Econ. Rev.* **2010**, *100*, 35–69. [CrossRef]
28. Chen, Q. *Advanced Econometrics and Stata Application*, 2nd ed.; Higher Education Press: Beijing, China, 2014.
29. Beyene, A.D.; Kassie, M. Speed of adoption of improved maize varieties in Tanzania: An application of duration analysis. *Technol. Forecast. Soc. Chang.* **2015**, *96*, 298–307. [CrossRef]
30. Mcqueen, G.; Thorley, S. Bubbles, Stock returns and duration dependence. *J. Financ. Quant. Anal.* **1994**, *29*, 379–401. [CrossRef]
31. Evans, G.W. Pitfalls in testing for explosive bubbles in asset prices. *Am. Econ. Rev.* **1991**, *81*, 922–930.
32. Ding, W. The Duration Dependence Test of Speculative Bubbles—Shanghai Composite Index as a Sample. *East China Econ. Manag.* **2005**, 121–125. [CrossRef]
33. Zhou, A.; Liu, Y. Analysis on the Rational Bubbles in Shenzhen Stock Market and the Distribution Characteristics of Industry Sectors. *J. Tianjin Univ. Financ. Econ.* **2009**, *29*, 52–56+60.

34. Jia, R.; Lu, Q. The impact of migrant workers and women's decision-making on farmers' collective action participation—Taking the survey data of farmers in Shaanxi, Gansu, and Ningxia as examples. *Agric. Technol. Econ.* **2019**, *286*, 122–134.
35. Xiao, J.; Tan, S.; Cheng, M. An empirical study on farmers' willingness to respond to conservation tillage. *China Land Sci.* **2012**, *26*, 57–63.

agriculture

MDPI

Article

Calibration of Collision Recovery Coefficient of Corn Seeds Based on High-Speed Photography and Sound Waveform Analysis

Xinping Li *, Wantong Zhang, Shendi Xu, Fuli Ma, Zhe Du, Yidong Ma and Jing Liu

College of Agricultural Equipment Engineering, Henan University of Science and Technology, Luoyang 471000, China; 210321041644@stu.haust.edu.cn (W.Z.); ztwtant@163.com (F.M.); lk35eee@163.com (Z.D.); qllzp4@163.com (Y.M.); 210321041647@stu.haust.edu.cn (J.L.)
* Correspondence: lwzyyxs@163.com; Tel.: +86-13592065522

Abstract: Compared with the movement of corn seeds in the sowing machine, the movement in the threshing machine is more intense. The collision between corn seeds and threshing parts and other corn seeds will not only change the movement path of corn seeds in the threshing clearance but also cause damage to corn seeds. Therefore, when using discrete element simulation technology to optimize the critical components of corn threshing machinery, it is necessary to measure corn seeds' accurate collision recovery coefficient. However, when measuring the collision recovery coefficient between corn seeds, there will be multi-point collisions between corn seeds, affecting the measurement results' accuracy. In order to solve this problem, this study combined high-speed photography and the sound waveform of corn seed collision to eliminate the interference of the multi-point collision phenomenon and improve the accuracy of measurement results. According to the above test method, the contact parameters of corn seeds were measured. Finally, the corn–corn rolling friction coefficient and corn–PMMA rolling friction coefficient were 0.0784 and 0.0934, respectively. The corn–corn static friction coefficient was 0.32, and the corn–PMMA static friction coefficient was 0.445. The corn–corn collision recovery and corn–PMMA collision recovery coefficients were 0.28 and 0.62, respectively. After that, the method's reliability and the measurement results' accuracy were verified by the plane collision test and repose angle test.

Keywords: corn seeds; collision recovery coefficient; static friction coefficient; mult-point collision; sound waveform; characteristic dimensions

Citation: Li, X.; Zhang, W.; Xu, S.; Ma, F.; Du, Z.; Ma, Y.; Liu, J. Calibration of Collision Recovery Coefficient of Corn Seeds Based on High-Speed Photography and Sound Waveform Analysis. *Agriculture* **2023**, *13*, 1677. https://doi.org/10.3390/agriculture13091677

Academic Editors: Muhammad Sultan, Redmond R. Shamshiri, Md Shamim Ahamed and Muhammad Farooq

Received: 31 July 2023
Revised: 22 August 2023
Accepted: 23 August 2023
Published: 25 August 2023

1. Introduction

Corn is grown in many parts of the world and has the highest planting area and yield in China. According to China's National Bureau of Statistics, the country's corn output in 2021 was 272.55 million tons, and the sown area was 433.24 million hectares. The sowing and harvesting of corn are the two most critical links in the mechanization of corn production [1–4]. However, due to the complex external environment, it is impossible to clearly observe the movement and stress of corn in the above machinery, which can not provide favorable information for the optimization of sowing or harvesting machinery. Discrete element simulation technology can simulate the stress and movement of corn in the above machinery and provide a theoretical reference for the structural design and optimization of critical components [5–9].

The most commonly used software is EDEM when using discrete element technology to optimize corn sowing and harvesting machinery. When using EDEM for simulation, it is necessary to set corn seeds' intrinsic and contact parameters [10–13]. The intrinsic parameters of corn are consistent with those obtained by physical experiments. However, due to the difference between the shape of the generated corn seed model and the actual corn seed, there are errors between the contact parameters of corn seeds and the actual measured values, so it is necessary to calibrate the contact parameters of corn seeds. According to the energy conservation theory and high-speed photography technology, Cui et al. [14] calculated the rolling friction coefficient between corn seeds and polymethyl methacrylate, steel, and corn seeds. Han et al. [15] glued rice onto a plate that could adjust the angle and made a particle plate. Then, the coefficient of static friction between rice was measured using the inclined plane method. Wang et al. [16] calibrated the rolling and static friction coefficients between corn seeds by combining physical and simulation tests. Based on the repose angle test data, a mathematical regression model obtained the rolling and static friction coefficients between corn seeds. Li et al. [17] used the Hertz–Mindlin with JKR Cohesion contact model to calibrate the relevant parameters of the clayey black soil with two moisture contents through the repose angle simulation test. By combining physical tests and simulation tests, Yu et al. [18] obtained the collision recovery, static, and rolling friction coefficients between panax notoginseng seeds. Through physical tests, Xing et al. [19] obtained several contact parameters of latosol particles in Hainan hot areas. Then, they applied the Plackett–Burman and Box–Behnken tests to obtain the best combination of parameters. When the response surface optimization method is used to calibrate the particle contact parameters, there is a problem in that the calibration parameters are distorted due to improper selection of factor zero level. To solve this problem, Zhang et al. [20] established a linear function of the zero level of variable range and the measured value during factor calibration. They obtained the best combination of static and rolling friction coefficients between corn seeds through the repose angle simulation test. When calibrating the contact parameters of camellia seeds, Ding et al. [21] adopted a BP artificial neural network based on a genetic algorithm to optimize the contact parameters.

Currently, most research on the discrete element parameter calibration of corn seeds is to optimize the critical components of corn sowing machinery through discrete element simulation technology. In sowing machinery, the movement of corn seeds is mainly rolling and sliding. The lifting method is the most commonly used in measuring the repose angle of corn seeds. In the process of using the lifting method, the motion of corn seeds is also mainly rolling and sliding, and the collision between corn seeds is relatively rare. Therefore, the rolling and static friction coefficients of corn seeds significantly influence the repose angle. When the optimum contact parameters of corn seeds obtained using the above method are applied to the discrete element simulation of the sowing machinery, the simulation results can maintain a high consistency with the physical tests. However, in the working process of the threshing machine, the movement of corn seeds is more intense, and the corn seeds will collide with the threshing parts and other corn seeds when it falls off the corn cob. These collisions will not only affect the movement of corn seeds in the threshing clearance but also cause damage to corn seeds. Therefore, in the study of low-damage threshing machinery, it is a very effective means to use discrete element simulation technology to analyze the stress and movement of corn seeds in threshing machinery to optimize its critical components. So, due to the above phenomena, the collision recovery coefficient of corn seeds will significantly impact the simulation results in the simulation process. Therefore, we need to measure corn seeds' exact collision recovery coefficient to improve the simulation results' accuracy.

The methods to measure the collision recovery coefficient of corn seeds include the inclined plane (plane) impact method and the single pendulum impact method. When the collision recovery coefficient between corn seeds is measured using the inclined plane (plane) collision method, the multi-point collision occurs due to the irregular surfaces of corn seeds, which reduces the accuracy of the measurement results. When the single pendulum collision method is used to measure the collision recovery coefficient between corn seeds, it is necessary to fix two corn seeds on the rope, then release one corn seed from a certain height to make it collide with another corn seed, and finally calculate the collision recovery coefficient between corn seeds according to the position changes of the two corn seeds. However, during the experiment, the tension of the rope on the corn seed will affect the measurement results, and the collision point of the two corn seeds will be artificially interfered with, which reduces the randomness of the experiment to a certain extent.

In order to solve the above problems, this study randomly selected a certain number of corn seeds. Then, these corn seeds were divided into three categories according to their surfaces and defined and measured the characteristic dimensions of these three categories of corn seeds. After measuring the characteristic dimensions of these three categories of corn seeds, the classification and statistical results of corn seeds are modified according to the relationship between the characteristic dimensions of corn seeds in order to establish an accurate discrete element simulation model. Then, the inclined plane method measured corn seeds' static friction coefficient. The energy conservation method based on high-speed photography measured corn seeds' rolling friction coefficient. The inclined plane collision method measured corn seeds' collision recovery coefficient. When measuring the collision recovery coefficient of corn seeds, the multi-point collision of corn seeds is removed by combining the high-speed photography technology and the sound waveform of the collision to improve the accuracy of the measurement results. Finally, the reliability of the measurement method and the accuracy of the measurement results are verified by the simulation and physical test of the plane collision test and repose angle test of corn seeds.

2. Materials and Methods

2.1. Materials

This study used the corn variety Zhengdan 958 to measure the contact parameters. The measurement of corn seed contact parameters includes the contact parameters between corn seeds and the contact parameters between the corn seeds and working parts. The material of the working part is polymethyl methacrylate (PMMA), so the contact parameters to be measured are the corn–corn static friction coefficient, the corn–corn rolling friction coefficient, the corn–corn collision recovery coefficient, the corn–PMMA static friction coefficient, the corn–PMMA rolling friction coefficient, and the corn–PMMA collision recovery coefficient. When measuring the contact parameters between corn seeds, it is necessary to use a particle plate made of corn seeds (as shown in Figure 1). When testing the contact parameters between corn seeds, the particle plate is installed on the test device to determine the contact parameters.

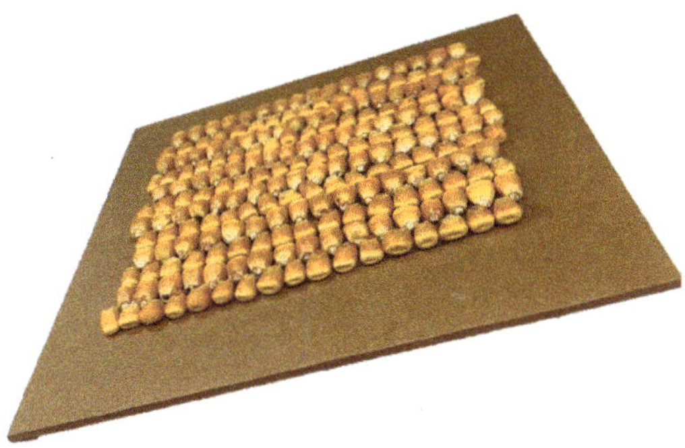

Figure 1. Particle plate.

2.2. Methods

In this study, the proposed measurement method's reliability and the measurement results' accuracy were verified by the plane collision test and repose angle test.

In the plane collision test, the particle plate made in advance was placed on the table, and a corn seed was selected to fall freely from a certain height. The high-speed camera was used to shoot the falling process of the corn seed and record the maximum height of the corn seed after the first collision (Figure 2a). The lifting method was used to measure the repose angle of corn seeds, and the test device is shown in Figure 2b. Place the bottomless PMMA cylinder on the horizontally fixed PMMA bottom plate, select six hundred corn seeds into the cylinder, and lift the PMMA cylinder at a speed of 0.04 m·s^{-1}.

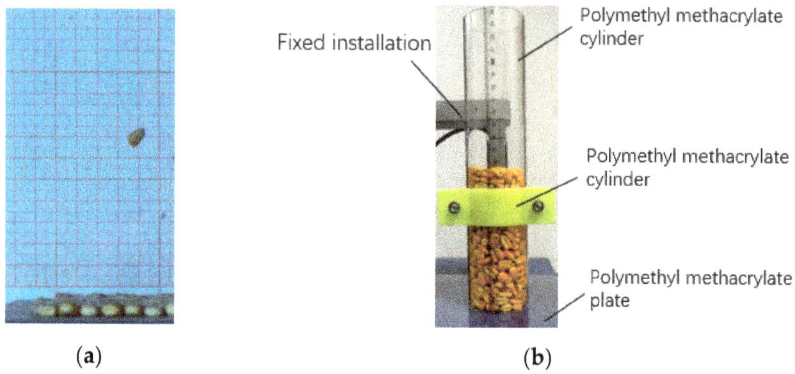

Figure 2. Validation test: (**a**) Plane collision test; (**b**) Test device for repose angle test.

3. Results and Discussion

3.1. Determination of Intrinsic Parameters of Corn Seeds

3.1.1. Density and Moisture Content of Corn Seeds

The corn seed variety was Zhengdan 958, which was selected for the experiment, and the density of the corn seeds was 1197 kg·m^{-3}, measured via the drainage method. The moisture content is measured according to the material moisture measurement method specified in the national standard GB/T3543.6. One hundred grams of corn seeds are randomly selected and dried in a constant temperature drying oven. The weight of the corn seeds after drying is measured, and the moisture content of the corn seeds is calculated by formula (1):

$$\omega = \frac{m_1 - m_2}{m_1} \times 100\% \quad (1)$$

where ω is the moisture content; m_1 is the mass of the corn seeds before drying and m_2 is the mass of the corn seeds after drying.

The above process was repeated five times, and the average moisture content of the three groups of corn seeds was 11.7%.

3.1.2. Determination of Poisson's Ratio and Shear Modulus

Poisson's ratio of corn seed was 0.4, referring to the literature [22]. The shear modulus of corn seeds is measured using a universal testing machine and calculated according to the following formula:

$$\begin{cases} E = \frac{FL_1}{S(L_2 - L_1)} \\ K = \frac{E}{2(1+\mu)} \end{cases} \quad (2)$$

where E, K, and μ are the elastic modulus of corn seed, shear modulus of corn seed, and Poisson's ratio of corn seed; F and S are the maximum bearing capacity of corn seed at elastic deformation stage and the cross-sectional area at the center of the corn seed; L_1 and L_2 are the lengths of the corn seeds before and after compression.

A uniaxial compression test was carried out on corn seeds using a universal testing machine, and the impact location of corn seeds is shown in Figure 3. Repeat the experiment three times and calculate the average. The shear modulus of corn seed is 1.36×10^8 Pa.

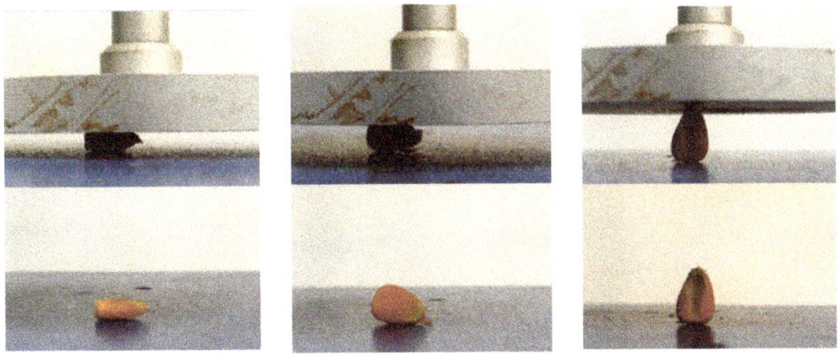

(a) (b) (c)

Figure 3. Impact locations of corn seeds: (**a**) impact the ventral part of the corn seed; (**b**) impact the side of the corn seed; (**c**) impact the top surface of the corn seed.

3.1.3. Determination of Characteristic Dimensions of Corn Seeds

Because the surfaces of corn seeds significantly influence the contact parameters of corn seeds [23], the selected corn seeds are divided into flat, quasi-conical, and quasi-cylindrical shapes according to their surfaces, and the characteristic dimensions of these three types of corn seeds are measured. The definition method of corn seed characteristic dimensions is shown in Figure 4.

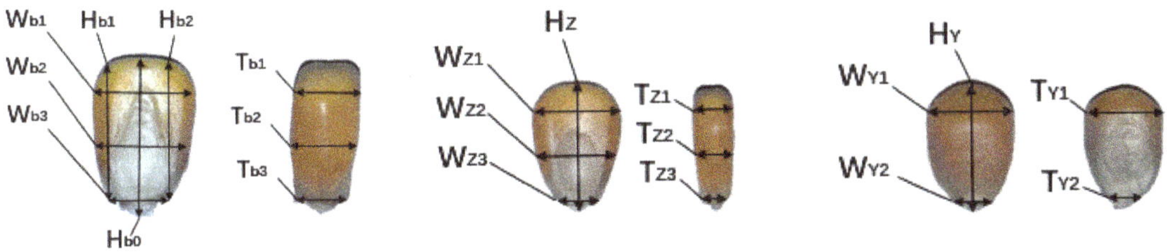

(a) (b) (c)

Figure 4. Characteristic dimensions definition diagram of corn seed: (**a**) flat; (**b**) quasi-conical; (**c**) quasi-cylindrical.

Three thousand corn seeds were randomly selected and classified according to the above shapes. The proportion of flat, quasi-conical, and quasi-cylindrical corn seeds was 75.7%, 17.6%, and 6.7%, respectively. The corn seeds were measured according to the outline dimensions of the corn seeds defined in Figure 4. Two hundred flat and two hundred quasi-conical corn seeds were randomly selected for measurement. All quasi-cylindrical corn seeds were measured, and the measurement results are shown in Table 1.

As can be seen from the measurement results in Table 1, the ratio between W_{b3} and W_{b2} of flat corn seeds is 0.553, and the ratio between W_{Z3} and W_{Z2} of quasi-conical corn seeds is 0.485. This study takes 0.5 as the standard value. The corn seed is flat when the ratio between the width of the bottom end of the corn seed and the width of the middle position is greater than or equal to 0.5. Otherwise, it is quasi-conical. According to the above conclusions, the corn seeds were classified again, and the proportion of each type of corn seed was counted. The proportion of flat, quasi-conical, and quasi-cylindrical corn seeds was 77.9%, 15.4%, and 6.7%, respectively. The modification results of characteristic dimensions of corn seeds are shown in Table 2.

Table 1. Summary of corn seed characteristic dimensions measurement results.

Shape	Characteristic Dimensions	Mean Value (mm)
Flat	W_{b1}	8.21
	W_{b2}	7.77
	W_{b3}	4.30
	H_{b0}	12.77
	H_{b1}	11.56
	H_{b2}	11.50
	T_{b1}	4.17
	T_{b2}	4.29
	T_{b3}	3.76
Quasi-conical	W_{Z1}	7.80
	W_{Z2}	7.21
	W_{Z3}	3.50
	H_Z	5.07
	T_{Z1}	4.83
	T_{Z2}	3.10
	T_{Z3}	12.23
Quasi-cylindrical	W_{Y1}	7.05
	W_{Y2}	6.54
	H_Y	4.08
	T_{Y1}	3.62
	T_{Y2}	10.28

Table 2. Modification results of characteristic dimensions of corn seeds.

Shape	Characteristic Dimensions	Mean Value (mm)
Flat	W_{b1}	8.14
	W_{b2}	7.83
	W_{b3}	4.37
	H_{b0}	12.71
	H_{b1}	11.57
	H_{b2}	11.45
	T_{b1}	4.14
	T_{b2}	4.29
	T_{b3}	3.78
Quasi-conical	W_{Z1}	7.76
	W_{Z2}	7.20
	W_{Z3}	3.32
	H_Z	5.02
	T_{Z1}	4.80
	T_{Z2}	3.04
	T_{Z3}	12.18
Quasi-cylindrical	W_{Y1}	7.05
	W_{Y2}	6.54
	H_Y	4.08
	T_{Y1}	3.62
	T_{Y2}	10.28

3.2. Determination of Static and Rolling Friction Coefficient of Corn Seeds

3.2.1. Static Friction Coefficient

The static friction coefficient is measured using the inclined plane sliding method [24,25], and the measurement principle is shown in Figure 5. Place the corn seed on a test plate in a horizontal position, slowly increase the angle between the test plate and the horizontal plane and record the angle when the corn seed appears to slide. When the corn seed first appears to slide, the force balance equation on the test plate is as follows:

$$\begin{cases} F_1 = G\sin\alpha = f \\ F_2 = G\cos\alpha = N \\ f - \mu F_2 \end{cases} \tag{3}$$

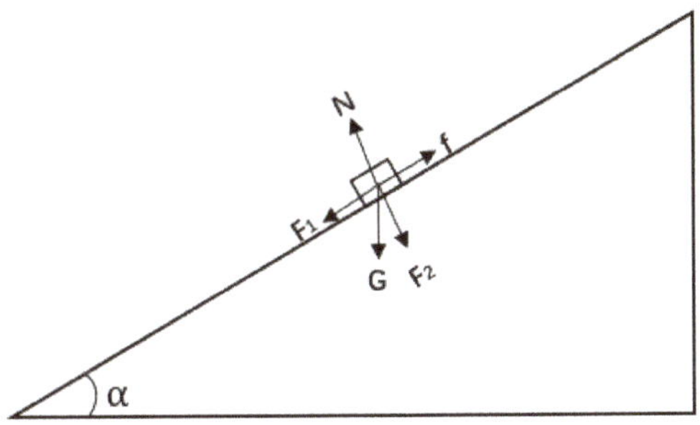

Figure 5. Principle diagram of static friction coefficient test.

The formula for calculating the static friction coefficient of corn seeds on the test plate is

$$\mu_1 = \frac{f}{F_2} = \frac{F_1}{F_2} = \frac{mg\sin\alpha}{mg\cos\alpha} = \tan\alpha \tag{4}$$

where F_1, F_2, f, and N are tensile force, the pressure of the corn seed on the test plate, static friction force between the corn seed and the test plate and support force of test plate on corn seed; μ_1 is coefficient of static friction between corn seed and test plate; α is angle between test plate and horizontal plane.

Figure 6 shows the contact parameter measuring device, which can be installed with different sizes and materials of the test plate. In this study, the test plate is a PMMA plate with a side length of 200 mm and a thickness of 3 mm. Place the corn seed in the center of the test plate and place the angle meter on the right side of the test plate. Slowly turn the test plate, use a high-speed camera to record the image of the moment when the corn seed appears to slide, and record the angle shown by the angle meter.

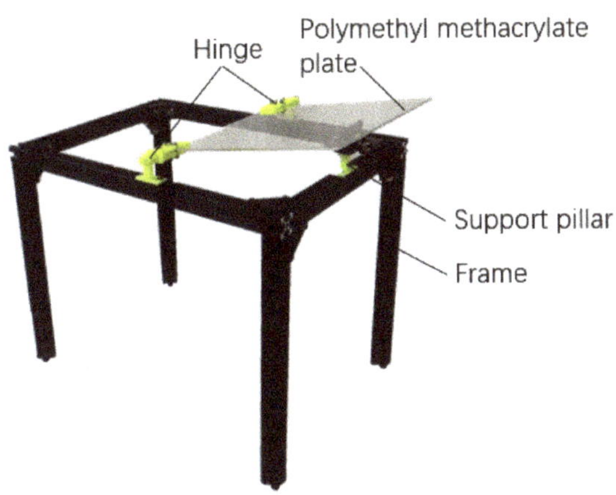

Figure 6. Contact parameter measuring device.

The static friction coefficient of corn–PMMA was calculated according to formula (4). When the coefficient of static friction between corn seeds is measured, the test plate is replaced with a prepared particle plate, and the test process is repeated. The two tests were repeated fifteen times, respectively, and the average value was obtained: the corn–PMMA static friction coefficient was 0.445, and the corn–corn static friction coefficient was 0.32.

3.2.2. Rolling Friction Coefficient

According to the energy conservation method, the rolling friction coefficient was measured using the contact parameter measuring device in Figure 6, and the measurement principle was shown in Figure 7.

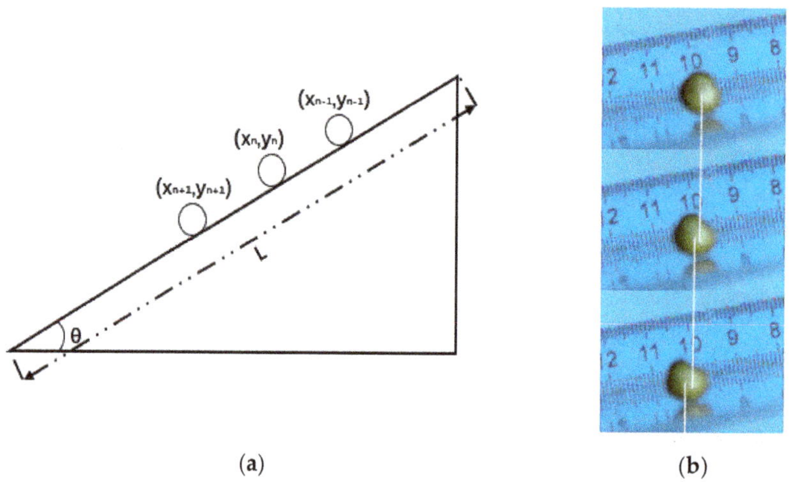

Figure 7. Rolling friction coefficient measurement method: (**a**) Principle diagram of rolling friction coefficient test; (**b**) The motion tracking of quasi-cylindrical corn seeds: Under the influence of the outer surface, flat and quasi-conical corn seeds display infrequent rolling. In contrast, quasi-cylindrical corn seeds are more prone to rolling. Therefore, experiments were conducted using quasi-cylindrical corn seeds. The X coordinate and Y coordinate of the corn seed are the cosine value and the sine value of the scale value of the position of the corn seed.

Fix the ruler on the test plate, and then turn the test plate to a certain angle to fix it. After that, the corn seed was placed on the test plate to make them naturally fall, and the rolling process of the corn seed was recorded with a high-speed camera (The shooting frame rate of the high-speed camera was set to 1000). The speed of the corn seed at the end of rolling is calculated according to the rolling image of the corn seed captured. Assuming that the rolling of corn seed ends at "n" frames, the time interval between "n" frames and the previous frame is 0.001 s, and the movement of corn seeds during this time interval is regarded as uniform rolling. The following formula can calculate the corn seed's speed at the rolling end:

$$V_n = \frac{k(\sqrt{(x_n - x_{n-1})^2 + (y_n - y_{n-1})^2} + \sqrt{(x_{n+1} - x_n)^2 + (y_{n+1} - y_n)^2})}{2\Delta t} \quad (5)$$

where V_n, k, and Δt are the speed at which the corn seed rolls at the end, the ratio factor of the actual size to the image size, and the time interval between two frames of the images.

According to the law of conservation of energy, the formula for calculating the proportion of energy lost by corn seed in the rolling process is as follows:

$$C_{f1} = \frac{U - E_k}{U} = k_1 \cot\theta_1 \quad (6)$$

where C_{f1}, U, and E_k are the proportion of energy loss caused by rolling friction to total energy, gravitational potential energy at the initial position of corn seed, and kinetic energy at the end of the rolling process of corn seed. It can be seen from Formula (6) that C_{f1} and $\cot\theta_1$ have a linear relationship.

The values of C_{f1} and $\cot\theta_1$ are calculated by selecting ten angles from 20°~42° according to Formulas (5) and (6). Each angle is repeated fifteen times, and the average value is taken. The calculated values of C_{f1} and $\cot\theta_1$ are shown in Table 3.

Table 3. The proportion of the energy loss caused by rolling corn seeds on the PMMA plate.

Angle (°)	$\cot\theta_1$	C_{f1}
20	2.747	0.273
22	2.475	0.251
25	2.145	0.257
27	1.963	0.181
30	1.732	0.241
32	1.600	0.165
35	1.428	0.151
37	1.327	0.131
40	1.192	0.135
42	1.111	0.137

According to the data in Table 3, the scatter plot of the proportion of energy loss during the rolling of corn seeds was drawn (Figure 8).

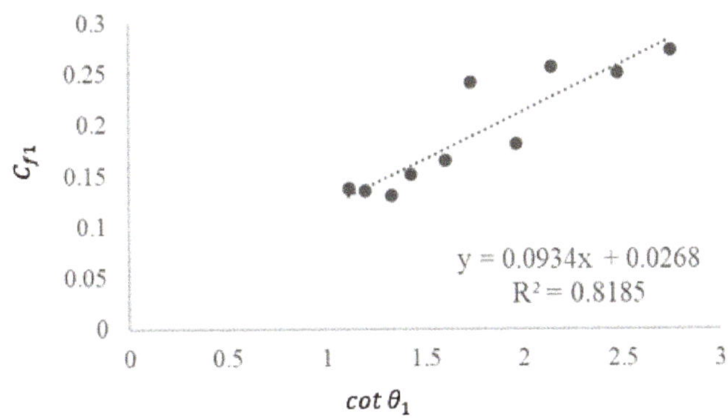

Figure 8. Corn–PMMA rolling friction coefficient.

The energy loss ratio C_{f1} of corn seeds rolling on the PMMA plate is fitted with the cotangent value of the angle between the test plate and the horizontal plane, and the equation is obtained as follows:

$$y = 0.0934x + 0.0268\left(R^2 = 0.8185\right) \quad (7)$$

The rolling friction coefficient between corn seeds was measured by replacing the test plate with a prepared particle plate. The same procedure was followed to obtain the energy loss ratio table (Table 4) and the energy loss scatter diagram (Figure 9) for rolling corn seeds on the particle plate.

Table 4. The proportion of the energy loss caused by rolling corn seeds on the particle plate.

Angle (°)	$\cot\theta_2$	C_{f2}
20	2.747	0.681
22	2.475	0.647
25	2.145	0.622
27	1.963	0.645
30	1.732	0.633
32	1.600	0.629
35	1.428	0.612
37	1.327	0.593
40	1.192	0.553
42	1.111	0.5

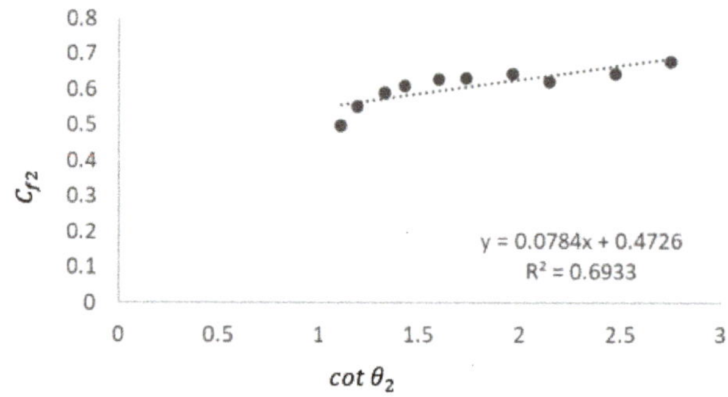

Figure 9. Corn–corn rolling friction coefficient.

After fitting the data, the equation is obtained as follows:

$$y = 0.0784x + 0.4726\left(R^2 = 0.6933\right) \quad (8)$$

3.3. *Determination of Collision Recovery Coefficient of Corn Seeds*

3.3.1. Multi-Point Collision of Corn Seeds

When measuring the corn–corn collision recovery coefficient, multi-point collision will occur between corn seeds due to the influence of the irregular surface of corn seeds and the angle of the collision of corn seeds. Figure 10 shows the process of single-point collision of corn seeds. Corn seeds bounce immediately after the collision (Figure 10a).

Figure 11 shows the process of the multi-point collision of corn seeds. Affected by the collision angle, the corn seed will swing or rotate after the first collision (Figure 11a), and the second collision will occur (Figure 11b). By comparing Figures 10e and 11e, it can be seen that multi-point collision between corn seeds will consume part of the kinetic energy of corn seed, thus reducing the height of corn seed's rebound and changing the motion trajectory of corn seed, reducing the accuracy of measurement results.

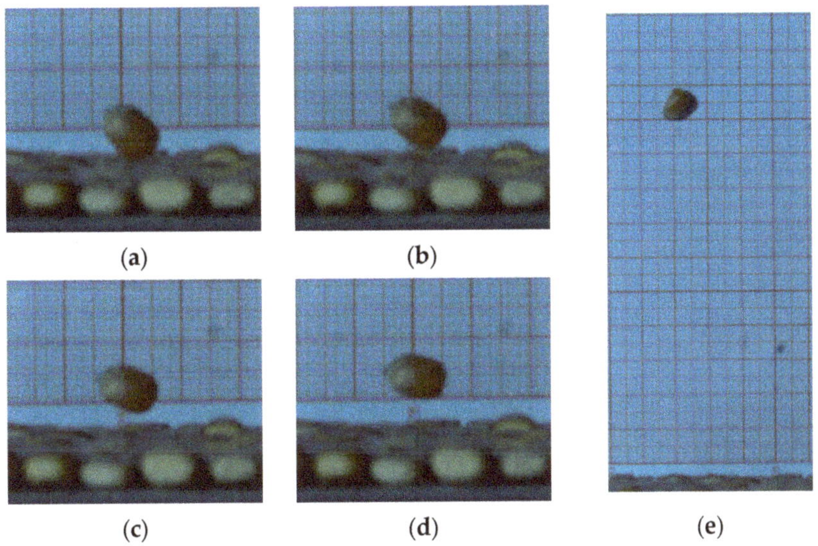

Figure 10. Single-point collision of corn seeds: (**a**) 1.232 s; (**b**) 1.233 s; (**c**) 1.234 s; (**d**) 1.235 s; (**e**) 1.380 s.

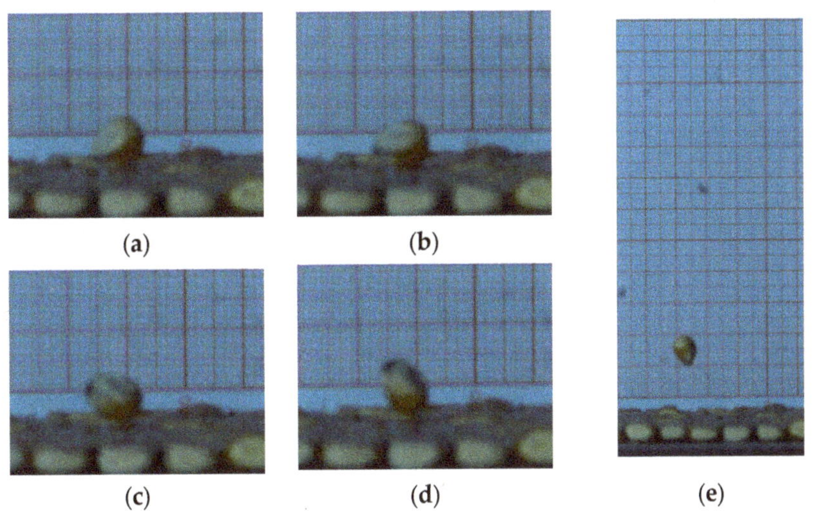

Figure 11. Multi-point collision of corn seeds: (**a**) 1.379 s; (**b**) 1.380 s; (**c**) 1.381 s; (**d**) 1.382 s; (**e**) 1.432 s.

In most cases, the multi-point collision can be identified in the pictures of collisions between corn seeds captured by high-speed cameras. However, due to the limitations of the structure of the test device, shooting angle, light, and the intensity of multi-point collision, some multi-point collision situations cannot be identified via high-speed photography. In order to solve this problem, this study combined high-speed photography and sound waveform to eliminate multi-point collision tests and improve the accuracy of measurement results. In the experiment, high-speed cameras were used to capture the images of corn seeds when they collided, and the sound of the collision of corn seeds was recorded. The recorded sound was then imported into Audacity, the sampling rate was selected according to the sampling device, and the waveform was enlarged. Then, the time node of the multi-point collision was found by the recorded sound waveform, and the multi-point collision was identified according to the change of sound waveform and the captured corn seed collision pictures. Figure 12 shows the high-speed photographic images of the single-point collision of corn seeds and the waveform of the collision sound. Figure 12a is the collision between corn seeds, corresponding to the sound waveform (The collision sound of corn seeds was processed using Audacity) in the time interval of Δt_1 in Figure 12c; Figure 12b shows the rebound of corn seeds after the collision, corresponding to the sound waveform in the time interval of Δt_2 in Figure 12c.

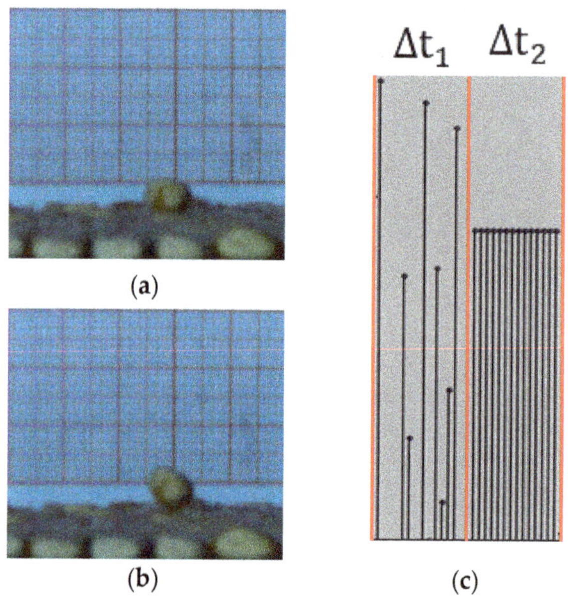

Figure 12. The sound waveform of the single-point collision: (**a**) Single-point collision; (**b**) the rebound of corn seeds after the collision; (**c**) sound waveforms generated by the collision between corn seeds.

Figure 13 shows the high-speed photographic images of the multi-point collision of corn seeds and the waveform of the collision sound. Figure 13a,b are two collisions of corn seeds, corresponding to the sound waveforms in Figure 13d's two-time intervals of Δt_1 and Δt_2 respectively. Figure 13c shows the rebound of corn seed after the second collision, corresponding to the sound waveform in the Δt_3 in Figure 13d.

In combination with Figures 12c and 13d, it can be seen that no matter the single-point or multi-point collision of corn seeds, each collision will cause changes in sound waveform. In the case of multi-point collision, the peak values of sound waveforms generated by different collisions were less different. After the single-point collision of corn seeds, the peak value of the sound waveform was positive and maintained the same value for a specific time interval. After the multi-point collision of corn seeds, the peak value of the sound waveform was zero and maintained the same value for a specific time interval.

The combination of high-speed photography and sound waveforms can not only identify multi-point collisions that are difficult to observe. Moreover, the time node of the first collision of corn seeds can be found faster through the sound waveform, thus reducing the time cost and improving the test efficiency.

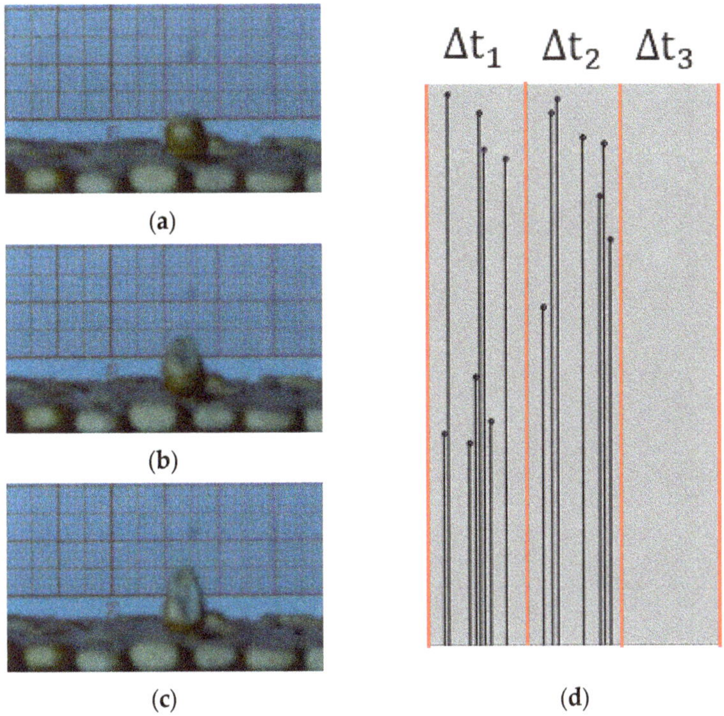

Figure 13. The sound waveform of the multi-point collision: (**a**) First collision; (**b**) Second collision; (**c**) The rebound of corn seeds after the second collision; (**d**) Sound waveforms generated by collisions between corn seeds.

3.3.2. Test Principle of Collision Recovery Coefficient

The collision recovery coefficient is measured using the inclined plane collision method [26], and the measurement principle is shown in Figure 14.

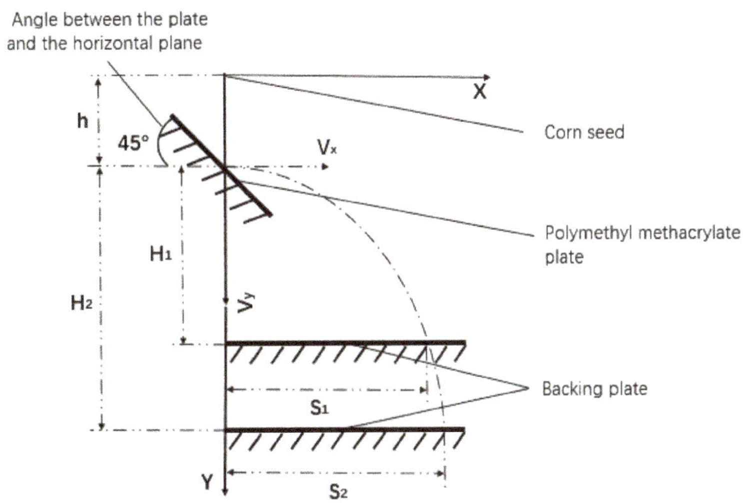

Figure 14. Principle diagram of inclined plate collision test.

The test plate is placed at an angle, and the corn seed falls from a certain height, collides with the test plate, and finally falls to the backing plate. The oblique motion of corn seed can be decomposed into uniform motion along the X direction and uniform motion with acceleration g along the Y direction. The motion equation is as follows:

$$\begin{cases} S_i = V_x t \\ H_i = V_y t + \frac{1}{2} g t^2 \end{cases} \tag{9}$$

where S_i, H_i (i= 1, 2) are the displacement in the X direction and Y direction after the collision between the corn seed and the test plate; V_x, V_y are the velocities in the X direction and Y direction after the collision between the corn seed and the test plate; g is the acceleration of gravity; t is the time of the corn seed's oblique throwing motion.

During the test, the height of the backing plate H_1 and H_2 were changed, and the displacement S_1 and S_2 of the corn seeds corresponding to the two heights were measured, respectively, in the X direction, and the following formula could be obtained:

$$\begin{cases} V_x = \sqrt{\dfrac{g S_1 S_2 (S_1 - S_2)}{2(H_1 S_2 - H_2 S_1)}} \\ V_y = \frac{H_1 V_x}{S_1} - \frac{g S_1}{2 V_x} \end{cases} \tag{10}$$

where S_1, S_2, H_1, and H_2 are the X and Y displacements corresponding to the two tests after a collision between corn seed and test plate, respectively.

According to Formulas (9) and (10), the calculation formula of the collision recovery coefficient of corn seed can be obtained as follows:

$$\begin{cases} C_r = \frac{V_n'}{V_n} = \dfrac{\sqrt{(V_x{}^2 + V_y{}^2)} \cos\left[45^\circ + \tan^{-1}\left(\frac{V_y}{V_x}\right)\right]}{V_0 \sin 45^\circ} \\ V_0 = \sqrt{2gh} \end{cases} \tag{11}$$

where C_r is collision recovery coefficient; $V_n{}'$ and V_n are the normal rebound velocity after collision and the normal approach velocity before collision.

According to the above principle, corn seeds' collision recovery coefficient was determined using a contact parameter measuring device (The test device is shown in Figure 15). The material of the test plate and the backing plate was PMMA. The angle between the PMMA plate and the horizontal plane is fixed at 45°, the backing plate is fixed at the lower part of the frame, and the corn seed falling point height is 5 cm from the test plate. Replace the test plate with a prepared particle plate when measuring the corn–corn collision recovery coefficient.

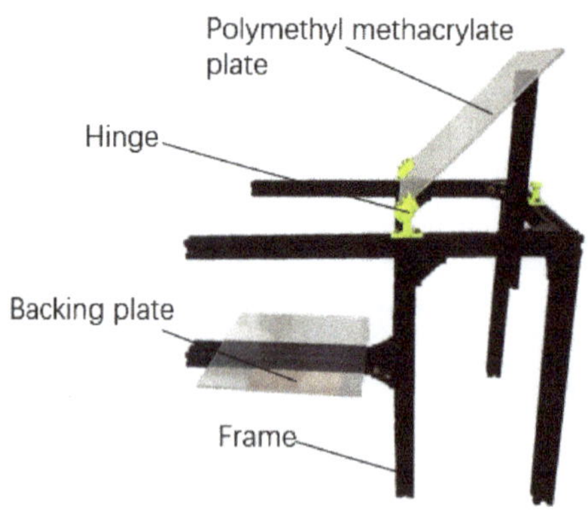

Figure 15. Collision recovery coefficient measuring device.

According to the above process, the two tests were repeated fifteen times, respectively, and the average value was obtained: the corn–PMMA collision recovery coefficient was 0.62, and the corn–corn collision recovery coefficient was 0.28.

3.4. Validation Test

3.4.1. Parameter Selection of Simulation Test

According to the classification of corn seeds in Section 3.1.3 and the measurement results of characteristic dimensions of these three types of corn seeds, the 3D model was established. The method of particle aggregate was used for modeling in EDEM2018 (Figure 16), and the Hertz–Mindlin with Bonding model was used for the contact model [27–29].

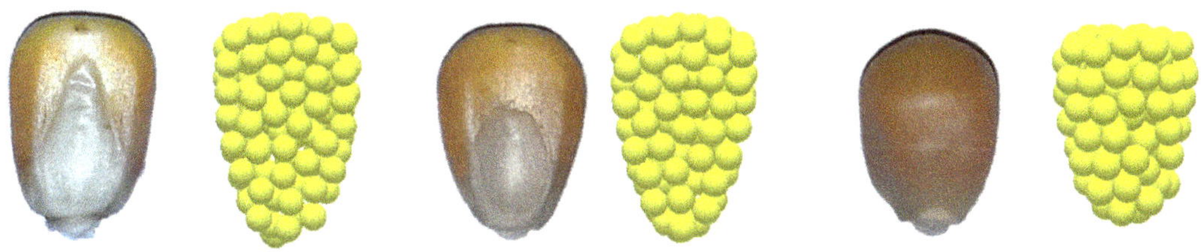

Figure 16. Simulation model of corn seeds: (**a**) flat corn seeds; (**b**) quasi-conical corn seeds; (**c**) quasi-cylindrical corn seeds.

Based on the above test results, the parameters required by the simulation test are shown in Table 5. All parameters in the table, except those obtained through experimental investigation in this study, were sourced from literature references [22].

Table 5. Parameters in DEM simulation.

Parameters	Value
Density of corn seed/(kg·m^{-3})	1197
Poisson's ratio of corn seed	0.4
Shear modulus of corn seed/Pa	1.36×10^8
Normal Stiffness of corn seed/(N·m^{-3})	3.54×10^9
Shear stiffness of corn seed/(N·m^{-3})	2.53×10^9
Critical normal stress of corn seed/Pa	1.1×10^7
Critical shear stress of corn seed/Pa	4.1×10^6
Bonded disk radius/mm	1
Density of PMMA/(kg·m^{-3})	1200
Poisson's ratio of PMMA	0.35
Shear modulus of PMMA/Pa	1.30×10^9
Corn–corn rolling friction coefficient	0.0784
Corn–PMMA rolling friction coefficient	0.0934
Corn–corn collision recovery coefficient	0.28
Corn–PMMA collision recovery coefficient	0.62
Corn–corn static friction coefficient	0.32
Corn–PMMA static friction coefficient	0.445

3.4.2. Plane Collision Test

In order to reduce the simulation test time, the plane collision method was used to carry out the verification test. In the simulation test, the contact parameters between the corn seed and the boundary are set to the above-measured contact parameters between the corn seeds. The results of the physical and simulation tests are shown in Figure 17. The experiment was divided into three groups, and each selected a corn seed type for the plane impact test. Release the corn seed 15 cm from the particle plate and record the maximum height at which the corn seed bounces back after the first collision (Using the method in Section 3.3.1 to eliminate the effect of multiple impacts on the test results). The results of the three groups of experiments were averaged, and the maximum height of the three types of corn seeds rebound was 5.95 cm (flat), 7.68 cm (quasi-conical), and 7.79 cm (quasi-cylindrical), respectively. The above process was repeated in EDEM2018, and the maximum rebound heights of three types of corn seeds were 6.17 cm (flat), 7.41 cm (quasi-conical), and 7.53 cm (quasi-cylindrical), respectively. The relative errors of the simulation and physical tests are 3.70%, 3.52%, and 3.34%, respectively. The relative errors of the three groups of tests are less than 5%, which proves that the simulation test is reliable.

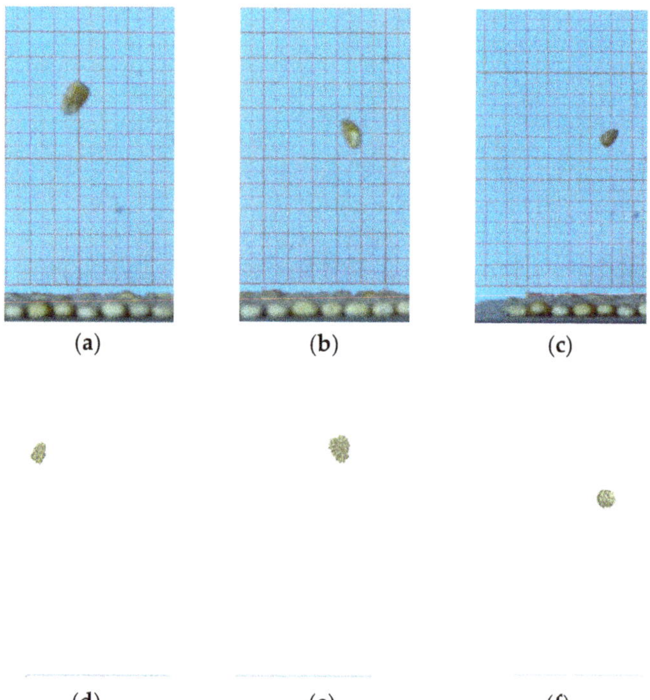

(a) (b) (c)

(d) (e) (f)

Figure 17. Plane collision test: (**a**) physical test of flat corn seed; (**b**) physical test of quasi-conical corn seed; (**c**) physical test of quasi-cylindrical corn seed; (**d**) simulation test of flat corn seed; (**e**) simulation test of quasi-conical corn seed; (**f**) simulation test of quasi-cylindrical corn seed.

3.4.3. Repose Angle Test

The repose angle test was carried out according to the method in Section 2.2. The particle pile formed by corn seeds was photographed (Figure 18a), then the photos were processed using Matlab, and the angle of the particle pile was obtained by linear fitting [30,31]. The above process was repeated fifteen times and obtained the average value. The repose angle of corn seeds was 31.38°.

(a) (b)

Figure 18. Repose angle test: (**a**) physical test; (**b**) simulation test.

After that, the three-dimensional model of the test device was imported into EDEM2018 software, and the parameters in Table 5 were substituted into EDEM for the simulation test. During the test, the number of corn seeds generated for each type was generated according to the final ratio in Section 3.1.3, and a total of six hundred corn seeds were generated. The simulation test was repeated fifteen times, and the average value was taken to obtain that the repose angle of corn seeds was 30.67°, the relative error between the simulated repose angle and the actual repose angle was 2.26%, and the relative error was less than 5%, indicating that the simulation test was reliable.

3.5. Discussion

(1) The multi-point collision of corn seeds will reduce the rebound height of corn seeds, but the rebound height of corn seeds is affected by various factors, such as the collision angle between corn seeds and the falling posture of corn seeds. Therefore, the rebound height of corn seeds cannot be used as the basis for judging the multi-point collision between corn seeds.

(2) After the collision between corn seeds, the more turns the corn seeds spin in the air, the lower the height of the corn seeds rebound.

(3) When using the lifting method to conduct the repose angle test of corn seeds, the high-speed camera is used to shoot the test process. At the beginning of the experiment, when the cylinder was lifted, the corn seeds were dispersed in all directions after losing the barrier of the cylinder. Because no other object was around to block the corn seeds, they directly collided with the bottom surface to a small degree (Figure 19a). As the experiment continued, the corn seeds piled up on the bottom surface, and the seeds inside the cylinder came into contact with the fallen seeds. When the cylinder is lifted, the corn seeds in the original cylinder will mainly slide and roll due to the obstruction of the seeds below (Figure 19b).

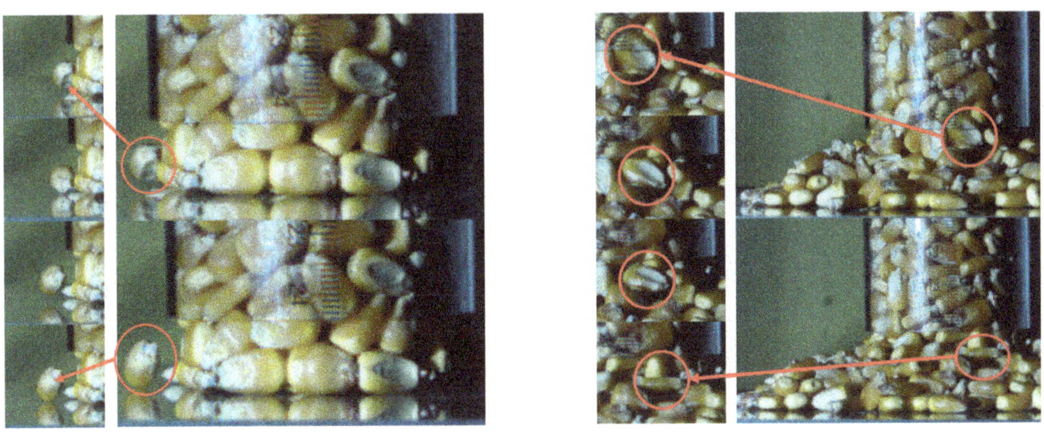

(a) (b)

Figure 19. Movement of corn seeds in repose angle test: (**a**) Collision: The movement of corn seed in the red circle is shown on the left side of the picture; (**b**) Slide: The movement of corn seed in the red circle is shown on the left side of the picture.

4. Conclusions

In this study, corn seeds were first divided into three categories according to their surfaces, and the proportion of each type of corn seed was counted. Then, these three types of seeds were measured according to the defined characteristic dimensions. The relationship between the characteristic dimensions of corn seeds modified the statistical and measurement results. Then, corn seeds' static and rolling friction coefficients were measured using the inclined plane and energy conservation methods, respectively. When measuring the collision recovery coefficient of corn seeds, the multi-point collision of corn seeds is identified by combining high-speed photography and sound waveform to improve the accuracy of measurement results. The results are as follows:

(1) According to the shape of the surfaces of corn seeds, they can be divided into three categories: flat, quasi-conical, and quasi-cylindrical. The number of flat, quasi-conical, and quasi-cylindrical corn seeds accounted for 77.9%, 15.4%, and 6.7%, respectively.
(2) The sound waveform's peak value after the corn seeds' single-point collision is positive and remains stable in a specific time interval. The peak value of the sound waveform after the multi-point collision of corn seeds is zero and maintains a stable value within a specific time interval.
(3) Through physical tests, the corn–corn rolling friction coefficient and corn–PMMA rolling friction coefficient were 0.0784 and 0.0934, respectively. The corn–corn static friction coefficient and corn–PMMA static friction coefficient were 0.32 and 0.445, respectively. The corn–corn collision recovery and corn–PMMA collision recovery coefficients were 0.28 and 0.62, respectively.
(4) The measurements are verified by plane collision and repose angle tests. The relative errors between the simulation test and physical test of the two verification methods are less than 5%, which proves that the technique combining high-speed photography and sound waveform is reliable.

Author Contributions: Conceptualization, X.L. and W.Z.; methodology, X.L. and W.Z.; investigation, S.X., F.M., Z.D. and J.L.; resources, Y.M.; data curation, J.L.; writing, original draft preparation, W.Z. and X.L. All authors have read and agreed to the published version of the manuscript.

Funding: This research was funded by the National Natural Science Foundation of China (52275245) and Henan Science and Technology Research Program (222103810041).

Institutional Review Board Statement: Not applicable.

Data Availability Statement: The data used to support the findings of this study are available from the corresponding author upon request.

Acknowledgments: The authors would like to thank their college and the laboratory, as well as gratefully appreciate the reviewers and editors who provided helpful suggestions for this manuscript.

Conflicts of Interest: The authors declare no conflict of interest.

References

1. Wang, K.; Xie, R.; Xue, J.; Sun, L.; Li, S. Comparison and analysis of maize grain commodity quality and mechanical harvest quality between China and the United States. *Int. J. Agric. Biol. Eng.* **2022**, *15*, 55–61. [CrossRef]
2. Li, S.; Wang, K.; Xie, R.; Hou, P.; Ming, B.; Yang, X.; Han, D.; Wang, Y. Implementing higher population and full mechanization technologies to achieve high yield and high efficiency in maize production. *Crops* **2016**, *4*, 1–6.
3. Ou, Y.; Fan, C.; Zhao, H.; Dong, J.; Jiao, Y. Present status and equipment research progress of maize full-mechanized production. *J. Chin. Agric. Mech.* **2022**, *43*, 207–214.
4. Yang, L.; Yan, B.; Zhang, D.; Zhang, T.; Wang, Y.; Cui, T. Research Progress on Precision Planting Technology of Maize. *Trans. Chin. Soc. Agric. Mach.* **2016**, *47*, 38–48.
5. Xu, Y.; Sun, Q.; Zhang, L.; Huang, W. Advances in discrete element methods for particulate materials. *Adv. Mech.* **2003**, *33*, 251–260.
6. Zhu, H.; Zhou, Z.; Yang, R.; Yu, A. Discrete particle simulation of particulate systems: A review of major applications and findings. *Chem. Eng. Sci.* **2008**, *63*, 5728–5770. [CrossRef]

7. Fleissner, F.; Gaugele, T.; Eberhard, P. Applications of the discrete element method in mechanical engineering. *Multibody Syst. Dyn.* **2007**, *18*, 81–94. [CrossRef]
8. Zeng, Z.; Ma, X.; Cao, X.; Li, Z.; Wang, X. Critical Review of Applications of Discrete Element Method in Agricultural Engineering. *Trans. Chin. Soc. Agric. Mach.* **2021**, *52*, 1–20.
9. Fu, H.; Wang, C.; Jin, C.; Yu, J. Development of digital design software platform AgriDEM for agricultural machinery parts. *Trans. Chin. Soc. Agric. Eng.* **2017**, *33*, 1–9.
10. Martin, C.; Bouvard, D.; Shima, S. Study of particle rearrangement during powder compaction by the discrete element method. *J. Mech. Phys. Solids* **2003**, *51*, 667–693. [CrossRef]
11. Wei, H.; Nie, H.; Li, Y.; Saxén, H.; He, Z.; Yu, Y. Measurement and simulation validation of DEM parameters of pellet, sinter and coke particles. *Powder Technol.* **2020**, *364*, 593–603. [CrossRef]
12. Liu, C.; Chen, X.; Zhang, W.; Wang, S. Study on the calibration process of parallel bonding meso-structure parameter in PFC numerical simulation. *Value Eng.* **2017**, *36*, 204–207.
13. Ma, Z.; Li, Y.; Xu, L. Summarize of particle movements research in agricultural engineering realm. *Trans. Chin. Soc. Agric. Mach.* **2013**, *44*, 22–29.
14. Cui, T.; Liu, J.; Yang, L.; Zhang, D.; Zhang, R.; Lan, W. Experiment and simulation of rolling friction characteristic of corn seed based on high-speed photography. *Trans. Chin. Soc. Agric. Eng.* **2013**, *29*, 34–41.
15. Han, Y.; Jia, F.; Tang, Y.; Liu, Y.; Zhang, Q. Influence of granular coefficient of rolling friction on accumulation characteristics. *Acta Phys. Sin.* **2014**, *63*, 4501–4507.
16. Wang, Y.; Liang, Z.; Zhang, D.; Cui, T.; Shi, S.; Li, K.; Yang, L. Calibration method of contact characteristic parameters for corn seeds based on EDEM. *Trans. Chin. Soc. Agric. Eng.* **2016**, *32*, 36–42.
17. Li, J.; Tong, J.; Hu, B.; Wang, H.; Mao, C.; Ma, Y. Calibration of parameters of interaction between clayey black soil with different moisture content and soil-engaging component in northeast China. *Trans. Chin. Soc. Agric. Eng.* **2019**, *35*, 130–140.
18. Yu, Q.; Liu, Y.; Chen, X.; Sun, K.; Lai, Q. Calibration and experiment of simulation parameters for Panax notoginseng seeds based on DEM. *Trans. Chin. Soc. Agric. Eng.* **2020**, *51*, 123–132.
19. Xing, J.; Zhang, R.; Wu, P.; Zhang, X.; Dong, X.; Chen, Y.; Ru, S. Parameter calibration of discrete element simulation model for latosol particles in hot areas of Hainan Province. *Trans. Chin. Soc. Agric. Eng.* **2020**, *36*, 158–166.
20. Zhang, R.; Zhou, J.; Liu, H.; Shi, S.; Wei, G.; He, T. Determination of Interspecific Contact Parameters of Corn and Simulation Calibration of Discrete Element. *Trans. Chin. Soc. Agric. Mach.* **2022**, *53*, 69–77.
21. Ding, X.; Li, K.; Hao, W.; Yang, Q.; Yan, F.; Cui, Y. Calibration of Simulation Parameters of Camellia oleifera Seeds Based on RSM and GA-BP-GA Optimization. *Trans. Chin. Soc. Agric. Mach.* **2023**, *54*, 139–150.
22. Li, F.; Wang, P.; Wang, Y.; Ge, Y.; Tang, K.; Li, D. Calibration of Discrete Element Parameters of Corn Coated Seeds Based on Stacking Test. *J. Agric. Sci. Technol.* **2022**, *24*, 97–107.
23. Joanna, W.; Marek, M.; Józef, H.; Jin, Y. Influence of grain shape and intergranular friction on material behavior in uniaxial compression: Experimental and DEM modeling. *Powder Technol. Int. J. Sci. Technol. Wet Dry Part. Syst.* **2012**, *217*, 435–442.
24. Liu, W.; He, J.; Li, H.; Li, X.; Zheng, K.; Wei, Z. Calibration of simulation parameters for potato minituber based on EDEM. *Trans. Chin. Soc. Agric. Mach.* **2018**, *49*, 125–135.
25. Zhong, J.; Tao, L.; Li, S.; Zhang, B.; Wang, J.; He, Y. Determination and interpretation of parameters of double-bud sugarcane model based on discrete element. *Comput. Electron. Agric.* **2022**, *203*, 107428. [CrossRef]
26. Teng, G.; Jia, Z.; Zhou, K. A theoretical model for the coefficient of restitution calculation of point impact. *Mach. Des. Res.* **2007**, *23*, 14–15.
27. Salot, C.; Gotteland, P.; Villard, P. Influence of relative density on granular materials behavior: DEM simulations of triaxial tests. *Granul. Matter* **2009**, *11*, 221–236. [CrossRef]
28. Kozhar, S.; Dosta, M.; Antonyuk, S.; Heinrich, S.; Bröckel, U. DEM simulations of amorphous irregular shaped micrometer-sized titania agglomerates at compression. *Adv. Powder Technol.* **2015**, *26*, 767–777. [CrossRef]
29. Wang, M.; Wang, W.; Yang, L.; Hou, M. Research of discrete element modeling method of maizekernel based on EDEM. *J. Henan Agric. Univ.* **2018**, *52*, 80–84.
30. Jia, X.; Chen, X.; Li, X. Experiment research on instantaneous natural repose angle of granular slope under different granularity and stack form. *Mod. Min.* **2015**, *31*, 25–27.
31. Feng, J.; Lin, J.; Li, S.; Zhou, J.; Zhou, Z. Calibration of discrete element parameters of particle in rotary solid state fermenters. *Trans. Chin. Soc. Agric. Mach.* **2015**, *46*, 208–213.

agriculture

Article

Development of Heat Stress Forecasting System in Mechanically Ventilated Broiler House Using Dynamic Energy Simulation

Jeong-Hwa Cho [1], In-Bok Lee [2,3,*], Sang-Yeon Lee [1], Se-Jun Park [1], Deuk-Young Jeong [1], Cristina Decano-Valentin [1,4], Jun-Gyu Kim [1], Young-Bae Choi [1], Hyo-Hyeog Jeong [1], Uk-Hyeon Yeo [5] and Seung-Jae Lee [6]

1 Research Institute for Agriculture and Life Sciences, College of Agriculture and Life Sciences, Seoul National University, 1 Gwanak-ro, Gwanak-gu, Seoul 08826, Korea
2 Department of Rural Systems Engineering, Research Institute for Agriculture and Life Sciences, Global Smart Farm Convergence Major, College of Agriculture and Life Sciences, Seoul National University, 1 Gwanak-ro, Gwanak-gu, Seoul 08826, Korea
3 Research Institute of Green Eco Engineering, Institute of Green Bio Science and Technology, Seoul National University, 1 Gwanak-ro, Gwanak-gu, Seoul 08826, Korea
4 Department of Agricultural and Biosystems Engineering, College of Engineering, Mariano Marcos State University, Batac 2906, Philippines
5 Agriculture, Animal & Aquaculture Intelligence Research Centre, Electronics and Telecommunications Research Institute, 218 Gajeong-ro, Yuseong-gu, Daejeon 34129, Korea
6 National Center for AgroMeteorology (NCAM), Seoul National University, 1 Gwanak-ro, Gwanak-gu, Seoul 08826, Korea
* Correspondence: iblee@snu.ac.kr

Citation: Cho, J.-H.; Lee, I.-B.; Lee, S.-Y.; Park, S.-J.; Jeong, D.-Y.; Decano-Valentin, C.; Kim, J.-G.; Choi, Y.-B.; Jeong, H.-H.; Yeo, U.-H.; et al. Development of Heat Stress Forecasting System in Mechanically Ventilated Broiler House Using Dynamic Energy Simulation. *Agriculture* **2022**, *12*, 1666. https://doi.org/10.3390/agriculture12101666

Academic Editors: Muhammad Sultan, Redmond R. Shamshiri, Md Shamim Ahamed and Muhammad Farooq

Received: 26 August 2022
Accepted: 7 October 2022
Published: 11 October 2022

Abstract: The internal rearing environment of livestock houses has become an important issue in the last few years due to the rapid increase in meat consumption. As the number of days of heat waves increase continuously, problems caused by abnormal weather changes steadily occurred. Thus, the main goal of this study is to develop a technology that can automatically calculate heat stress for livestock by considering weather forecast data. Specifically, a web-based heat stress forecasting system for the evaluation of heat stress in broilers was developed. The field experiments were carried out at the selected broiler house to measure and analyze the external weather, the internal environment, and the ventilation flow rate of fans used in tunnel ventilation. The developed model was validated by comparing the field and simulated thermal environment values. Based on a reliable model, Land-Atmosphere Modeling Package (LAMP) weather forecast data was used to show the stress index on the internal rearing environment with a heat stress index suitable for South Korea. When the users input the farm location, structure and equipment, and rearing information, users responded after receiving heat stress from the broiler raised in a mechanically ventilated broiler house.

Keywords: broiler house; building energy simulation; forecasting system; heat stress index; temperature humidity index (THI)

1. Introduction

Due to the increase in meat consumption, domestic livestock production continues to increase from KRW 19.1 trillion (USD 13.2 billion) in 2015 to KRW 20.3 trillion (USD 14.0 billion) in 2020, and the proportion of livestock production in agriculture and forestry also increased from 37.6% to 39.0% [1]. The poultry industry is divided into the broiler, a chicken raised for meat, and the laying hens, a chicken raised to produce edible eggs. The recent record showed that in Korea, the broiler industry accounts for 10.0% of the total livestock production in 2020. Moreover, in the same year, the per capita chicken consumption in Korea continuously increased from 13.4 kg in 2015 to 14.7 kg [2]. As the consumption of poultry products has steadily increased, the number of rearing birds per unit area has increased compared to the number of farm households. However, the current broiler houses lacked the production application of cutting-edge technologies such as ICT and lack of production. In these situations, the birds tend to have a higher mortality rate due to infections such as Avian Influenza (AI) during the cold season, and heat waves during the hot season.

The climate change scenario RCP (Representative Concentration Pathways) 8.5, which was a case where carbon emission reduction was not implemented as in the current trend, predicted that the heat wave in South Korea will increase by about three times in 2050. In fact, the temperature analysis reported Korea Meteorological Administration (KMA) in 2020 showed that the average annual temperature in the 2010s continuously increased to 13 °C compared to the 1980s (12.2 °C), the 1990s (12.6 °C) and the 2000s (12.8 °C) [3]. This trend reduces immunity to livestock and increases the incidence of diseases, increasing mortality rate and a decrease in productivity. A heat wave, by definition, refers to abnormal climates causing very hot weather. In South Korea, the KMA issued a 'heat wave warning' if the daily maximum temperature is predicted to last more than 33 °C for two days or more, and a 'heat wave alert' if it is predicted to last more than 35 °C. The number of days where heatwaves occurred also tends to increase along with the rise in average temperature in South Korea, and in particular, the average number of heatwave days has changed from 10 to 15.5 as of 2010. The highest temperature in 2018 in South Korea was 41 °C, and the average number of days of heatwave was 31.5 days [4].

So, a response to climate change is necessary for the livestock industry. Whereas, the average number of heat wave days in the Jeollabuk-do region was found to be 39.9 days, which accounts for 27.3% of domestic meat production [5]. The Jeollabuk-do region, where a lot of livestock were raised, has a high mortality rate of 42% compared to other regions. As an example, over 420,000 (chickens account for 97%, ducks 2.5%, and pigs 0.5%) livestock died in a 5 days continuous heat wave in July 2018 [6]. Most of this livestock was raised in a very dense internal rearing environment, so they have weak basic immunity and are vulnerable to stress, which often leads to death when the temperature rises. Specifically, the temperature adaptability of chickens is very low compared to other livestock, so the problem of heat stress in the summer poses a serious threat to the poultry industry. Thus, scientific actions against climate change are necessary to reduce the production risk of chickens.

In the situation where the climate change problem persists, a web-based heat stress forecasting simulator was developed for the evaluation of heat stress in the broiler house.

There are recent studies that continue to develop a heat stress index for poultry directly affected by high temperatures. Usually, most studies consider regional characteristics and the thermal status of chickens by correcting the coefficient of previously developed indices. In addition to considering the air temperature and humidity inside the house, activity [7], body weight gain [8], black glove temperature [9], and feed requirement [10] were additionally considered. Ha et al. predicted a heat stress index using external weather conditions such as air temperature and enthalpy [11]. These latest studies may better represent the thermal environment of chickens, but they have the disadvantage of not being able to preemptively respond because they need to understand the heat stress of chickens through real-time monitoring. In the study, through field experiments, factors causing heat stress in livestock were analyzed, and the model was verified using internal and external thermal environment values. Based on a reliable model, weather forecast data were used to show the degree of stress on the internal rearing environment with a heat stress index suitable for South Korea, and a system development study was conducted to provide information to users by web service. Previously, related users responded after receiving heat stress from chickens, but this study is meaningful in that it can automatically calculate the heat stress of livestock according to the weather forecast and provide it to farmers in advance.

2. Materials and Methods

The overall process of establishing a heat stress forecasting system for broilers to improve the internal rearing environment exposed to abnormal climates is shown in Figure 1. First, a broiler house in Jeollabuk-do was selected as the experimental site, and various energy-related factors were collected through field experiments. In addition, the internal environment, according to local weather data and actual fan ventilation rate were measured. The structure of the house, the type of envelope, and the growth environment conditions were also investigated to design the energy model of the house. The model was designed to simulate the internal and external flow of energy according to the time step by considering the sensible heat and latent heat generated by the chickens. The accuracy of the model was secured through statistical indicators using the previously measured field test data.

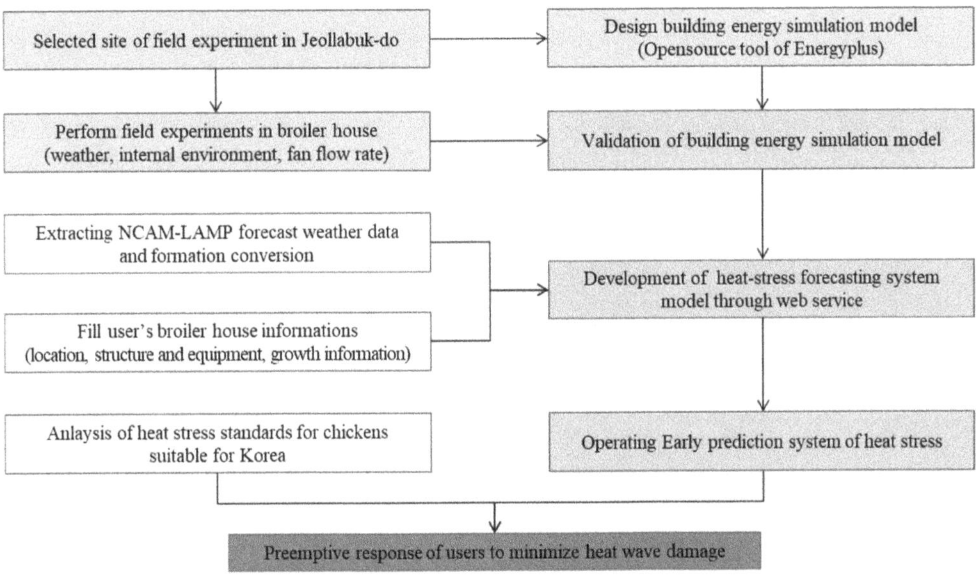

Figure 1. Research flow for heat stress energy model and forecasting system development procedure.

Next, to establish a service that automatically provides chicken heat stress information to users, a forecasting system was developed to calculate an energy model according to the user's farm information by receiving high-resolution weather forecast data from National Center for AgroMeteorolgy—Land Atmosphere Modelling Package (NCAM-LAMP). The input variables, such as livestock structure, environmental facility conditions, and livestock growth information, are needed for the energy model. And heat stress index was calculated one day and two days after the forecast. Among the heat stress index developed by researchers in the past, the heat stress index suitable for South Korea was selected.

2.1. Research Site

The target broiler house is located in Deokho-ri, Yeonggwang-eup, Yeonggwang-gun, South Korea (Latitude: 35.277, Longitude: 126.511, Elevation 85 m). The building was 18 m in width, 100 m in length, 3 m in side height, and 6 m in ridge height, and 34,000 broilers were raised every rearing cycle. The duration of one rearing cycle was about one month, and the average initial and shipment weights were 45 g and 1.7 kg, respectively. The average daily growth rate during the rearing cycle was about 50 g to 60 g.

The target broiler house used negative pressure forced ventilation where cross ventilation was applied in winter, tunnel ventilation in summer, and another ventilation was used by changing the two ventilators in order to remove the heat remaining in the house in the changing season when the outside air is lower than in summer. Cross ventilation is a method in which an exhaust fan was installed on one side along the side wall of the house, and an inlet was installed on the other side so that the air flows in the width direction of the house. On the other hand, in the tunnel ventilation, exhaust fans were installed on one side along the length of the house and an inlet is installed on the other side. Due to the high ventilation flow rate, it was mainly applied as a ventilation method in summer due to the effect of lowering the effective temperature. In this ventilation method, the temperature difference between the air inflow side and the discharge side was relatively large. In tunnel ventilation, the inlet for external air with a dimension of 23.5 m × 1.25 m was installed at the side wall. The tunnel exhaust fan (Euroemme EM50; Munters, Sweden) has a propeller diameter of 1.27 m and a maximum flow rate of 42,125 m^3/h. A windbreak was installed 3.0 to 4.0 m in front of the tunnel exhaust fan to prevent dust and odors emitted outside the facility from spreading to nearby villages or damaging crops. Figures 2 and 3 showed the external view and internal structure of the target farm.

Figure 2. Outside and inside the field test facility for experiment (Latitude: 35.277, Longitude: 126.511, Yeonggwang-gun, Jeollabuk-do, Republic of Korea).

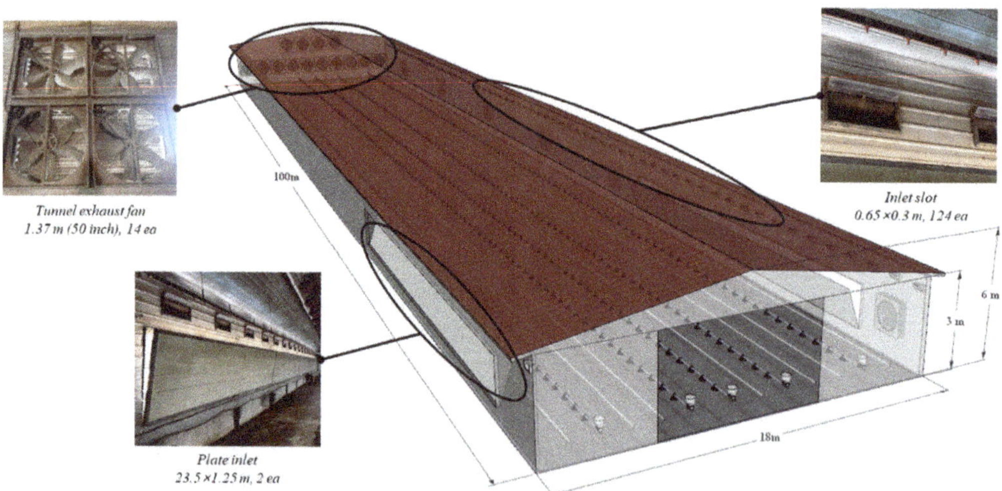

Figure 3. Test target facility size and internal facility status.

2.2. *NCAM-LAMP Weather Forecast Data*

In order to evaluate the heat stress of livestock, it is necessary to calculate the internal environment through dynamic energy simulations. In accordance with ISO 13790 process

(2008), the meteorological data was used in advance to check the local climatic condition of the building before entering the building information [12]. Buildings, including livestock houses, verified the performance from the planning stage through the energy performance evaluation program. The weather conditions had the greatest influence on the calculation of energy loads aimed at appropriate indoor conditions. The heating and cooling demand (kWh/m^{2}a) may change depending on the outdoor temperature, wind speed, wind direction, and precipitation, which has a significant impact on heat loss due to the outer wall, roof, floor, window, and door of the building. These factors were found to have a direct effect on heat loss according to the amount of ventilation. In addition, the amount of solar radiation was directly related to the amount of energy that the building can acquire and acted as an important variable in determining the energy load. Therefore, for precise building energy analysis, energy analysis should be performed based on meteorological data within the target area. The meteorological data mentioned here was basic input data for the evaluation of building energy and building environmental performance and was various data of 8760 h—365 days.

The NCAM developed the agricultural and forestry support numerical forecasting model based on the Weather Research and Forecasting (WRF)/Noah-MP Model, which regularly produces and provides weather, vegetation, and soil forecasting information with a high resolution of less than 1 km in South Korea. This agricultural support numerical forecast model was built in the Land Atmosphere Modelling Package (LAMP) (Figure 4) [13]. To support customized weather forecasting, medium to long-range weather/climate information in agriculture, forestry, and livestock industries are produced, and high-resolution decision-making information on the target area is provided to local governments and individual workers (e.g., Hong et al., 2021 [14]). LAMP produces forecast data not only for traditional meteorological variables such as air pressure, temperature, wind, humidity, and precipitation but also for various variables such as soil temperature and soil moisture, carbon dioxide, PM_{10}, and $PM_{2.5}$. Therefore, it is a practical option to use the LAMP meteorological data in order to inform users of the possibility of heat damage in advance and to respond preemptively.

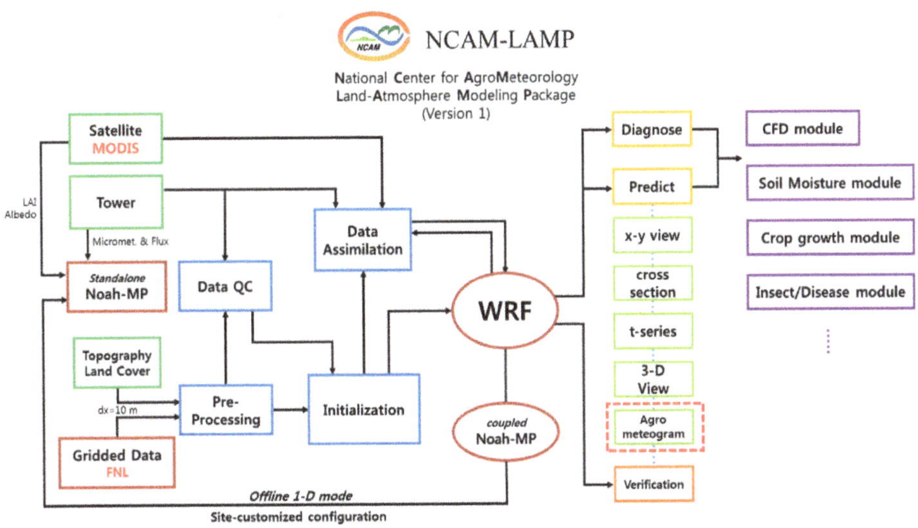

Figure 4. Framework of the medium-range weather prediction model implemented in the National Center for AgroMeteorology-Land-Atmosphere Modeling Package (NCAM-LAMP). The LAMP diagnosis and prediction data are used as high-resolution meteorological input data for poultry heat stress model in this study.

The LAMP was validated for the temperature and relative humidity at a height of 2 m and wind speed at a height of 10 m in the Jeollabuk-do region. Data from 19 AWS branches located in the region were collected, and model grid data were found nearby, and data were performed at 1-h intervals from 23 August 2020 to 19 September 2020. The Root Mean Square Error (*RSME*) of temperature, relative humidity, and wind speed were analyzed as 2.52, 11.0, and 3.54, respectively. The LAMP is now produced twice a week, and the forecast period per episode is 12 days corresponding to the medium-range weather forecast. In addition, the temporal resolution is 30 min, and the spatial resolution is about 800 m.

2.3. Building Energy Simulation (BES)

Quantitative evaluation of building energy load is required to prepare measures for the energy demand and efficient use of buildings. In particular, in the case of livestock facilities, the insulation properties are significantly lower than that of general residential houses. The internal environment may be changed by a heat source having a heat capacity, such as wall insulation material, and the heat transfer process reacts sensitively to weather data. Therefore, the building energy simulation (BES) was used to calculate the thermal energy flow of the broiler house using a numerical analysis method through the dynamic load calculation method that changes with time. The BES can evaluate the overall performance of various design proposals in the architectural field. In addition, it can be applied to the improvement of the system because it is possible to continuously evaluate the energy performance according to the operation and maintenance of the building after its construction.

Studies have been performed to analyze the internal environment of livestock like pigs and broilers using dynamic energy simulation [15–21]. In previous studies, indoor environmental conditions were analyzed, but thermal loads were additionally analyzed [15,16,18,19], and energy consumption was calculated [20,21]. Panagakis and Axaopoulous analyzed animal stress indexes to determine fogging strategies for pig rearing [17]. Mogharbel et al. analyzed environmental control as a stress index to properly control the thermal comfort of sow farrowing rooms [20]. Among the previous dynamic energy model studies, most house types were studied mainly for pig houses. And studies that analyzed the heat stress evaluation did not verify the model using the results of field experiments [22–24]. Moreover, a few studies simply focused on case analysis by factor. Therefore, it is necessary to evaluate the thermal environment inside the house using reliable results of the verified model. Moreover, if it is provided to users in connection with the service, a better rearing environment can be achieved.

In this study, Energyplus (Version 9.5.0, DOE, USA) was used to analyze the energy load and energy flow in the house. This simulation tool is a dynamic energy analysis engine created by the US Department of Energy by combining the strengths of BLAST, DOE-2, and COMIS. Energyplus has input items for various building and HVAC system components, and in particular, it performs calculations simultaneously through feedback between the building, air conditioning system, and heat source equipment. Among the programs verified according to the ASHRAE Standard 140-2007, the Energyplus software was used mainly due to open source and flexible source code modification, so it is easy to connect with the interface [25]. Existing commercial programs are difficult to frequently perform modeling and simulation according to field needs, and their licenses are limited, making them inappropriate to extend to services provided for general users. Therefore, using Energyplus, a model was developed to analyze the energy flow of the house building and determine the high-temperature stress of chickens through the air environment inside.

2.4. Field Experimental and Energy Model Developed Procedure

The heat stress calculation model for each environmental condition of broilers was developed through the following process (Figure 5). First, a broiler house was selected for the experiment, and meteorological data of the area where the broiler house was located was collected through a field experiment. In order to design the corresponding broiler house on the model, the structure and physical properties of the envelope were investigated. In addition, the summer growth environment conditions of the broiler were measured and reflected in the model. Next, to develop a dynamic numerical analysis model, the amount of sensible and latent heat generated by broilers and the ventilation operation plan was considered. In order to simulate the flow of energy inside and outside the house, the Air Heat Balance (AHB) algorithms adopted by Energyplus were used to calculate the air temperature and humidity over time. In addition, the reliability of the model was secured by comparing the values measured in the field with the results of model calculations using statistical indicators. The heat stress index of broilers according to the internal environmental conditions was expressed using the calculated air temperature and humidity values inside the house as variables. To prevent the death of livestock during the summer by using this energy model, a system was built that could provide a warning to users in case of an emergency by using weather forecast data as input data.

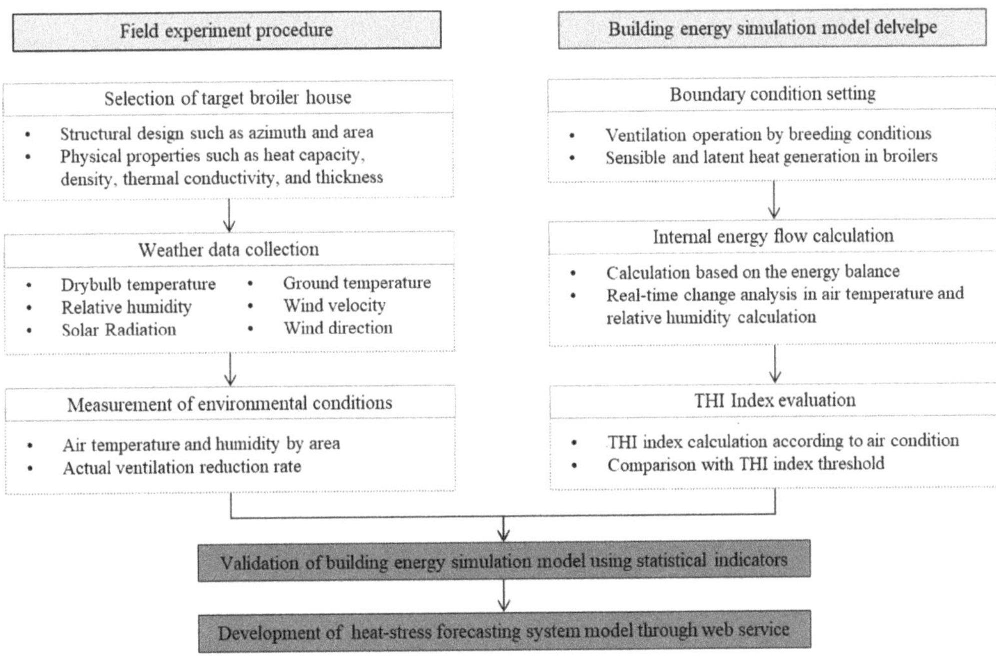

Figure 5. Research flow for heat stress energy model development procedure.

2.4.1. Field Experiment

The duration of the field experiment was done when the chicks weighed 50 g (25 September) until broiler weighed approximately 1.6 kg (28 October). External weather such as temperature, humidity, wind speed, etc. acts as major factors affecting the growing environment inside the facility. Therefore, the external weather at the target farm was measured at 5 min intervals, and Automated Synoptic Observing System (ASOS) data provided by the adjacent meteorological observatory where the target facility is located were additionally analyzed. For the analysis of the air environment in the broiler house, three rows of temperature and humidity sensors (a total of 12 sensors were installed at intervals of 20 m) were installed in the 18 m × 100 m house, and logging was performed every 5 min.

Since the actual air volume of the ventilation fan installed in the livestock facility may vary depending on the ventilation load inside the facility, such as the inlet conditions, the actual air volume of the ventilation fan installed in the target broiler house was measured using a micro-pressure measuring instrument (TSI DP-CALC Micro manometer, TSI Incorporated, Shoreview, MN, USA) (Figure 6). To measure the air volume, a prefabricated T-shaped (+) aero-flow meter with a pressure hole, and a duct of the same size as the tunnel fan was designed and manufactured so that the exhaust air can pass through the air volume meter. There are pitot tubes in the air volume meter, and the number and location of pitot tubes follow ANSI/AMCA Standard 210. According to the Bernoulli theorem, the dynamic pressure at the corresponding position was measured through a pitot tube and a pressure gauge through the difference between the voltage and the static pressure, and the flow rate was calculated. Assuming that the Bernoulli equation was followed, Equation (1) can be considered Equation (2) because P is equal to each other at the same atmospheric pressure and is zero or negligibly small in a streamlined flow. Therefore, the experiment was conducted by measuring the dynamic pressure for 1 min according to the operation of the ventilation system using a dynamic pressure measuring device and calculating the wind speed value under the relevant conditions. In this experiment, the difference between air flow rate and static pressure was measured while changing the number of tunnel fans operated, and the fan performance curve of the actual ventilation flow rate was derived.

$$P + \frac{\rho v^2}{2} + \rho g h = P + q + \rho g h = constant \tag{1}$$

$$q = \frac{\rho v^2}{2} \tag{2}$$

The target experimental broiler house has 14 tunnel exhaust fans installed at the end wall. In theory, when forced ventilation is performed through the exhaust fan, the static pressure inside the facility is kept lower than the outside, so the difference between the external and the internal static pressure was measured. The capacity of a fan is generally determined based on the maximum flow rate, but when applied in the field, the flow rate of the fan varies depending on the structural characteristics of the surrounding environment such as ducts and dampers. As the resistance of airflow increases due to structural changes around the fan, the difference in static pressure between the inlet and the outlet increases, and the flow rate decreases. In other words, the difference in static pressure between the outlet and the inlet of the fan corresponds to a pressure to overcome the resistance of flow due to the surrounding structure. The fan performance curve refers to the static pressure difference upstream and downstream of the fan. The performance curve is a unique characteristic of a fan that remains the same when the fan shape, power required, and rotation speed are constant. These factors are an important consideration when designing and installing a fan. The manufacturer of the fan generally provides a design performance curve for the fan measured according to the corresponding criteria. However, it is known that the fan performance curve changes due to pressure drop (ΔP) due to surrounding structures or aging of the fan, and the fan performance curve at the maximum rotational speed can be approximated by a quadratic polynomial Equation (3).

The design fan performance curves typically show performance at standard air conditions corresponding to 101.3 kPa barometric pressure, 20 °C temperature, 50% relative humidity, and 1.20 kg/m^3 density (ρ). To accurately estimate the performance in the field, it is necessary to perform the correction of the flow rate through the air temperature and density (ρ_d) through the following Equation (4) [26]. The fan performance curve is a relational expression between static pressure and flow rate and is a converted formula considering the area of the target tunnel exhaust fan with a diameter of 1.4 m, as shown in Equation (5). The orifice formula is a formula commonly used to estimate the flow of fluid through the inlet and can be derived from the Bernoulli equation under the assumption of incompressible steady flow. When the cross-sectional area (A) changes due to the throttle, which rapidly changes the cross-section during fluid flow, Equations (6) and (7) are established by the Bernoulli equation and the continuity equation. By linking the two equations, the flow rate (v_1) and the flow rate passing through the orifice (v_2) are calculated as shown in the following Equation (8). Energy loss occurs due to the contraction and friction of the cross-section of the fluid passing through the orifice.

$$\Delta P = c_0 + c_1 Q + c_2 Q^2 \tag{3}$$

$$\Delta P = \frac{\rho}{\rho_d}\left(c_0 + c_1 Q + c_2 Q^2\right) \tag{4}$$

$$\Delta P = -67.553 + 61.877v - 7.3758v^2 \tag{5}$$

$$\frac{1}{2}\rho {v_1}^2 + p_1 = \frac{1}{2}\rho {v_2}^2 + p_2 \tag{6}$$

$$Q = A_1 v_1 = A_2 v_2 \tag{7}$$

$$Q = A_2 v_2 = A_2 \times \sqrt{\frac{2(p_1 - p_2)}{\rho\left[1 - (A_2/A_1)^2\right]}} \tag{8}$$

$$Q = C_D A \sqrt{\frac{2\Delta P}{\rho}} \tag{9}$$

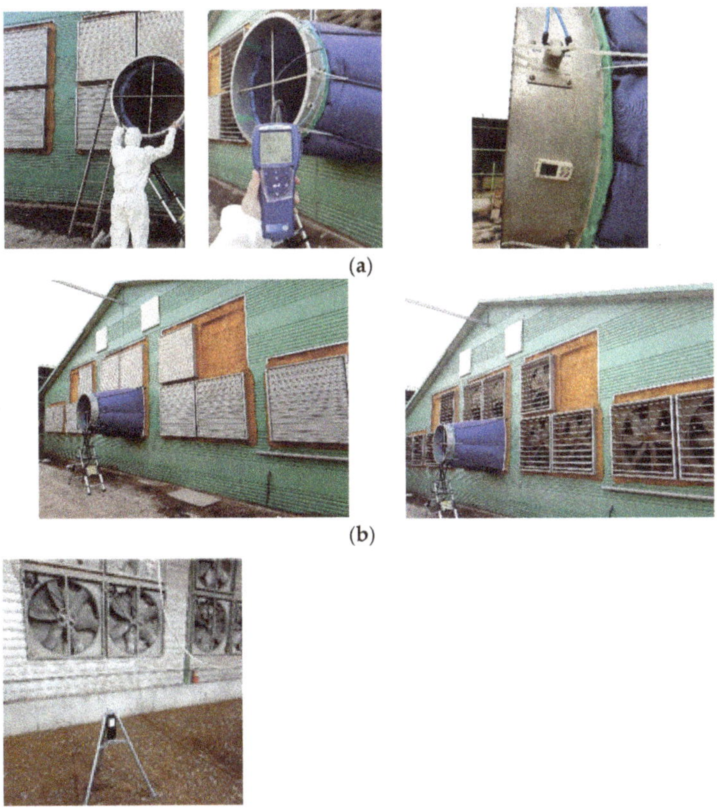

(a) (b) (c)

Figure 6. Measurement the ventilation flow rate of the broiler house of exhaust fan: (**a**) Installation of an air flow meter in the tunnel exhaust fan; (**b**) Ventilation flow rate change according to the number of fans; (**c**) Internal static pressure of exhaust fan.

2.4.2. Building Energy Model Design

To develop a dynamic numerical analysis model, the structure of the target broiler house was first modeled. The size of the house was 18 m in width, 3 m in inside height, 6m in height, and 100 m in total length. As there was no drawing of the house, reference was made to the standard design drawing for broilers [27]. On the floor of the house, it is suggested to install two layers of moisture-proof film on the 150 mm rubble compaction, and install 50 mm discarded concrete and 200 mm reinforced concrete slabs on it. 150 mm and 100 mm sandwich panels are presented for the roof and wall surfaces, respectively. The broiler house analysis model to be used in the study was designed based on the broiler house construction method announced in the standard design drawing, and the structural properties of the model used as input data are as follows. (Table 1) The envelope input of the energy model was implemented using the "Material" and "BuildingSurface:Detailed" classes and the modeling used Sketchup 2019 to shape the house structure.

Table 1. Physical properties input to the broiler house structural model.

Materials	Thermal Conductivity (kJ/hmK)	Thermal Capacity (kJ/kgK)	Density (kg/m^3)
sandwich panel (100 mm Thickness)	0.1368	1.5	100
sandwich panel (150 mm Thickness)	0.1404	1.5	73.33
reinforced concrete	7.92	0.84	2800
concrete	6.3	0.84	2000
gravel	7.2	1.0	1800
polyethylene film	0.8792	2.3037	0.96

Ventilation methods applied to livestock houses can be divided into natural and forced ventilation. In the case of natural ventilation, the environment may be maintained by introducing external air without using additional energy. Meanwhile, in hot summer, a large amount of cooler external air was introduced using a forced ventilation tunnel fan, and in the case of high external temperature, air enters the opening where the cooling pad was used. The "ZoneVentilation:DesignFlowRate" class was used to implement the ventilation operation recommendations according to the age and internal air temperature recommended by the broiler house on the model. The actual ventilation measured experimentally was entered into the design flow rate factor, and the exhaust option was selected as the ventilation type.

In order to simulate the effect of the cooling pad, which is a cooling method that can lower the heat stress index by removing latent heat of the outside air introduced into the house, the cooling efficiency was confirmed through a literature survey. In summary, factors affecting the cooling pad efficiency include air flow rate, pad type and thickness, and circulation water amount. The airflow rate and the amount of water have a significant impact on the cooling efficiency of the cooling pad, and the type and thickness of the pad have less impact on the efficiency [28]. As the control period is long and the flow rate becomes larger, the temperature difference between the inlet and the outlet increases, and the air resistance increases when water is supplied [29]. In addition, the air that passed through the cooling pad was mixed vertically at the side of the house and then moved to the center, and the wind speed of the central aisle was higher than that of the side passage, showing a difference in ventilation rate [30]. The widely used design wind speed of about 1.5m/s of air passing through the cooling pad to adjust the internal and external static pressure difference of negative pressure ventilation. In addition, the type of pad used in the calculation is CELdek7090 (Munters, Sweden), with 60 L per minute of water sprayed along the cross-sectional area of the upper part of the pad [31]. If the cooling efficiency of the pads differs depending on the outside air conditions, but on average, the cooling efficiency was 15 to 40% depending on the pad thickness [32]. The cooling efficiency of the cooling pad on the energy model was implemented using the "Evaporative Cooler: Direct: CelDekPad" class. As an input to the energy model, the direct pad area was 35 m^2, the direct pad depth was 150T, and the recirculation water pump power consumption was 0.7 kW

Next, it is necessary to simulate the energy generated by the broiler as an internal heat source in the house and reflect it as a boundary condition. Through previous studies, the empirical equations for the amount of sensible and latent heat generated according to the surrounding environment, weight, and feed efficiency of the broiler were determined. The heat and moisture generated by the broiler reflected the sensible and latent heat of the broiler suggested by Perdersen and Salvik (2002) [33]. In addition, in order to calculate the weight of the broiler, the empirical Equation (10) suggested by Yoo (2009) was used for the weight of the broiler by age. Perderson and Salvik (2002) presented a correction equation for the total amount of energy generated from the broiler according to the temperature change when 1000 W of unit energy was generated from the broiler, as shown in Equation (11) [34]. Whereas, Equations (12) and (13) refer to the amount of sensible and latent heat generated, respectively.

$$m = 1.1678 \times d^2 + 11.137 \times d + 35.753 \quad (10)$$

$$Q_{tot} = \frac{1}{1000} \times 10.62 \times m^{0.75} \times [1000 + 200 \times (20 - T)] \quad (11)$$

$$Q_{sen} = \frac{1}{1000} \times 10.62 \times m^{0.75} \times \left\{0.61 \times [1000 + 200 \times (20 - T)] - 0.228 \times T^2\right\} \quad (12)$$

$$Q_{lat} = Q_{tot} - Q_{sen} \quad (13)$$

The amount of heat generated by the broiler can be calculated by entering the input variables corresponding to Equations (10)–(13) as simulation results or schedule values through the Energy Management System (EMS) built into Energyplus. In Table 2, parameters in the group that drives the EMS function were written. "Variable" refers to a simulation result item collected through "Sensor", and the internal air temperature and relative humidity required for calculation were obtained as variables calculated as the result value according to the timestep during the Energyplus simulation process, and the number of rearing days and boilers was input through the "Schedule". Since the heat energy of the broiler affects the load inside the house, the value calculated using the sensor variable was set as "InternalGain:OtherEquipment" through the actuator and applied as an internal gain. In addition, the "EnergyManagementSystem:Program" class was used to simulate the high-temperature stress index and the amount of heat generated by the broiler as the result of simulation calculations. The location where the ERL (Energyplus Runtime Language) was written in the program was also determined. For the high-temperature stress index, "EndOfZoneTimestepBeforeZoneReporting" was selected, and the program was executed immediately before the output result was generated at every zone timestep. This calling point was used because it is useful when creating a custom output variable. Next, the amount of sensible and latent heat generation was executed at the beginning of every timestep by selecting "BeginTimestepBeforePredictor" and before calculating the load of the zone. It is useful when used by components related to internal and external heat acquisition (lighting, shading device, set temperature) based on the results of previous timesteps and current weather data [35].

Table 2. EMS objects to consider to heat the production of broilers.

Parameter	Unit	Physical Meaning	EMS Object	Variable	Program Calling Manager
Tin	°C	internal air temperature	Sensor	Calculation	-
Hin	%	internal air relative humidity	Sensor	Calculation	-
Day	day	rearing days	Sensor	Schedule value	-
Head	head	rearing scale	Sensor	Schedule value	-
Qbroiler	W/head	heat production per head	Actuator	-	-
THIindex	-	Temperature Humidity Index	Program	-	EndOfZoneTimestepBefore ZoneReporting
Qsen	W	sensible heat generation of broiler	Program	-	BeginTimestepBeforePredictor
Qlat	W	latent het generation of broiler	Program	-	BeginTimestepBeforePredictor

2.4.3. Energy Model Validation

When developing a building energy simulation model and deriving results, it is compared with field measurement data to determine the accuracy of the model. Through comparison, the reliability of the results through the building energy simulation model can be secured by verifying Measurement and Verification (DOE, 2015) through a statistical method that indicates the error of whether the thermal behavior and energy consumption are similar [36]. However, differences between data calculated inside the actual buildings and simulation models occurred due to errors and information uncertainty in design values, heat loss in each part of construction and equipment, and schedule differences in internal heat gains (occupant, lighting, electric equipment, etc.). The parameters used in the building energy simulation model were adjusted through the trial-and-error method and included the simulation results within the allowable range. The energy model implementing thermal zones was validated using data such as external weather, indoor air temperature and humidity, ventilation amount, and rearing information measured by the broiler house. The guides used in this study are ASHRAE (American Society of Heating, Refrigerating and Air-conditioning Engineers) and FEMP (Federal Energy Management Program), and *NMBE* (Normalized Mean Based Error), *Cv* (*RSME*), and R^2 which are applicable indicators as performance indicators are as follow Equations (14)–(17). MBE is a statistical indicator of the overall behavior of the model prediction value compared to the measurement data. The positive value means that the model underpredicts the measured data, and the negative value overpredicts. This index checks *Cv* (*RSME*) together due to a cancellation error that the sum of positive and negative numbers can reduce MBE. *Cv* (*RSME*) represents the relative error variability as a percentage, and the smaller the coefficient, the closer it is to the average value. Usually, the smaller the value of the two error indicators, the more reliable the model is, and *NMBE* and *Cv* (*RSME*) are ±5 and 15 when analyzed monthly, respectively, and ±10 and 30 when analyzed on a collimated basis. In this validation, the time step was set to 1 min, which is the same as the experimental data, and R^2 indicates how close the calculated value is to the regression line of the measured value. IPMVP and ASHRAE handbooks recommend that the coefficient of determination is not less than 0.75.

$$NMBE(\%) = \frac{\Sigma(S-M)}{\sum M} \times 100 \tag{14}$$

$$RMSE = \sqrt{\frac{\Sigma(S-M)^2}{N}} \tag{15}$$

$$C_{v\ (RMSE)} = \frac{RMSE}{M_{avg}} \times 100 \quad (16)$$

$$R^2 = \left(\frac{\sum_{i=1}^{n} (R_i - \overline{R_i})(C_i - \overline{C_i})}{\sqrt{\sum_{i=1}^{n} (R_i - \overline{R_i})^2 \times \sum_{i=1}^{n} (C_i - \overline{C_i})^2}} \right)^2 \quad (17)$$

2.5. Heat Stress Forecasting System Developed Procedure

The process of establishing a heat stress forecasting system service is as follows. (Figure 7) This study was conducted to receive NCAM-LAMP data, a high-resolution weather prediction site with 810 m spatial resolution for Jeollabuk-do, and to calculate the validated energy model according to chicken rearing conditions and provide it to users using web and text services. To establish a service that automatically provides risk information to users, first, weather data of the area were extracted with farm location information and constructed as input data for energy models. Next, the variables of the energy model were updated for livestock structure, environmental facility conditions, and livestock growth information, and the forecasted heat stress index (the next day and the day after) was calculated from the forecast time to prevent heat wave damage in advance.

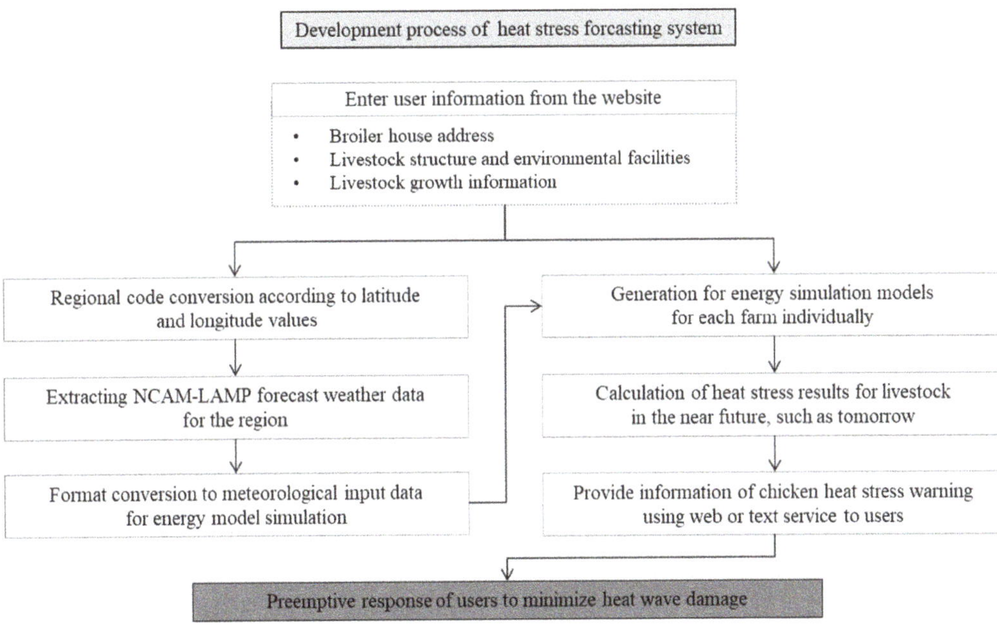

Figure 7. Research flow for heat stress forecasting system development procedure.

2.5.1. Heat Stress Modelling for Construction of LAMP

Using a web service site and a mobile phone, users will receive various information such as farm addresses, livestock structures, environmental facilities, and livestock growth information (Table 3). Among the input information, building dimensions (length, width, ridge, eave height, etc.), wall/roof insulation information (thickness and type), cooling pad, number of fans, fan size corresponds to livestock structure and environmental facilities, and the number of birds and days corresponding to livestock growth information were identified. The latitude and longitude of the house can be determined using the input farm address. Based on the open Application Programming Interface (API) grid of the Korea Meteorological Administration, the district code can be divided using data with maps of latitude and longitude grids per 1 km in Jeollabuk-do. For example, the code "45790" presented city and county, the first two digits "45" referred to Jeollabuk-do, while the last three digits "790" meant Gochang-gun. Weather data from the area where the input farm address is located is required as input data for the livestock heat stress model. Since Energyplus weather data can be calculated only when there is a full period of one year, only the data of the LAMP forecast period was updated from the original Energyplus weather data and a new file of Energyplus weather data for one year was created every time it is calculated. The LAMP file format is NetCDF (.nc), and the weather data file format to be entered into the Energyplus model was EnergyplusWeather (.epw), which analyses the file characteristics and internal variables of each data. First of all, since a nc file contains data only for one time in all regions, the file was generated by time and is provided at about 100 MB per file, so if the predicted and provided file is merged into one, a very large file is generated and hard disk capacity is required. The epw file was annual data of latitude and longitude as long as it corresponds to a house in one file, and is about 1.5 MB. Among the variables that require values per unit time in the epw file, variables that affect the calculation of the internal model include dry bulb temperature, dew point temperature, relative humidity, atmospheric pressure, horizontal infrared radiation, direct vertical radiation, diffusion horizontal radiation, wind direction, wind speed, cloud cover, opaque cloud, snow depth, precipitation, etc. LAMP weather data also has several variables, but none of the weather variables required for epw were added. The variables such as 10 m vertical wind, relative humidity, 10 m wind direction and speed, sea level pressure, lower cloud cover, total cloud cover, direct solar radiation, normal direct solar radiation, diffuse solar radiation, snow depth, horizontal longwave radiation intensity, etc. were added. The lower cloud cover was generated based on 850 hPa isostatic surface cloud cover, the total cloud volume was generated based on the maximum vertical cloud cover, and the horizontal radiation intensity was calculated by referring to Walton (1983), Clark and Allen (1978) literature [37,38]. In addition, the units that have been the same in the past have been unified. Since the nc file has weather data of one region in the entire Jeollabuk-do region, it was coded using jupyter notebook so that only the value of the required area where the farmhouse is located among the forecasted weather data can be obtained. The latitude and longitude values in the header of the original epw file and the target latitude and longitude values in the nc file must be exactly the same to extract the data of the region corresponding to the latitude and longitude. The date of the period to be used and the original/merged file name was modified according to the generated file. In addition, if month and day are single digits, two digits must be created by adding a leading zero to them. In addition, nc files are created from 00:00 to 23:00 in one day, and the following period is used as a standard of one day. The newly created epw file, including the LAMP forecast data, was shown through Algorithm 1.

Table 3. Variables the user enters into the system.

Type	Input Variable	Unit	Example	Input	Condition
location	latitude, longitude	-	127°35′	address	conversion to district code format
structure and equipment	farm width	m	18	constant	min 0, max 50
	farm length	m	80	constant	min 0, max 120
	farm side height	m	4	constant	min 0, max 10
	farm ridge height	m	6.4	constant	min 0, max 10
	wall insulation thickness	mm	100	constant	min 0, max 1000
	roof insulation thickness	mm	150	constant	min 0, max 1000
	insulation type	-	EPS	option	EPS (styropor, Neopor), XPS
	cooling pad	-	O	schedule	0 or 1
	number of fans	ea	14	constant	min 0, max 20
	fan size	inch	50	option	30 or 50
growth information	rearing days	day	5	constant	min 0, max 60
	rearing scale	head	34,000	constant	-

Algorithm 1. Pseudocode for merged data and converted LAMP format.

$Input:$
D_{today} : *Start date to replace weather data*
EPW_{in} : *Annual weather data for a specific latitude and longitude (.epw file)*
NC_H^D : *National weather data for H hour on date D* $(D_{today} \leq D < D_{today} + 7, 0 \leq H < 24,$ *.nc file*)
$Output:$
EPW_{out} : *Annual weather data with replacement completed (.epw file)*
Algorithm :
READ first line of EPW_{in} *(first line of .epw file contains latitude, longitude information)*
Lat := *Latitude information of* EPW_{in}
Lng := *Longitude information of* EPW_{in}
$Rows$:= *Hourly weather data list for one week at latitude Lat and longitude Lng location*
FOR $i := D_{today}$ *to* $(D_{today} + 6)$:
FOR $j := 0$ *to* 23 :
READ NC_j^i
EXTRACT weather parameters where the latitude and longitude values correspond to Lat and Lng
EXTRACT only the necessary parameters to compose .epw file
RELOCATE the extracted parameters to fit the format of .epw file (make one row)
ADD a row in the Rows
FOR Line in EPW_{in} : *(each Line contains weather data at a specific time for a specific date)*
D_{Line} := *Line's date information*
H_{Line} := *Line's hour information*
IF Line's date NOT IN Rows' dates :
WRITE Line to EPW_{out}
ELSE :
Line := Among the row stored in Rows, the one extracted from $NC_{H_{Line}}^{D_{Line}}$
WRITE Line to EPW_{out}

2.5.2. Web-Based Forecasting System Development

An Energyplus Input File (.idf) file developed using the Energyplus program was created for the broiler energy model. The livestock building structure, environmental equipment, and broiler growth information values were manually entered by the users at the corresponding coding line. The heat stress calculation time for 3 days takes about 2 s. T The calculation period was the next day and the day after the forecast day, and since the result varied depending on the initial value, it was calculated from the day before the forecast day. When the user selects the type of insulation material such as Styrofoam, Neopole, Isopink, or rigid urethane, the corresponding heat capacity, density, specific heat, and absorption rate were inputted and calculated. The schedule was divided according to whether the cooling pad was used or not. Specifically, when the cooling pad was used, it was "always on continuous" condition while when it was not in use, "always off discrete" is input. The cooling pad length was inputted as the actual house length, but for the user who does not know the exact length, the default value was set to 20% of the length of the house. According to the standard design of the broiler specified by Nonghyup (2016, 2019) [27], the cooling pad is 30 m when the house length is 114.3 m, in the year 2106 and when it is 112.0 m, it is specified to be 24 m, in the year 2019. According to the standard design, the length of the cooling pad compared to the length of the barn was about 20%, which was brought and used as the model default.

Next, the evaluation criteria for broilers suitable for South Korea were identified through a literature review on heat stress. A representative broiler's heat stress index was studied, and among the indexes, the summer weather condition in South Korea and the type of broiler variety were compared to each other. In addition, the improvements were analyzed by comparing the case of the existing heat stress forecasting system of livestock.

3. Results and Discussions

3.1. Weather and Internal Environment

The closest meteorological station was located in Gunseo-myeon, Yeonggwang-gun, about 4 km away from the broiler house, which provides Automated Synoptic Observing System (ASOS). The measured average outside air temperature outside the target building during the experimental period was 16.5 °C and the average relative humidity was 70.1%, respectively. A comparison of the field-measured external weather environment using a portable weather station with the ASOS showed that the difference between the highest temperature during the day and the lowest temperature at night was up to 4 °C and at least 1.5 °C, respectively. This was believed to be due to radiation from the concrete floor as the location where the simple station was installed is on the roof of an office building that does not block sunlight. In terms of dominant wind direction, it was confirmed that the highest wind frequency occurred from the south and southwest directions. The southwest wind was dominant during the day while the south wind prevailed during the night.

The air temperature change at different zones inside the poultry house until the broilers were introduced and shipped as shown in Figure 8. During the experiment and data collection period, the broiler growth environment was set to an appropriate level through appropriate adjustment of ventilation. Results showed that the relative humidity of the outside air increased to more than 90% at night and decreased during the day. From this, it can be noted that the relative humidity inside the broiler house is highly affected by the change in the outside relative humidity. In addition, it was also noted that the relative humidity is lower than the appropriate growth condition during the day and higher at night. The optimum humidity range was 70% at 1 week of the rearing period and 60% after 3 weeks of rearing period, and the actual humidity measured inside the broiler house was relatively similar to the appropriate growing humidity. The increase in humidity is dominated by the outside air, and there seems to be no tendency to increase during the entire growth period. In addition, the increase in moisture content of floor litter showed no significant effect on humidity generation as the litter used were newly replaced during the experimental period. Moreover, it was observed that the variation in the internal environment during the first 1 to 5 days of rearing age seems to be largely affected by the operation of the hot air blower. Specifically, the temperature during the early days of rearing was 0.07 °C higher than the appropriate rearing temperature of 35 °C. The use of hot air blower also showed a significant decrease in humidity, with 5.25% lower than the appropriate humidity of 70%. During the entire rearing period, the measured air temperature was almost similar to the allowable temperature value suggested by the Rural Development Administration (2007) [39]. When the broiler grew to the days of 22nd, the weight increased, and the amount of heat generated in the body increased, so on days when the outside temperature was high, negative pressure ventilation was performed using the tunnel exhaust fan during the day. The use of ventilation to lower the internal air temperature during this period showed that the air quality and internal humidity inside the broiler house were also improved.

3.2. Ventilation Flow Rate Depending on the Number of Fans

By analyzing the flow rate of the target tunnel exhaust fan measured through field experiments and the difference in static pressure inside and outside the house, the field fan performance curve was derived. The results showed that a high error was observed when the static pressure difference was 10 Pa or less. In this study, the analysis was performed using only the measured values of the static pressure difference of 10 Pa or more. (Figure 9).

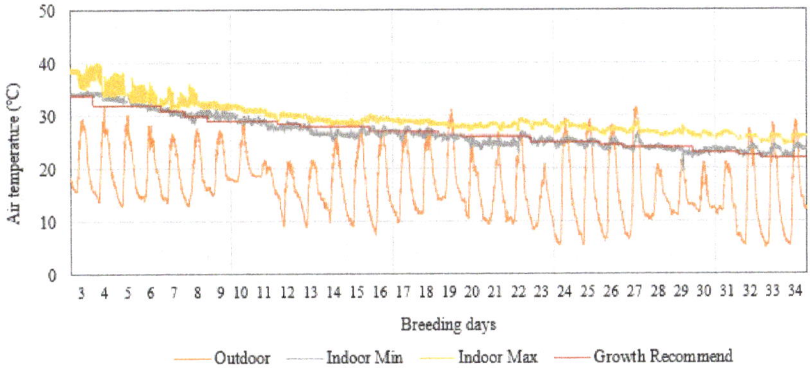

(**a**)

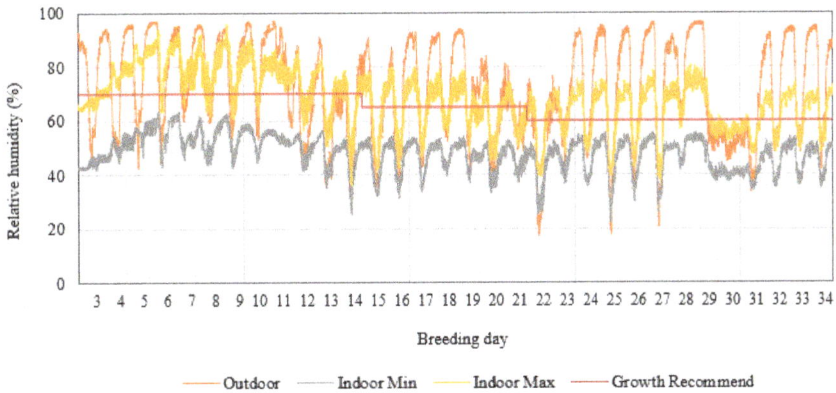

(**b**)

Figure 8. Measured internal environment the broiler house during the rearing period (25 September, 28 October 2020): (**a**) Measured and optimum growth of air temperature; (**b**) Measured and optimum growth of air relative humidity.

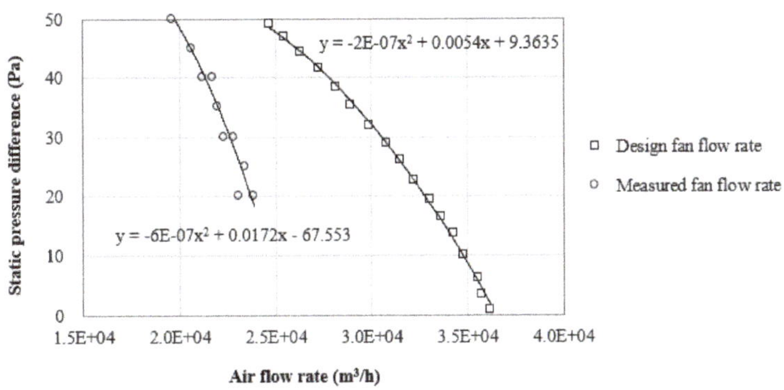

Figure 9. Fan performance curve results in the broiler house.

Compared to the design maximum fan flow rate provided by the manufacturer, the actual airflow rate measured during the field experiment was found to be only 70% of the designed flow rate. This may be caused by the difference between the evaluation method of the fan performance curve and the field measurement environment. For instance, in the approved fan performance test method, ducts are installed at the inlet and outlet of the target fan. Thus, the static pressure difference between the inlet and outlet of the fan is kept constant, making it easy to measure in the duct. However, in the case of the mechanically ventilated broiler house, the exhaust fans were installed in an open wall, not a duct. So, as the distance from the exhaust fan increased, the airflow rate inside the house decreased, and the static pressure increased. Other reasons for the decrease in performance of the target fan may include the installation of shutters to prevent infiltration through the exhaust fan, accumulation of dust around the fan, and aging of the fan belt due to long-term use. Therefore, in order to accurately estimate the ventilation rate in a mechanically ventilated broiler house, it is necessary to evaluate the fan performance change at the broiler house site. The experimental condition in which the minimum flow rate of the target tunnel exhaust fan was measured was that all 14 installed tunnel exhaust fans were operated and the slot opening was partially opened ($\theta = 13°$).

When tunnel ventilation is used in mechanically ventilated broiler houses, the average static pressure difference inside and outside the house is from 2.5–30 Pa [40]. Under this condition the measured flow rate is reduced by 25 to 30% when compared to the flow rate predicted through the design fan performance curve. The experimental values of the flow rate reduction rate according to the static pressure difference are shown in Table 4. Park et al. (2018) was concerned about the exhaust fan failure because excessive negative pressure is formed when the static pressure difference is greater than 40 Pa [41]. The reason for the large static pressure difference measured in this study is the difference in the inlet area. When the ventilation rate was used on the calculation model according to the fan design data, heat stress of broiler in house may be underestimated. Therefore, it is more appropriate to input the ventilation rate in the energy model in consideration of the fan performance decrease.

Table 4. Flow rate reduction rate of design fan curve and field fan performance curve according to static pressure difference.

Static Pressure Difference (Pa)	Tunnel Fan Flow (m^3/h)		Reduction Rate (%)
	Design	Measured	
20	32,934	23,061	30.0
30	30,428	22,048	27.5
40	27,861	20,880	25.1
50	24,713	19,451	21.3

3.3. Validatdion of BES Computed Results

The accuracy of the BES model in the broiler house used in this study was validated by comparing the data of the air temperature and relative humidity inside the house measured in the field experiments with those calculated using the BES model. In the validation, the field measured and calculated data from 4 to 10 October and was compared. A 5 min average value was used for the field experiment value, and the time interval of the energy simulation was also set to 5 min. Among the energy analysis method, it is recommended that the energy analysis calculation time step is 5 min or 10 min [42]. This is because the change in energy flow due to the equipment operation or internal heat gain is too long for one hour, and the one-minute interval takes too much computation time. Figure 10 shows the graph comparing the minimum, average, and maximum values of the internal air temperature and humidity measured in the field with the calculated internal air results.

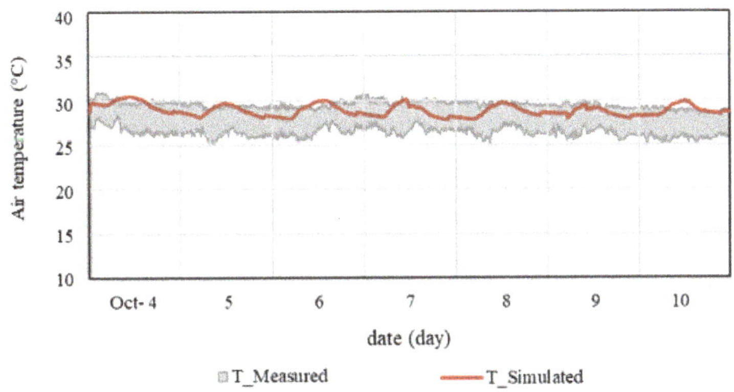

(**a**)

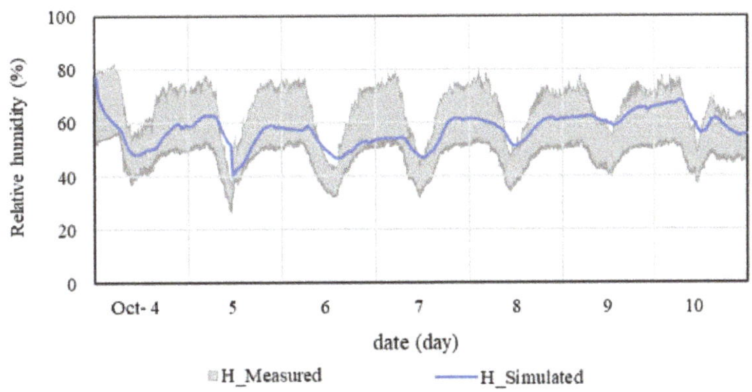

(**b**)

Figure 10. Comparison of internal air temperature and relative humidity for model validation: (**a**) Simulated value of air temperature compared to field measured data; (**b**) Simulated value of relative humidity compared to field measured data.

The validation result showed that the model error of air temperature and humidity inside the house fall within 10%. (Table 5) It means that the error range of the simulated results is within 10% based on the average values of the experimentally measured temperature and humidity results. Since relative humidity is a value that can be easily changed by various variables other than the air temperature inside the house, the error rate is higher than the temperature, but the measured value also has a deviation of more than 20%. The American Society of Heating, Refrigeration, and Air-conditioning Engineers (ASHRAE) suggests reliable standards for the error rate of energy analysis models in the M&V Guideline. According to this document, when the *RSME* is within 30% of the absolute value, it can be assumed that the model has high reliability and is suitable.

Table 5. Statistical analysis by air temperature and humidity for model validation.

	R^2	*RSME*	MAPE
Indoor air temperature	0.94	1.55	4.66
Indoor relative humidity	0.89	6.16	6.69

3.4. Heat Stress Index Evaluation Criteria for South Korea

Heat stress is the result of a heat load in which an animal cannot maintain homeostasis. In poultry, when the Effective Environmental Temperature (EET) is within the Thermo Neutral Zone (TNZ), which is within the upper and lower critical temperatures, the body temperature of the adult chicken is maintained at 41.2 °C [43]. When the EET rises above the upper critical temperature, the defense mechanism acts as though feed and water intake decreases and the evaporation rate increases. Heat balance formation is complex and includes several climatic factors (ambient temperature, relative humidity, wind speed, etc.), livestock factors (age, reproductive status, physical activity, adaptability, health status, metabolism, genotype, etc.), management factors (housing, shade provision, fans and sprinklers, nutrition management, etc.). In order to alleviate the heat load by using these factors, studies on high-temperature stress in livestock have been conducted for a long time.

Table 6 shows the heat stress index studied for broilers. The most commonly used index is the Temperature Humidity Index (THI), which uses air temperature and humidity data to show the effect of high-temperature stress on livestock and is useful for predicting livestock productivity simply. DeShazer and Beck (1988) analyzed whether water was sprayed or not, according to the region, the heat of vaporization was additionally calculated in the latent heat term [44]. Measured with time series data, the extreme weather of the top 1% of US weather data was also analyzed separately.

Tao and Xin (2003) developed thermal stress when exposed to high-temperature environments [45]. In addition, the Temperature Humidity Velocity Index (THVI) was developed considering the relative importance of air flow rate when growing with tunnel ventilation in summer. The experiment was conducted with male broilers 46 ± 3 days of age under 18 conditions divided into 3 stages (35, 38, 41 °C) of dry bulb temperature, 2 stages (19.4, 26.1 °C) of dew point temperature, and 3 stages (0.2, 0.7, 1.2 m/s) of air flow rate, and the body temperature increase of male broilers measured for 30 min was measured. In the case of broilers, the weight ratio of sensible heat and latent heat was divided into 85:15, and the mortality rate with respect to air flow rate was not linear, so an equation was derived in the form of an exponential function.

Chepete et al. (2005) conducted a study to derive the THI index of broilers according to the growth stage [46]. In the experimental method, about 1700 broilers were raised inside the naturally ventilated broiler house in summer and winter over 6 weeks. Here, the index was calculated by analyzing feed intake, average weight, and mortality according to the number of rearing weeks. Moraes et al. (2008) evaluated whether the high-temperature stress index for broilers proposed by the THI index was consistent with other regions [47]. In this study, they collected local weather data and confirmed that the internal environment was properly expressed as THI.

The various forms of THI index have been widely used as livestock stress index, but there are some obvious limitations. The use of this index requires the assumption that livestock responds to environmental stressors in exactly the same way, but it is impossible. For example, it is common that cows with low production (20 L of milk per day) are less affected by heat stress than cows with high production (45 L). In addition, changes in the response of livestock to climate variables have been established according to various literature. Moreover, the THI index does not consider thermal radiation (solar and long wave), wind speed, duration of exposure, or differences in age or genotype. Xin et al. (1994) reported that broilers were more sensitive to heat during the first 3 weeks of growth and that males were more affected by heat than females, Brown-Brandl et al. (1997) found age responses related to reaction response and reaction time in turkeys exposed to high heat loads [48,49], but no established relationship to respiration rate or rectal temperature was determined. Although there are recognized deficiencies, in this study, in order to analyze the most important factors of internal and external heat flow through field experiments at the house, the simulation results of inside air temperature and humidity are the most universal and have the greatest effect on heat stress, so only THI index representing only the air environment was used. Among the various THI indexes shown in Table 6, it was deemed necessary to calculate the high-temperature stress of livestock suitable for Korea's livestock and weather conditions.

Table 6. Heat stress index model for broilers.

Number	Equation	Reference	
(18)	$(1.8 \times T + 32) - [(0.55 - 0.0055 \times RH) \times (1.8 \times T - 26.8)]$	NRC (1971)	[50]
(19)	$0.6 \times T_{DB} + 0.4 \times T_{WB}$	DeShazer and Beck (1988)	[44]
(20)	$0.85 \times T_{DB} + 0.15 \times T_{WB}$	Tao and Xin (2003)	[45]
(21)	$(0.85 \times T_{DB} + 0.15 \times T_{WB}) \times V^{-0.058}$		
(22)	3~4 Weeks: $0.62 \times T_{DB} + 0.38 \times T_{WB}$, 5~6 Weeks: $0.71 \times T_{DB} + 0.29 \times T_{WB}$	Chepete et al. (2005)	[46]
(23)	$0.85 \times T_{DB} + \frac{RH(T_{DB} - 14.3)}{100} + 46.3$	Moraes et al. (2008)	[47]

Equation (18) is not only for poultry, but also for various livestock animals such as cows and pigs, and it is a level that suggests different standards for each livestock species through the threshold value of the model [50]. Equation (19) showed only the critical value by approximating the model developed for laying hen by DeShazer and Beck (1988) to a broiler. Equations (20) and (21) are models developed by Tao and Xin (2003) and are divided into equations expressed by the dry bulb and wet bulb temperatures of the ambient air around the broiler, and equations that consider the wind speed related to the effective temperature. At Iowa State University, a heat stress index model was developed by moving broilers to a control room with temperature and humidity control and monitoring the broilers. The broiler breed is a crossbreed of ROSS and ROSS, and the experiment was conducted similarly to the conditions of many rearing methods in Korea. The heat stress level criterion was expressed by dividing it into four levels which include Normal, Alert, Danger, and Emergency, and the critical values were calculated as 33.14, 36.08, 39.02 in Equation (20), and 33.90, 37.74, and 41.59 in Equation (21). Equation (22) can be seen as similar to the shipping period in Korea by making models every 3~4 weeks and 5~6 weeks using straight-run Cobb 500 broilers, which are the second most frequently raised broilers in Korea. Although the index was developed by measuring the temperature and humidity inside the broiler house without environmental control, the biggest limitation seems to be that it is very different from the weather conditions in Korea and does not provide a step-by-step threshold of the heat stress index. Equation (23) is an index that experiments with broilers and suggests thresholds as a model developed for dairy cows using Brazilian climate data. In order to evaluate the heat stress index for Korean broilers relatively similarly, Equation (20) is judged to be the most suitable considering the weather and the species of broiler during the experiment.

3.5. Improvements to the Web-Based Forecasting System

In this study, the internal environment was calculated through energy simulation by considering the NCAM-LAMP weather forecast data, user's basic information, calculation time, and the number of rearing broilers, and the stress level was evaluated using heat stress standards suitable for Korea. A system was developed to automatically predict all computational processing processes using Open Source energy simulation software. In order to create an automatically computable platform, the system was built to compute from ubuntu to the Command Language Infrastructure (CLI). On the developed web service site (https://poultryheatstress.co.kr, accessed on 1 August 2022) (Figure 11), the internal air temperature, humidity, and heat stress can be checked by time. It is available on the web and in text, and the color is different according to the risk of heat stress so that the user can easily understand the information through the chart. With the calculation time at intervals of 10 min, efforts were made to predict in more detail the time with a high risk of damage through the heat stress index.

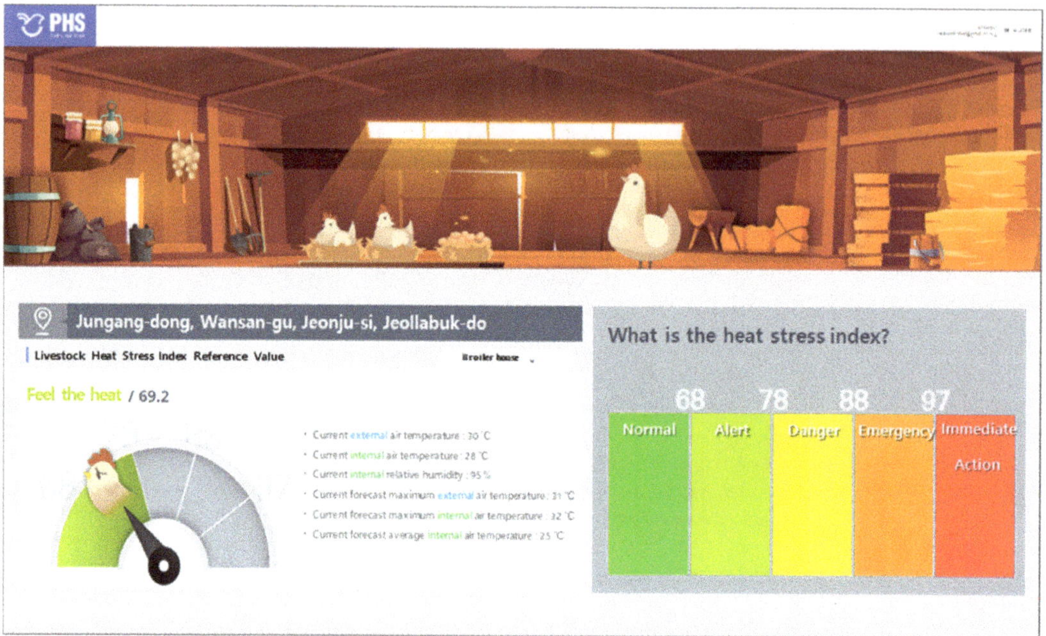

Figure 11. A sample image showing the web-based forecasting system for broiler heat stress.

There have been cases where real-time monitoring and forecasting systems for high-temperature stress in livestock were attempted at home and abroad using weather forecast data [51,52]. The National Livestock Science Institute of Rural Development Administration of Korea has developed a livestock-rearing weather information system to reduce damage to livestock from heat waves [53]. In connection with meteorological observation data from the Korea Meteorological Administration, a function was implemented to provide livestock farmers with real-time weather (temperature, humidity, etc.), heat stress index, and risks of meteorological disasters (heat waves, tropical nights, etc.) by region. In addition, the heat stress index customized livestock and livestock management guidelines for each type of livestock were provided so that farms could be equipped to reduce the productivity of livestock by responding to high-temperature stress. However, in the case of Korean cattle, natural ventilation is mainly performed, but in the case of pigs and chickens raised in windowless or semi-windowless facilities, the temperature and humidity index were calculated only by changes in external temperature and humidity. The internal temperature of most houses shows a tendency to increase and decrease similarly to the change in the outdoor temperature, but there is a limitation in that the change in the micro-weather environment in the facility according to the external weather is not simulated. To overcome this problem, this study developed an energy model that analyzes the micro-weather environment inside the facility by inputting external weather data into the model to quantify the high-temperature stress that occurs when livestock is exposed to heat waves.

The Livestock Science Institute of the Rural Development Administration applies the external weather as it is to the index, and the calculation result changes only by the weather (Figure 12). The high-temperature stress index through model calculation takes into account the age, rearing density, and ventilation rate, so it shows a value that gradually increases from the beginning to the latter half of rearing similar to the actual value, but it can be seen that the high-temperature stress increases significantly when incorrect ventilation is operated Therefore, by considering the recommended operating ventilation by internal air temperature for broilers, the high-temperature stress index does not increase significantly even when the broilers reach adulthood in the second half, and it is considered that the most absolute variable is the ventilation rate as it operates well similar to the critical point.

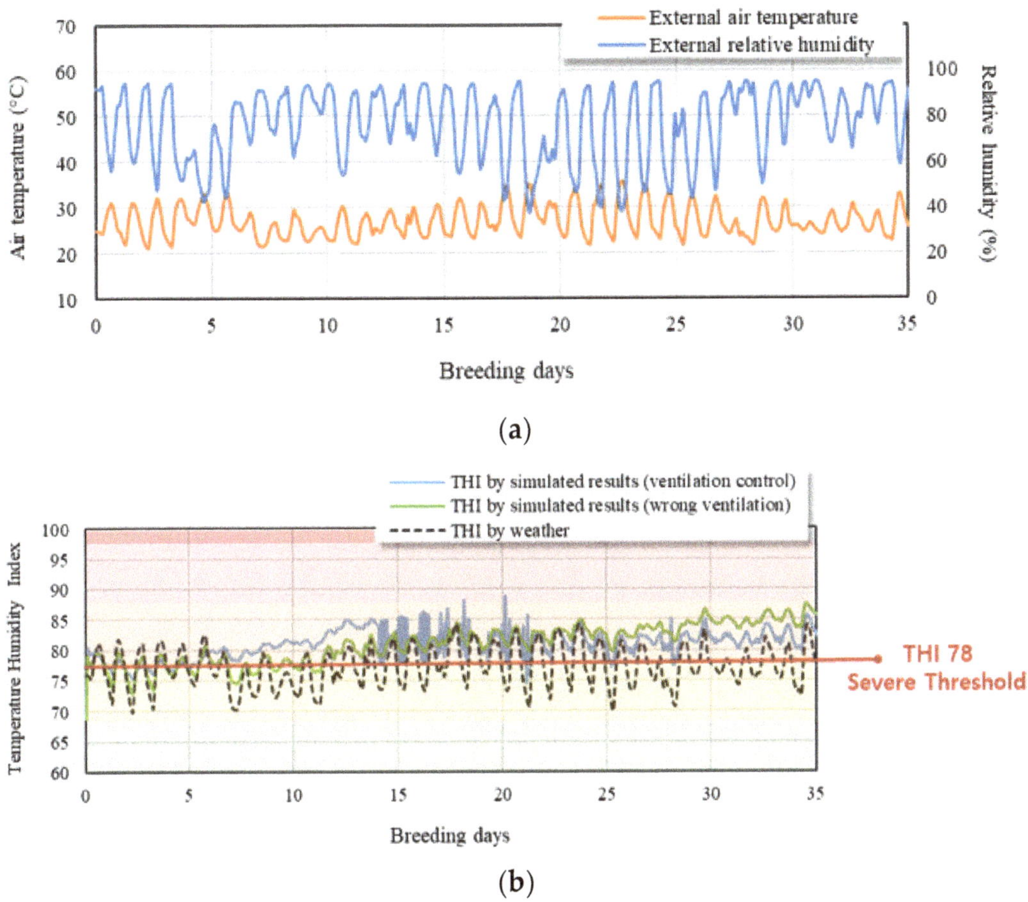

Figure 12. Comparison of heat stress index for the Rural Development Administration system predictive warning and this research model: (**a**) Jeonllabuk-do Weather Data (15 July–19 August); (**b**) Heat stress index analysis using energy model calculation results and experimental data.

4. Conclusions

This study attempted to develop a forecasting system that enables users to preemptively respond to heat stress evaluation of broilers before danger occurs. To do this, several processes were conducted, including field experiments at the selected broiler house to measure and analyze the external weather, the internal environment, and the ventilation flow rate during tunnel ventilation. Air temperature and relative humidity were analyzed according to the length and width using time series data for rearing days broilers. Field-measured data analysis showed that the actual ventilation rate of the tunnel fan was reduced by 30% compared to the design ventilation rate, specified by the manufacturers of the fans. Since the ventilation flow rate is a factor that directly affects the air thermal environment inside the broiler house, it is important to input the actual ventilation amount to accurately predict the internal environment. Then, a dynamic energy model that calculates the internal environment of the house was developed and validated using the field-measured data. Compared with the measured results, the simulated results were found to have a 1.55 and 4.66 for *RSME* and MAPE of internal air temperature, and 6.16 and 6.69 for *RSME* and MAPE of internal relative humidity, respectively. Using the LAMP weather forecast data, the broiler stress index for the rearing environment was expressed as the heat stress index suitable for South Korea, and the system was developed to provide information to users by turning it into a web service. When the users input the farm location, structure and equipment, and rearing information, the heat stress index of the broilers of the previous day and the day before the forecast was calculated from the forecast time.

Author Contributions: Conceptualization, J.-H.C. and I.-B.L.; methodology, J.-H.C. and S.-Y.L.; validation, J.-H.C. and S.-Y.L.; investigation, S.-J.P. and J.-G.K.; writing—original draft preparation, J.-H.C. and I.-B.L.; writing—review and editing, J.-H.C., D.-Y.J., C.D.-V., Y.-B.C., U.-H.Y. and S.-J.L.; visualization, H.-H.J.; All authors have read and agreed to the published version of the manuscript.

Funding: This work was supported by the Korea Institute of Planning and Evaluation for Technology in Food, Agriculture and Forestry (IPET) through Living Lab Project for Rural Issues, funded by Ministry of Agriculture, Food and Rural Affairs (MAFRA) (120099-03).

Conflicts of Interest: The authors declare no conflict of interest.

Nomenclature

Parameters

P	the static pressure at a point (Pa)
q	the dynamic pressure at a point (Pa)
v	the flow velocity at a point in the line (m/s)
ρ	the density of the fluid (kg/m^3)
g	gravitational acceleration (m/s^2)
h	the height of the point relative to the reference plane (m)
ΔP	the difference in the static pressure between the inlet and outlet of the fan (Pa)
Q	the flow rate of the conveying air of the fan (m^3/s)
c_0, c_1, c_2	fan performance curve coefficients
ρ_d	the density of air at the site
p_i	the static pressure in section *i*
v_i	the flow velocity in section *i*
A_i	the cross-sectional area of the section *i*
C_D	discharge coefficient
m	body mass of broiler (kg)
d	number of breeding days (day)
Q_{tot}	the total heat production of broiler (W/head)
T	indoor air temperature (°C)
Q_{sen}	the sensible heat production of broiler (W/head)
Q_{lat}	the latent heat production of broiler (W/head)
THI	temperature humidity index
RH	relative humidity (%)
T_{DB}	dry bulb temperature (°C)
T_{WB}	wet bulb temperature (°C)
$THVI$	temperature humidity velocity index
V	air velocity (m/s)

References

1. National Institute of Animal Science. Rural Development Administration (NIAS RDA) Livestock Statistics 30. 2021. Available online: www.nias.go.kr/30/2104.pdf (accessed on 1 May 2022).
2. Korea Rural Economic Institute (KREI). *Current Status and Countermeasures of Heat Wave Damage in Rural Areas;* Han Design Corporation Publisher: Tokyo, Japan, 2021.
3. Korea Meteorological Administration (KMA). *2019 Abnormal Climate Report;* Design Pickup Publisher: Seoul, Korea, 2020.
4. Korea Environment Institute (KEI); Climate Change Risk Research Center; Korea Environmental Policy and Assessment Institute (Chae Y. et al.) *2020 Heat Wave Impact Report*, 2020.

5. Korean Statistical Information Service (KOSIS). Available online: https://kosis.kr/index/index.do (accessed on 5 March 2022).
6. Korea Meteorological Administration (KMA). *2018 Abnormal Climate Report*; Design Pickup Publisher: Seoul, Korea, 2019.
7. Purswell, J.; Dozier, W.; Olanrewaju, H.; Davis, J.D.; Xin, H.; Gates, R. *Effect of Temperature-Humidity Index on Live Performance in Broiler Chickens Grown from 49 to 63 Days of Age*; American Society of Agricultural and Biological Engineers: St. Joseph, MI, USA, 2012.
8. Moura, D.; Vercellino, R.; Santos, J.; Vale, M. *Heat Stress Impact on Weight Gain in Broiler Chickens: A Meta-Analytical Study of Environmental Factor that Impact Production Losses*; American Society of Agricultural and Biological Engineers: St. Joseph, MI, USA, 2015.
9. Cândido, M.; Tinôco, I.; Pinto, F.; Santos, N.; Roberti, R. Determination of thermal comfort zone for early-stage broilers. *Eng. Agrícola* **2016**, *36*, 760–767. [CrossRef]
10. Lin, C.-Y.; Hsieh, K.-W.; Tsai, Y.-C.; Kuo, Y.-F. *Monitoring Chicken Heat Stress Using Deep Convolutional Neural Networks*, 2018.
11. Ha, J.-W.; Chang, H.-H.; Cha, K.-J.; Song, Y.-H. Verification of Thermal Environment and THI Prediction Equation of Hen House by Field Measurement and Model-based Simulation. *J. Korean Inst. Archit. Sustain. Environ. Build. Syst.* **2020**, *14*, 208–219.
12. European Committee for Standardization (CEN). *EN ISO 13790*; Energy Performance of Buildings: Calculation of Energy Use for Space Heating and Cooling. CEN: Brussels, Belgium, 2008.
13. Lee, S.-J.; Song, J.; Kim, Y.-J. The NCAM Land-Atmosphere Modeling Package (LAMP) Version 1: Implementation and Evaluation. *Korean J. Agric. For. Meteorol.* **2016**, *18*, 307–319. [CrossRef]
14. Hong, M.; Lee, S.-H.; Lee, S.-J.; Choi, J.-Y. Application of high-resolution meteorological data from NCAM-WRF to characterize agricultural drought in small-scale farmlands based on soil moisture deficit. *Agric. Water Manag.* **2021**, *243*, 106494. [CrossRef]
15. Daskalov, P.I. Prediction of Temperature and Humidity in a Naturally Ventilated Pig Building. *J. Agric. Eng. Res.* **1997**, *68*, 329–339. [CrossRef]
16. Daskalov, P.I.; Arvanitis, K.G.; Pasgianos, G.D.; Sigrimis, N.A. Non-linear Adaptive Temperature and Humidity Control in Animal Buildings. *Biosyst. Eng.* **2006**, *93*, 1–24. [CrossRef]
17. Panagakis, P.; Axaopoulos, P. Comparing fogging strategies for pig rearing using simulations to determine apparent heat-stress indices. *Biosyst. Eng.* **2008**, *99*, 112–118. [CrossRef]
18. Seo, I.-H.; Lee, I.-B.; Moon, O.-K.; Kim, H.-T.; Hwang, H.-S.; Hong, S.-W.; Bitog, J.; Yoo, J.-I.; Kwon, K.-S.; Kim, Y.-H.; et al. Improvement of the ventilation system of a naturally ventilated broiler house in the cold season using computational simulations. *Biosyst. Eng.* **2009**, *104*, 106–117. [CrossRef]
19. El Mogharbel, O.; Ghali, K.; Ghaddar, N.; Abiad, M.G. Simulation of a localized heating system for broiler brooding to improve energy performance. *Int. J. Energy Res.* **2014**, *38*, 125–138. [CrossRef]
20. Fabrizio, E.; Airoldi, G.; Chiabrando, R. Study of the Environmental Control of Sow Farrowing Rooms by Means of Dynamic Simulation. In Proceedings of the 8th International Symposium on Heating, Ventilation and Air Conditioning, Xi'an, China, 19–21 October 2013; Springer: Berlin/Heidelberg, Germany, 2014; Volume 263, pp. 3–11.
21. Zhou, Y.; Bidarmaghz, A.; Narsilio, G.; Aye, L. Heating and Cooling Loads of a Poultry Shed in Central Coast, NSW, Australia. In Proceedings of the World Sustainable Built Environment Conference, Hong Kong, China, 5–7 June 2017.
22. Ha, T.; Kwon, K.; Hong, S.; Choi, H.; Lee, J.; Yeo, U. Estimation of THI Index to Evaluate Thermal Stress of Animal-Occupied Zone in A Broiler House Using BES Method. *Korean Soc. Agric. Eng.* **2018**, *60*, 75–84.
23. Ha, T.; Kwon, K.; Lee, I.; Kim, R.; Yeo, U.; Lee, S. Estimation of THI index to evaluate thermal stress of piglets in summer season. *Korean Soc. Agric. Eng.* **2018**, *60*, 113–122.
24. Lee, S.Y.; Lee, I.B.; Kim, R.W.; Yeo, U.H.; Kim, J.G.; Kwon, K.S. Dynamic energy modelling for analysis of the thermal and hygroscopic environment in a mechanically ventilated duck house. *Biosyst. Eng.* **2020**, *200*, 431–449. [CrossRef]
25. American Society of Heating, Refrigerating and Air-Conditioning Engineers (ASHRAE). *ASHRAE Handbook—Fundamentals*; American Society of Heating, Refrigerating and Air-Conditioning Engineers (ASHRAE): Atlanta, GA, USA, 2017.
26. Liu, G.; Liu, M. Development of simplified in-situ fan curve measurement method using the manufacturers fan curve. *Build. Environ.* **2012**, *48*, 77–83. [CrossRef]
27. Ministry of Agriculture, Food and Rural Affairs (MAFRA) and Nonghyup (NH). Standard Design of the Broiler House Released Copy 2016. 2019. Available online: https://livestock.nonghyup.com/dtar/blupr.do (accessed on 18 July 2022).
28. Laknizi, A.; Ben Abdellah, A.; Mahdaoui, M.; Anoune, K. Application of Taguchi and ANOVA methods in the optimisation of a direct evaporative cooling pad. *Int. J. Sustain. Eng.* **2021**, *14*, 1218–1228. [CrossRef]
29. Rong, L.; Pedersen, P.; Jensen, T.L.; Morsing, S.; Zhang, G. Dynamic performance of an evaporative cooling pad investigated in a wind tunnel for application in hot and arid climate. *Biosyst. Eng.* **2017**, *156*, 173–182. [CrossRef]
30. Xue, H.; Qiang, Z.; Ni, J.Q.; Baoming, L.; Zhengxiang, S.; Shumei, Z.; Yu, W. Effect of cooling pad installation on indoor airflow distribution in a tunnel-ventilated laying-hen house. *Int. J. Agric. Biol. Eng.* **2016**, *9*, 169–177.
31. Lee, J.H.; Munters Korea Co. Essential facilities to improve productivity in hot weather—Misconceptions, truths, and effects about cooling pads (2). *Korean Poult. J.* **2018**, 156–158.
32. Choi, H.C. *Ventilation Method and Cooling Pad Technology in the Heat Season*; National Institute of Animal Science, Rural Development Administration (NIAS RDA): Jeonju, Korea, 2015.
33. Perdersen, S.; Sällvik, K. *Heat and Moisture Production at Animal and House Levels 4th Report of Working Group on Climatization of Animal Houses*; CIGR: Horsens, Denmark, 2002.

34. Yoo, J.S. New Feeding and Management for the Production of Poultry Farming. 2009.
35. Lee, G.H. *Energy Plus: Modeling Techniques with Air Conditioning Theory*; Hansol Academy: Lalitpur, Nepal, 2018; ISBN 979-11-5656-627-4-13530.
36. U.S. Department of Energy Federal Energy Management Program. M&V Guidelines: Measurement and Verification for Performance-Based Contracts Version 4.0. DOE/EE-1287-0286; November 2015. Available online: femp.energy.gov (accessed on 15 July 2022).
37. Walton, G.N. *Thermal Analysis Research Program Reference Manual*; National Bureau of Standards: Washington, DC, USA, 1983.
38. Clark, G.; Allen, C. The Estimation of Atmospheric Radiation for Clear and Cloudy Skies. In Proceedings of the 2nd National Passive Solar Conference (AS/ISES), Philadelphia, PA, USA, 16–18 March 1978; pp. 675–678.
39. National Institute of Animal Science, Rural Development Administration (NIAS RDA). *Korean Feeding Standard for Poultry*; Evergreen Publisher: Suwon, Korea, 2007.
40. Casey, K.D.; Gates, R.S.; Wheeler, E.F.; Xin, H.; Liang, Y.; Pescatore, A.J.; Ford, M.J. On-Farm Ventilation Fan Performance Evaluations and Implications. *J. Appl. Poult. Res.* **2008**, *17*, 283–295. [CrossRef]
41. Park, G.-Y.; Lee, I.-B.; Yeo, U.-H.; Ha, T.-H.; Kim, R.-W.; Lee, S.-Y. Ventilation rate formula for mechanically ventilated broiler houses considering aerodynamics and ventilation operating conditions. *Biosyst. Eng.* **2018**, *175*, 82–95. [CrossRef]
42. U.S. Department of Energy. EnergyPlus Version 9.5.0 Documentation Input Output Reference. 2021. Available online: https://energyplus.net/assets/nrel_custom/pdfs/pdfs_v9.5.0/InputOutputReference.pdf (accessed on 1 May 2022).
43. Mount, L.E. Thermal neutrality. In *Heat Loss from Animals and Man: Assessment and Control*; Monteith, J.L., Mount, L.E., Eds.; Butterworths: London, UK, 1974; pp. 425–439.
44. DeShazer, J.A.; Beck, M.M. *University of Nebraska Report for Northeast Regional Poultry Project NE-127*; Agricultural Research Division, University of Nebraska: Lincoln, NE, USA, 1988.
45. Tao, X.; Xin, H. Acute synergistic effects of air temperature, humidity, and velocity on homeostasis of market-size broilers. *Trans. ASAE* **2003**, *46*, 491.
46. Chepete, J.; Chimbombi, E.; Tsheko, R. Production performance and temperature-humidity index of Cobb 500 broilers reared in open-sided naturally ventilated houses in Botswana. In Proceedings of the Seventh International Symposium, Livestock Environment VII, Beijing, China, 18–20 May 2005.
47. De Moraes, S.R.P.; Yanagi, J.; De Oliveira, A.L.R.; Yanagi, S.; Café, M.B. Classification of the temperature and humidity index (THI), aptitude of the region, and conditions of comfort for broilers and layer hens in Brazil. In Proceedings of the International Conference of Agricultural Engineering, XXXVll Brazilian Congress of Agricultural Engineering, International Livestock Environment Symposium-ILES VIII, Iguassu Falls City, Brazil, 31 August–4 September 2008.
48. Xin, H.; Berry, I.L.; Barton, T.L.; Tabler, G.T. Feed and water consumption, growth, and mortality of male broilers. *Poult. Sci.* **1994**, *73*, 610–616. [CrossRef]
49. Brown-Brandl, T.M.; Beck, M.M.; Schulte, D.D.; Parkhurst, A.M.; DeShazer, J.A. Temperature Humidity Index for Growing Tom Turkeys. *Trans. ASAE* **1997**, *40*, 203–209. [CrossRef]
50. National Research Council (NRC). *A Guide to Environmental Research on Animals*; National Academic Science: Washington, DC, USA, 1971.
51. U.S. Department of Agriculture. Heat Stress Forecast Maps. Available online: https://www.ars.usda.gov/plains-area/clay-center-ne/marc/documents/heat-stress/main/ (accessed on 1 August 2022).
52. Kansas State University. Animal Comfort. Available online: https://mesonet.k-state.edu/agriculture/animal/#tab=download-tab (accessed on 1 August 2022).
53. National Institute of Animal Science. Livestock Breeding Weather Information System. Available online: https://chuksaro.nias.go.kr/lwis/gis/tpIdex/tpIdex.do (accessed on 1 August 2022).

agriculture

Article

Research on the Flexible Heating Model of an Air-Source Heat Pump System in Nursery Pig Houses

Hua Wang, Jijun Liu, Zhonghong Wu, Jia Liu, Lu Yi, Yixue Li, Siqi Li and Meizhi Wang *

College of Animal Science and Technology, China Agricultural University, Beijing 100193, China; wanghuatzlm@163.com (H.W.); liujijun@cau.edu.cn (J.L.); wuzhh@cau.edu.cn (Z.W.); liujiafg@163.com (J.L.); 18315154236@163.com (L.Y.); lyx200120@163.com (Y.L.); 15162358059@163.com (S.L.)
* Correspondence: meizhiwang@cau.edu.cn

Abstract: Maximizing the utilization of renewable energy for heating is crucial for reducing energy consumption in pig houses and enhancing energy efficiency. However, the mismatch between peak solar radiation and peak heat load demand in nursery pig houses results in energy waste. Therefore, we investigated a flexible air-source heat pump system (F-ASHP) based on the hourly-scale energy transfer of solar energy. A theoretical calculation model for F-ASHPs in pig houses in the heating areas of northern China has been established through on-site testing and Simulink. This study investigated the heat storage and release of four energy storage materials in pens and the variation in heat load in the house, validating the accuracy of the model. The results show that the F-ASHP can effectively match the peak solar heat and peak heat load in the house. Among the four energy storage materials in pens, the magnesium oxide heat storage brick material performed the best. During intermittent solar periods, it released 3319.20 kJ of heat, reducing the heat load in the pig house by 10.1% compared with that by the air-source heat pump (ASHP). This study provides a theoretical model for flexible heating calculations in pig houses in northern China and aims to serve as a valuable resource for selecting energy-storage pens.

Keywords: energy saving; F-ASHP model; nursery pig; storage pen; Simulink

Citation: Wang, H.; Liu, J.; Wu, Z.; Liu, J.; Yi, L.; Li, Y.; Li, S.; Wang, M. Research on the Flexible Heating Model of an Air-Source Heat Pump System in Nursery Pig Houses. *Agriculture* **2023**, *13*, 1059. https://doi.org/10.3390/agriculture13051059

Academic Editors: Muhammad Sultan, Redmond R. Shamshiri, Md Shamim Ahamed and Muhammad Farooq

Received: 24 April 2023
Revised: 9 May 2023
Accepted: 11 May 2023
Published: 15 May 2023

1. Introduction

With the proposal of China's carbon peaking and carbon neutrality development goals, reducing energy consumption and increasing energy utilization efficiency in buildings have become increasingly important research topics across various fields [1–4]. Cold stress can significantly affect animal health and breeding, and an appropriate temperature is crucial for healthy nursery pig farming [5–8]. Nursery pigs require higher indoor temperatures, necessitating additional heating equipment in pig houses [9,10]. In recent years, the energy-saving and emission-reduction effects of pig houses have improved with the enhancement of the insulation performance of the enclosure structure and the application of clean energy [11,12]. However, in accordance with the "dual carbon" goal, building energy systems face higher development requirements to satisfy the need for further innovative low-carbon goals based on energy-saving measures. Therefore, researching the transition from energy-saving to low-carbon livestock houses, reducing heat load, improving heating efficiency in piglet houses, and promoting a flexible heating system that fully utilizes renewable energy to satisfy the fluctuating energy demand in livestock houses are significant for achieving low-carbon development in piglet houses and fostering a sustainable livestock industry.

Air-source heat pumps (ASHPs) are widely used for heating livestock houses, and their heating capacity is influenced by the external environmental temperature [13]. As the environmental temperature decreases, the evaporation temperature and pressure decrease, leading to an increase in the compressor pressure ratio and compressor power consumption and decreases in heating capacity and coefficient of performance (COP) [14,15].

Due to the intermittent and unstable nature of solar energy, the peak heating capacity of solar energy does not align with the peak heat load of livestock houses within a day [16]. In recent years, researchers have explored two main approaches to address these issues. The first approach involves combining solar energy with ASHP for heating purposes. Kim [17] designed a combined solar energy and ASHP system and conducted performance testing and optimization analysis, resulting in a 5.1% increase in the system's COP. Besagni [18] focused on optimizing the solar energy-coupled ASHP system by incorporating additional components, such as photovoltaic panels and inverters, into the collector, thereby creating a solar photovoltaic-thermal-coupled heat pump system. The results demonstrated a COP of up to 3.8, leading to improved thermal efficiency. Furthermore, Sezen [19] investigated the impact of ambient temperature on solar energy-coupled heat pumps and identified the optimal system COP for heating performance across different ambient temperature ranges.

The second approach involves the utilization of heat storage within the building envelope during peak solar periods and subsequent heat release during intermittent solar periods. Choi [20] conducted a study on a residential solar-air heating system that utilized foundation concrete as the heat storage material. The results indicated that the heat accumulated in the foundation concrete during the day could be effectively utilized during non-collection periods. By incorporating energy storage materials into building components, we can entirely exploit their heat storage and release characteristics, minimize heat waste during periods of non-utilization, and significantly enhance indoor thermal comfort.

Furthermore, Park [21] conducted experimental investigations on the thermal performance of phase-change bricks doped with phase-change materials in both summer and winter environments, comparing their effectiveness in controlling indoor temperatures with different roof materials. The results revealed that, during winter, the phase-change cold roof significantly improved the insulation performance of the roof structure, thereby reducing indoor heat loss. In another study, Ye [22] implemented the storage of cold energy within a building wall using a summer jet attachment. This approach enabled the utilization of stored cold energy at night, resulting in a reduction of the building's cooling load during summer daytime hours.

Heat pumps are highly efficient devices commonly used for cooling and heating purposes that can be integrated with renewable energy systems [23]. Simulink, an essential component of MATLAB, provides an integrated environment for modeling, simulating, and conducting comprehensive analyses of dynamic systems. Simulink has demonstrated excellent results in solving various time-varying system differential equations and can be employed for simulating the thermal environment within buildings [24,25].

The optimization of ASHP performance and building energy storage research is primarily based on residential and public buildings. The architectural structure of pig houses is different from that of residential buildings owing to the widespread use of special production processes, such as individual stalls, slatted floors, and exhaust fans. To date, research on storing renewable energy in energy storage pens through heating systems has not been conducted in pig houses for conservation. This study proposes a flexible air-source heat pump system (F-ASHP) model based on the hourly scale energy transfer of solar energy in a nursery pig house. Figure 1 shows a schematic diagram of the F-ASHP model. During the peak period of solar radiation P1 (12:00–16:00), the required temperature inside the pig houses for conservation increased, and the solar peak heat was stored in the energy storage pen. During the solar intermittent period P2 (16:00 to 12:00, the next day), the energy storage pen released heat, thereby reducing the heat load demand in the pig houses for conservation, transferring solar energy on an hourly scale, and achieving the goal of a flexible heat load demand in pig houses for conservation.

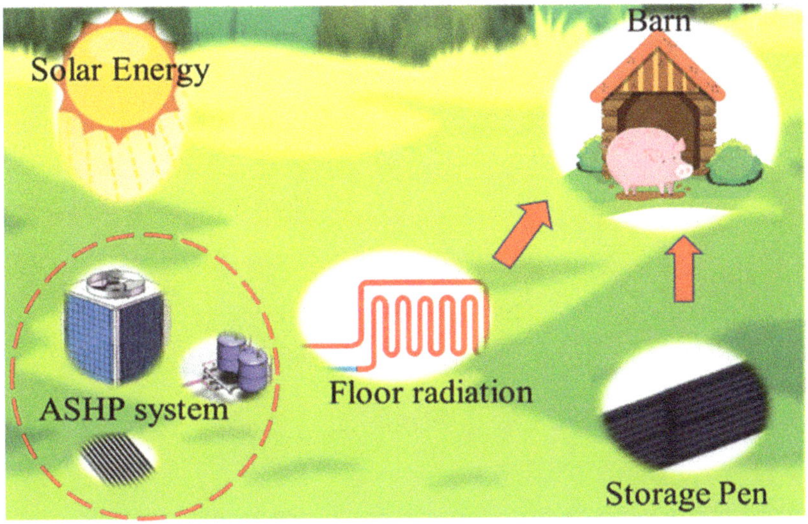

Figure 1. F-ASHP model.

This study constructs a mathematical model of the F-ASHP by coupling the ASHP, energy-storage pen, and solar collectors. The dynamic changes in heat storage, release time, and release amount in energy storage pens composed of different materials, as well as the dynamic change of heat load in pig houses for conservation, were studied with on-site testing and Simulink. Our findings provide the theoretical and technical foundation for reducing energy consumption and carbon emissions during the winter.

2. Materials and Methods

2.1. Geometric Model

The ASHP heating system was installed in a nursery pig house in Shunyi District, Beijing, China. The experimental pig house has a north–south orientation, and its peripheral dimensions are 42 m long and 9.3 m wide, as shown in Figure 2. There are 14 pens (3 m width and 7.5 m length each) in the pig house, with a 2 m × 3 m slatted floor area in each pen. The slat is 80 mm wide, and the opening is 20 mm wide. The pig house has two units, and each unit holds seven pens. The pig house is naturally ventilated. The wall of the pig house is made up of 0.37 m thick bricks and 90 mm thick extruded polystyrene board (for external insulation). The eave of the pig house is 2.9 m high, and the double pitch roof's material is color steel laminboard with a thickness of 100 mm. The windows are made of plastic and steel with two glass panes. There are 13 windows (2.00 × 1.45 m^2) and 1 window (1.15 × 1.15 m^2) in the southern wall, and 14 windows (1.00 × 0.80 m^2) in the northern wall. There is one window with a size of 1.18 × 1.07 m^2 and one window with a size of 0.74 × 1.17 m^2 in the eastern wall. There is one wooden door (0.90 × 1.77 m^2) in the western wall. There were 442 piglets in the two units of the pig house, and the age of the piglets was 61 d. The physical parameters of the building are shown in Table 1.

2.2. ASHP System

The thermal load of the pig house was 12.0 kW when the outdoor design temperature for heating was −7.6 °C in Beijing. Considering the efficiency of pipeline transportation and the safety factor of 1.3, the thermal load of the pig house needs to be configured at 15.6 kW. For heating the pig house, experimental heating equipment based on enhanced vapor injection technology, a type of low-temperature ASHP, was utilized. The outdoor extreme minimum temperature in Beijing can fall below −15 °C; thus, the heating capacity of the ASHP was 22.4 kW when the dry ball temperature was −7 °C and 20.2 kW when the dry ball temperature was −15 °C (parameters from the equipment factory). The ASHP utilized R410A refrigerant and was installed in a water tank with a thermal storage capacity of approximately 1000 L. The hot water stored in the tank was pumped to heat the pig house. In the event that the water temperature in the tank dropped below the set temperature, the ASHP would heat the water in the circulating system to supply hot water to the tank. When the water temperature in the tank was higher than the setting temperature, the ASHP would stop working, but the water in the tank would still be pumped to heat the pig house. The ASHP heating system is shown in Figure 3.

Figure 2. Experimental pig barn.

Table 1. Structure information.

Structure	Direction	Area (m^2)	Heat Transfer Coefficient ($W/m^2{\cdot}K$)
Door	West	1.59	4.7
Window	East	1.26	1.92
	South	39.02	4.7
	West	0.87	4.7
	North	11.2	4.7
Wall	East	36.52	0.35
	South	83.79	0.35
	West	35.32	0.35
	North	11.62	0.35
Floor	/	396.52	0.22
Roof	/	380.52	0.47

(**a**) Radiant floor (**b**) ASHP

Figure 3. ASHP systems.

2.3. Mathematical Models

The assumption was based on the heat losses in the ASHP system from the piping and heating water tank, and this study was based on a certain stocking density for 20 kg pigs. Figure 4 is a schematic diagram of the heat balance equations for the F-ASHP model, including the heat supply of the ASHP, heat supply of solar collectors, heat loss of the water tank, heat transfer through the enclosure structure of pig houses, pig heat production, ventilation heat loss, and heat release from the energy storage pen. Equations (1)–(7) represent the heat balance equations for each component of the F-ASHP model. A flexible dynamic simulation model of the F-ASHP was built based on MATLAB/Simulink (Figure 5), and the relative tolerance was less than 10^{-6}. Reference [12] describes the changes in the outdoor temperature and COP during the test period of the ASHP. During the test period, the heat storage water tank temperatures of 40 °C and 38 °C were used as the starting and stopping temperatures for the air-source heat pump, respectively.

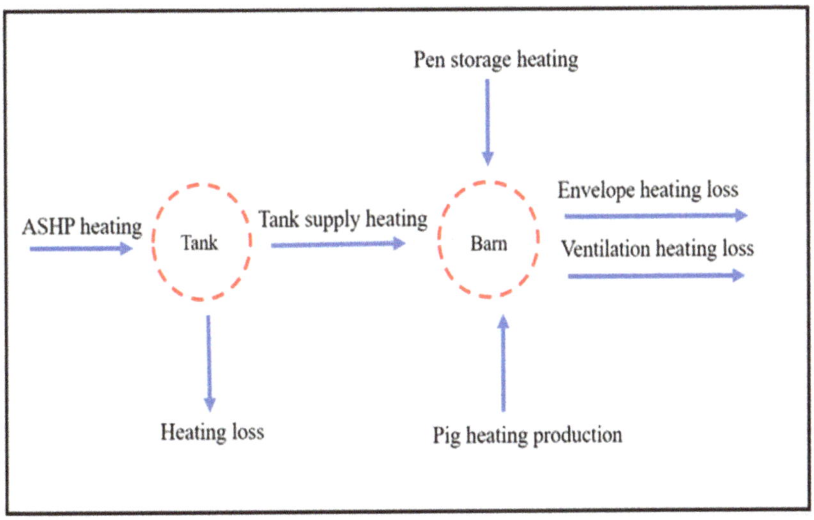

Figure 4. Heat balance schematic.

The major equations that were implemented in the subsystem of the Simulink model for F-ASHP were Equations (1) and (7), respectively [26].

$$C_p \times M_{TANK} \times \frac{dt}{d\tau} = Q_s + Q_{ASHP} - G \times C_p \times (T_{out} - T_{in}) - Q_{Loss} \quad (1)$$

where C_p is the water of specific heat capacity, J/kg·K; M_{TANK} is the quality of water tank, kg; Q_s is the solar energy, W; Q_{ASHP} is the heat of the ASHP, W; G is the quality flow rate, kg/s; T_{out} is the supply temperature of the tank, °C; T_{in} is the return temperature of the tank, °C; and Q_{LOSS} is the heat loss of the tank, W.

$$G \times C_p \times (T_{out} - T_{in}) = K_i \times A_b \times T_{barn} - T_{out} + Q_{ventilation} - Q_{pig} + Q_{source} \quad (2)$$

where G is the quality flow rate, kg/s; C_p is the water of specific heat capacity, J/kg·K; T_{out} is the supply temperature of the tank, °C; T_{in} is the return temperature of the tank, °C; K_i is the heat transfer coefficient of enclosure, W/m^2·K; A_b is the area of the barn, m^2; $Q_{ventilation}$ is ventilation heat loss, W; Q_{pig} is the pig heat production, W; and Q_{source} is the energy storage pen of heat, W;

$$T_a = T_{a,p} + (T_{a,min} - T_{a,p}) \times cos\left[\frac{2\pi}{N}(\tau - 3)\right] \quad (3)$$

where T_a is the ambient temperature, °C; $T_{a,min}$ is the min ambient temperature, °C; $T_{a,p}$; is the average ambient temperature, °C; and N is time, h.

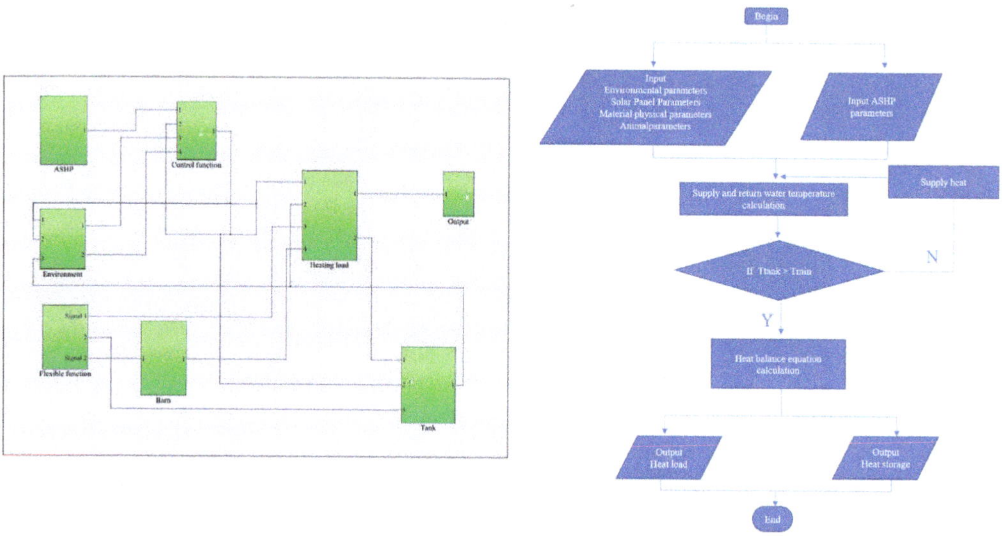

Figure 5. F-ASHP of the Simulink model.

$$\frac{dT_b}{d_\tau} = F_R \times [(S - U_L) \times \left(T_{f,i} - T_a\right)] + G \times C_p \times \left(T_{f,i} - T_{f,o}\right) \quad (4)$$

where $T_{f,i}$ is the heat collector inlet temperature, °C; $T_{f,o}$ is the heat collector outlet temperature, °C; T_a is the average ambient temperature, °C; U_L is the heating loss, W/m^2·°C; S is the solar radiation intensity, W/m^2; and F_R is the heat transfer factor.

$$Q_{source} = C_{sp} \times M_S \times \frac{dt}{d\tau} = K \times A_p \times (T_s - T_{barn}) \quad (5)$$

where C_{sp} is the energy storage pen of Specific heat capacity, J/kg·K; M_s is the quality of energy storage pen, kg; A_b is the Area of pen, m^2; T_s is the average temperature of energy storage pens, °C; and K is the heat transfer coefficient of energy storage pen, W/m^2·K.

$$Q_{pig} = \frac{1000 + 12 \times (20 - T_a)}{1000}\left\{5.09 \times m^{0.75} + [1 - (0.47 + 0.003 \times m)] \times \left[n \times 5.09m^{0.66} - 5.09m^{0.66}\right]\right\} \quad (6)$$

where m is the quality of nursery pig, kg; n is the ratio of energy obtained to the maintenance energy of the nursery pig; and T_a represents air temperature, °C.

$$Q_{ventilation} = C_{pa} \times V \times \rho \times T_{barn} - T_{out} \quad (7)$$

where C_{pa} is the air of specific heat capacity, J/kg·K; ρ is air density, kg/m^3; and V is the ventilation rate of the nursery pig, m^3/h.

2.4. Boundary Conditions

This study proposed an F-ASHP model in a nursery pig house, and 15 cases were used to simulate the heat storage and release, heat storage and release times, and variation in pig house heat load for different energy storage pen materials under 40 °C water tank set temperatures. The four storage pen materials are concrete, gray sand brick, magnesium oxide heat storage brick, and clay. The pen dimensions are 4.8 m in length, 2.4 m in width, and 0.03 m in thickness. The storage pen's materials and simulation conditions are listed in Tables 2–4.

Table 2. Storage pen materials.

Materials	Density (kg/m^3)	Specific Heat Capacity (J/kg·K)	Heat Exchange Area (m^2)	Quality (kg)
Concrete brick	2300	920	1.92	132.48
Gray sand brick	1900	1050	1.92	109.44
Clay brick	1842	1850	1.92	106.10
Magnesium oxide Storage brick	2500	1140	1.92	144.00

Table 3. Model parameters.

Parameter	Value	Parameter	Value
Water tank volume	1000 L	Number of pigs	448
Water specific heat capacity	4.2 kJ/kg·°C	Ventilation rate	5.8 m^3/h
Air specific heat capacity	1.005 kJ/kg·k	Collector area	6 m^2
Air density	1.29 kg/m^3	Energy utilization rate	0.8
Water density	1000 kg/m^3	Heat transfer factor	0.9
Nursery pig weight	13.6 kg	Barn set temperature	25
Average daily solar radiation	9.85 MJ/d	Average outdoor temperature	−7.9
Lowest temperature	−18.3 °C	Flow rate	110 L/h

Table 4. Simulation conditions.

Case	Tank Setting Temperature (°C)	Pen Material	Solar Energy	P1 Setting Temperature (°C)
1	40	Hollow plastic	/	26
2				27
3				28
4		Clay brick	Working	26
5				27
6				28
7		Gray sand brick	Working	26
8				27
9				28
10		Magnesium oxide Storage brick	Working	26
11				27
12				28
13		Concrete brick	Working	26
14				27
15				28

2.5. ASHP System Measuring Points

The ASHP heating system and indoor environment were monitored during the period between 15 November 2016 and 16 April 2017. The water temperatures were monitored by Pt100 temperature sensors (KZW/P-231, measuring range: 0~70 °C, accuracy: ±0.15 °C, KunLunZhongDa Sensors Co., Ltd., Beijing., China). All the data on water temperature and flow were displayed and recorded by a PLC system.

Air temperature and RH indoors and outdoors were monitored by a temperature and humidity automatic recorder (Apresys, 179A-TH, measuring range, RH: 0%~100%, temperature: −40~100 °C, accuracy: ±2% and ±0.2 °C, San Ramon, CA, USA). Ground temperature was monitored by ground thermometers (JTR09, measuring range: −20~85 °C, accuracy: <±4%, Tian Jian Hua Yi Co., Ltd., Beijing., China).

3. Results and Discussion

3.1. Model Validation

We referenced [11], monitored, and analyzed the ASHP supply and return water temperatures and the indoor environment of pig houses. A higher relative humidity will raise the dew-point temperature of the air, and air condensation will frost on the surface of the ASHP evaporator, affecting heat exchange. The lowest temperature in the Beijing area in the last 30 years was −11.8 °C. A defrosting device was activated after the evaporator surface reached the dew-point temperature, and condensation could occur in colder climate zones, which would affect the ASHP system.

To verify the accuracy of the simulation model, we selected Case 1. Figure 6 illustrates the measured and simulated values of the ASHP system's supply and return water temperatures on 4 January. The simulation results for the supply and return water temperatures of the ASHP heating system in the pig houses exhibited a similar changing trend to the measured values. During the noon period, the supply and return water temperatures were lower than those in other time periods. These values were attributed to the higher outdoor temperature during noon, which reduced the heat transfer from the pig houses to the outdoors and consequently decreased the heat load required by the pig houses. The maximum relative error between the simulated and measured values of the supply and return water temperatures was 5.1%. This fluctuation was attributed to short-term fluctuations in the outdoor temperature between 15:00 and 16:00, which deviated from the change trajectory of the outdoor temperature wave function. Overall, this mathematical model can be effectively utilized for numerical heat transfer research in pig houses for conservation purposes.

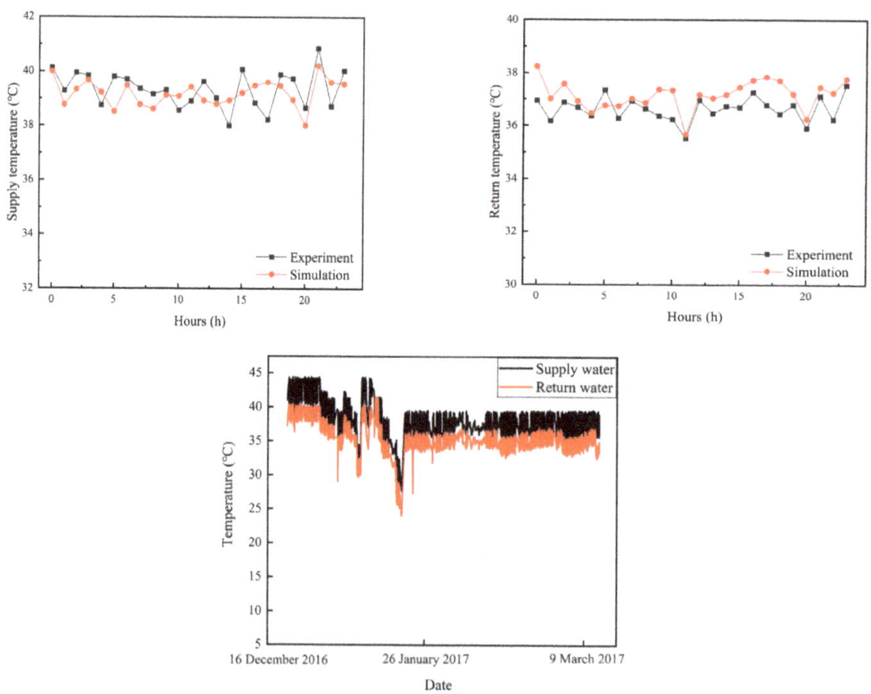

Figure 6. Experimental data and simulation validation of the model.

3.2. Analysis of Heat Storage and Release of the Pen

3.2.1. Analysis of Heat Storage

Cases 5, 8, 11, and 14 were selected to analyze the heat storage in the F-ASHP model. Figure 7 shows the heat storage of the four types of energy storage pens in operating hours, which demonstrated the same trend. The heat storage time required for the brick pen to increase by 2 °C was 14.4 h, whereas the heat storage time for the clay pen to rise by 2 °C was 15.0 h. The temperature change rate of the gray sand brick storage pen was the fastest, owing to its smallest specific heat capacity. The temperature rise rate of the storage pen of the four materials increased rapidly in the first 3.5 h, which could be explained by the larger initial temperature difference leading to a larger heat transfer rate. At 4 h of heat storage time, the average temperatures of the four types of material pens were 26.94 °C, 26.92 °C, 26.73 °C, and 26.82 °C, respectively.

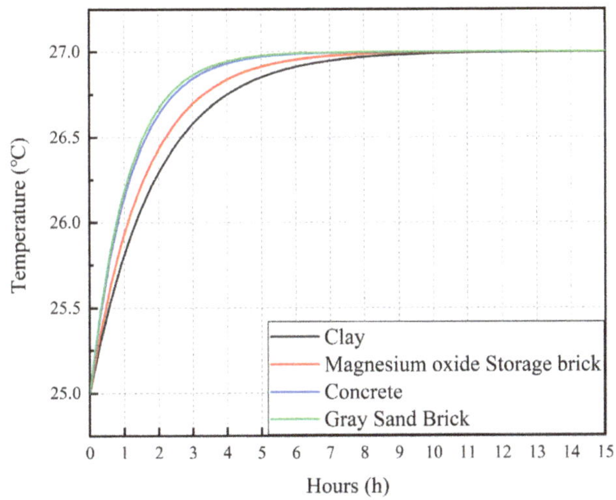

Figure 7. The heat storage of four pen materials.

3.2.2. Analysis of Heat Release

Figure 8 shows the heat release of the energy storage pens composed of four different materials (Cases 5, 8, 11, and 14) during the solar intermittent period. In the first 30 min of heat release, the gray sand brick storage pen released the most heat. As the heat release time increased, the heat release of the clay pen was higher than those of the other three materials, which was attributed to the highest specific heat capacity of the clay energy storage pig pen. During the solar intermittent period, the total heat release of individual clay energy storage pig pens, magnesium oxide storage brick pens, concrete brick energy storage pens, and gray sand brick energy storage pens was 374.44 kJ, 331.92 kJ, 270.72 kJ, and 260.64 kJ, respectively. This phenomenon indicates that the energy storage pen can effectively store energy during the solar peak period and achieve energy transfer on an hourly scale.

However, while the clay brick energy storage pig pen offered the best heat release effect, it can be adversely impacted by certain factors, such as cleaning and disinfection in the pig nursery. Therefore, we decided to use the magnesium oxide storage brick pen in the pig nursery.

Previous studies [27] have shown that pen attachment ventilation (PAV) for convective heating has better effects than ASHP floor radiant heating. PAV can directly perform convective heat exchange with the energy storage pen, and the combination of PAV and the energy storage pen may be increasingly conducive to energy storage during the solar peak period.

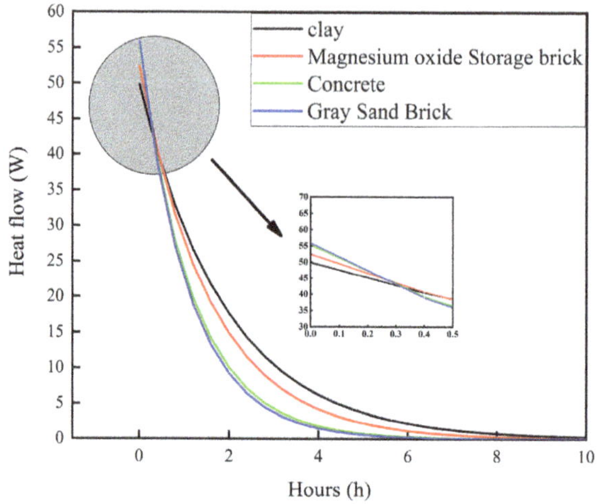

Figure 8. The heat release of four pen materials.

3.3. *The Analysis of F-ASHP Heat Load*

Figure 9 illustrates the heat load of the nursery pig house for Cases 1, 8, and 11. In Case 1, no flexible heat load regulation was observed, and the barn setting temperature was maintained at 25 °C. In Cases 8 and 11, the barn setting temperature increased during the P1 stage. The heating load in all three cases decreased during the P1 stage due to the rising outdoor temperature, which reduced the heat transfer within the enclosure. Cases 8 and 11 had higher heat loads during the P1 stage compared with that in Case 1 because they utilized solar energy for pig pen heat storage. During the P2 stage, the peak-valley heat loads for Cases 1, 8, and 11 were 7.80 kW, 7.31 kW, and 7.01 kW, respectively. Both Case 2 and Case 11 had lower heat loads than Case 1, indicating that the energy storage pen effectively facilitated hourly energy transfer, reducing the indoor heat load demand during periods of solar intermittency. The heat loads of Cases 8 and 11 in the P2 stage exhibited a similar trend. However, Case 8 had a higher heat load during the P2 stage compared with that in Case 11, attributable to the lower heat capacity of the energy storage pen in Case 8, which stored less heat during the P1 stage.

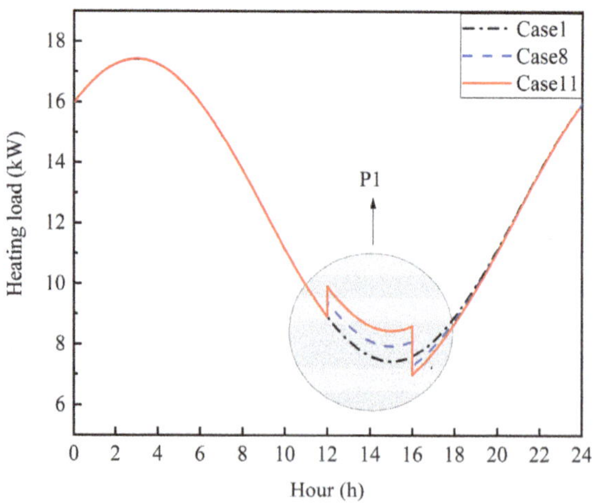

Figure 9. The heat load curve under different working conditions.

The decrease in barn heat load was highly pronounced during the early stages of P2, as the temperature of the energy storage pig pen was higher than the indoor temperature, resulting in a faster heat transfer rate. Concurrently, the outdoor temperature gradually decreased with the setting sun, increasing the heat transfer within the indoor enclosure. Compared with that in Case 1, the use of a magnesium oxide energy storage pen with a 2-degree increase during the P1 stage reduced the heat load of the ASHP by 10.1% in the P2 stage, highlighting the elasticity in heat load demand achieved by the F-ASHP model in the pig nursery. However, not all the heat added in the P1 stage was transferred to the P2 stage, as some of it was lost through the enclosure, and the insulation effect of the enclosure affected the energy storage during the P1 stage. In addition, the energy storage pen and F-ASHP were suitable for regions with abundant sunlight, whereas cloudy conditions could affect the energy storage and transfer efficiency of the energy storage pig pen.

4. Conclusions

In this study, a flexible heating model combining an ASHP and energy storage pig pens was proposed. Using a nursery pig house in Beijing, simulations were conducted to analyze different material energy storage pens, heat storage capacity, heat storage time, and variations in barn heat load using on-site testing and Simulink. The main conclusions are summarized as follows:

(1) The F-ASHP model for a nursery pig house can accurately predict the heat load demand with a relative error of 5.1%. This model can be used to evaluate the feasibility of F-ASHP in different regions with abundant solar energy.
(2) The energy storage pen effectively stores heat during solar peak times and releases it during solar intermittency, achieving hour-scale energy transfer. Compared with the ASHP system, the magnesium oxide energy storage pen reduced the heat load demand in the pig house by 10.1%, resulting in a highly flexible barn heat load demand.
(3) Among the four energy storage pens, the magnesium oxide storage brick demonstrated the best performance, with a total accumulated heat storage of 3319.20 kJ.

5. Limitations

This study proposed that the F-ASHP heating model could achieve hour-scale energy transfer of solar energy and reduce the internal heat load of nursery pig houses during solar intermittency through the use of an energy storage pen. However, this study still has some limitations. Principally, it investigated the F-ASHP model using theoretical calculations, where differences in its actual application in pig production may be present. Additionally, energy storage pens are made of solid materials, and their heat storage capacity is affected by the specific heat capacity of the specific materials used. Therefore, future studies should further monitor and analyze the indoor thermal environment of the F-ASHP heating system and investigate a flexible heating model of phase-change energy storage for pig pens coupled with a PAV.

Author Contributions: Data curation, L.Y., J.L. (Jia Liu) and S.L.; investigation, H.W.; methodology, H.W., J.L. (Jijun Liu) and Z.W.; project ad-ministration, M.W.; software, H.W.; supervision, M.W.; writing—original draft, H.W.; writing—review and editing, H.W., Y.L., S.L. and M.W. All authors have read and agreed to the published version of the manuscript.

Funding: This research is supported by the Science and Technology Innovation 2023—"New Generation Artificial Intelligence" Major Project of China (2022ZD0115704).

Institutional Review Board Statement: Not applicable.

Data Availability Statement: All datasets from the current study are available from the corresponding author upon reasonable request.

Conflicts of Interest: The authors declare that they have no known competing financial interests or personal relationships that could have appeared to influence the work reported in this paper.

Nomenclature

Symbols	
A_b	Area of barn, m^2
A_p	Area of pen, m^2
C_{pa}	Air of specific heat capacity, J/kg·K
C_p	Water of specific heat capacity, J/kg·K;
C_{sp}	Energy storage pen of specific heat capacity, J/kg·K
F_R	Heat transfer factor
G	Quality flow rate, kg/s
K	Heat transfer coefficient of energy storage pen, $W/m^2{\cdot}K$
K_i	Heat transfer coefficient of enclosure, $W/m^2{\cdot}K$
m	Quality of nursery pig, kg
M_s	Quality of energy storage pen, kg
M_{TANK}	Quality of water tank, kg
Q_{ASHP}	Heat of the ASHP, W
Q_{Loss}	Heat loss of the tank, W
Q_{pig}	Pig heat production, W
Q_s	Solar energy, W
Q_{source}	Energy storage pen of heat, W
$Q_{ventilation}$	Heat loss of Ventilation, W
S	Solar radiation intensity, W/m^2
T_a	Ambient temperature, °C
$T_{a,p}$	Average ambient temperature, °C
$T_{a,min}$	Min ambient temperature °C
T_{barn}	Supply air temperature, °C
$T_{f,i}$	Heat collector inlet temperature, °C
$T_{f,o}$	Heat collector outlet temperature, °C
T_{in}	Return temperature of the tank, °C
T_{out}	Supply temperature of the tank, °C
T_s	Average temperature of energy storage pens, °C [2]
U_L	Heating loss, $W/m^2{\cdot}°C$
V	Ventilation rate of nursery pig, m^3/h.
ρ	Air density, kg m^{-3}
Abbreviations	
ASHP	Air-source heat pump
F-ASHP	Flexible air-source heat pump system
COP	Coefficient of performance
PAV	Pen-attached ventilation

References

1. Goel, A.; Ganesh, L.S.; Kaur, A. Sustainability integration in the management of construction projects: A morphological analysis of over two decades' research literature. *J. Clean. Prod.* **2019**, *236*, 117676. [CrossRef]
2. Zhang, Y.; Jiang, X.; Cui, C.; Skitmore, M. BIM-based approach for the integrated assessment of life cycle carbon emission intensity and life cycle costs. *Build. Environ.* **2022**, *226*, 109691. [CrossRef]
3. Yang, X.; Hu, M.; Wu, J.; Zhao, B. Building-information-modeling enabled life cycle assessment, a case study on carbon footprint accounting for a residential building in China. *J. Clean. Prod.* **2018**, *183*, 729–743. [CrossRef]
4. Wang, Y.; Liu, J.; Zhao, Z.; Ren, J.; Chen, X. Research on carbon emission reduction effect of China's regional digital trade under the "double carbon" target—Combination of the regulatory role of industrial agglomeration and carbon emissions trading mechanism. *J. Clean. Prod.* **2023**, *405*, 137049. [CrossRef]
5. Bao, J.; Xie, Q. Artificial intelligence in animal farming: A systematic literature review. *J. Clean. Prod.* **2022**, *331*, 129956. [CrossRef]
6. Yeo, U.-H.; Lee, I.-B.; Kim, R.-W.; Lee, S.-Y.; Kim, J.-G. Computational fluid dynamics evaluation of pig house ventilation systems for improving the internal rearing environment. *Biosyst. Eng.* **2019**, *186*, 259–278. [CrossRef]
7. Zheng, P.; Zhang, J.; Liu, H.; Bao, J.; Xie, Q.; Teng, X. A wireless intelligent thermal control and management system for piglet in large-scale pig farms. *Inf. Process. Agric.* **2021**, *8*, 341–349. [CrossRef]
8. Gautam, K.R.; Rong, L.; Zhang, G.; Bjerg, B.S. Temperature distribution in a finisher pig building with hybrid ventilation. *Biosyst. Eng.* **2020**, *200*, 123–137. [CrossRef]

9. Vasdal, G.; Møgedal, I.; Bøe, K.E.; Kirkden, R.; Andersen, I.L. Piglet preference for infrared temperature and flooring. *Appl. Anim. Behav. Sci.* **2010**, *122*, 92–97. [CrossRef]
10. Requejo, J.M.; Garrido-Izard, M.; Correa, E.C.; Villarroel, M.; Diezma, B. Pig ear skin temperature and feed efficiency: Using the phase space to estimate thermoregulatory effort. *Biosyst. Eng.* **2018**, *174*, 80–88. [CrossRef]
11. Wang, H.; Yi, L.; Wu, Z.; Liu, J.; Wang, M. Feasibility of the air source heat pump system for heating wine houses in Beijing. *Trans. CSAE* **2021**, *37*, 236–242.
12. Wang, M.; Yi, L.; Liu, J.; Yang, L.; Chen, Z.; Wu, Z. Study on heating effect of air source heat pump for floor heating of nursery pig house in Beijing. *Trans. CSAE* **2019**, *35*, 203–210.
13. Tu, Q.; Zhang, L.; Li, L.; Deng, C.; Wang, B.; Gu, B.; Dai, Z. Comparison of Application Effects of Capillary Radiation Heat Pump and Electric Heating Wire in Greenhouse Seedling Cultivation. *Agriculture* **2022**, *12*, 1401. [CrossRef]
14. Ma, J.; Kim, D.; Braun, J.E.; Horton, W.T. Development and validation of a dynamic modeling framework for air-source heat pumps under cycling of frosting and reverse-cycle defrosting. *Energy* **2023**, *272*, 127030. [CrossRef]
15. Bansal, P.; Mohabir, A.; Miller, W. A novel method to determine air leakage in heat pump clothes dryers. *Energy* **2016**, *96*, 1–7. [CrossRef]
16. Song, B.; Bai, L.; Yang, L. Analysis of the long-term effects of solar radiation on the indoor thermal comfort in office buildings. *Energy* **2022**, *247*, 123499. [CrossRef]
17. Kim, T.; Choi, B.-I.; Han, Y.-S.; Do, K.H. A comparative investigation of solar-assisted heat pumps with solar thermal collectors for a hot water supply system. *Energy Convers. Manag.* **2018**, *172*, 472–484. [CrossRef]
18. Besagni, G.; Croci, L.; Nesa, R.; Molinaroli, L. Field study of a novel solar-assisted dual-source multifunctional heat pump. *Renew. Energy* **2019**, *132*, 1185–1215. [CrossRef]
19. Sezen, K.; Tuncer, A.D.; Akyuz, A.O.; Gungor, A. Effects of ambient conditions on solar assisted heat pump systems: A review. *Sci. Total Environ.* **2021**, *778*, 146362. [CrossRef]
20. Choi, Y.; Takase, K.; Mae, M. System performance of a residential building using the air-based solar heating system. *Solar Energy* **2018**, *171*, 47–63. [CrossRef]
21. Chung, M.H.; Park, J.C. Development of PCM cool roof system to control urban heat island considering temperate climatic conditions. *Energy Build.* **2016**, *116*, 341–348. [CrossRef]
22. Ye, X.; Kang, Y.; Yang, X.; Zhong, K. Temperature distribution and energy consumption in impinging jet and mixing ventilation heating rooms with intermittent cold outside air invasion. *Energy Build.* **2018**, *158*, 1510–1522. [CrossRef]
23. El Haj Assad, M.; Nazari, M.A.; Ehyaei, M.A.; Rosen, M.A. Chapter 11—Heat pumps and absorption chillers. In *Design and Performance Optimization of Renewable Energy Systems*; Assad, M.E.H., Rosen, M.A., Eds.; Academic Press: Cambridge, MA, USA, 2021; pp. 163–180.
24. Wang, R.; Feng, W.; Xue, H.; Gerber, D.; Li, Y.; Hao, B.; Wang, Y. Simulation and power quality analysis of a Loose-Coupled bipolar DC microgrid in an office building. *Appl. Energy* **2021**, *303*, 117606. [CrossRef]
25. Zhu, M.; Pan, Y.; Wu, Z.; Xie, J.; Huang, Z.; Kosonen, R. An occupant-centric air-conditioning system for occupant thermal preference recognition control in personal micro-environment. *Build. Environ.* **2021**, *196*, 107749. [CrossRef]
26. Wang, X.; Cao, M.; Hu, F.; Yi, Q.; Amon, T.; Janke, D.; Xie, T.; Zhang, G.; Wang, K. Effect of Fans' Placement on the Indoor Thermal Environment of Typical Tunnel-Ventilated Multi-Floor Pig Buildings Using Numerical Simulation. *Agriculture* **2022**, *12*, 891. [CrossRef]
27. Wang, H.; Liu, J.; Wu, Z.; Feng, G.; Shen, Z.; Wang, M. Airflow characteristics of attachment ventilation in a nursery pig house under heating mode. *Biosyst. Eng.* **2022**, *224*, 346–360. [CrossRef]

agriculture

MDPI

Article

Smart Operation of Climatic Systems in a Greenhouse

Aurora González-Vidal [1,2], José Mendoza-Bernal [1], Alfonso P. Ramallo [1,*], Miguel Ángel Zamora [1], Vicente Martínez [3] and Antonio F. Skarmeta [1]

1 Department of Information and Communication Engineering, University of Murcia, 30100 Murcia, Spain
2 Information Technologies Institute, The Centre for Research & Technology, Hellas (ITI-CERTH), 57001 Thessaloniki, Greece
3 Department of Vegetal Nutrition, Centro de Edafología y Biología Aplicada del Segura del Consejo Superior de Investigaciones Científicas (CEBAS-CSIC), 30100 Murcia, Spain
* Correspondence: alfonsop.ramallo@um.es

Abstract: The purpose of our work is to leverage the use of artificial intelligence for the emergence of smart greenhouses. Greenhouse agriculture is a sustainable solution for food crises and therefore data-based decision-support mechanisms are needed to optimally use them. Our study anticipates how the combination of climatic systems will affect the temperature and humidity of the greenhouse. More specifically, our methodology anticipates if a set-point will be reached in a given time by a combination of climatic systems and estimates the humidity at that time. We performed exhaustive data analytics processing that includes the interpolation of missing values and data augmentation, and tested several classification and regression algorithms. Our method can predict with a 90% accuracy if, under current conditions, a combination of climatic systems will reach a fixed temperature set-point, and it is also able to estimate the humidity with a 2.83% CVRMSE. We integrated our methodology on a three-layer holistic IoT platform that is able to collect, fuse and analyze real data in a seamless way.

Keywords: smart agriculture; greenhouse technologies; artificial intelligence

Citation: González-Vidal, A.; Mendoza-Bernal, J.; Ramallo, A.P.; Zamora, M.Á.; Martínez, V.; Skarmeta, A.F. Smart Operation of Climatic Systems in a Greenhouse. *Agriculture* **2022**, *12*, 1729. https://doi.org/10.3390/agriculture12101729

Academic Editors: Muhammad Sultan, Redmond R. Shamshiri, Md Shamim Ahamed and Muhammad Farooq

Received: 9 September 2022
Accepted: 7 October 2022
Published: 19 October 2022

1. Introduction

Traditional agriculture can no longer adequately absorb the world's growing demand for food. Water scarcity and climate change seriously threaten global food security for present and future generations. In this context, smart agriculture emerges as a new approach that focuses on optimizing production efficiency, increasing quality, minimizing environmental impact and reducing the use of resources (energy, water and fertilizers) [1].

The emergence of smart agriculture implies high productivity ratios thanks to the modernization of irrigation and climate systems and the application of advanced control systems. Greenhouses are widespread within Europe, with an area of nearly 405,000 dedicated hectares [2], and the total greenhouse agricultural area in Spain exceeded 73 thousand hectares in 2021 [3]. In addition, awareness of the potential environmental impact of these intensive systems and, above all, the existence of more sustainable methods and technologies in agriculture is permeating into the population, which causes changes in behaviors and new regulations for farmers and technology providers.

Among the advances applied in protected crops with strong growth potential is the use of soil-less crops, which allow for increases in productivity and a more efficient use of water and fertilizers. Initially, most soil-less farming systems were open, where drainage is released into the environment. This leads to a high water consumption, and approximately 31% of nitrates and 48% of potassium applied during the crop cycle are discharged into the environment, causing the contamination of aquifers and possibly environmental problems of eutrophication also [4]. However, in recent years, European policies have been aimed at making agricultural practices more sustainable, with initiatives such as the Green Deal [5]. In the Mediterranean area, where more than 60% of its production is under greenhouses, the implementation of closed systems in crops without soil would have a significant environmental impact, since it would notably reduce discharges of nitrates and phosphates into the environment. In addition, it would mean significant savings in fertilizers and water, these being scarce resources in the Mediterranean area.

An important challenge in smart agriculture applied to greenhouses is the high consumption of energy when one wants to carry out a certain degree of automation of the production system. The use of renewable energies and energy efficiency applied to the different climatic and fertilization machinery makes these new agricultural systems more profitable than traditional agriculture and reduces their carbon footprint. This is because the solar resource tends to be high in agricultural sites.

The introduction of artificial intelligence (AI) and massive data processing (big data) as tools in the field of smart agriculture and greenhouses opens up a range of possibilities. These new techniques help to address the challenges of this new type of agriculture, such as the optimization of production with respect to resources, pest control, rapid reaction to adverse situations, predictions of extreme events, etc., in order to make this type of agriculture a sustainable and profitable industrial sector in the long term.

In this work, we present an ICT platform that is composed of three different technological layers. The solution has the latest innovations, such as the use of big data processing and machine learning as a service applied to a closed-cycle hydroponic greenhouse in order to estimate the effects of choosing a specific combination of climatic systems. The authors believe that this is a relevant research question as the piping of the hydroponic system and the fact that the roots are not embedded in the soil contribute to substantial inertia, making the thermodynamics of the environment affecting the plant truly different than those of a conventional greenhouse.

The specific objectives of this work are as follows:

- To create a ICT platform that is able to gather data, fuse it, analyze it and provide greenhouse-related services;
- To create a methodology that anticipates the thermal response of a greenhouse according to real-time variables using artificial intelligence:
  - Gather and prepare the data to analyze the operation of different climatic systems and their effect on temperature and humidity;
  - Create classification models that anticipate if a combination of climatic systems will reach the desired temperature in a greenhouse in a certain time;
  - Create regression models to estimate greenhouse humidity;
- To integrate the created methodology into the created platform.

This will serve as a tool to optimize the use of different climatic systems attending the energy consumption in order to increase the efficiency of the installation.

This paper is organized as follows: Section 2 contains the review of the literature on the use of AI in greenhouses. Section 3 provides an overview of the ICT system architecture (explaining the three layers) and Section 4 delves into the analytical proposal, including the data processing and the description of the classification and regression models. Section 5 shows the experimental set-up and how our methodology can be integrated into any platform, and discusses the novelty and originality together with some practical applications of the work. Section 6 concludes the work.

2. State of the Art

Agriculture has a primary role in meeting global nutrition needs. In recent decades, the food and agriculture sectors have passed through substantial changes [6]. Improvements in agricultural productivity are expected to be increased in order to ensure planned development, improvements and maintenance that may cover the needs of growing world populations [7]. The excessive use of phytosanitary products, soil tillage techniques and production must necessarily change if we want to meet current challenges. Climate change, the depletion of fossil resources, environmental issues (protection of biodiversity, fight against pollution, etc.) and health standards require a change in model to one that tends towards a more sustainable use of the soil [8].

Protected cropping is defined as growing within, under or sheltered by structures such as cover material, shade cloths, plastic tunnels or greenhouses. The level of protection and control can range from an inexpensive canopy (e.g., fabric/cloth) in a field to greenhouses or complete controlled environment horticulture systems [9]. In recent years, soil-less cultivation has become increasingly important as a promising strategy for growing a variety of crops. It is difficult to cultivate crops in horticulture greenhouses under hot and arid climate conditions. The main challenge is to provide a suitable greenhouse indoor environment with sufficiently low costs and with low environmental impacts. Once in place, greenhouses are highly non-linear, complex, multi-input and multi-output (MIMO) systems when looked at as thermodynamic enclosures that respond to changes in morphology and operations with changes in temperature and humidity [10]. Sustainable agriculture cultivation in greenhouses is constantly evolving thanks to new technologies and methodologies able to improve the crop yield and solve the common concerns that occur in protected environments [11]. Due to climate change, variations in weather have occurred in the last decades, and the introduction of new methodologies and technologies has become essential in supporting agriculture and optimizing greenhouse productions [12].

Soil-less cultivation (hydroponics) allows for a higher control of plant growth thanks to the better control of the feeding medium. This leads to a high productivity and better product quality, as well as a very high efficiency of water and fertilizer use [13]. Especially in regions with limited water or soil resources, hydroponic cultivation techniques can open up new approaches to food production. Not all systems are equally efficient, nor can they be applied in all areas and locations. Nowadays, the cultivation of horticultural crops, including leafy and fruiting vegetables and medicinal herbs with pharmaceutical value, are commercially grown in recycled (i.e., recirculating) hydroponics under controlled environments [14]. Recirculating hydroponics has great implications in practice under controlled environment agriculture toward economic considerations and environmental sustainability. Even more controlled agriculture is generally used in indoor farming plant factories for producing a range of high-value crops such as leafy and fruiting vegetables and medicinal plants under artificial light organized vertically [15]. In addition to the lighting and nutrient supply, the numerous systems, applications, substrates, organisms and their economic viability and sustainability have to be considered in order to obtain an understanding of whether and when it is worthwhile to use soil-less techniques [16].

Hydroponics could solve a considerable weakness of Mediterranean greenhouses. This is the large amount of energy required to maintain optimal environmental conditions for crop growth [17], which limits the operation period to approximately 9 months due to very high summer temperatures. The major challenge in Mediterranean greenhouses is to find ways to improve the yield per drop of water and unit of energy. An increased investment is required and needs to be considered in terms of the return on investments [18]. Greenhouses and indoor farming systems play an important role in providing fresh food, such as fruits and vegetables being high in vitamins and minerals. Controlled-environment greenhouses combine a high crop production per unit area with a high water use efficiency per unit of product [19], but at the cost of a high energy demand [20] and high investments. Increasing agricultural systems' resource efficiency is a key action for producing adequate food quantities in semi-arid Mediterranean regions while coping with water scarcity, environmental constraints and economic issues. There have been several studies on the high consumption of energy that some greenhouse solutions may have [21]. Recently, Soussi et al. (2022) provided an updated literature review of the climate control methods and cooling systems, with a particular focus on their reliability under hot and arid climate conditions. The main criteria of the performance evaluation are the effectiveness of these systems in generating suitable climate conditions for crops, and their reduction in the use of energy and water. Due to their lightweight construction and inefficient insulation, greenhouses are considered as one of the most energy-intensive agents of the agricultural industry. According to the literature, they consume more fossil-fuel-generated energy than that used by buildings of a similar size [22]. Smart agriculture applied to greenhouses is considerably interesting from an energy point of view. The growing request for the quality and production of fruits and vegetables imposes on the producer the equipping of greenhouses with suitable heating and cooling units to regulate the temperature, humidity and lighting, driving a healthy and controlled production. In smart applications, in general, it is mandatory to select proper high-efficiency technical solutions to reduce the energy consumption and, as a consequence, the production costs and environmental impact [12].

The future smart agriculture will be based on advanced data acquisition combined with several technologies. All of these outlooks involve data management in a certain way. With the development of open-source and big data, different techniques have emerged to remedy the limitations of traditional decision-making systems [23]. Sustainable agricultural development is a significant solution with fast population development through the use of information and communication (ICT) in precision agriculture, and has produced new methods for making cultivation more productive, proficient and well-regulated while not contributing more to climate change. Big data (machine learning, deep learning, etc.) is among the vital technologies of ICT, employed in smart agriculture for their rather large data analytic capabilities to abstract significant information and to assist agricultural professionals in comprehending full-farming practices on all of their facets, helping them to take precise decisions [24].

3. System Architecture

The future of smart agriculture will be based on advanced data acquisition combined with several technologies. All of these outlooks involve data management in a certain way. With the development of open-source and big data, different techniques have emerged to remedy the limitations of traditional decision-making systems [23]. Sustainable agricultural development is a significant solution to fast population development through the use of information and communication (ICT) in precision agriculture, which has produced new methods for making cultivation more productive, proficient and well-regulated while not contributing more to climate change. Big data (machine learning, deep learning, etc.) is among the vital technologies of ICT, employed in smart agriculture for their rather large data analytic capabilities to abstract significant information and to assist agricultural professionals in comprehending full-farming practices on all of their facets, helping to take precise decisions [24].

The data sources of the lower layers are fed into an NGSI-LD-based platform (an evolution from the European FIWARE initiative, based on the ETSI standards institute), enabling massive data management and providing support for the integration of machine learning and data analytic tools. This third layer (services layer) allocates the database with all of the information about the status of the crop and parameters, and it represents the interface with users. Moreover, the FIWARE-based platform employs standardized IoT-data protocols to facilitate the acquisition, integration and exchange of massive data from CPS and gateways, and also provides an interface to the big data techniques. The overall architecture scheme can be seen in Figure 1.

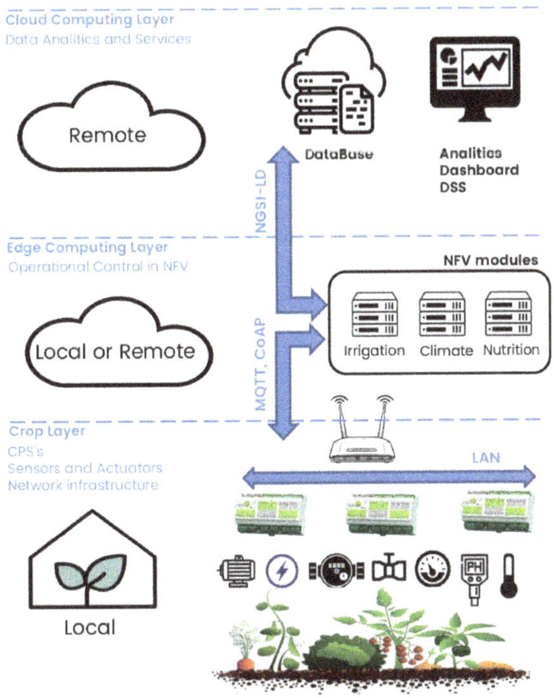

Figure 1. Architecture of the control system in the greenhouse.

4. Materials and Methods

The greenhouse under study had a set of equipment mainly deployed inside a machinery room beside the crops within the greenhouse. These facilities were composed of several hardware units; the climate unit is the one that this work focuses on.

The greenhouse had an IPv6 over Low Power Wireless Personal Area Network (6LoWPAN) deployment, where sensors are directly interconnected using Internet protocols. The sensors measure temperature, humidity, solar radiation and CO_2. A CPS device was provided with peripherals dedicated to managing the different climatic devices. Although the platform is capable of managing up to thirteen climatic systems, the physical facilities of the prototype greenhouse have the following hardware available:

- Ventilation system with four overhead motorized windows.
- Thermal-shade screen system on the roof, with an electro-mechanical traction system for opening and closing. It receives 48% of shadow and 55% of energy saving.
- Air cooling system, which consists of a humidifier module and three helical extractors. The extractors have a flow of 38,000 m^3/h.
- Air fog system used to humidify and cool the greenhouse through water evaporation. It has a pressure pump of 1 HP and 25 L, and an air compressor of 4 HP and 50 L.

4.1. Data Preprocessing

The greenhouse sensors provide climatological data on temperature, humidity, radiation and photosynthetically active radiation (PAR). The latter discriminates the spectral range of solar radiation from 400 to 700 nanometers, which photosynthetic organisms are able to use in the process of photosynthesis. The data were collected between 1st February and 31st August, including, therefore, the hottest period in Murcia, Spain, when the true effect of the climate systems in lowering the temperature and controlling the humidity in the desired ranges can be seen. For some variables, such as temperature and humidity, there were two sensors placed at 1.5 m (edge of the plants) and 3 m (above the plants) and places within the greenhouse. For analytical purposes, we used the average value for both.

While the sampling frequency of the sensors was 60 s, the climate systems can be switched on or off for less time, and also their sampling may not coincide in time. In order to homogenize the databases, we upsampled the time series to 10 s. For this purpose, a simple linear interpolation was performed by taking the values provided by the sensors of two consecutive samples ($v[i]$ and $v[i+1]$), dividing the time interval between both measurements into equal parts and assigning to each of them the value:

$$v(k) = v[i] + \frac{v[i+1] - v[i]}{6} \cdot k, \qquad \forall k \in \{1, \ldots, 6\}, \tag{1}$$

where $i = \{1, \ldots, N-1\}$.

$$v'(6 \cdot i + k) = v[i] + \frac{v[i+1] - v[i]}{6} \cdot k \tag{2}$$

where $i = \{0, \ldots, N-1\}$ and $k = \{0, \ldots, 6\}$.

Categorical variables were stored only every time a change in their status occurs. For example, windows return their percentage of opening and climatic systems, for which, data are available for the date and time of switching on and off; they were replicated in a similar way to the numerical variables, but taking into account that the intermediate values to be assigned between two consecutive measurements $w[i]$ and $w[j]$ are:

$$w'(k) = w[i], \quad \forall k \in \{i+1, \ldots, j-1\} \tag{3}$$

In addition to the sensors inside the greenhouse, we integrated the data from an external weather station that is next to the greenhouse. The weather station provides data on temperature, relative humidity, radiation, wind direction, wind speed and precipitation. The sampling frequency of the sensors in the station was five minutes. The same linear interpolation was applied for 10 s of upsampling to homogenize this data also.

The so-called vapor pressure deficit (vpd) was calculated inside and outside the greenhouse. The vapor-pressure deficit is the difference (deficit) between the amount of moisture in the air and how much moisture the air can hold when it is saturated. This variable is closely related to crop development and yield [25,26], and could be useful for estimating which climate systems should be switched on at which time. The vpd was calculated as follows in Equation (4).

$$vpd(t, rh) = svp(t) - \frac{rh}{100 \cdot svp(t)} \tag{4}$$

where t is the temperature in Celsius degrees (°C), rh is the relative humidity (percentage) and $svp(t)$ is the saturation vapor pressure (kPa), and is calculated by the Tetens formula [27] that can be seen in Equation (5).

$$svp(t) = 0.6108 \cdot e^{\frac{17.27 \cdot t}{t + 237.3}} \tag{5}$$

Once the data were generated, we looked for missing values, errors in the sensor measurements or inconsistencies. There was a period of time (between 8 June and 23 July) when one of the humidity sensors in the greenhouse did not work properly. To compensate for these small gaps, two regression models were built as a function of the other variables and taking into consideration the climate systems. The regression models were radial support vector machines (RSVM) and a random forest (RF). These two models were chosen because they are well known and perform well in regression problems. The rest of the variables were within reasonable parameters and no missing values were found.

The regression models were trained and evaluated with the data, where the humidity inside the greenhouse was known. A total of 255,745 samples were used for training and

63,936 observations for testing, which is an 80–20 setup. To evaluate the performance of the models during training, 10 repeats of five-fold cross-validation were performed. The final values used for the RSVM model were $\sigma = 0.12$ and $C = 2.27$, and the final values used for the RF model were min.node.size = 2, mtry = 6 and splitrule = "extratrees". The test results are shown in Table 1:

Table 1. Combinations of climate systems, number of occurrences observed and average switch-on times.

Combination	Occurrences	Avg. Time in Seconds
Two recirculators	2469	65
One fan and two recirculators	52	26
Two fans	83	28
Two fans and two recirculators	2407	56
Two fans, two recirculators and cooling system	49	28
Three fans and two recirculators	2438	45
Three fans, two recirculators and cooling system	1257	5050

RF is better than RSVM in all metrics. The coefficients of determination R^2 are close to 1, so the models fit the data perfectly. As the two models were trained and evaluated on the same datasets, statistical inference can be performed to determine if there are differences between using one or the other. Using Student's *t*-test method, the differences between the two models were found to be significant at 95% confidence (*p*-value adjustment by Bonferroni's method = $0.12 > \alpha = 0.05$). Therefore, the RF model was chosen to estimate the missing points of the humidity sensor.

For each combination of climate systems, we created a different table that includes the values of the above variables every 10 s.

4.2. Climatic Systems Combinations

Among the systems in place to modify the thermal conditions inside the greenhouse are: three fans, two recirculators, a ventilation system with four motorized upper windows, a thermal-shade screen system and the air cooling system. The ventilation system and the thermal screen system on the roof were used less than 1% of the time during this study, so the analysis of variables showed that these variables have a variance close to zero and were discarded. Therefore, only the three fans, the two recirculators and the air cooling system were used to modify the greenhouse thermal conditions.

Of the systems that were actually used to modify the conditions inside the greenhouse, seven combinations of them were found. Table 2 shows the combinations of climate systems that were found during this study, the number of occurrences of each of them and the average time of switching on.

Table 2. Test accuracy results of the four models for the different combinations of climate systems.

Climatic Systems	Model	Accuracy
2 fans + recirculators	Random forest	89.2%
	Linear SVM	74.5%
	Radial SVM	81.9%
	Logistic regressor	78.7%
3 fans + recirculators	Random forest	89.9%
	Linear SVM	75.4%
	Radial SVM	84.8%
	Logistic regressor	79.4%
3 fans + recirculators + cooling	Random forest	98.4%
	Linear SVM	88.8%
	Radial SVM	86.5%
	Logistic regressor	88.8%

The statistical analysis carried out for each combination of climate systems found shows that the average switch-on times differ from one system to another, as can be seen in Table 2.

Considering that the two recirculators alone were used to homogenize the greenhouse conditions and were switched on periodically, this combination of systems was not taken into account when deciding which systems should be switched on to achieve the ideal climatic conditions inside the greenhouse. The statistical analysis showed that this combination of systems was only able to reduce the temperature slightly when the outside temperature decreased.

The statistical analysis showed that there were three combinations of climate systems (one fan and two recirculators, two fans, and two fans, two recirculators and the cooling system) that were switched on together very few times during this study. The average switch-on times of these systems were less than 30 s, and they did not change the temperature inside the greenhouse in any case. Therefore, these systems were discarded due to insufficient quality information.

In summary, three combinations of systems were used to decrease the greenhouse temperature: (1) two fans and two recirculators, (2) three fans and two recirculators and (3) three fans, two recirculators and a cooling system.

4.3. Methodology: Binary Classification for Climatic Systems and Regression for Humidity Prediction

In this work, we created a methodology that consists of the sequential application of two algorithms. The first one will output a logical value that indicates if a certain combination of climatic systems under the current conditions would be able to reach the desired temperature in the desired time, and it was formulated as a binary classification model. The second one will output the final humidity reached when using such a combination of climatic systems under the current conditions after the established time. These models are going to be trained using a machine learning (ML) approach as in the data interpolation.

ML is a part of the AI paradigm that studies the mathematical algorithms that can learn patterns using examples or previously collected data and then make a prediction or classification over new data [28].

For the first part, we are examining only binary classification, $Y = 0$ being the target variable when the required temperature is not reached using a particular combination of climatic systems and $Y = 1$ when it is reached. Several inputs were used related to the weather outside the greenhouse and meteorological conditions within the greenhouse. This is a form of supervised learning because the training and testing datasets contain a response label and the algorithm observes the input vector and attempts to learn a probability distribution to predict y given x [29]. The objective of the algorithm is to learn a set of weights on a subset of the data that minimizes the error or loss between the ground truth and predicted value in order to precisely classify the input to the associated label [30].

Logistic regression and support vector machines (SVMs) are popular binary classification algorithms and a random forest is also very well-known in ML practices. We chose these models, including linear and radial kernels for SVMs for our purpose.

With a similar approach but from a regression point of view, we used the same strategy, including the same algorithms since they were valid for both classification and regression, for predicting the humidity.

4.4. Manual Data Augmentation

The data collected from the greenhouse could have been seen as insufficient in obtaining robust classification models. Additionally, the data were unbalanced after being partitioned between train/validation and test sets, as can be seen in Figure 2, so the models were biased toward the majority classes. Due to this, a manual data augmentation process was applied to the training data of the three combinations of climate systems that were successfully lowering the temperature inside the greenhouse. This data augmentation trained the models with balanced data and helped to better predict new unseen examples (increases in the generalizability of the models). For example, from an observation where the target temperature was reached by turning on certain climatic systems for S seconds, we can generate new observations with the same conditions as before, but, with turn-on times shorter than S seconds, we found that those climatic systems could not reach the target. This process generates new examples by applying common sense and adds value to the data.

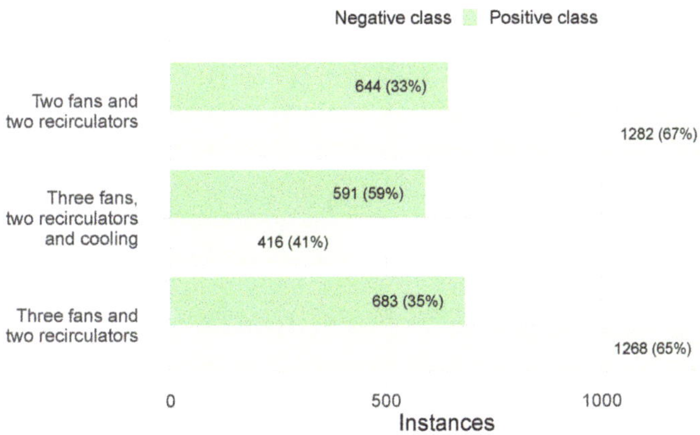

Figure 2. Distribution of training classes before manual data augmentation.

Manual data augmentation was applied to the training data to add new observations, from both the positive and negative classes. For the first case, new examples were generated by modifying the on-time of the systems, indicating that, if the target was reached in S seconds with certain internal and external conditions, then, for on-times in the interval $[S + 10, S + 120]$ seconds, the target would also be reached for those same conditions. The number of new examples generated from the positive class depended on the distribution of the classes in the different combinations of weather systems. It was seven new examples for each observation where the target was reached in the case of "two fans and two recirculators" and "three fans and two recirculators", and five new examples for each observation where the target was reached in the case of "three fans, recirculators and cooling". On the other hand, if given certain conditions, where the turn-on of the systems failed to reach the target for a certain time S, then, for turn-on times $0 \leq S' < S$ seconds, the targets would not be reached either. To generate new examples, the interval $[0, S - 10]$ was partitioned into three equal parts and new instances were created, keeping the rest of the conditions and adding the new turn-on times. Finally, another three new examples were generated for each instance of the initial positive class, indicating that, if a temperature T was reached in a given time, then temperatures of $T - 1$, $T - 2$ and $T - 3$ degrees Celsius could not be reached in the same time.

In the case of the two fans and the two recirculators, 2407 observations were available, 1602 (67%) of them failed to lower the temperature inside the greenhouse and 805 were successful (33%). The stratified random split between training/validation and the test (80–20%) provided 1926 observations for the training and validation of the models and 481 for the test. After applying the data augmentation, a training set of 12,140 instances, 6988 of the negative class (58%) and 5152 of the positive class (42%), became available.

For the three fans and the two recirculators, there were 2438 observations, of which, 1585 (65%) of them failed to lower the temperature inside the greenhouse and 853 were successful (35%). The stratified random split between training/validation and the test (80–20%) provided 1951 observations for the training and validation of the models and 487 for the test. After applying the data augmentation, a training set of 12,539 instances, 7075 from the negative class (56%) and 5464 from the positive class (44%), became available.

For the three fans, the two recirculators and the cooling, 1257 observations were available, 519 (41%) of them failed to lower the temperature inside the greenhouse and 738 were successful (59%). When performing the stratified random division between training/validation and the test (80–20%), 1007 observations were available for the training and validation of the models and 250 for the test. After applying the data augmentation, a training set of 6959 instances, 3413 of the negative class (49%) and 3546 of the positive class (51%), became available.

The above process increased the number of instances available to train the models, balanced the distribution of classes and added new unseen examples to increase the generalizability of the models, as can be seen in Figure 3.

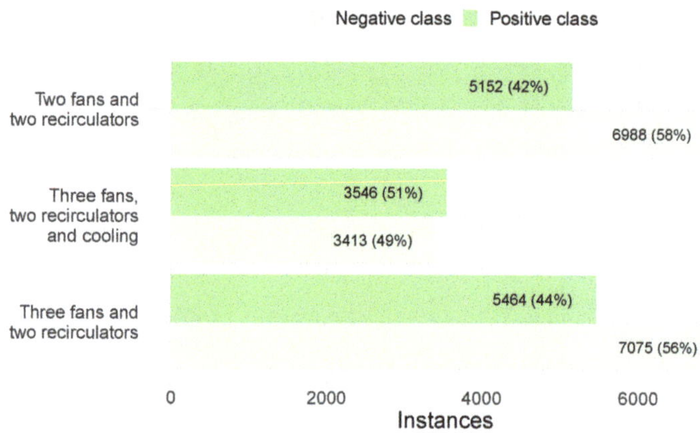

Figure 3. Distribution of training classes after manual data augmentation.

5. Results and Discussion

5.1. Variable Selection

During the experiment, three combinations of climate systems were detected that successfully lowered the temperature inside the greenhouse (the other combinations were not applied or did not lower the temperature). These combinations were: (1) two fans and two recirculators, (2) three fans and two recirculators and (3) three fans, two recirculators and the cooling system. Therefore, we will focus on these combinations of climate systems to estimate whether they will achieve the target temperature and to obtain the final humidity inside the greenhouse that would be reached after switching on for S seconds. For the analysis at hand, it was firstly important to identify which were the variables that were going to influence the biggest decrease in the temperature of the greenhouse. With this aim, the first task performed was a variable selection using random forests with five-fold cross-validation. Variable selection is necessary because most models do not deal well with a large number of irrelevant variables. These variables will only introduce noise into the model, or, worse, they might lead to over-fitting. Therefore, variable selection serves two purposes: firstly, it helps to determine all of the variables that are related to the outcome, which makes the model complete and accurate. Secondly, it helps in selecting a model with few variables by eliminating irrelevant variables that decrease the precision out of the areas of training and increase the complexity of the model. Ultimately, the variable selection provides a balance between simplicity and fit [31]. This method dropped the variables related to greenhouse internal radiation, wind (speed and direction), rainfall and the use of windows due to the low usage during data collection.

5.2. Predictive Models

The initial dataset was divided into three subsets, one for each combination of climate systems. This was carried out to simplify the problem and to try to achieve maximum accuracy. Each subset was divided, by stratified random partitioning, into 80% of the data for the training and validation of the models and the remaining 20% for testing. All data were centred and scaled to have a mean of 0 and a standard deviation of 1. This is important when operating with variables at different scales, as it ensures that they all contribute equally to the result so that no bias is introduced. This ensures that the criteria for finding combinations of predictors is based on how much variation they explain and, therefore, improves numerical stability.

Four classification models were trained: random forest, linear SVM, radial SVM and multiple logistic regression. Ten-fold cross-validation repeated five times was performed to perform an exhaustive hyperparameter search, using 80% of the training/validation data for training and the remaining 20% for model validation. Once the best hyperparameters were obtained, the four models were trained with all of the training data (together with the data generated by data augmentation) and their performance on the test set was evaluated.

All models were trained and evaluated on the same dataset, which makes it possible to compare them and see which one is better. As we can see in Table 3, the random forest model is the most accurate in all cases, significantly improving the results of the other models. If we compare the results obtained with the initial class distribution (see Figure 2), we can see that, in the case of the two fans and the recirculators, the accuracy obtained is improved by 22% if we had classified all of the instances as belonging to the most frequent class (negative class), whereas, for the other two combinations of climatic systems, it is improved by more than 30%. This shows that the achieved models are reasonably good.

Table 3. Results of the tests to estimate the humidity inside the greenhouse at each instant.

Metrics	Radial SVM	Random Forest
RMSE	3.55	0.16
R^2	0.96	0.99
CVRMSE	4.42	0.20
MAPE	0.03	0.001
MAE	2.33	0.08

As all models were trained and evaluated on the same datasets, statistical inference could be performed to determine if there are differences between using one model or another. Using the Student's t-test method, the random forest model was the best in all cases and there were significant differences with the other models.

If we analyze in more detail the best model for each combination of climatic systems, we can obtain the following:

- Two fans and two recirculators: The accuracy is 89.2%, as we can see in Figure 4, so the model correctly classifies approximately 90% of the examples. The precision of the model is 92.7%, so it will only be wrong approximately 7% of the time when it predicts that this combination of systems will reach the target. The recall is 90.9%, so the model is able to correctly classify more than 90% of the examples where this combination of systems reaches the target temperature. Regarding the reliability of the model (kappa index), following Landis and Koch's standards [32], we can say that it is substantial and the results are not due to random chance.
- Three fans and two recirculators: The results obtained by this model are very similar to the previous one (see Figure 5).
- Three fans, two recirculators and cooling: This model has an accuracy of 98.4%, as we can see in Figure 6. The accuracy of the model is 100%, so it is absolutely perfect when predicting that this combination of systems will reach the target. The recall is 96.1%, so the model is able to correctly classify more than 96% of the examples where this combination of systems reaches the target temperature. Concerning the reliability of the model, following Landis and Koch's standards [32], we can say that it is almost perfect as the kappa index is very close to 1.

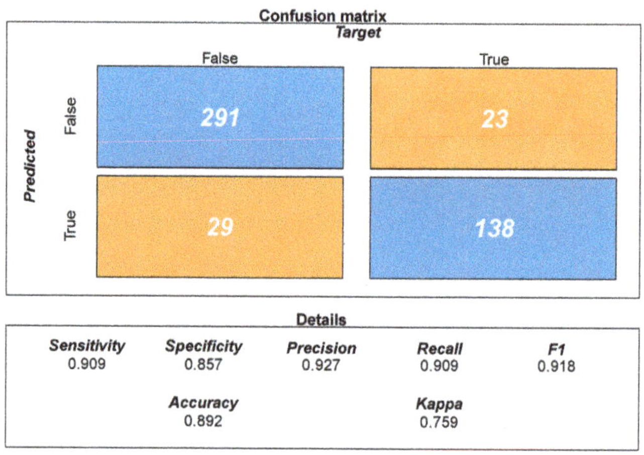

Figure 4. Test results of the random forest classification model for two fans and two recirculators.

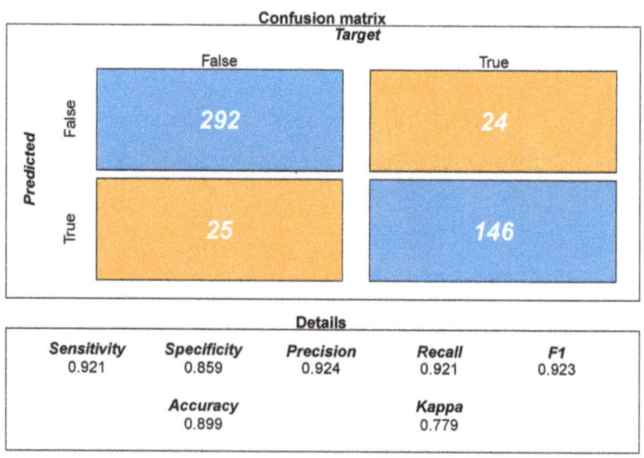

Figure 5. Test results of the random forest classification model for three fans and two recirculators.

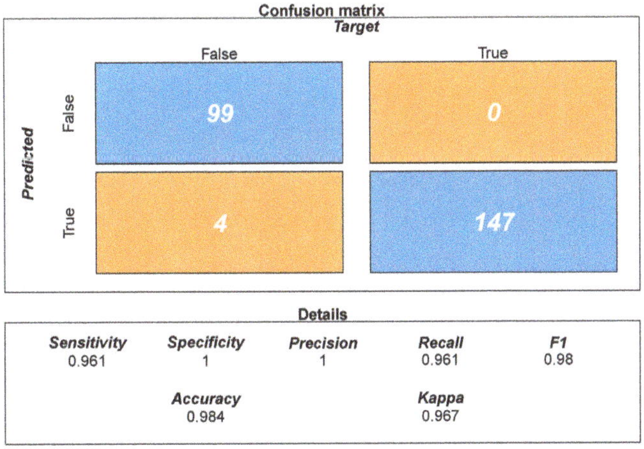

Figure 6. Test results of the random forest classification model for three fans, two recirculators and cooling system.

5.3. Modeling the Humidity

Another crucial aspect to control in greenhouses is the air humidity since it affects the nutrient uptake [33], water waste and, in general, the quality of the crop. A low humidity results in greater transpiration, water waste and problems in the development of the crops linked, amongst others, to calcium absorption. When conditions are too humid, it may promote the growth of mold and bacteria that cause plants to die and crops to fail, as well as conditions such as root or crown rot. Humid conditions also invite the presence of pests, such as fungus gnats, whose larvae feed on plant roots and thrive in moist soil. Because of this, an algorithm was created to predict the final humidity inside the greenhouse that would be reached when using a certain combination of climatic systems under certain environmental conditions.

For this task, the same steps were followed as in the previous case, but, this time, the four regression models were trained to try to predict the final humidity inside the greenhouse that would occur if the different climate systems were turned on for a certain time. The random forest model performed better in the test dataset than the other models for all evaluation metrics, as can be seen in Table 4. Due to the four models being trained and evaluated on the same datasets, statistical inference can be performed to determine if there are differences between using one or the other. Using the Student's *t*-test method, the differences between the random forest model and the other models were found to be highly significant at 95% confidence. Therefore, the random forest model was chosen to estimate the humidity inside the greenhouse.

Table 4. Test results for the estimation of inside final humidity.

Model	R^2	RMSE	CVRMSE	MAE	MAPE
Random forest	0.96	1.81	2.83	1.18	0.02
Radial SVM	0.94	2.20	3.45	1.50	0.02
Linear SVM	0.89	3.02	4.72	2.31	0.04
Log. regressor	0.89	2.99	4.68	2.32	0.04

In Figure 7, we can see the differences between the predictions of the four models and the expected value in the first twenty observations of the test set.

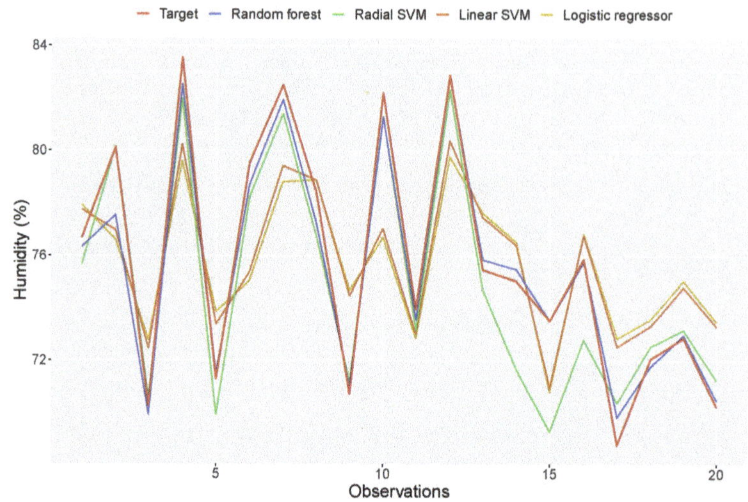

Figure 7. Estimation of the final greenhouse indoor humidity for the first twenty observations of the test set using the four regression models together with the true value.

Considering that the final indoor humidity varies between 41.5% and 85% in the test dataset, and the complexity of the problem, an error of 2.83% could be considered as acceptable. These results allow us to predict whether, given current climatic conditions and future predictions for a given time, a combination of climatic systems will be able to achieve the objective or not, at least with a 90% efficiency, which will allow for the automation of the process and the saving of energy and time in which the crops are subjected to undesired conditions, which will increase the yield of the crops. This is a major improvement on the primitive techniques previously used, which were based on trial and error methods that added new systems if the desired conditions were not achieved after a certain period of time. Therefore, when the climatic conditions of the greenhouse are not fulfilled, we can use the three models to decide which of them would reach the target temperature in the shortest time and directly use that combination of climate systems to reduce energy consumption.

5.4. Not Observed Cases/Extreme Cases

Evaluating the performance of the models, it was observed that they were able to predict examples similar to the ones that they were trained on reasonably well. However, when testing new examples with "extreme" conditions or conditions that were not similar to the previous ones, it was observed that the models no longer performed as well as expected. For example, when the initial and final temperatures were similar, all of the models predicted that the target could be reached, regardless of the on-time of the systems (even when the on-time was zero). It was also observed that, if a combination of systems was able to reach a temperature T in S seconds when a number of seconds S' (with $S' > S$) outside the usual range of on-time that they had been trained on was introduced, they predicted that the target could not be reached, which was erroneous. Due to these observations, the data augmentation described in Section 4.4 was performed, with the aim of adding new examples and correcting such erroneous predictions.

Figure 8 shows an example of an erroneous prediction of the three fans, recirculators and cooling model, trained without data augmentation, versus the same model trained with data augmentation.

```
> getPred.rf.3FansRecirculatorsCoolingNoAugmented(27, 25, 70, 22, 21, 65, 60, 414, 477, 0)
  Target  Humidity
1 TRUE        70
> getPred.rf.3FansRecirculatorsCooling(27, 25, 70, 22, 21, 65, 60, 414, 477, 0)
  Target  Humidity
1 FALSE       70
```

Figure 8. Example of misclassification of a model trained without data augmentation together with the prediction of the same model trained with data augmentation.

5.5. Optimized Operation: Machine Learning as a Service

The algorithms here presented can be of help for greenhouse management, especially for the optimization of the yield and for the prevention of catastrophic effects of extreme events such as heat waves and cold snaps. To make sure that the algorithms have an impact and that they can be used by the majority of professionals, providing them as a cloud service that can be accessed by all types of agricultural management solutions was considered. The implementation of them on the cloud was under the use of OpenCPU. OpenCPU is a framework for embedded scientific computing and reproducible research. The OpenCPU server provides a reliable and interoperable HTTP API for data analysis based on R. We used OpenCPU so that our methodology can be integrated into any platform, since all-state in OpenCPU is managed by controlling objects in sessions on a server [34]. For this purpose, an R package was constructed that performs the classification and prediction processes and supports a set of URLs in which the data are stored. The package's functions are the following:

- getPred.rf.2FansRecirculators();
- getPred.rf.3FansRecirculators();
- getPred.rf.3FansRecirculatorsCooling().

All of the package's functions use the following parameters:

- innerTemp0: Initial indoor temperature.
- innerTempF: Final indoor temperature.
- innerHum0: Initial indoor humidity.
- extTemp0: Initial external temperature.
- extTempF: Final external temperature.
- extHum0: Initial external humidity.
- extHumF: Final external humidity.
- extRad0: Initial external radiation.
- extRadF: Final external radiation.
- secondsON: Expected seconds of operation.

In order to access the functionality, we can use OpenCPU in two ways: graphically and by means of a request using curl. Figure 9 shows the graphic user interface of openCPU when making a request with certain inputs, and Figure 10 shows its output. The output consists of a logical value that tells you if, under the input conditions, the desired temperature will be achieved (TRUE in our case), and the final indoor humidity that will be achieved by the selected combination of climatic systems.

Figure 11 shows the same results from a request using curl. This type of request is integrated into our platform in order to make the results accessible to anyone.

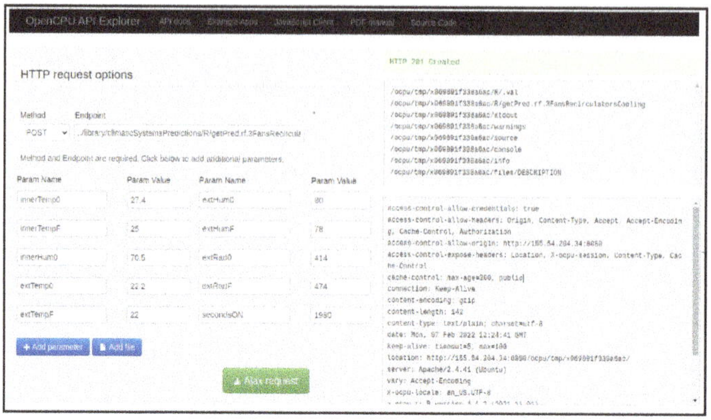

Figure 9. Example of a query to the three fans, two recirculators and cooling model to check if it achieves the target under the given conditions using the API.

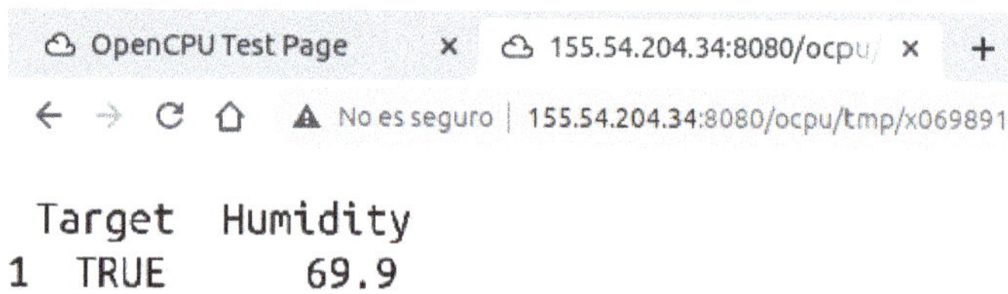

Figure 10. Result of query Figure 9.

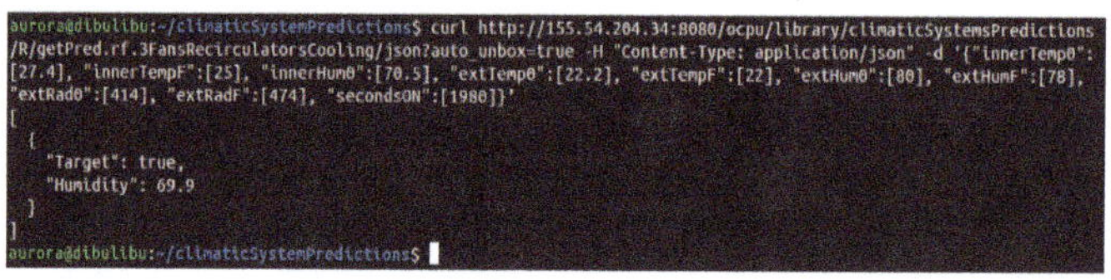

Figure 11. Example of a query to the three fans, two recirculators and cooling model to check if it achieves the target under the given conditions using the terminal.

5.6. Novelty and Practical Applications

The research here presented has two facets that make it novel. On the one hand, the methodology used is formed by a comprehensive platform that takes into consideration a large number of data streams, and that offers the possibility of implementing complex AI algorithms. In that respect, we used, for the first time, the support vector machine for the modeling of conditioning systems on greenhouses, which produced outstanding results. On the other hand, we implemented the modeling of temperature and humidity. Although moisture has been seen as a key environmental parameter for the health and yield of crops, humidity had not been included in past studies like the one we present here. We believe that having the model of humidity has made the work highly relevant and realistic for its application in the real world.

Two of the most important environmental parameters that need to be controlled for an optimal greenhouse climate are temperature and relative humidity. Temperature is the most important single parameter in greenhouse controls as temperature has a significant role in plant growth and development. In addition to optimizing the greenhouse temperature, humidity control is of vital importance as optimal plant growth can only be achieved within a certain humidity range.

Modern greenhouses have control systems for these parameters; however, the programming is far from optimizing their operation, saving energy and establishing the optimal set points for crop development. This work contributes significantly to obtaining a accurate control of temperature and humidity in order to minimize the energy consumption and to establish optimal conditions for plant growth. The results are already being applied to a greenhouse where blueberries are grown commercially.

6. Conclusions and Future Work

This paper presents an AI-based mechanism for the smart operation of climatic systems in a greenhouse. We show that we can predict with an accuracy equal to 90% if, under certain conditions, a climatic system will reach a fixed temperature set-point. We are also able to estimate the humidity with a 2.83% CVRMSE. These two machine learning models are crucial for the automatic choice of a combination of the climatic system in order to maintain certain conditions inside a greenhouse and, also, in order to select those that consume less so as to be more efficient. Our mechanism is part of an IoT platform and can be easily integrated into other frameworks since we have developed, through OpenCPU, a package that can be accessed as an API.

For future work, we are currently expanding the climatic systems to be considered. Precisely, we will include those that help to modify the humidity, such as fog machines. In addition, we are analyzing the simplest and more efficient way to lower the temperature in a greenhouse during winter, which is window opening. We are also collecting data from other greenhouses in order to assess how robust our model is with regard to transference, and are testing mechanisms to carry that out.

Author Contributions: Conceptualization, A.G.-V., J.M.-B. and A.P.R.; methodology, A.G.-V. and J.M.-B.; software, J.M.-B.; validation, A.G.-V., A.P.R. and M.Á.Z.; formal analysis, A.G.-V.; investigation, A.G.-V., J.M.-B. and V.M.; resources M.Á.Z., V.M. and A.F.S.; data curation, J.M.-B.; writing—original draft preparation, A.G.-V., J.M.-B. and A.P.R.; writing—review and editing, M.Á.Z. and V.M.; visualization, J.M.-B.; supervision, V.M. and A.F.S.; project administration, M.Á.Z. and V.M.; funding acquisition, M.Á.Z., V.M. and A.F.S. All authors have read and agreed to the published version of the manuscript.

Funding: This work was sponsored by the European Commission through the H2020 DEMETER (g.a. 857202) project and by PRIMA Foundation and the EU Grant Agreement number: 1821 WATERMED4.0 -Call 2018 Section 1 Water. It was also co-financed by the Spanish Ministry of Universities by means of the Margarita Salas linked to the European Union through the NextGenerationEU programme. The publication is also part of the project PERSEO PDC2021-121561-100, funded by MCIN/AEI/10.13039/501100011033 and by the European Union "NextGenerationEU"/"PRTR", and by Berries 4.0: Invernaderos 4.0 para la producción sostenible de superalimentos (2I18SAE00060).

Institutional Review Board Statement: Not applicable.

Informed Consent Statement: Not applicable.

Data Availability Statement: The data presented in this study are available on request from the corresponding author. The data are not publicly available due to privacy restrictions.

Conflicts of Interest: The authors declare no conflict of interest.

References

1. Kakraliya, S.; Jat, H.; Singh, I.; Gora, M.; Kakraliya, M.; Bijarniya, D.; Sharma, P.; Jat, M. Energy and economic efficiency of climate-smart agriculture practices in a rice—Wheat cropping system of India. *Sci. Rep.* **2022**, *12*, 8731. [CrossRef] [PubMed]
2. Baudoin, W.; Nersisyan, A.; Shamilov, A.; Hodder, A.; Gutierrez, D.; Pascale, S.D.; Nicola, S.; Gruda, N.; Urban, L.; Tanny, J.; et al. *Good Agricultural Practices for Greenhouse Vegetable Production in the South East European Countries-Principles for Sustainable Intensification of Smallholder Farms*; FAO: Rome, Italy, 2017; Volume 230.
3. Statista. Area of Greenhouse in Spain from 2015 to 2021. 2021. Available online: https://www.statista.com/statistics/1218861/greenhouse-total-area-spain/ (accessed on 12 October 2012).
4. Zamora-Izquierdo, M.A.; Santa, J.; Martínez, J.A.; Martínez, V.; Skarmeta, A.F. Smart farming IoT platform based on edge and cloud computing. *Biosyst. Eng.* **2019**, *177*, 4–17. [CrossRef]
5. Fetting, C. *The European Green Deal*; ESDN Report; ESDN Office: Vienna, Austria, 2022.
6. FAO. The future of food and agriculture–Trends and challenges. *Annu. Rep.* **2017**, *296*, 1–180.

7. Lugonja, D.; Jurišić, M.; Plaščak, I.; Zbukvić, I.; Glavica-Tominić, D.; Krušelj, I.; Radočaj, D. Smart Agriculture Development and Its Contribution to the Sustainable Digital Transformation of the Agri-Food Sector. *Teh. Glas.* **2022**, *16*, 264–267. [CrossRef]
8. Ranganathan, J.; Waite, R.; Searchinger, T.; Hanson, C. *How to Sustainably Feed 10 Billion People by 2050, in 21 Charts*; World Resources Institute: Washington, DC, USA, 2018.
9. Geilfus, C.M. *Controlled Environment Horticulture*; Springer: Berlin/Heidelberg, Germany, 2019.
10. Huang, Y.H.; Te Lin, T. High-throughput image analysis framework for fruit detection, localization and measurement from video streams. In Proceedings of the 2019 ASABE Annual International Meeting. American Society of Agricultural and Biological Engineers, Boston, MA, USA, 7–10 July 2019; p. 1.
11. Aznar-Sánchez, J.A.; Velasco-Muñoz, J.F.; López-Felices, B.; Román-Sánchez, I.M. An analysis of global research trends on greenhouse technology: Towards a sustainable agriculture. *Int. J. Environ. Res. Public Health* **2020**, *17*, 664. [CrossRef] [PubMed]
12. Bersani, C.; Fossa, M.; Priarone, A.; Sacile, R.; Zero, E. Model Predictive Control versus Traditional Relay Control in a High Energy Efficiency Greenhouse. *Energies* **2021**, *14*, 3353. [CrossRef]
13. Gonnella, M.; Renna, M. The Evolution of soilless systems towards ecological sustainability in the perspective of a circular economy. Is it really the opposite of organic agriculture? *Agronomy* **2021**, *11*, 950. [CrossRef]
14. Son, J.E.; Kim, H.J.; Ahn, T.I. Hydroponic systems. In *Plant Factory*; Elsevier: Amsterdam, The Netherlands, 2020; pp. 273–283.
15. Asaduzzaman, M.; Niu, G.; Asao, T. Nutrients Recycling in Hydroponics: Opportunities and Challenges Toward Sustainable Crop Production Under Controlled Environment Agriculture. *Front. Plant Sci.* **2022**, *13*, 403. [CrossRef]
16. Fussy, A.; Papenbrock, J. An Overview of Soil and Soilless Cultivation Techniques—Chances, Challenges and the Neglected Question of Sustainability. *Plants* **2022**, *11*, 1153. [CrossRef]
17. Gázquez, J.; López, J.; Pérez-Parra, J.; Baeza, E.; Saéz, M.; Parra, A. Greenhouse cooling strategies for mediterranean climate areas. In Proceedings of the International Symposium on High Technology for Greenhouse System Management: Greensys2007, Naples, Italy, 4–6 October 2007; Volume 801, pp. 425–432.
18. Kittas, C.; Katsoulas, C.; Bartzanas, T.; Bakker, S. Greenhouse climate control and energy use. In *Good Agricultural Practices for Greenhouse Vegetable Crops*; FAO: Rome, Italy, 2013; p. 63.
19. Stanghellini, C. Horticultural production in greenhouses: Efficient use of water. In Proceedings of the International Symposium on Growing Media and Soilless Cultivation, Leiden, The Netherlands, 17–21 June 2013.
20. Graamans, L.; Baeza, E.; Van Den Dobbelsteen, A.; Tsafaras, I.; Stanghellini, C. Plant factories versus greenhouses: Comparison of resource use efficiency. *Agric. Syst.* **2018**, *160*, 31–43. [CrossRef]
21. Nikolaou, G.; Neocleous, D.; Christou, A.; Polycarpou, P.; Kitta, E.; Katsoulas, N. Energy and Water Related Parameters in Tomato and Cucumber Greenhouse Crops in Semiarid Mediterranean Regions. A Review, Part I: Increasing Energy Efficiency. *Horticulturae* **2021**, *7*, 521. [CrossRef]
22. Iddio, E.; Wang, L.; Thomas, Y.; McMorrow, G.; Denzer, A. Energy efficient operation and modeling for greenhouses: A literature review. *Renew. Sustain. Energy Rev.* **2020**, *117*, 109480. [CrossRef]
23. Saadane, R.; Chehri, A.; Jeon, S.; Ouafiq, E.M. AI-based modeling and data-driven evaluation for smart farming-oriented big data architecture using IoT with energy harvesting capabilities. *Sustain. Energy Technol. Assess.* **2022**, *52*, 102093.
24. Bhat, S.A.; Huang, N.F.; Sofi, I.B.; Sultan, M. Agriculture-Food Supply Chain Management Based on Blockchain and IoT: A Narrative on Enterprise Blockchain Interoperability. *Agriculture* **2021**, *12*, 40. [CrossRef]
25. Boulard, T.; Wang, S. Greenhouse crop transpiration simulation from external climate conditions. *Agric. For. Meteorol.* **2000**, *100*, 25–34. [CrossRef]
26. LShamshiri, R.; Jones, J.W.; Thorp, K.R.; Ahmad, D.; Che Man, H.; Taheri, S. Review of optimum temperature, humidity, and vapour pressure deficit for microclimate evaluation and control in greenhouse cultivation of tomato: A review. *Sci. Hortic.* **2018**, *32*, 287–302.
27. Tetens, O. Über Einige Meteorologische Begriffe. *Z. Geophys.* **1930**, *6*, 297–309.
28. Mitchell, T.M.; Mitchell, T.M. *Machine Learning*; McGraw-Hill: New York, NY, USA, 1997; Volume 1.
29. Goodfellow, I.; Bengio, Y.; Courville, A. *Deep Learning*; MIT Press: Cambridge, MA, USA, 2016.
30. Kirasich, K.; Smith, T.; Sadler, B. Random forest vs logistic regression: Binary classification for heterogeneous datasets. *SMU Data Sci. Rev.* **2018**, *1*, 9.
31. Chowdhury, M.Z.I.; Turin, T.C. Variable selection strategies and its importance in clinical prediction modelling. *Fam. Med. Community Health* **2020**, *8*, e000262. [CrossRef]
32. Landis, J.R.; Koch, G.G. The measurement of observer agreement for categorical data. *Biometrics* **1977**, *33*, 159–174. [CrossRef]
33. Gisleröd, H.; Selmer-Olsen, A.; Mortensen, L. The effect of air humidity on nutrient uptake of some greenhouse plants. *Plant Soil* **1987**, *102*, 193–196. [CrossRef]
34. Zafeiropoulos, A.; Fotopoulou, E.; González-Vidal, A.; Skarmeta, A. Detaching the design, development and execution of big data analysis processes: A case study based on energy and behavioral analytics. In Proceedings of the 2018 Global Internet of Things Summit (GIoTS), Bilbao, Spain, 4–7 June 2018; IEEE: Piscataway, NJ, USA, 2018; pp. 1–6.

agriculture

Article

Effective Laser Fly Control with Modulated UV-A Light Trapping for Mushroom Fungus Gnats (Diptera: Sciaridae)

Sumesh Nair [1], Yvonne Yuling Hu [2], Ching-Chieh Su [1], Ming-Jeh Chien [3] and Shean-Jen Chen [1,*]

1 College of Photonics, National Yang Ming Chiao Tung University, Tainan 711, Taiwan; nairsumesh008@gmail.com (S.N.); jingjie-su@nctu.edu.tw (C.-C.S.)
2 Department of Photonics, National Cheng Kung University, Tainan 701, Taiwan; la.heureu@gmail.com
3 Lohas Biotech Development Corporation, Taipei 114, Taiwan; ming@fic.com.tw
* Correspondence: sheanjen@nycu.edu.tw

Abstract: Fungus gnats (Sciaridae) are one of the most devastating pests on mushroom farms. Generally, they are controlled using relatively inefficient physical means, like sticky or pheromone traps, or with chemical pesticides. Here, we have proposed an integrated fungus gnat control system combining a UV-A LED source at 365 nm and a high-power laser diode at 445 nm. The 365 nm UV-A LED serves a light trap, since previous studies have concluded that fungus gnats show maximum attraction in the range of 365–390 nm. The UV-A LED is also modulated at different frequencies, and the response of the gnats corresponding to these different frequencies was observed. We utilized an Arduino Uno microcontroller to run the integrated device, and a BASLER USB camera was used to capture the images. Our experiments indicated that a frequency of 40 Hz is the optimal choice for attracting the gnats most rapidly. Within 20 s of exposure, the UV-A LED operated at 40 Hz was found to trap approximately 80% of the gnats. In a restricted trapping zone measuring $2.5 \times 2.5 \times 3$ cm^3, our integrated module, which includes a 40 Hz modulated UV-A LED and a laser, resulted in a survival rate of only 50% for the total number of gnats. This outcome was accomplished through periodic 200 ms long exposures, amounting to a total duration of 2 min for a group of 100 gnats.

Keywords: UV-A LED; laser fly control; light trap; fungus gnats; mushroom pests

Citation: Nair, S.; Hu, Y.Y.; Su, C.-C.; Chien, M.-J.; Chen, S.-J. Effective Laser Fly Control with Modulated UV-A Light Trapping for Mushroom Fungus Gnats (Diptera: Sciaridae). *Agriculture* **2023**, *13*, 1574. https://doi.org/10.3390/agriculture13081574

Academic Editors: Muhammad Sultan, Redmond R. Shamshiri, Md Shamim Ahamed and Muhammad Farooq

Received: 19 June 2023
Revised: 3 August 2023
Accepted: 5 August 2023
Published: 7 August 2023

1. Introduction

Fungus gnats (Insecta: Diptera: Sciaridae) are an omnipresent pest species in a wide variety of plants around the world. They can be found in small gardens and crop plantations. Although generally considered to be not much of a nuisance as a pest, some species can attack and feed on important cash crops like potatoes, mushrooms, ornamental plants (such as begonias and young orchids), and a handful of others crops as well [1]. While the adult fly does not cause any damage to the plants and is generally a hindrance to the workers, the larval stage causes the most significant damage of all growth stages. During this stage, the larvae can aggressively feed on the roots and the absorbent hairs of the roots, thereby stunting the growth of the plant by cutting off or reducing the nutrition intake from the soil or substrate [1,2]. If seedlings, especially delicate ones, such as legume seedlings, are affected by fungus gnat larvae, it leads to wilting, thus reducing the efficiency and quantity of produce [1,3]. Also, this rampant feeding can cause injuries to plants, which serves as a gateway for soil borne bacteria and other disease-causing agents to enter the plants, thereby shortening the life span and the possible utility of the plants [3]. Even when not feeding on the plants directly, they can consume the organic matter in the compost and the growing substrate, leaving fewer nutrients for the plants themselves [4]. The larvae also feed invasively on the mycelium of mushrooms, thereby causing significant damage to mushroom plantations [2–6]. Even when not feeding on the plants directly, they can consume the organic matter in the compost and the growing substrate, leaving fewer nutrients for the plants themselves [4]. The environmental conditions used for mushroom cultivation are perfect for their rampant reproduction [7]. This unhindered attack on the mycelium layers negatively affects the quality and quantity of the mushrooms, thereby rendering them worthless for trade and human consumption. If left uncontrolled, these gnats can wreak havoc, leading to decreased yield and significant losses to the growers [8,9]. Current control methods utilize physical, chemical and biological applications such as sticky traps, pesticides, and insecticides, or the introduction of microbial agents to control the reproduction of the fungus gnat itself [10–13]. But these methods have their own problems or shortcomings. Physical methods can tend to be tedious and time consuming, ending up being rather ineffective when taking into account the effort involved and the resulting effectivity of the method. Chemical methods, such as pesticides and insecticides, have come under the scrutiny of the scientific community in recent times, with the possible ill effects on health of the consumers resulting from improper application of the same. There have been studies observing the residues of the common pesticides such as benomyl, parathion, malathion, *beta*-cypermethrin, diflubenzuron, and pyriproxyfen, among others [14–16]. These studies have indicated that residues can be found on mushrooms for up to 7-10 days after spraying them with the aforementioned insecticides. These chemical residues can accumulate in the body over time, and there has been research which indicates that these pesticides, especially parathion and malathion, can lead to carcinogenesis of the breast tissue [17]. There is also a likelihood of the gnats developing resistance through generations to these pesticides, necessitating increased dosage and frequent application, thus increasing the associated risks [18,19]. Use of microbial or viral agents or other biological means, such as nematodes or killer mites, for the control of fungus gnats can be effective, but generally the effectiveness depends on the time of application, with earlier application being better [20]. These bacterial or viral agents can perhaps hinder the other beneficial organisms in the substrate, affecting the balance of the symbiotic system.

Recently, there have been a few innovative ways utilized to trap the fungus gnats by using LEDs as attraction traps. As an example, broad spectrum LEDs with UV peak emission, or fluorescent black light sources with UV and blue emission have been used as appropriate sources for light trap in mushroom plantations and farms [21,22]. In other studies, focusing on the response of fungus gnats to different wavelength stimuli, it has been concluded that wavelengths in the broad band UV range and green range (365-390 nm and 525 nm) elicit the highest attraction response from the gnats [23–26]. There have been studies where LEDs traps have been used in congruence with the existing physical methods to improve the efficiency of these physical methods of fly trapping. Yellow glue traps along with 530 nm peak emission LEDs have been utilized resulting in double trapping efficiency when compared to just the glue trap for multiple greenhouse pests, such as dark winged fungus gnats (Sciaridae), sweet potato whiteflies (Aleyrodidae), western flower thrips (Thripidae), and leaf hoppers (Cicadellidae) [20]. Use of LEDs seems like the right direction for application as standalone traps or for enhancing the efficacy of existing traps since there has been evidence of color preferences and color-based flight behavior in a variety of flies [27]. In general, these light traps, when operated alone, leave the flies to die after trapping them, with no active mechanism involved in the eradication of the pests. This can result in the escape of a significant percentage of the trapped flies, thereby reducing the applicability. Use of lasers as a means of instant elimination of these trapped flies is an interesting approach. There have been a few studies with respect to the use of lasers as a means to effective pest control systems, but none have been used synergistically along with another light-based trapping mechanism to our knowledge. The effects of 532 nm and 1064 nm laser exposure on *Drosophila melanogaster* larvae have been studied thoroughly; 532 nm was found to be the better than 1,064 nm at eradicating the larvae [28]. Experiments on adult *D. melanogaster* samples and found that exposure of 650 nm semiconductor laser beam for 1282 s at 60 mW power or more resulted in a 99% mortality rate [29]. In another study, it was demonstrated that using a 532 nm, 500 mW laser for 2.5 s caused lasting damage to aphids within 3.5 s of exposure, achieving a 100% kill rate. [30]. Also, the effect of different frequencies on the behavior of various animals and insects have been a topic of interest. These studies have concluded from the experiments that there is a strong positive correlation between body size and metabolic rate on temporal perception, as measured by critical flicker fusion frequency, which is a lowest flashing frequency at which the flickering is perceived as constant [31].

In the study, we have developed an all-optical system, including a light trap, which employs a modulated UV-A lamp with 365 nm to attract the fungus gnats to an assigned small area, and a cheap engraving laser at 445 nm, which works as the elimination laser. Initially, we discuss the experimental procedure used to study the response of the gnats to different modulated frequencies and the resulting observations. Forty Hz frequency from the UV-A lamp was observed to elicit the maximum response of the gnats, suggesting it to be the optimum modulation correlating to the most efficient attraction. The aforementioned trapping area was then made subject to a few-watt blue laser irradiation of millisecond time duration, thereby incapacitating the gnats in the trap effectively. These experiments and their corresponding results are discussed in the latter part of the article.

2. Materials and Methods

2.1. Fungus Gnat Rearing

To facilitate our research, we sourced adult fungus gnats from a cooperative mushroom farm plantation (Lohas Biotech, Tainan, Taiwan). The specific genus that we have used for our study is the dark winged fungus gnat (Diptera: Sciaridae: *Bradysia* sp.). Here, we utilized a laboratory rearing method to study the behavior and life cycle of these pests in a controlled environment, based on this group's approach [32]. Peat moss was procured from the same local horticultural supplier to the mushroom farm and was chosen as the substrate for the rearing process. In the laboratory, we used plastic containers measuring $15 \times 10 \times 3$ cm^3 to house the fungus gnats. Each container was filled with 100 g of peat moss, which was moistened with 30 milliliters of water. To provide nutrition for the larvae, raw potato shavings and 2×2 cm^2 potato disks were added to the substrate. This ensured that the newly hatched larvae had a suitable food source to support their growth and development. To maintain optimal conditions for the fungus gnats, the containers were kept under controlled environmental conditions. To mimic natural conditions found on mushroom farms and plantations, a 16 h daylight cycle, a constant temperature of 26 °C, and a humidity level of 85–95% were maintained. Refreshing the substrate every three days involved adding 30 milliliters of water to maintain the desired moisture level and replacing the potato shavings and disks to ensure a constant supply of food for the larvae. As the potato disks began to sprout in the growing medium, the effects of the larvae's aggressive feeding on the plant roots were quite clear. When compared to new potato disc growth in substrate without fungus gnats, the affected plants were visibly stunted and were observed to wilt within 7–10 days of sprouting. Based on the previous literature [33,34] and our observations throughout the fungus gnat rearing, this seems to be a telltale sign of damage to the absorbent hairs and, eventually, its roots. Thus, there seems to be an urgent need for effective fly control measures.

2.2. Modulated Frequency Testing

To investigate the behavior of fungus gnats in response to different frequencies of LED light exposure, we constructed a transparent arena using glass slides and acrylic sheet materials. The dimensions of the arena were $150 \times 25 \times 3$ mm^3. By utilizing a Basler USB camera controlled by pylon viewer software and an Arduino Uno microcontroller, we achieved precise monitoring of the gnats' movement within the arena. The setup involved modulating LED lamps across frequencies ranging from 10 Hz to 100 Hz for a duration of 40 s. Figure 1A depicts the experimental setup for modulated frequency testing. To optimize the LED output, a buck-boost circuit was employed, regulating the voltage and current supplied to the LEDs. This ensured consistent and efficient operation throughout the experiments. The LED lamps were powered by a 12V-3A external power supply, while the Basler camera and Arduino Uno microcontroller received power from their respective notebook ports. The LED illumination and image acquisition processes were synchronized using an external trigger generated by the Arduino Uno microcontroller. A simple code was written in the Arduino IDE environment to achieve the synchronization. This seamless coordination minimized noise in the data and ensured accurate tracking of the gnats' behavior. Images were captured at 200 ms intervals, resulting in a total of 200 images for post-processing and analysis. Figure 1B shows the acquired image of the transparent arena with the gnats inside.

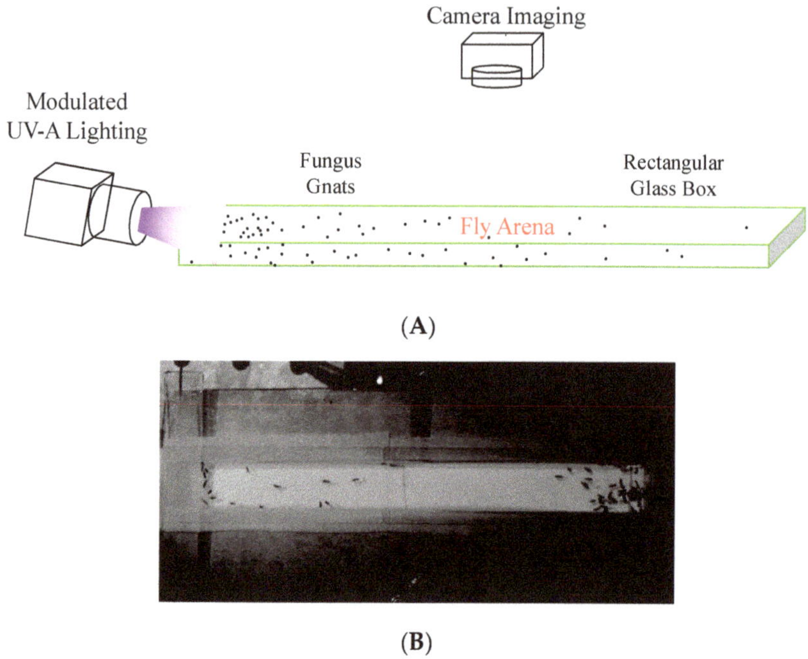

Figure 1. (**A**) Depicts the schematic for the modulated frequency experiment. (**B**) UV-illuminated arena with sample flies.

2.3. Image Processing for Fly Counting

To analyze the acquired images, we employed MATLAB software. The acquired images underwent background subtraction to eliminate static elements and focus on the gnats' movement within the arena. This process involved utilizing an averaged background image derived from a set of 200 background images. After background subtraction, the resulting images were cropped to isolate the arena area. The left half of the images, which indicated the gnats' movement towards the light when viewed sequentially, was extracted. Figure 2A portrays the schematic of the image processing pipeline, while Figure 2B shows the result of the background subtraction that isolates the arena for further fly counting. Various image processing techniques, including distance transform and watershed functions, were applied to segment and track the gnats' movement and count them accurately. Video S1 in the supplementary materials shows the result of this processing in real time. The processed images revealed the trajectories and positions of the gnats over time, providing valuable data for analysis. By counting the number of gnats at different time instances during the LED light exposure for each modulated frequency, we obtained quantitative information about their response to the light stimuli. Origin software was used to plot and visualize the fungus gnat count data, facilitating further analysis and interpretation of the results.

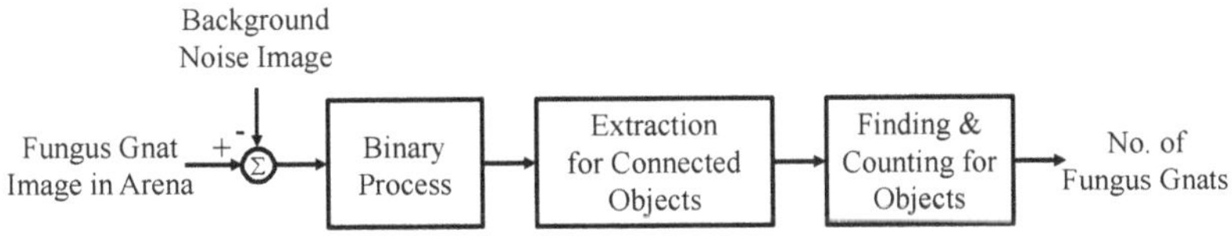

(A)

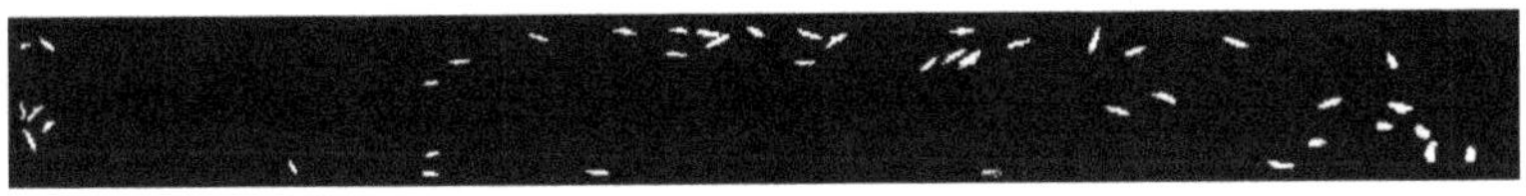

(B)

Figure 2. (**A**) Image process diagram for automatic fly number counting. (**B**) Cropped binary image (see Video S1).

Thus, the experimental setup for modulated frequency testing and image processing techniques allowed us to gain insights into the behavior of fungus gnats to different frequency of light stimuli and develop an effective fly control solution using LED light exposure. At the same time, the experimental setup and image analysis enabled precise tracking and counting of the same.

3. Experimental Results

3.1. Light Power and Wavelength

The impact of varying intensity on the response of fungus gnats has been a relatively unexplored area of study, prompting us to investigate this aspect in our research. We considered three intensities based on three possible outputs of the UV-A LED, which were 0.020 W, 0.190 W, and 0.380 W. There was a positive response when higher intensities were applied at elevated powers, consistent with the findings of a previous study conducted by Cloyd, R. A., and his group [1]. As a control, the attraction of fungus gnats to normal ambient fluorescent lights were studied. There was no correlation, as shown in Figure 3D. Based on our observations and the aforementioned work, the maximum output of the UV-A LED was chosen for further experiments. Regarding the wavelength parameter, we found that the 365 nm UV-A LED light demonstrated the most effective response from the gnats. This aligns with previous observations and emphasizes the preference of fungus gnats for this specific wavelength. The attraction response to the UV-A LED light further supports the importance of considering wavelength as a crucial parameter in designing effective light traps for attracting and controlling fungus gnats. By utilizing the optimal wavelength, our laser-based fly control device can precisely target and lure these gnats, enhancing the overall efficiency of the system.

3.2. Optimal Modulated Frequency

In addition to investigating the impact of intensity and wavelength, we conducted a comprehensive frequency study to determine the response of fungus gnats to different frequencies of LED light exposure. This study aimed to identify the optimal modulation frequency that elicits the highest and quickest attraction response from the gnats. To cover a wide spectrum of frequencies, experiments were conducted using frequencies ranging from 10 Hz to 100 Hz. Remarkably, our results revealed that the gnats exhibited the highest attraction response at a frequency modulation of 40 Hz, followed by 50 Hz and then 100 Hz. These frequencies consistently elicited rapid and significant attraction responses from the fungus gnats. These results were subject to Pearson's correlation coefficient to statistically analyze the presence of any discernable correlation between exposure time and increase in the number of fungus gnats being attracted to the illumination spot. For each of the frequencies as well as the control light, observations were recorded at 200 milliseconds intervals for a time duration of 40 s, resulting in a total of 200 observations for each frequency as well as control light. It was observed that for 40 Hz, 50 Hz, and 100 Hz, the Pearson's correlation coefficients were 0.53, 0.76, and 0.64, respectively, indicating moderate to strong positive correlation. When compared to control, the Pearson correlation coefficient was 0.38, indicating weak correlation. The Pearson correlation for each of the frequencies and control light were statistically significant at p value < 0.001 (two-tailed test). Thus, the UV-light trap works positively to attract the fungus gnats, compared to ambient fluorescent light. At the 40 Hz modulation frequency, approximately 33 out of 40 gnats in the test set were attracted to the light source within a remarkably short duration of 14.4 s. Similarly, at 50 Hz and 100 Hz, 34 gnats were attracted in 21.6 s and 35 gnats were attracted in 22.8 s, respectively. These observations were subject to sigmoidal fitting to see if there was any appreciable statistical significance. Figure 3A–C represents the results of sigmoidal fitting. Figure 3D represents the graph of fungus gnat response to ambient fluorescent light. It can be seen that 40 Hz shows the fastest attraction compared to the other frequencies, statistically as well. These findings highlight the significance of frequency modulation in influencing the response behavior of fungus gnats towards the LED light source. The observed average attraction rate of 82% across all tested frequencies underscores the effectiveness of the LED light trap in attracting the gnats. The optimal frequency modulation of the LED light trap creates an effective trap for the gnats, ensuring a rapid and efficient trapping process.

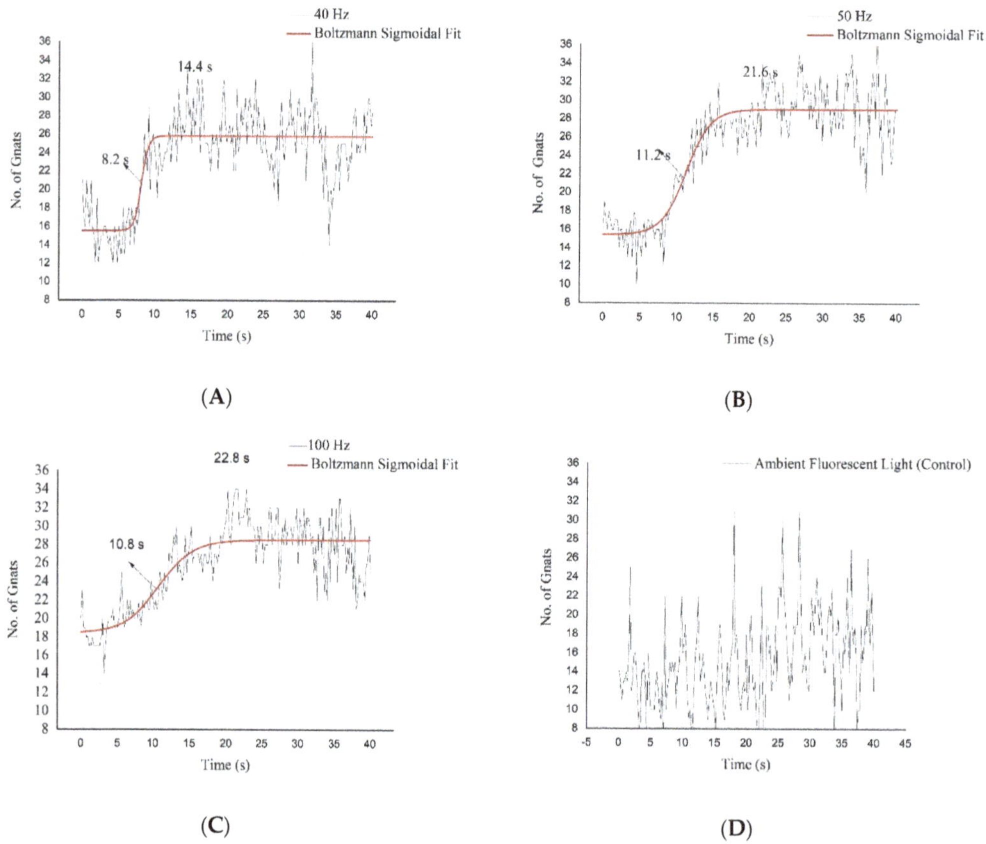

Figure 3. Attraction responses to (**A**) 40 Hz, (**B**) 50 Hz, and (**C**) 100 Hz modulations and (**D**) ambient fluorescent light (Control).

3.3. Integrated Laser Fly Control with UV Light Trap

Building upon the insights gained from the frequency study, we proceeded to integrate the UV-A LED light trap with a 445 nm engraving blue laser, creating a novel laser fly control system with an integrated LED light trap. This integration aimed to enhance the efficiency and effectiveness of fungus gnat control by combining the attractive properties of the LED light trap with the lethal capability of the laser. For the study, a conical trap area with a smaller triangular volume to trap the gnats was made out of acrylic sheet. The gnats were lured into a trapping area measuring $2.5 \times 2.5 \times 3$ cm^3, and the light trap operated at a 40 Hz modulation frequency based on the optimal frequency determined from our previous experiments. Figure 4A depicts the overall system schematic mentioned above. The introduction of 100 gnats into the laser trap and the activation of the laser beam every 10 s for a duration of 200 milliseconds allowed for sufficient time to capture the gnats within the trap. By limiting the exposure duration to just 200 milliseconds, we ensured that the structural integrity of the trap, made of acrylic material, would not be compromised by potential melting due to high local temperatures. This approach minimized the risk of damaging the trap while effectively eliminating the gnats. In total, 100 gnats were included in the experiment, and after the laser exposure, only 50 gnats remained alive, resulting in an elimination rate of approximately 50% within just 2 min. These promising results demonstrate the effectiveness of the integrated laser and UV light trap system in significantly reducing the population of fungus gnats in our experiment. The presence of deceased gnats within the arena, as shown in Figure 4B compared to Figure 4C, depicts the results of the integrated system in achieving targeted fungus gnat control. While successful within lab tests, work is ongoing to further study and understand the real-time applicability on mushroom farms.

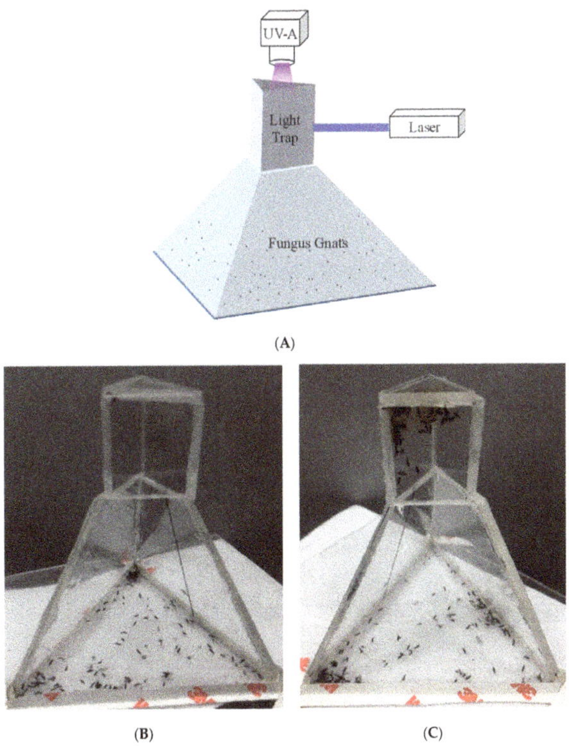

Figure 4. (**A**) The schematic for the final system. (**B**) Flies in trap before laser exposure. (**C**) Remaining flies with eliminated flies on top (Video S2).

4. Discussions

The presence of fungus gnats can have detrimental effects on mushroom plantations, as these pests proliferate rapidly and can disrupt the delicate balance that is required for profitable cultivation. Also, by rearing the fungus gnats in the laboratory, we gained valuable insights on the larvae's most damaging phase for agricultural crops. Although there has been an increase in the use of chemical means such insecticides in the last four decades, the absolute economical value of crop losses due to pests and the proportion of crop losses on the whole has increased [35]. Moreover, traditional approaches, such as chemical-based pesticide application, present additional challenges and risks, including environmental concerns, rising resistance to the pesticide, and potential health hazards. Considering other conventional control methods that often rely on the application of sticky traps or pheromone traps, our laser-based system can alleviate the effort required on the part of mushroom cultivators by providing them with a prototype plug-and-play system, lightening their burden of having to place physical traps. Our approach can be considered as a step towards incorporating automated systems in mushroom plantations that require minimum supervision while assisting the conventional methods, which can lead to high quality produce without the need for overuse of insecticides.

By studying the response of fungus gnats to different frequency modulations of a light trap within the UV-A range (approximately 365 nm), 40 Hz frequency was found to elicit the fastest response from the gnats. With this knowledge, we engineered a UV-light-trap-based laser fly control system that provides a possible substitute to the conventional methods. This targeted approach can enhance the sustainability and ecological balance of mushroom farms, allowing for the coexistence of fungus gnat control measures and natural biological processes. Also, 365 nm is attractive to fungus gnats the most, while not eliciting response from beneficial insects like bees, thereby minimizing the impact on beneficial insects and possibly extending its application to other fruit or vegetable orchards as well. For instance, there have been studies conducted on naive honey bee foragers (Hymenoptera: Apidae) and bumble bees (Hymenoptera: Apidae) which indicate that these insects show maximum attraction to 410 nm and 430–480 nm, respectively [36–38].

Thus, our approach utilizing the precise frequency of 40 Hz of 365 nm UV-A light creates an irresistible lure for the gnats, effectively containing them in the trap. Furthermore, utilizing the 445 nm integrated laser to specifically target the gnats trapped in the container enhances the efficiency of our system by instantly killing the gnats, providing an additional mechanism to terminate captured gnats swiftly and effectively and thereby effectively contributing to reducing or controlling the fungus gnats in a mushroom plantation.

5. Conclusions

Our research and development efforts have demonstrated the potential of laser-based fly control as a promising choice for managing fungus gnats on mushroom farms. By leveraging the advantages of optimal frequency modulation and UV light attraction, our system offers a reliable and efficient means of controlling these gnats. The approach of integrating a laser with a UV light trap presents a breakthrough in the field of fly control, providing a comprehensive solution that combines attraction, trapping, and termination of fungus gnats. We believe that our findings contribute to the advancement of sustainable and environmentally friendly practices within the agricultural industry. The successful development of our laser-based fly control device opens up new possibilities for pest management in various farming systems. The application of light-based control methods can be extended beyond mushroom farms, potentially benefiting other agricultural sectors facing similar pest challenges. The optimal frequency modulation and targeted attraction approach can be adapted to target different fly species, offering customized solutions for specific farming needs. The combination of optimal frequency modulation, UV light attraction, and integrated laser termination offers an efficient, environmentally friendly, and economically viable solution. By employing our laser-based fly control device, farmers can effectively manage fungus gnat populations, ensuring healthier crops and reducing potential yield losses. This innovative technology paves the way for a more sustainable and ecologically conscious approach to pest management in agriculture, benefiting both farmers and the environment alike.

Supplementary Materials: The following supporting information can be downloaded at: https://www.mdpi.com/article/10.3390/agriculture13081574/s1, Video S1: The image processing for real-time fly counting; Video S2: The laser fly control processing with UV light trap.

Author Contributions: Conceptualization, S.N., Y.Y.H., C.-C.S., M.-J.C. and S.-J.C.; methodology, S.N., C.-C.S., Y.Y.H. and S.-J.C.; software, S.N. and C.-C.S.; formal analysis, S.N.; investigation, S.N., C.-C.S. and Y.Y.H.; writing—original draft preparation, S.N.; writing—review and editing, S.N., Y.Y.H. and S.-J.C. All authors have read and agreed to the published version of the manuscript.

Funding: This research received no external funding.

Institutional Review Board Statement: Not applicable.

Data Availability Statement: All data supporting the conclusions of this article are included in this article.

Acknowledgments: This work was supported by the National Science and Technology Council (NSTC) in Taiwan with the grant numbers NSTC 112-2622-8-A49 -009 -TE3 and 111-2622-E-A49 -015. Also, we are thankful to Lohas Biotech Development Corp. for providing the initial fungus gnats and peat soil for rearing of the same.

Conflicts of Interest: The authors declare no conflict of interest.

References

1. Cloyd, R.A.; Dickinson, A.; Larson, R.A.; Marley, K.A. Phototaxis of fungus gnat, *Bradysia* sp. *nr coprophila* (Lintner) (Diptera: Sciaridae) adults to different light intensities. *HortScience* **2007**, *42*, 1217–1220. [CrossRef]
2. Shin, S.G.; Lee, H.S.; Lee, S. Dark winged fungus gnats (Diptera: Sciaridae) collected from shiitake mushroom in Korea. *J. Asia. Pac. Entomol.* **2012**, *5*, 174–181. [CrossRef]
3. Cloyd, R.A. Ecology of fungus gnats (*Bradysia* spp.) in greenhouse production systems associated with disease-interactions and alternative management strategies. *Insects* **2015**, *6*, 325–332. [CrossRef] [PubMed]
4. Binns, E.S. Field and laboratory observations on the substrates of the mushroom fungus gnat *Lycoriella auripila* (Diptera: Sciaridae). *Ann. Appl. Biol.* **1980**, *96*, 143–152. [CrossRef]
5. Cantelo, W.W. *Lycoriella mali*: Control in mushroom compost by incorporation of insecticides into compost. *J. Econ. Entomol.* **1979**, *72*, 703–705. [CrossRef]
6. Lewandowski, M.; Sznyk, A.; Bednarek, A. Biology and morphometry of *Lycoriella ingenua* (*Diptera: Sciaridae*). *Biol. Lett.* **2004**, *41*, 41–50.
7. Stanton, G.; Sanders, A. *Ball Identification Guide to Greenhouse Pests and Beneficials*; Ball Publishing (Grower Talks Bookshelf): Batavia, IL, USA, 1998.
8. Erler, F.; Polat, E.; Demir, H.; Catal, M.; Tuna, G. Control of mushroom sciarid fly *Lycoriella ingenua* populations with insect growth regulators applied by soil drench. *J. Econ. Entomol.* **2011**, *104*, 839–844. [CrossRef]
9. Park, I.K.; Choi, K.S.; Kim, D.H.; Choi, I.H.; Kim, L.S.; Bak, W.C.; Choi, J.W.; Shin, S.C. Fumigant activity of plant essential oils and components from horseradish (*Armoracia rusticana*), anise (*Pimpinella anisum*) and garlic (*Allium sativum*) oils against *Lycoriella ingenua* (Diptera: Sciaridae). *Pest Manag. Sci.* **2006**, *62*, 723–728. [CrossRef]
10. Kecskeméti, S.; Szelényi, M.O.; Erdei, A.L.; Geösel, A.; Fail, J.; Molnár, B.P. Fungal volatiles as olfactory cues for female fungus gnat, *Lycoriella ingenua* in the avoidance of mycelia colonized compost. *J. Chem. Ecol.* **2020**, *46*, 917–926. [CrossRef]
11. Anderson, V.M.; Sward, G.F.H.; Ranger, C.M.; Reding, M.E.; Canas, L. Microbial control agents for fungus gnats (Diptera: Sciaridae: *Lycoriella*) affecting the production of oyster mushrooms, *Pleurotus* spp. *Insects* **2021**, *9*, 786. [CrossRef]
12. Chen, C.; Shan, T.; Liu, Y.; Wang, C.; Shi, X.; Gao, X. Identification and functional analysis of a cytochrome P450 gene involved in imidacloprid resistance in *Bradysia odoriphaga* Yang et Zhang. *Pestic. Biochem. Physiol.* **2019**, *153*, 129–135. [CrossRef]
13. Chen, C.; Shi, X.; Desneux, N.; Han, P.; Gao, X. Detection of insecticide resistance in *Bradysia odoriphaga Yang et Zhang* (Diptera: Sciaridae) in China. *Ecotoxicology* **2017**, *26*, 868–875. [CrossRef] [PubMed]
14. Navickiene, S.; Possidônio de Amarante, O.; Brito, N.M.; Graciolli, L.A.; Ribeiro, M.L. Determination of benomyl residues in shiitake mushrooms (*Lentinula edodes*) by liquid chromatography with UV detection. *J. Chromatogr. Sci.* **2007**, *45*, 340–344. [CrossRef]
15. Pigatti, A.; Do Amaral, J.F. The determination of residues in mushrooms treated with insecticides. *Arq. Inst. Biol.* **1960**, *27*, 35–41.
16. Du, P.; Wu, X.; Xu, J.; Dong, F.; Shi, Y.; Li, Y.; Liu, X.; Zheng, Y. Different residue behaviors of four pesticides in mushroom using two different application methods. *Environ. Sci. Pollut. Res.* **2018**, *25*, 8377–8387. [CrossRef] [PubMed]
17. Calaf, G.M. Role of organophosphorous pesticides and acetylcholine in breast carcinogenesis. *Semin. Cancer Biol.* **2021**, *76*, 206–217. [CrossRef]

18. Bartlett, G.R.; Keil, C.B.O. Identification and characterization of a permethrin resistance mechanism in populations of the fungus gnat *Lycoriella Mali*(Fitch) (Diptera: Sciaridae). *Pestic. Biochem. Physiol.* **1997**, *58*, 173–181. [CrossRef]
19. Brewer, K.K.; Keil, C.B. A mixed function oxidase factor contributing to permethrin and dichlorvos resistance in *Lycoriella mali* (Fitch) (Diptera: Sciaridae). *Pestic. Sci.* **1989**, *26*, 29–39. [CrossRef]
20. Chen, T.Y.; Chu, C.; Henneberry, T.J.; Umeda, K. Monitoring and trapping insects on poinsettia with yellow sticky card traps equipped with light-emitting diodes. *HortTechnology* **2004**, *14*, 337–341. [CrossRef]
21. Otieno, J.A.; Stukenberg, N.; Weller, J.; Poehling, H.M. Efficacy of LED-enhanced blue sticky traps combined with the synthetic lure Lurem-TR for trapping of western flower thrips (*Frankliniella occidentalis*). *J. Pest Sci.* **2018**, *91*, 1301–1314. [CrossRef]
22. Shimoda, M.; Honda, K. Insect reactions to light and its applications to pest management. *Appl. Entomol. Zool.* **2013**, *48*, 413–421. [CrossRef]
23. Sanbu, I.; Chiba-Ken, F.E.S.; Gotoh, T.; Kawasaki, T. Development of sticky light trap and attractiveness to mushroom-infesting sciarids, *Lycoriella mali* (Fitch) and *Bradysia paupera Tuomikoski* (Diptera: Sciaridae). *Jpn. J. Appl. Entomol. Zool.* **1997**, *41*, 141–146.
24. Jess, S.; Bingham, J.F.W. The spectral specific responses of *Lycoriella ingenua* and *Megaselia halterata* during mushroom cultivation. *J. Agric. Sci.* **2004**, *142*, 421–430. [CrossRef]
25. Stukenberg, N.; Ahrens, N.; Poehling, H.M. Visual orientation of the black fungus gnat, *Bradysia difformis*, explored using LEDs. *Entomol. Exp. Appl.* **2018**, *166*, 113–123. [CrossRef]
26. Kecskeméti, S.; Geösel, A.; Fail, J.; Egri, Á. In search of the spectral composition of an effective light trap for the mushroom pest *Lycoriella ingenua* (Diptera: Sciaridae). *Sci. Rep.* **2021**, *11*, 12770. [CrossRef]
27. Keller, M.; Leahy, D.; Norton, B. Laser induced mortality of *Anopheles stephensi* mosquitoes. *Sci. Rep.* **2016**, *6*, 20936. [CrossRef] [PubMed]
28. Obayashi, K.; Sato, K.; Ito, N.; Wang, X.L.; Takagi, S. Physical pest control of *Drosophila* using laser (part 1). *J. Jpn. Soc. Agric. Mach.* **2005**, *67*, 93–100.
29. Ren, K.; Tu, K.; Li, H.W. Control effects of semiconductor laser on *Drosophila melanogaster*. *Zhongguo Jiguang/Chin. J. Lasers* **2006**, *33*, 1148–1152.
30. Dix, P.D.; Flippo, D.K.; Zukoff, S.N.; McCornack, B.; Frink, E.; Sharda, A. *Use of Lasers as an Alternative to Pesticide*; American Society of Agricultural and Biological Engineers: St. Joseph, MI, USA, 2016. Available online: https://elibrary.asabe.org/abstract.asp?aid=47242 (accessed on 2 July 2023).
31. Healy, K.; McNally, L.; Ruxton, G.D.; Cooper, N.; Jackson, A.L. Metabolic rate and body size are linked with perception of temporal information. *Anim. Behav.* **2013**, *86*, 685–696. [CrossRef]
32. Ju, J.H.; Heon, Y.; Ri, K.H.; Man, Y.Y.; Nam, Y.Y. Mass-rearing method of the fungus gnat, *Bradysia difformis* (Sciaridae. Diptera) in Laboratory. *Korean J. Appl. Entomol.* **2018**, *57*, 57–64.
33. Zhang, D.J.; Yang, Y.J.; Liu, C.Y.; Zhang, F.; Wu, Q.S. Root hair growth and development in response to nutrients and phytohormones. In *Root Biology*; Springer: Cham, Switzerland, 2018; pp. 65–84.
34. Iqbal, S. Insect, pest and disease management in rice. In *Rice Production Knowledge and Practices for Ensuring Food Security*; Austin Publishing Group: Jersey City, NJ, USA, 2010.
35. Oerke, E.C. Crop losses to pests. *J. Agric. Sci.* **2006**, *144*, 31–43. [CrossRef]
36. Gumbert, A. Color choices by bumble bees (*Bombus terrestris*): Innate preferences and generalization after learning. *Behav. Ecol. Sociobiol.* **2000**, *48*, 36–43. [CrossRef]
37. Giurfa, M.; Núñez, J.; Chittka, L. Colour preferences of flower-naive honeybees. *J. Comp. Physiol. A* **1995**, *177*, 247–259. [CrossRef]
38. Ostroverkhova, O.; Galindo, G.; Lande, C.; Kirby, J.; Scherr, M.; Hoffman, G.; Rao, S. Understanding innate preferences of wild bee species: Responses to wavelength-dependent selective excitation of blue and green photoreceptor types. *J. Comp. Physiol.* **2018**, *7*, 667–675. [CrossRef]

agriculture

Article

Design of Wireless Sensor Network for Agricultural Greenhouse Based on Improved Zigbee Protocol

Ruipeng Tang *, Narendra Kumar Aridas and Mohamad Sofian Abu Talip

Department of Electrical Engineering, Faculty of Engineering, University of Malaya, Kuala Lumpur 50603, Malaysia; narendra.k@um.edu.my (N.K.A.); sofian_abutalip@um.edu.my (M.S.A.T.)
* Correspondence: 22057874@siswa.um.edu.my; Tel.: +60-11-3339-1886

Abstract: Greenhouse cultivation technology has greatly contributed to the development of agriculture in Malaysia. Understanding how to monitor the greenhouse environment with high efficiency and low power consumption is particularly important. In this research, a wireless sensor network for agricultural greenhouses based on the improved Zigbee protocol is designed. Its hardware consists of various sensors and Zigbee nodes commonly used in agricultural greenhouses. On the basis of this hardware, this research designed the network topology of WMN (Wireless mesh network) by comparing the advantages and disadvantages of various topologies, and combined with this structure, proposed an improved ZigBee routing protocol EMP-ZBR to solve the question regarding energy loss and the network congestion of wireless networks. After testing EMP-ZBR and traditional Zigbee routing protocols, the improved EMP-ZBR protocol is superior to traditional Zigbee routing in terms of the end-to-end average delay, packet delivery rate, routing control overhead and routing discovery frequency, which were optimized about 1.1%, 15.2%, 15.2%, 8.1 ms in different mobile pause times, and 9.8%, 19.3%, 15.7% and 121 ms in different packet sending rates. The agreement proves that EMP-ZBR can more effectively alleviate the impact of congestion and improve the overall performance of the data monitoring system for agricultural greenhouses.

Keywords: agricultural sensor network; EMP-ZBR; network topology; routing transmission algorithm

Citation: Tang, R.; Aridas, N.K.; Abu Talip, M.S. Design of Wireless Sensor Network for Agricultural Greenhouse Based on Improved Zigbee Protocol. *Agriculture* **2023**, *13*, 1518. https://doi.org/10.3390/agriculture13081518

Academic Editors: Muhammad Sultan, Redmond R. hamshiri, Md Shamim Ahamed and Muhammad Farooq

Received: 22 June 2023
Revised: 20 July 2023
Accepted: 27 July 2023
Published: 29 July 2023

1. Introduction

In the early stages, agricultural data monitoring relied mainly on manual statistical methods and instrument-based monitoring [1]. However, in recent years, various information technologies have begun to be applied to the data monitoring of agricultural greenhouses [2]. Since agricultural greenhouses usually occupy a large area and require many monitoring nodes, and agricultural greenhouses are generally distributed in the suburbs [3], this requires high networking performance and the power consumption of data collection equipment. ZigBee technology is a wireless two-way communication technology based on the IEE802.15.4 protocol standard. It has the characteristics of a low frequency band, low power consumption and low cost [4], which can meet the needs of agricultural greenhouse monitoring; some scholars have carried out research on Zigbee or other types of technology in agricultural greenhouse monitoring and achieved some results.

Kang BJ and other scholars [5] developed an automated system for greenhouses, which stores information on greenhouse temperature, moisture, leaf temperature and leaf moisture in a database. The system is designed using Zigbee to enable the collection and automated control of plant information within the system. T.Veeramani kandasamy and other scholars [6] used GSM and Zigbee to build an automatic system for monitoring agricultural water resources and crop growth, so as to achieve the purpose of increasing food production. Zhang Hui and other scholars [7] proposed a wireless communication data acquisition system based on Zigbee and 4G, which meets the short-distance communication and long-distance data transmission. Yang Wei and other scholars [8] proposed a wireless-based real-time monitoring environment solution in accordance with the environmental monitoring requirements. It is designed to monitor environmental parameters using Zigbee. Xiao Xiao and other scholars [9] improved the Zigbee routing algorithm according to the actual application environment, and optimized the data transmission process and node energy usage according to the characteristics of different systems. Li Yan and other scholars [10] improved the ZBR routing algorithm and clustered the entire network. When the algorithm selects the cluster head, it has various factors around the cluster head, the life cycle of cluster head and the remaining energy of nodes. Niu Yugang and other scholars [11] addressed the congestion problem in the Zigbee network by considering node congestion avoidance and energy consumption balance, but the algorithm did not consider the node load and energy issues. Shan Chenggang and other scholars [12] designed a multi-path selection algorithm SMSA. This algorithm calculates the frequency of data packet collisions in the transmission process, and selects the least collisions as the reference target, which can effectively reduce the transmission delay, but it also does not consider the remaining energy of the nodes in the transmission path.

In the above studies, although researchers in various countries have conducted extensive research and made some improvements to Zigbee routing algorithms, most studies are aimed at general Zigbee algorithm improvements or the application of Zigbee in some specific environments, and few studies have combined the above two aspects. At the same time, most studies on Zigbee are mainly for outdoor crops and there are few studies on greenhouse vegetables. Although there are new technologies similar to Zigbee, for example, in May 2015, Huawei and Qualcomm jointly researched a solution called NB-CIoT (NarrowBand Cellular IoT) [13,14], but since the new technology has only been developed for a few years, it is far less mature than Zigbee.

Therefore, this research uses Zigbee as the research object; a greenhouse monitoring system based on the improved Zigbee algorithm is designed, which forms the wireless sensor network through various sensor nodes, and improves the network topology and routing algorithm of Zigbee to achieve high-efficiency and the low-consumption monitoring of environmental parameters. Compared with the previous research, this research innovatively combines the wireless topology structure with the improved Zigbee routing algorithm; it not only optimizes Zigbee itself, but also optimizes the wireless network composed of Zigbee, and is mainly oriented to the vegetable and fruit scenes in greenhouses, making the research more realistic.

2. Design and Implementation of the Wireless Sensor Network

2.1. Topology Design of the Wireless Sensor Network

In this research, the data required for crop monitoring were collected using various sensors commonly used in greenhouses and Zigbee in battery self-powered mode to form a wireless sensor network [15]. Due to the large number of Zigbee nodes required in actual production, a good network topology is particularly important for the fast communication between the network nodes. There are three types of Zigbee networks, namely, star topology, tree topology and WMN topology [16]. By comparing the advantages and disadvantages of various topologies, this research designed a WMN network topology; it uses direct data communication between routers, which can prevent the potential failure of the entire network system in the star model in case a problem with the coordinator is caused by a single link (Figure 1 shows the Zigbee wireless network topology).

The coordinator, router, and terminal are interconnected in layers, which expand the monitoring area and the establishment and later maintenance of the network. The devices communicate with one another in the Zigbee wireless network, and their own addresses are unique. According to the IEEE802.15.4 standard [17], the address of the devices consists of a 64-bit physical address and a 16-bit network address. However, it is worth noting that the physical address of all Zigbee devices is standardized at 64 bits, with nothing required to be carried out to the physical address or the Zigbee network structure; thus, the allocation of the network addresses is indeed crucial in the Zigbee network. The distribution of Zigbee network addresses is determined by the relationship between parent nodes and child nodes within the network. The parent node in the lower layer can establish connections with child nodes. On the other hand, child nodes can only connect to the parent nodes in the upper layer. L_m, C_m and R_m are three essential values in the Zigbee network, which, respectively, represent the Zigbee network level, the number of child nodes that can be connected to the parent nodes and the number of routers in the network [18,19]. The address interval can be calculated. The address interval can be calculated according to Equation (1):

$$Cskip(d) = \begin{cases} 1 + C_m \times (L_m - d - 1) & R_m = 1 \\ \left(1 + C_m - R_m - C_m \times R_m^{L_m - d - 1}\right) / (1 - R_m) & \end{cases} \quad (1)$$

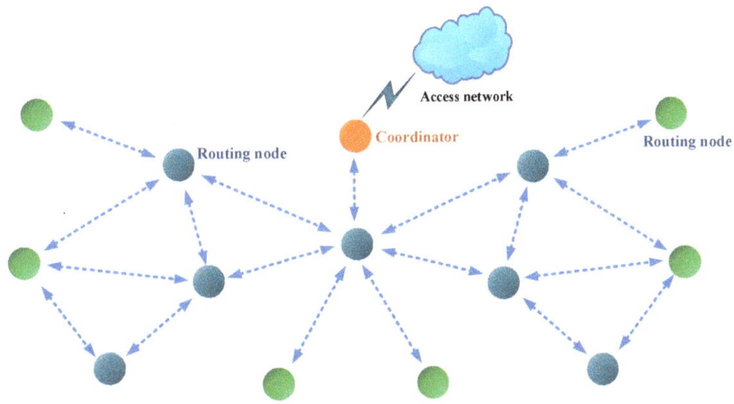

Figure 1. Zigbee wireless network topology.

In a Zigbee wireless network, the network topology works and the network address of each device can be calculated, as shown in Figure 2.

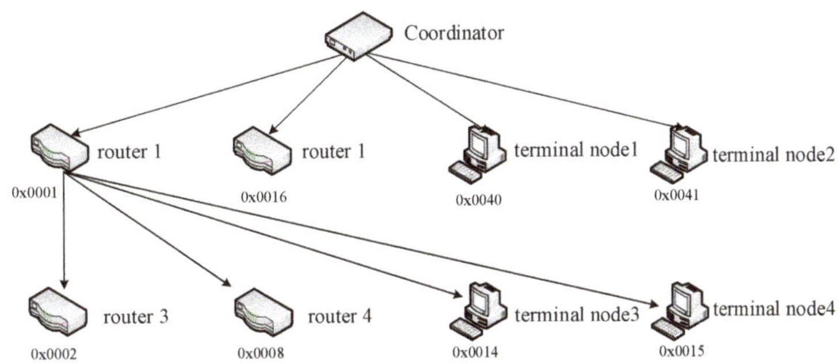

Figure 2. Network address distribution of each device in the WMN network.

Determining the network address of each device is of great importance for wireless sensor networks. If a single device is abnormal or faulty, the network address can quickly determine the faulty node, which can greatly facilitate the operation of wireless sensor networks.

2.2. Routing Protocol Design

The transmission delay and the successful reception rate of data packets in the ZigBee network will be affected by the routing protocol to a large extent. In order to make the network more stable, this study proposes an improved ZigBee-based routing protocol EMP-ZBR after combining the designed WMN network topology, thereby reducing energy loss and relieving network congestion.

First, the cross-layer mechanism is introduced on the basis of the original Zigbee routing protocol, in which the routing layer can obtain the queue buffer packets in the MAC (Media Acccss Control Address) layer and use the length of the cache queue in the node to determine the load of the node, defined as the queue cache group occupancy, according to the protocol built on an agricultural information monitoring system; the energy consumption of the node is determined against the total energy in the node and the energy consumed [16], and is defined as the energy consumption occupancy ratio. The algorithm description: Q_{cr} and E_{cr} is defined to represent the cache packet occupation ratio and energy consumption ratio of the node, respectively. Q_{cr} and E_{cr} values are defined as Equations (2) and (3):

$$Q_{cr}= Q_{le}/Q_{max} \tag{2}$$

$$Q_{cr}= \left(E_{int} - E_{left}\right)/E_{int} \tag{3}$$

Among them, Q_{le} represents the length of the cache queue at the moment of the node, and Q_{max} refers to the maximum cache queue length that the node can accommodate. E_{int} represents the total energy of the node at the beginning, E_{left} represents the remaining energy of the node after working for a period of time and E_{int} E_{left} represents the total amount of energy consumed by the node during work.

The routing update criterion for the Zigbee routing protocol was then improved accordingly. In the original Zigbee protocol, the shortest hop count is applied as the routing update criterion; while MP-ZBR (Multi-Link Zone Border Router) is the routing protocol, the new routing update criterion cost is applied. The value of cost will be calculated following certain rules through the energy consumption ratio of intermediate nodes, cache packet occupation ratio and link quality. When the RREQ (Route Request Packet) packet arrives at the intermediate node [20], the E_{cr} (the energy consumption ratio in the node), Q_{cr} (cache packet occupancy ratio) and the link quality LQI value are first calculated. Q_{cr} and E_{cr} are calculated in Equations (4) and (5). The LQI value can be obtained by calculating the received signal strength RSSI value, which is an integer from 0 to 255. The larger the LQI value, the better the link quality. The Q_{cr}, E_{cr} and LQI values are added to Q_{crsum} (the sum of the node load ratio), Q_{ersum} (the sum of the energy consumption ratio) and L_sum (link quality sum). To unify the rules, the L_sum value is defined as the reciprocal of the sum of the actual link quality, and the cost value is calculated with the following rules:

$$Q_{crsum} = \sum Q_{cr} \tag{4}$$

$$E_{crsum} = \sum E_{cr} \tag{5}$$

$$\text{L\_sum} = 1/\sum LQI \tag{6}$$

$$\cos t = \alpha \frac{Q_{crsum}}{hop\_count} + \beta \frac{E_{crsum}}{hop\_count} + \lambda \frac{\text{L\_sum}}{hop\_count} \tag{7}$$

where α, β and λ are the weight value and satisfy α + β + λ =1. The greater the weight value, the greater the influence of the node energy condition or the node's buffer queue length and link quality condition on the routing update criterion. Since this research studies the impact of congestion on the network, after multiple experimental verifications, α the value of α is 0.4 and the values of β and λ are 0.3. Q_{crsum} is the sum of the proportions of all node cached packets experienced by the RREQ packet from the source node to the receiving or forwarding node, and E_{crsum} is the sum of the energy consumption proportions of all nodes experienced by the RREQ packet from the source node to the receiving or forwarding node. Also, L_sum is the sum of the link quality of all nodes experienced by the RREQ packet from the source node to the receiving or forwarding node, and Hop_count is the total number of hops experienced by the RREQ packet from the source node to the receiving or forwarding node [21]. $\cos t$ is the energy consumption, queue buffer score, and link quality of the RREQ packet from the source node to the receiving or forwarding node. The smaller the average energy consumption ratio and the average cache packet occupation ratio in the path are, the higher the link quality is, and the lower the $\cos t$ value is, indicating a better comprehensive performance of the path. At this time, the established path will be relatively stable, and can effectively alleviate network congestion.

Finally, this research improves the path establishment method of the Zigbee routing protocol, and finds an effective path for the source node to reach the destination node by RREQ packets during the routing discovery process [22]. Upon receipt of a RREQ packet, the destination node initiates the cache timer and sends the RREP (Residential Real Estate Project) packet back to the source node in the order of their arrival [23]. After the first RREP packet hits the source node, the source node will cache all the RREP packets that arrived during this period before the timer expires, and sorts these packets according to certain priority rules. The path with the highest priority is defined as the primary path, and the path with the second priority is defined as the backup path. When the main path is successfully established, the data are preferentially transmitted to the destination node through this path. When the primary path cannot continue to transfer data due to congestion, the previously established backup path is selected to successfully transmit the data to the destination node. The improved algorithm first compares the comprehensive performance index cost value of each feasible path in the path establishment process, and selects the path with the lowest cost value as the main path during data transmission [24]. As maintaining multiple backup paths would consume more network resources, this article will select only one backup path from other paths. Figure 3 shows the specific flow of the algorithm.

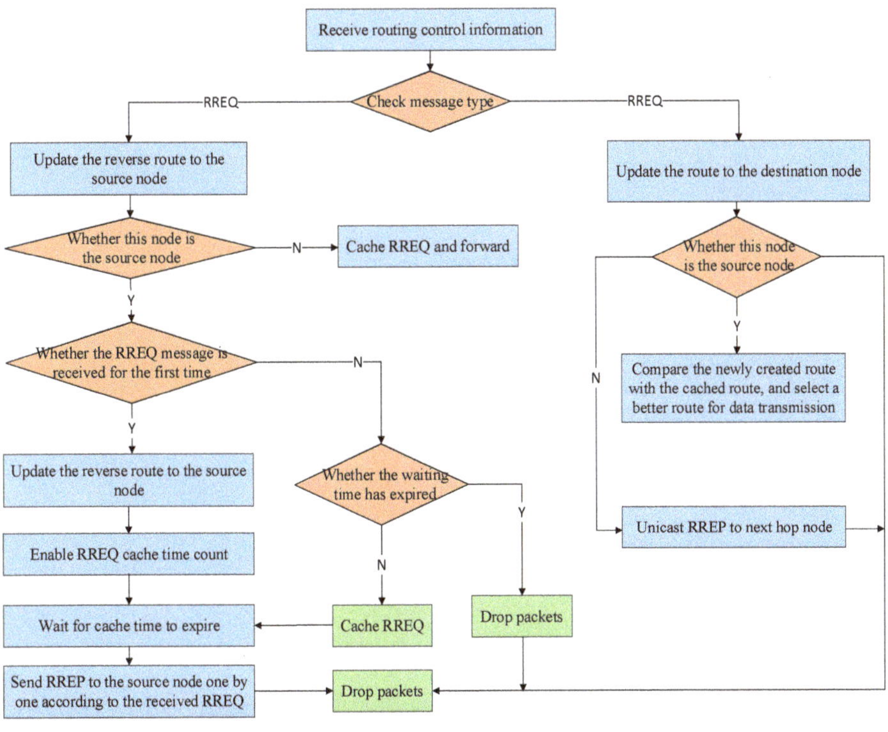

Figure 3. The specific flow of the algorithm.

3. Experimental Design and Results

3.1. Experimental Design

NS2 (Network Simulator Version 2) is a professional simulation software specially used for network simulation [25]. It has very rich module components, its configuration is flexible and its scalability is relatively strong. It is favored by many scholars who study network technology. To this end, this research will randomly arrange a large number of sensor nodes in a certain area on the NS2 platform, imitating the actual system application situation, and set the node energy and packet sending rate in the network. Finally, in the same scenario, the proposed EMP-ZBR routing protocol and the traditional ZigBee routing protocol are simulated separately, and the performance of the two protocols is evaluated in four aspects: packet delivery rate, route discovery frequency, route control overhead, and end-to-end average delay.

3.2. Experimental Indicators

The indicators to measure the performance of the Zigbee routing protocol are the end-to-average delay and packet delivery rate. This research will use the following four indicators for the comparative analysis of the original ZigBee routing protocol and EMP-ZBR routing protocol:

(1) Packet delivery rate: This indicates the ratio of the number of data packets successfully reaching the destination node to the number of data packets sent by the source node. The larger the packet delivery rate, the better the reliability of the transmitted data [26]. Equation (8) is as follows:

$$Packet\ delivery\ rate = \frac{\begin{array}{c}The\ number\ of\ data\ packets\ successfully\\ reaching\ the\ destination\ node\end{array}}{The\ number\ of\ data\ packets\ sent\ by\ the\ source\ node} \tag{8}$$

(2) Route discovery frequency: This indicates the number of route requests initiated by the source node within a unit time. The higher the frequency, the greater the loss of network resources and energy [27]. Equation (9) is as follows:

$$Route\ discovery\ frequency = \frac{\begin{array}{c}The\ total\ number\ of\ routing\ requests\\ initiated\ by\ the\ source\ node\end{array}}{Simulation\ time} \tag{9}$$

(3) Routing control overhead: This indicates the ratio of routing control packets to received data packets. The more routing control packets, the more energy consumed by the network [28]. Equation (10) is as follows:

$$Routing\ control\ overhead = \frac{Total\ routing\ control\ messages}{Total\ number\ of\ packets\ received} \tag{10}$$

(4) End-to-end average delay: This represents the average time from the source node sending out the data packet to the destination node receiving the data packet in the whole process [28]. Equation (11) is as follows:

$$End - to - end\ average\ delay = \frac{Packet\ Received\ Time - Packet\ Sent\ Time}{Total\ number\ of\ packets\ received} \tag{11}$$

3.3. Experimental Environment Settings

In this study, in order to better simulate the real scene, a simulation area with a length of 10 and width of 30 ft × 90 ft is selected. A total of 60 nodes are randomly arranged in it, and 40 nodes are set to be in a static state. These node simulation systems have coordinator nodes and sensor nodes such as air temperature and moisture, soil temperature and moisture. The remaining 20 nodes are set as random mobile nodes. The random movement of nodes will make the network topology change more frequently so that the performance of the two protocols under different topological conditions can be studied and the final simulation results obtained are more authentic. Table 1 shows the simulation parameters.

Table 1. The simulation parameters.

Parameters	Value
Channel type	Channel/Wireless Channel
Network Interface	Phy/WirelessPhy/802_15_4
MAC type	Mac/802_15_4
Wireless communication model	Antenna/OmniAntenna
Transfer model	TwoRayGround
Topological size	30 ft × 90 ft
Number of nodes	60
Packet size	512 bytes
Simulation time	400 s
Energy model	EnergyModel
Node initial energy	80 J
Transmit power	0.85 W
Received power	0.49 W
Sleep power	0.15 W

3.4. Experimental Results

3.4.1. Result Analysis under Different Pause Times of Mobile Nodes

In this research, the CBR sending rate is always three packets/s, and the pause time of the mobile node is changed to 0 s, 50 s, 100 s, 150 s, 200 s, 250 s, 300 s, 350 s, 400 s. The node pause time will affect the network topology changes. The mobility of nodes in the network will weaken with the increase in the pause time, and the topology changes will gradually become stable [29].

(1) Packet delivery rate

Figure 4 shows the performance difference of the two routing protocols in the packet delivery rate under different mobile node pause times. Under the same circumstances, the packet delivery rate of the EMP-ZBR routing protocol is always greater than that of the ZigBee routing protocol, which increases the packet delivery rate by 1.1% on average, and improves the network congestion to a certain extent [30].

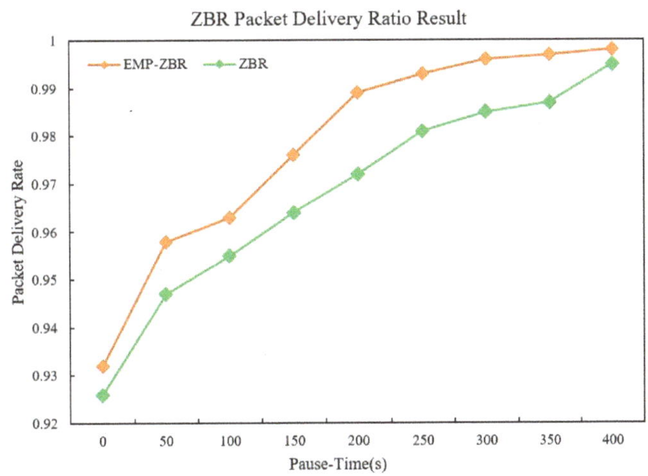

Figure 4. Result of the packet delivery rate at different pause times.

(2) Route discovery frequency

Figure 5 shows the difference in route discovery frequency of the two routing protocols under different mobile node pause times. Under the same pause time, the routing discovery frequency of the EMP-ZBR routing protocol is significantly lower than that of the ZigBee routing protocol, which slows down the routing discovery frequency by 15.2% on average, indicating that EMP-ZBR reduces the number of times the source node routing requests are initiated and that the network is more stable.

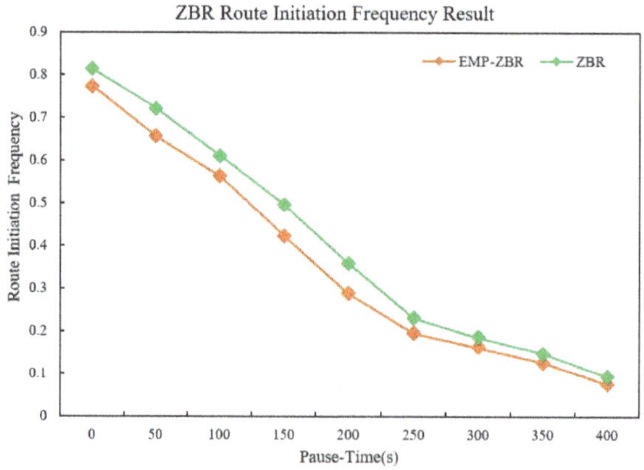

Figure 5. Route discovery frequency result with different pause times.

(3) Routing control overhead

Figure 6 shows the relationship between the routing control overhead and mobile node pause time. Under the same pause time, the EMP-ZBR routing protocol selects nodes with sufficient energy and a large remaining queue buffer length to establish paths, and fewer messages are used for routing initiation and routing maintenance. Compared with the traditional ZigBee protocol, the average reduction of 21% of the routing control overhead more effectively alleviates the impact of path congestion and link breaks on the network.

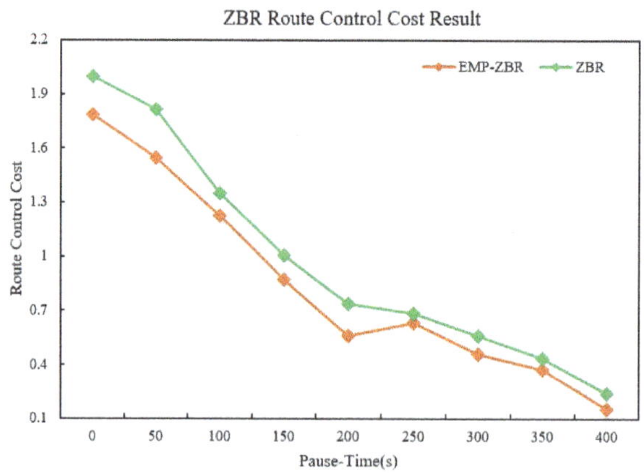

Figure 6. Routing control overhead results for different pause times.

(14) End-to-end average delay

Figure 7 shows the relationship between the end-to-end average delay and mobile node pause time. Under the same pause time, the average delay of the EMP-ZBR routing protocol is significantly lower than that of the ZigBee routing protocol, and it will not pass through congested nodes during data transmission, which reduces the average delay in the network [31]. Compared with the traditional ZigBee protocol, the EMP-ZBR routing protocol reduces the end-to-end average delay by 8.1 ms on average, which plays a great role in alleviating congestion.

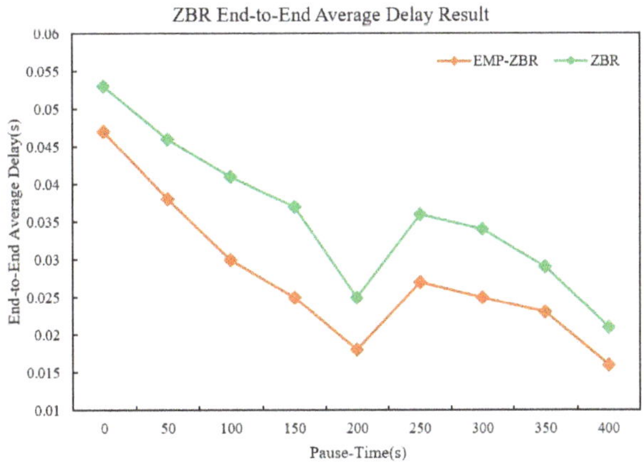

Figure 7. End-to-end average delay results for different pause times.

3.4.2. Analysis of Results at Different Packet Sending Rates

In this research, controlling the pause time of the mobile node remains unchanged (0 s), changing the CBR sending rate to 3, 6, 9, 12, 15, 18, 21, and 24, respectively. It reflects the load level of nodes in the network. The faster the packet sending rate, the stronger the node load, and the more prone to congestion.

(1) Packet delivery rate

Figure 8 shows the variation in the packet delivery rate of the two routing protocols under the background of different CBR packet sending rates [32]. As the packet sending rate increases, the packet delivery rates of the two routing protocols show a downward trend; however, the EMP-ZBR routing protocol always increases the packet delivery rate by an average of 9.8% compared to the traditional ZigBee protocol because it forwards the information criterion to avoid packet loss caused by some nodes due to heavy load, and improves network congestion.

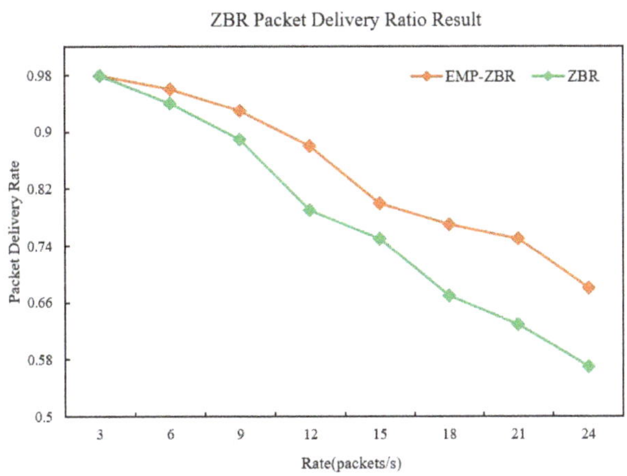

Figure 8. Packet delivery rate results at different packet sending rates.

(2) Route discovery frequency

Figure 9 shows the performance difference between the two routing protocols in route discovery frequency under different CBR sending rates. As the packet sending rate of nodes increases, the route discovery frequency of the two routing protocols gradually increases. However, the route discovery frequency of the EMP-ZBR routing protocol is always lower than that of the ZigBee routing protocol, which slows down the route discovery frequency by 19.3% on average [33]; this is because the IMP and ZBR routing protocol caches the backup path while establishing the main path. It can then use the backup path for data transmission, reducing the number of routing initiations, thereby reducing the energy consumption and network load.

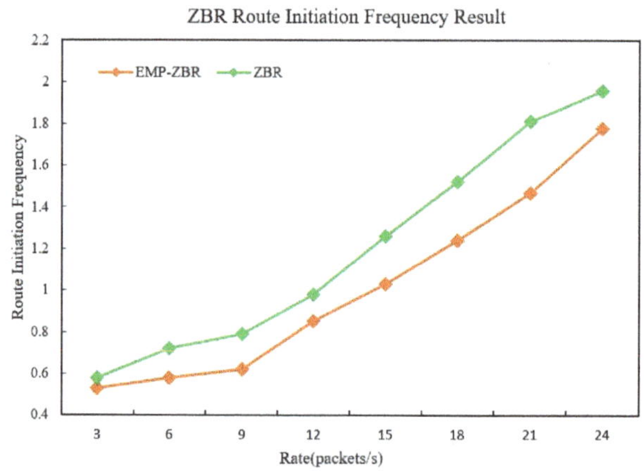

Figure 9. Route discovery frequency results at different packet sending rates.

(3) Routing control overhead

Figure 10 shows the difference in the routing control overhead performance of the two routing protocols at different CBR sending rates. The EMP-ZBR routing protocol improves the forwarding mechanism of the intermediate nodes. When forwarding data packets, it fully considers the energy of the intermediate nodes and the remaining status of the cache queue [34]. The frequency of route discovery and route maintenance saves the route control overhead; compared with the traditional ZigBee protocol, the route control cost is reduced by 15.7% on average.

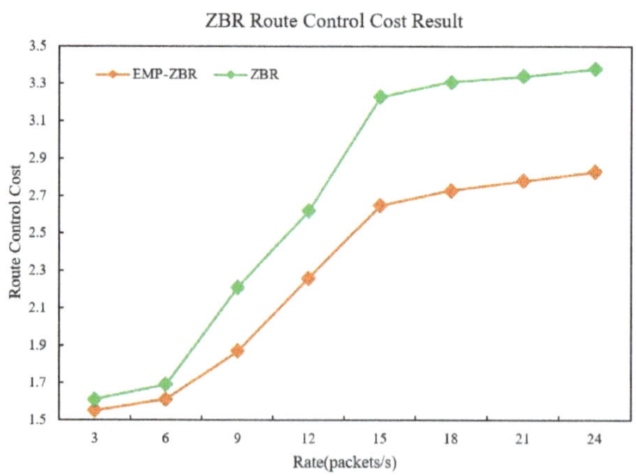

Figure 10. Routing control overhead results at different packet sending rates.

(4) End-to-end average delay

Figure 11 shows the differences in the end-to-end average delays of the two routing protocols at different CBR sending rates. As the CBR packet sending rate increases, the average network delay of the two routing protocols increases, but the delay of the EMP-ZBR routing protocol is always lower than that of the ZigBee routing protocol, reducing the average end-to-end average delay by 121 ms [35]. This is because it selects the effective path with the smallest average energy-occupied ratio, the average queue cache packet, and the largest link quality, thereby reducing the possibility of congestion.

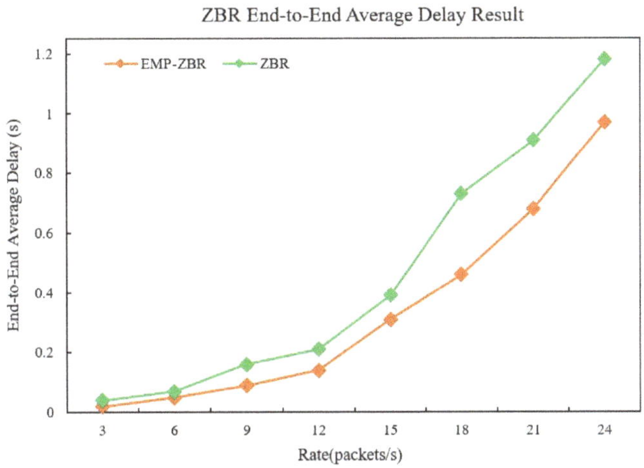

Figure 11. End-to-end average delay results of different packet sending rates.

4. Conclusions

In this research, a wireless sensor network for agricultural greenhouses based on the improved Zigbee protocol is designed. This research uses a variety of commonly used sensors and Zigbee to form an agricultural greenhouse data acquisition network, and improves the topology networking mode and routing algorithm of Zigbee nodes. After NS2 simulation experiments, the improved Zigbee sensor network shows lower network delay. The average end-to-end delay, high packet delivery rate, small routing control overhead and low routing discovery frequency prove that this research can effectively improve the network speed and reduce congestion, improving the overall performance of the agricultural information monitoring system. In the future, we envisage combining the wireless sensor network with 5G technology, connecting the 5G router through the Zigbee coordinator, and transmitting the collected sensor data to the remote terminal faster. In addition, we envisage combining the Zigbee network with the agricultural greenhouse control equipment so that the agricultural greenhouse control equipment can perform regular operations according to the environmental parameters of the Zigbee sensor network, so as to realize automatic production and remote terminal control.

Author Contributions: Conceptualization, R.T.; Methodology, R.T.; Software, R.T.; Validation, R.T.; Formal analysis, R.T.; Investigation, R.T.; Resources, N.K.A.; Data curation, R.T.; Writing—original draft, R.T.; Writing—review & editing, R.T., N.K.A. and M.S.A.T.; Visualization, R.T.; Supervision, N.K.A. and M.S.A.T.; Project administration, M.S.A.T.; Funding acquisition, N.K.A. and M.S.A.T. All authors have read and agreed to the published version of the manuscript.

Funding: This research received no external funding.

Institutional Review Board Statement: Not applicable.

Informed Consent Statement: Not applicable.

Data Availability Statement: Not applicable.

Conflicts of Interest: The authors declare no conflict of interest.

References

1. Chunjiang, Z. *Precision Agriculture Research and Practice;* Science and Technology Press: Beijing, China, 2016; pp. 56–58.
2. Dongmin, H. Thoughts on Meteorology Serving Disaster Prevention and Mitigation in Modern Agriculture. *China Agric. Meteorol.* **2009**, *S2*, 4.
3. Chi, T.; Chen, M.; Gao, Q. Implementation and study of a greenhouse environment surveillance system based on wireless sensor network. In Proceedings of the 2008 International Conference on Embedded Software and Systems Symposia, Chengdu, China, 29–31 July 2008; IEEE: Manhattan, NY, USA; pp. 287–291.
4. Li, Z. Smart agriculture based on Zigbee and NB-IoT wireless sensor network. *Hebei Agric. Mach.* **2021**, *15*, 143–144.
5. Kang, B.J.; Park, D.H.; Cho, K.R.; Shin, C.S.; Cho, S.E.; Park, J.W. A Study on the Greenhouse Auto Control System Based on Wireless Sensor Network. In Proceedings of the International Conference on Security Technology, Hainan, China, 22 December 2008; IEEE: Manhattan, NY, USA; pp. 41–44.
6. Veeramani kandasamy, T.; Rajendran, K.; Sambath, K. Remote monitoring and closed loop control system for social modernization in agricultural system using GSM and Zigbee technology. In Proceedings of the 2014 International Conference on Advances in Electrical Engineering (ICAEE), Vellore, India, 9–11 January 2014.
7. Hui, Z.; Gang, H. Remote Wireless Soil Parameter Monitoring System Based on SCM and 3G. *Microprocessor* **2015**, *37*, 92–96.
8. Wei, Y.; Minzan, L.; Xiu, W. Current status and research progress of farmland information transmission methods. *J. Agric. Eng.* **2013**, 230–297.
9. Xiao, X.; Jiabin, Y.; Xue, W. Improved Zigbee Routing Algorithm for Airport Noise Perception. *Small Micro Comput. Syst.* **2016**, *37*, 361–365.
10. Yan, L.; Anna, Y.; Xue, L. A weight cluster-based hybrid routing algorithm of Zigbeenetwork. *Int. J. Future Gener. Commun. Netw.* **2014**, *7*, 65–72.
11. Yugang, N.; Wenguang, C. A Grid-based WSN Routing Algorithm Considering Congestion Avoidance and Energy Consumption Balance. *Control. Decis. Mak.* **2016**, *31*, 1985–1990.
12. Chenggang, S.; Jie, M.; Wei, Z. An Improvement of AODV Algorithmin a Large- scale Outdoor Area for Zigbee Networks. In Proceedings of the Fifth International Conference on Instrumentation & Measurement, Computer, Communication, and Control(IMCCC), Qinhuangdao, China, 18–20 September 2015; pp. 1679–1684.
13. Yunzhou, X. The evolution of NB-IoT standard system and the development of the Internet of Things industry. *J. Internet Things* **2018**, *2*, 76–87.
14. Zhu, J.; Yang, B.; Chen, P. Research on Wireless Access Technology of Internet of Things. *J. Internet Things* **2018**, *2*, 73–84.
15. Gupta, K.; Vohra, R.; Sawhney, R.S. Envisaging Performance Metrics of Zigbee WirelessSensors by Topology Variations. *Int. J. Comput. Appl.* **2015**, *121*, 975–987.
16. Lee, T.J.; Lee, O.; Kim, J.; Shterengas, L. CO_2 sensor with data logger system. In Proceedings of the IEEE Long Island Systems, Applications and Technology Conference (LISAT), Farmingdale, NY, USA, 4 May 2012; pp. 1–7.
17. Dai, J. Application practice of Philips horticultural LED in greenhouse production. *Agric. Eng. Technol.* **2017**, *37*, 28–32.
18. Adumbabu, I.; Selvakumar, K. Energy Efficient Routing and Dynamic Cluster Head Selection Using Enhanced Optimization Algorithms for Wireless Sensor Networks. *Energies* **2022**, *15*, 8016. [CrossRef]
19. Patnaik, S.; Kountchev, R.; Jain, V. (Eds.) *Smart and Sustainable Technologies: Rural and Tribal Development Using IoT and Cloud Computing: Proceedings of ICSST 2021*; Springer: Cham, Switzerland, 2022.
20. Faghihniya, M.J.; Hosseini, S.M.; Tahmasebi, M. Security upgrade against RREQ flooding attack by using balance index on vehicular ad hoc network. *Wirel. Netw.* **2017**, *23*, 1863–1874. [CrossRef]
21. Luo, L.; Sheng, L.; Yu, H.; Sun, G. Intersection-Based V2X Routing via Reinforcement Learning in Vehicular Ad Hoc Networks. In *IEEE Transactions on Intelligent Transportation Systems*; IEEE: Manhattan, NY, USA, 2022; Volume 23, pp. 5446–5459. [CrossRef]
22. Kusumamba, S.; Kumar, S.M.D. A reliable cross layer routing scheme (CL-RS) for wireless sensor networks to prolong network lifetime. In Proceedings of the 2015 IEEE International Advance Computing Conference (IACC), Banglore, India, 12–13 June 2015; pp. 1050–1055. [CrossRef]
23. Wang, X.; Meng, X. UAV Online Path Planning Based on Improved Genetic Algorithm. In Proceedings of the 2019 Chinese Control Conference (CCC), Guangzhou, China, 27–30 July 2019; pp. 4101–4106. [CrossRef]
24. Shipu, X.; Yunsheng, W.; Yong, L.; Weixiong, R.; Mingzhou, M.; Jingyin, Z.; Chenxi, Z. Research on wsn routing algorithm for vegetable greenhouse. In Proceedings of the 2018 2nd High Performance Computing and Cluster Technologies Conference, Beijing, China, 22–24 June 2018; pp. 37–42.
25. Zhang, D.; Cui, Y.; Zhang, T. New quantum-genetic based OLSR protocol (QG-OLSR) for mobile ad hoc network. *Appl. Soft Comput.* **2019**, *80*, 285–296. [CrossRef]
26. Tao, J.; Bai, G.; Shen, H.; Cao, L. ECBRP: An efficient Cluster-based routing protocol for real-time multimedia streaming in MANETs. *Wirel. Pers. Commun.* **2011**, *61*, 283–302. [CrossRef]

27. Marwaha, S.; Tham, C.K.; Srinivasan, D. Mobile agents based routing protocol for mobile ad hoc networks. In Proceedings of the Global Telecommunications Conference, 2002. GLOBECOM '02, Taipei, Taiwan, 17–21 November 2002; Volume 1, pp. 163–167. [CrossRef]
28. Sun, Y.; Sun, J.; Zhao, F.; Hu, Z. Delay constraint multipath routing for wireless multimedia ad hoc networks. *Int. J. Commun. Syst.* **2016**, *29*, 210–225. [CrossRef]
29. Gui-kai, L.; Chun-li, S.; Gang, W.; Hong-jiang, W. Subarea Tree Routing in multi-hop wireless ad hoc networks. In Proceedings of the 2008 11th IEEE Singapore International Conference on Communication Systems, Guangzhou, China, 19–21 November 2008; pp. 1695–1699. [CrossRef]
30. Zhang, D.; Gong, C.; Zhang, T.; Zhang, J.; Piao, M. A new algorithm of clustering AODV based on edge computing strategy in IOV. *Wirel. Netw.* **2021**, *27*, 2891–2908. [CrossRef]
31. Wang, Y.; Vuran, M.C.; Goddard, S. Cross-layer analysis of the end-to-end delay distribution in wireless sensor networks. *IEEE/ACM Trans. Netw.* **2011**, *20*, 305–318. [CrossRef]
32. Chatterjee, S.; Ladia, R.; Misra, S. Dynamic Optimal Pricing for Heterogeneous Service-Oriented Architecture of Sensor-Cloud Infrastructure. In *IEEE Transactions on Services Computing*; IEEE: Manhattan, NY, USA, 2017; Volume 10, pp. 203–216. [CrossRef]
33. Jia, L. Distributed energy balance routing algorithm for wireless sensor network based on multi-attribute decision-making. *Sustain. Energy Technol. Assess.* **2021**, *45*, 101192.
34. de Olivera, S.; Shiroma, P.M.; Ilídio, R.S.; Pereira, M.A.; Silva, C.M. Extending the Lifetime of Wireless Sensor Networks using a Genetic Model for defining Mobile Sinks. *Int. Robot. Autom. J.* **2017**, *2*, 00040.
35. Ahn, G.-S.; Campbell, A.T.; Veres, A.; Sun, L.-H. SWAN: Service differentiation in stateless wireless ad hoc networks. In Proceedings of the Twenty-First Annual Joint Conference of the IEEE Computer and Communications Societies, New York, NY, USA, 23–27 June 2002; Volume 2, pp. 457–466. [CrossRef]

agriculture

Article

Research on Optimization of the Bulb Form of the Bulb Tubular Pump Device for a Low-Head Agricultural Irrigation Pumping Station

Hongyin Zhang [1], Jianlong Liu [2], Jinxin Wu [1], Weixuan Jiao [1,*], Li Cheng [1] and Mingbin Yuan [1]

1 College of Hydraulic Science and Engineering, Yangzhou University, Yangzhou 225100, China; chengli@yzu.edu.cn (L.C.)
2 Management Division of Qinhuai River Hydraulic Engineering of Jiangsu Province, Nanjing 210022, China
* Correspondence: jiaowx@yzu.edu.cn

Abstract: A bulb tubular pump unit is a horizontal tubular pump unit composed of a water pump and bulb with an electric motor installed. Electric motors, transmission equipment, and bearings are usually placed in the bulb. The bulb is located in the flow channel and has a relatively narrow space. Therefore, the shape of the bulb has a significant influence on the flow pattern and pump efficiency in the flow channel. In this study, the CFX 19.2 software was employed to optimize the bulb hydraulically according to its geometry and parameters. The research results indicate that the flow pattern at the tail of the elliptical bulb was better, the hydraulic loss at the bulb section was small, and the device efficiency was higher than that at the tail, which was round. The streamlined support had small flow resistance, minimal hydraulic loss, and a high pump unit head and efficiency. Nine schemes were selected, and the geometrical characteristics and parameters of the bulb were determined as follows: the shape of the tail of the bulb was oval, the bulb ratio was 0.96, and the shape of the support parts was streamlined. The results hold important reference significance to improve efficiency and broaden the operating conditions of bulb tubular pump devices.

Keywords: bulb tubular pump; bulb form; optimization design; numerical simulation; hydraulic performance

Citation: Zhang, H.; Liu, J.; Wu, J.; Jiao, W.; Cheng, L.; Yuan, M. Research on Optimization of the Bulb Form of the Bulb Tubular Pump Device for a Low-Head Agricultural Irrigation Pumping Station. *Agriculture* **2023**, *13*, 1698. https://doi.org/10.3390/agriculture13091698

Academic Editors: Redmond R. Shamshiri, Muhammad Sultan, Md Shamim Ahamed and Muhammad Farooq

Received: 11 August 2023
Revised: 22 August 2023
Accepted: 25 August 2023
Published: 28 August 2023

1. Introduction

Agricultural drought leads to a reduction in food production and is a serious natural disaster phenomenon. Water conservancy is an important piece of infrastructure of the national economy, and pumping stations are an important component of water conservancy projects, which are key to protecting and developing food production. In particular, large pumping stations bear the heavy responsibility of regional flood control, irrigation, water diversion, and water supply. The benefits of irrigation pumping stations are mainly reflected in the reduction in the effects agricultural drought [1,2]. By increasing water used for irrigation, agricultural drought is alleviated, and food production is guaranteed.

Low-head pump stations are the main type of pump stations in the middle and lower reaches of the Yangtze River and the Pearl River Deltas in China. Due to the low and flat terrain in these areas, low-head pumping stations are often used. The construction of low-head pumping stations has promoted the construction of agricultural production bases with stable and high yields in China, ensuring economic development and people's productivity and survival. The bulb tubular pump device is one of the main types of low-head pumping stations along the Yangtze River and is also an important component of high-efficiency irrigation systems. The bulb tubular pump device has the advantages of smooth inlet and outlet channels, low hydraulic loss, and high channel efficiency, making it the most reasonable device type in theory for large-flow and low-head pumping stations. This economical pump type combines the technical characteristics of a diving motor and through-flow pump. With the rapid development of water conservancy in China and the extensive implementation of water diversion projects, the use of submersible tubular pumps has been consistently increasing and fully applied, which has played a significant role in promoting the development of the national economy.

The pump device is the core component of the pumping station. At present, research on pump devices mainly focuses on hydraulic optimization [3–5] and hydraulic stability [6–8]. The research results provide a reference for the design and optimization of a bulb tubular pump device. Numerical simulation based on computational fluid dynamics (CFD) technology is the main method for hydraulic optimization research of pump devices [9–11]. Zhou et al. [12] adopted the RNG k–ε turbulence model and SIMPLEC algorithm, based on the multi-rotation coordinate system model, and calculated the steady flow of two typical forms of post-installed tubular pumps, as well as a traditional cylindrical bulb and a spin-cone slender bulb with a total length of 10.72 D (D is the impeller diameter), and obtained the overall flow field results. Zhang et al. [13] combined the working condition's regulation mode, transmission mode, and overall structure and found that bulb tubular pumps with different structures were safe and reliable. Sun et al. [14] used the method of combining numerical simulation and experimental testing to analyze the differences between the external and internal characteristics of the two schemes involving a front-mounted bulb and a rear-mounted bulb. This research can provide reference for the design and form selection of a submersible tubular pump device, which has great engineering significance. Although some experts have conducted some work on the hydraulic optimization of bulb tubular pump devices in the early stage, with the development of CFD technology and pump device theory, the original research cannot meet the current needs.

The hydraulic stability of the pump device is also an important factor affecting the efficiency of the pumping station [15–17]. Stall [18,19], cavitation [20,21], and inlet vortices [22,23] are unstable water phenomena that occur during the operation of the pump unit. At present, experts and scholars mainly conduct research on the formation mechanism and suppression measures of hydraulic instability in pump devices and have achieved certain research results, which can provide guidance for the safe and stable operation of pump stations.

To sum up, however, little research has been conducted on the shape of the bulb. Therefore, in this study, we used CFX 19.2 software to optimize the shape of the bulb, which holds important reference significance to improve efficiency and broaden the operating conditions of bulb tubular pump devices.

2. Numerical Simulation

2.1. Three-Dimensional Geometric Model

In this study, the single-unit design flow rate of the tubular pump was 64 m^3/S, the single power was 3550 kW, the speed was 85.7 r/min, the head size was 3.15 m, and the impeller diameter range D was 5.14 m. As shown in Figure 1, the bulb tubular pump device consists of components such as the inlet duct, impeller, guide vane, bulb body, and outlet duct. The bulb tubular pump device is a rear-mounted bulb tubular pump device, and the bulb body is placed in the outlet duct. Unlike other types of pumping stations, placing the bulb body inside the pump device will have an impact on the flow of water in the duct and increase hydraulic losses. Therefore, it is necessary to conduct research on the key structural parameters of the bulb body.

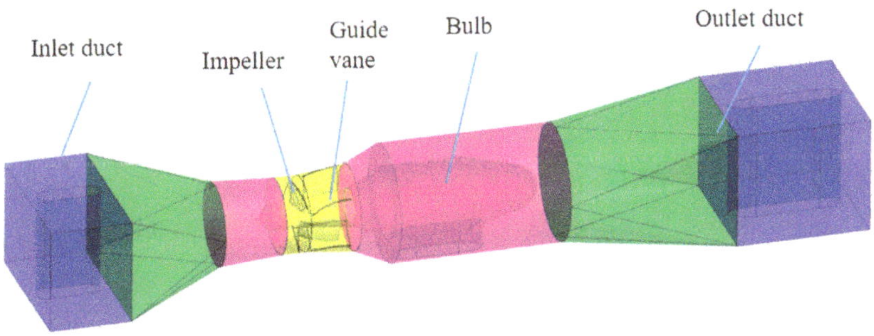

Figure 1. Schematic diagram of bulb tubular pump.

2.2. Numerical Simulation

2.2.1. Calculation Method

In order to ensure the accuracy of the calculation, this article divides the overall pump device into block grids as shown in Figure 2. ANSYS 19.2 Turbo Grid software was used for grid division of the impeller and guide vane calculation domain, and ANSYS 19.2 Mesh software was used for grid division of other areas. Hexahedral grids were used for each part of the grid to ensure the orthogonality and high quality of the calculation grid, and the grid quality of each part was above 0.85. At the same time, in order to accurately simulate the real situation, this paper conducts mesh refinement processing in various calculations and boundary layers. After this grid was partitioned into blocks, each block was output separately and then each part was merged during the CFX pretreatment to establish the interface area and solution conditions for calculation.

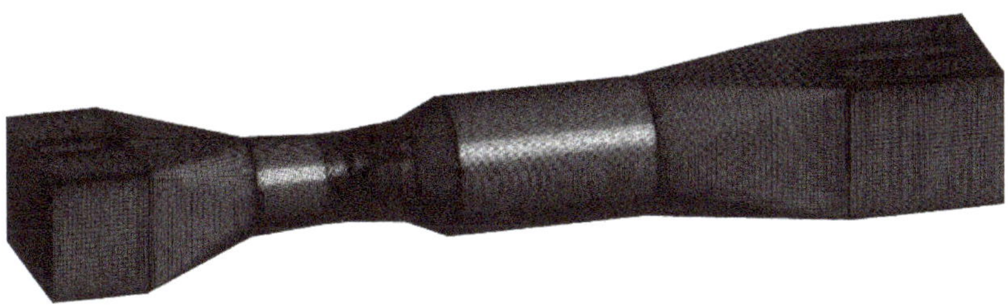

Figure 2. Mesh layout.

2.2.2. Computational Grid Independence Analysis

The quality and quantity of the grid have a significant influence on the accuracy and reliability of the numerical results. Turbo-Grid provides a self-checking function of grid quality, which generally makes the surface shape of the blades smooth and also generates better topology and grid quality. In terms of the number of grids, in theory, the more grids the model has, the smaller the solution error caused by the grid is. As the number of grids increases, so too does the requirement for computer configuration and calculation speed. Therefore, on the basis of ensuring the solution accuracy to a certain extent, the number of grids should not be too large. Before the numerical calculation, the number of grids should be analyzed to achieve the purpose of grid independence. Therefore, we conducted a grid independence analysis for a bulb tubular pump impeller (see Table 1).

Table 1. Number of impeller domain grids.

Grid	Impeller Grid/Million	Guide Vane Grid/Million	Passage Inlet Grid/Million	Passage Outlet (Including Bulb) Grid/Million	Total Grid/Million
1	1.40	1.80	2.60	3.00	8.80
2	1.50	1.80	2.60	3.00	8.90
3	1.60	1.80	2.60	3.00	9.00
4	1.70	1.80	2.60	3.00	9.10
5	1.80	1.80	2.60	3.00	9.20
6	1.90	1.80	2.60	3.00	9.30

A grid independence analysis was conducted according to the calculation formula of efficiency and head. The head H and the efficiency η of the bulb tubular pump device were used as the main evaluation indices. As the number of grids increased, the head and efficiency of the device decreased gradually. It is worth noting that the relative error between the fifth and sixth scheme is within 1%, and it can be considered that the number of grids has already met the independence requirement at this time, as shown in Figure 3.

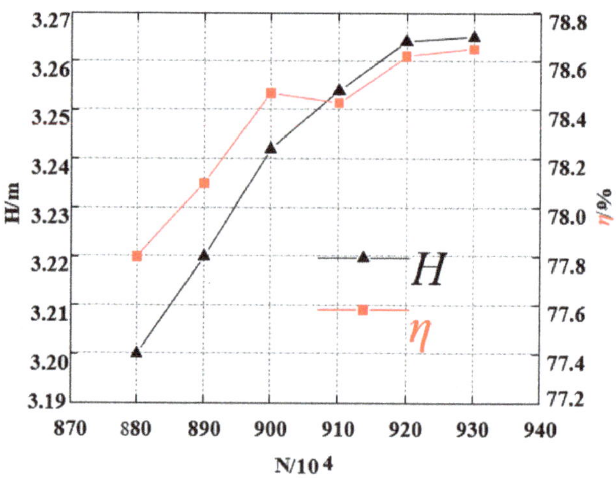

Figure 3. Grid independence analysis.

2.2.3. Calculation Parameters and Boundary Condition Setting

This paper uses the continuity equation based on incompressible fluid and the time-averaged Reynolds equation (RANS equation), and the SST k–ω model is used for the turbulence model. Hydraulic calculation based on the averaged Reynolds N-S equation is the prime method of numerical mathematics. Therefore, it is necessary to supplement the turbulence model to close the equation, and the selection of the turbulence model has a decisive influence on the accuracy of numerical simulation. Currently, the commonly used turbulence models include standard k–ε, SST k–ω, and others. Compared with the standard k–ε turbulence model, the SST k–ω turbulence model has the following advantages: the model can fully adapt to various physical phenomena, such as back pressure changes, and can be applied to the dense inner layer. The boundary layer phenomenon can be accurately simulated by applying the wall function without using the thick attenuation function, which is more easily distorted. Therefore, the SST k–ω turbulence model was applied to the numerical simulation of this pump device. The inlet of the device model was set as the mass flow inlet, which was initially set as 64,000 kg/s. The outlet was the average static pressure outlet, and the setting reference pressure was 1 atm. The interface was set to transfer the values between the components. The fixed wall was set as the nonslip boundary condition, the speed was initially set to 85.7 r/min, and the calculated residual control value was set to 10^{-4}.

2.2.4. Entropy Production Dissipation Theory

According to the second law of thermodynamics, the loss of mechanical energy is irreversibly converted into thermodynamic energy, a thermodynamic process that ultimately increases entropy production dissipation. In the interior of a pumping device, turbulent motion inevitably generates energy dissipation. Therefore, using the entropy production dissipation theory is appropriate to study the mechanism of hydraulic loss and energy dissipation inside the pumping device.

For turbulent flow, the entropy production dissipation based on the Reynolds time average can be divided into two parts: the viscous entropy production dissipation caused by fluid viscosity $\Delta S_{pro,\overline{D}}$ and the turbulent entropy production dissipation caused by turbulent pulsation $\Delta S_{pro,D'}$. In addition, the wall entropy production dissipation due to wall friction loss is $\Delta S_{pro,w}$.

The total entropy production dissipation of the whole device can be obtained from Equation (1) ΔS_{pro}.

$$\Delta S_{pro} = \Delta S_{pro,\overline{D}} + \Delta S_{pro,D'} + \Delta S_{pro,w} \tag{1}$$

The entropy production dissipation due to the time-averaged velocity can be calculated as follows:

$$\dot{S}'''_{\overline{D}} = \frac{\mu_{eff}}{T}\left\{2\left[\left(\frac{\partial\overline{u}}{\partial x}\right)^2+\left(\frac{\partial\overline{v}}{\partial y}\right)^2+\left(\frac{\partial\overline{w}}{\partial z}\right)^2\right]+\left[\left(\frac{\partial\overline{v}}{\partial x}+\frac{\partial\overline{u}}{\partial y}\right)^2+\left(\frac{\partial\overline{w}}{\partial x}+\frac{\partial\overline{u}}{\partial z}\right)^2+\left(\frac{\partial\overline{v}}{\partial z}+\frac{\partial\overline{w}}{\partial y}\right)^2\right]\right\} \tag{2}$$

where μ_{eff} is the dynamic viscosity (Pa-s) and $\overline{u}$, $\overline{v}$, and $\overline{w}$ are the time-averaged velocities in the direction of x, y, z components, (m/s).

The entropy production dissipation of turbulent dissipation due to velocity fluctuations can be calculated using the following Equation (3):

$$\dot{S}'''_{D'} = \beta\frac{\rho\omega k}{T} \tag{3}$$

where β = 0.09, ω is the turbulent eddy frequency (s^{-1}), and k is the turbulence energy (m^2/s^2).

The local entropy production dissipation integral of the computational domain is shown in Equations (4) and (5).

$$\Delta S_{pro,\overline{D}} = \int_V \dot{S}'''_{\overline{D}}\,dV \tag{4}$$

$$\Delta S_{pro,D'} = \int_V \dot{S}'''_{D'}\,dV \tag{5}$$

The wall entropy production dissipation can be calculated using the following equation:

$$\Delta S_{pro,w} = \int_A \frac{\vec{\tau}\cdot\vec{v}}{T}\,dA \tag{6}$$

where $\vec{\tau}$ is the wall shear stress (Pa), A is the area (m^2), and $\underset{V}{\rightarrow}$ denotes the first grid velocity near the wall (m/s).

The analysis of energy loss in the bulb area of the pump device by means of entropy generation theory can intuitively show the size and distribution of energy dissipation in the pump device. Therefore, when optimizing the structure of the pump device, we can focus on the intensive area of entropy generation and dissipation in the pump.

3. Analysis of Calculation Results

3.1. Analysis on Influencing Factors of Hydraulic Performance of Bulb Section

Table 2 shows that the hydraulic loss of the bulb was the largest in the whole plant. Because the bulb was in the key position of the flow channel, it affected the flow pattern of water in the flow channel, thus greatly affecting the efficiency of the whole device [21]. Further analysis showed that under the designed flow condition, the hydraulic loss (except impeller) of the whole unit was the smallest; however, under the small-flow condition, the hydraulic loss was larger, and under the large-flow condition, the hydraulic loss was smaller than that under the small-flow condition. Under the condition of a large flow rate, the hydraulic loss also gradually increased with an increase in flow rate. The change rule of hydraulic loss of the bulb under different flow conditions was consistent with that of the whole plant. Therefore, an in-depth study on the bulb structure of the bulb tubular pump device would be helpful for a solid modeling and optimum design of the pump device.

Table 2. Proportion of hydraulic losses (%).

Flow Rate	Hydraulic Loss/M	Passage Inlet	Guide Vane	Bulb	Passage Outlet
$0.9Q_d$	0.663	3.66	43.06	39.30	13.98
$1.0Q_d$	0.258	15.43	28.29	35.43	20.85
$1.1Q_d$	0.343	9.89	25.73	42.42	21.96
$1.2Q_d$	0.484	5.95	28.97	47.43	17.65

Figures 4 and 5 respectively show the velocity streamline and velocity contour of the central axis of the bulb section. According to the diagram, under the condition of a small flow rate deviating from the design flow, the flow line at the rear of the bulb was chaotic and created a serious reflow phenomenon. As the flow rate increased, the reflow phenomenon at the rear of the bulb was significantly improved, and the flow line around the bulb was smoother. This result corresponded to the device efficiency values given in Table 3. At 1.2 times the design flow, the device efficiency value reached the highest level, which was 83.28%. Figure 4 shows a cloud image of the axial velocity distribution in the bulb segment. According to the diagram, under the small-flow condition, a large range of a low-speed zone appeared at the rear of the bulb and the front of the support. With an increase in flow, the range of the low-speed zone gradually decreased.

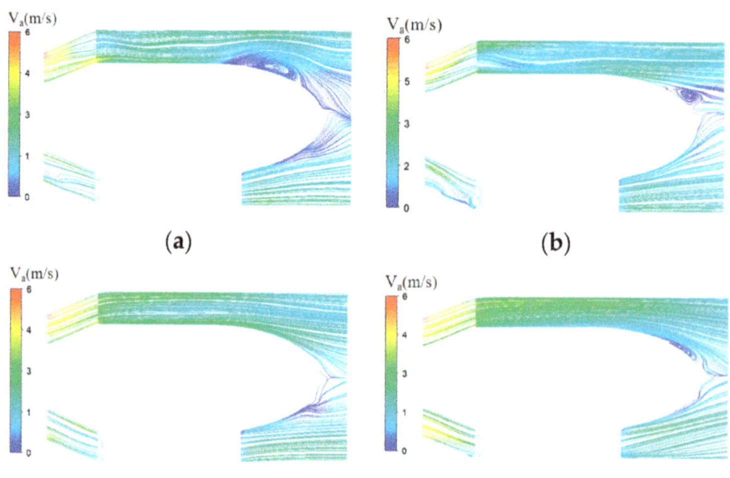

Figure 4. Streamline cloud diagram of axial velocity in bulb section: (**a**) $0.9Q_d$; (**b**) $1.0Q_d$; (**c**) $1.1Q_d$; (**d**) $1.2Q_d$.

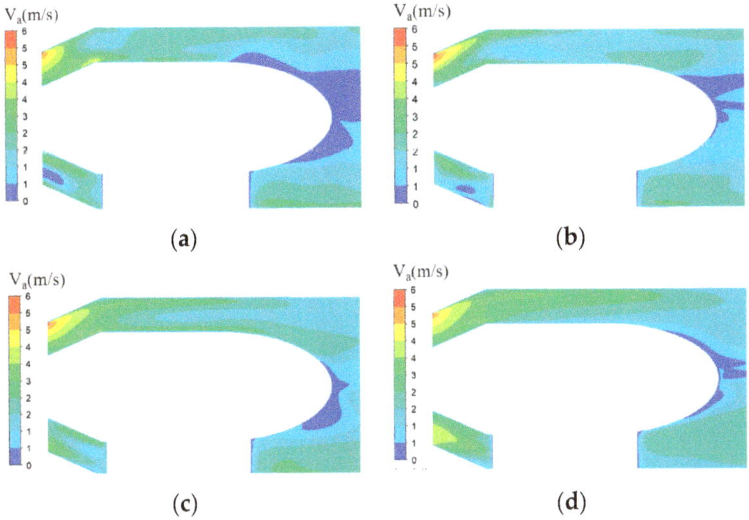

Figure 5. Contour image of axial velocity distribution in bulb section: (**a**) $0.9Q_d$; (**b**) $1.0Q_d$; (**c**) $1.1Q_d$; (**d**) $1.2Q_d$.

Table 3. Quantitative analysis of hydraulic performance of unit.

Flow	$\mathbf{0.9}Q_d$	$\mathbf{1.0}Q_d$	$\mathbf{1.1}Q_d$	$\mathbf{1.2}Q_d$
Head/m	3.56	3.27	2.89	2.38
Efficiency/%	73.87	78.75	82.17	83.28

3.2. Influence of Bulb Tail Shape on Hydraulic Performance

The shapes of the tail of the bulb studied were elongated, round, and oval (see Table 4). Different tail shapes have different effects on the flow pattern and thus on the performance of the whole device [19]. Therefore, in the design of the bulb structure, the shape of the tail of the bulb must be considered as an important research object.

Table 4. Research schemes 1–3.

Scheme	**Bulb Tail Form**	**Three-Dimensional Model**
1	First straight-line progressive section, then semi-ellipsoid; tail contraction angle is 27 degrees	

Table 4. *Cont.*

Scheme	Bulb Tail Form	Three-Dimensional Model
2	Hemisphere	
3	Semi-ellipsoid	

Figure 6 shows the head flow curve and the efficiency flow curve of the device with different bulb tail shapes. The diagram in scheme 3 clearly shows that the elliptical device at the tail of the bulb had a slightly higher efficiency than the other two schemes under the small flow rate from the highest efficiency point to the left. Under the condition of the high flow rate from the highest efficiency point to the right, the device efficiency was slightly lower than in scheme 1. Furthermore, the device head of scheme 3 was slightly higher than that of the other two schemes. This was due to the different shape and structure of the rear of the bulb and the different hydraulic losses of the bulb section. For the tail of the oval bulb, the low-speed range of the tail was reduced, the area where the vortex formed was smaller, and the water flow also followed a gradually diffusing trend. This result was more in line with the flow law of water flow, and its efficiency and head were improved.

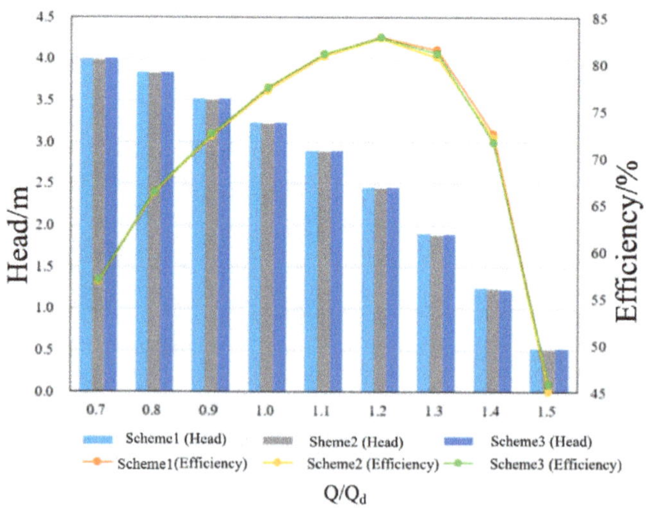

Figure 6. Device head flow curve and device efficiency flow curve with different bulb tail shapes.

Figure 7 shows the axial velocity streamline and velocity distributions in the bulb bodies with different tail shapes at design flow rate. As shown in the diagram, the tail back flow area of the bulb in scheme 2 was the largest, and the streamline was the most chaotic. In scheme 1, the streamline improved slightly because the tail of the bulb was extended slightly and the hydraulic loss in the bulb section was reduced. In scheme 3, the oval bulb tail body was used, and the water flow diffused smoothly. Therefore, the range of low-speed zone in the tail was the smallest, the area of vortex formation was the smallest, and the streamline was the smoothest.

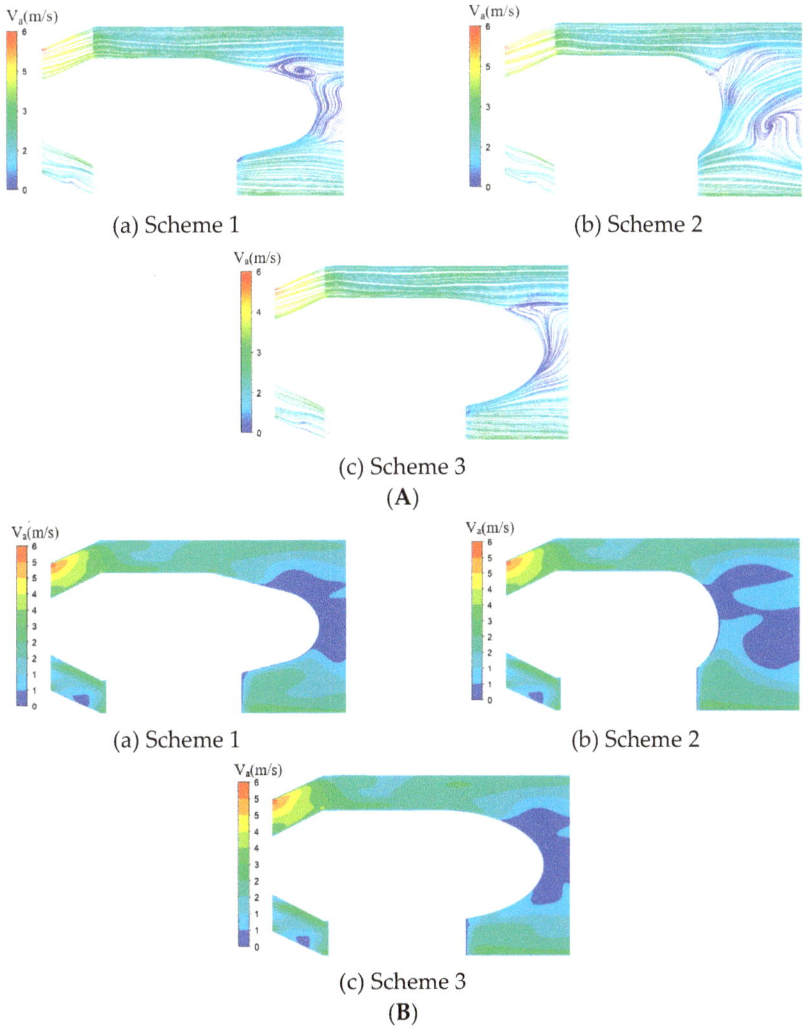

Figure 7. (**A**) Axial velocity streamline contour chart in the bulb bodies with different tail shapes at design flow rate; (**B**) Axial velocity contour chart in the bulb bodies with different tail shapes at design flow rate.

In order to further analyze the changes in pressure head caused by the changes in the structure of the bulb body, Figure 8 shows the static pressure contour chart at the outlet section of the bulb with different tail shapes. As shown in the diagram, the static pressure distribution at the outlet section of the bulb in scheme 3 was the most uniform, and the static pressure changed smoothly.

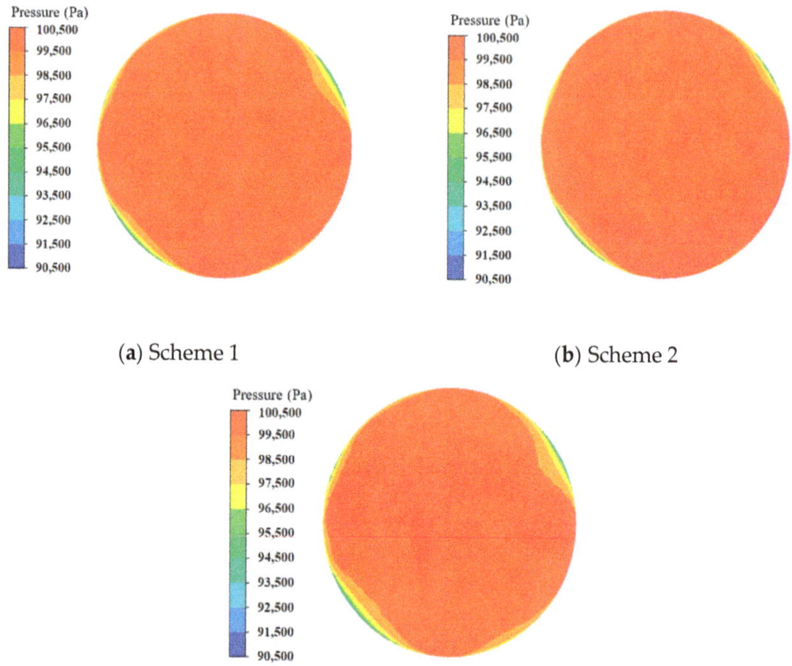

(**a**) Scheme 1 (**b**) Scheme 2 (**c**) Scheme 3

Figure 8. Static pressure contour chart.

The bulb structure in the rear-bulb tubular pump had a significant influence on hydraulic performance. In engineering design, it is recommended that the elliptical bulb body tail be used in the pumping station in combination with construction practice. To reduce the swirl area at the tail, the tail size of the bulb generally can be based on the total length of the bulb. Thus, the flow pattern distribution at the elliptical bulb body tail was improved to a certain extent by using the tail of the elliptical bulb. The hydraulic loss of the bulb section and the device performance also were improved compared with the round tail. For the pumping station, scheme 3 optimized the device and further reduced hydraulic losses by changing the bulb ratio.

3.3. *Influence of Bulb Ratio on Hydraulic Performance*

The bulb structure is used to place the electric motors and gearboxes in the tubular pump unit. The important dimension parameters are the bulb ratio and the ratio of the bulb diameter to the outer diameter of the impeller. The impeller diameter is generally a known parameter when designing the bulb section structure, whereas the bulb is determined according to its internal structure size, that is, according to the type and size of the motor and transmission equipment used in the device. Therefore, even in a pump unit with a same impeller, the bulb ratio can vary depending on the size and diameter of the bulb.

In practical design, the bulb diameter is usually expressed by the product of the bulb ratio and the impeller's outer diameter. The bulb diameter generally depends on the internal structural dimensions, such as the motor and transmission equipment. By referring to the data, we could determine when it would be necessary to use a small bulb and submerged asynchronous motor for the direct drive. The minimum bulb size could be arranged using a high-speed motor and gear box, and the bulb diameter was 0.76 D. When it was necessary to use the headlight body, according to the current manufacturing level of a high-voltage, high-speed motor and planetary gear reducer, it was reasonable to use a 0.9 D bulb diameter. If the synchronous motor was used in a direct connection, the diameter of the core was larger because the synchronous motor had to match the excitation system. Therefore, the diameter of the bulb should be more than 1.03 D. In the East Route Project of South-to-North Water Transfer, the bulb ratio of Jinhu Pump Station is 1.03 (i.e., the diameter of the bulb is 1.03 D). The bulb ratio of Linjiaba Pumping Station is 0.98 (i.e., the diameter of bulb is 0.98 D). The bulb ratio of Shandong Pumping Station with a water pump manufactured by the Japanese Yokogawa Company is 1.04 (i.e., the bulb diameter is 1.04 D) (Table 5).

Table 5. Research schemes 4–6.

Scheme	Bulb Ratio ($\frac{d_1}{D}$)	Three-Dimensional Diagram
4	0.88	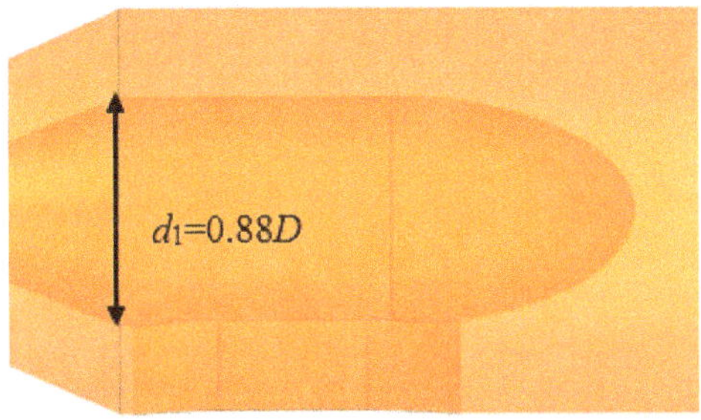
5	0.96	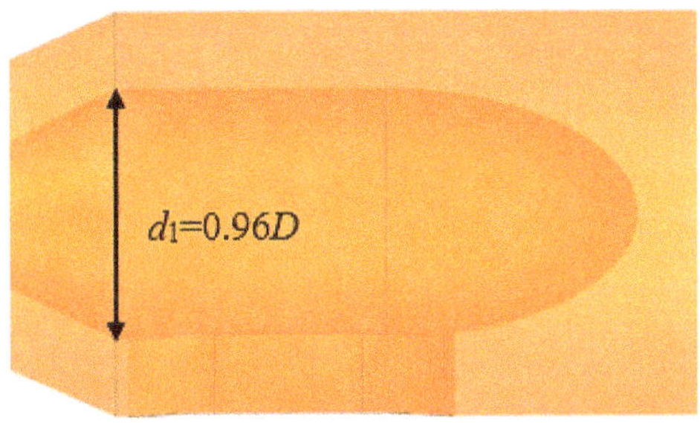
6	1.04	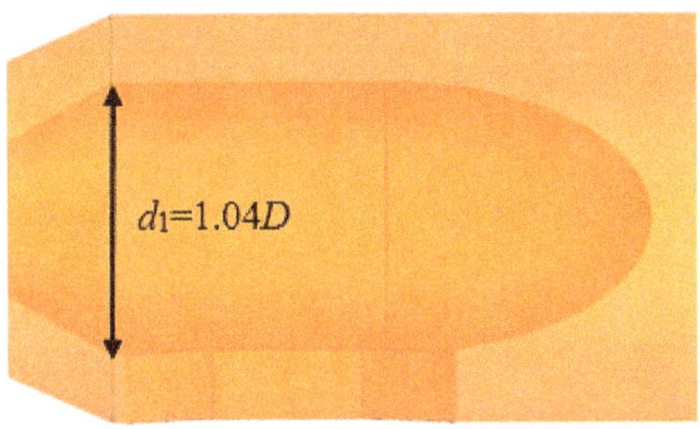

The device head curves and device efficiency curves for different bulb ratios are shown in Figure 9. According to this analysis, the device head was not affected by the change in the bulb ratio. The efficiency curves of the three bulb ratio schemes basically coincided with each other under the condition of a small flow rate from the highest efficiency point to the left, with little change. The device efficiency increased with the decrease in the bulb ratio under the condition of large flow to the right from the highest efficiency point, and the variation was obvious. This result was due to the small bulb ratio and the large flow area, creating a smoother flow pattern and higher efficiency in the bulb section.

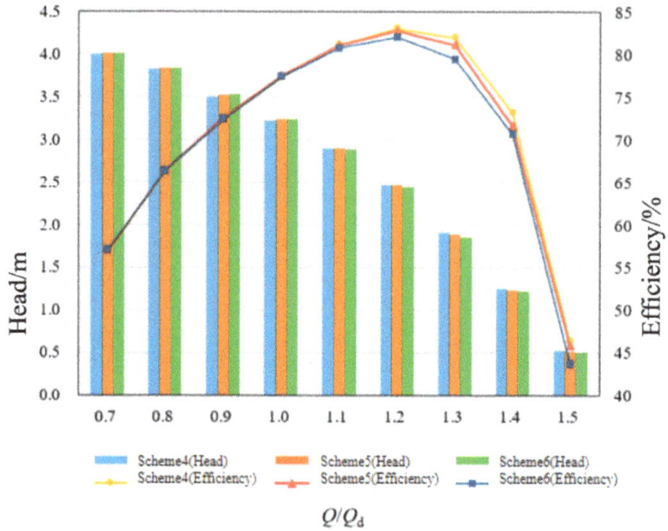

Figure 9. Head flow curve and efficiency flow curve of different bulb ratios.

The axial two-dimensional velocity streamline distributions in bulbs with different bulb ratios at the design flow rate are shown in Figure 10A. According to the diagram for scheme 6, the flow area of the bulb was reduced because of the large bulb ratio, the tail streamline was disordered, and the reflow phenomenon was serious. The tail streamlines of schemes 4 and 5 were relatively smooth. As shown in Figure 10B, we found that the loss of the entropy generation at the tail of scheme 6 was relatively large, whereas the loss of the entropy generation at the tails of schemes 4 and 5 had little change, which indicated that the hydraulic loss at the tail of the bulb of scheme 4 and 5 was small and the efficiency of the whole plant improved.

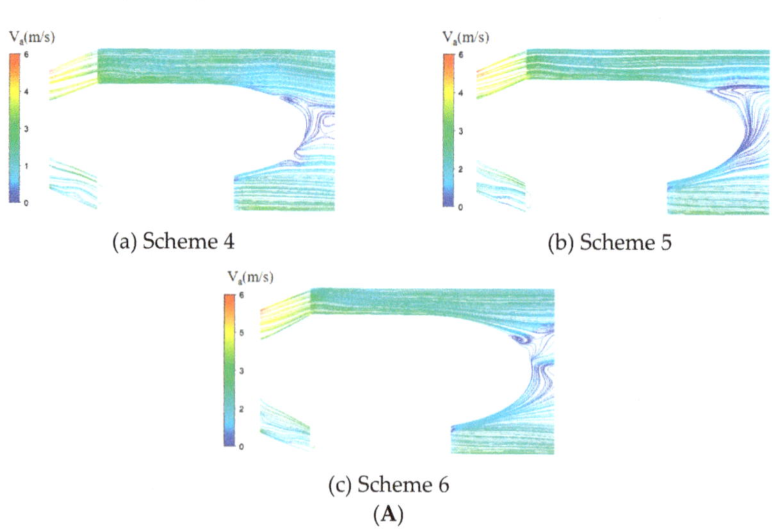

Figure 10. *Cont.*

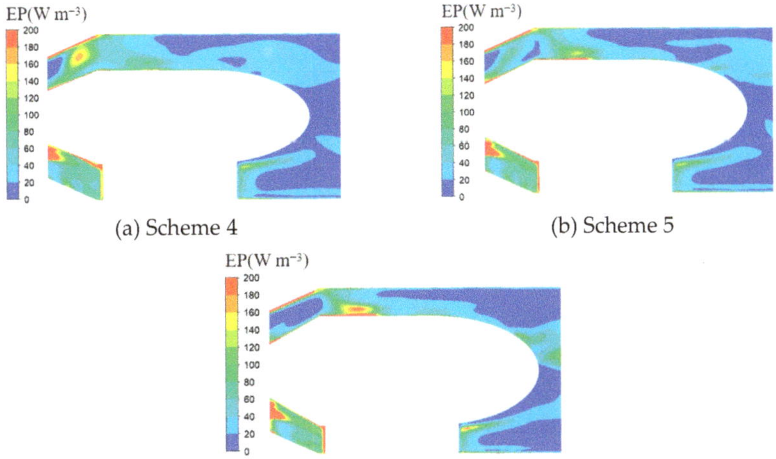

(a) Scheme 4 (b) Scheme 5

(c) Scheme 6

(B)

Figure 10. (**A**) Axial velocity streamline distribution in bulb bodies with different bulb ratios at design flow rate; (**B**) Axial surface entropic production distribution cloud in bulb bodies with different bulb ratios at design flow rate.

Schemes 4 and 5 had a small tail entropy output value, small hydraulic loss, and high efficiency. Considering that the pump station was directly connected by the synchronous motor, to facilitate the installation of the motor, we selected scheme 5 for optimization. By changing the shape of the support parts, the hydraulic loss of the bulb was further reduced, and the device efficiency was improved.

3.4. Influence of Support Shape on Hydraulic Performance

Bulb support is an important internal flow-passing part of the tubular pump unit. In this section, we used the CFD method to simulate the internal flow field of the bulb tubular pump unit and compared, analyzed, and optimized the shapes of the different supports. This provided a certain reference basis for the design and manufacture of the bulb tubular pump [20]. We used three schemes to compare the shapes of different bulb supports, as shown in Table 6. Among these supports, scheme 7 was the original support, and the support included one bottom support. In scheme 8, one bottom support in scheme 1 was divided into two parts, and the total width, length, and dimensions of the two supports remained unchanged. The distance between the two supports was 0.5 m. Scheme 9 streamlined the bottom support of the outer surface in scheme 1.

Figure 11 shows the flow head curve and flow efficiency curve of the bulb tubular pump unit for three schemes. From the flow efficiency curve, the differences among the three schemes were obvious. The efficiency of the pump unit of scheme 7 was about 0.4% higher than that of scheme 8 and that of scheme 9 was about 0.8% higher than that of scheme 8 at design flow rate, which indicated that the hydraulic loss caused by bulb support of tubular pump had a more significant influence on the hydraulic performance of the pump unit. From the flow head curve, the performance of scheme 7 was better than that of scheme 8, and scheme 9 was better than that of scheme 7, with a difference of about 0.4%. This result showed that different supports had an influence on the unit head, and this trend was consistent with hydraulic loss. The improved support structure in scheme 7 reduced the wet circumference of water flow and thus the friction loss of water. The shape of the streamlined support in scheme 9 followed the principle of minimum resistance of fluid movement, and the flow resistance of water flow was the smallest. Thus, the hydraulic loss was the smallest, and the pump unit head and efficiency were high.

Table 6. Research schemes 7–9.

Scheme	Support Shape	Three-Dimensional Model
7	Original scheme	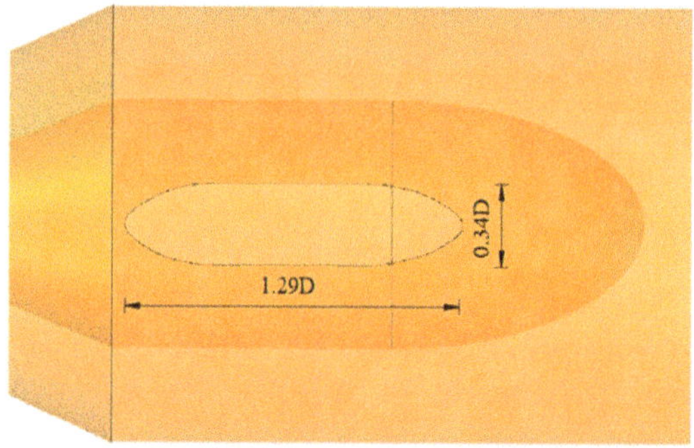
8	Split into two supports	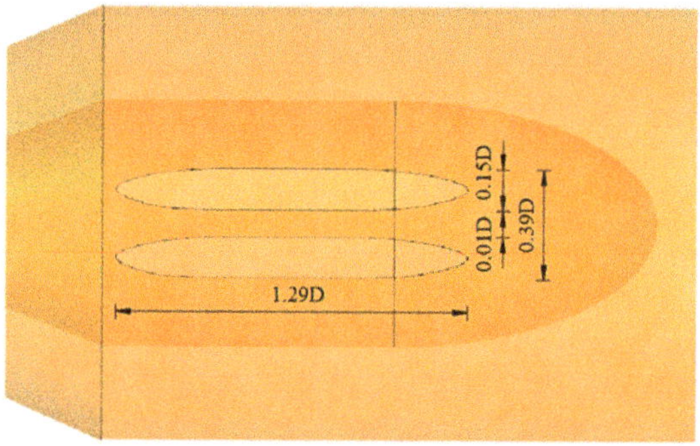
9	Streamline design	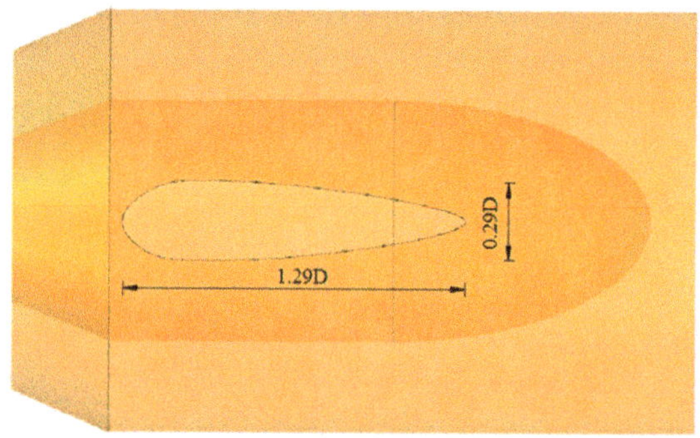

In Scheme 8, the support consisted of two bottom supports for the bulb. Because of the small distance between the two supports, there was little water flowing through this area, making it a dead-water zone. As shown in Figure 12, the water flow behind the support and the tail of the bulb was disordered, and the flow pattern was poor. The tail of the bulb also had a low-speed zone and a small range of back flow. In scheme 7, because the bottom support was combined into one, the flow area was increased and the flow resistance was reduced. As a result, the low-speed area and the return area at the rear of the bulb were significantly reduced. In scheme 9, the bottom support was changed into a streamlined one. The calculation results showed that the original undesirable flow pattern was significantly improved and the flow was relatively smooth, which was beneficial to further improve the efficiency of the pump unit.

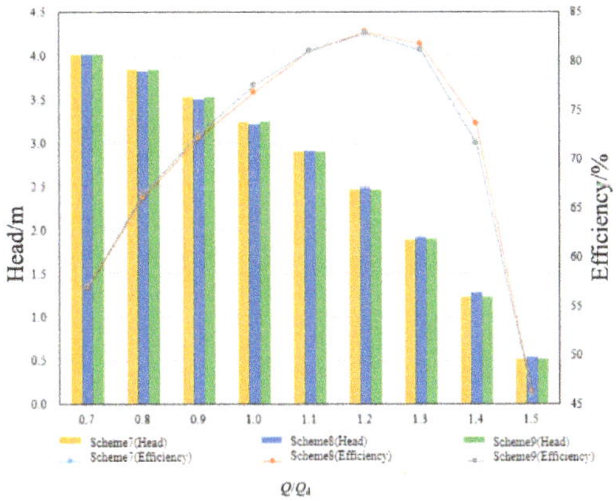

Figure 11. Head flow curve and device efficiency flow curve of different bulb support shapes.

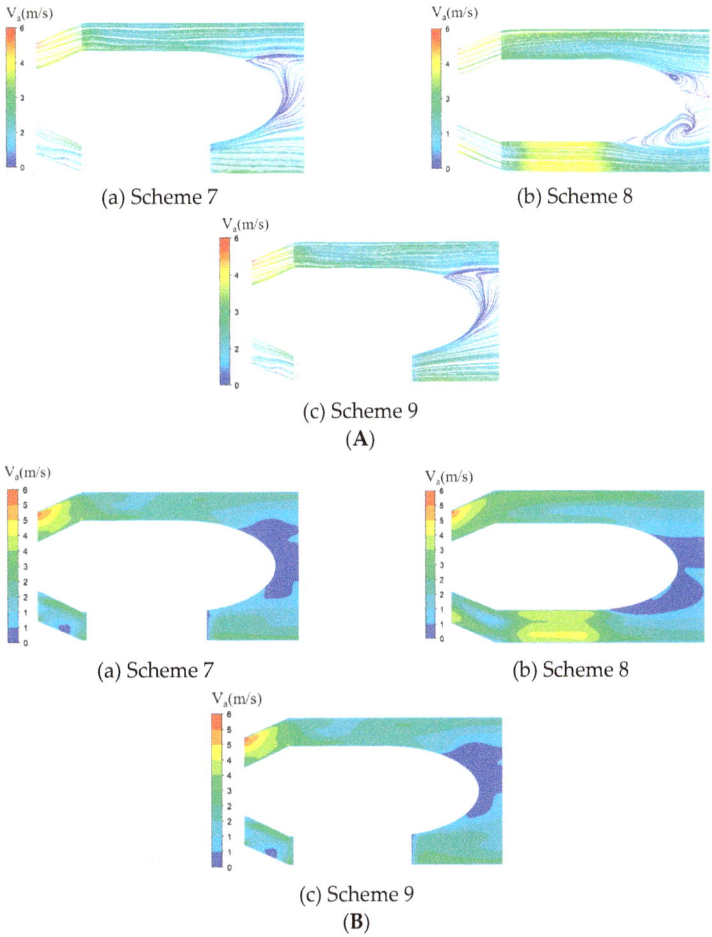

Figure 12. (**A**) Axial velocity streamline contour chart in the bulb bodies with different bulb support shapes at design flow rate; (**B**) Axial velocity contour chart in the bulb bodies with different bulb support shapes at design flow rate.

The variations in the hydraulic loss of the bulb section with flow rate under different schemes are shown in Figure 13. The hydraulic loss corresponding to the high-efficiency point was the smallest, and the hydraulic loss on the left side deviating from the high-efficiency point increased with a decrease in the flow rate. The hydraulic loss on the right side of the high-efficiency point increased with an increase in the flow rate. The increase in hydraulic loss in the small-flow area was mainly due to the increase in the pump outlet circulation caused by secondary back flow in the pump, which led to an increase in loss in the bulb. We found that merging the bottom supports reduced the hydraulic loss of the bulb section, and the water loss was also reduced after streamlining, thus improving the device efficiency.

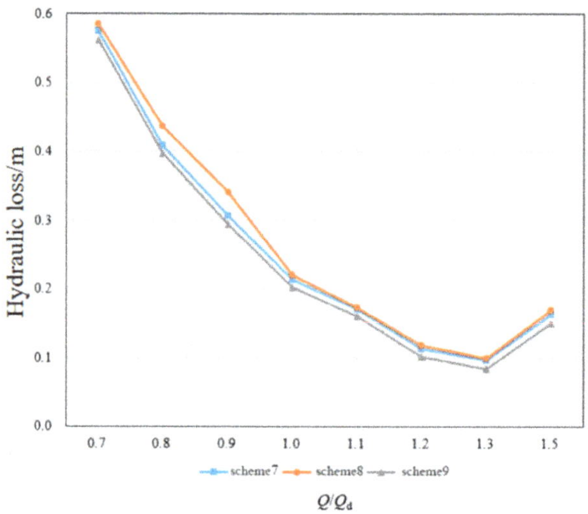

Figure 13. Hydraulic loss curve of bulb section.

In conclusion, the shape of the bottom support had a clear influence on the hydraulic loss and flow pattern of the pump unit. A reasonable shape of the support could improve the flow pattern, reduce water circulation, and improve the efficiency of the pump unit. Scheme 9's supports had better hydraulic performance, improved device efficiency, and achieved the purpose of optimization. Therefore, we selected scheme 9 as the bulb optimization scheme. Its geometric characteristics and parameters were as follows: the shape of the bulb tail was oval, the bulb ratio was 0.96, and the shape of the supports was streamlined.

4. Conclusions

In this study, the CFX 19.2 software was used to hydraulically optimize the bulb in accordance with the geometry and parameters. Through the calculation analysis and scheme comparison, we drew the optimized scheme of the bulb section. The conclusions are as follows:

(1) Because the bulb was in the key position of the flow channel, it affected the flow state in the flow channel and had a significant influence on the efficiency of the entire device. The hydraulic loss of the bulb part is about 40% of the overall hydraulic loss of the pump device, which accounts for the largest proportion of the entire pump device.
(2) Different tail shapes of the bulb had different effects on the flow pattern and the performance of the entire device. The flow pattern distribution at the tail of the bulb was improved to a certain extent by using the elliptical bulb tail. This reduces the

hydraulic loss of the bulb section and increases the efficiency of the device by about 5% compared to the hemisphere bulb structure.

(3) The bulb ratio of the bulb with a semi-ellipsoid tail shape was small, the flow area was large, the flow pattern was smoother, and the bulb section had higher efficiency. Schemes 4 and 5 had a small tail entropy output value, small hydraulic loss, and high efficiency. Considering that the pump station was directly connected by the synchronous motor, we selected scheme 5 to facilitate the installation of the motor.

(4) The bulb support is an important internal flow component of the tubular pump unit. A change in its shape had a significant influence on the flow pattern of the water in the pump unit and also on the hydraulic performance of the pump unit. Reasonable shape design of the support parts according to minimum resistance requirements effectively improved the flow pattern, reduced hydraulic losses, and improved the efficiency of the pump unit. Scheme 9's supports had better hydraulic performance, improved the device efficiency, and achieved the optimization purpose. Therefore, we selected scheme 9 as the bulb optimization scheme. Its geometric characteristics and parameters were as follows: the shape of the bulb tail was oval, the bulb ratio was 0.96, and the shape of the supports was streamlined.

Author Contributions: Software, W.J., H.Z. and J.W.; Validation, L.C. and W.J.; Writing—original draft, H.Z., M.Y. and J.W.; Formal analysis, H.Z. and J.W.; Writing—review and editing, W.J., J.L. and L.C.; Methodology, J.L. and L.C.; Supervision, W.J., L.C. and J.L. All authors have read and agreed to the published version of the manuscript.

Funding: This research was funded by the National Natural Science Foundation of China (grant no. 52279091), the Project Funded by the Priority Academic Program Development of Jiangsu Higher Education Institutions (PAPD), the Key Project of Water Conservancy in Jiangsu Province (Grant No: 2022017, 2022055), the Natural Science Foundation of the Jiangsu Higher Education Institutions of China (Grant No:23KJB570003), the Jiangsu Province College Students' Innovation and Entrepreneurship Training Program (202311117041Z), and the Open Project Program of Engineering Research Center of High-efficiency and Energy-saving Large Axial Flow Pumping device, Jiangsu Province, Yangzhou University (Grant No:ECHEAP025).

Institutional Review Board Statement: Not applicable.

Data Availability Statement: Data on the analysis and reporting results during the study can be obtained by contacting the authors.

Conflicts of Interest: The authors declare no conflict of interest.

References

1. Li, W.; Huang, Y.; Ji, L.; Ma, L.; Agarwal, R.K.; Awais, M. Prediction model for energy conversion characteristics during transient processes in a mixed-flow pump. *Energy* **2023**, *271*, 127082. [CrossRef]
2. Jiao, W.; Chen, H.; Cheng, L.; Zhang, B.; Gu, Y. Energy loss and pressure fluctuation characteristics of coastal two-way channel pumping stations under the ultra-low head condition. *Energy* **2023**, *278*, 127953. [CrossRef]
3. Sun, L.; Pan, Q.; Zhang, D.; Zhao, R.; van Esch, B.B. Numerical study of the energy loss in the bulb tubular pump system focusing on the off-design conditions based on combined energy analysis methods. *Energy* **2022**, *258*, 124794. [CrossRef]
4. Zhang, B.; Cheng, L.; Jiao, W.; Zhang, D. Experimental and statistical analysis of the flap gate energy loss and pressure fluctuation spatiotemporal characteristics of a mixed-flow pump device. *Energy* **2023**, *272*, 127117. [CrossRef]
5. Jia, X.; Lv, H.; Zhu, Z. Research on the influence of impeller tip clearance on the internal flow loss of axial circulating pump under unpowered driven condition. *ASME J. Fluids Eng.* **2023**, *145*, 021202. [CrossRef]
6. Zhang, B.; Yang, A.; Cheng, L.; Jiao, W.; Chen, Y.; Zhao, H. Spatial-temporal evolution and pressure fluctuation characteristics of the combined submerged vortex in a closed pump sump. *Phys. Fluids* **2023**, *35*, 065140.
7. Jiao, W.; Chen, H.; Cheng, L.; Zhang, B.; Yang, Y.; Luo, C. Experimental study on flow evolution and pressure fluctuation characteristics of the underwater suction vortex of water jet propulsion pump unit in shallow water. *Ocean. Eng.* **2022**, *266*, 112569. [CrossRef]
8. Song, X.J.; Luo, Y.Y.; Wang, Z.W. Numerical prediction of the influence of free surface vortex air- entrainment on pump unit performance. *Ocean. Eng.* **2022**, *256*, 111503. [CrossRef]
9. Pan, Q.; Shi, W.; Zhang, D.; Van Esch, B.P. Analysis, Design, and Validation of a Vaned Diffuser for Improved Fish Friendliness. *ASME J. Fluids Eng.* **2022**, *144*, 051502. [CrossRef]

10. Jiao, W.; Li, Z.; Cheng, L.; Wang, Y.; Zhang, B. Study on the Hydraulic and Energy Loss Characteristics of the Agricultural Pumping Station Caused by Hydraulic Structures. *Agriculture* **2022**, *12*, 1770. [CrossRef]
11. Ji, L.; Li, W.; Shi, W.; Tian, F.; Agarwal, R. Effect of Blade Thickness on Rotating Stall of Mixed-Flow Pump Using Entropy Generation Analysis. *Energy* **2021**, *236*, 121381. [CrossRef]
12. Zhou, J.R.; Cheng, L.; Liu, C.; Tang, F.P. Numerical simulation of bulb form of rear bulb penetration device. *J. Disch. Irrig. Mech. Eng.* **2011**, *29*, 72–76.
13. Zhang, R.T.; Shan, H.C.; Bu, G.; Zhou, W.; Zhu, H.G.; Yao, L.B. Structural characteristics of bulb tubular pump for Phase I project of East Route of South-to-North Water Transfer. *J. Drain. Irrig. Mech. Eng.* **2016**, *34*, 774–782+789.
14. Sun, Z.; Yu, J.; Tang, F. The Influence of Bulb Position on Hydraulic Performance of Submersible Tubular Pump Device. *J. Mar. Sci. Eng.* **2021**, *9*, 831. [CrossRef]
15. Wang, Z.; Zheng, Y.; Wang, F.; Lin, Z.; Huan, C.; Li, Y. Experimental Study on Energy Consumption and Hydraulic Stability for Distributed Pumping System. *Arab. J. Sci. Eng.* **2014**, *39*, 6883–6894. [CrossRef]
16. Zhang, X.; Tang, F.; Jiang, Y. Experimental and numerical study of reverse power generation in coastal axial flow pump system using coastal water. *Ocean. Eng.* **2023**, *271*, 113805. [CrossRef]
17. Liu, A.; Zhang, C.; Zhang, Y.; Zhang, Y. Analysis of hydraulic characteristics of fluid multiplier fish pump based on CFD simulation. *Ocean. Eng.* **2023**, *272*, 113854. [CrossRef]
18. Yang, Y.; Chen, X.; Bai, L.; Yao, Y.; Wang, H.; Ji, L.; He, Z.; Song, X.; Zhou, L. Quantification and investigation of pressure fluctuation intensity in a multistage electric submersible pump. *Phys. Fluids* **2023**, *35*, 035122. [CrossRef]
19. Yang, Y.; Li, W.; Shi, W.; Ping, Y.; Yang, Y.; Wang, L. Numerical investigation on the unstable flow at off-design condition in a mixed-flow pump. *Proc. Inst. Mech. Eng. Part A J. Power Energy* **2019**, *233*, 849–865. [CrossRef]
20. Huang, R.; Wang, Y.; Du, T.; Luo, X.; Zhang, W.; Dai, Y. Mechanism analyses of the unsteady vortical cavitation behaviors for a waterjet pump in a non-uniform inflow. *Ocean. Eng.* **2021**, *233*, 108798. [CrossRef]
21. Yang, Y.; Zhou, L.; Bai, L.; Xu, H.; Lv, W.; Shi, W.; Wang, H. Numerical Investigation of Tip Clearance Effects on the Performance and Flow Pattern within a Sewage Pump. *J. Fluids Eng. Trans. ASME* **2022**, *144*, 081202.
22. Jiao, W.; Zhao, H.; Cheng, L.; Yang, Y.; Li, Z.; Wang, C. Nonlinear dynamic characteristics of suction-vortex-induced pressure fluctuations based on chaos theory for a water jet pump unit. *Ocean. Eng.* **2023**, *268*, 113429.
23. Song, X.J.; Liu, C. Experiment study of the floor-attached vortices in pump sump using V3V. *Renew. Energy* **2021**, *164*, 752–766. [CrossRef]

agriculture

Communication

Predicting Models for Plant Metabolites Based on PLSR, AdaBoost, XGBoost, and LightGBM Algorithms Using Hyperspectral Imaging of *Brassica juncea*

Hyo In Yoon [1], Hyein Lee [1], Jung-Seok Yang [1], Jae-Hyeong Choi [1,2], Dae-Hyun Jung [1,3], Yun Ji Park [1], Jai-Eok Park [1], Sang Min Kim [1,2] and Soo Hyun Park [1,*]

[1] Smart Farm Research Center, Korea Institute of Science and Technology (KIST), Saimdang-ro 679, Gangneung 25451, Republic of Korea; yoonhi@snu.ac.kr (H.I.Y.); leehi@kist.re.kr (H.L.); inenviron@kist.re.kr (J.-S.Y.); choi_jh@kist.re.kr (J.-H.C.); daehyun@khu.ac.kr (D.-H.J.); pyj@kist.re.kr (Y.J.P.); j-park@kist.re.kr (J.-E.P.); kimsm@kist.re.kr (S.M.K.)
[2] Department of Bio-Medical Science & Technology, University of Science and Technology, Seoul 02792, Republic of Korea
[3] Department of Smart Farm Science, Kyung Hee University, Yongin 17104, Republic of Korea
[*] Correspondence: ecoloves@kist.re.kr; Tel.: +82-33-3661; Fax: +82-33-650-3429

Abstract: The integration of hyperspectral imaging with machine learning algorithms has presented a promising strategy for the non-invasive and rapid detection of plant metabolites. For this study, we developed prediction models using partial least squares regression (PLSR) and boosting algorithms (such as AdaBoost, XGBoost, and LightGBM) for five metabolites in Brassica juncea leaves: total chlorophyll, phenolics, flavonoids, glucosinolates, and anthocyanins. To enhance the model performance, we employed several spectral data preprocessing methods and feature-selection algorithms. Our results showed that the boosting algorithms generally outperformed the PLSR models in terms of prediction accuracy. In particular, the LightGBM model for chlorophyll and the AdaBoost model for flavonoids improved the prediction performance, with R2p = 0.71–0.74, com-pared to the PLSR models (R2p = 0.53–0.58). The final models for the glucosinolates and anthocya-nins performed sufficiently for practical uses such as screening, with R2p = 0.82–0.85 and RPD = 2.4–2.6. Our findings indicate that the application of a single preprocessing method is more effective than utilizing multiple techniques. Additionally, the boosting algorithms with feature selection ex-hibited superior performance compared to the PLSR models in the majority of cases. These results highlight the potential of hyperspectral imaging and machine learning algorithms for the non-destructive and rapid detection of plant metabolites, which could have significant implications for the field of smart agriculture.

Keywords: prediction models; hyperspectral image; PLSR model; AdaBoost; XGBoost; LightGBM

Citation: Yoon, H.I.; Lee, H.; Yang, J.-S.; Choi, J.-H.; Jung, D.-H.; Park, Y.J.; Park, J.-E.; Kim, S.M.; Park, S.H. Predicting Models for Plant Metabolites Based on PLSR, AdaBoost, XGBoost, and LightGBM Algorithms Using Hyperspectral Imaging of *Brassica juncea*. *Agriculture* **2023**, *13*, 1477. https://doi.org/10.3390/agriculture13081477

Academic Editors: Muhammad Sultan, Redmond R. Shamshiri, Md Shamim Ahamed and Muhammad Farooq

Received: 24 May 2023
Revised: 29 June 2023
Accepted: 29 June 2023
Published: 26 July 2023

1. Introduction

Mustard (*Brassica juncea*), also commonly known as Chinese mustard, brown mustard, leaf mustard, vegetable mustard, and oriental mustard, is an annual plant that belongs to the Brassicaceae family [1]. Mustard contains bioactive components such as glucosinolates and their degradation products; polyphenols (flavonoids and anthocyanins); and large amounts of dietary fiber, chlorophyll, β-carotene, ascorbic acid, minerals, and volatile components [2]. Mustard is used as a spice because of its pungent taste. It also has important uses in medicine; its leaves are used as a diuretic, stimulant, and expectorant in folk medicine. Previous studies have found that *B. juncea* has bactericidal properties, can reduce the risk of atherosclerosis, and has antioxidant- and peroxynitrite-scavenging effects [1,3]. Additionally, *B. juncea* has exhibited antibacterial and antitumor properties and has been shown to improve various metabolic disorders [4]. Despite its excellent bioactivity, its industrial use as a raw material for medicine is limited by traditional analytical techniques, which are time-consuming and destructive. Therefore, a non-destructive method of determining bioactive compound contents should be developed for quality control in the production stage.

Recently, hyperspectral imaging has been used for the assessment of the biophysical traits of plants. Spectral information from hyperspectral images can be combined with various data processing and mining tools to ensure fast, non-destructive, and highly accurate detection of functional component contents [5]. Preprocessing of spectral data is an important step for suppressing the undesired effects of measurement conditions and enhancing relevant features, which commonly contain normalization, derivatives, and smoothing [6]. Partial least squares regression (PLSR) is a widely used method to analyze large amounts of hyperspectral data and predict functional components in plants, such as chlorophyll and carotenoids in spinach [7] or total polyphenols in cocoa beans [8]. In our previous study, we aimed to develop a predictive model for the functional components of mustard plants using a PLSR prediction model based on hyperspectral images and preprocessing techniques [9]. In that study, we found that a preprocessing combination of SNV transformation and 1st-Der with spectral data resulted in high-performance prediction models for the total chlorophyll, carotenoid, and glucosinolate contents, while a preprocessing combination of the S.G. filter and SNV transformation gave the highest prediction rate for the total phenolics. However, the accuracy of this model was limited because the amount of data was relatively small and it was only applied in an indoor environment.

Machine learning techniques, combined with hyperspectral imaging, have been extensively used for the determination of food quality [10], such as identifying contaminants in food [11]. Among them, boosting methods in ensemble learning are attracting attention for their outstanding performance and have paved the way for data analysis. Boosting algorithms, such as those for adaptive boosting (AdaBoost), extreme gradient boosting (XGBoost), and the light gradient-boosting model (LightGBM), have performed well in hyperspectral imaging-based data classification tasks [12,13]. Effective training of machine learning models usually requires abundant data for a more accurate predictive model [14]. To train the model and improve the accuracy of the PLSR prediction model for functional components such as chlorophyll, phenolics, flavonoids, glucosinolates, and anthocyanins in mustard plants, we first acquired more hyperspectral imaging data of plant leaves. For this study, we aimed to develop a model with excellent predictive performance by adding enough training data to apply boosting algorithms and applying a combination of data processing methods. This analysis has expanded upon the previous study by including the prediction of the total phenolic components, which was not previously considered. However, the prediction of the total carotenoids was excluded from the current study due to its poor performance in the previous study. To apply the developed model and predict functional components in the growing environment, hyperspectral images were measured from various angles.

2. Materials and Methods

2.1. Training Data Acquisition

The plant growth conditions and analysis methods used for this study were the same as those described in detail in the previous study [9]. Briefly, mustard plants (*B. juncea* L. Czern.) were cultivated in three different environments. Plants in an indoor farm were hydroponically grown under mixed LEDs and with Hoagland nutrient solution. Plants in a greenhouse and an open field were grown in pots filled with commercial soil and fertilizer. Fifteen plants from each cultivation environment were harvested for 4 weeks after the transplant to ensure variation in growth stage and leaf color. A total of 122 fully expanded leaves were collected for analysis.

As with the experimental setup in the previous study [9], the hyperspectral imaging system consisted of a hyperspectral imaging camera (MicroHSI 410 SHARK; Corning Inc., Corning, NY, USA) and eight 15 W halogen lamps. A total of 112 hyperspectral images were acquired, with 1408 spatial pixels and 150 spectral bands in the range of 400–1000 nm. After the hyperspectral imaging data were obtained, the leaves were freeze-dried for 4 days and powdered for component analysis [9,15]. The powder obtained using pulverization after freeze-drying was subdivided into 3 repetitions of 20 mg each and used for the analysis of 5 functional components. Briefly, the previous methods were used for the determination of the total chlorophyll content [16], total phenolic content [17], total flavonoid content [18], total glucosinolate content [19], and total anthocyanin content [20]. As a result of the component analysis, analysis values with high degrees of variation in content were excluded, and the average value of the rest was used as the component value for model development.

2.2. Data Processing and Prediction Models

The average of the spectral data was extracted from hyperspectral images within predefined regions of interest, as in the previous method [9]. The average spectral data of 150 bands for each of the 112 hyperspectral images were obtained [9]. The preprocessing methods, used alone or in combination (Table S1), included normalization, logarithmic transformation, a Savitzky–Golay filter, the 1st and 2nd derivative after SG filtering, multiplicative scatter correction (MSC), and standard normal variate (SNV) transformation. The SG filter was applied with a three-order polynomial fit with five data points, using the SciPy package in Python 3.9. A total of 36 preprocessing combinations were used to prepare the spectral data for the development of the predictive models.

Partial least squares regression (PLSR), adaptive boosting (AdaBoost), extreme gradient boosting (XGboost), and light gradient boosting model (LightGBM) algorithms were applied to predict the content of each metabolite in the plants. PLSR is a method that is commonly used to predict metabolite content from hyperspectral data. It works by extracting latent variables (LVs), which are linear combinations of original predictor variables that capture the maximum variation in data. The number of LVs is chosen based on the optimal performance of the relevant model, which is typically determined through cross-validation.

AdaBoost, XGboost, and LightGBM are boosting algorithms that are also commonly used for regression tasks [21–23]. Boosting algorithms combine multiple weak learners (e.g., decision trees) into a strong learner, which improves the accuracy of predictions. In this study, boosting algorithms were used for both feature selection and regression.

To reduce redundant information in the hyperspectral data, feature selection based on the importance of boosting was used. Only bands with a feature importance value greater than 1.25 times the average value were selected. The implementation of model development was programmed using the Scikit-learn, XGboost, and LightGBM packages in Python 3.9.

The preprocessing and feature-selection methods were determined, and the parameters for all of the algorithms were optimized after tenfold cross-validation based on the training dataset, corresponding to 80% of the data. After hyperparameter tuning, the performance of the final model was tested with an independent validation dataset that corresponded to 20% of the data. The model performance was evaluated based on the coefficient of determination (R^2) and the root mean square error ($RMSE$), as follows:

$$R^2 = 1 - \frac{\sum_{i=1}^{n}\left(\hat{y}_i - y_i\right)^2}{\sum_{i=1}^{n}\left(y_i - \overline{y}\right)^2} \quad (1)$$

$$RMSE = \sqrt{\frac{1}{n}\sum_{i=1}^{n}\left(\hat{y}_i - y_i\right)^2} \quad (2)$$

where y_i is the measured value of the component analysis; $\hat{y}_i$ is the value predicted by the model; $\overline{y}$ is the mean value of the component analysis; and n is the number of samples.

3. Results and Discussion

3.1. Development of a Prediction Model Based on Hyperspectral Imaging with the PLSR, AdaBoost, XGboost, and LightGBM Algorithms

The total chlorophyll, phenolic, flavonoid, glucosinolate, and anthocyanin contents in the *B. juncea* plants are summarized in Table 1. The reflectance spectra of 112 leaves were obtained by averaging the hyperspectral data, followed by preprocessing, as shown in Figure 1. In the spectra of the *B. juncea* leaves, the green and red regions were relatively low and high, respectively, compared to those of a typical green leaf. These different spectra could be caused by the absolute contents and ratio of chlorophyll and anthocyanin in the leaf [24]. The plant used in this study was a red mustard cultivar with purple–green leaves and a high anthocyanin content (Table 1).

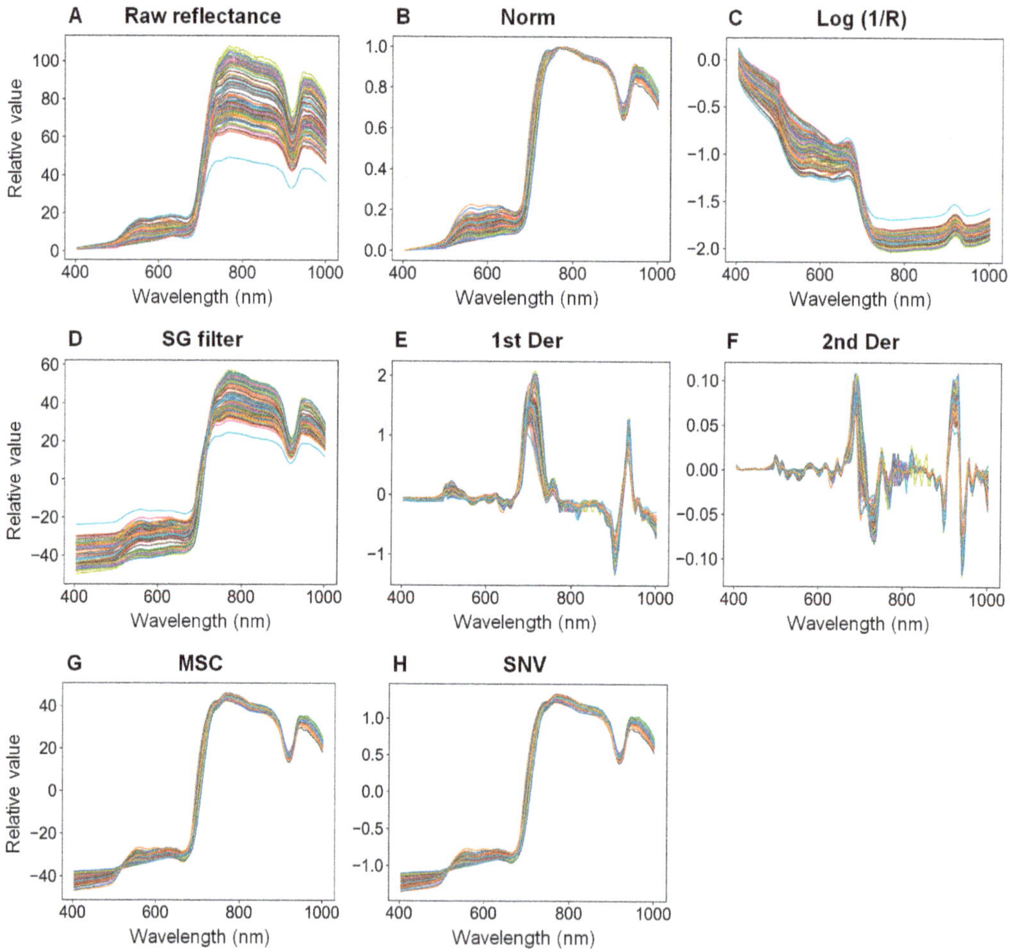

Figure 1. Single preprocessing method for hyperspectral data of *B. juncea* plants: raw reflectance (**A**), normalization (**B**), logarithmic transformation (**C**), Savitzky–Golay filter (**D**), first and second derivative after SG filtering (**E**,**F**), multiplicative scatter correction (**G**), and standard normal variate transformation (**H**). Colored lines represent different leaf samples. The combination of preprocessing methods refers to Table S1.

Table 1. Statistical summary of the five components in *B. juncea* plants.

Metabolites	Min	Max	Mean	Standard Deviation
Chlorophyll (mg g^{-1} DW)	2.54	12.17	7.24	2.28
Phenolics (mg g^{-1} DW)	2.13	11.28	6.22	2.33
Flavonoids (mg g^{-1} DW)	3.00	13.59	7.52	2.51
Glucosinolates (μmol g^{-1} DW)	11.88	55.08	31.36	11.25
Anthocyanins (mg g^{-1} DW)	0.00	33.80	3.23	5.35

PLSR models for five metabolites were developed using 36 preprocessing methods (Table S1). The optimal combination of preprocessing methods for each of the five PLSR models, as well as the optimal number of latent variables (LVs) for each component, was determined based on the low root mean square error of cross-validation (RMSECV) values, as shown in Table 2. Spectral preprocessing is an essential step in order to avoid undesirable scattering effects and reveal signals that correspond to chemical components [25]. The appropriate preprocessing method will depend on various factors, including the wavelength range and interval, the prediction model, the target compound, and the plant organs used, such as leaves and fruits. In previous studies, the performance of the PLSR model in detecting total phenolic content using a VIS-NIR hyperspectral imaging system was improved with the normalization method in apple fruits [26] and with the SG filter and derivative transformation in Arabidopsis leaves [27]. Derivative transforms emphasize spectral features but also emphasize the noise of data. The first and second derivatives removed an additive and a linear baseline, respectively. Logarithmic transformation can be employed to address a non-linear problem. In a previous study using the SWIR hyperspectral imaging system, the logarithmic transformation Log (1/R) improved the performance of the PLSR model for the ABA content in zucchini leaves [28]. MSC and SNV transformation are useful in reducing spectral variability due to scattering and baseline shifts. To further improve model performance, spectral preprocessing methods can be used in combination [9]. In this study, the prediction performance of the PLSR model was higher with the single preprocessing methods than with combinations of multiple methods (Table 2).

The AdaBoost, XGboost, and LightGBM prediction models were also developed using 36 preprocessing methods. The best preprocessing method for each algorithm and metabolite was determined based on low RMSECV values (Table S2). After that, the prediction models were compared according to the selection of three features (bands) based on the feature importance in the boosting algorithms (Table 3). The spectral bands that were reduced by the algorithms made the performances of several models better compared with the full bands. Hyperspectral data require band selection due to the large amount of highly correlated and redundant information. Reducing the number of features, even to less than 20% of the total band, can enhance the performance of a regression algorithm [28,29]. Combinations of different feature-selection and regression algorithms can improve model accuracy [30].

The importance values for selecting the features of each best performance model are given in Figure 2. For chlorophyll prediction, the highest importance was at 480.57 nm, followed by 916.83 nm, among 17 bands selected based on the XGBoost algorithm with 1st Der processing data. For phenolic prediction, the feature importance was the highest at 904.82 nm, followed by 760.73 nm, among 28 bands selected based on the LightGBM algorithm with Norm processing data. The selected features were distributed in the ranges of 488.57–544.61 nm and 672.68–992.87 nm, respectively. For flavonoid prediction, the feature importance was the highest at 692.69 nm, followed by 608.64 nm, among 33 bands selected based on the AdaBoost algorithm with 2nd Der processing data. The feature importance for glucosinolate prediction was concentrated in the range of 870–900 nm. The highest importance values were, in order, at 872.80, 896.81, 880.8, and 628.66 nm among 28 bands selected based on the AdaBoost algorithm with SNV processing data. For anthocyanin prediction, the feature importance was the highest at 924.83 nm among 37 bands selected with the LightGBM algorithm with Log (1/R), 1st Der, and MSC processing data. The features were selected depending on the spectral data preprocessing method as well as the selection algorithm.

Table 2. Performances of five PLSR models based on best preprocessing methods and optimal latent variables (LVs) for each component of *B. juncea* plants.

Metabolites	Preprocessing Method	Optimal LVs	Calibration		Cross-Validation		Prediction	
			R^2_C	*RMSEC*	R^2_{CV}	*RMSECV*	R^2_P	*RMSEP*
Chlorophyll	Log (1/R) + 1st Der + MSC	5	0.667	1.332	0.407	1.777	0.567	1.349
	Log (1/R) + 1st Der + SNV	5	0.667	1.332	0.405	1.779	0.553	1.371
	Raw reflectance	8	0.619	1.425	0.393	1.798	0.530	1.405
	SG filter	2	0.431	1.740	0.388	1.806	0.526	1.411
	1st Der	**5**	**0.659**	**1.347**	**0.384**	**1.811**	**0.575**	**1.337**
Phenolics	**SG filter**	**11**	**0.731**	**1.206**	**0.433**	**1.751**	**0.558**	**1.487**
	Norm + SNV	6	0.566	1.532	0.425	1.761	0.458	1.648
	SNV	6	0.566	1.532	0.425	1.761	0.458	1.648
	Norm + SG filter + SNV	6	0.554	1.552	0.424	1.763	0.440	1.675
	SG filter + SNV	6	0.554	1.552	0.424	1.763	0.440	1.675
Flavonoids	1st Der	1	0.452	1.814	0.406	1.888	0.507	1.845
	SG filter	1	0.442	1.830	0.404	1.892	0.477	1.902
	Raw reflectance	**2**	**0.448**	**1.820**	**0.398**	**1.901**	**0.531**	**1.802**
	2nd Der	1	0.450	1.818	0.382	1.926	0.412	2.016
	Log (1/R) + 1st Der + MSC	3	0.509	1.716	0.365	1.953	0.233	2.303
Glucosinolates	Raw reflectance	8	0.783	5.229	0.647	6.667	0.725	5.804
	Norm + SG filter	7	0.746	5.651	0.647	6.668	0.662	6.435
	SG filter	**11**	**0.807**	**4.922**	**0.646**	**6.679**	**0.759**	**5.436**
	Norm + SG filter + MSC	8	0.753	5.581	0.633	6.794	0.646	6.591
	SG filter + MSC	8	0.753	5.581	0.633	6.799	0.645	6.592
Anthocyanins	Log (1/R) + SNV	8	0.854	2.109	0.746	2.780	0.836	1.808
	Log (1/R) + SG filter + SNV	9	0.855	2.104	0.745	2.786	0.849	1.737
	Log (1/R)	11	0.868	2.002	0.745	2.787	0.837	1.801
	Log (1/R) + MSC	**9**	**0.853**	**2.115**	**0.744**	**2.791**	**0.861**	**1.664**
	Log (1/R) + SG filter + MSC	9	0.850	2.137	0.743	2.794	0.850	1.728

R^2: coefficient of determination; RMSEC, RMSECV, and RMSEP: root mean square errors of calibration, cross-validation, and prediction, respectively. Bold indicates the best performance based on the RMSEP for each component.

Table 3. Performance of AdaBoost, XGboost, and LightGBM prediction models for five metabolites in *B. juncea* plants according to feature-selection algorithms after determination of preprocessing and hyperparameter tuning.

Prediction Model	Preprocessing Method	Feature Selection		Calibration		Cross-Validation		Prediction	
		Method	Feature No.	R^2_C	*RMSEC*	R^2_{CV}	*RMSECV*	R^2_P	*RMSEP*
Total Chlorophyll									
AdaBoost	Log (1/R) + 2nd Der + MSC	Full band	150	0.878	0.807	0.448	1.714	0.594	1.307
		AdaBoost	28	0.926	0.628	0.573	1.507	0.476	1.483
		XGboost	13	0.868	0.838	0.463	1.690	0.348	1.656
		LightGBM	35	0.929	0.616	0.541	1.563	0.502	1.447
XGboost	Log (1/R) + 2nd Der + MSC	Full band	150	0.996	0.137	0.519	1.600	0.476	1.484
		AdaBoost	28	0.997	0.120	0.594	1.471	0.545	1.382
		XGboost	13	0.891	0.763	0.488	1.651	0.455	1.514
		LightGBM	35	1.000	0.033	0.628	1.407	0.576	1.334

Table 3. *Cont.*

Prediction Model	Preprocessing Method	Feature Selection		Calibration		Cross-Validation		Prediction	
		Method	Feature No.	R^2_C	*RMSEC*	R^2_{CV}	*RMSECV*	R^2_P	*RMSEP*
LightGBM	**1st Der**	Full band	150	0.945	0.543	0.414	1.766	0.695	1.133
		AdaBoost	31	0.829	0.954	0.463	1.691	0.648	1.217
		XGboost	**17**	**0.743**	**1.170**	**0.388**	**1.805**	**0.737**	**1.052**
		LightGBM	35	0.960	0.462	0.551	1.547	0.657	1.201
Total Phenolics									
AdaBoost	**Norm**	Full band	150	0.924	0.642	0.581	1.505	0.521	1.549
		AdaBoost	37	0.931	0.611	0.641	1.393	0.517	1.554
		XGboost	16	0.921	0.652	0.646	1.382	0.512	1.562
		LightGBM	**28**	**0.925**	**0.637**	**0.618**	**1.437**	**0.594**	**1.426**
XGboost	Norm + SG filter	Full band	150	1.000	0.027	0.627	1.419	0.390	1.748
		AdaBoost	34	0.974	0.372	0.573	1.518	0.354	1.798
		XGboost	15	0.969	0.406	0.605	1.461	0.406	1.724
		LightGBM	30	0.933	0.601	0.557	1.546	0.378	1.765
LightGBM	1st Der	Full band	150	0.882	0.798	0.538	1.580	0.559	1.486
		AdaBoost	36	0.862	0.864	0.572	1.520	0.517	1.556
		XGboost	10	0.770	1.115	0.565	1.532	0.386	1.753
		LightGBM	38	0.942	0.558	0.602	1.467	0.499	1.583
Total Flavonoids									
AdaBoost	**2nd Der**	Full band	150	0.872	0.878	0.512	1.712	0.704	1.429
		AdaBoost	**33**	**0.827**	**1.018**	**0.551**	**1.642**	**0.709**	**1.417**
		XGboost	12	0.913	0.724	0.572	1.602	0.575	1.714
		LightGBM	34	0.847	0.958	0.538	1.666	0.623	1.615
XGboost	1st Der	Full band	150	0.972	0.409	0.516	1.705	0.586	1.692
		AdaBoost	36	0.986	0.286	0.586	1.577	0.564	1.736
		XGboost	7	0.932	0.640	0.545	1.653	0.644	1.569
		LightGBM	46	0.997	0.138	0.580	1.588	0.568	1.728
LightGBM	1st Der	Full band	150	0.874	0.868	0.483	1.761	0.585	1.693
		AdaBoost	36	0.905	0.754	0.543	1.657	0.519	1.823
		XGboost	7	0.651	1.448	0.531	1.678	0.594	1.676
		LightGBM	46	0.955	0.518	0.548	1.648	0.503	1.854
Total Glucosinolates									
AdaBoost	**SNV**	Full band	150	0.935	2.868	0.666	6.481	0.768	5.333
		AdaBoost	**28**	**0.907**	**3.417**	**0.674**	**6.401**	**0.816**	**4.744**
		XGboost	14	0.913	3.301	0.699	6.157	0.782	5.169
		LightGBM	34	0.935	2.852	0.677	6.372	0.730	5.748
XGboost	SG filter + SNV	Full band	150	0.997	0.644	0.670	6.445	0.751	5.521
		AdaBoost	29	0.996	0.670	0.676	6.382	0.763	5.389
		XGboost	12	0.993	0.928	0.715	5.987	0.778	5.211
		LightGBM	41	1.000	0.233	0.707	6.071	0.776	5.238
LightGBM	Log (1/R) + 1st Der + SNV	Full band	150	0.875	3.959	0.702	6.122	0.675	6.308
		AdaBoost	30	0.962	2.183	0.741	5.709	0.662	6.435
		XGboost	8	0.901	3.538	0.744	5.678	0.613	6.890
		LightGBM	51	0.985	1.386	0.739	5.729	0.665	6.411
Total Anthocyanins									
AdaBoost	**Log (1/R) + 1st Der**	Full band	150	0.975	0.865	0.834	2.246	0.714	2.390
		AdaBoost	26	0.968	0.986	0.819	2.349	0.639	2.682
		XGboost	11	0.969	0.973	0.735	2.839	0.519	3.097
		LightGBM	**37**	**0.976**	**0.851**	**0.822**	**2.329**	**0.824**	**1.876**

Table 3. *Cont.*

Prediction Model	Preprocessing Method	Feature Selection		Calibration		Cross-Validation		Prediction	
		Method	Feature No.	R^2_C	*RMSEC*	R^2_{CV}	*RMSECV*	R^2_P	*RMSEP*
XGboost	1st Der	Full band	150	1.000	0.003	0.724	2.899	0.265	3.830
		AdaBoost	20	1.000	0.041	0.725	2.892	0.742	2.271
		XGboost	11	0.987	0.625	0.738	2.826	0.251	3.865
		LightGBM	40	0.997	0.297	0.664	3.198	0.389	3.492
LightGBM	Log (1/R) + 1st Der	Full band	150	0.899	1.756	0.685	3.097	0.743	2.264
		AdaBoost	24	0.918	1.575	0.699	3.028	0.485	3.204
		XGboost	9	0.826	2.303	0.687	3.089	0.314	3.699
		LightGBM	39	0.933	1.430	0.742	2.804	0.717	2.375

R^2: coefficient of determination; RMSEC, RMSECV, and RMSEP: root mean square errors of calibration, cross-validation, and prediction, respectively. Bold indicates the best performance based on the RMSEP for each component.

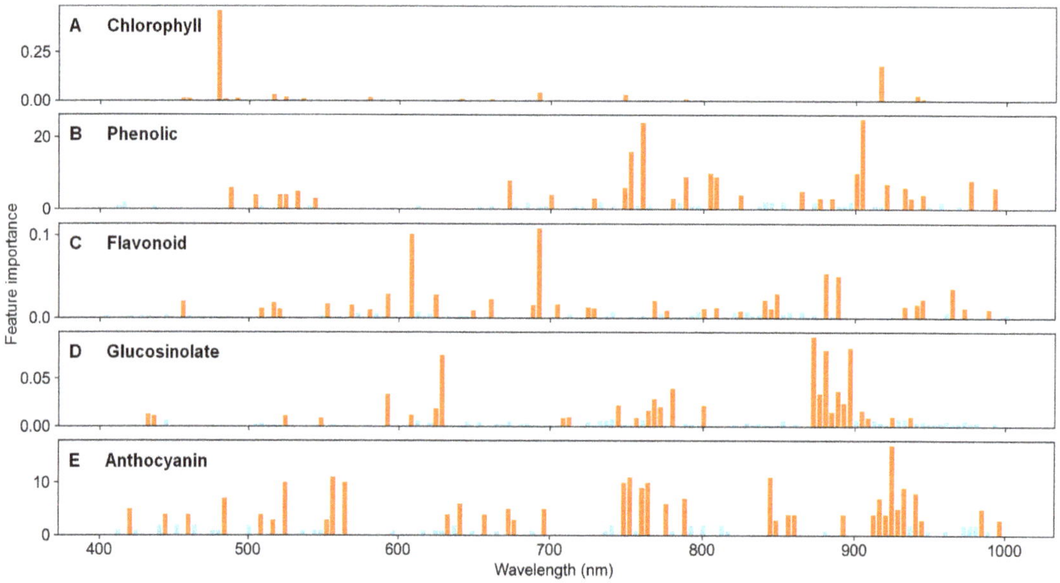

Figure 2. Feature importance values used to determine the best prediction models for total chlorophyll (**A**), phenolics (**B**), flavonoids (**C**), glucosinolates (**D**), and anthocyanins (**E**) in *B. juncea* plants. Orange bars represent selected features, and light blue bars represent unselected features, i.e., those not used in the prediction model. The best prediction models are documented in Table 3.

Overall, the boosting algorithms showed better prediction performances compared to the PLSR models (Tables 2 and 3). Specifically, the LightGBM model was found to be the best for predicting chlorophyll, while the AdaBoost model was the best for predicting phenolics, flavonoids, glucosinolates, and anthocyanins. The boosting models performed better than the best PLSR models, except when it came to anthocyanins, where the PLSR models showed better performances. The performances of the best prediction models for five metabolites are given in Figure 3. The best model for chlorophyll was the 1st Der processing–XGBoost selection–LightGBM prediction model with 17 bands selected (R^2_P = 0.737, RMSEP = 1.052). The best model for phenolics was the Norm processing–LightGBM selection–AdaBoost prediction model with 28 bands selected (R^2_P = 0.594, RMSEP = 1.426). For flavonoids, the 2nd Der processing–AdaBoost selection–AdaBoost prediction model with 33 bands selected performed the best (R^2_P = 0.709, RMSEP = 1.417).

For glucosinolates, the SNV processing–AdaBoost selection–AdaBoost prediction model with 28 bands selected was best ($R^2{}_P$ = 0.816, RMSEP = 4.744). The best boosting model for anthocyanins was the Log (1/R)–1st Der-MSC processing–LightGBM selection–AdaBoost prediction model with 37 bands selected ($R^2{}_P$ = 0.824, RMSEP = 1.876). The best PLSR model for anthocyanins was that with 9 LVs and using Log (1/R)–SG filter–MSC processing data, which had the highest performance of all of the models ($R^2{}_P$ = 0.850, RMSEP = 1.728). The ratio of prediction to deviation (RPD) value of the final AdaBoost models for glucosinolates and anthocyanins was 2.4, which indicates that these models are sufficient for practical screening applications [31,32]. However, the best PLSR model for anthocyanins performed even better, with an RPD value of 2.6. It is worth noting that the selection of the best model depended on various factors, including the spectral data preprocessing method and the selection algorithm used.

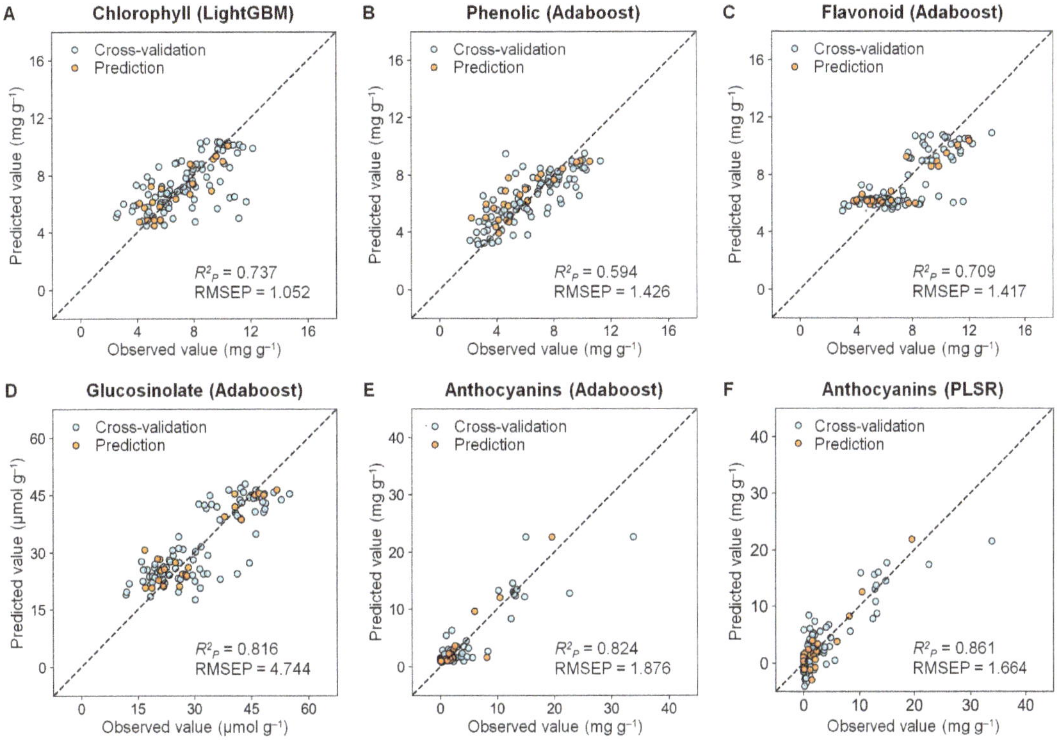

Figure 3. The optimal models for predicting the concentrations of the total chlorophyll (**A**), phenolics (**B**), flavonoids (**C**), glucosinolates (**D**), and anthocyanins (**E**,**F**) in *B. juncea* plants, as presented in Table 3. $R^2{}_P$ and RMSEP indicate coefficient of determination and root mean square error of prediction, respectively.

3.2. Application of the Functional Component Prediction Model with Visualization

Prediction models based on hyperspectral imaging are used to predict content at a single-pixel level and generate compound distribution maps. The prediction model developed here was applied to actual plants that were grown and utilized the spectrum of every pixel. The spatial distribution of five metabolites was found to be uneven across the leaf area (Figure 4). Yuan et al. (2021) visualized the distribution of SPAD values, which indicate chlorophyll content in pepper leaves [29]. The distribution of the total phenolics has been visualized using hyperspectral imaging and modeling in Arabidopsis plants [27] and shelled cocoa beans [8]. Hence, by employing a hyperspectral imaging system and the necessary software to run the algorithm, we could non-destructively and continuously monitor the compound distribution. This phytochemical monitoring will aid in making cultivation decisions to effectively control the quality of functional plants.

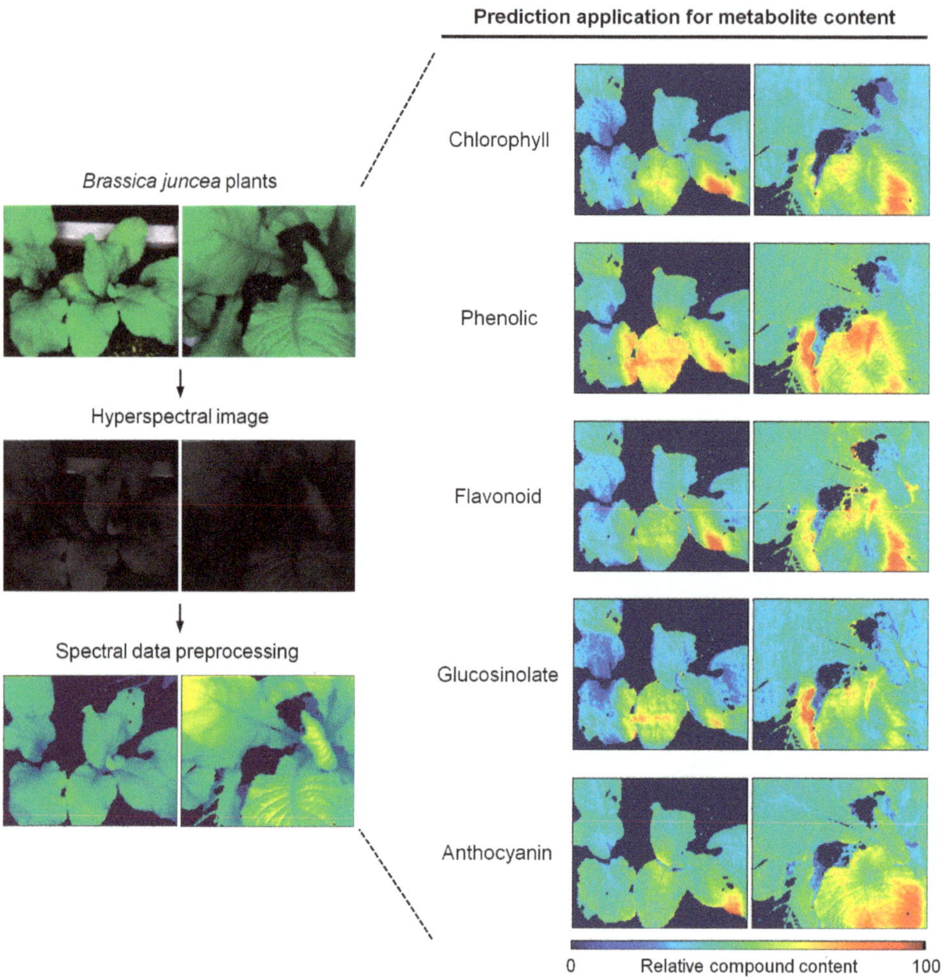

Figure 4. Distribution map of five metabolites, described by an application of hyperspectral image-based prediction in a growing environment: total chlorophyll, phenolics, flavonoids, glucosinolates, and anthocyanins in *B. juncea* plants.

4. Conclusions

A prediction model using hyperspectral imaging was developed based on PLSR and boosting algorithms such as AdaBoost, XGboost, and LightGBM to predict five metabolites in *B. juncea*: total chlorophyll, phenolics, flavonoids, glucosinolates, and anthocyanins. To improve the model performance, various spectral data preprocessing methods and feature-selection algorithms were adopted. The prediction performance was higher with the single preprocessing methods than with combinations of multiple PLSR- and boosting-model methods. Feature selection based on boosting algorithms could improve prediction performance. The cross-validation and prediction performances were better in the boosting algorithms than in the PLSR models, except regarding anthocyanin prediction. The final models for glucosinolates and anthocyanins especially performed sufficiently for practical use such as screening, as $R^2{}_p$ = 0.82–0.85 and RPD = 2.4–2.6. This research presents a promising approach for the rapid and accurate prediction of metabolites in plants using hyperspectral imaging, which can contribute to the development of precision agriculture and plant breeding.

Overall, our results showed that boosting algorithms can be applied to predict the functional components of medicinal plants. Many studies have compared spectral data preprocessing methods and tried to improve prediction performance. We have confirmed that prediction performance can be improved by reducing spectral bands with a feature-selection algorithm. To develop faster and more accurate prediction techniques, it is necessary to continuously introduce the latest algorithms and data processing methods. Based on hyperspectral images, non-destructive monitoring techniques of functional components can be used as tools for quality control in the field of smart agriculture, including in the medicinal plant industry.

Supplementary Materials: The following supporting information can be downloaded at: https://www.mdpi.com/article/10.3390/agriculture13081477/s1, Table S1: List of preprocessing methods for hyperspectral data of *B. juncea* plants; Table S2: Determination of preprocessing methods for AdaBoost, XGBoost, and LightGBM prediction algorithms for five metabolites in *B. juncea* plants.

Author Contributions: Conceptualization, S.H.P. and S.M.K.; funding acquisition, S.M.K.; investigation, S.H.P. and H.I.Y.; methodology, H.I.Y., J.-H.C., D.-H.J., J.-E.P. and Y.J.P.; project administration, S.H.P.; software, H.I.Y. and H.L.; validation, H.I.Y. and S.H.P.; writing—original draft, H.I.Y. and S.H.P.; review and editing, H.I.Y., J.-S.Y. and S.H.P. All authors have read and agreed to the published version of the manuscript.

Funding: This work was supported by the Korean Institute of Planning and Evaluation for Technology in Food, Agriculture and Forestry (IPET) and by the Korean Smart Farm R&D Foundation (KosFarm) through the Smart Farm Innovation Technology Development Program, funded by the Ministry of Agriculture, Food and Rural Affairs (MAFRA), the Ministry of Science and ICT (MSIT), and the Rural Development Administration (RDA) (421034-04).

Institutional Review Board Statement: Not applicable.

Data Availability Statement: The original contributions presented in this study are included in the article and Supplementary Materials; further inquiries can be directed to the corresponding authors.

Conflicts of Interest: The authors declare that this research was conducted in the absence of any commercial or financial relationships that could be construed as potential conflicts of interest.

References

1. Szőllősi, R. Indian Mustard (*Brassica juncea* L.) Seeds in Health. In *Nuts and Seeds in Health and Disease Prevention*; Academic Press: Cambridge, MA, USA, 2020; pp. 357–364.
2. Tian, Y.; Deng, F. Phytochemistry and Biological Activity of Mustard (*Brassica juncea*): A Review. *CyTA—J. Food* **2020**, *18*, 704–718. [CrossRef]
3. Kumar, V.; Kumar Thakur, A.; Dev Barothia, N.; Chatterjee, S.S. Therapeutic Potentials of *Brassica juncea*: An Overview. *CellMed* **2011**, *1*, e2. [CrossRef]
4. Park, C.H.; Park, Y.E.; Yeo, H.J.; Kim, J.K.; Park, S.U. Effects of Light-Emitting Diodes on the Accumulation of Phenolic Compounds and Glucosinolates in *Brassica juncea* Sprouts. *Horticulturae* **2020**, *6*, 77. [CrossRef]
5. Sarić, R.; Nguyen, V.D.; Burge, T.; Berkowitz, O.; Trtílek, M.; Whelan, J.; Lewsey, M.G.; Čustović, E. Applications of Hyperspectral Imaging in Plant Phenotyping. *Trends Plant Sci.* **2022**, *27*, 301–315. [CrossRef] [PubMed]
6. Beć, K.B.; Grabska, J.; Bonn, G.K.; Popp, M.; Huck, C.W. Principles and Applications of Vibrational Spectroscopic Imaging in Plant Science: A Review. *Front. Plant Sci.* **2020**, *11*, 1226. [CrossRef]
7. Zhang, C.; Wang, Q.; Liu, F.; He, Y.; Xiao, Y. Rapid and Non-Destructive Measurement of Spinach Pigments Content during Storage Using Hyperspectral Imaging with Chemometrics. *Measurement* **2017**, *97*, 149–155. [CrossRef]
8. Caporaso, N.; Whitworth, M.B.; Fowler, M.S.; Fisk, I.D. Hyperspectral Imaging for Non-Destructive Prediction of Fermentation Index, Polyphenol Content and Antioxidant Activity in Single Cocoa Beans. *Food Chem.* **2018**, *258*, 343–351. [CrossRef]
9. Choi, J.-H.; Park, S.H.; Jung, D.-H.; Park, Y.J.; Yang, J.-S.; Park, J.-E.; Lee, H.; Kim, S.M. Hyperspectral Imaging-Based Multiple Predicting Models for Functional Component Contents in *Brassica juncea*. *Agriculture* **2022**, *12*, 1515. [CrossRef]

10. Saha, D.; Manickavasagan, A. Machine Learning Techniques for Analysis of Hyperspectral Images to Determine Quality of Food Products: A Review. *Curr. Res. Food Sci.* **2021**, *4*, 28–44. [CrossRef]
11. Bonifazi, G.; Capobianco, G.; Gasbarrone, R.; Serranti, S. Contaminant Detection in Pistachio Nuts by Different Classification Methods Applied to Short-Wave Infrared Hyperspectral Images. *Food Control* **2021**, *130*, 108202. [CrossRef]
12. Jafarzadeh, H.; Mahdianpari, M.; Gill, E.; Mohammadimanesh, F.; Homayouni, S. Bagging and Boosting Ensemble Classifiers for Classification of Multispectral, Hyperspectral and PolSAR Data: A Comparative Evaluation. *Remote Sens.* **2021**, *13*, 4405. [CrossRef]
13. Weksler, S.; Rozenstein, O.; Haish, N.; Moshelion, M.; Wallach, R.; Ben-Dor, E. Detection of Potassium Deficiency and Momentary Transpiration Rate Estimation at Early Growth Stages Using Proximal Hyperspectral Imaging and Extreme Gradient Boosting. *Sensors* **2021**, *21*, 958. [CrossRef]
14. Sha, W.; Guo, Y.; Yuan, Q.; Tang, S.; Zhang, X.; Lu, S.; Guo, X.; Cao, Y.-C.; Cheng, S. Artificial Intelligence to Power the Future of Materials Science and Engineering. *Adv. Intell. Syst.* **2020**, *2*, 1900143. [CrossRef]
15. Park, Y.J.; Park, J.-E.; Truong, T.Q.; Koo, S.Y.; Choi, J.-H.; Kim, S.M. Effect of Chlorella Vulgaris on the Growth and Phytochemical Contents of "Red Russian" Kale (*Brassica napus* Var. *Pabularia*). *Agronomy* **2022**, *12*, 2138. [CrossRef]
16. Lichtenthaler, H.K.; Buschmann, C. Chlorophylls and Carotenoids: Measurement and Characterization by UV-VIS Spectroscopy. *Curr. Protoc. Food Anal. Chem.* **2001**, *1*, F4.3.1–F4.3.8. [CrossRef]
17. Thomas, M.; Badr, A.; Desjardins, Y.; Gosselin, A.; Angers, P. Characterization of Industrial Broccoli Discards (*Brassica oleracea* Var. *Italica*) for Their Glucosinolate, Polyphenol and Flavonoid Contents Using UPLC MS/MS and Spectrophotometric Methods. *Food Chem.* **2018**, *245*, 1204–1211. [CrossRef]
18. Dewanto, V.; Xianzhong, W.; Adom, K.K.; Liu, R.H. Thermal Processing Enhances the Nutritional Value of Tomatoes by Increasing Total Antioxidant Activity. *J. Agric. Food Chem.* **2002**, *50*, 3010–3014. [CrossRef]
19. Mawlong, I.; Sujith Kumar, M.S.; Gurung, B.; Singh, K.H.; Singh, D. A Simple Spectrophotometric Method for Estimating Total Glucosinolates in Mustard De-Oiled Cake. *Int. J. Food Prop.* **2017**, *20*, 3274–3281. [CrossRef]
20. Yang, Y.-C.; Sun, D.-W.; Pu, H.; Wang, N.-N.; Zhu, Z. Rapid Detection of Anthocyanin Content in Lychee Pericarp during Storage Using Hyperspectral Imaging Coupled with Model Fusion. *Postharvest Biol. Technol.* **2015**, *103*, 55–65. [CrossRef]
21. Freund, Y.; Schapire, R.E. A Short Introduction to Boosting. *J. Jpn. Soc. Artif. Intell.* **1999**, *14*, 771–780.
22. Chen, T.; Guestrin, C. Xgboost: A Scalable Tree Boosting System. In Proceedings of the 22nd ACM SIGKDD International Conference on Knowledge Discovery and Data Mining, San Francisco, CA, USA, 13–17 August 2016; ACM: New York, NY, USA, 2016; pp. 785–794.
23. Ke, G.; Meng, Q.; Finley, T.; Wang, T.; Chen, W.; Ma, W.; Ye, Q.; Liu, T.-Y. LightGBM: A Highly Efficient Gradient Boosting Decision Tree. *Adv. Neural Inf. Process Syst.* **2017**, *30*, 3146–3154.
24. Gitelson, A.A.; Merzlyak, M.N.; Chivkunova, O.B. Optical Properties and Nondestructive Estimation of Anthocyanin Content in Plant Leaves. *Photochem. Photobiol.* **2001**, *74*, 38–45. [CrossRef] [PubMed]
25. Mishra, P.; Lohumi, S.; Khan, H.A.; Nordon, A. Close-Range Hyperspectral Imaging of Whole Plants for Digital Phenotyping: Recent Applications and Illumination Correction Approaches. *Comput. Electron. Agric.* **2020**, *178*, 105780. [CrossRef]
26. Hasanzadeh, B.; Abbaspour-Gilandeh, Y.; Soltani-Nazarloo, A.; Hernández-Hernández, M.; Gallardo-Bernal, I.; Hernández-Hernández, J.L. Non-Destructive Detection of Fruit Quality Parameters Using Hyperspectral Imaging, Multiple Regression Analysis and Artificial Intelligence. *Horticulturae* **2022**, *8*, 598. [CrossRef]
27. Jayapal, P.K.; Joshi, R.; Sathasivam, R.; Van Nguyen, B.; Faqeerzada, M.A.; Park, S.U.; Sandanam, D.; Cho, B.-K. Non-Destructive Measurement of Total Phenolic Compounds in Arabidopsis under Various Stress Conditions. *Front. Plant Sci.* **2022**, *13*, 982247. [CrossRef]
28. Burnett, A.C.; Serbin, S.P.; Davidson, K.J.; Ely, K.S.; Rogers, A. Detection of the Metabolic Response to Drought Stress Using Hyperspectral Reflectance. *J. Exp. Bot.* **2021**, *72*, 6474–6489. [CrossRef]
29. Yuan, Z.; Ye, Y.; Wei, L.; Yang, X.; Huang, C. Study on the Optimization of Hyperspectral Characteristic Bands Combined with Monitoring and Visualization of Pepper Leaf SPAD Value. *Sensors* **2021**, *22*, 183. [CrossRef]
30. Luo, M.; Wang, Y.; Xie, Y.; Zhou, L.; Qiao, J.; Qiu, S.; Sun, Y. Combination of Feature Selection and CatBoost for Prediction: The First Application to the Estimation of Aboveground Biomass. *Forests* **2021**, *12*, 216. [CrossRef]
31. Bellon-Maurel, V.; Fernandez-Ahumada, E.; Palagos, B.; Roger, J.-M.; McBratney, A. Critical Review of Chemometric Indicators Commonly Used for Assessing the Quality of the Prediction of Soil Attributes by NIR Spectroscopy. *TrAC Trends Anal. Chem.* **2010**, *29*, 1073–1081. [CrossRef]
32. Heil, K.; Schmidhalter, U. An Evaluation of Different NIR-Spectral Pre-Treatments to Derive the Soil Parameters C and N of a Humus-Clay-Rich Soil. *Sensors* **2021**, *21*, 1423. [CrossRef]

agriculture

Article

Tea Bud Detection and 3D Pose Estimation in the Field with a Depth Camera Based on Improved YOLOv5 and the Optimal Pose-Vertices Search Method

Zhiwei Chen [1], Jianneng Chen [1,*], Yang Li [2], Zhiyong Gui [1,2] and Taojie Yu [1]

[1] School of Mechanical Engineering, Zhejiang Sci-Tech University, Hangzhou 310018, China
[2] Tea Research Institute, Chinese Academy of Agricultural Sciences, Hangzhou 310008, China
* Correspondence: jiannengchen@zstu.edu.cn

Abstract: The precise detection and positioning of tea buds are among the major issues in tea picking automation. In this study, a novel algorithm for detecting tea buds and estimating their poses in a field environment was proposed by using a depth camera. This algorithm introduces some improvements to the YOLOv5l architecture. A Coordinate Attention Mechanism (CAM) was inserted into the neck part to accurately position the elements of interest, a BiFPN was used to enhance the small object detection ability, and a GhostConv module replaced the original Conv module in the backbone to reduce the model size and speed up model inference. After testing, the proposed detection model achieved an mAP of 85.2%, a speed of 87.71 FPS, a parameter number of 29.25 M, and a FLOPs value of 59.8 G, which are all better than those achieved with the original model. Next, an optimal pose-vertices search method (OPVSM) was developed to estimate the pose of tea by constructing a graph model to fit the pointcloud. This method could accurately estimate the poses of tea buds, with an overall accuracy of 90%, and it was more flexible and adaptive to the variations in tea buds in terms of size, color, and shape features. Additionally, the experiments demonstrated that the OPVSM could correctly establish the pose of tea buds through pointcloud downsampling by using voxel filtering with a 2 mm × 2 mm × 1 mm grid, and this process could effectively reduce the size of the pointcloud to smaller than 800 to ensure that the algorithm could be run within 0.2 s. The results demonstrate the effectiveness of the proposed algorithm for tea bud detection and pose estimation in a field setting. Furthermore, the proposed algorithm has the potential to be used in tea picking robots and also can be extended to other crops and objects, making it a valuable tool for precision agriculture and robotic applications.

Keywords: tea bud detection; YOLOv5; depth camera; pose estimation; CAM; OPVSM

Citation: Chen, Z.; Chen, J.; Li, Y.; Gui, Z.; Yu, T. Tea Bud Detection and 3D Pose Estimation in the Field with a Depth Camera Based on Improved YOLOv5 and the Optimal Pose-Vertices Search Method. *Agriculture* **2023**, *13*, 1405. https://doi.org/10.3390/agriculture13071405

Academic Editors: Redmond R. Shamshiri, Muhammad Sultan, Md Shamim Ahamed and Muhammad Farooq

Received: 6 June 2023
Revised: 12 July 2023
Accepted: 13 July 2023
Published: 14 July 2023

1. Introduction

Tea is a popular beverage consumed worldwide and also a vital economic crop for many countries. In comparison to the high demand and output of tea, the traditional method of harvesting tea leaves remains quite rudimentary, particularly for famous teas, which are often picked manually. This picking process is labor-intensive, time-consuming, and physically demanding, which may result in strains and injuries among laborers. Instead of picking manually, an automatic tea picking system has the potential to lower the requirement of labor and ease labor intensity [1]. In order to develop this automatic system, detecting and positioning the tea leaves in 3D space is an essential process, which allows the system to accurately perceive the leaves in the field for picking. Moreover, the algorithm for detecting and positioning tea leaves can be used to enhance production, facilitate research, and monitor and preserve the environment. Research into tea detection and positioning algorithms plays a crucial role in automatic picking systems, which have the potential to significantly improve the efficiency and productivity of tea production in the future [2].

The development of tea bud detectors has been sustained for several years. Researchers have extracted and combined various features, such as shape, color, and texture, from tea bud images to design advanced features using complicated processes. For example, Zhang et al. utilized R-B factors to convert the image to grayscale and performed threshold segmentation using the Otsu method. They applied area filtering, erosion, and expansion algorithms to generate a binary image of the tea buds. Finally, the center of mass method was employed to determine the position of the tea buds within the binary image [3]. Zhang et al. applied a Gaussian filter to reduce image noise and split the images into their respective R, G, and B components. Image operations were performed to obtain G-B′ components. The minimum error method was utilized to determine the most suitable adaptation thresholds, which underwent a piecewise linear transformation to enhance the distinction between the tea buds and the background. Binarization and the application of the Canny operator were then employed to detect edges in the binary image. Finally, the unknown area was calculated and marked, and the segmentation process was completed using the watershed function [4]. Wu et al. analyzed the color information of G and G-B components of tea buds and leaves. They calculated segmentation thresholds using an improved Otsu method to recognize tea buds [5]. While these methods have demonstrated effectiveness under specific conditions, their limited generalizability and lack of intelligence restrict their applicability in production scenarios.

With the progression of deep learning methods, the object detection field has shifted the focus toward acquiring larger labeled datasets and refining the network architecture, rather than emphasizing image features. Many studies on object detection have used large public datasets to train and test detectors based on convolutional networks, with subjects such as cars [6], traffic signs [7] and pedestrians [8]. Similarly, many studies on object detection based on deep learning in agriculture have also been conducted [9]. Zeng et al. proposed using a lightweight YOLOv5 model to efficiently detect the location and ripeness of tomato fruits in real time, and the improvement reduced the model size by 51.1% while maintaining a 93% true detection rate [10]. Ma et al. integrated a coordinate attention (CA) module with YOLOv5 and trained and tested a YOLOv5-lotus model to effectively detect overripe lotus seedpods in a natural environment [11]. Wang et al. introduced an apple fruitlet detection method that utilizes a channel-pruned YOLOv5s deep learning algorithm, achieving rapid and accurate detection, and the compact model size of 1.4 MB facilitated the development of portable mobile fruit thinning terminals [12]. Sozzi et al. evaluated six versions of YOLO (YOLOv3, YOLOv3-tiny, YOLOv4, YOLOv4-tiny, YOLOv5x, and YOLOv5s) for real-time bunch detection and counting in grapes. Finally, YOLOv4-tiny emerged as the best choice due to its optimal balance between accuracy and speed. This study provided valuable insights into the performance of different YOLO versions and their applicability in the grape industry [13]. Cardellicchio et al. used single-stage detectors based on YOLOv5 to effectively identify nodes, fruit, and flowers on a challenging dataset acquired during a stress experiment conducted on multiple tomato genotypes, and achieved relatively high scores [14].

In particular, studies on the detection of tea buds based on deep learning have also made some significant advancements. Murthi and Thangavel employed the active contour model to identify potential tea bud targets in images, then utilized a DCNN for recognition [15]. Chen et al. introduced a new fresh tea sprout detection method, FTSD-IEFSSD, which integrates image enhancement and a fusion single-shot detector. This method separately inputs the original and enhanced images into a ResNet50 subnetwork and improves detection accuracy through score fusion. This method achieved an AP of 92.8% [16]. Xu et al. combined YOLOv3 with DenseNet201 to quickly and accurately detect and classify tea buds, achieving 82.58% precision on the top-shot dataset and 99.28% precision on the side-shot dataset [17]. Gui et al. enhanced and reduced the weight of the YOLOv5 network for tea bud detection by replacing the standard convolution with a Ghost_conv module, adding a BAM into the backbone, applying MS-WFF in the neck, and switching to a CIOU loss function. After training and testing on a dataset of 1000 samples, the model achieved a frame rate of 29,509 FPS, an mAP of 92.66%, and a model size of 90 M [18]. Moreover, tea buds can also be segmented using deep learning. Hu et al. presented a discriminative pyramid (DP) network-based method with exceptional accuracy for the semantic segmentation of tea geometrids in natural scene images [19], while TS-SegNet, a novel deep convolutional encoder-decoder network method for tea sprout segmentation proposed by Qian et al., produced a good segmentation result [20].

Currently, research is primarily focused on detecting and segmenting tea buds, with very few studies proposing methods for identifying the specific points for picking. However, identifying the position of the picking point is crucial for efficient and effective tea bud harvesting. The Faster R-CNN was used to locate tea shoot picking points in the field, achieving a precision of 79% [21], and the Mask-RCNN was utilized to create a recognition model for tea buds, leaves, and picking points. The results showed an average detection accuracy of 93.95% and a recall rate of 92.48% [22]. Although these algorithms position the picking point of tea buds in the image, the lack of a depth sensor limits the acquisition of spatial coordinates for picking points. Thus, the robot would be unable to determine the target position in robot coordinates for end-effector movement. Li et al. used the YOLO network to detect the tea buds and acquired the position of the target in the 3D space by using an RGB-D camera [23]. This method calculates the center point of all points and represents the grab position using a vertical minimum circumscribed cylinder. Then, they inclined the minimum circumscribed cylinder according to the growth direction of tea buds calculated via the PCA method [24]. Chen et al. developed a robotics system for the intelligent picking of tea buds, training the YOLOv3 model to detect the tea buds from images and proposing an image algorithm to locate the picking point. While their system has a successful picking rate of 80%, the detection precision of this system could be improved, and the importance of addressing measurement errors and motion errors in the picking process should not be overlooked [25].

In conclusion, tea bud detection research has experienced good development, but some datasets are still quite simple and small-sized. Moreover, the pose of tea buds within a 3D space is important for the robot to grasp the target accurately, yet there are few studies that have talked about this issue. Therefore, the aim of this study is to detect tea buds using an improved YOLOv5 network and determine their 3D pose through the OPVSM. The contributions of this study are as follows:

1. The convolution (CBS) of YOLOv5 is replaced by GhostConv to reduce FLOPs and compress the model size by generating more feature maps from cheap operations [26].
2. The Coordinate Attention Mechanism (CAM) is integrated into the neck, which could accurately locate the region of interest by reserving the coordinates of elements in the feature map, leading to more precise detection results [27].
3. The conversion from PANet to BiFPN enhances feature aggregation and multi-scale feature fusion, resulting in improved detection efficiency [28].
4. The OPVSM is proposed to search the pose-vertices from the pointcloud to build the skeleton of tea buds.

2. Materials and Methods

2.1. Improvement of YOLOv5

2.1.1. YOLOv5

As the most popular network architecture for object detection, the YOLOv5 (You Only Look Once) network model [29] is widely used in detection and recognition systems in various fields, as shown in Figure 1. This model architecture has spawned many improved versions because of its good modifiability, portability and trainability. This model provides two hyperparameters, the depth-multiple and width-multiple, used to control the depth (number of bottlenecks) and width (number of channels) of the backbone in the model, respectively.

Figure 1 shows the structure of the original YOLOv5 model. This model can be divided into four parts, including the input, backbone, neck and head. During training, the input part reads the images and labels, enhances these images and puts them into the backbone part. The backbone network extracts the features from images by means of the CSPDarknet53, which is composed of CSP modules. In the neck, FPN and PANet are used to aggregate image features and output three layers of feature maps with different sizes. Finally, the head part predicts and outputs the detection results according to these feature maps.

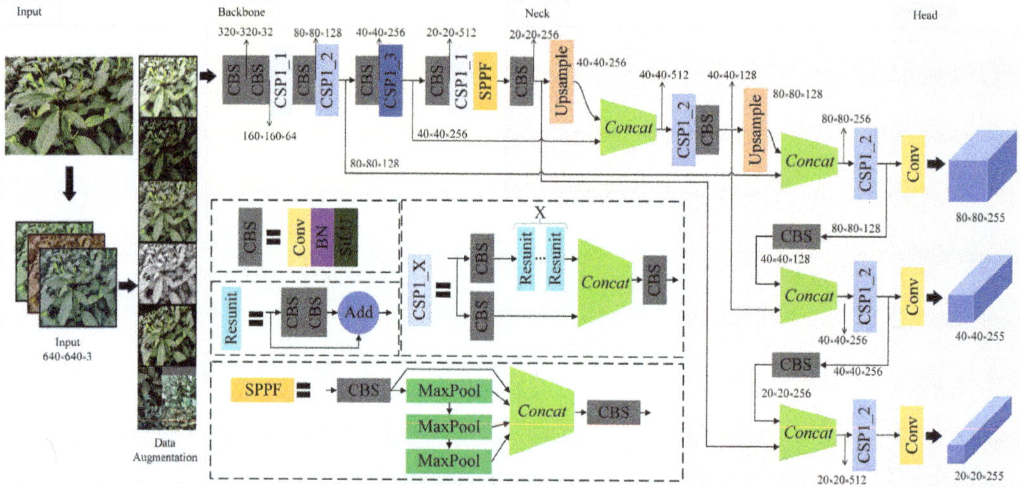

Figure 1. Structure of the YOLOv5 model.

2.1.2. Architecture of Improved YOLOv5

Although the YOLOv5 model has good precision and speed, its performance in small target detection is inadequate. Figure 2 shows the structure of the improved YOLOv5 model. The main modifications are made in the backbone and neck parts of the model. In Figure 2, the red circle 1 represents the Coordinate Attention Mechanism (CAM). It is incorporated into the neck section to precisely locate the region of interest by preserving the coordinates of elements in the feature map, thereby improving the accuracy of object detection. Meanwhile, the red circle 2 denotes the modified connection known as BiFPN. It replaces the PANet in the neck to enhance feature aggregation and multi-scale feature fusion. Additionally, the red circle 3 represents the GhostConv module utilized in the backbone. It replaces the standard convolution (CBS) in the backbone, which reduces the computation cost by generating more feature maps from cheap operations. These modifications improve the small target detection performance of the YOLOv5 model and make it more efficient.

2.1.3. Coordinate Attention Mechanism

At the beginning of the neck section, the Coordinate Attention Mechanism (CAM) shown in Figure 3 is added. Firstly, the variables with a size of $H \times W \times C$ are shrunk in the X and Y directions using separate adaptive average pooling layers. Here, the X direction corresponds to the width of the feature map, while the Y direction corresponds to its height. Two feature maps with different sizes of $H \times 1 \times C$ and $1 \times W \times C$ are then obtained.

Next, the Y-pooling feature map is permuted and aggregated with the X-pooling feature map to create a new feature map with a size of $(H + W) \times 1 \times C$. This aggregated feature is then processed using a CBS consisting of a convolution layer, batch normalization layers, and the SiLU activation function. The SiLU function applies the sigmoid function to the input value x and multiplies the result with the input itself, providing smoothness and non-linearity, and preserving certain linearity characteristics suitable for deep learning models.

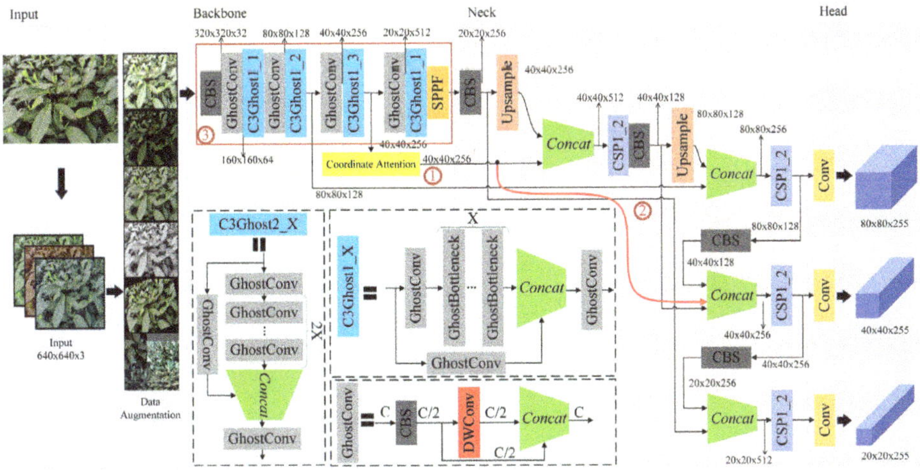

Figure 2. Architecture of the improved YOLOv5.

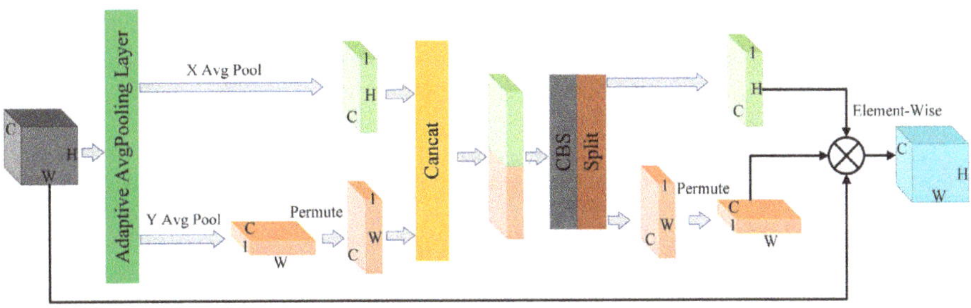

Figure 3. Structure of the Coordinate Attention Mechanism.

Following this, the feature map is split back into two feature maps of the same size as before the aggregation step. The previously permuted feature map is restored to its original size. Finally, the two feature maps are combined element-wise with the original feature map to generate a new feature map with a size of H × W × C, incorporating aggregated contextual information.

The Coordinate Attention Mechanism utilizes a separate global pooling operation in the X and Y directions, respectively, so that the attention module can retain the element position of the precise coordinates in the feature map. The formula for the separate global pooling operation is as follows:

$$\begin{cases} z_c^h(h) = \frac{1}{W} \sum\limits_{0 \le i < W} x_c(h, i) \\ z_c^w(w) = \frac{1}{H} \sum\limits_{0 \le j < H} x_c(j, w) \end{cases} \quad (1)$$

More precisely, the above pooling operation is an information embedding process in the attention module. The subsequent matrix transformations such as Concat and CBS are used to aggregate the contextual information. After these operations, the region of interest can be obtained on the basis of the features generated by the pooling. In brief, this structure is more conducive to accurately positioning the elements of interest.

The Coordinate Attention Mechanism (CAM) captures and leverages spatial information by utilizing separate global pooling operations in the X and Y directions. This allows the CAM to emphasize or suppress specific spatial locations based on their importance. Unlike other attention mechanisms such as Efficient Channel Attention (ECA), Squeeze-and-Excitation (SE), or Convolutional Block Attention Module (CBAM), which primarily focus on channel-wise attention, the CAM specifically targets spatial reasoning. By incorporating CAM, this model gains the ability to perform precise localization and spatial reasoning, which are particularly valuable in tasks requiring accurate object detection and a good understanding of spatial relationships. Additionally, CAM exhibits high effectiveness in localizing important features within an input feature map. By applying separate global pooling in the X and Y directions, the CAM attends to specific spatial locations, enabling the model to capture fine-grained details and accurately localize objects. In contrast, other attention mechanisms like ECA, SE, or CBAM may not explicitly prioritize precise localization, limiting their performance in tasks where precise object localization is crucial. Furthermore, the separate global pooling operations in the X and Y directions enable efficient computation, as they do not require extensive calculations across channels or intricate transformations. This leads to improved efficiency, reduced memory consumption, and faster inference times, making CAM particularly suitable for real-time applications. Overall, compared to other attention mechanisms, CAM demonstrates stronger spatial reasoning ability, higher localization accuracy, and lower computational complexity. It is particularly suitable for tea bud detection.

2.1.4. BiFPN

The bidirectional feature pyramid network (BiFPN) improves the FPN by incorporating a bidirectional feature fusion approach. It deletes the feature maps with single inputs, adds the extra connection between original and final feature maps, and down-samples the original feature map to make the network more accurate and efficient. In this study, the BiFPN is adapted and ported based on the PANet in the neck part to improve the detection efficiency and the muti-scale feature fusion. As shown in Figure 4, the relationship drawn by the red dotted arrow is the new improved connection. This new connection is utilized to aggregate more features in the middle scale layer for enhancing the small object detection ability without increasing the computational cost too much.

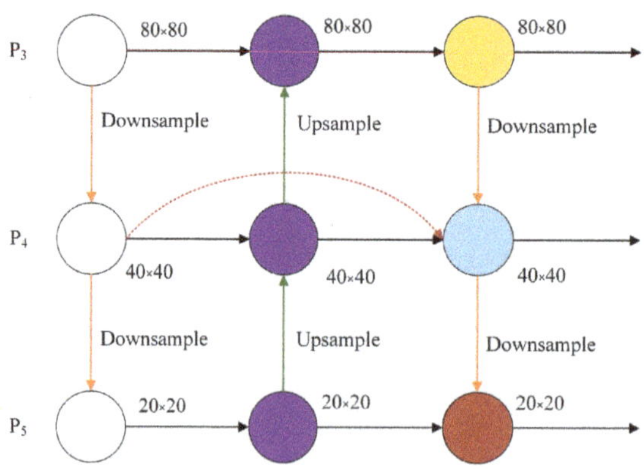

Figure 4. Structure of the BiFPN.

Specifically, unlike the unidirectional flow of information in the original FPN, BiFPN enables bidirectional information flow, allowing it to propagate in both bottom-up and top-down directions. Moreover, it efficiently combines features from different levels of the pyramid, incorporating both top-down and bottom-up pathways as well as lateral connections. This multi-scale fusion process enables effective feature capture at various scales, facilitating the network's ability to handle objects of different sizes. Additionally, BiFPN is designed to be computationally efficient by reducing overheads through shared computations across multiple feature levels, distinguishing it from other fusion approaches. Furthermore, BiFPN addresses FPN's limitations by introducing extra connections between adjacent levels of the pyramid, facilitating improved information propagation across scales for the transmission of fine-grained details between higher-level and lower-level feature maps. Therefore, in this study, BiFPN enhances the expressive power, adaptability, and efficiency of the feature fusion process, improving the model's ability to detect small objects. This improvement is particularly valuable for tea bud detection, reducing the rate of missed detections.

2.1.5. GhostConv

GhostConv can be viewed as a factorization of the CBS block into two parts, where one part has a large number of filters but fewer channels, and the other part has fewer filters but more channels. This factorization allows for a reduction in the number of parameters and computational cost, while maintaining a good balance between representation power and efficiency. As shown in Figure 5, the CBS block convolves the feature maps from C channels to C/2 channels by reducing the number of kernels. Then, the DWConv (depth-wise convolution) block is applied to each channel of the input feature maps independently, rather than being applied across all channels as in regular convolution, which is computationally efficient as it reduces the number of parameters. Unfortunately, the disadvantage of GhostConv is that it increases the number of layers of the model due to the depth-wise convolution. Therefore, GhostConv should only be used in the backbone in order to limit the growth in the number of layers.

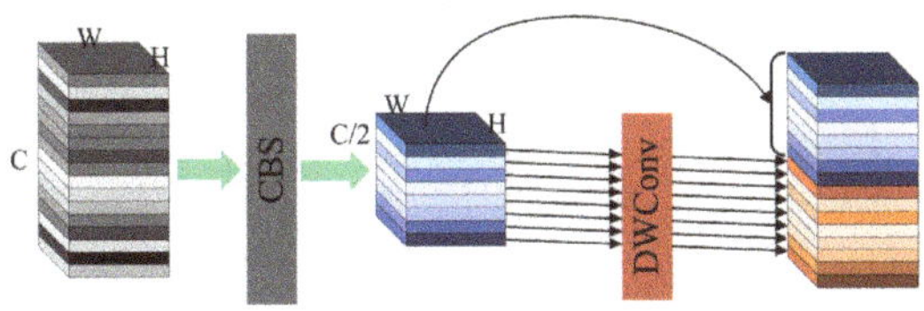

Figure 5. The GhostConv block.

According to the above explanation of GhostConv, the module applied in this research offers several advantages. By adopting a split-transform-merge strategy, GhostConv reduces computation by dividing a larger convolutional kernel into a primary path and a secondary "ghost" path. The primary path conducts a standard convolution on a subset of input channels, while the ghost path applies a lightweight convolution to the remaining channels. Despite its reduced computation, GhostConv achieves a larger receptive field by utilizing a larger kernel size in the primary path, enabling the network to capture extensive context and long-range dependencies. Balancing computation and receptive field size, GhostConv strikes an effective trade-off between efficiency and expressive power. It employs a fusion operation to merge features from the primary and ghost paths, facilitating information flow and enhancing the representation learning process. By leveraging both paths, GhostConv efficiently captures and propagates features throughout the network. The ghost path in GhostConv serves as a regularization mechanism, promoting the learning of more robust and discriminative features by introducing sparsity in the computation. GhostConv exhibits scalability and can be seamlessly integrated into various network architectures, making it suitable for diverse resource constraints and deployment scenarios. Furthermore, GhostConv has exhibited excellent generalization capabilities across diverse datasets and tasks, consistently demonstrating performance improvements. These advantages enable the tea bud detection model to reduce model size and computation cost while increasing detection speed, making it beneficial for porting to embedded platforms.

2.2. Three-Dimensional Pose Estimation

The improved YOLOv5 model can detect tea buds using bounding boxes in the images. However, a 6-DOF picking robot always requires the 6-DOF pose of the tea buds within the 3D space to pick them precisely with an ingenious end-effector under field conditions. In this paper, a pose estimation algorithm based on the optimal pose-vertices search method (OPVSM) is proposed in order to find the best vertices from the tea buds' pointcloud to form a graph structure for fitting their pose.

2.2.1. Tea Bud Pointcloud Extraction

After detection using the improved YOLOv5 model, the depth camera can acquire the pointcloud from the bounding box, which refers to a tetrahedron in the 3D space, as shown in Figure 6.

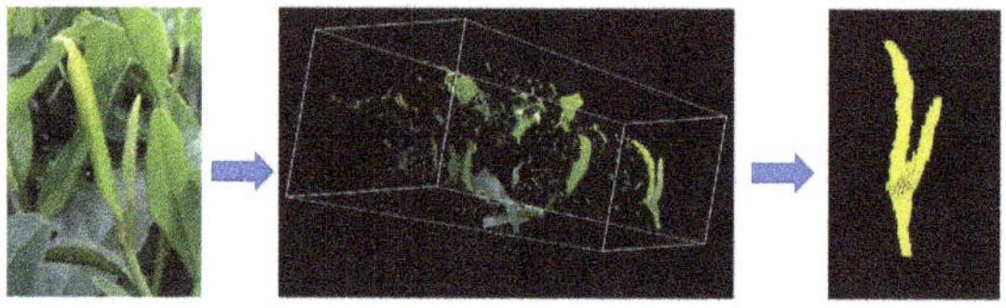

(a) (b) (c)

Figure 6. Obtaining the pointcloud of tea bud from the depth camera. (**a**) Tea bud detected in image; (**b**) Pointcloud extracted from a depth camera using a passthrough filter; (**c**) Pointcloud of a tea bud extracted from pervious pointcloud using DBSCAN.

Figure 6a shows the area of a bounding box in the image obtained by the tea bud detector, and the tetrahedron shown in Figure 6b is obtained from the depth camera via a passthrough filter, which includes the tea bud points, background points and many outliers. These outliers can be removed by the radius filter, and the points of the tea bud can be extracted from this pointcloud using the Density-Based Spatial Clustering of Applications with Noise (DBSCAN) method [30]. The DBSCAN algorithm starts by selecting a random point and finding all points within its neighborhood. If the number of these points is greater than the minimum, a new cluster will be formed. Otherwise, this point will be marked as noise. This process will be repeated for all points within the cluster. After that, the algorithm then moves to the next unprocessed data point and repeats the process until all data points have been processed. Finally, the points of tea buds can be extracted from the original pointcloud, as shown in Figure 6c.

2.2.2. Tea Bud Graph Structure

The tea buds are diverse because of their biodiversity. Their size, color, and shape features such as the opening and closing angles of the bud and the side leaf are not consistent for each tea bud. These features of the tea bud are mainly affected by many factors such as the growth phase, climate, field conditions, and tea variety. It is difficult to design a standard template for all tea buds. Therefore, the traditional template matching method cannot easily be directly applied to the pose detection of tea buds.

By contrast, the geometric features of tea buds are much clearer, and these features are highly abstracted and obtained as shown in Figure 7. The geometric features of tea buds can be divided into three parts: the first part is the stem of the tea bud near the ground, which is highly abstracted into a vertex; the second part is the bifurcation of the tea bud, which is also abstracted as a vertex; the third part is each bud or leaf, and these are abstracted as a vertex, respectively. The bud vertex, leaf vertices and stem vertex are all connected to the bifurcation vertex through an edge. Thus, a finite, simple and undirected graph is established.

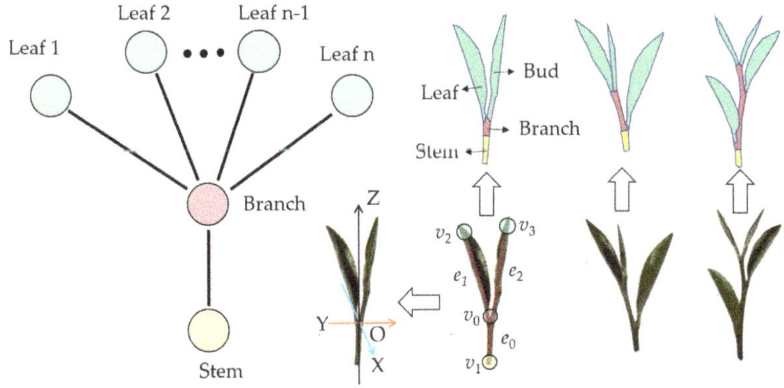

Figure 7. The geometry features of the tea buds.

In this paper, we take one bud and one leaf (BL1) as an example, because the BL1 is a standard that appears most widely in famous and high-quality tea. The vertex v_0 is the bifurcation of BL1, v_1 is the stem, and v_2 and v_3 are the bud and leaf. e_0, e_1, e_2 are the edges that connect v_0 to other vertices. According to these elements of this graph, the space coordinates of this BL1 shown in Figure 7 can be established. The coordinate origin is the v_0 vertex, the Z-axis that refers to the growth direction is the unit vector of v_1v_0, the X-axis is the normal vector of the plane $v_0v_2v_3$, and the Y-axis can be obtained according to the right-handed rule.

2.2.3. Optimal Pose-Vertices Search Method

In this study, the OPVSM is proposed to find the best vertices from the pointcloud to build a graph for fitting the pose of the tea bud. The flow of this algorithm is shown in Figure 8.

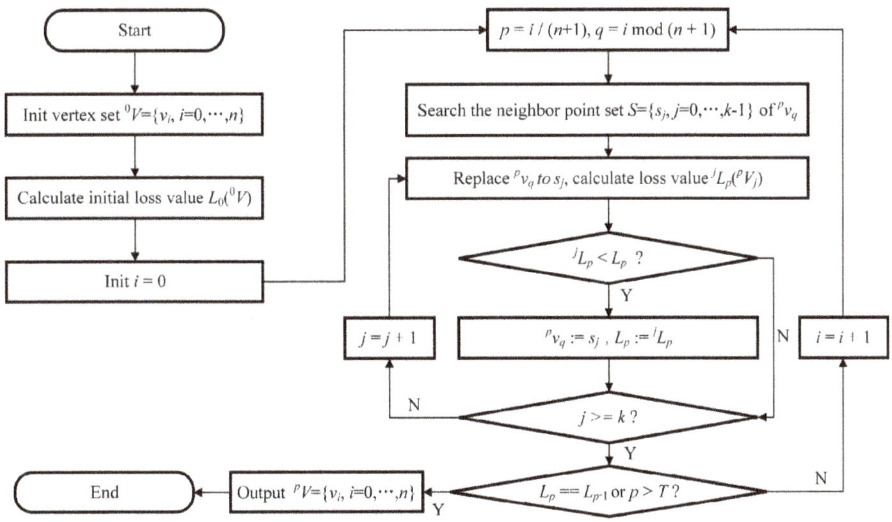

Figure 8. Flowchart of the vertex search algorithm.

The vertex set V has $n + 1$ vertices. v_0 refers to the bifurcation, and the other vertex includes a stem, a bud, and $n - 2$ leaves. Firstly, some vertices are selected from the pointcloud at equal spacing to initialize the set V, named ^0V, where the number 0 represents the number of iterations V. Then, the initial loss value L_0 can be calculated according to the loss function. This loss function refers to the fit between the graph model and the pointcloud, which will be described in the next section in more detail. In brief, the better the fit is, the lower the loss value is, which means that the graph model can better reflect the shape features of the tea bud. Finally, we set the i equal to 0, and the initialization is finished. The loop stage of this algorithm starts and continues to run to iterate the vertex set V_p until the loss value L_p cannot be decreased or the number of iterations p is bigger than the threshold T.

In each iteration, to start with, the nearest k points in the pointcloud of the query vertex ${}^{p}v_q$ are searched and arranged from near to far to obtain the nearest neighbor points set $S = \{s_j, j = 0, ..., k - 1\}$. Then, in the order of the index of set S, s_j is used to replace the query vertex ${}^{p}v_q$ to obtain a candidate set ${}^{p}V_j$. The loss value ${}^{j}L_p$ of this candidate set ${}^{p}V_j$ is calculated for comparison with the current L_p obtained from the last iteration. If the loss value is decreased, this query vertex is updated by this point, and the loss value is replaced by the new value. Otherwise, this point would be overlooked. After traversing set S, the query vertex ${}^{p}v_q$ is replaced by each of the elements in set S in sequence, and the new vertex set ${}^{p}V$ with the smallest loss value L_p is obtained. If the loss value is changed and the number of iterations p is smaller than T, a new iteration is carried out.

Along with the running of the algorithm, each vertex in set V is considered to replace its neighbors in the pointcloud according to the loss value. Finally, the optimal vertex set V can be obtained to build the graph model for fitting tea buds.

2.2.4. Loss Function

It is important to design a loss function for measuring the fit between the model and the pointcloud of a tea bud. On the basis of the vertex set V, a weighted undirected graph $G(V, E)$ can be established as follows:

$$G(V,E), V = \{v_i, i = 0, \ldots, n\}, E = \{e_j, j = 0, \ldots, n-1\} \tag{2}$$

where V is the vertex set that is already acquired and E refers to the edge set that consists of all edges in this model. The edge e_j in E is the connection between v_{j+1} and v_0, and the weight of this edge is the positional relationship between v_{j+1} and v_0.

The loss function is given as follows:

$$L(G) = (1+\eta)(L_s + L_v + L_e) \tag{3}$$

where L_s is the loss of the bifurcation v_0, L_v is the loss of other vertices, L_e is the loss of edges, and η is the structural parameter of this model to punish some incorrect structures.

To begin with, point sets are created for each element in the graph to hold the points closest to it in the pointcloud. These point sets are denoted as P_s, which corresponds to v_0, ${}^{i}P_v$, which corresponds to v_i, and ${}^{j}P_e$, which corresponds to e_j. The mean and standard deviation of the distance between the point and element in these point sets are then calculated as follows:

$$\mu = \frac{1}{n}\sum_{i=0}^{n-1} d_i, \sigma = \sqrt{\frac{1}{n}\sum_{i=0}^{n-1} (d_i - \mu)^2} \tag{4}$$

where n represents the number of elements in the point set and d_i denotes the distance between the element q_i in point set P and the corresponding element in the graph. Ideally, the graph elements should be located in the center of their point set, with the elements in the point set uniformly distributed around them. Therefore, the small mean distance of the point set indicates that it is close to its graph element, while the small standard deviation of the distance in the point set indicates that it is both centralized and uniform. With these considerations in mind, the forms of L_s, L_v, and L_e are shown below:

$$\begin{cases} L_s = \mu_s + \sigma_s \\ L_v = \sum\limits_{i=0}^{n} {}^{i}\lambda_v({}^{i}\mu_v + {}^{i}\sigma_v) \\ L_e = \sum\limits_{j=0}^{n-1} {}^{j}\lambda_e({}^{j}\mu_e + {}^{j}\sigma_e) \end{cases}, \tag{5}$$

where i is the index of the vertex set V and j is the index of the edge set E. As mentioned previously, the edge e_j in E is the connection between v_{j+1} and v_0. Therefore, the parameter ${}^{i}\lambda_v$ is related to the point set ${}^{i}P_v$, which corresponds to v_i, and the parameter ${}^{j}\lambda_e$ is related to the point set ${}^{j}P_e$, which corresponds to e_j. In general, the bifurcation vertex v_0 should be positioned at the center of the pointcloud. The other vertices, representing the stem, bud and leaves, should extend to the border of the pointcloud, and their point sets should be small as they only contain points near the border. The point sets of the edges should mainly include points from the stem, bud, and leaves in the pointcloud. Based on these considerations, the parameters ${}^{i}\lambda_v$ and ${}^{j}\lambda_e$ of L_v and L_s are determined as follows:

$$\begin{cases} {}^{i}\lambda_v = \frac{1}{Z}(|{}^{i}P_v|)^2 \\ {}^{j}\lambda_e = (1 + \frac{1}{Z}|{}^{j}P_e|)^2 \end{cases}, \tag{6}$$

where $|{}^{i}P_v|$ and $|{}^{j}P_e|$ represent the number of elements in each point set, and Z is the size of the pointcloud.

Moreover, to prevent an irrational situation, the loss function includes a penalty factor denoted as η, which is introduced in the following form:

$$\eta = \eta_{angle} + \eta_{length}$$
$$\begin{cases} \eta_{angle} = \begin{cases} 0.01 & N_{acute} < 0.35N_{angle} \\ 0.1 & 0.35N_{angle} <= N_{acute} < 0.7N_{angle} \\ 5 & N_{acute} >= 0.7N_{angle} \quad or \quad \alpha_{\min} < \pi/18 \end{cases} \\ \eta_{length} = \begin{cases} 0 & Len_{min}/Len_{avg} > 0.2 \\ 5 & Len_{min}/Len_{avg} <= 0.2 \end{cases} \end{cases} \tag{7}$$

The penalty factor η is the sum of two parameters, η_{angle} and η_{length}. N_{angle} represents the total number of angles between edges, while N_{acute} is the number of acute angles among those angles between edges. If the ratio of N_{acute} to N_{angle} is too high, η_{angle} will be increased to prevent edges from being too close. α_{min} is the minimum angle among these angles. If α_{min} is too small, indicating that two edges are too close, the parameter will be assigned a large value to punish them. Len_{min} represents the shortest edge, and Len_{avg} represents the average length of all edges. If the ratio of Len_{min} to Len_{avg} is too small, η_{length} will be increased to prevent the shortest edge from vanishing during the search. The specific values of the constants in Formula (7) were chosen based on experiential knowledge and are used as penalty terms without further elaboration.

Figure 9 illustrates a flowchart summarizing the loss calculation process. Firstly, the original pointcloud of a tea bud, denoted as P, is divided into multiple point sets based on the shortest distance from the elements in the graph. These sets include P_s, relative to vertex v_0, ${}^{i}P_v$, relative to vertex v_i, and ${}^{j}P_e$, relative to edge e_j. Subsequently, the parameters μ, σ, and λ can be computed from these sub point sets using Formulas (4) and (6). Following that, the sub-losses in the loss function, namely L_s, L_v, and L_e, are derived using Formula

(5). The penalty factor η is determined based on the vertex set and edge set in the graph using Formula (7). Finally, the loss value $L(G)$ can be computed using Formula (3).

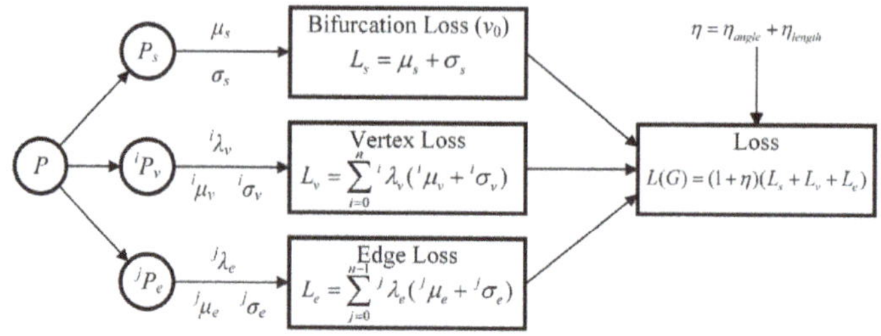

Figure 9. Flowchart of the loss function calculation process.

3. Results and Discussion

3.1. Experiments with Improved YOLOv5

3.1.1. Preparation

The deep learning method utilizes a known dataset with a homologous pattern to build an approximate model for predicting unknown data. The deep neural network can be considered as an approximation of the feature distribution. Therefore, the foundation of deep learning is the quality of the dataset, as only high-quality data can ensure the reliability of the model. There are three important properties that determine the quality of data, namely authenticity, multiplicity, and correctness. The authenticity of the data requires that the samples accurately reflect the application scenarios in the real world. The multiplicity of the data requires that the dataset contains different samples under varying conditions such as weather, illumination, shooting angle, camera parameters, and background. The correctness of the data requires that the labels of the dataset are accurate and precise for labeling the objects in the images. Moreover, a larger volume of data typically leads to better model performance.

As mentioned above, a large tea bud image dataset was established over a period of three years, from 2020 to 2022. It includes tens of thousands of samples acquired with different cameras, during different seasons and weather conditions and from different tea plant varieties. A selection of these samples can be seen in Figure 10.

Figure 10. Variety of samples in the dataset.

In this paper, tea buds from longjing43 tea plants were selected as the detection object. All of the images in the dataset were of the longjing43 variety. The variations in plantation, seasons, and shooting time account for the distinct appearances observed in these samples.

A total of 20,859 samples of the longjing43 variety were acquired, and the remaining samples were excluded. From these samples, we randomly selected 16,000 samples for both the training and validation datasets. In each training epoch, the 16,000 samples were split randomly into training and validation datasets in a 7:3 ratio. The remaining 4859 samples formed the test set, which was not used during the training stage, but rather only for testing after training to eliminate any potential bias.

This study aimed to detect tea buds that conform to the "one bud and one leaf" (BL1) standard specified in the tea industry guidelines. They were labeled into two different classes, namely ST (Spring Tea) and AT (Autumn Tea), due to their significant differences. The dataset included a total of 198,227 objects, out of which 122,762 were Spring Tea and 75,465 were Autumn Tea. The labeling process employed the Labelimg software, where tea buds in the images were selected using rectangular frames and assigned corresponding class labels. The resulting labeling data were saved in a .txt file format, adhering to the specified format $(c, x_{center}, y_{center}, w, h)$. Here, c represents the class number, (x_{center}, y_{center}) represents the normalized central coordinates of the rectangle, and (w, h) indicates the normalized length and width of the rectangle.

Data augmentation is a technique that enhances the diversity and variability of the training set, resulting in a more robust model with reduced possibility of overfitting. It promotes better generalization by exposing the model to various variations of the same data points. Data augmentation expands the dataset without the need for additional labeled examples, thereby providing the model with more instances to learn from and reducing the risk of overfitting to noise or outliers. It also serves as a regularization technique by imposing constraints and introducing random perturbations to mitigate the model's sensitivity to individual training examples. By applying data augmentation with other techniques, such as dropout, batch normalization and weight decay, the model can avoid overfitting and achieve optimal results [31].

Therefore, some of the images in the dataset were pre-processed before training. This process may involve various techniques such as translation, rotation, cropping, noise addition, mirroring, contrast, color, and brightness transformation. All of these augmentation methods are shown in Figure 11.

Figure 11. Data augmentation.

This research applied specific constraints to the parameters of various data augmentation methods to strike a balance between introducing meaningful variations and avoiding excessive distortions. The translation distance was limited to a maximum of 0.1 times the image width horizontally and 0.1 times the image height vertically. Rotation augmentation was constrained to a maximum angle of 30 degrees in both clockwise and counterclockwise directions. For cropping augmentation, the cropping parameter was set to 0.8, indicating that the cropped region covered 80% of the original image size. Gaussian noise augmentation employed a variance parameter of 0.02. Mirroring augmentation involved horizontal flipping with a 50% chance. Contrast, color, and brightness transformations were limited to a magnitude range of −0.2 to 0.2. These specific parameter constraints were selected based on empirical observations and domain knowledge to ensure reasonable variations while preserving image quality and information integrity.

In addition, the mosaic [32] process was used to improve the robustness and generalization of the model in this study. As shown in Figure 12, the mosaic image was created by randomly selecting four images from the training dataset and placing them together in a square. The images were resized and padded to fit into the square, with random flips and rotations applied to each individual image. This technique exposes the model to a wide range of object arrangements and backgrounds in a single image, which helps to improve the performance of object detection models.

Figure 12. An example of a mosaic image.

3.1.2. Results

To evaluate the performance of the improved YOLOv5, the improved model was compared with the original YOLOv5 using several evaluation metrics, including P (precision), R (recall), mAP (mean average precision), Params (parameters), FLOPs (floating point operations) and speed (FPS). Precision is defined as the ratio of true positive predictions to the total number of positive predictions made by the model. It is calculated as follows:

$$\mathrm{P} = \frac{TP}{TP + FP}, \quad (8)$$

where TP represents the number of correctly predicted positive instances and FP represents the number of instances that were predicted as positive but are actually negative. Recall is defined as the ratio of true positive predictions to the total number of positive instances in the dataset. It can be expressed as:

$$\mathrm{R} = \frac{TP}{TP + FN}, \quad (9)$$

where FN represents the number of instances that were predicted as negative but are actually positive. Recall measures the ability of the model to correctly identify positive instances. AP is a measure of the precision–recall curve for a given class, where precision represents the fraction of true positive detections among all predicted detections at a certain recall level. mAP is calculated by computing the average precision (AP) for each class in the dataset and then taking the mean of all AP values, defined as follows:

$$mAP = \frac{\sum_{i=1}^{k} AP_i}{k}, \quad (10)$$

where k is the number of classes of detection object; in this dataset, k is equal to 2. mAP is widely used to evaluate detectors because it can consider both precision and recall, providing a more comprehensive evaluation of the model's performance. Params refers to the number of learnable weights and biases in a neural network model. The number of parameters in the model determines its complexity and capacity to learn from the data. Models with a large number of parameters require more computational resources and longer training times, but may also achieve higher accuracy. The FLOPs value represents the number of floating-point arithmetic operations that the model performs during inference or training. A large FLOPs value directly reflects the massive computational resources required by the model. The speed measures the processing speed of a deep learning model during inference. It is usually represented by FPS (frames per second), which indicates how many frames or data samples the model can process per second.

To ensure optimal performance of the proposed model, it was essential to determine its depth and width prior to training. In order to accomplish this, various YOLOv5 models with differing depths (number of bottlenecks) and widths (number of channels) of the backbone were trained using the training data, and subsequently evaluated using the testing data. The evaluation metrics obtained from this process are presented in Table 1.

Table 1. The performance of different original Yolov5 models for our data.

Model	Depth Multiple	Width Multiple	mAP (%)	Params (M)	FLOPs (G)	Speed (FPS) GPU	Speed (FPS) CPU	Size (M)
YOLOv5x	1.33	1.25	81.1	86.18	203.8	51.54	1.69	165.10
YOLOv5l	1	1	82.2	46.11	107.7	82.64	3.8	88.55
YOLOv5m	0.67	0.75	81.4	20.86	47.9	90.09	5.84	40.25
YOLOv5s	0.33	0.5	78.1	7.02	15.8	112.36	13.35	13.75

Despite YOLOv5l having only 75% of the depth and 80% of the width of YOLOv5x, it achieved the highest mAP of 82.2%, which is even better than YOLOv5x's 81.1% mAP due to the overfitting problem. However, the large number of parameters in YOLOv5l makes its size much larger than models with shallower and narrower networks. Additionally, the FLOPs value, which represents the magnitude of computation, is also very high in YOLOv5l compared to smaller models, which limits the detection speed of the model. Therefore, the proposed model should be implemented based on the YOLOv5l to reduce its size and FLOPs without compromising its detection effectiveness.

The improved YOLOv5 was developed using Python and based on the original YOLOv5l architecture. The model was trained and tested on Ubuntu 18.04 LTS using an Intel i7-10700 processor, an NVIDIA GeForce RTX 3090, CUDA Version 11.7, PyTorch version 1.8.0, and Python version 3.8. The training process consisted of 700 epochs with a batch size of 64 and an SGD optimizer. The training results are shown in Figure 13. The proposed model achieved its best performance in epoch 473, with a precision of 76.2%, a recall of 77.6%, and an mAP of 82%.

Furthermore, the model was tested on a previously unseen test dataset. The results for precision, recall, and mAP at different confidence levels are shown in Figure 14, with the best mAP of 85.2% achieved at a class ST of 84.8% and a class AT of 85.6%.

Following the training of the proposed model, a comparison was made between its testing results and those of the YOLOv5l model, as detailed in Table 2. The proposed model showed improved performance over YOLOv5l, with higher precision, recall, and mAP scores. Specifically, the mAP of all classes in the proposed model was found to be 3.2% higher than that of YOLOv5l. The ST class of the proposed model showed a 3.7% increase in mAP compared to YOLOv5l, while the AT class of the proposed model showed a 2.7% increase. Moreover, the proposed model had only 60% of the parameters and size of YOLOv5l. The number of FLOPs in the proposed model was only 55% that in YOLOv5l. Furthermore, despite our model having more layers than YOLOv5l, which could limit its performance, the speed of our model on GPU and CPU devices was higher than that of YOLOv5l, and was close to that of YOLOv5m.

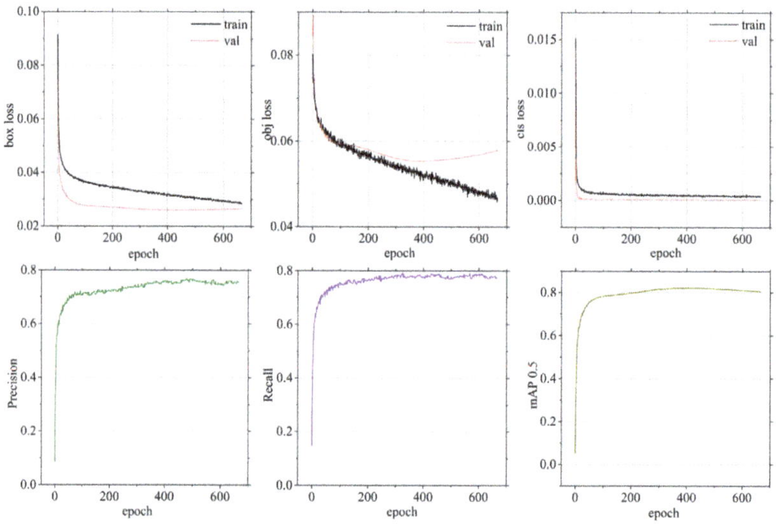

Figure 13. Training results of the proposed model.

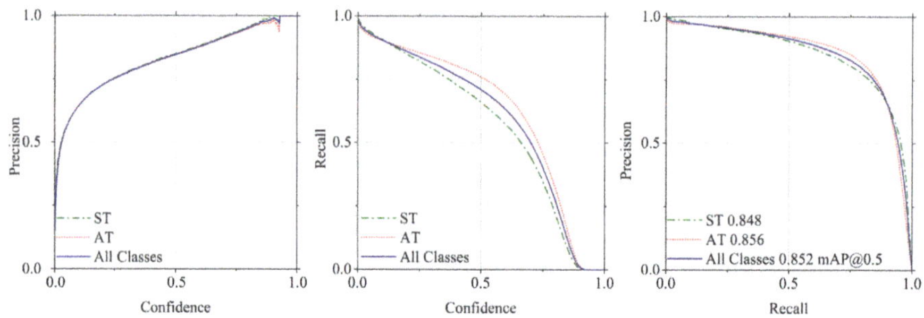

Figure 14. Testing results of the proposed model.

Table 2. Comparative experiments between improved and original models.

Model	Class	P (%)	R (%)	mAP (%)	Params (M)	FLOPs (G)	Speed (FPS) GPU	Speed (FPS) CPU	Size (M)	Layers
YOLOv5l	ST	76.3	76.3	81.1	46.11	107.7	82.64	3.8	184.44	367
	AT	76.6	80.7	82.9						
	All	76.5	78.5	82						
Ours	ST	78.4	78.7	84.8	29.25	59.8	87.71	6.88	112.05	624
	AT	78	83.5	85.6						
	All	78.2	81.1	85.2						

Some sample detections performed using both the original and improved models are shown in Figure 15. A yellow circle indicates that the detector missed this target, while a red circle indicates that the detector detected this non-target. It is shown that both the proposed and original methods rarely detected non-targets, and the improved model could detect more small targets in the same scene, demonstrating higher recall.

Figure 15. Comparison of detection results on images between YOLOv5l and the proposed model.

3.1.3. Ablation Study

An ablation study was conducted to investigate the contribution of individual components or features to the overall performance of the improved model. By systematically removing or disabling different parts of the model and evaluating the resulting performance, insights into the relative importance of each component can be gained, including CAM, BiFPN and GhostConv.

The results of the ablation study are presented and compared in Table 3. The Coordinate Attention Mechanism was found to improve the mAP of the model without significantly increasing the number of parameters and FLOPs. The BiFPN improved the mAP of the model by utilizing multi-scale feature fusion, but it also increased the number of layers and FLOPs. GhostConv reduced the number of parameters and model size to 63% that of the previous version, as well as reducing the number of FLOPs to 55% that of the original. Despite the increase in the number of layers due to DWConv in the GhostConv block, it improved the speed of the model when running on both GPU and CPU devices.

Table 3. Comparison of ablation study results.

Model	mAP (%)	Params (M)	FLOPs (G)	Speed (FPS) GPU	Speed (FPS) CPU	Size (M)	Layers
YOLOv5l	82	46.11	107.7	82.64	3.8	184.44	367
YOLOv5l + CAM	84	46.13	107.7	81.3	3.77	184.52	377
YOLOv5l + BiFPN	83	46.37	108.5	81.3	3.72	185.48	367
YOLOv5l + CAM + BiFPN	84.4	46.4	108.6	77.5	3.29	185.6	377
YOLOv5l + CAM + BiFPN + GhostConv	85.2	29.25	59.8	87.71	6.88	112.05	624

3.1.4. Experimental Comparison with Different Detection Methods

The improved model presented in this paper is compared with other state-of-the-art detection models in Table 4. The results show that the proposed algorithm performed extremely well in terms of detection speed, model size, and computational amount while ensuring the highest mAP. The SSD model, which uses Vgg16 as the backbone network, had a fast detection speed, low computational complexity, and small model size, but low precision. The Faster RCNN model, which uses Resnet50 as the backbone network, had higher accuracy, but the largest model size and computational amount, and the slowest detection speed. The YOLO series models, including YOLOv3, YOLOv4, YOLOv5, YOLOv6, YOLO7, and YOLOv8l, performed worse than the proposed improved model in terms of mAP, model size, and detection speed. The YOLOv8m and YOLOX models have a model size similar to the proposed algorithm, and these two models performed well in GPU detection speed, both above 90 FPS, but their performance on CPU devices was worse, and their precision scores were lower than that of the proposed model. Overall, the proposed improved object detection model for tea buds had the best performance among those state-of-the-art detection models.

Table 4. Comparison between different object detection methods.

Model	mAP (%)	Params (M)	FLOPs (G)	Speed (FPS) GPU	Speed (FPS) CPU	Size (M)
Vgg16-SSD	67.18	26.15	62.8	84.26	4.82	104.6
Resnet50-Faster-RCNN	80.05	137.07	370.41	12.09	1.03	548.28
YOLOv3-SPP	82.4	62.55	155.4	76.92	3.08	250.2
CSP-YOLOv4-Mish	80.1	126.67	177.7	49.75	1.91	242.67
YOLOv5	82	46.11	107.7	78.74	2.77	184.44
YOLOR	74.6	52.49	119.3	73.53	2.89	100.61
YOLOv6	79.2	59.54	150.5	44.82	3.43	114.19
YOLOv7	76.6	36.48	103.2	75.19	2.63	71.35
YOLOv8m	81.8	25.84	78.7	93	5.66	103.36
YOLOv8l	81.1	43.61	164.8	76.34	3.05	174.44
YOLOX	82.3	23.27	62.1	95.23	5.35	93.08
Ours	85.2	29.25	59.8	87.71	6.88	112.05

3.2. *Experiments on OPVSM*

3.2.1. Preparation

In this paper, the target was set as BL1 (one bud and one leaf). A total of 50 pointclouds including the BL1 target were acquired from the field using a depth camera. As shown in Figure 16, according to the method in Section 2.2.1, each pointcloud was detected using the improved model, filtered by a passthrough filter and a radius filter, clustered using the DBSCAN algorithm, and finally estimated via the OPVSM to build the coordinate system of tea buds.

Figure 16. The tea bud pose estimation process.

3.2.2. Results

All pointclouds collected in the field for testing were processed, and the results of the model construction were examined and analyzed. Some of the test results are shown in Figure 17.

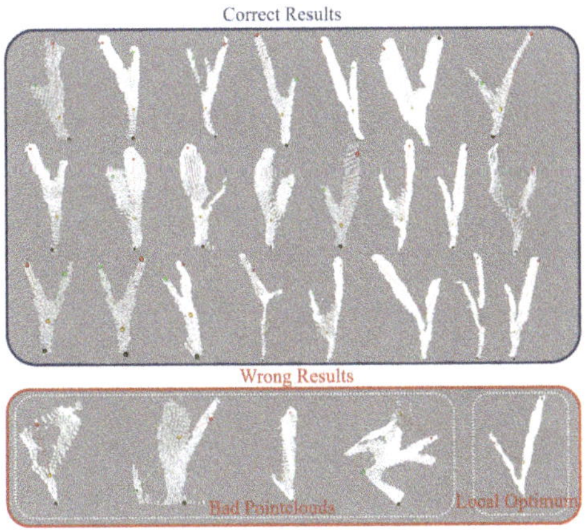

Figure 17. Samples of algorithm-processed images.

After testing, the OPVSM successfully fitted the actual poses of 45 tea buds out of the total samples. The accuracy of this method was around 90%. The wrong results were primarily due to the bad pointclouds and local optima. As shown in Figure 17, the imprecise bounding boxes and measurement instability of the depth camera led to the acquisition of bad pointclouds. Additionally, the search process may have been affected by the initial parameters of the algorithm, resulting in local optima.

Next, in order to evaluate the performance of the OPVSM, various metrics were recorded during the process, including the pointcloud size, number of iterations, and processing time. To increase the size of test samples, each pointcloud was downsampled multiple times, resulting in a total of 619 test samples, as shown in Figure 18.

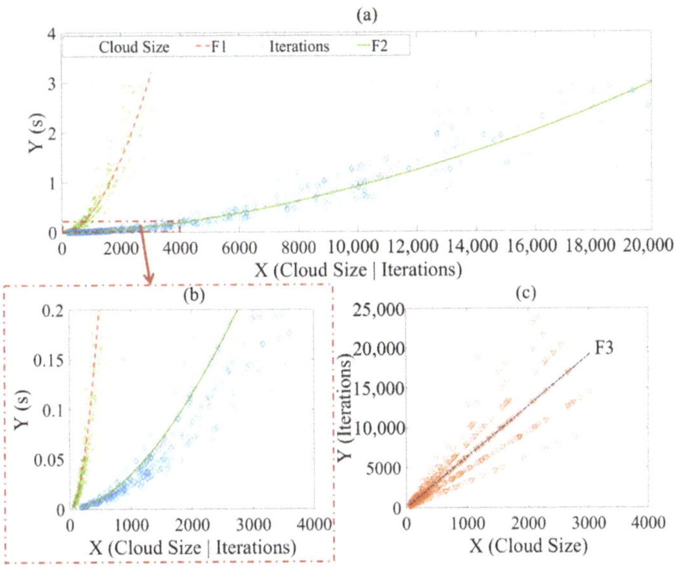

Figure 18. The performance plots of the proposed algorithm. (**a**) The impact graph of cloud size and iteration times on processing time; (**b**) The impact graph of cloud size and iteration times on processing time within 0.2s; (**c**) The impact graph of cloud size on iteration times.

As the pointcloud size and number of iterations increase, the processing time also increases exponentially. F1 in Figure 18a represents the fitting curve between pointcloud size and processing times, built using the Levenberg–Marquardt method with a function of $f(x) = ax^b$. Similarly, F2 in Figure 18b represents the fitting curve between number of iterations and processing times, built using the same method and function as F1.

Furthermore, Figure 18c shows the proportional relationship between pointcloud size and number of iterations, which could be fitted as a proportional curve (F3). The parameters and evaluation metrics of these fitting curves are shown in Table 5.

Table 5. Parameters and metrics of fitting curves.

Curve	Function	a	b	SSE	R-Square	RMSE
F1	$f(x) = ax^b$	1.71×10^{-6}	1.889	0.8841	0.9992	0.03785
F2	$f(x) = ax^b$	2.45×10^{-7}	1.717	0.7245	0.9994	0.03427
F3	$f(x) = ax$	6.375	\	8,052,000	0.9995	114.1

The primary independent variable is the cloud size. As the cloud size increases, the number of iterations also increases proportionally, while the processing time increases exponentially. The proportional function (F3) was identified between the cloud size and number of iterations, with a coefficient a of 6.375. Although the SSE and RMSE of this fitting curve were large, indicating that the data points in Figure 18c were scattered, the fitting curve had strong explanatory power and a good fitting result due to a high R-square value of 0.9995.

Notably, the two exponent coefficients, b, in F1 and F2 were close to each other. Additionally, the difference between the two coefficients a in F1 and F2 was 6.98 times, which was close to the coefficient a in F3. The time complexity $O(n)$ of this algorithm is was nearly equal to $n^{1.7\sim1.9}$, where n is the size of the pointcloud.

To achieve real-time processing, the process of the proposed algorithm had to be completed within 0.2 s. Thus, in Figure 18b, the plot with processing times of 0 s to 0.2 s was examined. The cloud size should be no larger than 800, and the number of iterations should be no greater than 4000, allowing for a processing time of less than 0.2 s. To reduce the pointcloud size, the voxel filter was used to down-sample the pointcloud to a size of 800. The voxel filter replaced all points within a voxel grid cell with the point closest to the center of the cell, effectively reducing the number of 3D points in the pointcloud while preserving the overall shape of the object being represented. The algorithm results after filtering are shown in Figure 19.

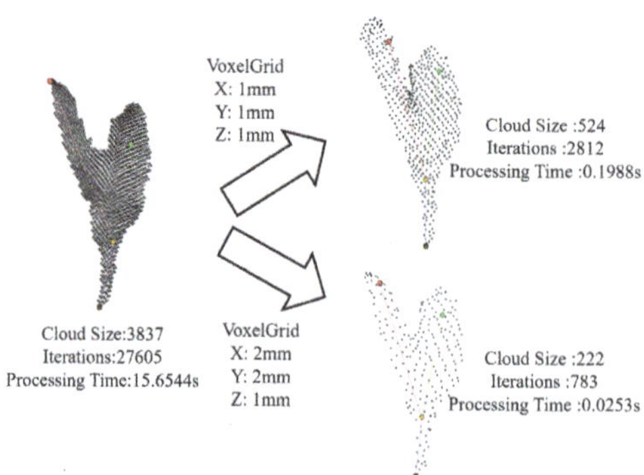

Figure 19. Test results after voxel filtering.

The processing time in Figure 19 includes the voxel filtering process. After voxel filtering with a voxel grid 1 mm × 1 mm × 1 mm in size, the cloud size was reduced to 1/6, the number of iterations was reduced to 1/10, and the processing time was reduced to 1/78, while the pose of the tea buds was correctly built. Furthermore, considering the sparse distribution of the pointcloud along the z axis due to the acquisition principle of the depth camera, we set the voxel grid with a size of 2 mm × 2 mm × 1 mm as the voxel filter, and the cloud size was reduced to 1/17, the number of iterations was reduced to 1/35, the processing time was reduced to 1/618, and the pose of the tea bud was correctly built. This test demonstrated that this method can be applied in real time.

3.3. Discussion

After analyzing the errors in the test dataset, it was found that blurred objects in the image often confused the detector. This was especially true when the camera captured tea buds from a side angle, as the distant buds were also captured in the image. Due to the limited depth of field of the camera, these tea buds were often blurred, making it difficult to manually label all tea buds accurately. As a result, the model was more prone to errors when it encountered these blurred targets. Additionally, the occlusion between the tea buds and illumination (especially overexposure) also led to some mistakes in these detection results. All of these possible factors reduced the precision and recall of this detection model.

There are several ways to address these problems. Firstly, improving the image quality can be achieved by using a better camera or adjusting the illumination conditions to avoid overexposure. Secondly, developing an image processing algorithm combined with a depth camera to segment the distant background and remove blurred targets far away from the camera can be helpful. Additionally, training the model on datasets with images captured from multiple angles may improve its ability to detect objects from different perspectives, thereby reducing the impact of occlusions and improving detection accuracy.

The OPVSM's performance is affected by bad pointclouds and local optima. The imprecise bounding boxes and measurement instability of the depth camera result in incomplete pointclouds for the tea buds. While some of these incomplete pointclouds still retain the shape of the tea buds, others have their shapes destroyed, making it difficult for the algorithm to operate effectively and increasing the likelihood of it getting stuck in local minima. Combining multiple sensors to fill in missing data or preprocessing the pointclouds to classify and remove low-quality pointclouds may help to solve this problem.

Overall, this research work has several limitations that need to be acknowledged. Firstly, the dataset used in our research, although extensively acquired, annotated, and curated over a period of three years, may not fully capture the diversity of tea buds in real-world scenarios. This limitation could potentially impact the generalizability of our results to unseen data or different settings. To overcome this limitation, future studies could focus on incorporating larger and more diverse datasets to enhance the robustness and applicability of the proposed methods. Secondly, while OPSVM is based on breadth-first search and does not require learning from big data, it is important to validate the method on a larger and more diverse test set to enhance the generalizability of our results. This validation will be conducted in future studies. Furthermore, the proposed methodology and algorithms are subject to computational constraints, and the scalability and computational efficiency of the model may pose challenges when applied in real time or in resource-constrained environments. Addressing these limitations could involve exploring optimization techniques or alternative architectures that improve efficiency without compromising performance. Lastly, this study primarily focused on tea bud detection in the field as a specific application domain. The effectiveness of the proposed approach in other locations, such as laboratory or greenhouse settings, or for detecting other small targets, remains an open question. Further investigations and experiments are necessary to evaluate the generalizability of our findings and determine the suitability of the proposed approach in diverse contexts.

4. Conclusions

In conclusion, an algorithm for detecting and estimating the pose of tea buds using a depth camera was proposed in this paper. The improved YOLOv5l model with CAM, BiFPN and GhostConv components demonstrated superior performance in detecting tea buds, achieving a higher mAP of 85.2%, a faster speed of 87.71 FPS on GPU devices and 6.88 FPS on CPU devices, a lower parameter value of 29.25 M, and a lower FLOP value of 59.8 G compared with other models. Moreover, the datasets were collected under different conditions and were augmented to enhance the model's generalization ability under complex scenes. The OPVSM achieved results with 90% accuracy during testing. It builds a graph model that fits the tea buds by iteratively searching for the best vertices from the point cloud guided by a loss function. The algorithm gradually improves the fitness of the model until it reaches a local minimum. The optimal vertex set V acquired could build a graph model that accurately represents the pose of the tea bud. This approach is adaptive to variations in size, color, and shape features. The algorithm's time complexity $O(n)$ is $n^{1.7 \sim 1.9}$, and it can be completed within 0.2 s by using voxel filtering to compress the pointcloud to around 800.

The combination of the proposed detection model and the OPVSM resulted in a reliable and efficient approach for tea bud detection and pose estimation. This study has the potential to be used in tea picking automation and can be extended to other crops and objects for precision agriculture and robotic applications. Future work will aim to improve the model performance and inference speed via data cleaning and expansion, model pruning and compression, and other methods, and parallel processing will be used to accelerate the OPVSM.

Author Contributions: Conceptualization, Z.C., J.C., Y.L., Z.G. and T.Y.; methodology, Z.C.; software, Z.C.; validation, Z.C., Z.G. and T.Y.; formal analysis, Z.C.; investigation, Z.C., Z.G. and T.Y.; resources, Z.C., J.C. and Y.L.; data curation, Z.C. and J.C.; writing—original draft preparation, Z.C.; writing—review and editing, Z.C. and Y.L.; visualization, Z.C.; supervision, J.C.; project administration, J.C.; funding acquisition, J.C. All authors have read and agreed to the published version of the manuscript.

Funding: This research was funded by "Leading Goose" R&D Program of Zhejiang (grant number: 2022C02052) and the National Natural Science Foundation of China (Grant No. 51975537).

Institutional Review Board Statement: Not applicable.

Informed Consent Statement: Not applicable.

Data Availability Statement: The data presented in this study are available on request from the corresponding author. The data are not publicly available due to confidentiality restrictions.

Acknowledgments: The authors deeply appreciate the Tea Research Institute, Chinese Academy of Agricultural Sciences for their assistance in conducting field experiments.

Conflicts of Interest: The authors declare no conflict of interest.

References

1. Chen, J.; Chen, Y.; Jin, X.; Che, J.; Gao, F.; Li, N. *Research on a Parallel Robot for Tea Flushes Plucking*; Atlantis Press: Amsterdam, The Netherlands, 2015; pp. 22–26.
2. Yang, H.; Chen, L.; Ma, Z.; Chen, M.; Zhong, Y.; Deng, F.; Li, M. Computer Vision-Based High-Quality Tea Automatic Plucking Robot Using Delta Parallel Manipulator. *Comput. Electron. Agric.* **2021**, *181*, 105946. [CrossRef]
3. Zhang, H.; Chen, Y.; Wang, W.; Zhang, G. Positioning Method for Tea Picking Using Active Computer Vision. *Trans. Chin. Soc. Agric. Mach.* **2014**, *45*, 61–65+78.
4. Zhang, L.; Zou, L.; Wu, C.; Jia, J.; Chen, J. Method of Famous Tea Sprout Identification and Segmentation Based on Improved Watershed Algorithm. *Comput. Electron. Agric.* **2021**, *184*, 106108. [CrossRef]
5. Wu, X.; Zhang, F.; Lv, J. Research on Recognition of Tea Tender Leaf Based on Image Color Information. *J. Tea Sci.* **2013**, *33*, 584–589. [CrossRef]
6. Lyu, S.; Chang, M.-C.; Du, D.; Li, W.; Wei, Y.; Coco, M.D.; Carcagnì, P.; Schumann, A.; Munjal, B.; Dang, D.-Q.-T.; et al. UA-DETRAC 2018: Report of AVSS2018 & IWT4S Challenge on Advanced Traffic Monitoring. In Proceedings of the 2018 15th IEEE International Conference on Advanced Video and Signal Based Surveillance (AVSS), Auckland, New Zealand, 27–30 November 2018; pp. 1–6.
7. Zhang, J.; Huang, M.; Jin, X.; Li, X. A Real-Time Chinese Traffic Sign Detection Algorithm Based on Modified YOLOv2. *Algorithms* **2017**, *10*, 127. [CrossRef]
8. Zhang, S.; Benenson, R.; Schiele, B. CityPersons: A Diverse Dataset for Pedestrian Detection. *arXiv* **2017**, arXiv:1702.05693.

9. Cui, M.; Lou, Y.; Ge, Y.; Wang, K. LES-YOLO: A Lightweight Pinecone Detection Algorithm Based on Improved YOLOv4-Tiny Network. *Comput. Electron. Agric.* **2023**, *205*, 107613. [CrossRef]
10. Zeng, T.; Li, S.; Song, Q.; Zhong, F.; Wei, X. Lightweight Tomato Real-Time Detection Method Based on Improved YOLO and Mobile Deployment. *Comput. Electron. Agric.* **2023**, *205*, 107625. [CrossRef]
11. Ma, J.; Lu, A.; Chen, C.; Ma, X.; Ma, Q. YOLOv5-Lotus an Efficient Object Detection Method for Lotus Seedpod in a Natural Environment. *Comput. Electron. Agric.* **2023**, *206*, 107635. [CrossRef]
12. Wang, D. Channel Pruned YOLO V5s-Based Deep Learning Approach for Rapid and Accurate Apple Fruitlet Detection before Fruit Thinning. *Biosyst. Eng.* **2021**, *210*, 271–281. [CrossRef]
13. Sozzi, M.; Cantalamessa, S.; Cogato, A.; Kayad, A.; Marinello, F. Automatic Bunch Detection in White Grape Varieties Using YOLOv3, YOLOv4, and YOLOv5 Deep Learning Algorithms. *Agronomy* **2022**, *12*, 319. [CrossRef]
14. Cardellicchio, A.; Solimani, F.; Dimauro, G.; Petrozza, A.; Summerer, S.; Cellini, F.; Renò, V. Detection of Tomato Plant Phenotyping Traits Using YOLOv5-Based Single Stage Detectors. *Comput. Electron. Agric.* **2023**, *207*, 107757. [CrossRef]
15. Murthi, M.; Thangavel, S.K. *A Semi-Automated System for Smart Harvesting of Tea Leaves*; IEEE: Banglore, India, 2017; pp. 1–10.
16. Chen, B.; Yan, J.; Wang, K. Fresh Tea Sprouts Detection via Image Enhancement and Fusion SSD. *J. Control Sci. Eng.* **2021**, *2021*, 6614672. [CrossRef]
17. Xu, W.; Zhao, L.; Li, J.; Shang, S.; Ding, X.; Wang, T. Detection and Classification of Tea Buds Based on Deep Learning. *Comput. Electron. Agric.* **2022**, *192*, 106547. [CrossRef]
18. Gui, Z.; Chen, J.; Li, Y.; Chen, Z.; Wu, C.; Dong, C. A Lightweight Tea Bud Detection Model Based on Yolov5. *Comput. Electron. Agric.* **2023**, *205*, 107636. [CrossRef]
19. Hu, G.; Li, S.; Wan, M.; Bao, W. Semantic Segmentation of Tea Geometrid in Natural Scene Images Using Discriminative Pyramid Network. *Appl. Soft Comput.* **2021**, *113*, 107984. [CrossRef]
20. Qian, C.; Li, M.; Ren, Y. Tea Sprouts Segmentation via Improved Deep Convolutional Encoder-Decoder Network. *IEICE Trans. Inf. Syst.* **2020**, *103*, 476–479. [CrossRef]
21. Chen, Y.-T.; Chen, S.-F. Localizing Plucking Points of Tea Leaves Using Deep Convolutional Neural Networks. *Comput. Electron. Agric.* **2020**, *171*, 105298. [CrossRef]
22. Wang, T.; Zhang, K.; Zhang, W.; Wang, R.; Wan, S.; Rao, Y.; Jiang, Z.; Gu, L. Tea Picking Point Detection and Location Based on Mask-RCNN. *Inf. Process. Agric.* **2021**, *10*, 267–275. [CrossRef]
23. Li, Y.; He, L.; Jia, J.; Lv, J.; Chen, J.; Qiao, X.; Wu, C. In-Field Tea Shoot Detection and 3D Localization Using an RGB-D Camera. *Comput. Electron. Agric.* **2021**, *185*, 106149. [CrossRef]
24. Li, Y.; Wu, S.; He, L.; Tong, J.; Zhao, R.; Jia, J.; Chen, J.; Wu, C. Development and Field Evaluation of a Robotic Harvesting System for Plucking High-Quality Tea. *Comput. Electron. Agric.* **2023**, *206*, 107659. [CrossRef]
25. Chen, C.; Lu, J.; Zhou, M.; Yi, J.; Liao, M.; Gao, Z. A YOLOv3-Based Computer Vision System for Identification of Tea Buds and the Picking Point. *Comput. Electron. Agric.* **2022**, *198*, 107116. [CrossRef]
26. Han, K.; Wang, Y.; Tian, Q.; Guo, J.; Xu, C.; Xu, C. GhostNet: More Features from Cheap Operations. *arXiv* **2020**, arXiv:1911.11907.
27. Hou, Q.; Zhou, D.; Feng, J. Coordinate Attention for Efficient Mobile Network Design. *arXiv* **2021**, arXiv:2103.02907.
28. Tan, M.; Pang, R.; Le, Q.V. EfficientDet: Scalable and Efficient Object Detection. In Proceedings of the 2020 IEEE/CVF Conference on Computer Vision and Pattern Recognition (CVPR), Seattle, WA, USA, 13–19 June 2020; pp. 10778–10787.
29. Jocher, G.; Chaurasia, A.; Stoken, A.; Borovec, J.; NanoCode012; Kwon, Y.; Michael, K.; Xie, T.; Fang, J.; Imyhxy; et al. Ultralytics/Yolov5: V7.0—YOLOv5 SOTA Realtime Instance Segmentation. *Zenodo* **2022**. [CrossRef]
30. Ester, M.; Kriegel, H.-P.; Xu, X. A Density-Based Algorithm for Discovering Clusters in Large Spatial Databases with Noise. *Kdd-96 Proceeding*. 1996, pp. 226–231. Available online: https://cdn.aaai.org/KDD/1996/KDD96-037.pdf (accessed on 5 June 2023).
31. Shorten, C.; Khoshgoftaar, T.M. A Survey on Image Data Augmentation for Deep Learning. *J. Big Data* **2019**, *6*, 60. [CrossRef]
32. Bochkovskiy, A.; Wang, C.; Liao, H.M. YOLOv4: Optimal Speed and Accuracy of Object Detection. *arXiv* **2020**, arXiv:2004.10934.

agriculture

MDPI

Article

Effect of Traditional Spices on the Quality and Antioxidant Potential of Paneer Prepared from Buffalo Milk

Tahir Mahmood Qureshi [1], Muhammad Nadeem [2,*], Javeria Iftikhar [2], Salim-ur-Rehman [3], Sobhy M. Ibrahim [4], Faizan Majeed [5,6] and Muhammad Sultan [6,*]

1 Department of Food Sciences, Cholistan University of Veterinary & Animal Sciences, Bahawalpur 63100, Pakistan
2 Institute of Food Science and Nutrition, University of Sargodha, Sargodha 40100, Pakistan
3 Department of Food Science and Technology, Faisalabad Campus, Riphah International University, Faisalabad 38000, Pakistan
4 Department of Biochemistry, College of Science, King Saud University, P.O. Box 2455, Riyadh 11451, Saudi Arabia
5 Department of Agricultural and Biosystems Engineering, University of Kassel, 37213 Witzenhausen, Germany
6 Department of Agricultural Engineering, Bahauddin Zakariya University, Multan 60800, Pakistan
* Correspondence: nadeem.abdul@uos.edu.pk (M.N.); muhammadsultan@bzu.edu.pk (M.S.)

Abstract: This study aims to evaluate the effect of different spices (black pepper, cumin, clove, nigella, cardamom, and cinnamon) on the physicochemical characteristics and microbial quality, as well as antioxidant potential, of paneer during storage. Different types of spices were incorporated into the paneer at different levels (0.2 and 0.3%). In addition to paneer, the antioxidant potential of spices was also investigated. The results concerning total plate counts (TPC) or yeast and molds (Y & M) (log_{10} CFU/g) of all treatments were substantially ($p < 0.05$) increased during storage. Generally, all freshly prepared spicy paneer and control had higher sensory scores for all the sensory characteristics which declined during subsequent storage. All the paneer samples having 0.3% spices showed very slight variations (nonsignificant) in sensory score of all the attributes of their relative samples containing 0.2% spice. The incorporation of spices into the paneer matrix also showed promising results concerning all the above-mentioned attributes revealing antioxidant potential. There was significant ($p < 0.05$) effect of treatments and storage days on antioxidant potential of paneer. The freshly prepared control paneer (P_0) showed the lowest total phenol (TP), total flavonoids (TF), 2, 2-diphenyl-1-picrylhydrazyl (DPPH), reducing power (RP), and total antioxidant capacity (TAC) values compared to all the spicy paneer (treatments). The freshly prepared control paneer (P_0) showed the lowest antioxidant potential compared to all the spicy paneer (treatments). The maximum antioxidant potential was observed in the paneer having 0.3% clove (P_6). All the spicy paneer showed increasing trend of all the attributes, showing antioxidant potential up to 6 days of storage, but afterwards, the activities were slightly decreased. It may be concluded that spicy paneer would be considered as a functional dairy product with enhanced sensory and antioxidant properties, and shelf stability.

Keywords: spices; paneer; storage; proximate; composition; sensory evaluation; antioxidant potential

Citation: Qureshi, T.M.; Nadeem, M.; Iftikhar, J.; Salim-ur-Rehman; Ibrahim, S.M.; Majeed, F.; Sultan, M. Effect of Traditional Spices on the Quality and Antioxidant Potential of Paneer Prepared from Buffalo Milk. *Agriculture* **2023**, *13*, 491. https://doi.org/10.3390/agriculture13020491

Academic Editor: Massimo Malacarne

Received: 17 January 2023
Revised: 7 February 2023
Accepted: 15 February 2023
Published: 19 February 2023

1. Introduction

Among the dairy products, paneer is considered a conventional soft cheese variety, produced by heat and acid coagulation of the milk [1]. It is used in the preparation of many sweet products such as rasogolla, rasamalai, and sandesh, and it is also used in culinary dishes and snacks, thereby deemed as a rich source of fat, minerals, vitamins, and high-quality proteins.

Spices are reported to have been used for culinary and medicinal purposes for centuries. In addition to improving the taste and color of food as well as beverages, they also offer protection from both acute and chronic diseases [2]. It is obvious that regular use of spicy foods is associated with a minimal risk of death from ischemic heart diseases and cancer [2]. The health aspects ascribed to spices include antimicrobial, antioxidant, anti-inflammatory, anti-Type 2 diabetes, etc. [3,4].

Among the spices, black pepper is a popular spice and is locally identified as "kali mirch". Owning to its carminative property, immune enhancer ability, and antimicrobial activity, it is known as "Black Gold" or "The King of Spices" [5]. Gülçin et al. [6] found strong antioxidant and radical scavenging action of the extracts of black pepper seeds due to presence of phenolic compounds. According to the findings of [7], adding black pepper and piperine to a diet can lessen the oxidative stress that a high-fat diet causes in the cells of rats.

Clove (*Syzygium aromaticum* L.) is locally known as "long" and is reported to have strong antioxidant activity [8,9] owning to existence of tocopherol, ascorbates, and phenolic compounds. Moreover, clove oil contains eugenol followed by eugenyl acetate, β-caryophyllene, gallic acid, caryophyllene, and α-humulene [10–12], and eugenol is considered to be the main bioactive compound of clove oil [13]. Cumin (*Cuminum cyminum* L.) is locally known as "zeera" and is widely used in cooking due to its unique fragrance [14].

The *Nigella sativa* (locally called kalonji) is an annual herb whose essential oil could be used as natural antifungal agent in soft cheese [15]. Moreover, *N. sativa* oil is also reported to have marked antioxidant activity and anti-inflammatory properties due to presence of a high level of thymoquinone [16].

Cardamom (*Elettaria cardamomum*) is locally known as "ilaichi". It comes from a perennial herbaceous plant and is reported to contain predominant compounds including α-terpineol, α-terpinyl acetate, 1,8-cineole, β-linalool, and sabinene [17,18]. Owing to the existence of such compounds in crucial oil of cardamom, it is reported to have antioxidant potential [17,19,20].

The bark of cinnamon, locally known as "darchini", is typically employed as a flavoring agent and a spice. Many studies confirmed that cinnamon contains many properties including antidiabetic, antioxidant, antimicrobial, and anticancer, and inhibits cardiovascular diseases [21].

Since the extracts from spices are reported to have antioxidant activities [8,16,20,22–24], it would be of great interest for the food industries to incorporate such kinds of spices (powders or their extracts) into their products for their value addition.

Consumption of different spices for different dishes is a well-established trend in our society. Therefore, to induce the trend of paneer consumption, it would be enticing to incorporate spices in the matrix of paneer. In addition, it would be a good idea to use products having negligible quantities of fat as the people with hypertensive conditions may consume the aforementioned type of products. In the present study, skimmed milk was used for the preparation of paneer, which would be good for the people with hypertensive conditions.

It may be anticipated that the consumption of paneer would be increased after making it somewhat spicy. Moreover, owing to the presence of health aspects of spices as well as paneer, the spicy paneer would have more beneficial effects regarding health. Different previous studies focused on antioxidant activity of whey cheese [25], burfi (milk-based confection) [26], and paneer [27] through incorporation or application of black cumin (*Nigella damascena* L.), clove bud essential oil, and clove essential oil, respectively. It is also reported that spices have been incorporated in different dairy products such as cheese, butter, ghee, and ice-cream [28], but so far, no comprehensive studies have been carried out concerning monitoring of antioxidant potential of paneer through incorporation of powders from different traditionally used spices at different concentrations. Therefore, this study aimed to monitor the physicochemical and microbiological quality and antioxidant potential of spicy paneer during storage. Moreover, sensory evaluation was also performed to check the suitability of different spicy paneer.

2. Materials and Methods

2.1. Procurement of Raw Material

Buffalo milk used for paneer manufacturing was taken from a local farm at Sargodha city (Pakistan), and spices, i.e., black pepper, cumin, clove, nigella, cardamom, and cinnamon, were acquired from the city's local market. All the spices were ground separately to make powders. Fresh buffalo skimmed (9.34% solids not fat (SNF), 0.5% fat) milk was used for the manufacturing of paneer. Aluminum foil, plastic sheet, and muslin cloth were purchased from a local store in Sargodha.

2.2. Production of Spicy Paneer

Paneer was produced from fresh raw skimmed buffalo milk as suggested by Khan et al. [29], with a few variations. First of all, pH of milk was computed employing a pH meter as suggested by Ardö [30], whereas fat was explored by the Gerber–van Gulik method of Ardö and Polychroniadou [30]. For the preparation of one batch of paneer, 40 L of skimmed buffalo milk was coagulated using diluted lemon juice at 75 °C after its pasteurization (82 °C for 5 min). The whey was drained off using a muslin cloth. The coagulum acquired after whey drainage was firstly pressed for 5 min. The paneer matrix obtained was crushed and made into grains. The whole paneer matrix was separated into 13 equal parts and spices were added in different ratios (see treatment plan, Table 1).

Table 1. Treatment plan of the present study.

Treatments	Spices (% by Weight of Expected Yield) Addition into Paneer Matrix
P_0	Control, without spices
P_1	Black pepper powder (0.2%)
P_2	Black pepper powder (0.3%)
P_3	Cumin powder (0.2%)
P_4	Cumin powder (0.3%)
P_5	Clove powder (0.2%)
P_6	Clove powder (0.3%)
P_7	Nigella powder (0.2%)
P_8	Nigella powder (0.3%)
P_9	Cardamom powder (0.2%)
P_{10}	Cardamom powder (0.3%)
P_{11}	Cinnamon powder (0.2%)
P_{12}	Cinnamon powder (0.3%)

Then, the paneer of each treatment was separately pressed for 60 min. The manufacturing of paneer was carried out separately in the same manner three times. In our preliminary trials, spices were added into the paneer matrix at different ratios, but, finally, two levels were selected, i.e., 0.2% and 0.3%, based on sensory perception. By increasing the quantities of spices into paneer, irritating taste was being developed as assessed by the panel of assessors.

2.3. Sampling of Paneer

The paneer from each treatment was cut into four identical pieces and stored for 3, 6, and 9 days at refrigerated temperature. The remaining one piece was frozen instantly (0 day) till evaluated. All paneer's pieces were enveloped in polyethylene films with an outer covering of aluminum foil. In our preliminary trials, the spicy paneer samples were also stored for up to 12 days, but their outer appearance and color were not liked by the panel of assessors due to slight mold growth over the surfaces of paneer. Therefore, paneer samples were stored for only up to 9 days for all kind of analysis.

2.4. Preparation of Water Soluble Extract (WSE) of Spicy Paneer and Spices

The WSEs of paneer were prepared according to the procedure described by [31] with a few changes. Briefly, 20 g of paneer was included into 60 mL of distilled water and stirred for one hour at room temperature. The pH was adjusted at 4.6 and centrifuged at 10 kg for 10 min at 4 °C. The supernatant was designated as WSE and filtered further via Whatman filter paper. These were immediately frozen at −20 °C. The WSEs of spices were also prepared. Approximately 0.01 kg powder of each spice was added into 200 mL distilled water and mixed for 180 min at room temperature. Then, the mixture was centrifuged at 10 kg for 10 min at 4 °C. The mixture was filtered, and the filtrate was immediately frozen at −20 °C.

2.5. Physicochemical Characteristics of Spicy Paneer

The pH of paneer was examined according to the method of Ardö and Polychroniadou [30] using a pH meter, whereas the acidity of paneer samples was explored by AOAC method [32]. The IDF standard 4/ISO 5534 was employed for determining the paneer moisture content (MC) [33]. The paneer fat contents were computed by the method of Ardö and Polychroniadou [30]. Total nitrogen (%) was computed by the well-known IDF standard 20B [34]. Additionally, protein contents (%) were computed by multiplying total nitrogen TN (%) by 6.38.

2.6. Microbiological Analysis of Spicy Paneer

Total plate count (TPC, log CFU/g of paneer) and yeast and mold (Y & M, log CFU/g of paneer) counts were carried out as described by [35]. Out of each treatment, 10 g of crushed paneer sample was homogenized in a domestic blender for 5 min by inserting 90 mL of sterilized sodium citrate (2%, pH 7.5) water. Different dilutions of the above suspension were made up to10^{-4} and then 1 mL of diluted samples were plated on plate count agar media (Titan Biotech Ltd., Rajasthan, India). The TPC was computed after incubating the plates for 2 days at 37 °C. The counts of Y & M were also carried out in a similar way (after 48 h at 30 °C), employing potato dextrose agar (HiMedia, Mumbai, Maharashtra 400086, India).

2.7. Sensory Evaluation of Spicy Paneer

Sensory evaluated of paneer was performed using a 9-point hedonic scale for every attribute including appearance, flavor, texture, color, and overall acceptability [36]. A group of 15 people, including faculty and students from our institute, were selected for evaluation of the paneer.

2.8. Determination of Total Phenolic Contents (TP)

The TP of paneer were explored using the method given by Reis et al. [37]. A 1 mL aliquot of WSE of both spices and paneer was used in the assay, and gallic acid (in ethanol) was employed as a standard. One mL of Folin–Ciocalteu reagent (10%) was added into each sample. After being vortexed, 2 mL of sodium carbonate (20%) solution was included. The absorbance was evaluated at 760 nm using a spectrophotometer (Halo DB-20, UV–Vis double beam, Dynamica Scientific Ltd., Livingston, UK) after incubating the mixture for 60 min at 30 °C.

2.9. Determination of Total Flavonoid Contents (TF)

The TF were also explored via the spectrophotometric method given by Jia et al. [38]. A 1 mL aliquot of WSE of both spices and paneer was used in the assay, and the results were articulated as mg catechin equivalent (CE) per 100 g of paneer or spices. Approximately 75 µL of sodium nitrite (5%) solution was added into each sample. After vortexing for 1 min, 150 µL of aluminum chloride (10%) solution was also inserted. The absorbance was evaluated at 510 nm, employing the spectrophotometer after inserting 500 µL of 1 M NaOH into the above mixture. All calculations were performed in triplicate.

2.10. Determination of Reducing Power (RP) or Ferricyanide/Prussian Blue Assay

The RP of WSE of paneer was computed via method as given by Reis et al. [37] with slight changes. A 1 mL aliquot of WSE of both spices and paneer was used in the assay, and calculation was performed employing the standard curve of Trolox as mg of Trolox per 100 g of paneer or spices. The sample was mixed into 500 µL sodium phosphate (Na_3PO_4) buffer (0.2 M, pH 6.6) and potassium ferricyanide ($K_3Fe(CN_6)$). After incubation of the mixture for 20 min at 50 °C, 500 µL of trichloroacetic acid (%) was included and vortexed. The mixture was centrifuged (Hermle Labortechnik GmbH Siemensstr-25 D-78564 Wehingen, Germany) at 3000× g for 10 min at 4 °C in order to obtain clear supernatant. Then, 150 µL of ferric chloride (0.1%) was inserted to the supernatant and the absorbance was computed employing spectrophotometer (Halo DB-20, UV–Vis double beam, Dynamica Scientific Ltd., Livingston, UK) at 700 nm.

2.11. Determination of DPPH Radical Scavenging Activity Assay

The potential for WSE of paneer to scavenge 2, 2-diphenyl-1-picrylhydrazyl radical (DPPH) was acquired via the method of Yi et al. [39] with some modifications. A 1 mL aliquot of WSE of both spices and paneer was used in the assay, and DPPH radical scavenging activity was calculated from the Trolox standard (TE). Approximately 2 mL of DPPH (60 µM in absolute ethanol) solution was added into each sample. The mixture was vortexed and then incubated for 30 min at room temperature in the dark. The absorbance was measured at 517 nm, employing the spectrophotometer. All calculations were performed in triplicate.

2.12. Determination of Total Antioxidant Capacity (TAC) Assay

The TAC of paneer was determined employing a similar method to that given by Prieto et al. [40]. A 1 mL aliquot of WSE of both spices and paneer was used in the assay and TAC was calculated using the Trolox standard. Approximately 4 mL of reagent (0.6 M sulfuric acid, 28 mM Na_3PO_4, and 4 mM ammonium molybdate) solution was added into each sample. After incubating the mixture for 95 min at 90 °C, the absorbance was computed at 695 nm using a spectrophotometer. All calculations were assessed in triplicate, while experiments were performed in duplicate.

2.13. Statistical Analysis

Statistical software (Minitab 16, Minitab, LLC, State College, PA, USA) was employed for statistical analysis using two-way ANOVA and Tukey's test for pairwise assessment at a level of $p < 0.05$. The normal distribution of the data was tested by Shapiro–Wilk test, and the normality assumptions were observed to be satisfied.

3. Results

3.1. Physicochemical Characteristics of Spicy Paneer

Data regarding physicochemical characteristics of all the paneer samples (control and spicy paneer) made in the current study are presented in Table 2. The initial pH of all the samples (treatments) was in the range 5.65–5.69 and significantly ($p < 0.05$) decreased during storage. That decrease in pH was slight after subsequent storage period. Such kind of decreasing trend might be due to the activities of contaminated microorganisms resulting in the accumulation of more and more acids during storage. The most significant decreasing trend was observed in the control paneer, reaching 5.32 at the end of the storage period (9 days). The initial MCs of all the treatments were around 58%, but spicy paneer showed slightly lower MCs compared to the control treatment, which might be due to slightly less water residing in the paneer matrix. The MCs of all the paneer were significantly decreased during storage which may have been caused by moisture loss from the paneer matrix during storage. The fat contents of fresh paneer samples were less than 4%, which slightly increased during subsequent storage. There were no great differences in the fat contents among all treatments. The skimmed milk was used for the preparation of paneer in all the treatments; therefore, the paneer showed very low quantities of fat. Similarly, all the freshly prepared paneer (control and spicy) showed around 32% of protein. Even though there was significant effect of treatments and days of protein contents of all the paneer samples, that increase was slight during successive storage period. The increased contents of protein during storage might be due to increase in dry matter during storage. A similar trend was also experienced in the ash contents of all the paneer samples during storage. Although there was significant effect of storage on the protein contents of paneer, that increasing trend was very slight. All the prepared paneer showed ash contents in the range 2.42–3.00%. The maximum fat to dry matter ratio (fat/DM) was observed in freshly prepared control paneer (8.83%) followed by freshly prepared paneer incorporated with cardamom (8.71–8.79%). The increased contents of fat/DM of the control paneer might be due to lower contents of DM compared to other paneer samples having spice powders. The MCs have an inverse relationship with dry matter. Most of the paneer samples showed increasing trend regarding fat/DM. In general, it was observed that all the paneer samples having 0.3% spices showed nonsignificant variations concerning physicochemical characteristics of their relative samples containing 0.2% spice.

3.2. Microbiological Quality of Spicy Paneer

Figure 1 represents the data depicting the microbiological quality of spicy paneer during storage. The counts of microbes in paneer may be affected due to the quality of milk, heat treatment of milk, hygienic practices during handling and manufacturing, and post-manufacture conditions. The TPC values of freshly prepared control treatment were 3.04 ($\log_{10}$ CFU/g), whereas all other freshly prepared spicy paneer depicted slightly lower values (2.78–3.00 $\log_{10}$ CFU/g). The results concerning TPC of all treatments were significantly ($p < 0.05$) increased during storage, which showed that the counts steadily increased after each successive storage. At the end of the storage period, the control treatment showed the maximum TPC values up to 5.58 ($\log_{10}$ CFU/g), whilst most of the treatments showed significant ($p < 0.05$) variations in comparison to the control. All the spicy paneer treatments showed TPC values around 5.30 ($\log_{10}$ CFU/g) at the end of the investigated storage period (9 days). All the paneer samples having 0.3% spices showed very slight variations (nonsignificant) in TPC counts of their relative samples containing 0.2% spice. A similar trend was also seen considering Y & M ($\log_{10}$) counts. The Y & M values ($\log_{10}$) of freshly prepared control treatment showed 2.60 ($\log_{10}$ CFU/g), whereas all other freshly prepared spicy paneer showed slightly lower values (nonsignificant). The values of Y & M were significantly increased during storage in control paneer, reaching up to 3.45 ($\log_{10}$ CFU/g), which was the highest among all other treatments. All the paneer samples having 0.3% spices showed very slight variations (nonsignificant) in Y & M counts of their relative samples containing 0.2% spice. Thus, microbial counts raised with the progress of storage in all treatments having spices, but to a lesser extent as compared to control paneer. It was suspected that fungus might began to grow very slightly on the paneers' surfaces that were stored for 12 days. The taste of 12-days-stored spicy paneer was slightly changed.

Table 2. Proximal characteristics (means ± SD) of paneer supplemented with different spices during storage.

Treatments	Time (days)	pH	Moisture (%)	Fat (%)	Protein (%)	Ash (%)	Fat/DM
Control paneer (P_0)	0	5.67 ± 0.06 a	58.33 ± 1.65 a	3.68 ± 0.56 a	31.03 ± 0.98 f	2.53 ± 0.03 h–j	8.83 ± 0.23 a
	3	5.48 ± 0.05 a–i	56.93 ± 1.33 a–e	3.88 ± 0.43 a	32.34 ± 1.13 a–f	2.70 ± 0.04 c–j	9.02 ± 0.41 a–d
	6	5.37 ± 0.04 g–i	56.96 ± 1.84 a–e	3.88 ± 0.63 a	32.56 ± 1.19 a–f	2.66 ± 0.06 b–j	9.04 ± 0.54 a–c
	9	5.32 ± 0.06 i	55.26 ± 1.36 d–k	3.91 ± 0.71 a	32.93 ± 0.92 a–e	2.84 ± 0.04 b–g	8.75 ± 0.14 b–j
Black pepper (0.2%) paneer (P_1)	0	5.68 ± 0.06 a	56.80 ± 1.52 a–f	3.71 ± 0.59 a	32.43 ± 0.87 a–f	2.83 ± 0.07 b–h	8.58 ± 0.29 a–f
	3	5.56 ± 0.07 a–g	55.80 ± 1.47 b–k	3.80 ± 0.60 a	32.63 ± 0.95 a–f	2.98 ± 0.03 a–c	8.59 ± 0.26 a–i
	6	5.48 ± 0.09 b–i	55.06 ± 1.32 d–k	3.86 ± 0.75 a	33.40 ± 0.84 a–d	2.99 ± 0.04 a–c	8.61 ± 0.49 b–m
	9	5.45 ± 0.07 b–i	54.33 ± 1.21 h–k	3.90 ± 0.64 a	33.76 ± 0.90 ab	3.16 ± 0.05 a	8.54 ± 0.39 e–n
Black pepper (0.3%) paneer (P_2)	0	5.67 ± 0.05 a	56.73 ± 1.89 a–g	3.68 ± 0.69 a	31.96 ± 1.21 b–f	2.78 ± 0.06 b–i	8.51 ± 0.26 a–g
	3	5.47 ± 0.04 b–i	55.63 ± 1.56 c–k	3.76 ± 0.59 a	32.76 ± 0.86 a–f	2.90 ± 0.04 a–f	8.49 ± 0.24 b–k
	6	5.47 ± 0.09 b–i	54.56 ± 1.69 f–k	3.86 ± 0.62 a	33.46 ± 0.81 a–c	2.92 ± 0.05 a–f	8.50 ± 0.19 d–n
	9	5.40 ± 0.04 f–i	54.50 ± 1.59 f–k	3.92 ± 0.81 a	33.90 ± 0.69 a	2.91 ± 0.03 a–c	8.62 ± 0.14 c–n
Cumin (0.2%) paneer (P_3)	0	5.69 ± 0.03 a	56.30 ± 1.48 a–j	2.80 ± 0.66 fg	32.33 ± 0.78 a–f	2.42 ± 0.03 j	6.41 ± 0.10 i–p
	3	5.58 ± 0.07 a–f	55.93 ± 1.55 b–j	2.89 ± 0.79 e–g	32.76 ± 0.95 a–f	2.51 ± 0.05 i	6.56 ± 0.28 i–p
	6	5.51 ± 0.09 a–i	55.23 ± 1.84 d–k	2.92 ± 0.55 d–g	32.33 ± 1.17 a–f	2.55 ± 0.07 ij	6.53 ± 0.23 l–p
	9	5.47 ± 0.06 b–i	54.03 ± 1.57 j–k	2.98 ± 0.68 c–g	32.83 ± 0.85 a–f	2.65 ± 0.08 e–j	6.49 ± 0.41 o–p
Cumin (0.3%) paneer (P_4)	0	5.67 ± 0.05 a	57.06 ± 1.69 a–d	2.78 ± 0.73 fg	31.56 ± 0.88 d–f	2.52 ± 0.09 h–j	6.47 ± 0.47 f–o
	3	5.55 ± 0.09 a–h	55.46 ± 1.39 c–k	2.84 ± 0.83 fg	32.16 ± 0.96 a–f	2.53 ± 0.05 g–j	6.39 ± 0.56 l–p
	6	5.45 ± 0.04 b–i	54.16 ± 1.49 i–k	2.93 ± 0.81 d–g	33.16 ± 0.91 a–e	2.61 ± 0.04 e–j	6.40 ± 0.26 o–p
	9	5.41 ± 0.05 e–i	53.46 ± 1.44 k	2.98 ± 0.79 c–g	32.83 ± 1.16 a–f	2.81 ± 0.03 b–i	6.41 ± 0.37 p
Clove (0.2%) paneer (P_5)	0	5.67 ± 0.04 a	56.60 ± 1.77 a–h	2.77 ± 0.52 g	31.70 ± 1.17 c–f	2.77 ± 0.06 b–i	6.38 ± 0.36 h–p
	3	5.52 ± 0.08 a–h	55.63 ± 1.79 c–k	2.82 ± 0.67 f–g	32.73 ± 1.11 a–f	2.84 ± 0.05 b–g	6.38 ± 0.27 k–p
	6	5.51 ± 0.06 a–i	55.23 ± 1.68 d–k	2.89 ± 0.56 e–g	32.73 ± 0.97 a–f	2.95 ± 0.06 a–e	6.46 ± 0.38 m–p
	9	5.46 ± 0.03 b–i	54.63 ± 1.59 e–k	2.97 ± 0.62 c–g	33.46 ± 0.85 a–c	2.97 ± 0.04 a–d	6.54 ± 0.36 n–p
Clove (0.3%) paneer (P_6)	0	5.68 ± 0.09 a	56.66 ± 1.66 a–h	2.85 ± 0.69 fg	32.40 ± 0.98 a–f	2.80 ± 0.06 b–i	6.58 ± 0.22 f–o
	3	5.47 ± 0.06 b–i	56.10 ± 1.97 a–j	3.03 ± 0.76 b–f	32.46 ± 0.91 a–f	2.94 ± 0.03 a–f	6.90 ± 0.38 g–o
	6	5.52 ± 0.07 a–h	55.53 ± 1.69 c–k	3.15 ± 0.83 b–d	33.40 ± 0.93 a–d	2.96 ± 0.04 a–e	7.09 ± 0.27 h–p
	9	5.36 ± 0.05 h–i	54.53 ± 1.74 f–k	3.19 ± 0.91 bc	33.33 ± 1.19 a–e	3.12 ± 0.07 ab	7.02 ± 0.42 l–p
Nigella (0.2%) paneer (P_7)	0	5.67 ± 0.04 a	56.43 ± 1.59 a–i	2.86 ± 0.63 e–g	32.06 ± 1.15 a–f	2.94 ± 0.07 a–f	6.57 ± 0.51 h–p
	3	5.56 ± 0.09 a–g	55.60 ± 1.56 c–k	3.03 ± 0.74 b–f	32.36 ± 0.89 a–f	3.03 ± 0.08 a–c	6.83 ± 0.15 i–p
	6	5.51 ± 0.04 a–i	55.06 ± 1.48 d–k	3.10 ± 0.85 b–e	32.90 ± 0.87 a–e	3.07 ± 0.04 ab	6.90 ± 0.32 j–p
	9	5.47 ± 0.07 b–i	54.23 ± 1.46 i–k	3.22 ± 0.96 bc	33.46 ± 0.90 a–c	3.08 ± 0.06 ab	7.04 ± 0.55 m–p

Table 2. *Cont.*

Treatments	Time (days)	pH	Moisture (%)	Fat (%)	Protein (%)	Ash (%)	Fat/DM
Nigella (0.3%) paneer (P_8)	0	5.66 ± 0.08 a	56.00 ± 1.38 a–j	2.97 ± 0.91 c–g	32.96 ± 1.15 a–e	2.94 ± 0.08 a–f	6.75 ± 0.23 h–p
	3	5.59 ± 0.05 a–f	55.06 ± 1.11 d–k	3.00 ± 0.59 c–g	33.06 ± 0.96 a–e	3.05 ± 0.09 ab	6.69 ± 0.27 l–p
	6	5.55 ± 0.03 a–h	54.43 ± 1.72 g–k	3.15 ± 0.66 b–d	33.43 ± 0.99 a–c	3.07 ± 0.05 ab	6.92 ± 0.16 m–p
	9	5.48 ± 0.08 b–i	54.13 ± 1.38 i–k	3.26 ± 0.78 b	33.66 ± 0.93 ab	3.17 ± 0.03 a	7.11 ± 0.37 m–p
Cardamom (0.2%) paneer (P_9)	0	5.68 ± 0.04 a	57.80 ± 1.92 a–c	3.71 ± 0.98 a	31.96 ± 1.18 b–f	2.83 ± 0.07 b–h	8.79 ± 0.27 ab
	3	5.52 ± 0.08 a–h	56.80 ± 1.68 a–f	3.80 ± 0.91 a	32.63 ± 1.18 a–f	2.98 ± 0.04 a–c	8.80 ± 0.48 a–e
	6	5.44 ± 0.03 c–i	56.06 ± 1.62 a–j	3.86 ± 0.87 a	33.40 ± 1.17 a–d	3.02 ± 0.06 a–c	8.76 ± 0.15 a–h
	9	5.41 ± 0.06 e–i	55.33 ± 1.83 d–k	3.90 ± 0.68 a	33.76 ± 1.16 ab	3.16 ± 0.08 a	8.72 ± 0.12 b–j
Cardamom (0.3%) paneer (P_{10})	0	5.67 ± 0.05 a	57.73 ± 1.59 a–c	3.68 ± 0.69 a	31.96 ± 0.98 b–f	2.78 ± 0.05 b–i	8.71 ± 0.25 ab
	3	5.43 ± 0.08 c–i	56.63 ± 1.69 a–h	3.76 ± 0.88 a	32.76 ± 0.89 a–f	2.90 ± 0.03 a–f	8.65 ± 0.28 a–f
	6	5.43 ± 0.05 c–i	55.56 ± 1.60 c–k	3.86 ± 0.69 a	33.46 ± 0.86 a–c	2.92 ± 0.06 a–f	8.68 ± 0.16 a–j
	9	5.36 ± 0.09 hi	55.50 ± 1.89 c–k	3.92 ± 0.74 a	33.90 ± 0.97 a	3.04 ± 0.08 a–c	8.82 ± 0.14 a–i
Cinnamon (0.2%) paneer (P_{11})	0	5.68 ± 0.05 a	57.30 ± 1.58 a–d	2.80 ± 0.82 fg	32.33 ± 1.17 a–f	2.42 ± 0.06 j	6.55 ± 0.29 e–n
	3	5.54 ± 0.06 a–h	56.93 ± 1.47 a–e	2.89 ± 0.85 e–g	32.76 ± 1.12 a–f	2.51 ± 0.09 ij	6.72 ± 0.51 e–n
	6	5.47 ± 0.09 b–i	56.23 ± 1.84 a–j	2.92 ± 0.77 d–g	32.33 ± 1.15 a–f	2.55 ± 0.04 ij	6.69 ± 0.31 h–p
	9	5.43 ± 0.05 d–i	55.03 ± 1.42 d–k	2.98 ± 0.71 c–g	32.83 ± 0.93 a–f	2.65 ± 0.07 e–j	6.65 ± 0.13 l–p
Cinnamon (0.3%) paneer (P_{12})	0	5.68 ± 0.08 a	58.06 ± 1.27 ab	2.78 ± 0.73 fg	31.50 ± 0.85 e–f	2.52 ± 0.05 h–j	6.64 ± 0.32 b–l
	3	5.51 ± 0.04 a–i	56.46 ± 1.28 a–i	2.84 ± 0.96 fg	32.16 ± 0.99 a–f	2.53 ± 0.07 g–j	6.53 ± 0.44 h–p
	6	5.41 ± 0.08 f–i	55.16 ± 1.85 d–k	2.93 ± 0.85 d–g	33.16 ± 0.94 a–e	2.63 ± 0.08 f–j	6.54 ± 0.41 l–p
	9	5.37 ± 0.09 g–i	54.46 ± 1.89 f–k	2.98 ± 0.74 c–g	32.83 ± 1.14 a–f	2.81 ± 0.04 b–i	6.55 ± 0.11 n–p

Means with different letters in the same column present significant ($p < 0.05$) differences among treatments and storage.

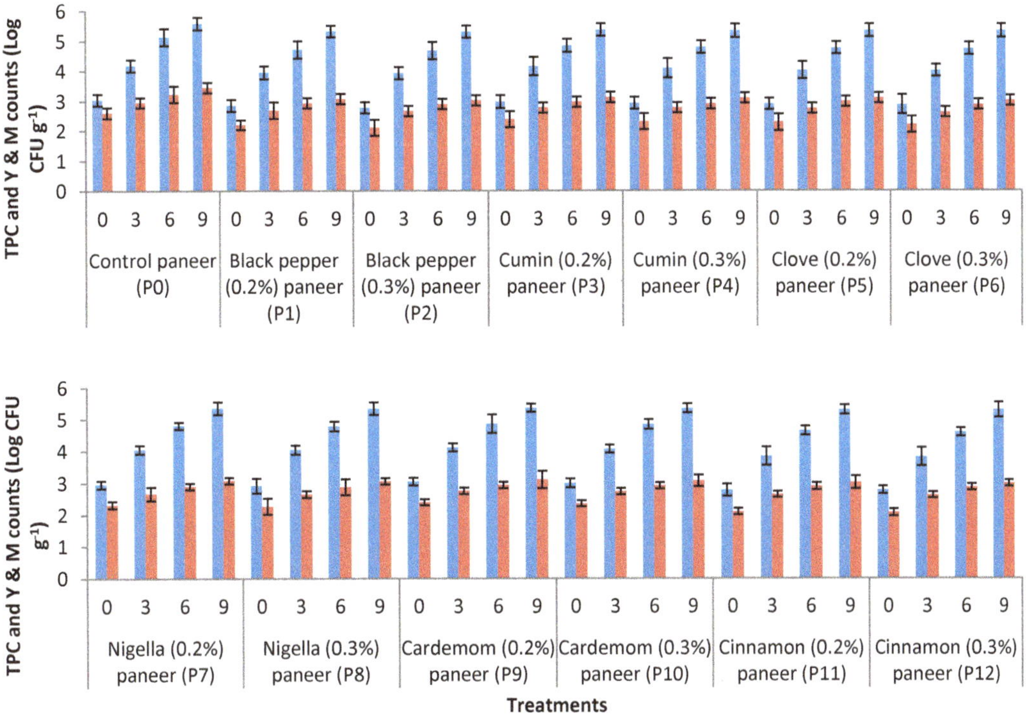

Figure 1. TPC (means ± SD, blue bars) and Y & M (means ± SD, red bars) of spicy paneer during storage at 5 °C.

3.3. Sensorial Quality of Prepared Paneer

Figure 2 shows the data regarding sensorial quality of paneer formulated in this study during storage. Different sensory characteristics, i.e., color, texture, appearance, flavor, and overall acceptability, were used for evaluating sensorial quality of experimentally prepared paneer. These sensory attributes illustrate the acceptance of the prepared paneer. Generally, all freshly prepared spicy paneer and control treatment had higher sensory scores for all the sensory characteristics, which declined during subsequent storage. Moreover, all the spicy paneer had higher sensory scores than control paneer at each storage period. All the paneer samples having 0.3% spices showed very slight variations (nonsignificant) in sensory score of all the attributes of their relative samples containing 0.2% spice. Among all the spicy paneer, paneer having black pepper (0.2%) was much liked by the people, followed by paneer with cumin (0.3%). The control paneer stored for 12 days showed visible mold growth on the surfaces. The taste of spicy paneer was also changed a little bit. Therefore, all the paneers stored for 12 days were excluded in the current study.

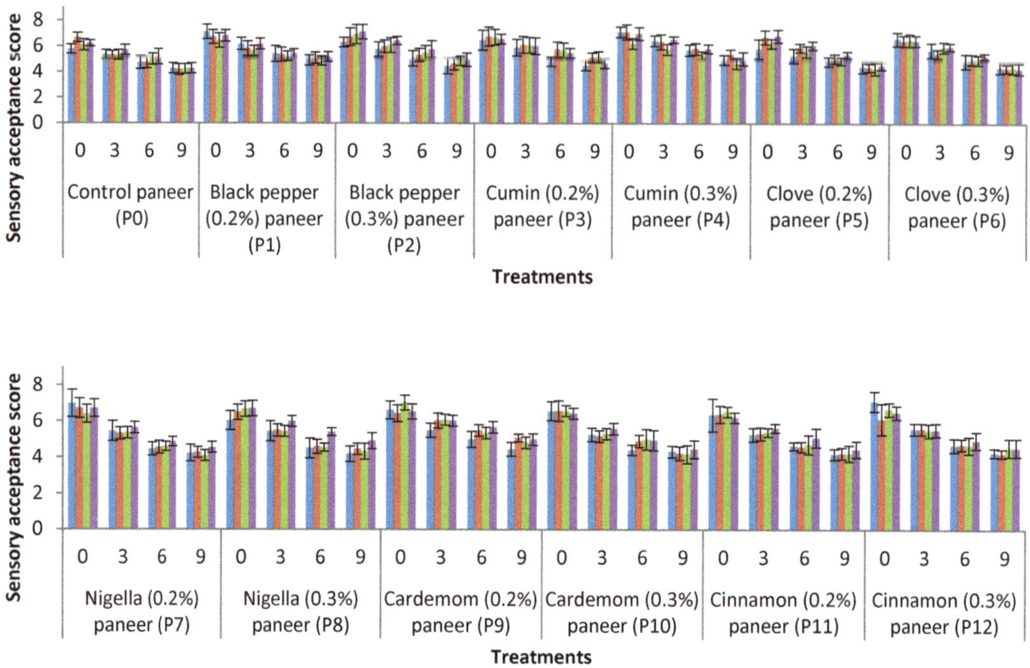

Figure 2. Sensory acceptance scores regarding appearance and color (means ± SD, blue bars), flavor (means ± SD, red bars), texture (means ± SD, green bars), and overall acceptability (means ± SD, purple bars) of spicy paneer during storage at 5 °C.

3.4. Antioxidant Potential of Spices and Spicy Paneer

Figure 3 shows the results regarding antioxidant potential of spices, whereas Figures 4 and 5 depict data regarding the antioxidant potential of paneer formulated in this study. There was a significant ($p < 0.05$) effect of spices on the values of TP, TF, RP, TAC, and DPPH radical scavenging activity. The maximum TP values were shown by cumin (354.60 mg GAE/100 g), followed by clove (142.53 mg GAE/100 g) and black pepper (109.00 mg GAE/100 g). The maximum TF values were also shown by cumin (526.40 mg CE/100 g), followed by clove (247.15 mg CE/100 g) and cinnamon (144.52 mg CE/100 g). Regarding RP, the maximum RP values were shown by cumin (512.04 mg TE/100 g), followed by cinnamon (162.66 mg TE/100 g) and clove (153.79 mg TE/100 g). The highest values of DPPH radical scavenging activity were shown by cumin (18.89 mM TE/g), followed by Nigella (16.59 mM TE/g) and cinnamon (11.18 mM TE/g). The maximum values of TAC were shown by cumin (791.45 mg TE/100 g), followed by clove (539.61 mg TE/100 g). All other spices also showed moderate values in connection to total antioxidant capacity.

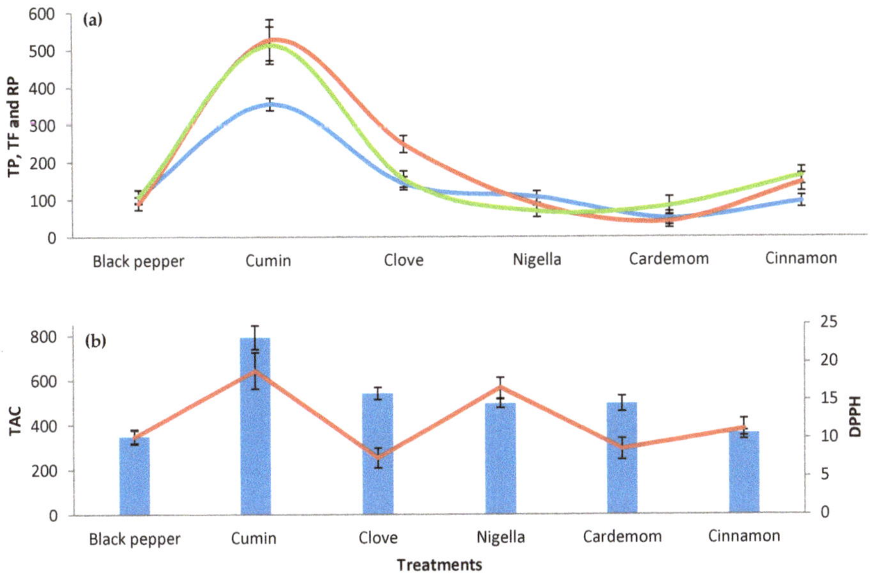

Figure 3. (**a**) Total phenolics (TP, means ± SD, mg GAE/100 g) (blue line), total flavonoids (TF, means ± SD, mg CE/100 g) (red line), reducing power (RP, means ± SD, mg TE/100 g) (green line), (**b**) total antioxidant activity (TAC, means ± SD) (blue bar, mg TE/100 g), and DPPH radical scavenging activity (means ± SD, red line, mM TE/g) of spices used for incorporation into paneer.

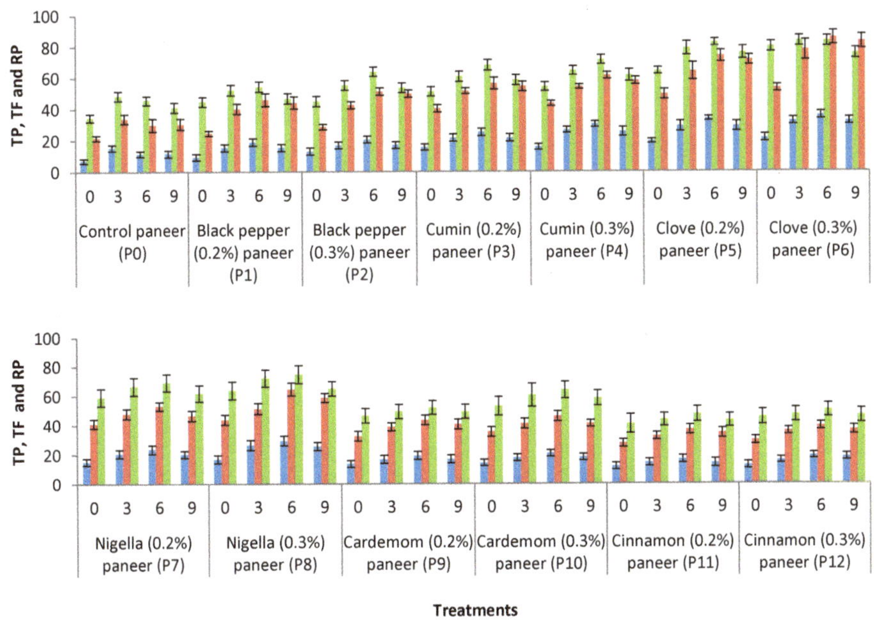

Figure 4. Total phenolics (TP, means ± SD, mg GAE/100 g) (blue bar), total flavonoids (TF, means ± SD, mg CE/100 g) (red bar), and reducing power (RP, means ± SD, mg TE/100 g) (green bar) of spicy paneer during storage at 5 °C.

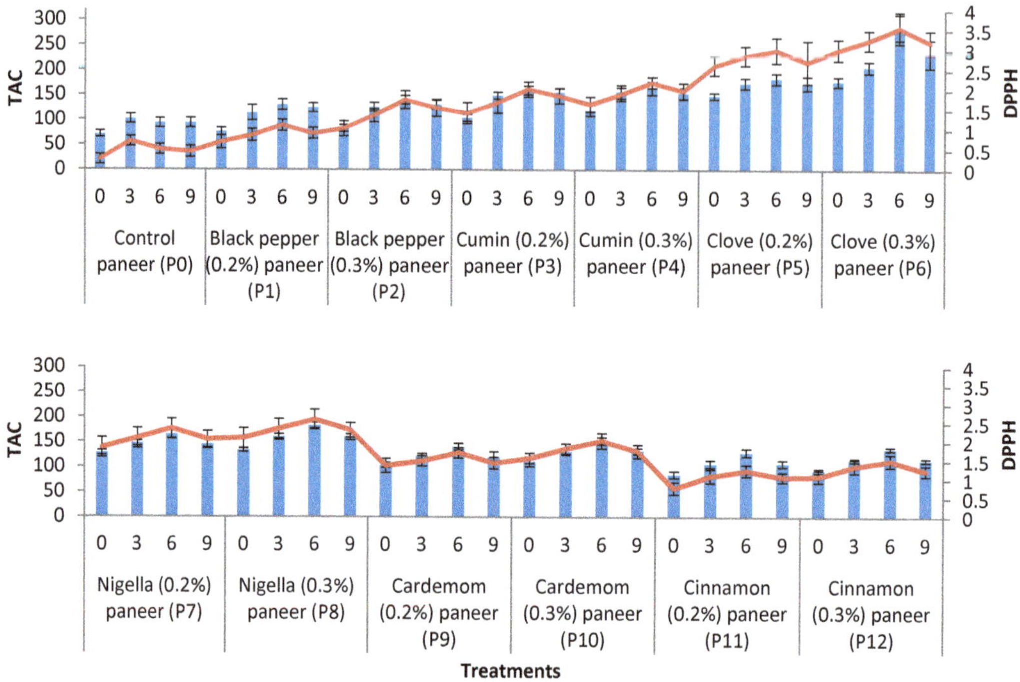

Figure 5. Total antioxidant activity (TAC, means ± SD) (blue bar, mg TE/100 g) and DPPH radical scavenging activity (means ± SD, red line, mM TE/g) of spicy paneer during storage at 5 °C.

There was a significant effect of treatments and storage days on the values of TP (Figure 4) of paneer. The freshly prepared control paneer (P_0) showed the lowest TP mean values (7.10 mg GAE/100 g) compared to the values of all other respective spicy paneer (treatments), whereas the highest value (21.36 mg GAE/100 g) was obtained by freshly prepared paneer containing 0.3% clove (P_6). All the spicy paneer showed increasing trend of TP values up to 6 days of storage, but afterwards, TP values were slightly decreased.

Regarding TF values, there was a significant effect of treatments and storage days (Figure 4). The freshly prepared control paneer (P_0) showed the lowest TF mean values (21.66 mg CE/100 g) compared to the values of all other respective spicy paneer (treatments), whereas the highest value (53.06 mg CE/100 g) was obtained by freshly prepared paneer containing 0.3% clove (P_6). All the spicy paneer showed increasing trend of TF values up to 6 days of storage, but afterwards, TF values were slightly decreased.

There was significant effect of treatments and storage days on RP of paneer (Figure 4). The freshly prepared control paneer (P_0) showed the lowest RP mean values (34.52 mg TE/100 g) compared to the values of all other respective spicy paneer (treatments), whereas the highest value (79.74 mg TE/100 g) was obtained by freshly prepared paneer having 0.3% clove (P_6). All the spicy paneer showed increasing trend of RP values up to 6 days of storage, but afterwards, RP values were slightly decreased.

There was significant effect of treatments and storage days on DPPH radical scavenging activity of paneer (Figure 5). The freshly prepared control paneer (P_0) showed the lowest DPPH mean values (0.25 mM TE/g) compared to the values of all other respective spicy paneer (treatments), whereas the highest value (3.01 mM TE/g) was obtained by freshly prepared paneer having 0.3% clove (P_6). All the spicy paneer showed increasing trend of DPPH values up to 6 days of storage, but afterwards, DPPH values were slightly decreased.

There was a significant ($p < 0.05$) effect of treatments and storage days on total antioxidant capacity of paneer (Figure 5). The freshly prepared control paneer (P_0) showed the lowest TAC mean values (70.52 mg TE/100 g) compared to the values of all other respective spicy paneer (treatments), whereas the highest value (177.27 mg TE/100 g) was obtained by freshly prepared paneer having 0.3% clove (P_6). All the spicy paneer showed increasing trend of antioxidant capacity up to 6 days of storage, but afterwards, TAC values were slightly decreased.

In general, it was observed that there was significant variation in the values of TP, TF, RP, DPPH, TAC, and RP among the paneer having different concentrations of the same spice. The control paneer (P_0) showed lower values of all these assays compared to all other treatments at all stages during the storage period.

4. Discussion

The pH at the time of milk coagulation during the manufacturing of paneer was observed to be more or less the same (around 5.65) in all the treatments, which was in agreement with the results (5.63) obtained by [26]. The more significant decreasing trend in pH of control paneer at the end of the storage period might be because of more activities of microorganisms compared to spicy paneer. Such kinds of conditions may cause more accumulation of acids during the storage period, thereby leading to more acidic conditions. The spices may cause a delay in the activities of pathogenic microorganisms, thereby keeping the freshness of paneer until 9 days of storage. The control paneer lost its freshness after 9 days. Regarding the decreasing trend of pH of paneer, the current study outcomes are in good agreement with the findings of [41,42]. The MCs (~58%) of fresh paneer satisfied the Bureau of Indian Standards [43]. A slight decrease in MCs of paneer during storage could be because of removal of MCs from paneer. Khatkar et al. [27] also observed the same kind of decreasing trend of MCs during the storage of paneer, as observed in this study. The fat contents of paneer treatments were very low because paneer was prepared from skimmed milk. On the other hand, protein contents were slightly increased during storage in the paneer matrix because of increased dry matter during storage. Moreover, protein constitutes most of the dry matter of paneer due to very low quantities of fat. The higher ash contents of all the treatments of paneer might also be due to lower quantities of fat, thereby constituting more of the dry matter. The slight increasing trend of ash contents during storage might be due to the increase in dry matter during storage.

Usually, bacteria as well as Y & M are mostly damaged during heating of milk, but these microorganisms may reinfect the paneer matrix during post-manufacture conditions, thereby leading to increased counts during storage. The rising trend of TPC and Y & M counts were inconsistent with the observations of [44,45]. Moreover, Rani et al. [46] also observed a rising trend of TPC and Y & M counts of paneer incorporated with cumin and black pepper during storage (up to 8 days). It was also observed that, due to presence of spices in the paneer matrix, there was an inhibitory effect against bacteria and yeast and molds. Therefore, microbial counts did not increase to such an extent as observed in the control paneer. In this study, it was suspected that all the spicy paneer might have slight fungus on the surface of paneer stored for 12 days due to its slight change in taste. On the other hand, the control paneer showed visible growth of fungus over its surface after 12 days of storage. Das et al. [47] also reported visible reddish brown discoloration on the surface of 12-days-stored paneer. As the paneer does not contain any starter or probiotic bacteria which cause inhibition of harmful microbes, deterioration happened in paneer with the advancement of storage period due to increase in the number of microbes.

The decreasing score of all the sensory characteristics of paneer in this study was concurrent with the results obtained by [45,47]. Our results were concurrent with the results obtained by [27], who observed decreased sensory acceptance score of clove-treated paneer during storage. The control paneer was accepted for up to 6 days on the basis of sensory acceptance score, which was in accordance with the results obtained by [27], who reported 5-day-old paneer as acceptable. The control paneer showed some signs of mold growth after 9 days due to its flavor. The growth of fungus on the surfaces of control paneer after 12 days of storage caused exclusion of such paneer for sensory evaluation. All the spicy paneer after 12 days of storage also showed some signs of mold growth due to a slight change in taste. Khatkar et al. [27] observed the shelf life of clove-treated paneer up to 10 days. In addition, some people liked the taste of paneer having 0.2% spices, and some people liked the paneer containing 0.3% spices; therefore, paneer with both percentages of each spice were included in the study.

The results of DPPH and TAC of all the spicy paneers were very promising, which might be due to the presence of enormous quantities of TP and TF of spices. Our results regarding TP of spicy paneer that incorporated 0.3% cardamom (14.25 mg GAE/100 g) were lower in comparison with the results (47.2 mg GAE/100 g) obtained by [26], who incorporated mixture of black pepper (0.25%) and cardamom (0.50%). These discrepancies might be due to variations in raw materials, i.e., milk and quantity and source of spices. The hydroxy groups of flavonoids have the potential to donate hydrogen or electrons to DPPH free radicals, thereby leading to the termination of reactions of free radicals [48]. The RP of spices, as well as spicy paneer, may exhibit potent antioxidant activity due to the existence of antioxidants. Such compounds caused the diminution of ferricyanide complex to ferrocyanide, which subsequently reacts with ferric chloride to form ferric ferrous complex, exhibiting maximum absorption at 700 nm [49]. All the parameters showing antioxidant potential of spices (depicting the highest values) were many folds compared to spicy paneer (depicted the highest values). In this context, TP values of spices were 17 folds compared to the values of spicy paneer. Similarly, TF values of spices were 10 folds compared with the values of spicy paneer. Moreover, RP and DPPH values of spices were 6 folds compared with the values of spicy paneer. Such kinds of manyfold antioxidant potential of spices revealed that these contained abundant quantities of bioactive compounds (phenolics, flavonoids, and many other compounds) relevant to antioxidant activities. For instance, the pericarp of black pepper contained significant quantities of phenolic and flavonoid compounds [22] as well as some other compounds such as ascorbic acid, β-carotene, camphene, carvacrol, eugenol, piperine, and ubiquinone [5], which are considered to be strong antioxidants. Similarly, clove is reported to have strong antioxidant activity [8,9] due to the presence of tocopherol, ascorbates, and phenolic compounds [23]. It also contains some health-promoting components such as volatile oils, acids, cymene, pinene, cuminaldehyde, terpinene, thymol, and oleoresin [14,45]. Chaudhry et al. [50] also reported that the extracts of *Nigella sativa* contained thymol and thymoquinone which exhibit strong antioxidant and antibacterial potential. Cardamom is reported to contain predominant compounds including α-terpineol, α-terpinyl acetate, 1,8-cineole, β-linalool, and sabinene [17,18]. Owing to the existence of such compounds in essential oil of cardamom, it was reported to have antioxidant potential [17,19,20]. The bark of cinnamon contains phytoconstituents, such as flavonoids, phenolics, and carotenoids, which are considered to be antioxidants [51]. Ghosh et al. [52] evaluated the antioxidant potential of pectic polysaccharides extracted from its bark. They found that the extracted compounds responsible for the antioxidant potential include arabinogalactan, uronic acid, and glucan. It was also reported that cinnamon had antioxidant potential [53,54] due to the presence of major constituents such as (*E*)-cinnamaldehyde, linalool, *β*-caryophyllene, eucalyptol, eugenol, benzyl benzoate, α-felandrene, α-pinene, and cinnamaldehyde acetate [24,51]. Since the spices are reported to have antioxidant activities, it would be of great interest for the food industries to incorporate such kinds of spices (powders or their extracts) into their products for their value addition.

5. Conclusions

Based on the results of the present study, it may be concluded that all the treatments concerning spicy paneer (varying percentages of spices) were sensorially acceptable. Similarly, it may be concluded that spicy paneer had more antioxidant potential compared to the control paneer due to the presence of more phenolics and flavonoids and many other compounds. Moreover, paneer having 0.3% clove (P_6) showed the maximum values regarding TP, TF, RP, DPPH, and TAC. In this way, it may be recommended that spicy paneer with very-low-fat contents should be produced by the dairy industry as it would be more nutritious with noteworthy worth due to the presence of more antioxidant compounds as mentioned above in all the spices.

Author Contributions: Conceptualization, T.M.Q., M.N. and M.S.; data curation, M.N. and J.I.; formal analysis, T.M.Q., M.N. and J.I.; funding acquisition, M.N. and S.M.I.; investigation, T.M.Q., J.I., S.M.I., F.M. and M.S.; methodology, T.M.Q., M.N., S.-u.-R. and J.I.; project administration, M.N. and S.M.I.; resources, M.N.; software, T.M.Q. and J.I.; validation, M.N.; visualization, T.M.Q., J.I., S.-u.-R., S.M.I., F.M. and M.S.; writing—original draft, T.M.Q. and M.N.; writing—review and editing, J.I., S.M.I., F.M. and M.S. All authors have read and agreed to the published version of the manuscript.

Funding: This work was supported by Researchers Supporting Project number (RSP2023R100), King Saud University, Riyadh, Saudi Arabia.

Institutional Review Board Statement: Not applicable.

Informed Consent Statement: Not applicable.

Data Availability Statement: The data are contained within the article.

Acknowledgments: This work was supported by Researchers Supporting Project number (RSP2023R100), King Saud University, Riyadh, Saudi Arabia. The authors acknowledge the financial support of the Institute of Food Science and Nutrition, University of Sargodha, Pakistan through utilization of the Institutional budget for carrying out all research activities.

Conflicts of Interest: The authors declare no conflict of interest.

References

1. Qureshi, T.M.; Amjad, A.; Nadeem, M.; Murtaza, M.A.; Munir, M. Antioxidant potential of a soft cheese (paneer) supplemented with the extracts of date (*Phoenix dactylifera* L.) cultivars and its whey. *Asian-Australas. J. Anim. Sci.* **2019**, *32*, 1591–1602. [CrossRef] [PubMed]
2. Jiang, T.A. Health Benefits of Culinary Herbs and Spices. *J. AOAC Int.* **2019**, *102*, 395–411. [CrossRef]
3. Vázquez-Fresno, R.; Rosana, A.R.R.; Sajed, T.; Onookome-Okome, T.; Wishart, N.A.; Wishart, D.S. Herbs and Spices- Biomarkers of Intake Based on Human Intervention Studies—A Systematic Review. *Genes Nutr.* **2019**, *14*, 18. [CrossRef] [PubMed]
4. Li, Y.-X.; Erhunmwunsee, F.; Liu, M.; Yang, K.; Zheng, W.; Tian, J. Antimicrobial mechanisms of spice essential oils and application in food industry. *Food Chem.* **2022**, *382*, 132312. [CrossRef] [PubMed]
5. Meghwal, M.; Goswami, T.K. Chemical Composition, Nutritional, Medicinal and Functional Properties of Black Pepper: A Review. *Open Access Sci. Rep.* **2012**, *1*, 1–5. [CrossRef]
6. Gülçin, I. The antioxidant and radical scavenging activities of black pepper (*Piper nigrum*) seeds. *Int. J. Food Sci. Nutr.* **2005**, *56*, 491–499. [CrossRef] [PubMed]
7. Vijayakumar, R.S.; Surya, D.; Nalini, N. Antioxidant efficacy of black pepper (*Piper nigrum* L.) and piperine in rats with high fat diet induced oxidative stress. *Redox Rep.* **2004**, *9*, 105–110. [CrossRef] [PubMed]
8. Alfikri, F.N.; Pujiarti, R.; Wibisono, M.G.; Hardiyanto, E.B. Yield, Quality, and Antioxidant Activity of Clove (*Syzygium aromaticum* L.) Bud Oil at the Different Phenological Stages in Young and Mature Trees. *Scientifica* **2020**, *2020*, 9701701. [CrossRef]
9. Dua, A.; Singh, A.; Mahajan, R. Antioxidants of Clove (*Syzygium aromaticum*) Prevent Metal Induced Oxidative Damage of Biomolecules. *Int. Res. J. Pharm.* **2015**, *6*, 273–278. [CrossRef]
10. Selles, S.M.A.; Kouidri, M.; Belhamiti, B.T.; Amrane, A.A. Chemical composition, in-vitro antibacterial and antioxidant activities of *Syzygium aromaticum* essential oil. *J. Food Meas. Charact.* **2020**, *14*, 2352–2358. [CrossRef]
11. Cortés-Rojas, D.F.; de Souza, C.R.F.; Oliveira, W.P. Clove (*Syzygium aromaticum*): A precious spice. *Asian Pac. J. Trop. Biomed.* **2014**, *4*, 90–96. [CrossRef]
12. Sulistyoningrum, A.S.; Saepudin, E.; Cahyana, A.H.; Rahayu, D.U.C.; Amelia, B.; Haib, J. Chemical Profiling of Clove Bud Oil (*Syzygium aromaticum*) from Toli-Toli and Bali by GC-MS Analysis. *AIP Conf. Proc.* **2017**, *1862*, 030089-1–030089-6.
13. Ghadermazi, R.; Keramat, J.; Goli, S.A.H. Antioxidant Activity of Clove (*Eugenia caryophyllata* Thunb), Oregano (*Oringanum vulgare* L.) and Sage (*Salvia officinalis* L.) Essential Oils in Various Model Systems. *Int. Food Res. J.* **2017**, *24*, 1628.
14. Fatima, T.; Beenish, N.B.; Gani, G.; Qadri, T.; Bhat, T.A. Antioxidant Potential and Health Benefits of Cumin. *J. Med. Plants Stud.* **2018**, *6*, 232–236.
15. Abdel-Latif, E.F.; Abbas, K.A.; Abdelmontaleb, H.S.; Hamdy, S.M. *Nigella sativa* oil: A promising prospective antifungal agent in the manufacture of low-salt soft cheese. *Ital. J. Food Saf.* **2021**, *10*, 9862. [CrossRef] [PubMed]

16. Bordoni, L.; Fedeli, D.; Nasuti, C.; Maggi, F.; Papa, F.; Wabitsch, M.; De Caterina, R.; Gabbianelli, R. Antioxidant and Anti-Inflammatory Properties of *Nigella sativa* Oil in Human Pre-Adipocytes. *Antioxidants* **2019**, *8*, 51. [CrossRef] [PubMed]
17. Alam, A.; Jawaid, T.; Alam, P. *In vitro* antioxidant and anti-inflammatory activities of green cardamom essential oil and *in silico* molecular docking of its major bioactives. *J. Taibah Univ. Sci.* **2021**, *15*, 757–768. [CrossRef]
18. Ashokkumar, K.; Vellaikumar, S.; Murugan, M.; Dhanya, M.K.; Ariharasutharsan, G.; Aiswarya, S.; Akilan, M.; Warkentin, T.D.; Karthikeyan, A. Essential Oil Profile Diversity in Cardamom Accessions from Southern India. *Front. Sustain. Food Syst.* **2021**, *5*, 639619. [CrossRef]
19. Amma, K.P.P.; Sasidharan, I.; Sreekumar, M.M.; Sumathykutty, M.A.; Arumughan, C. Total Antioxidant Capacity and Change in Phytochemicals of Four Major Varieties of Cardamom Oils During Decortication. *Int. J. Food Prop.* **2015**, *18*, 1317–1325. [CrossRef]
20. Ben Nouri, A.; Dhifi, W.; Bellili, S.; Ghazghazi, H.; Aouadhi, C.; Chérif, A.; Hammami, M.; Mnif, W. Chemical Composition, Antioxidant Potential, and Antibacterial Activity of Essential Oil Cones of Tunisian *Cupressus sempervirens*. *J. Chem.* **2015**, *2015*, 538929. [CrossRef]
21. Rao, P.V.; Gan, S.H. Cinnamon: A Multifaceted Medicinal Plant. *Evid.-Based Complement. Altern. Med.* **2014**, *2014*, 642942. [CrossRef]
22. Lee, J.-G.; Chae, Y.; Shin, Y.; Kim, Y.-J. Chemical composition and antioxidant capacity of black pepper pericarp. *Appl. Biol. Chem.* **2020**, *63*, 1–9. [CrossRef]
23. Rebey, I.B.; Zakhama, N.; Karoui, I.J.; Marzouk, B. Polyphenol Composition and Antioxidant Activity of Cumin (*Cuminum cyminum* L.) Seed Extract Under Drought. *J. Food Sci.* **2012**, *77*, C734–C739. [CrossRef] [PubMed]
24. Ribeiro, P.R.E.; Montero, I.F.; Saravia, S.A.M.; Ferraz, V.P.; Santos, R.A.; Marcía, J.A.F.; Linhares, B.M. Chemical composition and antioxidant activity in the essential oil of *Cinnamomum zeylanicum* Nees with medicinal interest. *J. Med. Plants Res.* **2020**, *14*, 326–330. [CrossRef]
25. Akan, E.; Yerlikaya, O.; Akpinar, A.; Karagozlu, C.; Kinik, O.; Uysal, H.R. The effect of various herbs and packaging material on antioxidant activity and colour parameters of whey (Lor) cheese. *Int. J. Dairy Technol.* **2021**, *74*, 554–563. [CrossRef]
26. Badola, R.; Danish, M.; Kumar, S.; Fahad, M.; Kanade, P.P.; Upadhayay, S.; Kohli, D.; Rautela, I. Effect of Incorporation of Black Pepper and Cardamom on Quality Characteristics of Paneer. *Int. J. Appl. Sci. Eng.* **2018**, *6*, 121–127. [CrossRef]
27. Ray, A.B.K.A.; Kaur, A. Studies on Shelf Life Extension of Paneer with the Addition of Plant Essential Oil and Different Packaging Materials. *Int. J. Curr. Microbiol. Appl. Sci.* **2017**, *6*, 376–389. [CrossRef]
28. El-Sayed, S.M.; Youssef, A.M. Potential application of herbs and spices and their effects in functional dairy products. *Heliyon* **2019**, *5*, e01989. [CrossRef]
29. Khan, S.U.; Pal, M.A.; Malik, A.H.; Sofi, A.H. Process Optimization for Paneer Production from Milk Powder. *Int. J. Food Nutr. Saf.* **2012**, *2*, 62–71.
30. Ardö, Y.; Polychroniadou, A. *Laboratory Manual for Chemical Analysis of Cheese: Improvement of the Quality of the Production of Raw Milk Cheeses*; Office for Official Publications of the European Communities: Luxembourg, 1999.
31. Gupta, A.; Mann, B.; Kumar, R.; Sangwan, R.B. ACE-Inhibitory Activity of Cheddar Cheeses Made with Adjunct Cultures at Different Stages of Ripening. *Adv. Dairy Res.* **2013**, *1*, 1000102. [CrossRef]
32. AOAC. *Official Methods of Analysis*, 17th ed.; The Association of Official Analytical Chemists: Gaithersburg, MD, USA, 2000.
33. *ISO 5534:2004*; Cheese and Processed Cheese. Determination of the Total Solids Content (Reference Method). International Organization for Standardization: Geneva, Switzerland, 2004.
34. IDF. *Milk: Determination of Nitrogen Content*; IDF Standard 20B; International Dairy Federation: Brussels, Belgium, 1993.
35. Broadbent, J.; Brighton, C.; McMahon, D.; Farkye, N.; Johnson, M.; Steele, J. Microbiology of Cheddar cheese made with different fat contents using a *Lactococcus lactis* single-strain starter. *J. Dairy Sci.* **2013**, *96*, 4212–4222. [CrossRef]
36. Dongare, S.A.; Syed, H.M. Physicochemical and Sensory Properties of Paneer Prepared from Buffalo Milk. *J. Pharmacogn. Phytochem.* **2018**, *7*, 3852–3854.
37. Reis, F.S.; Stojković, D.; Soković, M.; Glamočlija, J.; Ćirić, A.; Barros, L.; Ferreira, I.C. Chemical characterization of *Agaricus bohusii*, antioxidant potential and antifungal preserving properties when incorporated in cream cheese. *Food Res. Int.* **2012**, *48*, 620–626. [CrossRef]
38. Zhishen, J.; Mengcheng, T.; Jianming, W. The determination of flavonoid contents in mulberry and their scavenging effects on superoxide radicals. *Food Chem.* **1999**, *64*, 555–559. [CrossRef]
39. Yi, Z.; Yu, Y.; Liang, Y.; Zeng, B. In vitro antioxidant and antimicrobial activities of the extract of *Pericarpium Citri Reticulatae* of a new Citrus cultivar and its main flavonoids. *Lwt* **2008**, *41*, 597–603. [CrossRef]
40. Prieto, P.; Pineda, M.; Aguilar, M. Spectrophotometric Quantitation of Antioxidant Capacity through the Formation of a Phosphomolybdenum Complex: Specific Application to the Determination of Vitamin E. *Anal. Biochem.* **1999**, *269*, 337–341. [CrossRef]
41. Shanaziya, A.S.F.; Mangalika, U.L.P.; Nayananjalie, W.A.D. Effect of Different Coagulants on the Quality of Paneer Made from Cow Milk. *Int. J. Sci. Res. Publ. IJSRP* **2018**, *8*, 189–194. [CrossRef]
42. Singh, R.R.; Singh, R.; Shakya, B.R. Impact of Turmeric Addition on the Properties of Paneer, Prepared from Different Types of Milk. *Int. J. Curr. Eng. Technol.* **2014**, *4*, 1874–1883.
43. *BIS IS: 10484*; Specification for Paneer. Bureau of Indian Standards: New Delhi, India, 1983.

44. Mishra, D.; Rao, K.J.; Bhardwaj, R.; Sutariya, H.; Kavitkar, R.S.; Subhash, W.S. Effect of PH on Sensory, Textural, Microbial Quality and Shelf-Life of Paneer. *Int. J. Food Ferment. Technol.* **2016**, *6*, 405. [CrossRef]
45. Singh, B.K.; Sen, D.C.; Kumar, R.; Rani, B.; Barman, A.K. Storage Study of Danadar Prepared from Cow Milk. *Chem. Sci. Rev. Lett.* **2017**, *6*, 1958–1969.
46. Rani, M.; Dabur, R.S.; Garg, S.R.; Jadhav, V. Preparation, storage and microbiological quality of ready-to-serve low cholesterol masala paneer. *Vet. World* **2014**, *7*, 443–447. [CrossRef]
47. Das, A.; Chauhan, G.; Agrawal, R.K.; Tomar, S.; Sirajuddin, S.K. Comparative Study on Evaluation of Refrigerated (4±1 C) Storage Stability of Paneer Incorporated with Crude Extract from Indian Curd, Nisin and Lactic Acid. *Int. J. Curr. Microbiol. Appl. Sci.* **2018**, *7*, 167–180. [CrossRef]
48. Zha, X.-Q.; Wang, J.-H.; Yang, X.-F.; Liang, H.; Zhao, L.-L.; Bao, S.-H.; Luo, J.-P.; Xu, Y.-Y.; Zhou, B.-B. Antioxidant properties of polysaccharide fractions with different molecular mass extracted with hot-water from rice bran. *Carbohydr. Polym.* **2009**, *78*, 570–575. [CrossRef]
49. Prabha, M.R.; Vasantha, K. Antioxidant, cytotoxicity and polyphenolic content of *Calotropis procera* (Ait.) R. Br. Flowers. *J. Appl. Pharm. Sci.* **2011**, *1*, 136–140.
50. Chaudhry, H.; Fatima, N.; Ahmad, I.Z. Evaluation of Antioxidant and Antibacterial Potentials of *Nigella sativa* L. Suspension Cultures under Elicitation. *BioMed Res. Int.* **2015**, *2015*, 708691. [CrossRef] [PubMed]
51. Behbahani, B.A.; Falah, F.; Arab, F.L.; Vasiee, M.; Yazdi, F.T. Chemical Composition and Antioxidant, Antimicrobial, and Antiproliferative Activities of *Cinnamomum zeylanicum* Bark Essential Oil. *Evid.-Based Complement. Altern. Med.* **2020**, *2020*, 5190603. [CrossRef]
52. Ghosh, T.; Basu, A.; Adhikari, D.; Roy, D.; Pal, A.K. Antioxidant activity and structural features of *Cinnamomum zeylanicum*. *3 Biotech* **2015**, *5*, 939–947. [CrossRef]
53. Shahid, M.Z.; Saima, H.; Yasmin, A.; Nadeem, M.T.; Imran, M.; Afzaal, M. Antioxidant capacity of cinnamon extract for palm oil stability. *Lipids Health Dis.* **2018**, *17*, 116. [CrossRef]
54. Pan, C.; Yang, K.; Erhunmwunsee, F.; Li, Y.-X.; Liu, M.; Pan, S.; Yang, D.; Lu, G.; Ma, D.; Tian, J. Inhibitory effect of cinnamaldehyde on *Fusarium solani* and its application in postharvest preservation of sweet potato. *Food Chem.* **2023**, *408*, 135213. [CrossRef]

agriculture MDPI

Article

Cavitation Reactor for Pretreatment of Liquid Agricultural Waste

Alexey Abdrashitov [1], Alexander Gavrilov [2], Evgeny Marfin [1], Vladimir Panchenko [3], Andrey Kovalev [4], Vadim Bolshev [4,*] and Julia Karaeva [1,*]

1 Institute of Power Engineering and Advanced Technologies, FRC Kazan Scientific Center, Russian Academy of Sciences, 420111 Kazan, Russia; abdary@mail.ru (A.A.); marfin76@mail.ru (E.M.)
2 Department of Radio Electronics, Kazan Federal University, 420008 Kazan, Russia; alexander.gavrilov@kpfu.ru
3 Department of Theoretical and Applied Mechanics, Russian University of Transport, 127994 Moscow, Russia; pancheska@mail.ru
4 Federal Scientific Agroengineering Center VIM, 109428 Moscow, Russia; kovalev_ana@mail.ru
* Correspondence: vadimbolshev@gmail.com (V.B.); julieenergy@list.ru (J.K.)

Abstract: One of the most well-known methods of intensifying the process of anaerobic digestion is the pretreatment of raw materials. For the first time, the use of a jet-driven Helmholtz oscillator for biomass pretreatment is proposed. The design of the device is optimal for creating hydraulic cavitation; however, in this case, acoustic oscillations are generated in the system and resonance occurs. In this study, the optimal design of this device was determined for the subsequent design of a cavitation reactor. The diameter of the resonant chamber was varied in the range from 28.3 to 47.5 mm, and its length from 6 to 14 mm; in addition, the diameter of the outlet was changed from 6.1 to 6.3 mm. Based on the experimental data obtained, it was found that the optimal ratio of the length of the resonator chamber to the diameter of the inlet nozzle is 1.73, and the inner diameter of the resonator chamber to the diameter of the inlet nozzle corresponds to 5.5. Improving the technology of agricultural waste disposal will ensure their maximum involvement in economic circulation, reduce the consumption of traditional fuel and energy resources, and improve the technological and machine-building base, which makes it possible to produce competitive cavitation reactors.

Keywords: agricultural waste; cavitation; pretreatment; jet-driven Helmholtz oscillator; anaerobic digestion

Citation: Abdrashitov, A.; Gavrilov, A.; Marfin, E.; Panchenko, V.; Kovalev, A.; Bolshev, V.; Karaeva, J. Cavitation Reactor for Pretreatment of Liquid Agricultural Waste. *Agriculture* **2023**, *13*, 1218. https://doi.org/10.3390/agriculture13061218

Academic Editors: Muhammad Sultan, Redmond R. Shamshiri, Md Shamim Ahamed, Muhammad Farooq and Na Duan

Received: 14 April 2023
Revised: 18 May 2023
Accepted: 8 June 2023
Published: 9 June 2023

1. Introduction

Anaerobic digestion (AD) technology is widely throughout the world. The unique advantage of this technology is the solution of both environmental and energy problems. Environmental protection and energy conservation are among the most important issues at present [1]. The efficient use of biomass resources, including agricultural waste, will increase the world's reserves of renewable energy resources and help achieve climate change mitigation goals [2]. It should be noted that 772.6 million tons of waste with a content of 228 million tons of dry matter is generated annually in the agro-industrial complex of Russia [3], and this is a constant and large-tonnage source of energy.

To increase the productivity of AD technology, it is necessary to pretreat the feedstock entering the reactor [4]. This allows the preparation of substrates for biological decomposition by microorganisms involved in the fermentation process, as well as increasing the yield of biogas [5]. All methods of pretreatment can be divided into the following categories: physical, thermal, chemical, biological, and a combination of these. A separate group is represented by more innovative methods [6]. Figure 1 shows a number of methods used to prepare raw materials. Innovative pretreatment technologies are based on extreme and non-classical processes, such as ultrasound [6], gamma rays [7], electron beam irradiation [8], pulsed electric field [9], high hydrostatic pressure [10], and high-pressure homogenization [11].

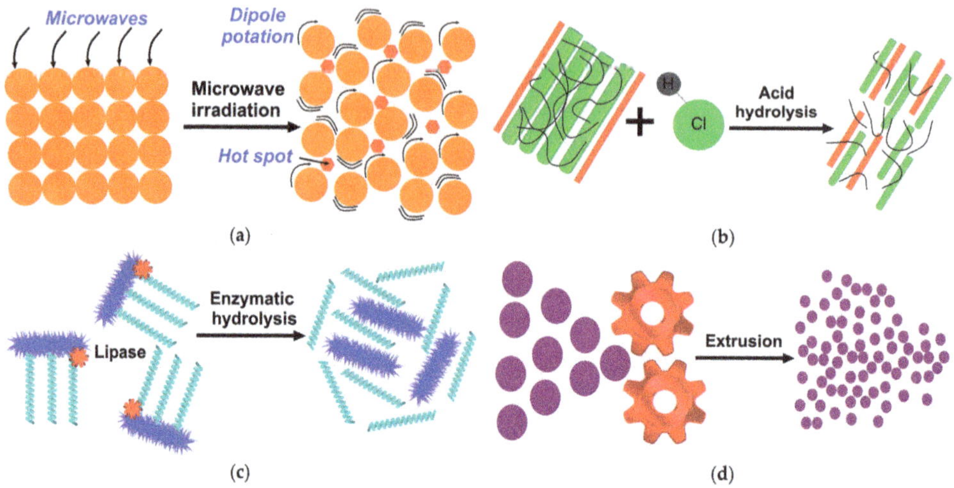

Figure 1. Examples of pretreatment: (**a**) thermal, (**b**) chemical, (**c**) biological, and (**d**) physical.

Of particular interest among innovative methods is the generation of cavitation, when vapor microbubbles in a liquid medium collapse due to sharp drops in local pressure, and a large amount of energy is released, as well as intense local heating (about 5000 °C) and high pressure (about 50 MPa) [12]. If the pressure decrease occurs due to high local velocities in the flow of a moving droplet liquid, then the cavitation is called hydrodynamic, and if it is due to the passage of acoustic waves it is known as acoustic. Under the action of directed and controlled cavitation in biomass, the destruction of complex fibrous bonds of organic substances at the molecular level occurs, the particle sizes decrease to 0.1–8 μm, as a result, the biogas yield increases by 30–50% [13].

The main advantages of integrating ultrasonic technologies that create acoustic cavitation into agricultural biogas plants (full-scale application) [14] are:

- the intensification of the AD process;
- increase in biogas output;
- increasing the content of methane in the resulting gas;
- the economy of the substrate;
- decrease in viscosity;
- reduction in the energy consumption of agitators and pumps.

Cavitation reactors have shown significant promise for application in biogas technologies due to advantages such as short processing times and higher degradation rates [15]. In practice, the energy consumption of high-intensity ultrasonic systems can be maintained at a fairly low level (<5 kWh/m^3) [14]. Thus, the use of high frequency sonic waves is an attractive pretreatment method, but experimental evaluation is still needed to weigh the potential benefits of sonication against additional energy requirements [16].

Hydrodynamic cavitation devices can be broadly classified into two categories [17]:

- those with moving parts, such as a rotor–stator [18–21];
- without moving parts, for example, a vortex-based device [22–25], swirling jet [26], venturi tube [27,28], hydrosonic pump [12], and orifice plate [29,30].

The hydrodynamic cavitation reactor is based on a stator and rotor and has been used for pretreatment of wheat straw, which made it possible to increase the yield of biogas by two times [18]. Hydrodynamic cavitation occurs in a flow system with a constriction, such as a venturi tube [27]. The efficiency of the pretreatment of lignocellulosic biomass using a combination of hydrodynamic cavitation and the introduction of sodium percarbonate was studied in [27]. It has been proven that the efficiency of the process can be increased by changing the geometry of the constriction, i.e., to achieve this, it is necessary to use a narrower neck. A hydrodynamic cavitation reactor, including a stator–rotor assembly, was used for the pretreatment of wheat straw, which allowed a two-fold increase in the biogas yield at AD [18]. The synergistic effect of the combined pretreatment can also be achieved by maximizing the release of extracellular polymeric substances from the biomass [31]. In a number of studies, the influence of temperature on hydrodynamic cavitation flows was determined in order to find optimal conditions that increase the intensity of cavitation treatment [32–36]. It should be noted that the effect of temperature on hydrodynamic cavitation characteristics was significant for lignocellulosic raw materials, since pretreatment at 70 °C resulted in greater enzymatic digestibility than at 40 °C [37].

In ref. [12], acoustic and hydrodynamic cavitation was compared during pretreatment. The same effect of these two methods on the anaerobic fermentation of cattle manure and slurry mixed with wheat straw was experimentally established. Thus, any pretreatment that generates cavitation is a promising solution due to low capital investment and ease of operation [38].

Acoustic cavitation in raw materials pre-treatment systems before anaerobic digestion is based on the use of an ultrasonic transducer, which converts high-frequency voltage into ultrasonic frequency mechanical vibrations. Hydrodynamic cavitation is artificially created by constriction in the channel where the liquid moves. A feature of the jet-driven Helmholtz oscillator (JDHO) is the presence of a constriction, as a result of which acoustic oscillations are generated in the system. The JDHO, which can continuously generate a spatial jet stream entirely based on its internal geometry without any moving parts [39]. The JDHO creates an oscillating jet with different frequencies without the influence of any external forces [40]. Thus, this cavitation device is very efficient [41]. The authors of [42] present the results of a study of the cavitation behavior of an oscillating jet flowing through a JDHO. They showed that such a design creates cavitation clouds in the liquid, which propagate over long distances. In the study [43], it was also demonstrated that such a cavitation device design is more efficient than the classical "organ pipe nozzle". A greater effect is achieved due to the resonant amplification of self-oscillations of the submerged jet [44]. The fluidic oscillators are widely used in the physical and chemical effects of acoustic cavitation, which are utilized in biotechnology [45,46], oil refining [47], and the oil industry [48], in particular, in drilling wells and intensifying oil production [49], green chemistry [50], the food industry [51], agriculture [52], etc. [53]. The JDHO is currently used in the oil industry and has never before been used for the pre-treatment of raw materials before anaerobic digestion. There is no description in the literature of studies on the use of a jet oscillator in the process of the pretreatment of biomass for the subsequent production of biogas. The aim is to determine the optimal design of the JDHO for the design of a cavitation reactor that allows the pretreatment of liquid agricultural waste before AD.

2. Object and Method of the Research

2.1. *Jet-Driven Helmholtz Oscillator*

The design of the considered device is based on the JDHO [41,44]. It is presented as a dual device that combines two independent elements in one housing: a jet generator of pressure oscillations and an acoustic resonator. The acoustic resonator consists of a cylindrical resonant chamber with two parallel covers (Figure 2a).

In the front cover (1), in the direction of flow, there is an inlet nozzle through which a working agent (gas, liquid) is fed into the resonator. The inlet nozzle forms a jet of the working agent. In the rear cover (2), there is an outlet with sharp edges through which the working agent is removed from the resonator. The inlet nozzle and the outlet are located on the axis of the resonant chamber (3). The configuration "nozzle–jet–hole" is a jet generator. It generates the hole tone. The liquid volume contained in the resonant chamber (3), the inlet nozzle in the front cover, and the outlet in the power housing is responsible for amplifying the harmonic of the hole tone at frequencies close to the natural frequency of the resonator, the value of which can be determined by the formula [49]:

$$f_0 = \frac{c_0}{2\pi D_C}\sqrt{\frac{1}{L_C}\left(\frac{d_1^2}{l_1 + 0.25\pi d_1} + \frac{d_2^2}{l_2 + 0.25\pi d_2}\right)} \tag{1}$$

where c_0 is the speed of sound in the working medium, (m/s); D_C is the inner diameter of the resonator chamber (m); L_C is the length of the resonator chamber (m); d_1 and d_2 are the diameters of the inlet nozzle and outlet (m); and ℓ_1 and ℓ_2 are the effective lengths of the inlet nozzle and outlet (m).

In this work, the object of study was the JDHO with the following geometric characteristics: an inlet nozzle with a diameter of d_1 = 5.5 mm and a length of ℓ_1 = 1 mm; outlet with diameters d_2 = 6.3, 6.2, and 6.1 mm and length ℓ_2 = 1 mm; and the diameter of the resonant chamber D_C varied in the range from 28.3 to 47.5 mm, and its length L_C varied from 6 to 14 mm. The thickness of the covers in which the inlet nozzle and the outlet are placed is 4 mm.

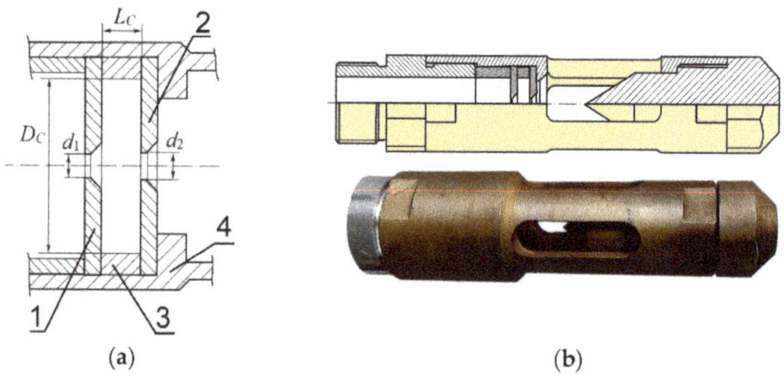

Figure 2. Jet-driven Helmholtz oscillator: (**a**) design diagram, where 1—front cover with an inlet nozzle, 2—rear cover with an outlet, 3—resonant chamber, and 4—power housing; (**b**) construction drawing and photography.

2.2. Experimental Setup

The working agent in this work was air. The results of experiments presented as the dependences of dimensionless similarity numbers (Strouhal number, Mach number) can be extended to other liquids, including incompressible ones.

As any oscillatory fluid–flow phenomena with fixed-geometry boundaries, jet-driven oscillators are governed by the Strouhal number, defined as [54]:

$$\mathrm{Sh_d} = \frac{f \cdot d}{W} \tag{2}$$

where f is the frequency of the oscillations (Hz); d is inlet nozzle diameter (m); and W bulk velocity in the exit of the inlet nozzle (m/s).

The definition employed for the Mach number was [55]:

$$M = \frac{W}{c_0} \tag{3}$$

To conduct experimental studies, a stand was developed, the scheme of which is shown in Figure 3.

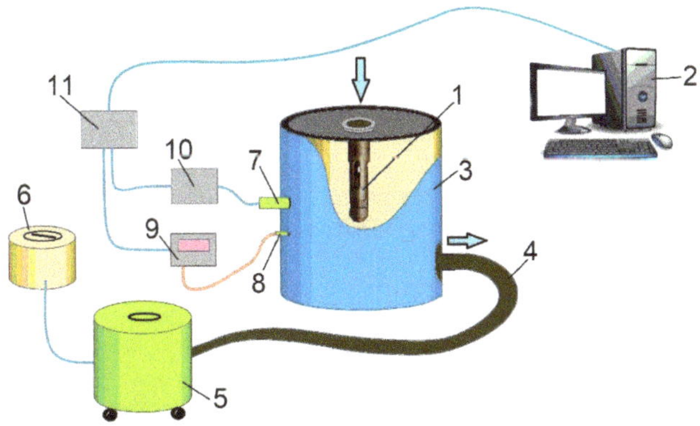

Figure 3. The scheme of the experimental setup: 1—emitter, 2—computer, 3—vacuum chamber, 4—pipeline for the removal of the working agent, 5—vacuum pump, 6—linear voltage converter, 7—measuring microphone, 8—fitting for measuring pressure in the vacuum chamber, 9—converter pressure sensor (pressure sensor), 10—microphone amplifier, and 11—ADC module.

As seen in Figure 3, the emitter (1) under study was located in the cover of the vacuum chamber. The formation of a jet in the resonant chamber of the device was carried out by sucking air from the vacuum chamber with the help of a vacuum pump (5) through a hose (4). Air was supplied to the inlet of the device from the surrounding space. With the help of voltage regulator (6), the suction pressure and, accordingly, the jet velocity were varied. The change in air pressure in the vacuum chamber and the generated pressure fluctuations were recorded on a computer through an external analog-to-digital converter (ADC) board with a sampling rate of 10 kHz. To measure these parameters, a measuring microphone and a pressure transducer were used.

The experimental data recorded on a computer were processed and the change in the pressure drop across the emitter (in time), the root-mean-square amplitude p, and the characteristic frequency of the generated oscillations f were determined. The pressure drop was used to calculate the jet velocity W in the resonant chamber of the device. According to the data obtained, the change in the mean square oscillation amplitude p/q reduced to the velocity head ($q = \rho \cdot W^2/2$) on the jet velocity was studied, similarly to [56].

An experimental study of the frequency of natural oscillations of the resonator was determined by the method of sounding the device with "white noise". For this, two measuring microphones were used: one of which recorded the incident sound wave and the second one recorded that which passed through the resonator. The ratio of the spectra of the transmitted signal to the incident signal makes it possible to obtain the resonant curve of the device.

2.3. Goal and Scope

The goal of this study was to determine the optimal design of the JDHO for the design of a cavitation reactor that allows the pre-treatment of organic waste. Based on the results of experimental studies, the following are determined: (a) the range of experimental data at which the amplitude of pressure pulsations is maximum, (b) the optimal ratio of the length of the resonator chamber to the diameter of the inlet pipe and the ratio of the inner diameter of the resonator chamber to the diameter of the inlet pipe, and (c) the optimal location of the JDHO in the design of the cavitation reactor depending on the Mach number and a dimensionless parameter that takes into account the atmospheric pressure of the medium and the operating pressure.

2.4. Assumptions and Limitations of the Study

When conducting research, the main assumptions and limitations should be noted:

- To determine the optimal design of the JDHO, which provides the maximum amplitude of pressure pulsations, the parameters d_2, D_C, and L_C were varied.
- During the operation of the JDHO, two cavitation mechanisms operate, namely: hydrodynamic and acoustic.
- The hydrodynamic mechanism is based on the fact that a local pressure drop occurs in the fluid flow when flowing through the nozzle. If the pressure in this area becomes lower than the pressure of saturated vapors or dissolved gases, then microbubbles are formed. Then, with an increase in local pressure and the collapse of microbubbles, cavitation occurs. This hydrodynamic mechanism also works during the development of vortex structures in the resonant chamber, since when the flow swirls, a region of low pressure is created in the center of the vortex.
- The acoustic mechanism is caused by the fact that pressure fluctuations created by the oscillator propagate in the environment and create elastic waves. During the passage of an elastic wave (in the half-cycle of the lower half-wave), a reduced pressure is created, which is lower than the pressure of the saturated vapors of a liquid or dissolved gases. This creates conditions for the formation of cavitation bubbles, which, when the pressure rises (in the half-cycle of the upper half-wave), collapse and create a cavitation effect.
- The shape of the cavitation bubbles, their size, and many other factors influence the collapse pressure. However, these issues are beyond the scope of this work and will be the subjects of future research.
- The JDHO is fundamentally new and has never before been used for pretreatment in AD technology.

3. Results of Experimental Studies

The study of natural oscillations of the emitter made it possible to obtain its resonance curve (Figure 4) and to determine the frequency of natural oscillations f_0. The obtained values with an error of less than 2% coincide with the theoretical value calculated by Equation (1).

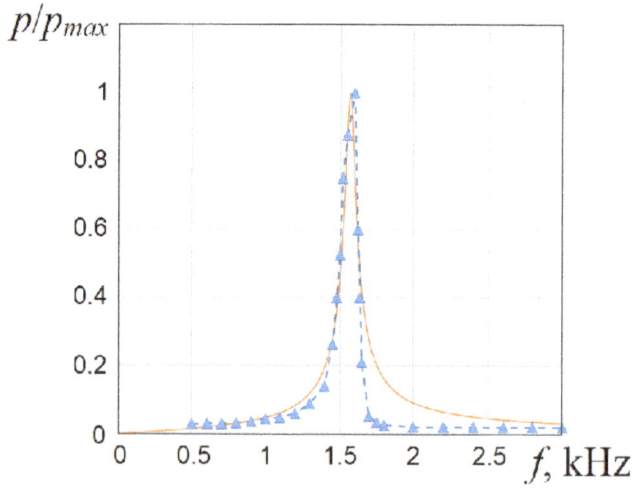

Figure 4. Resonance curve of natural oscillations of the JDHO: orange line—calculated data using formula (1), blue line with triangular markers—experimental points.

An increase in the pressure drop across the emitter leads to the formation of a jet, the speed of which is determined by the properties of the working agent (in our case, air) and the value of the pressure drop, which is measured by a strain gauge pressure difference sensor. At a certain value of the jet velocity, the generation of the tone of the hole begins.

Hole tone harmonics are amplified through resonance when their frequency coincides with the natural frequency of the radiator. The resonant amplification band is determined by the quality factor of the resonator (the width of the resonant curve).

The mechanism of generation of pressure oscillations by a JDHO is based on the resonant amplification of pressure fluctuations in a liquid (gas) flow passing through the emitter. The frequency of generated oscillations depends on the jet velocity and, at a certain liquid flow rate, coincides with the frequency of the natural oscillations of the resonator. From the point of view of the efficiency of its application, it is important that the operating mode of the device be resonant, which is why it is so important to determine the natural frequency of the emitter.

The mechanism observed in this work, due to the presence of the JDHO cavity, differs from the theoretical model proposed for the wedge tone and hole tone due to the presence of a closed (resonant) volume inside the chamber. The following sequence of events is assumed:

(1) The flow in the jet contains a low-frequency ordered axisymmetric variable component (and periodic volume flow fluctuations). When this component enters the outlet and the jet encounters various resistances in the outlet plane of the oscillator, periodic pressure pulses arise inside the chamber.
(2) These pulses are selectively amplified by the Helmholtz resonance mechanism and a pulsating pressure field is installed in the chamber.
(3) The pulsating pressure field causes flow rate pulsations at the chamber inlet, which leads to an effective amplification of the jet oscillations at the frequency of the ordered component.
(4) The viscous jet displacement layer (the expansion of the jet from its exit from the inlet nozzle to the start of collision with the outlet nozzle) responds to the amplification of jet oscillations in the range of its own acoustic frequencies and amplifies them. As a result, the ordered motion inside the jet is enhanced, vortex rings appear, and the circuit closes.

As the jet velocity increases, the amplitude of oscillations increases and, having reached the maximum value (it has its own for each version of the device geometry), begins to decrease. This is accompanied by an increase in the frequency of the generated oscillations. This pattern of changing the parameters of the generated oscillations is known as the mode (Mode). Each mode has its own Strouhal numbers. The first mode (Mode 1) is characterized by high jet velocities, and a further increase in velocity does not lead to the generation of harmonic oscillations at the resonant frequency. Figure 5 shows the dependences of the Strouhal numbers on the dimensionless length of the resonant chamber. The frequency of the generated oscillations and the jet velocity corresponded to the maximum value of the oscillation amplitude within each mode. These dependencies were approximated via the following linear function:

$$\mathrm{Sh_I} = 0.417 - 0.08\frac{L_C}{d_1} \tag{4}$$

$$\mathrm{Sh_{II}} = 0.609 - 0.087\frac{L_C}{d_1} \tag{5}$$

Within each mode, the characteristic Strouhal number practically does not change with increasing jet velocity. As the jet velocity increases, a jump from one mode to another (from Mode 2 to Mode 1) can occur. From the point of view of process hydrodynamics, the mode number represents the number of simultaneously moving vortex rings in the resonant chamber around the jet [41]. Inside the vortex ring, there is a zone of low pressure due to the circulation of the flow, and favorable conditions are created in this zone for the formation of cavitation clouds. In the study [42], photographs were taken using a high-speed camera, showing that the cavitation cloud is able to stretch over a length of 1 d_1 to 10 d_1, and at this distance, the cloud is continuous. The cavitation cloud grows in the radial direction until it reaches the maximum distance at which it collapses [57].

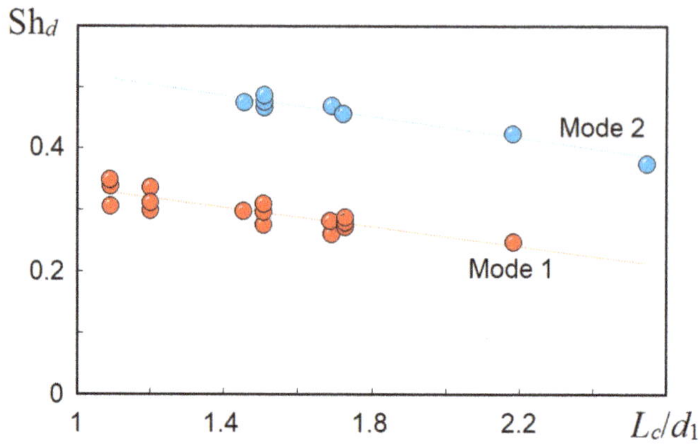

Figure 5. Dependences of the Strouhal number for two modes on the length of the resonant chamber.

Figure 6 shows the dependences of the maximum value of the root-mean-square amplitude of pressure fluctuations in the resonant chamber reduced to the velocity head on the length of the chamber. It can be seen that although the oscillations in the first mode are larger in absolute values than the oscillations in other modes, in relative terms, the amplitude is inferior to the second mode, which operates at lower jet velocities.

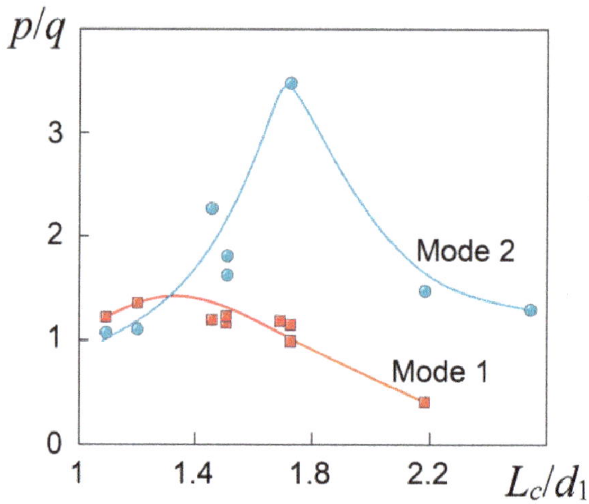

Figure 6. Dependences of the maximum root-mean-square oscillation amplitude reduced to the velocity head on the length of the resonant chamber.

Figure 7 shows the dependences of the relative root-mean-square amplitude of the generated oscillations on the jet velocity (Mach number) for various geometry options. So, for a chamber (D_C/d_1 = 5.5, L_C/d_1 = 1.5), the second mode is the strongest, while an increase in the resonant chamber diameter leads to the first mode becoming strong (red and blue lines). Another experiment with L_C/d_1 = 1.5 showed (Figure 7b) that a decrease in the resonant chamber diameter leads to a decrease in the oscillation generation efficiency—the relative amplitude of the second mode is two times lower. Moreover, the maximum value is shifted towards higher values of speed. There is a completely logical explanation for this, since a resonator with a smaller diameter of the resonant chamber has a higher frequency of natural oscillations and, in order to achieve the required value of the Strouhal number, large jet velocities are required. The hatching on the graphs means that in these ranges of

Mach numbers (jet velocity), the emitter operates not only in the hydrodynamic cavitation mode but also acoustic cavitation due to the generation of intense acoustic waves.

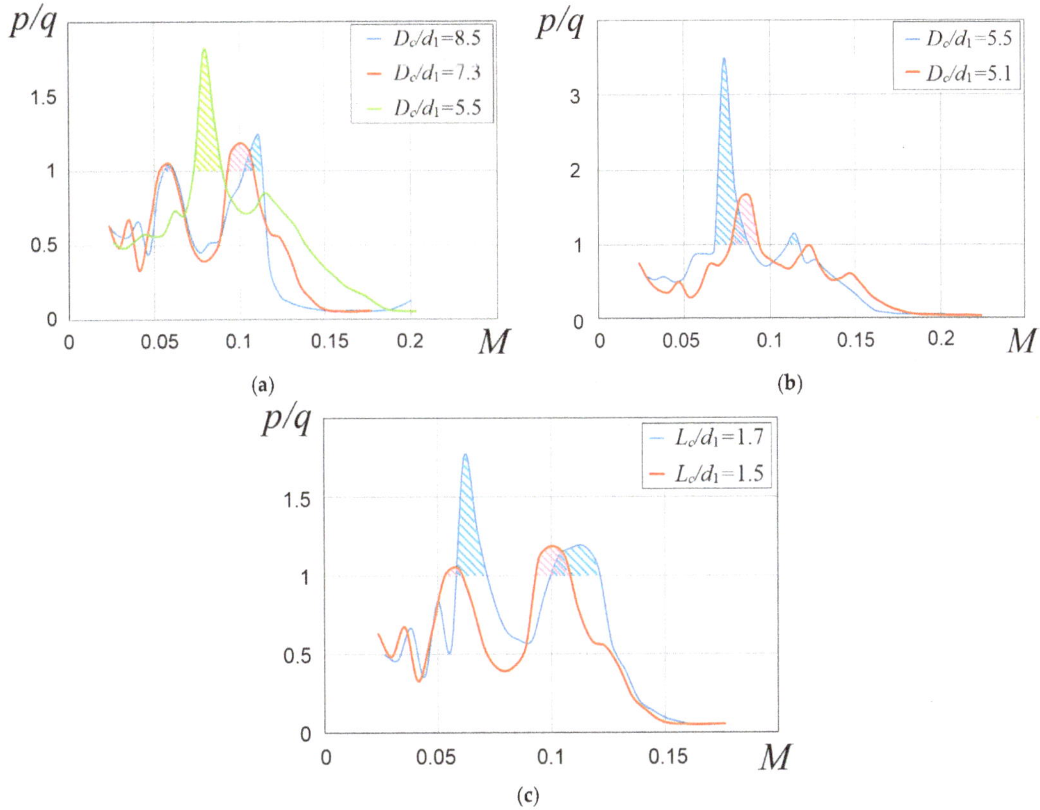

Figure 7. Dependences of the root-mean-square amplitude of oscillations in the chamber reduced to the velocity head on the jet velocity for various geometry options: (**a**) L_C/d_1 = 1.5; (**b**) L_C/d_1 = 1.73; (**c**) D_C/d_1 = 7.3.

Figure 6c shows a comparison of the curves for emitters with the same chamber diameter but different lengths. The oscillation amplitude in the second mode is much larger for a resonator with a larger chamber length, although resonance sets in at high velocities. Obviously, the length of the chamber determines the frequency of the hole-tone generation, and this is confirmed by the data in Figure 5.

The experiments performed show that the maximum value of the oscillation amplitude depends on the consistency of the regime parameters and the geometric dimensions of the emitter channel. Based on the experimental studies carried out, it can be argued that the optimal, from the point of view of the efficiency of converting the jet energy into oscillation energy, geometric relationships are L_C/d_1 = 1.73 and D_C/d_1 = 5.5.

4. Discussion and Future Research

The maximum values of the oscillation amplitude within each mode are reached at a frequency whose value is slightly (about 7%) higher than the frequency of natural oscillations of the resonator. This experimental fact was also observed by other researchers [41]. The obtained data on the values of the Strouhal number (Equations (4) and (5)) and the optimal ratios of the geometric parameters make it possible to calculate the required jet velocity, including for liquid, to achieve efficient generation. As an example, consider the calculation for the second mode. According to Equation (5) we obtain:

$$\mathrm{Sh_{II}} = 0.609 - 0.087\frac{L_C}{d_1} = 0.609 - 0.087 \cdot 1.73 = 0.458$$

on the other hand,

$$\mathrm{Sh_{II}} = \frac{f \cdot d_1}{W} = \frac{1.07 \cdot f_0 \cdot d_1}{W} \quad (6)$$

Expression (1) for the frequency of natural oscillations can be rewritten as

$$f_0 = \frac{c_0}{2\pi D_C}\sqrt{\frac{1}{L_C}\left(\frac{d_1^2}{l_1+0.25\pi d_1} + \frac{d_2^2}{l_2+0.25\pi d_2}\right)} = \\ = \frac{c_0}{2\pi(D_C/d_1)\sqrt{L_C}}\sqrt{\frac{1}{l_1+0.25\pi d_1} + \frac{(d_2/d_1)^2}{l_2+0.25\pi d_2}} \quad (7)$$

Substituting Expression (7) into (6) we obtain

$$\mathrm{Sh_{II}} = \frac{1.07}{(W/c_0)}\frac{d_1}{2\pi(D_C/d_1)\sqrt{L_C}}\sqrt{\frac{1}{l_1+0.25\pi d_1} + \frac{(d_2/d_1)^2}{l_2+0.25\pi d_2}} = \\ = \frac{1.07}{(W/c_0)}\frac{\sqrt{d_1}}{2\pi(D_C/d_1)\sqrt{L_C/d_1}}\sqrt{\frac{1}{l_1+0.25\pi d_1} + \frac{(d_2/d_1)^2}{l_2+0.25\pi d_2}} = \\ = \frac{1.07}{2\pi \cdot M \cdot (D_C/d_1)\sqrt{L_C/d_1}}\sqrt{\frac{1}{\frac{l_1}{d_1}+0.25\pi} + \frac{(d_2/d_1)^2}{\frac{l_2}{d_1}+\frac{0.25\pi d_2}{d_1}}} \quad (8)$$

Substituting into the resulting Expression (8) the optimal ratios of geometric parameters for the second mode, we obtain

$$M_{\mathrm{II}} = \frac{0.035}{\mathrm{Sh_{II}}} = 0.076 \quad (9)$$

The value was calculated similarly for the first mode and obtained $M_{\mathrm{I}} = 0.126$. It should be noted that the values obtained fall within the range of experimental data (Figure 8), for which the values of the amplitude of pressure fluctuations should be maximum.

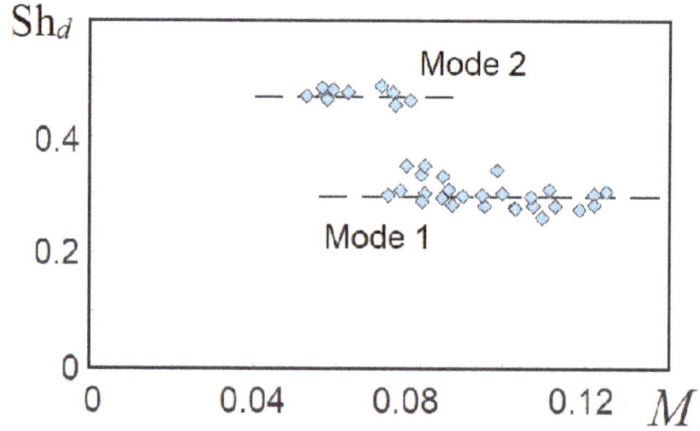

Figure 8. Relationship between the Strouhal number and the Mach number for two modes.

The cavitation intensity of a jet is affected by multiple factors, such as operating pressure, ambient pressure, and nozzle structure [42]. To evaluate the effect of operating pressure and ambient pressure, the dimensionless parameter P_r can be used.

$$P_r = \frac{P_{in}}{P_{out}} \rightarrow P_{in} - P_{out} = P_r P_{out} - P_{out} = P_{out}(P_r - 1) \tag{10}$$

where P_{out} is the ambient pressure (pressure at the JDHO outlet, Pa) and P_{in} is the working pressure (pressure in front of the JDHO, Pa).

The jet velocity in the JDHO is determined by the pressure drop and can be written as:

$$\Delta P = P_{in} - P_{out} = \rho\frac{W^2}{2} = \rho\frac{M^2\, c_0^2}{2} \tag{11}$$

Based on Equations (3), (10) and (11), we can write an expression for the dimensionless parameter:

$$\rho\frac{M^2\, c_0^2}{2} = P_{out}(P_r - 1) \rightarrow \rho\frac{M^2\, c_0^2}{2P_{out}} = P_r - 1 \tag{12}$$

$$P_r = \frac{\frac{M^2\, c_0^2}{2} + \frac{P_{atm}}{\rho} + gh}{\frac{P_{atm}}{\rho} + gh} \tag{13}$$

where P_{atm} is the atmospheric pressure (Pa), h is the JDHO location depth in the cavitation reactor of the anaerobic digestion system (m), g is the gravitational acceleration (9.81 m/s), and ρ is the fluid density (kg/m^3).

Equation (13) relates to the jet velocity, characterized by the Mach number; the depth of the JDHO installation in the tank; and the dimensionless parameter P_r (Figure 9). In ref. [42], the optimal value of the pressure coefficient is considered to be 25. This value corresponds to our experimental data for the second mode. The resulting dependence (13) allows the determination of the optimal location when designing a cavitation reactor.

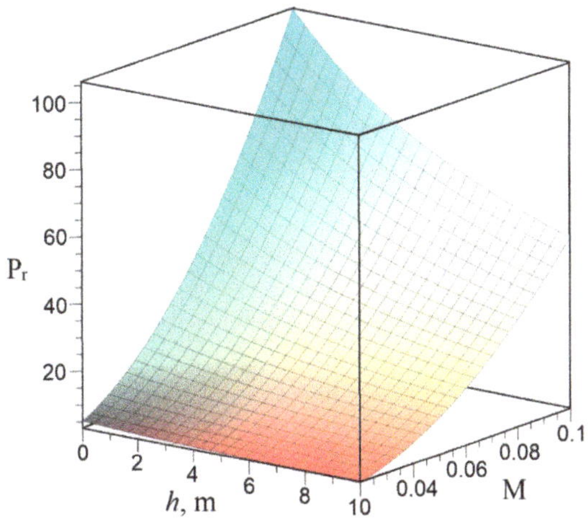

Figure 9. Dependence of P_r on the jet velocity and the depth of the JDHO in the liquid.

The choice of the substrate pretreatment method is a difficult step in the design of the AD process (Figure 10) [3]. It is important to rank the methods according to energy costs and the effect they give. The physical method of pretreatment of liquid agricultural waste is an important step for improving the efficiency of the bioconversion, compaction, and distribution of the particles, enzymatic availability, and overall conversion of feedstock to biogas without the formation of toxic by-products [58].

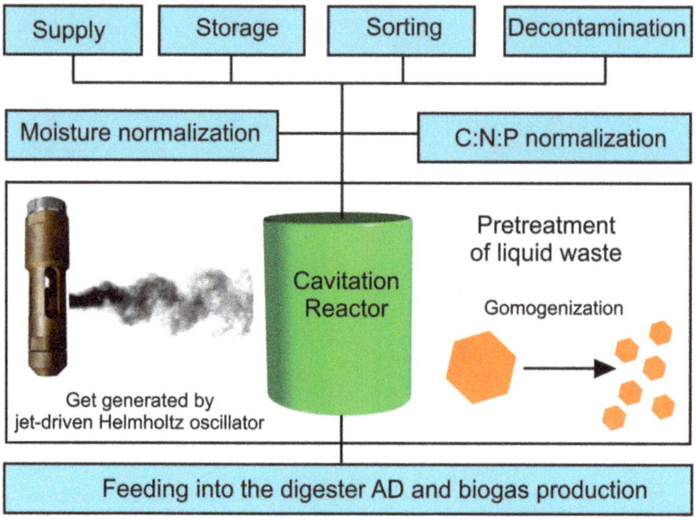

Figure 10. Application of cavitation reactor in AD technology.

The advantages of the proposed cavitation reactor is the simplicity of its design and operation, easy scalability, the possibility of obtaining stable emulsions from fat-containing waste, and a high degree of the homogenization of liquid raw materials [59]. This is confirmed by the presence of a number of commercial cavitation reactors. For example, Cavitation Technologies, Inc., owns the patented CTi Nano Neutralization technology, which includes a multi-stage hydrodynamic cavitation device [59]. The reaction system is flexible in scale and can be applied in the field of edible oil refining, algae oil extraction, renewable fuel production, biodiesel production, etc. A number of studies have shown that with a decrease in the particle size of the raw material supplied to the digester, the yield of biogas increases [56,60–63]. An increase in the surface area of the organic material provides a larger contact area for microorganisms, which leads to an intensification of the gas formation process [63].

Currently, the design and development of a laboratory–analytical complex for studying the processes of anaerobic digestion on a semi-industrial scale with a hydraulic mixing system [64], as well as subsequent thermochemical conversion of the digestate [65], is underway. It is planned to use a cavitation reactor for the pre-treatment of agricultural waste.

5. Conclusions

Acoustic cavitation is a promising approach that can be applied to biomass pretreatment to improve AD efficiency. The JDHO are easy to scale, and this is an advantage of their use. Based on the experimental data, it was found that the optimal ratio of the length of the resonator chamber to the diameter of the inlet nozzle is 1.73 and the ratio of the inner diameter of the resonator chamber to the diameter of the inlet nozzle is 5.5. The results obtained make it possible to improve existing jet vibration emitters for the implementation of acoustic impact and to expand the scope of their application in the economy.

Author Contributions: Conceptualization, A.A., A.G., E.M. and J.K.; methodology, A.A.; software, E.M.; validation, A.A., E.M. and A.K.; formal analysis, A.G., V.P., A.K. and V.B.; investigation, A.A., A.G., E.M., V.P., A.K., V.B. and J.K.; resources, A.A., V.P., A.K. and V.B.; data curation, A.A., A.G., E.M., V.P., V.B. and J.K.; writing—original draft preparation, E.M. and J.K.; writing—review and editing, E.M., J.K. and V.P.; visualization, A.A. and E.M.; supervision, V.P., A.K. and V.B.; project administration, J.K., V.P. and V.B.; funding acquisition, V.P., A.K. and V.B. All authors have read and agreed to the published version of the manuscript.

Funding: This research received no external funding.

Institutional Review Board Statement: Not applicable.

Data Availability Statement: The data that support the findings of this study are available within the article.

Conflicts of Interest: The authors declare no conflict of interest.

References

1. Su, X.; Shao, X.; Geng, Y.; Tian, S.; Huang, Y. Optimization of feedstock and insulating strategies to enhance biogas production of solar-assisted biodigester system. *Renew. Energy* **2022**, *197*, 59–68. [CrossRef]
2. Tiwari, P.; Wang, T.; Indlekofer, J.; Haddad, I.E.; Biollaz, S.; Prevot, A.S.H.; Lamkaddam, H. Online detection of trace volatile organic sulfur compounds in a complex biogas mixture with proton-transfer-reaction mass spectrometry. *Renew. Energy* **2022**, *196*, 1197–1203. [CrossRef]
3. Palvinskiy, V.V.; Vasilev, F.A.; Evteev, V.K. Methods of Pretreating Raw Materials Containing Organic Compounds before Anaerobic Digestion. *Eng. Technol. Syst.* **2022**, *32*, 10–27. [CrossRef]
4. Nguyen, V.K.; Chaudhary, D.K.; Dahal, R.H.; Trinh, N.H.; Kim, J.; Chang, S.W.; Hong, Y.; La, D.D.; Nguyen, C.X.; Ngo, H.H.; et al. Review on pretreatment techniques to improve anaerobic digestion of sewage sludge. *Fuel* **2021**, *285*, 119105. [CrossRef]
5. Atelge, M.R.; Atabani, A.E.; Banu, J.R.; Krisa, D.; Kaya, M.; Eskicioglu, C.; Kumar, G.; Lee, C.; Yildiz, Y.Ş.; Unalan, S.; et al. A critical review of pretreatment technologies to enhance anaerobic digestion and energy recovery. *Fuel* **2020**, *270*, 117494. [CrossRef]
6. Poddar, B.J.; Nakhate, S.P.; Gupta, R.K.; Chavan, A.R.; Singh, A.K.; Khardenavis, A.A.; Purohit, H.J. A comprehensive review on the pretreatment of lignocellulosic wastes for improved biogas production by anaerobic digestion. *Int. J. Environ. Sci. Technol.* **2022**, *19*, 3429–3456. [CrossRef]
7. Fei, X.; Chen, T.; Jia, W.; Shan, Q.; Hei, D.; Ling, Y.; Feng, J.; Feng, H. Enhancement effect of ionizing radiation pretreatment on biogas production from anaerobic fermentation of food waste. *Radiat. Phys. Chem.* **2020**, *168*, 108534. [CrossRef]
8. Chmielewski, A.G.; Sudlitz, M.; Han, B.; Pillai, S.D. Electron beam technology for biogas and biofertilizer generation at municipal resource recovery facilities. *Nukleonika* **2021**, *66*, 213–219. [CrossRef]
9. Kovačić, Đ.; Rupčić, S.; Kralik, D.; Jovičić, D.; Spajić, R.; Tišma, M. Pulsed electric field: An emerging pretreatment technology in a biogas production. *Waste Manag.* **2021**, *120*, 467–483. [CrossRef]
10. Bardi, M.J.; Mahmood, A.; Lippert, T.; Bandelin, J.; Koch, K. Stimulating effect of hydrostatic pressure on ultrasonic sewage sludge treatment for COD solubilization and methane production. *Bioresour. Technol.* **2022**, *348*, 126785. [CrossRef]
11. Nabi, M.; Gao, D.; Liang, J.; Cai, Y.; Zhang, P. Combining high pressure homogenization with free nitrous acid pretreatment to improve anaerobic digestion of sewage sludge. *J. Environ. Manag.* **2022**, *318*, 115635. [CrossRef] [PubMed]
12. Zieliński, M.; Dębowski, M.; Kisielewska, M.; Nowicka, A.; Rokicka, M.; Szwarc, K. Cavitation-based pretreatment strategies to enhance biogas production in a small-scale agricultural biogas plant. *Energy Sustain. Dev.* **2019**, *49*, 21–26. [CrossRef]
13. Chubur, V.; Danylov, D.; Chernysh, Y.; Plyatsuk, L.; Shtepa, V.; Haneklaus, N.; Roubik, H. Methods for Intensifying Biogas Production from Waste: A Scientometric Review of Cavitation and Electrolysis Treatments. *Fermentation* **2022**, *8*, 570. [CrossRef]
14. Neis, U. The use of power ultrasound for wastewater and biomass treatment. In *Power Ultrasonics: Applications of High-Intensity Ultrasound*; Gallego-Juarez, J.A., Graff, K.F., Eds.; Woodhead Publishers: Sawston, UK, 2014; pp. 974–996. [CrossRef]
15. Gogate, P.R.; Bhosale, G.S. Comparison of effectiveness of acoustic and hydrodynamic cavitation in combined treatment schemes for degradation of dye wastewaters. *Chem. Eng. Process. Process Intensif.* **2013**, *71*, 59–69. [CrossRef]
16. Buller, L.S.; Sganzerla, W.G.; Lima, M.N.; Muenchow, K.E.; Timko, M.T.; Forster-Carneiro, T. Ultrasonic pretreatment of brewers' spent grains for anaerobic digestion: Biogas production for a sustainable industrial development. *J. Clean. Prod.* **2022**, *355*, 131802. [CrossRef]
17. Nagarajan, S.; Ranade, V.V. Valorizing Waste Biomass via Hydrodynamic Cavitation and Anaerobic Digestion. *Ind. Eng. Chem. Res.* **2021**, *60*, 16577–16598. [CrossRef]
18. Patil, P.N.; Gogate, P.R. Csoka, Levente Agota Dregelyi-Kiss, Miklos Horvath. Intensification of biogas production using pretreatment based on hydrodynamic cavitation. *Ultrason. Sonochem.* **2016**, *30*, 79–86. [CrossRef]
19. Garuti, M.; Langone, M.; Fabbri, C.; Piccinini, S. Monitoring of full-scale hydrodynamic cavitation pretreatment in agricultural biogas plant. *Bioresour. Technol.* **2018**, *247*, 599–609. [CrossRef]
20. Zubrowska-Sudol, M.; Dzido, A.; Garlicka, A.; Krawczyk, P.; Stępień, M.; Umiejewska, K.; Walczak, J.; Wołowicz, M.; Sytek-Szmeichel, K. Innovative Hydrodynamic Disintegrator Adjusted to Agricultural Substrates Pre-treatment Aimed at Methane Production Intensification–CFD Modelling and Batch Tests. *Energies* **2020**, *13*, 4256. [CrossRef]
21. Garlicka, A.; Zubrowska-Sudol, M.; Umiejewska, K.; Roubinek, O.; Palige, J.; Chmielewski, A. Effects of Thickened Excess Sludge Pre-Treatment Using Hydrodynamic Cavitation for Anaerobic Digestion. *Energies* **2020**, *13*, 2483. [CrossRef]

22. Nagarajan, S.; Ranade, V.V. Pretreatment of Lignocellulosic Biomass Using Vortex-Based Devices for Cavitation: Influence on Biomethane Potential. *Ind. Eng. Chem. Res.* **2019**, *58*, 15975–15988. [CrossRef]
23. Nagarajan, S.; Ranade, V.V. Pre-treatment of distillery spent wash (vinasse) with vortex based cavitation and its influence on biogas generation. *Bioresour. Technol. Rep.* **2020**, *11*, 100480. [CrossRef]
24. Kovalev, A.A.; Kovalev, D.A.; Grigoriev, V.S. Energy Efficiency of Pretreatment of Digester Synthetic Substrate in a Vortex Layer Apparatus. *Eng. Technol. Syst.* **2020**, *30*, 92–110. [CrossRef]
25. Kovalev, A.A.; Kovalev, D.A.; Panchenko, V.A.; Zhuravleva, E.A.; Laikova, A.A.; Shekhurdina, S.V.; Vivekanand, V.; Litti, Y.V. Approbation of an innovative method of pretreatment of dark fermentation feedstocks. *Int. J. Hydrogen Energy* **2022**, *47*, 33272–33281. [CrossRef]
26. Langone, M.; Soldano, M.; Fabbri, C.; Pirozzi, F.; Andreottola, G. Anaerobic Digestion of Cattle Manure Influenced by Swirling Jet Induced Hydrodynamic Cavitation. *Appl. Biochem. Biotechnol.* **2018**, *184*, 1200–1218. [CrossRef]
27. Nakashima, K.; Ebi, Y.; Shibasaki-Kitakawa, N.; Soyama, H.; Yonemoto, T. Hydrodynamic Cavitation Reactor for Efficient Pretreatment of Lignocellulosic Biomass. *Ind. Eng. Chem. Res.* **2016**, *55*, 1866–1871. [CrossRef]
28. Saxena, S.; Saharan, V.K.; George, S. Modeling & simulation studies on batch anaerobic digestion of hydrodynamically cavitated tannery waste effluent for higher biogas yield. *Ultrason. Sonochem.* **2019**, *58*, 104692. [CrossRef]
29. Lee, I.; Han, J.-I. The effects of waste-activated sludge pretreatment using hydrodynamic cavitation for methane production. *Ultrason. Sonochem.* **2013**, *20*, 1450–1455. [CrossRef]
30. Jung, K.-W.; Hwang, M.-J.; Yun, Y.-M.; Cha, M.-J.; Ahn, K.-H. Development of a novel electric field-assisted modified hydrodynamic cavitation system for disintegration of waste activated sludge. *Ultrason. Sonochem.* **2014**, *21*, 1635–1640. [CrossRef]
31. Mancuso, G.; Langone, M.; Andreottola, G.; Bruni, L. Effects of hydrodynamic cavitation, low-level thermal and low-level alkaline pre-treatments on sludge solubilisation. *Ultrason. Sonochem.* **2019**, *59*, 104750. [CrossRef]
32. Ge, M.; Petkovšek, M.; Zhang, G.; Jacobs, D.; Coutier-Delgosha, O. Cavitation dynamics and thermodynamic effects at elevated temperatures in a small Venturi channel. *Int. J. Heat Mass Transf.* **2021**, *170*, 120970. [CrossRef]
33. Ge, M.; Manikkam, P.; Ghossein, J.; Subramanian, K.R.; Coutier-Delgosha, O.; Zhang, G. Dynamic mode decomposition to classify cavitating flow regimes induced by thermodynamic effects. *Energy* **2022**, *254*, 124426. [CrossRef]
34. Lanfranchi, A.; Tassinato, G.; Valentino, F.; Martinez, G.; Jones, E.; Gioia, C.; Bertin, L.; Cavinato, C. Hydrodynamic cavitation pre-treatment of urban waste: Integration with acidogenic fermentation, PHAs synthesis and anaerobic digestion processes. *Chemosphere* **2022**, *301*, 134624. [CrossRef] [PubMed]
35. Ge, M.; Sun, C.; Zhang, G.; Coutier-Delgosha, O.; Fan, D. Combined suppression effects on hydrodynamic cavitation performance in Venturi-type reactor for process intensification. *Ultrason. Sonochem.* **2022**, *86*, 106035. [CrossRef]
36. Ge, M.; Zhang, G.; Petkovšek, M.; Long, K.; Coutier-Delgosha, O. Intensity and regimes changing of hydrodynamic cavitation considering temperature effects. *J. Clean. Prod.* **2022**, *338*, 130470. [CrossRef]
37. Terán Hilares, R.; de Almeida, G.F.; Ahmed, M.A.; Antunes, F.A.F.; da Silva, S.S.; Han, J.-I.; dos Santos, J.C. Hydrodynamic cavitation as an efficient pretreatment method for lignocellulosic biomass: A parametric study. *Bioresour. Technol.* **2017**, *235*, 301–308. [CrossRef]
38. Marta, K.; Paulina, R.; Magda, D.; Anna, N.; Aleksandra, K.; Marcin, D.; Kazimierowicz, J.; Marcin, Z. Evaluation of Ultrasound Pretreatment for Enhanced Anaerobic Digestion of Sida hermaphrodita. *Bioenerg. Res.* **2020**, *13*, 824–832. [CrossRef]
39. Ziyan, L.; Kaiwen, Z.; Yingzheng, L.; Xin, W. Jet sweeping angle control by fluidic oscillators with master-slave designs. *Chin. J. Aeronaut.* **2021**, *34*, 145–162. [CrossRef]
40. Joulaei, A.; Nili-Ahmadabadi, M.; Kim, K.C. Parametric study of a fluidic oscillator for heat transfer enhancement of a hot plate impinged by a sweeping jet. *Appl. Therm. Eng.* **2022**, *205*, 118051. [CrossRef]
41. Morel, T. Experimental study of a jet-driven Helmholtz oscillator. *J. Fluids Eng. Trans. ASME* **1979**, *101*, 383–390. [CrossRef]
42. Wu, Q.; Wei, W.; Deng, B.; Jiang, P.; Li, D.; Zhang, M.; Fang, Z. Dynamic characteristics of the cavitation clouds of submerged Helmholtz self-sustained oscillation jets from high-speed photography. *J. Mech. Sci. Technol.* **2019**, *33*, 621–630. [CrossRef]
43. Fang, Z.L.; Wu, Q.; Jiang, P.; Wei, W. Numerical investigation on flow field characteristics of Helmholtz oscillator. *IOP Conf. Ser. Earth Environ. Sci.* **2019**, *240*, 062047. [CrossRef]
44. Liu, W.; Kang, Y.; Zhang, M.; Wang, X.; Li, D. Self-sustained oscillation and cavitation characteristics of a jet in a Helmholtz resonator. *Int. J. Heat Fluid Flow* **2017**, *68*, 158–172. [CrossRef]
45. Hiremath, L.; Nipun, S.; Sruti, O.; Kala, N.G.; Aishwarya, B.M. Sonochemistry: Applications in Biotechnology. In *Sonochemical Reactions*; Karakuş, S., Ed.; IntechOpen: London, UK, 2020.
46. Ghasemi, M.; Sivaloganathan, S. Effect of inertial acoustic cavitation on antibiotic efficacy in biofilms. *Appl. Math. Mech.* **2021**, *42*, 1397–1422. [CrossRef]
47. Stebeleva, O.P.; Minakov, A.V. Application of Cavitation in Oil Processing: An Overview of Mechanisms and Results of Treatment. *ACS Omega* **2021**, *6*, 31411–31420. [CrossRef]
48. Gensheng, L.; Zhonghou, S.; Changshan, Z.; Debin, Z.; Hongbing, C. Investigation and application of self-resonating cavitating water jet in petroleum engineering. *Pet. Sci. Technol.* **2005**, *23*, 1–15. [CrossRef]
49. Marfin, E.A.; Gataullin, R.N.; Abdrashitov, A.A. Acoustic stimulation of oil production by a downhole emitter based on a jet-driven Helmholtz oscillator. *J. Pet. Sci. Eng.* **2022**, *215*, 110705. [CrossRef]

50. Wu, Z.; Tagliapietra, S.; Giraudo, A.; Martina, K.; Cravotto, G. Harnessing cavitational effects for green process intensification. *Ultrason. Sonochem.* **2019**, *52*, 530–546. [CrossRef]
51. Strieder, M.M.; Silva, E.K.; Meireles, M.A.A. Advances and innovations associated with the use of acoustic energy in food processing: An updated review. *Innov. Food Sci. Emerg. Technol.* **2021**, *74*, 102863. [CrossRef]
52. Dyussenov, K.M.; Dyussenova, J.; Nedugov, I. The Using of Controlled Cavitation Processes in Some Engineering and Agricultural Applications. *Univers. J. Eng. Sci.* **2013**, *1*, 89–94. [CrossRef]
53. Smirnov, I.; Mikhailova, N. An Analysis of Acoustic Cavitation Thresholds of Water Based on the Incubation Time Criterion Approach. *Fluids* **2021**, *6*, 134. [CrossRef]
54. Sanmartın, P.C. Numerical Study of Fluidic Oscillators with Compressible Flow. Escola Tecnica Superior d'Enginyeries Industrial i Aeron Autica de Terrassa (ETSEIAT) Treball de fi de Grau Grau en Enginyeria en Tecnologies Aeroespacials. 2017. Available online: https://upcommons.upc.edu/bitstream/handle/2117/100392/REPORT_295.png (accessed on 23 December 2022).
55. Bergadà, J.M.; Baghaei, M.; Prakash, B.; Mellibovsky, F. Fluidic Oscillators, Feedback Channel Effect under Compressible Flow Conditions. *Sensors* **2021**, *21*, 5768. [CrossRef]
56. Olatunji, K.O.; Madyira, D.M.; Ahmed, N.A.; Ogunkunle, O. Experimental evaluation of the influence of combined particle size pretreatment and Fe_3O_4 additive on fuel yields of Arachis Hypogea shells. *Waste Manag. Res.* **2023**, *41*, 467–476. [CrossRef]
57. Gavaises, M.; Villa, F.; Koukouvinis, P.; Marengo, M.; Franc, J.-P. Visualisation and les simulation of cavitation cloud formation and collapse in an axisymmetric geometry. *Int. J. Multiph. Flow* **2015**, *68*, 14–26. [CrossRef]
58. Amin, F.R.; Khalid, H.; Zhang, H.; Rahman, S.; Zhang, R.; Liu, G.; Chen, C. Pretreatment methods of lignocellulosic biomass for anaerobic digestion. *AMB Express* **2017**, *7*, 72. [CrossRef]
59. Zheng, H.; Zheng, Y.; Zhu, J. Developments in Hydrodynamic Cavitation Reactors: Cavitation Mechanism, Reactor Design, and Applications. *Engineering* **2022**, *19*, 180–198. [CrossRef]
60. Armaha, E.K.; Chettya, M.; Deenadayalu, N. Effect of Particle Size on Biogas Generation from Sugarcane Bagasse and Corn Silage. *Chem. Eng. Trans.* **2019**, *76*, 1472–1476. [CrossRef]
61. Luo, L.; Qu, Y.; Gong, W.; Qin, L.; Li, W.; Sun, Y. Effect of Particle Size on the Aerobic and Anaerobic Digestion Characteristics of Whole Rice Straw. *Energies* **2021**, *14*, 3960. [CrossRef]
62. Olatunji, K.O.; Madyira, D.M.; Ahmed, N.A.; Jekayinfa, S.O.; Ogunkunle, O. Modelling the effects of particle size pretreatment method on biogas yield of groundnut shells. *Waste Manag. Res.* **2022**, *40*, 1176–1188. [CrossRef]
63. Heller, R.; Roth, P.; Hülsemann, B.; Böttinger, S.; Lemmer, A.; Oechsner, H. Effects of Pretreatment with a Ball Mill on Methane Yield of Horse Manure. *Waste Biomass Valorization* **2023**. [CrossRef]
64. Karaeva, J.V.; Khalitova, G.R.; Kovalev, D.A.; Trakhunova, I.A. Study of the process of hydraulic mixing in anaerobic digester of biogas plant. *Chem. Process Eng.* **2015**, *36*, 101–112. [CrossRef]
65. Karaeva, J.V.; Timofeeva, S.S.; Islamova, S.I.; Gerasimov, A.V. Pyrolysis kinetics of new bioenergy feedstock from anaerobic digestate of agro-waste by thermogravimetric analysis. *J. Environ. Chem. Eng.* **2022**, *10*, 3. [CrossRef]

agriculture

MDPI

Article

Dynamic Cutting Performance Test and Parameter Optimization of Longicorn Bionic Blade for Industrial Hemp Harvester

Kunpeng Tian [1,2], Bin Zhang [2], Cheng Shen [2], Haolu Liu [2], Jicheng Huang [2,*] and Aimin Ji [1,*]

1 College of Mechanical and Electrical Engineering, Hohai University, Changzhou 213022, China
2 Nanjing Institute of Agricultural Mechanization, Ministry of Agriculture and Rural Affairs, Nanjing 210014, China; xtsset@hotmail.com (B.Z.); shencheng1989@cau.edu.cn (C.S.); liuhaolu@caas.cn (H.L.)
* Correspondence: huangjicheng@caas.cn (J.H.); jiam@hhuc.edu.cn (A.J.)

Abstract: In response to the unclear issue of whether the dynamic cutting performance and structural parameters of an industrial hemp blade, which was developed earlier based on the bionic prototype of the batocera horsfieldi, can be optimized in actual working conditions, this paper analyzes the effective clamping conditions of a reciprocating double-acting cutting blade for stalks and the cutting motion. To investigate the effect of different structural and motion parameters, as well as their interactions, of the bionic blade on cutting energy consumption, bionic blades with different combinations of tooth pitch and tooth angle were designed. A Box–Behnken response surface method with three factors and three levels was used to design an experimental scheme. Utilizing rigid-flexible coupling numerical simulation technology, numerical simulation experiments were conducted to investigate the cutting performance of industrial hemp stalks using the blade. A regression model for cutting energy consumption was established, and variance analysis indicated that tooth angle, speed ratio, and the interaction between tooth angle and speed ratio had an extremely significant effect on the regression model. The primary and secondary orders of factors affecting cutting energy consumption were determined to be: speed ratio > tooth angle > tooth pitch. Through optimization, the optimal parameter combination was found to be a blade tooth pitch of 6.61 mm, a tooth angle of 30°, and a speed ratio of 1.62. Under these conditions, the cutting energy consumption was 3947.99 mJ. The optimized parameters were verified through numerical simulation cutting experiments, and the results showed that the error compared with the optimization results was only 8.16%. This indicates that the optimization results have high credibility and further verifies the reliability of the model. This study can provide a reference for the development of cutting devices for industrial hemp harvesters and the selection of motion parameters.

Keywords: industrial hemp; bionic blade; cutting energy consumption; numerical simulation; parameter optimization

Citation: Tian, K.; Zhang, B.; Shen, C.; Liu, H.; Huang, J.; Ji, A. Dynamic Cutting Performance Test and Parameter Optimization of Longicorn Bionic Blade for Industrial Hemp Harvester. *Agriculture* **2023**, *13*, 1074. https://doi.org/10.3390/agriculture13051074

Academic Editors: Muhammad Sultan, Redmond R. Shamshiri, Md Shamim Ahamed and Muhammad Farooq

Received: 29 April 2023
Revised: 15 May 2023
Accepted: 15 May 2023
Published: 17 May 2023

1. Introduction

Industrial hemp, also known as *Cannabis sativa* L., refers to a non-psychoactive variety of cannabis with a THC (tetrahydrocannabinol) content of less than 0.3% [1–3]. Due to its low THC content, it lacks drug-related value, but its fibers are highly resilient, moisture-absorbent, breathable, antibacterial, and radiation-resistant, so it is widely used in various fields, such as textiles, papermaking, environmental materials, and military equipment [4–7]. In recent years, with the medicinal value of industrial hemp extract cannabidiol (CBD) in anti-tumor, neuroprotection, metabolism, and immune regulation being discovered, more and more countries and regions have announced the legalization of industrial hemp, and the industrial hemp industry is rapidly developing [8,9], gradually becoming a new global industry hotspot [10].

Harvesting is a crucial step in the industrial hemp production process. However, due to the lack of mature harvesting equipment, current industrial hemp harvesting in China still relies on traditional manual methods, which are labor-intensive, inefficient, and costly, making it difficult to meet the demands of modern hemp production. In recent years, the research and development work of industrial hemp harvesters have been carried out by the author's team [11]. However, due to the lack of specialized cutting blades with high adaptability, the harvester can only use reciprocating blades from traditional rice and corn harvesters. Compared to crops, such as rice, wheat, corn, and flax, industrial hemp plants are taller and sturdier, with tough and strong hemp fiber and hard woody stalks. The traditional blades from rice and wheat harvesters are no longer suitable for the cutting requirements of hemp stalks, which has led to high cutting energy consumption, severe fiber tearing, low stalk cutting rate, and issues, such as fiber wrapping, stalk blockage, and high machine failure rate, during harvesting operations. Therefore, cutting has become a bottleneck issue that urgently needs to be addressed in the field of industrial hemp harvesting [12].

To address the bottleneck issue of poor cutting quality, the author utilized the biomimetic approach in the early stage, taking the batocera horsfieldi, a common pest with excellent biting and cutting performance in hemp fields, as the bio-inspired object. They designed industrial hemp harvester bionic blades based on the structural characteristics of the batocera horsfieldi's mouthpart palates. Comparative experiments were conducted between the bionic blades and conventional rice and wheat harvester blades for single stalk cutting performance. The results showed that compared to conventional blades, the average maximum cutting force and cutting power of the bionic blades were reduced by 7.4% and 8.0%, respectively, and the cutting residues were more even. The bionic blades demonstrated superior cutting performance [12]. However, this study only focuses on the horizontal comparison of cutting performance between two types of blades, bionic and ordinary, and the experimental method only involves single stalk cutting at low speed (loading speed of 25 mm/min) using an electronic universal testing machine. It is still unclear how the dynamic cutting performance of the blades will be in actual operating conditions, and whether the blade structure can be further optimized, which requires further research.

Cutting is a complex process of interaction between rigid blades and flexible stalks, which is difficult to study theoretically. The use of numerical simulation technology can establish a visual virtual cutting environment and provide guidance for the development of cutters [13]. Relevant studies are as follows: Huang H. et al. used Ansys/LS-Dyna to simulate and analyze the sugarcane cutting process, and they determined the maximum stress values and distribution positions of the blade and sugarcane [14]. Qiu M. et al. utilized numerical simulation techniques to investigate the blade motion and stress wave propagation patterns during the sugarcane cutting process, revealing the mechanism of sugarcane cutting [15]. Yang W. et al. simulated the forces acting on the surface of the sugarcane blade and root system using a finite element and SPH coupling method [16]. Xue Z. combined the constitutive model parameters of cassava stem and used ANSYS to carry out numerical simulation analysis of the circular cutter cutting cassava stem, analyzing the stress, deformation, and damage to the cassava stem under different cutter structural parameters, providing reference for the optimization design of the cutter [17]. Meng Y. et al. conducted an explicit dynamic simulation based on Ansys/LS-Dyna on the process of saw cutting mulberry branches, analyzed the interaction process between the saw blade and mulberry branches, and optimized the structural parameters of the saw blade through virtual cutting experiments [18]. Similar research has also been carried out in areas, such as sugarcane topping [19], soybean [20], tomato [21], cotton [22], chrysanthemum stalks [23,24], and wood cutting [25]. The above studies show that numerical simulation methods represented by the finite element method can intuitively and quantitatively analyze the dynamic cutting mechanical properties of the blade on the stalk, and can provide an effective method for the development of low-energy-consumption crop stalk blades.

Considering this, this study will use numerical simulation technology to conduct numerical simulation experiments on blade cutting under different blade structure parameters and motion parameters, in order to determine the optimal parameter combination for industrial hemp stalk cutting and provide a reference for the development of industrial hemp harvester cutters.

2. Overall Structure and Working Principle of Industrial Hemp Harvester

The industrial hemp harvester is mainly composed of components, such as a divider, a cutting platform frame, a lateral conveying device (with three layers of conveying chains), a cutting device, a longitudinal conveying device, a stalk collection bin, and a chassis. The overall structure is shown in Figure 1, and the cutting device is illustrated in Figure 2.

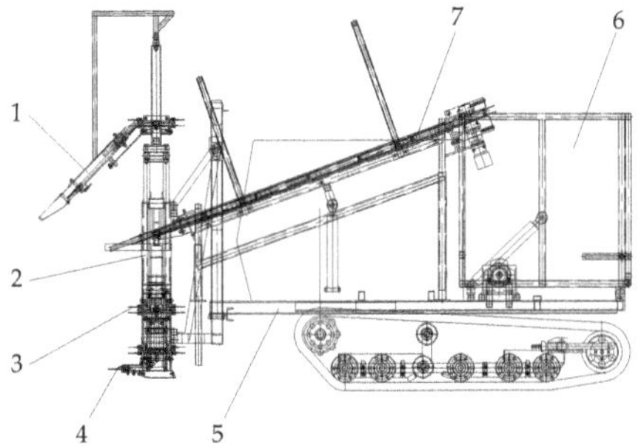

Figure 1. Overall structure and working principle of industrial hemp harvester. 1. Divider; 2. cutting platform frame; 3. lateral conveying device; 4. cutting device; 5. chassis; 6. stalk collection bin; 7. longitudinal conveying device.

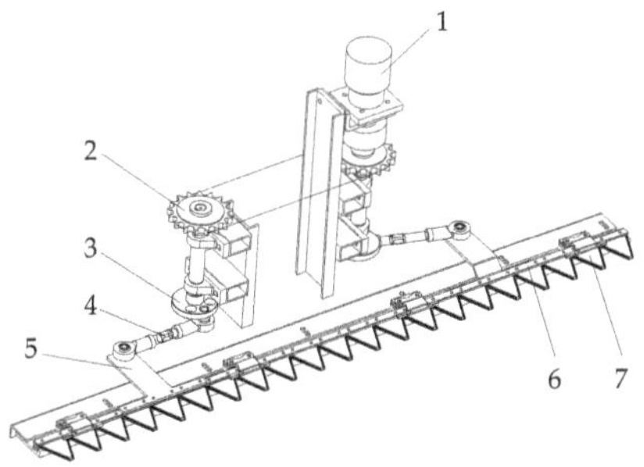

Figure 2. Structure diagram of the cutting device. 1. Hydraulic motor; 2. sprocket drive device; 3. eccentric disc; 4. articulated bearing components; 5. connecting arm; 6. blade-fixing rod; 7. blade.

The working principle is as follows: During operation, the divider separates the intertwined upper part of the hemp stalk and transfers it to the rear lateral conveying device. At the same time, the cutting device located in the lower part of the cutting platform cuts off the bottom part of the hemp stalk. The severed hemp stalk is pushed by the lateral conveying device towards one side of the cutting platform. When the stalk reaches the end of the cutting platform, it turns and enters the longitudinal conveying device. Finally, the longitudinal conveying device transports the hemp stalk to the stalk collection bin. When the stalk collection bin is full, it is manipulated to tilt and unload the collected hemp stalk. The cutting device is driven by a hydraulic motor, and power is transmitted through a sprocket drive to rotate an eccentric disc. The crank-connecting rod mechanism, composed of the left and right eccentric discs and articulated bearing components, converts the rotary motion into linear reciprocating motion, thereby achieving the reciprocating cutting of the blade.

3. Stalk Cutting Theoretical Analysis and Bionic Blade Design

3.1. Analysis of Cutting Force on Reciprocating Double-Acting Blade

For the reciprocating double-acting blade, the premise for ensuring effective cutting is that the blade should be able to firmly clamp the stalk and not slip. The force analysis of the blade when cutting hemp stalks is shown in Figure 3.

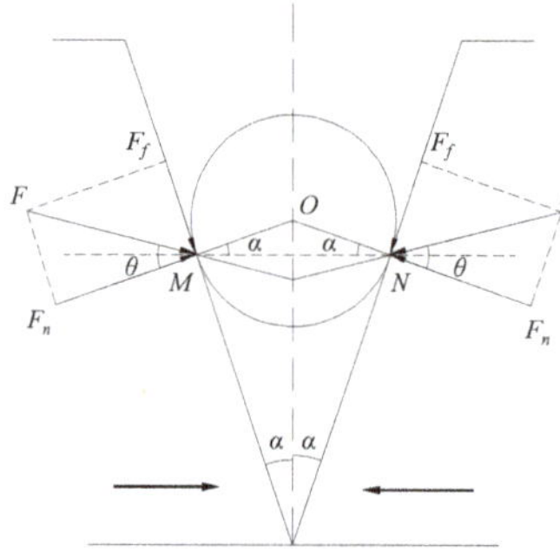

Figure 3. Analysis of cutting force. Note: an arrow indicates the direction of the blade's motion.

When the left and right blades simultaneously move towards the center for cutting, the hemp stalk tends to slide forward due to the squeezing force of the blade. At the contact points M and N between the blades and the stalk, the stalk is subjected to normal compression force F_n and frictional force F_f along the direction of the blade, and the resultant force is F. The angle between the normal force F_n and the resultant force F is the friction angle θ, which can be expressed by the following equation:

$$F_f = F_n \tan\theta \tag{1}$$

According to the force balance condition, the critical equilibrium condition for the stalk to be clamped and not slide is that the magnitude of the resultant force F at the contact points M and N between the blades and the stalk is equal and opposite, and they are on the same line. At this point, the cutting angle α of the blade is equal to the friction angle θ, which indicates that the condition for the stalk to be stably clamped by the blade is that the cutting angle of the blade is less than or equal to the friction angle, i.e.,

$$\alpha \leq \theta \tag{2}$$

The cutting angle α of the reciprocating cutting blade designed in this study is 19° [12]. It is known that the friction angle between the blade and the agricultural crop stalk is generally greater than 34° [26] (p. 913), which means that the blade satisfies the condition of effective clamping and no slippage.

3.2. Cutting Motion Analysis

The reciprocating cutter is driven by a crank-rocker mechanism. Compared with a single-acting cutter, the cutting stroke S of the double-acting blade can be reduced by half after one cutting, that is,

$$S = b/2 = 2r \tag{3}$$

where b is the blade width and r is the crank radius.

The movement of the blade of a reciprocating cutter driven by a crank-link mechanism can be approximated as intermittent motion. The displacement x, velocity v, and acceleration a of any point on the cutting edge are variable (as shown in Figure 2) and can be expressed by the following equations:

$$\begin{cases} x = -r\cos\omega t \\ v = r\omega\sin\omega t = \omega\sqrt{r^2 - x^2} \\ a = r\omega^2\cos\omega t = -\omega^2 x \end{cases} \tag{4}$$

In the equation, ω represents the angular velocity of the crankshaft, measured in rad/s, and t represents the time elapsed since the crankshaft passed through the left dead point, measured in s.

According to Equation (4), the following formula can be obtained:

$$\frac{v^2}{(r\omega)^2} + \frac{x^2}{r^2} = 1 \tag{5}$$

It is known that there is an elliptical equation between the movement speed v and the displacement x of the blade. When x = 0, the blade's movement speed is at its maximum, and at this time, v_{max} = $r\omega$. When $x = \pm r$, that is, when a certain point on the cutting edge is in the extreme left or right position of the blade, the blade's movement speed is at its minimum and the acceleration is at its maximum. At this time, v = 0 and v_{max} = $r\omega^2$.

A schematic diagram of the cutter movement is shown in Figure 4.

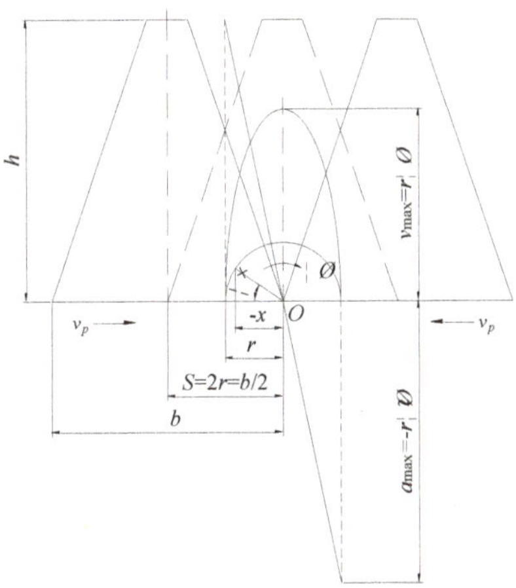

Figure 4. Schematic diagram of cutter movement.

The average cutting speed of the blade is generally expressed by the ratio of the cutting distance S to the time t taken to complete a cutting stroke, which is represented by v_p. The formula for v_p is:

$$v_p = \frac{nS}{30} \times 10^{-3} = \frac{rn}{15} \times 10^{-3} \tag{6}$$

Referring to the average cutting speed range of reciprocating blades for a ramie harvester [27], and taking its middle value, i.e., v_p = 1.5 m/s, the crank rotation speed n can be found to be 1181.1 r/min.

3.3. Bionic Blade Design

To facilitate the comparative study of the effects of different tooth pitch and tooth angle of bionic blades on cutting energy consumption, three-times the size of the biomimetic prototype (mouthparts palate of batocera horsfieldi) dimensions, i.e., 5.0 mm, 7.5 mm, and 10 mm, are used as tooth pitch sizes, and the mouthparts palate of batocera horsfieldi is shown in Figure 5. The biomimetic principle involves using MATLAB image processing software to extract boundary point data of the arc-shaped contour of the mouthparts palate of batocera horsfieldi. After the extraction of the boundary points, curve fitting is performed on these points, and the fitted curve is then used as the design for the blade edge of the cutting blade [12]. For each tooth pitch, three tooth angles of 0°, 15°, and 30° are set for the blades. As a result, there are nine different combinations of blades with different tooth pitches and tooth angles. The different types of blades are shown in Figure 6, with their respective tooth pitch and tooth angle parameters listed in Table 1.

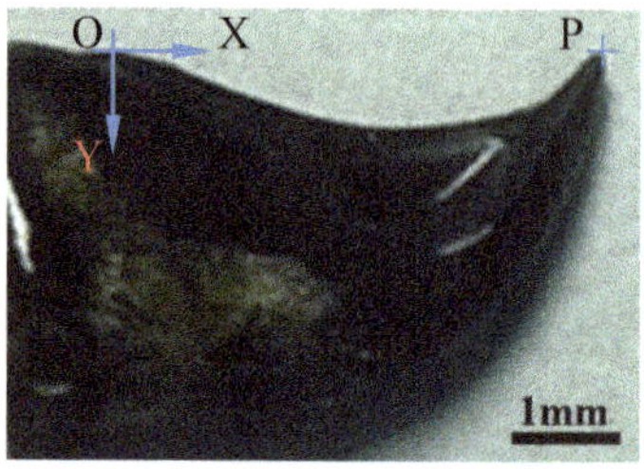

Figure 5. Mouthparts palate of batocera horsfieldi.

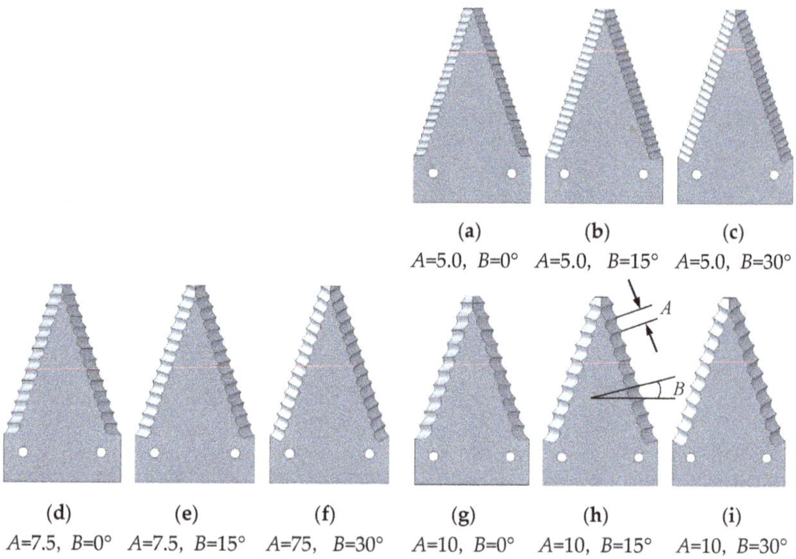

Figure 6. Bionic blades with different tooth pitches and tooth angles.

Table 1. Blade structure dimension parameter table.

Blade Code	Tooth Pitch *A*/mm	Tooth Angle *B*/°
a	5.0	0
b	5.0	15
c	5.0	30
d	7.5	0
e	7.5	15
f	7.5	30
g	10	0
h	10	15
i	10	30

Notes: *A* represents the tooth pitch; *B* represents the tooth angle.

4. Numerical Simulation and Analysis of Cutting Tests with Blade–Stalk Coupled Display of Dynamic Kinetics

4.1. Cutting Simulation Process and Determination of Material Parameters

The process of the numerical simulation of cutting experiments for the nonlinear display dynamics of the rigid–flexible coupling of the blade and hemp stalk is as follows: Based on the measured size of the industrial hemp stalk (stalk diameter of 18 mm), an industrial hemp stalk model is built in PTC Creo software. Different structural parameter bionic blade 3D models are designed and constructed, and the bionic blade and stalk models are assembled according to the actual cutting relationship. After determining the cutting relationship, the blade–hemp stalk 3D model is imported into the finite element analysis software ABAQUS through an intermediate format. According to the experimental plan, numerical simulation cutting experiments are conducted by defining material properties, assembling components, creating analysis steps, defining loads and boundary conditions, dividing the mesh, and submitting calculations. In defining the material properties of the model, the blade is set as a rigid body, while the stalk is set as a flexible body. According to measurements, the density of the stem is determined to be 9.5×10^{-9} t/mm^3, and the diameter is taken as the average value of the stem diameter, which is 18 mm. As the mechanical properties of industrial hemp stalks follow the transverse isotropy constitutive relationship, the engineering constants can be obtained from the literature [28]. Additionally, as the material characteristics of industrial hemp stalks are highly similar to wood, the anisotropic strength relationship of wood can be referenced to determine the anisotropic failure strength of industrial hemp stalks [29]. Therefore, the material property parameters of industrial hemp stalks based on the Hashin failure criterion can be obtained. The specific parameters are shown in Table 2.

Table 2. Mechanical property parameter table of stalk material of industrial hemp.

Mechanical parameter	E_X /MPa	E_Y /MPa	E_Z /MPa	G_{XY} /MPa	G_{XZ} /MPa	G_{YZ} /MPa	U_{XY}	U_{XZ}	U_{YZ}
Value	88	88	1743.50	33.85	31.99	31.99	0.3	0.02	0.02
Mechanical parameter	X_T /MPa	X_C /MPa	Y_T /MPa	Y_C /MPa	Z_T /MPa	Z_C /MPa	S_{XY} /MPa	S_{YZ} /MPa	S_{XZ} /MPa
Value	25	10	1	2	1	2	5	2	2

Note: E_X represents Young's modulus in X-direction; E_Y represents Young's modulus in Y-direction; E_Z represents Young's modulus in Z-direction; G_{XY} represents Shear modulus of plane XY; G_{XZ} represents Shear modulus of plane XZ; G_{YZ} represents Shear modulus of plane YZ; U_{XY} represents Poisson's ratio in plane XY; U_{XZ} represents Poisson's ratio in plane XZ; U_{YZ} represents Poisson's ratio in plane YZ; X_T represents Ultimate tensile stress in X-direction; X_C represents Ultimate compressive stress in X-direction; Y_T represents Ultimate tensile stress in Y-direction; Y_C represents Ultimate compressive stress in Y-direction; Z_T represents Ultimate tensile stress in Z-direction; Z_C represents Ultimate compressive stress in Z-direction; S_{XY} represents Ultimate shear stress in plane XY; S_{YZ} represents Ultimate shear stress in plane YZ; S_{XZ} represents Ultimate shear stress in plane XZ.

4.2. Experimental Design

Numerical simulation cutting experiments mainly investigated the effects of blade pitch, tooth angle, cutting speed ratio, and their interaction on blade cutting performance. In the case where the blade structure parameters are determined, it is necessary to determine the range of the speed ratio values. Referring to the speed ratio range of reciprocating cutters for stalk harvesters in the "Agricultural Machinery Design Manual", the speed ratio is set to be between 0.75 and 2 [26] (pp. 917). The experiment uses the three-factor three-level Box–Behnken response surface analysis method and cutting energy consumption as the index for orthogonal experimental design. The experimental factors and their level coding table are shown in Table 3.

The speed ratio coefficient is the ratio of the average cutting speed of the blade to the forward speed. As stated in Section 3.2, in this study, the value of v_p is 1500 mm/s. According to the formula for cutting speed and speed ratio $v_p = Cv_m$, the forward speed of the blade under different speed ratios can be calculated. When C = 0.75, the blade's forward speed v_m is 2000 mm/s; when C = 1.38, v_m is 1090 mm/s; when C = 2, v_m is 750 mm/s.

Table 3. Factors and level coding table.

Levels	Factors		
	Tooth Pitch ***A*****/mm**	**Tooth Angle** ***B*****/°**	**Speed Ratio** ***C***
−1	5.0	0	0.75
0	7.5	15	1.38
1	10.0	30	2.00

4.3. Test Index

The bionic blade has the advantage of reducing resistance and consumption. This study focuses on the effect of different structural parameters and motion parameters of the bionic blade on the energy consumption of cutting, and the test index is the energy consumption of single stalk cutting.

Numerical simulations using ABAQUS to measure cutting energy consumption were carried out as follows: two sets, set-1 and set-2, were established for the driving constraints of the left and right blades, respectively. In the time history output of the analysis step, the reaction forces for these two sets were output separately. After submitting the job, the force–time curves and data of the two driving reference points could be obtained from the post-processing ODB file. The cutting energy consumption was calculated according to Formula (7).

$$W = \sum_{i}^{n} (F_{i1} + F_{i2}) \cdot v_P \cdot \Delta t \tag{7}$$

In the equation, W represents the cutting energy consumption in mJ; F_{i1} and F_{i2}, respectively, represent the cutting reaction forces of the left and right blades at the i-th sampling point in N; v_p is the average cutting speed of the blade in mm/s; Δt is the sampling frequency interval time in s.

4.4. Experimental Results and Analysis

4.4.1. Regression Models and Analysis of Variance

The experimental design and analysis using Design-Expert software, experimental plan, and results are shown in Table 4. Taking experiment No. 13 as an example, the numerical simulation results, stubble cutting effect, and cutting force–time curve are shown in Figure 7.

Table 4. Experiment plan and results.

Test No.	Tooth Pitch *A*/mm	Tooth Angle *B*/°	Speed Ratio *C*	Cutting Energy *W*/mJ
1	−1	−1	0	6014.07
2	0	−1	1	4948.56
3	−1	0	1	4278.28
4	0	0	0	5531.58
5	1	1	0	5190.95
6	0	0	0	5531.58
7	0	0	0	5531.58
8	−1	0	−1	12,397.10
9	0	1	1	3972.02
10	1	0	1	3999.39
11	0	0	0	5531.58
12	−1	1	0	5424.18
13	1	−1	0	7201.73
14	0	−1	−1	16,033.50
15	0	0	0	5531.58
16	1	0	−1	11,869.80
17	0	1	−1	7343.06

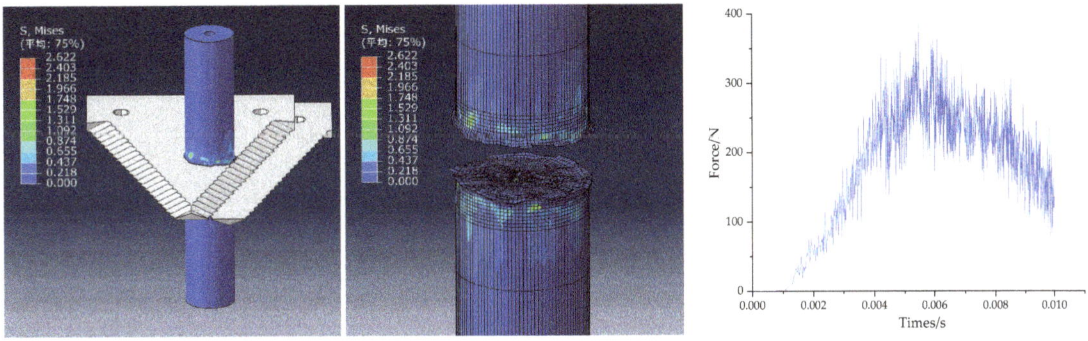

(**a**) A = 10 mm, B = 0°, C = 2 (**b**) Stubble cutting effect (**c**) Cutting force-time curve

Figure 7. Stalk cutting numerical simulation, stubble cutting effect, and cutting force–time curve. NOTE: " 平均" denotes average.

Through the Design-Expert, the response analysis of the experimental results was carried out using second-order regression, and the regression model equation for cutting energy consumption W was obtained as follows:

$$W = \begin{array}{c} 5531.58 + 18.53A - 1533.57B - 3805.77C - 355.22AB + \\ 62.10AC + 1928.24BC + 244.12A^2 + 182.03B^2 + 2360.44C^2 \end{array} \quad (8)$$

Further analysis of variance was conducted on the regression model of cutting energy consumption, and the results are shown in Table 5.

Table 5. Analysis of variance table for the regression model of cutting energy consumption.

Souces	Sum of Squares	*df*	Mean Square	*F*-Value	*p*-Value
Model	1.75×10^8	9	1.94×10^7	19.61	0.0004 **
A	2746.89	1	2746.89	0.0028	0.9594
B	1.88×10^7	1	1.88×10^7	19.02	0.0033 **
C	1.16×10^8	1	1.16×10^8	117.14	<0.0001 **
AB	5.05×10^5	1	5.05×10^5	0.5103	0.4981
AC	15,426.88	1	15,426.88	0.0156	0.9041
BC	1.49×10^7	1	1.49×10^7	15.04	0.0061 **
A^2	2.51×10^5	1	2.51×10^5	0.2537	0.6300
B^2	1.40×10^5	1	1.40×10^5	0.141	0.7184
C^2	2.35×10^7	1	2.35×10^7	23.72	0.0018 **
Residual	6.92×10^6	7	9.89×10^5		
Lack of Fit	6.92×10^6	3	2.31×10^6		
Cor Total	1.82×10^8	16			
R^2	0.9618				

Note: ** represents highly significant ($p < 0.01$).

The analysis of variance table above shows that the regression model for cutting energy consumption has a significance level p less than 0.01, indicating that the model has high significance. At the same time, the lack of a fit term in the model is not significant, and the fitting coefficient R^2 is 0.9618, indicating that the regression model obtained has a high degree of fit with the actual results, and the results are reliable. Therefore, the above model can be used to predict and analyze changes in cutting energy consumption.

Based on the above analysis of significance, for the cutting energy consumption indicator W, the factors of tooth angle B, speed ratio C, the interaction between B and C, and

C^2 have extremely significant impacts on the model, while other factors are not significant. The order of the main influences of each factor on cutting energy consumption is: $C > B > A$.

After eliminating the non-significant factors from the regression model of cutting energy consumption, the simplified regression model can be obtained as follows:

$$W = 5720.98 - 1533.57B - 3805.77C + 1928.24BC + 2384.12C^2 \quad (9)$$

4.4.2. Analysis of the Influence of Factor Interaction on Index

As can be seen from the variance analysis of the regression model mentioned above, the interaction between factors has a significant impact on the model. Therefore, the interaction between factors cannot be ignored. In this study, one factor was fixed at the middle level, and the interaction effects of the other two factors on the evaluation index were analyzed. Response surface plots were generated (as shown in Figure 8) to analyze the effects of the three factors, namely pitch, tooth angle, and speed ratio, on the cutting energy consumption index.

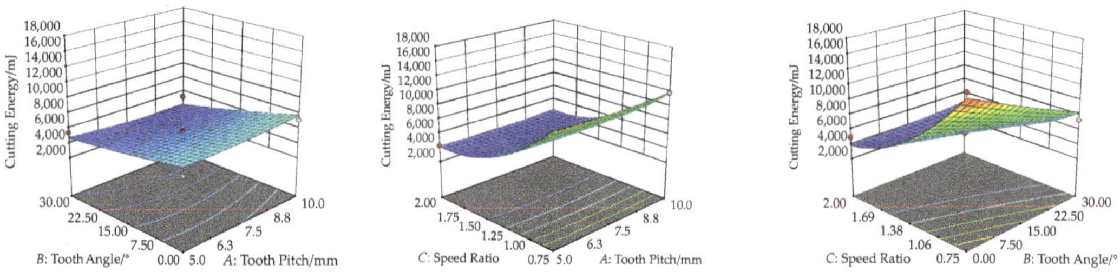

(**a**) $W = f(A, B, 1.38)$ (**b**) $W = f(A, 15, C)$ (**c**) $W = f(7.5, B, C)$

Figure 8. The interaction effect of two factors on cutting energy consumption.

Figure 8a shows the response surface of the interaction between tooth pitch A and tooth angle B on the cutting energy consumption when the speed ratio is at the middle level, i.e., $C = 1.38$. As can be seen from the figure, when the speed ratio is constant, the cutting energy consumption decreases as the tooth angle increases. Analysis shows that when the speed ratio is 1.38, the angle between the combined velocity vector of the blade's cutting speed and forward speed and the horizontal direction is 36° (calculated by arccot (1.38)). When the tooth angle B is 36°, the blade can cut into the stalk smoothly along the tooth groove, which is beneficial to reduce cutting resistance and energy consumption. Therefore, the closer the tooth angle is to 36 degrees, the lower the cutting energy consumption will be. This is shown in Figure 8a, where the cutting energy consumption decreases as the cutting rake angle increases. It can also be seen from Figure 8a that the change in cutting energy consumption with tooth pitch is not significant, indicating that cutting energy consumption is less affected by changes in tooth pitch.

Figure 8b shows the response surface plot of the interaction between the tooth pitch A and the speed ratio C on the cutting energy consumption when the tooth angle is set to the middle level of $B = 15°$. It can be seen from the plot that when the tooth angle is constant, the cutting energy consumption decreases rapidly with the increase in the speed ratio. Analysis indicates that to achieve the purpose of cutting with a parallel tooth system when the tooth angle is 15°, according to the calculation formula $\mathbf{C} = \cot B$, the speed ratio should be 3.73. Therefore, the cutting energy consumption in the plot decreases with the increase in the speed ratio. It can also be observed from the plot that the change in cutting energy consumption with tooth pitch is not significant, indicating that the variation in cutting energy consumption is less affected by the tooth pitch.

Figure 8c shows the response surface plot of the interaction between the tooth angle B and the speed ratio C on the cutting energy consumption when the tooth pitch is set to the middle level of $A = 7.5$ mm. It can be seen from the plot that when the tooth pitch is constant, the cutting energy consumption decreases with the increase in the tooth angle and the speed ratio, indicating that the interaction between the tooth angle and the speed ratio has a significant impact on the cutting energy consumption index. The reason is similar to the analysis in Figure 8a,b.

4.5. Parameter Optimization and Experimental Verification

To further obtain the optimal combination of factor parameters corresponding to the lowest cutting energy consumption, it is necessary to comprehensively optimize the blade structure parameters and motion parameters. Design-Expert software was used for optimization, with the objective and constraint conditions as follows:

$$\begin{cases} \min W \\ s.t. \begin{cases} 5.0mm \leq A \leq 10.0mm \\ 0^\circ \leq B \leq 30^\circ \\ 0.75 \leq C \leq 2 \end{cases} \end{cases} \tag{10}$$

Via optimization, the optimal combination of blade structure and motion parameters was obtained, with a blade pitch of 6.61 mm, a tooth angle of 30°, and a speed ratio of 1.62 (corresponding to a horizontal motion speed of 1.5 m/s and a feed speed of 0.93 m/s). At this point, the cutting energy consumption was 3947.99 mJ.

To verify the optimization results, a three-dimensional blade model with a pitch of 6.61 mm and a tooth angle of 30° was constructed, imported into the finite element analysis ABAQUS software, and the cutting simulation experiment was conducted again with the cutting speed ratio set to 1.62. The cutting energy consumption under this parameter was calculated to be 3625.84 mJ, with an error of only 8.16% compared to the optimization result. The experiment showed that the optimization results were highly reliable, further validating the model's reliability.

5. Conclusions

(1) An analysis was carried out on the critical force condition for effective clamping of the stalk of the reciprocating double-acting cutting blade, and the analysis showed that the designed reciprocating bio-inspired cutting blade satisfies the conditions of effective clamping and non-slip. The motion analysis of the reciprocating double-acting cutting blade was carried out, and the analysis showed that compared with the single-acting cutting tool, the cutting stroke of the double-acting cutting blade can be reduced by half to complete a single cut. The elliptic equation relationship between the motion speed and displacement of the blade was also determined.

(2) In order to investigate the effects of different structural and motion parameters and their interactions on cutting energy consumption of bionic blades, a combination of bionic blades with different tooth pitch and tooth angle was designed. Numerical simulation experiments were conducted using numerical simulation techniques to cut industrial hemp stalk with the blades. Based on the experimental results, a regression optimization model for cutting energy consumption was established, and the optimal parameter combination was determined through optimization as follows: blade tooth pitch of 6.61 mm, tooth angle of 30°, and speed ratio of 1.62. The cutting energy consumption under these conditions was 3948.99 mJ. The accuracy of the numerical simulation model was verified through validation experiments. This study can provide reference for the development of an industrial hemp harvester cutter and the matching of motion parameters.

Author Contributions: Conceptualization, K.T. and A.J.; methodology, J.H. and B.Z.; software, J.H.; validation, C.S.; investigation, H.L.; writing—original draft preparation, K.T.; writing—review and editing, A.J. All authors have read and agreed to the published version of the manuscript.

Funding: This research was funded by the National Natural Science Foundation of China (Grant No. 52005274), the Key Laboratory of Modern Agricultural Equipment and Technology (Jiangsu University) (Grant No. MAET202103), the Earmarked Fund for Mordern Agro-industry Technology Research System (Grant No. CARS-16-E20), and the Agricultural Science and Technology Innovation Program of Chinese Academy of Agricultural Sciences (ASTIP, CAAS).

Institutional Review Board Statement: Not applicable.

Informed Consent Statement: Not applicable.

Data Availability Statement: Not applicable.

Conflicts of Interest: The authors declare no conflict of interest.

References

1. Zuk-Golaszewska, K.; Golaszewski, J. *Cannabis sativa* L.–cultivation and quality of raw material. *J. Elem.* **2018**, *23*, 971–984. [CrossRef]
2. Zhang, X.; Sun, Y.; Cao, K.; Jiang, Y.; Hang, C.; Zhao, Y.; Han, X.; Wang, X. Status and Prospect of Industrial Hemp Breeding in Heilongjiang Province. *Crops* **2019**, *35*, 15–19. [CrossRef]
3. Guo, L.; Wang, M.; Wang, D.; Li, Z.; Che, Y.; Zhang, H. Research progress and prospect of comprehensive utilization of industrial hemp. *Heilongjiang Agric. Sci.* **2014**, *8*, 132–134.
4. Kymäläinen, H.; Sjöberg, A. Flax and hemp fibres as raw materials for thermal insulations. *Build. Environ.* **2008**, *43*, 1261–1269. [CrossRef]
5. Lü, J.; Long, C.; Ma, L.; Liu, J.; He, H. Design and experiment on decorticator of hemp fresh stem. *Trans. CSAE* **2014**, *30*, 298–307. [CrossRef]
6. Sepe, R.; Bollino, F.; Boccarusso, L.; Caputo, F. Influence of chemical treatments on mechanical properties of hemp fiber reinforced composites. *Compos. Part B Eng.* **2018**, *133*, 210–217. [CrossRef]
7. Cao, K.; Wang, X.; Sun, Y.; Han, C.; Zhang, X.; Zhao, Y.; Jiang, Y.; Han, X.; Guo, Y. The research progress on the breeding of industrial hemp varieties in china. *Plant Fiber Sci. China* **2019**, *41*, 187–192.
8. Guo, R.; Chen, X.; Guo, H. Review on pharmacological effects of tetrahydrocannabinol and cannabidiol. *Nat. Prod. Res. Dev.* **2017**, *29*, 1449–1453. [CrossRef]
9. Pisanti, S.; Malfitano, A.; Ciaglia, E.; Lamberti, A.; Ranieri, R.; Cuomo, G.; Abate, M.; Faggiana, G.; Bifulco, M. Cannabidiol: State of the art and new challenges for therapeutic applications. *Pharmacol. Ther.* **2017**, *175*, 133–150. [CrossRef]
10. Available online: https://www.cn-healthcare.com/articlewm/20190430/content-1053495.html (accessed on 5 April 2022).
11. Huang, J.; Shen, C.; Li, X.; Tian, K.; Chen, Q.; Zhang, B. Design and tests of hemp harvester. *Int. Agric. Eng. J.* **2017**, *26*, 117–127.
12. Tian, K.; Li, X.; Shen, C.; Zhang, B.; Huang, J.; Wang, J.; Zhou, Y. Design and test of cutting blade of cannabis harvester based on longicorn bionic principle. *Trans. Chin. Soc. Agric. Eng.* **2017**, *33*, 56–61. [CrossRef]
13. Ahmadi, E.; Barikloo, H.; Kashfi, M. Viscoelastic finite element analysis of the dynamic behavior of apple under impact loading with regard to its different layers. *Comput. Electron. Agric.* **2016**, *121*, 1–11. [CrossRef]
14. Huang, H.; Wang, Y.; Tang, Y.; Zhao, F.; Kong, X. Finite element simulation of sugarcane cutting. *Trans. CSAE* **2011**, *27*, 161–166.
15. Qiu, M.; Meng, Y.; Li, Y.; Shen, X. Sugarcane stem cut quality investigated by finite element simulation and experiment. *Biosyst. Eng.* **2021**, *206*, 135–149. [CrossRef]
16. Yang, W.; Zhao, W.; Liu, Y.; Chen, Y.; Yang, J. Simulation of forces acting on the cutter blade surfaces and root system of sugarcane using FEM and SPH coupled method. *Comput. Electron. Agric.* **2021**, *180*, 105893. [CrossRef]
17. Xue, Z. Cutting Mechanical Characteristics and Simulation Analysis of Cassava Stalk. Ph.D. Dissertation, Huazhong Agriculture University, Wuhan, China, 2018.
18. Meng, Y.; Wei, J.; Wei, J.; Chen, H.; Cui, Y. An ANSYS/LS-Dyna simulation and experimental study of circular saw blade cutting system of mulberry cutting machine. *Comput. Electron. Agric.* **2019**, *157*, 38–48. [CrossRef]
19. Xie, L.; Wang, J.; Cheng, S.; Zeng, B.; Yang, Z. Optimization and finite element simulation of the chopping process for chopper sugarcane harvesting. *Biosyst. Eng.* **2018**, *175*, 16–26. [CrossRef]
20. Dun, G.; Yang, Y.; Li, H.; Yu, C.; Du, J.; Zhang, J.; Gao, Z. Working parameters optimization of soybean serration rotary cutter by ANSYS/LS—DYNA. *J. Henan Agric. Univ.* **2019**, *53*, 739–744.
21. Guo, Q.; Zhang, X.; Xu, Y.; Li, P.; Chen, C. Mechanism and simulation analysis of efficient cutting for tomato straw. *J. Agric. Mech. Res.* **2017**, *39*, 11–24. [CrossRef]
22. Zheng, C.; Zhao, J.; Zhang, J.; Zhang, R.; Li, F. Construction of finite element model of cotton pole and calibration of cutting parameters. *J. Agric. Mech. Res.* **2021**, 43, 198–203. [CrossRef]

23. Wang, T.; Liu, Z.; Yan, X.; Mi, G.; Liu, S.; Chen, K.; Zhang, S.; Wang, X.; Zhang, S.; Wu, X. Finite Element Model Construction and Cutting Parameter Calibration of Wild Chrysanthemum Stem. *Agriculture* **2022**, *12*, 894. [CrossRef]
24. Liu, Z.; Wang, T.; Liu, S.; Yan, X.; Zhao, H.; Wu, X.; Zhang, S. Design and experimental study of a bionic blade for harvesting the wild chrysanthemum stem. *Agriculture* **2023**, *13*, 190. [CrossRef]
25. Xia, Z.; Yao, L.; Kan, J. The numerical simulation of rotating tool cutting soil and wood based on ANSYS/LS-DYNA. *For. Eng.* **2016**, *32*, 43–47. [CrossRef]
26. Chinese Academy of Agricultural Mechanization Science. *Agricultural Machinery Design Manual*, 1st ed.; China Agricultural Science and Technology Press: Beijing, China, 2007.
27. Shen, C.; Chen, Q.; Li, X.; Zhang, B.; Tian, K.; Huang, J. Design, analysis and test on cutting test bench of ramie stalk. *Int. Agric. Eng. J.* **2017**, *26*, 86–95.
28. Zhou, Y.; Li, X.; Shen, C.; Tian, K.; Zhang, B.; Huang, J. Experimental analysis on mechanical model of industrial hemp stalk. *Trans. CSAE* **2016**, *32*, 22–29. [CrossRef]
29. Available online: http://dec3.jlu.edu.cn/webcourse/T000185/T000199/files/bjjx/z8_2.html (accessed on 7 February 2023).

agriculture

MDPI

Article

Effect of Rotary Speed on Soil and Straw Throwing Process by Stubble-Crushing Blade for Strip Tillage Using DEM-CFD

Yiwen Yuan [1], Jiayi Wang [1], Xin Zhang [2] and Shuhong Zhao [1,*]

[1] College of Engineering, Northeast Agricultural University, Harbin 150030, China; yuanyw@neau.edu.cn (Y.Y.)
[2] Heilongjiang Academy of Agricultural Machinery Engineering Sciences, Harbin 150081, China
* Correspondence: shhzh@neau.edu.cn

Abstract: Strip tillage is a widely used land preparation approach for effective straw management in conservation agriculture. Understanding the dynamic throwing process during the stubble-crushing operation has important implications for seedbed preparation. However, the airflow generated by the high-speed rotation of stubble-crushing blades has yet to be considered. We established a coupled DEM-CFD simulation model and explored the dynamic motion of soil particles varied with their initial depth (at 0, 20, 40, 60, 80 mm depth) and surface straw under different blade rotary speeds (270, 540, 720, and 810 rpm) based on the model. The coupled model simulation results were proved to be well correlated with the field test results by the field high-speed camera test. The simulation results showed that airflow had a significant effect on the longitudinal displacement of straw ($p < 0.05$). An increase in rotary speed could increase the longitudinal and lateral throwing displacement of soil particles and straw and increase the blade–soil–straw interaction, while there was no directional effect on the vertical motion. The lateral movement of soil particles decreased with increasing soil particle depth. The stubble-crushing operation allowed the exchange of deep and shallow soil layers, as well as the burial of straw. Plain, straight stubble-crushing blades with a rotary speed of 540 rpm were able to form the optimal seeding row with a width of 80.86 mm. The simulation results were useful for assessing the design solutions of high-speed rotational tools and evaluating seedbed preparation practices.

Keywords: stubble-crushing; plain straight blade; DEM-CFD; dynamic motion; seedbed preparation; strip tillage

Citation: Yuan, Y.; Wang, J.; Zhang, X.; Zhao, S. Effect of Rotary Speed on Soil and Straw Throwing Process by Stubble-Crushing Blade for Strip Tillage Using DEM-CFD. *Agriculture* **2023**, *13*, 877. https://doi.org/10.3390/agriculture13040877

Academic Editors: Muhammad Sultan, Redmond R. Shamshiri, Md Shamim Ahamed and Muhammad Farooq

Received: 28 March 2023
Revised: 13 April 2023
Accepted: 14 April 2023
Published: 15 April 2023

1. Introduction

Conservation agriculture is an idea with expanding global adoption due to its potential for soil conservation and increased crop productivity [1–3]. The core principles of conservation agriculture include adequate retention of straws, minimum soil disturbance, and diversified crop rotation [4], the first two leading to increased soil surface straw loads [5]. In conservative agricultural systems, excessive straw remaining in the fields has been a yield-limiting problem. Therefore, mechanized technology for managing straws is necessary to ensure the quality of no and minimum tillage and straw returning operations in a conservation agriculture system. Northeastern China is the main crop-producing area of the country [6], and maize is one of the most important grain crops, which is cultivated in areas of 12.06×10^6 ha, while generating an enormous amount of straw each year.

Under the straw returning and mulching management model, the terrain is complex, with a large amount of straw and stubble. Therefore, treating straw is crucial after fall harvest or before planting in the spring, or to plough the land as little as possible. Strip tillage, the procedure of placing seed into narrow furrows and limiting soil inversion, is relatively new, having been first evaluated in the early 1990s [7]. This tillage mode, shown in Figure 1, confines tillage to a narrow strip where the crop will be planted and divides the cropping system into two distinct adjacent rows: tilled and untilled [8]. Tillage within the crop row can create a finer seedbed [9], and the untilled zone between rows maintains no-tilled states. Compared with traditional integrated tillage and rotational tillage, strip tillage has less soil disturbance, higher operational efficiency for soil water storage and moisture conservation, and improvement of the ground temperature of the seedbed, which could significantly improve the quality of subsequent seeding [10]. A high-quality strip crop row improves seed emergence rate, shortens emergence time, and then improves crop yield. Strip tillage, according to Al-Kaisi and Yin [11] and Gathala et al. [12], could help intensify dry land cropping by reducing land fallow and ensuring adequate yields, while also reducing greenhouse gas emissions and improving energy efficiency.

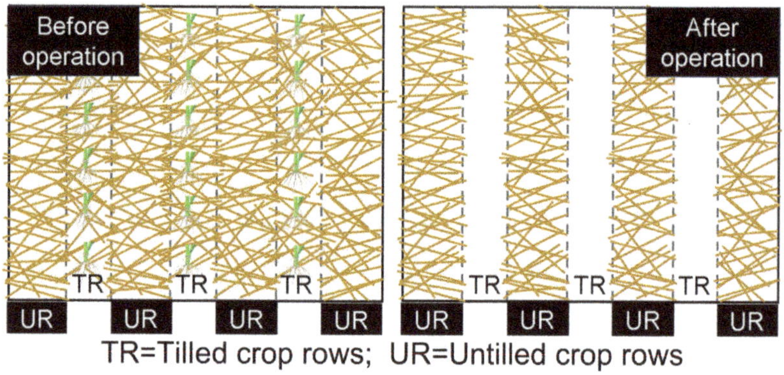

Figure 1. Effectiveness of maize straw mulch strip-till stubble-crushing operation under-conservation agriculture system.

Evidence indicates the critical soil engaging component is the blade [13]. In strip-till under-conservation agriculture system, stubble-crushing blades cut through crop straw and soil during the stubble-crushing operation. Soil and straw displacement are the highlights of the relevant literature. Fang et al. [14,15] analyzed the soil macroscopic and fine motion behavior and the displacement of rice and wheat straw in the rototilling process, which helped us to understand the interaction of the rotary blade, soil, and straw. To address the issue in the working process of strip tillage seeders, Zhao et al. [16] examined the effect of different edge–curve geometries on soil disturbance characteristics. Overthrowing, on the other hand, may degrade the quality of rotary strip tillage seedbed furrows. The effectiveness of seed germination was shown to be decreased when too much dirt was dumped out of the furrow, leading to a lack of soil cover over the seeds and a propensity to coat the furrow (resulting in a thin layer of compacted soil). In stubble-crushing operations, the high-speed rotation of the driven blades creates a non-negligible airflow around them. Airflow has a great influence on the motion of soil and straw, while airflow generated by high-speed rotational tools has not been considered. Under the condition of straw returning, the effect of soil and straw throwing separation affects the sowing zone environment and increases working resistance and power consumption. Hence, it is of great significance to explore the mechanism of throwing separation of soil and straw.

The numerical method has demonstrated considerable advantages over analytical and experimental methods in investigating soil–tool interaction problems, being less time-consuming and more capable of studying the complex geometry of soil-engaging tools [17]. It is becoming common practice to use the discrete element method (DEM) for numerical modeling in order to accurately predict soil dynamics [18–20]. A soil–tool–straw interaction DEM model was developed using the discrete element method to reproduce the soil bin test [21]. The DEM model was validated using experimental results by simulating the effect of narrow point opener geometry on soil disturbance and cutting [22]. Zhao et al. [16] used the discrete element method (DEM) to examine the effects of different edge–curve geometries on torque requirements and soil disturbance characteristics. Fang et al. [14]

studied the macro- and meso- movements of soil particles during rotary tillage with the help of DEM simulation. The flow field generated by the high-speed rotation of blades cannot be simulated by DEM softwares, despite the fact that it can be used to obtain some microscopic data that cannot be collected in a real test. Computational fluid dynamics (CFD) has also become a widely used and highly valued engineering tool among agricultural researchers for simulating fluid flow, the characterization of flow fields, and revealing fluid phase mechanisms for sustainable development [23]. To describe detailed dynamic information, including particle velocity, instantaneous forces acting on each particle by airflow, and interaction between gas–solid phases, discrete element method and computational fluid dynamics (DEM-CFD) techniques have been combined [24]. By the DEM-CFD coupling method, the purpose of forming a flow field by simulating the high-speed rotation of blades can be achieved.

Bent blade and plain straight blade are the two main types used for strip tillage. For a variety of reasons, the use of bent blades might result in an increased amount of soil being thrown out in untilled crop rows [13]. A plain straight blade is more compatible with the requirements of strip tillage to reduce soil disturbance. Therefore, the objective of this study was to investigate the effect of rotary speed and airflow on soil and straw throwing processes in plain straight blade operation based on the fluid–solid coupling simulation model of the strip-till under-conservation agriculture system.

2. Materials and Methods

2.1. Development of Coupling Contact Model

The discrete element method simulates the process of motion propagation in a collection of particles. The particle motion will inevitably cause mutual collisions between the particles, and the particles will inevitably generate force between them. A contact model describes how elements behave when they come into contact with each other. The discrete element method, which is based on contact mechanics and elastoplastic analysis of granular materials under quasi-static circumstances, relies heavily on the contact model. Particle trajectories are calculated analytically from forces and moments applied to them, which are then determined by the trajectory of those particles during testing. To increase the accuracy of the simulation findings, distinct contact types must be constructed for different simulation objects.

Hertz–Mindlin (no slip) is the default contact model added between soil and blade. In this model, both normal and tangential forces have damping components, where the damping coefficient is related to the recovery factor. The tangential friction follows Coulomb's law of friction, and the rolling friction is implemented as a directional constant torque model independent of the contact.

To describe the cohesion of agricultural soils, in addition to the Hertz–Mindlin (no slip) contact model between soil particles [25–27], considering the strong cohesion properties between clay particle, the Hertz–Mindlin with bonding (HMB) contact model was added [28–30]. This contact model is applied to bond the particles to simulate the surface straw as well as the root stubble model after the straw was crushed and returned to the field. In the HMB contact model, a "binder" is added between the particles to bond them together. This "binder" is capable of withstanding normal and shear displacements, and the bond is broken when normal and shear stresses exceed their limits. The moisture content is determined by the bonded radius and density under this model, shown in Figure 2, satisfying the following relationship:

$$\omega_s = \frac{\rho_W V_w}{\rho_s V_s} \tag{1}$$

$$V_s = \frac{4}{3}\pi R_P^3 \tag{2}$$

$$V_w = \frac{4}{3}\pi R_{\mathrm{B}}^3 - \frac{4}{3}\pi R_p^3 \tag{3}$$

where ω_s, V_s and ρ_s are, respectively, moisture content, volume (mm^3), and density (g·mm^{-3}) of soil particle. V_w and ρ_w represent the volume (mm^3) and density (g·mm^{-3}) of water. R_{B} and R_P are the bonded radius (mm) and soil particle radius (mm), respectively. Under the condition of known water density, the soil moisture content and soil density were measured, and the water volume was obtained by Equation (1). Furthermore, the soil particle radius was measured, and the bonded radius was calculated by Equation (3). Therefore, the definition of particle density and particle bonded radius in the HMB model indirectly reflected the particle moisture content.

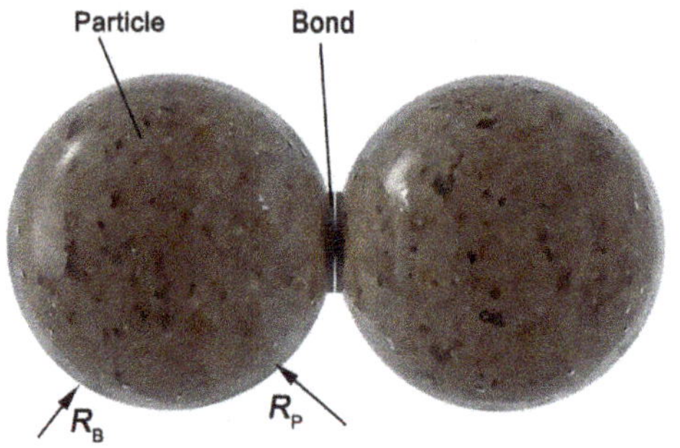

Figure 2. The bond relationship between the wet soil particles.

A Hertz–Mindlin with the JKR (Johnson–Kendall–Roberts) cohesion contact model, based on the Johnson–Kendall–Roberts theory [31], was added between the stubble and soil [32–36] to simulate crop growth.

The model was imported into the CFD software (Fluent 17.0, ANSYS, Canonsburg, PA, USA) module to establish the interaction between the stubble-crushing blade and airflow. A workflow of the CFD-DEM coupling process is presented in Figure 3. In the CFD-DEM coupling model, no complex energy transfer processes are involved, so only mass conservation and momentum conservation equations (Navier–Stokes equations) are used in this simulation to describe the continuous gas phase. The equations for the gas-phase mass and its momentum conservation, respectively, are given as follows:

$$\frac{\partial \varepsilon \rho}{\partial t} + \nabla \cdot \rho \varepsilon \vec{u} = 0 \tag{4}$$

$$\frac{\partial \varepsilon \rho \vec{u}}{\partial t} + \nabla \cdot (\rho \varepsilon \vec{u}\vec{u}) = -\varepsilon \nabla p + \nabla \cdot \left(\varepsilon \vec{\tau}\right) + \rho \varepsilon g - \vec{F}_{pf} \tag{5}$$

where ρ is the fluid density, ε is the fluid volume fraction without the CFD mesh, $\vec{u}$ is the fluid velocity, p is viscous stress tensor, and $\vec{F}_{pf}$ is total force density due to the presence of particles.

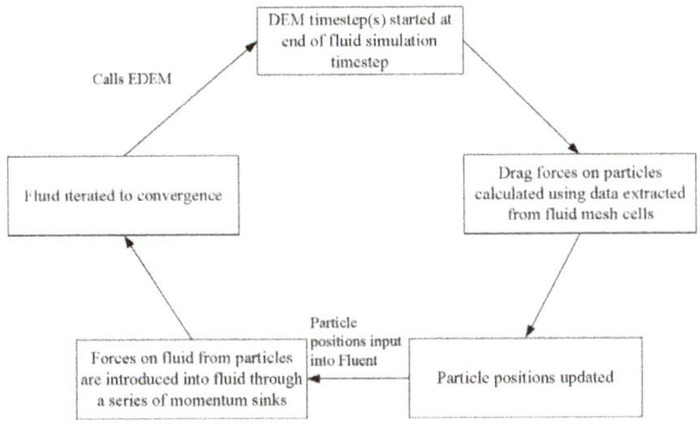

Figure 3. Flow chart of EDEM-FLUENT coupling processes and information exchange between two solvers.

The k-ε turbulence model was developed based on Boussinesq's assumption of isotropic eddy viscosity coefficient—Reynolds stress is proportional to the average velocity gradient. The working condition was turbulent, and the RNG (renormalization group) k-ε turbulent flow energy dissipation rate model was used [37–40]. The transport equation and turbulent viscosity equation were

$$\frac{\partial(\rho k)}{\partial t}+\frac{\partial(\rho k u_i)}{\partial x_i}=\frac{\partial}{\partial x_j}\left[\alpha_k\mu_{\text{eff}}\frac{\partial_k}{\partial x_j}\right]+G_k+\rho\varepsilon_t \tag{6}$$

$$\frac{\partial(\rho\varepsilon_t)}{\partial\varepsilon_t}+\frac{\partial(\rho\varepsilon_t)}{\partial x_i}=\frac{\partial}{\partial x_j}\left[\alpha_s\mu_{\text{eff}}\frac{\partial_c}{\partial x_j}\right]+\frac{C_{1c}\varepsilon_t}{k}G_k-C_{2c}\rho\frac{{\varepsilon_t}^2}{k} \tag{7}$$

$$\mu_{\text{eff}}=\mu+\mu_t \tag{8}$$

$$\mu_t=\rho C_\mu\frac{k^2}{\mathcal{E}_t} \tag{9}$$

$$C_{1\varepsilon}^*=C_{1\varepsilon}-\frac{\eta(1-\eta/\eta_0)}{1+\beta_{k-\varepsilon}\eta^3} \tag{10}$$

$$\eta=(2E_{ij}E_{ij})^{\frac{1}{2}}\frac{k}{\mathcal{E}_t} \tag{11}$$

$$E_{ij}=\frac{1}{2}\left(\frac{\partial u_i}{\partial x_j}+\frac{\partial u_j}{\partial x_i}\right) \tag{12}$$

where x_i, x_j are components of displacement in i and j directions; α_k α_ε are turbulent kinetic energy k and dissipation rate ε corresponding Prandtl number, $\alpha_k = \alpha_\varepsilon = 1.39$; u_{eff} represents correction parameters; u_i, u_j are components of velocity in i and j directions; ε_t represents turbulent dissipation rate; μ, μ_t are turbulent viscosity coefficient and viscosity; E_{ij} is mainstream time average strain rate, and η is strain rate; $C_{1\varepsilon}$, $C_{2\varepsilon}$, C_μ are empirical constant; $C_{1\varepsilon} = 1.42$, $C_{2\varepsilon} = 1.68$ and $C_\mu = 0.0845$; η_0, $\beta_{k-\varepsilon}$ represent constant.

2.2. Simulation Model Development and Parameters

To simulate the interaction between soil particles, maize straw, maize stubble, and plain straight blades, a simulation model was developed in EDEM 2018^{TM} software (DEM Solutions, Edinburgh, UK), which consisted of blades and a soil bin containing maize straw and stubble model composition.

In order to ensure the spatial adequacy, a virtual soil bin model of 1500 mm long × 400 mm wide × 150 mm deep was created using default circular particles, which were made up of 4 mm radius (selected based on available computation time), and soil particles were randomly generated and settled to reach an equilibrium state in soil bin. Black soil (clay loam) in Northeast China was used for this modeling. The maize straw piece hollow model was created using the 3D modeling software SolidWorks 2016 (Dassault Systemes, Suresnes, France), imported into EDEM 2018^{TM}, and the position of the straw model was adjusted by Reposition under Geometry. Straw models were filled by default circular particles of 1.5 mm in radius. Six straw pieces were placed on the soil surface as straw tracers. An additional 110 straw pieces, which were straws with lengths of 50, 75, and 100 mm and radii of 10, 12.5, and 15 mm, respectively (suggested by [28,34,41]), were randomly spread on the surface to achieve the same straw cover in mass as in the validation tests. The geometric model surface of stubble was designed as suggested by Wang et al. [42]. Four maize root stubbles, each with a spacing of 30 mm, were generated by combining the solid and virtual stubble geometry model and placed uniformly in the soil bin. The geometric models of the straw and stubble overlapped with the virtual surface, and the geometric model surface was deleted when the particles filled the geometric model (the soil bin, straw, and stubble models in Figure 4a). The plain straight rotary blade used for strip tillage was adopted in this study, as shown in Figure 5. A three-dimensional model of the stubble-crushing blade (density of 7865 kg·m^{-3}) was established in the Catia V5 software (Dassault Systemes, Suresnes, France), and then imported into the ICEM software for dynamic meshing, and the dynamic meshed stubble-crushing blade was imported into the EDEM 2018^{TM} to establish the interaction model of the coupling field. The EDEM 2018^{TM} was coupled with Fluent 17.0 using the udf coupling interface, and a flow field area of 1500 × 400 × 1000 mm in length × width × height was created in Fluent 17.0 by adding a natural wind of 3 m·s^{-1} in the direction opposite to that of the stubble-crushing blade movement (Figure 4b). Three cylindrical air layers of 230 mm radius and 6 mm thickness were added around the stubble-crushing blade so that the blades were placed in them. The parameters to be determined for the soil–straw–stubble–blade interaction model are shown in Table 1.

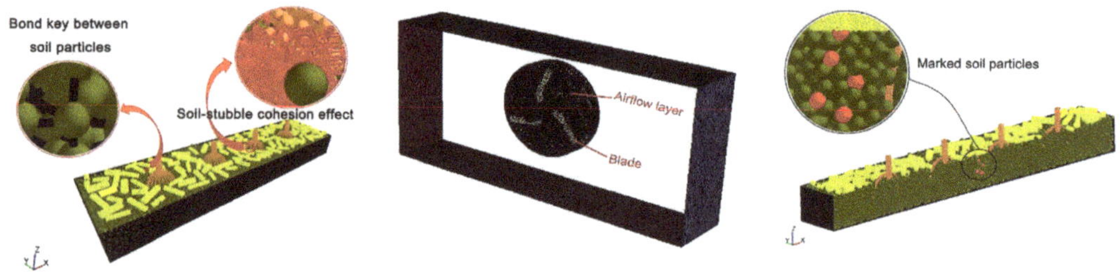

(a) (b) (c)

Figure 4. DEM-CFD coupling model: (**a**) soil, straw, and stubble model; (**b**) flow field model; (**c**) layout of marked soil particles single rental mark.

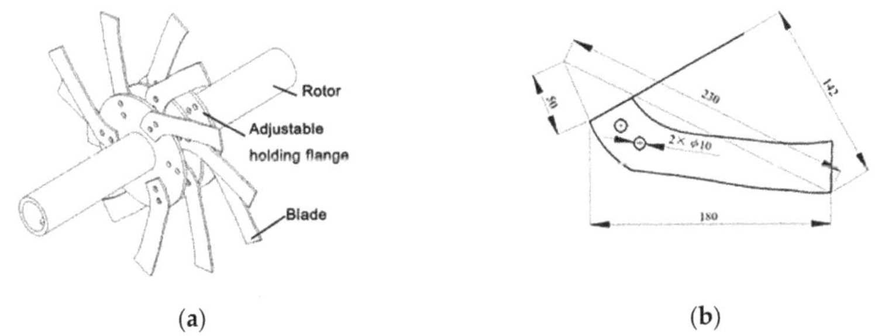

Figure 5. Soil-engaging tool of stubble-crushing tested: (**a**) model of the field validation experimental rotor with twelve blades fitted onto adjustable holding flanges; (**b**) geometric parameters of the plain straight stubble-crushing blade.

Table 1. Simulation model key parameters.

Objects	Parameters and Units	Values
Soil	Poisson's ratio	0.41
	Shear modulus (MPa)	1.24×10^6
	Density (kg·m^{-3})	2150
	Coefficient of rolling friction	0.2
	Coefficient of static friction	0.3
	Coefficient of restitution	0.6
	Critical normal stress (kPa)	200
	Critical shear stress (kPa)	68
	Normal contact bond stiffness (N·m^{-1})	3.4×10^8
	Shear contact bond stiffness (N·m^{-1})	1.5×10^8
	Moisture	20 ± 1%
Straw	Poisson's ratio	0.4
	Density (kg·m^{-3})	241
	Shear modulus (MPa)	1
	Shear modulus (MPa)	0.6
	Coefficient of static friction	0.01
	Coefficient of static friction	0.3
	Critical normal stress (kPa)	8.72×10^3
	Critical shear stress (kPa)	7.5×10^3
	Normal contact bond stiffness (N·m^{-1})	9.6×10^6
	Shear contact bond stiffness (N·m^{-1})	6.8×10^6
Stubble	Bonded radius (mm)	1.79
	Poisson's ratio	0.33
	Density (kg·m^{-3})	107.64
	Shear modulus (MPa)	6.293
	Coefficient of restitution	0.6
	Coefficient of rolling friction	0.21
	Coefficient of static friction	0.573
	Bond radius (mm)	1.7
	Critical normal stress (kPa)	500
	Critical shear stress (kPa)	500
	Normal contact bond stiffness (N·m^{-1})	1.034×10^6
	Shear contact bond stiffness (N·m^{-1})	1.034×10^6
Blade	Shear modulus (MPa)	7.9×10^4
	Poisson's ratio	0.3
Soil-stubble	Coefficient of restitution	0.60
	Coefficient of static friction	0.60
	Coefficient of rolling friction	0.02
	Surface energy (J·m^{-2})	10
Soil-blade	Coefficient of restitution	0.6
	Coefficient of static friction	0.313
	Coefficient of rolling friction	0.107

Table 1. *Cont.*

Objects	Parameters and Units	Values
Straw-blade	Coefficient of static friction Coefficient of rolling friction	0.3 0.01
Stubble-blade	Coefficient of static friction Coefficient of rolling friction	0.6 0.02

The time step is the amount of time between iterations in the EDEM 2018^{TM} (computational particle and contact between particles and geometry). In the process of soil particle generation, the fixed time step was 8.2×10^{-5} s (less than 20% of the Raileigh time step [28]). The Rayleigh time step is the time taken for a shear wave to propagate through a solid particle [43], which was set to 5.47×10^{-4} s. The soil particles generation time was set from 0 to 6 s, and the settling time was 1 s. The generation times of the maize straw pieces and root stubbles were 1 s, respectively. Based on this, the airflow–soil solid coupling field of the stubble-crushing blade operation was established.

2.3. Simulation Experiment

The two-factor experiment was designed as a randomized complete block with four replications [44,45]. The experimental factors were the rotary speed of the blade and the presence or absence of airflow. In addition, the blade rotary speeds were set to 270, 540, 720, and 810 rpm (including the high and low output shaft cutter roll speeds of the tractor). The airflow layer interacting with the blade (the airflow rate was directly output in the Fluent17.0 software) was set in the Fluent17.0 software, and the Coupling Server module in the EDEM 2018^{TM} was started for coupling simulation. At the same time, a separate simulation of the operation of the blade was performed in the EDEM 2018^{TM}. The experimental indicators were the lateral displacement of soil particles and straw particles (perpendicular to the driving direction of the tractor), vertical displacement, longitudinal displacement (opposite to the driving direction of the tractor), lateral velocity, longitudinal velocity, and vertical velocity. The average value of the indicators was used as the final experimental result of each group of experiments.

2.4. Model Monitoring

Throughout the simulation, soil particles were monitored by coloring them red. In the study of soil particles, since the operating depth of the stubble-crushing blade was 80 mm, the soil particles at each depth were marked in layers, which were 0, 20, 40, 60, and 80 mm, and four locations were randomly marked at each depth, and the straw model was randomly marked in four places on the surface, as shown in Figure 4c. The X, Y, and Z directions were defined as longitudinal (opposite to the working direction), lateral (perpendicular to the working direction), and vertical directions, respectively. The dynamic motion (displacement and velocity) of the marker in each of the three directions during the blade stabilization operation was output using the built-in post-processing module in the EDEM 2018^{TM}. The simulation process is shown in Figure 6.

(a) (b)

Figure 6. Simulation motion trajectory: (**a**) soil particles; (**b**) straw.

2.5. Field Experiment

In order to obtain the airflow velocity around working-blade and validate DEM-CFD model, the field test was conducted at the research farm (44°04′~46°40′ N and 125°42′~130°10′ E) of the Northeast Agricultural University located in the Heilongjiang Province of northeastern China in October 2020. This region has a typical temperate continental monsoon climate and shows a continental climate, with a short hot period during summer and a long cold period during winter. The mean annual precipitation is 569.1 mm and more than 60% of precipitation occurs from June to September. The average annual potential evapotranspiration is 1009 mm (1971–2000) and markedly exceeds annual precipitation. Soil in the northeast is typically black soil with a clay loam texture containing 36.0% sand, 24.5% silt, and 39.5% clay [46]. The average soil moisture content of the cultivation layer within 80 mm of the experimental site was 20 ± 1%, and the average moisture content of maize root stubble is 184.3% (d.b). Prior to the tillage tests, the field had been used for maize with a ridged culture under a conservation tillage system. The distance between two adjacent ridges was 650 mm. Before the test, straws were returned to the field. The average stubble height on the ground was 110 mm, and the straw coverage was 1.031 kg·m^{-2}. The average moisture content of soil, straw, and stubble was 20.5%, 19.4%, and 43.9%, respectively.

The field test is shown in Figure 7. A single test rotor was fitted with twelve blades and set to cut maize straws and stubbles on the ground. A double-row stubble-crushing implement fitted with two rotors assembly was pulled by a Benye 454 Tractor (45 hp, Ningbo Benye Tractor & Automobile Manufacturing Co., Ningbo, China). The forward travel speed was kept constant at 0.56 m·s^{-1}, as was the rotor rotational speed at 810 rpm (equivalent to the rotational speed of the high-grade output shaft of the tractor). The test was repeated four times. Each test area was divided into an experimental area of 10 m and a preparation area of 5 m. The equipment was driven into the experimental area when the tractor traveling speed was kept constant at 0.56 m·s^{-1} in the preparation area.

(a) (b) (c)

Figure 7. Apparatus of field experiment: (**a**) actual installation of air volume meter; (**b**) experimental devices; (**c**) electrical device connection.

Since the soil and straw thrown during the operation would hit the wind meter, we measured the wind speed data before the official test. A digital fractional wind meter (manufactured by Sigma Instruments Group Ltd., Hong Kong, with an accuracy of 0.001 m·s^{-1}) was installed on the stubble-crushing implement frame prior to the field experiment to measure and record the wind speed around the blades during the idling operation. During the operation of stubble-crushing blades at high-speed, blades can cut maize stubble, straws, and soil. The patterns of cutting and throwing are not possible to observe with the naked eye. Consequently, following Lee et al. [47] and Matin et al. [48], a high-velocity camcorder was used to catch symbolism-permitting representation of these examples. The video camera (Phantom V5.1, Vision Research Inc, Wayne, NJ, USA) was fitted to capture the images (1200 frames·s^{-1}, 1024 × 1024 dpi) of the process of throwing actions that help validate the simulation test results. The images were acquired on a laptop as RAW3 files using Phantom® Camera Control Application software (2.8, Vision Research Inc, Wayne,

NJ, USA) during the test runs. The images were analyzed with Promon Studio software (Vision Research Inc, Wayne, NJ, USA).

3. Results and Discussion

The dynamic displacement and velocity of soil particles with depths of 0, 20, 40, 60, and 80 mm and straw were extracted in EDEM 2018^{TM}, and the average velocity and displacement of four random soil particles at each depth and four random straws at the surface in different directions were taken for treatment. The processed experimental data were subjected to ANOVA, as shown in Table 2.

Table 2. Analysis of variance (F-test) of measured data.

Source of Variation	Degree of Freedom	Soil *F*-Value					
		X-Displacement	Y-Displacement	Z-Displacement	X-Velocity	Y-Velocity	Z-Velocity
Rotary speed, A	3	1.434 ^{ns}	0.153 ^{ns}	5.449 **	3.696 **	0.041 ^{ns}	0.126 ^{ns}
Airflow, B	1	1.818 ^{ns}	1.112 ^{ns}	0.337 ^{ns}	2.496 ^{ns}	0.177 ^{ns}	0.074 ^{ns}
AB	3	0.823 ^{ns}	0.096 ^{ns}	0.322 ^{ns}	0.078 ^{ns}	0.027 ^{ns}	0.101 ^{ns}
Error	32						
Source of Variation	**Degree of Freedom**	**Straw *F*-Value**					
		X-Displacement	**Y-Displacement**	**Z-Displacement**	**X-Velocity**	**Y-Velocity**	**Z-Velocity**
Rotary speed, A	3	1.811 ^{ns}	1.102 ^{ns}	0.028 ^{ns}	0.232 ^{ns}	0.042 ^{ns}	0.011 ^{ns}
Airflow, B	2	3.830 *	1.379 ^{ns}	0.113 ^{ns}	0.653 ^{ns}	0.228 ^{ns}	0.001 ^{ns}
AB	2	0.247 ^{ns}	0.003 ^{ns}	0.074 ^{ns}	0.088 ^{ns}	0.009 ^{ns}	0.027 ^{ns}
Error	152						

'*' means significantly different at $p < 0.05$; '**' means significantly different at $p < 0.01$; 'ns' means not significantly different at $p > 0.05$.

3.1. Testing Results and Model Validation

The field test airflow velocity was compared with the simulated airflow velocity. From the results of the field test, the airflow velocity around stubble-crushing blades was in the range of 0.602~4.988 m·s^{-1} during the operation, and the airflow velocity in the simulation test was in the range of 0.613~4.85 m·s^{-1}. The airflow velocity set in the simulation test was basically the same as the surrounding airflow velocity during the operation of the plain straight stubble-crushing blade in the field test.

As obtained from the shooting results taken from the field high-speed camera test, the soil particles and straw on the ground surface were mainly thrown in the longitudinal direction. The longitudinal displacement and velocity of soil and straw were obtained in the DEM and the DEM-CFD coupled field, respectively. The correlation between simulated and measured values is given in Figure 8. The results obtained from the coupled models all lie above y = x. This indicates that the airflow action can increase the longitudinal displacements and velocities of the soil and straw, especially straw, which will accelerate the throwing motion and lead to an increase in the throwing range. The results in Figure 8a,c,e,g show that a better correlation was obtained between the DEM-CFD simulated and measured longitudinal motions, while a fair correlation was obtained between the measured and DEM simulated longitudinal motions (shown in Figure 8b,d,f,h). The correlation between the DEM simulated and measured straw longitudinal velocity was extremely poor. This meant that airflow did affect the accuracy of simulation results. Overall, the simulation results of the DEM-CFD coupled model showed better correlation with the results of the field tests compared with those DEM simulation model. Thus, the coupling simulation test can simulate the soil and straw thrown trend during operation.

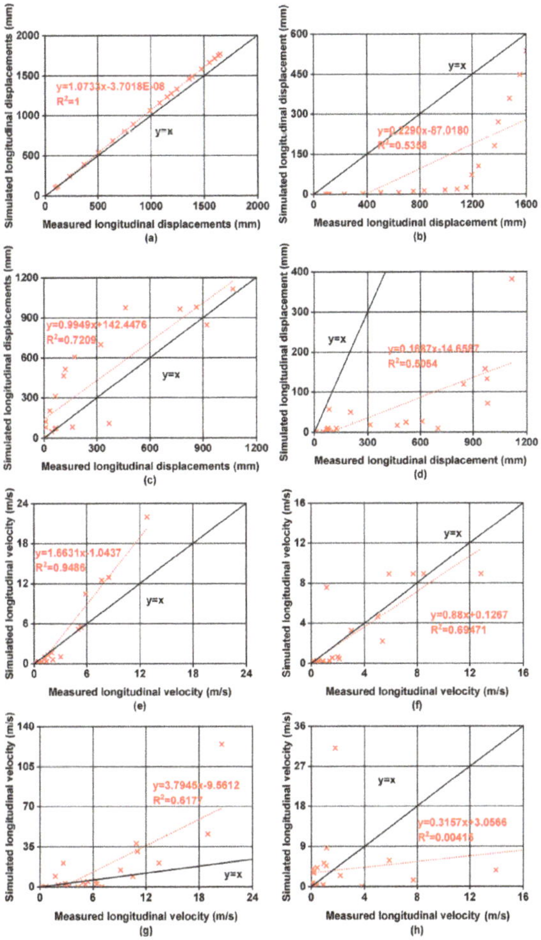

Figure 8. Correlation of longitudinal motion between the measured and simulated: (**a**,**b**) soil displacement in DEM-CFD and DEM model; (**c**,**d**) straw displacement in DEM-CFD and DEM model; (**e**,**f**) soil velocity in DEM-CFD and DEM model; (**g**,**h**) straw velocity in DEM-CFD and DEM model.

The overall determination coefficient (R^2) value of the longitudinal motion of straw is lower than that of soil due to the instability of straw throwing during stubble-crushing blade operation. The accuracy of the longitudinal motion results in the DEM-CFD simulation of the soil was higher than that of the straw, with an average relative error of 7.5% for the former and 17.8% for the latter. The results of the longitudinal displacement of soil and straw were more accurate, which was attributed to the fact that the extraction of velocity during the software processing was based on the change of displacement, which was the average velocity at the stage time instead of the instantaneous velocity. This had some deviation from the actual velocity variation, but it could reflect the velocity variation tendency of straw and soil and provide data support for the test results. On the whole, the model had an average relative error of 7.3% as compared to the testing results, which was lower than the 24.9% relative error of the analytical models in predicting soil forward displacement, as proposed by Fang et al. [14]. Therefore, the model was considered to be reasonably accurate in the simulation of soil–tool–straw interaction with airflow conditions.

3.2. Simulation of Soil and Straw Dynamics Displacements

3.2.1. Longitudinal Displacement

From the simulation test results (Figure 9), as the rotary speed of stubble-crushing blades increased from 270 to 540 rpm, the average longitudinal displacement of soil particles increased by 16.6% and that of straw by 89.9%. As the stubble-crushing blade speed increased from 540 to 720 rpm, the average longitudinal displacement of soil particles increased by 19.9%, while the average longitudinal displacement of straw increased by 106.8%. From 720 to 810 rpm, the average longitudinal displacement of soil particles increased by 17.3% and that of straw by 98.2%. Fang et al. [49] also observed that straw forward displacement was significantly larger than that of soil. Accordingly, increasing stubble-crushing blade rotary speed would increase the longitudinal throwing of soil and straw, which was the main throwing direction of soil and straw in stubble-crushing operations. According to Hendrick and Gill [50], the rotary speed of the blade influences soil throwing by affecting the soil acceleration. An increase in rotary speed results in a greater acceleration of soil particles, leading to more throwing.

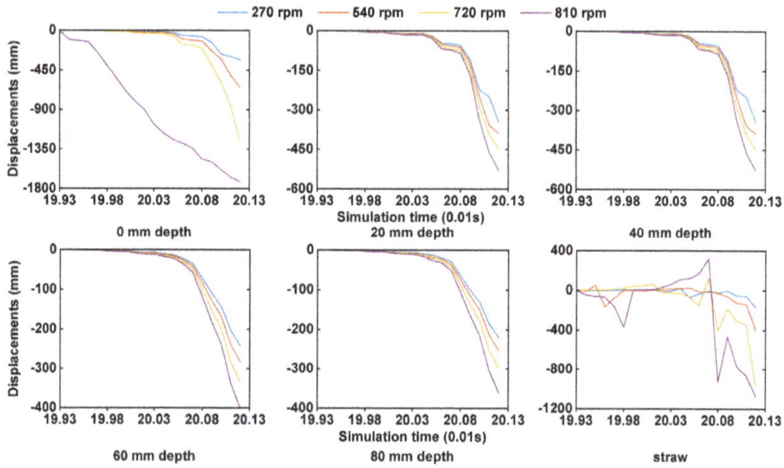

Figure 9. Longitudinal displacements of soil particles at different depths and surface straws at different rotary speeds.

The average longitudinal displacement of the soil particles decreased by 13% when the soil depth increased from 0 to 20 mm, by 13.7% when the soil depth increased from 20 mm to 40 mm, by 15.1% when the soil depth increased from 40 mm to 60 mm, and by 8.8% when the soil depth increased from 60 mm to 80 mm. Therefore, with the increase in soil depth, the longitudinal displacement of soil particles gradually decreased, and the longitudinal displacement of shallow soil was the largest, while the displacement of deep soil was the smallest. Therefore, most of the longitudinal soil thrown farther during the stubble-crushing operation was shallow soil.

As shown in Table 2, the presence or absence of airflow has a significant effect on straw displacement in X direction. Compared with no airflow, the longitudinal displacement of soil particles increased by 19.1% and the longitudinal displacement of straw increased by 335.9% under all of the working conditions with airflow. The effect of airflow on soil and straw throwing in stubble-crushing operation cannot be ignored. In summary, the lower the rotary speed of stubble-crushing blades, the smaller the loss of soil moisture, but this is not conducive to the separation of soil and straw throwing, so the maximum and minimum rotary speed were not the optimal choices.

3.2.2. Lateral Displacement

According to the simulation results (Figure 10), the average lateral displacement of soil particles increased by 31.5%, and the average lateral displacement of straw increased by 217.8% when the rotary speed of the stubble-crushing blades increased from 270 to 540 rpm. The average lateral displacement of soil particles increased by 14%, while the average lateral displacement of straw increased by 83.5%, when the rotary speed was raised from 540 to 720 rpm. The average lateral displacement of the soil particles increased by 20.3%, while the average lateral displacement of the straw increased by 61.3% when the rotary speed was raised from 720 to 810 rpm. Matin et al. [48] also reported an increasingly high amount of soil being thrown out of the furrow with greater rotary speed. Therefore, when the rotary speed rose, the lateral displacement of soil and straw also increased noticeably. This result showed that increasing rotary speed may greatly enhance the interaction between the blades, soil, and straw, aggravating the disturbance of soil and straw.

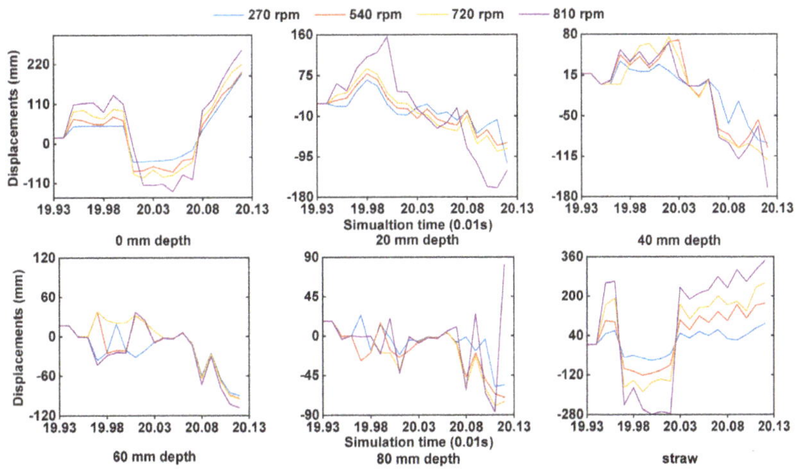

Figure 10. Lateral displacements of soil particles at different depths and surface straws at different rotary speeds.

The average lateral displacement of soil was reduced by 23.4% as the soil depth increased from 0 to 20 mm, and by 6.3% as the soil depth increased from 20 to 40 mm. The average lateral displacement of the soil particles also fell by 6.3% as the soil depth climbed from 40 to 60 mm. The average lateral displacement of the soil particles decreased by 32.9% when the soil depth was increased to 60 mm, and by 43.8% when it was increased from 60 mm to 80 mm. This is consistent with the findings of Fang et al. [14] that the lateral displacement of soil would be significantly reduced with an increase in soil depth. The lateral displacement of deep soil during the operation was minimally likely because the interaction between surface soil and lateral undisturbed soil was relatively weak, and as the soil depth increased, the force of lateral undisturbed soil on thrown soil increased. Therefore, the rotary speed should be suitably decreased to produce less lateral soil disturbance and lower power consumption.

3.2.3. Vertical Displacements

According to the test results (Figure 11), when the rotary speed of stubble-crushing blades was raised from 270 to 540 rpm, the average vertical displacement of the soil particles increased by 17.9% and that of straw decreased by 1.2%. However, when the rotary speed was raised from 540 to 720 rpm, the average vertical displacement of the soil particles decreased by 26.1% and that of the straw decreased by 1%. The average vertical displacement of the soil rose by 3.2% and that of the straw dropped by 2.1% when the rotary speed was raised from 720 to 810 rpm. The vertical displacement of soil particles was not significantly affected by the rotary speed of the stubble-crushing blades. This is different from the higher rotary speed soil being thrown higher, as reported by Matin et al. [51], and may be related to the different soil types and working conditions. The findings suggest that increasing the rotary speed of the stubble-crushing blades will result in the disorderly vertical distribution of soil particles, while the final position of straw will definitely be lower than in the initial position. Thus, stubble-crushing operation may accomplish a specified soil covering of straw on the ground surface.

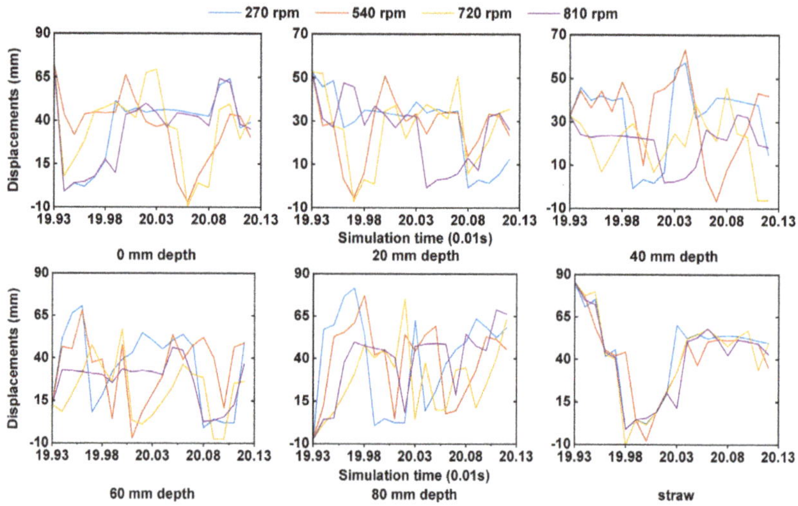

Figure 11. Vertical displacements of soil particles at different depths and surface straws at different rotary speeds.

When the soil depth increased from 0 to 20 mm, the average vertical displacement of the soil particles decreased by 24.2%. When the soil depth increased from 20 to 40 mm, the average vertical displacement of the soil particles decreased by 3.4%. When the soil depth increased from 40 to 60 mm, the average vertical displacement of the soil particles increased by 12.6%. Therefore, after the operation of stubble-crushing blades, the displacement of shallow soil will decrease and the displacement of deep soil will increase, resulting in the exchange of the deep and shallow soil layers and the change of soil layers. In conclusion, when the rotary speed of the stubble-crushing blades was 540 rpm with airflow action, the lateral displacement of the straw was analyzed. The statistics of the covered straw particles indicated that 12.5% of the total number of straws in the area were covered. Therefore, the majority of straws were disrupted in the lateral direction, and this operating condition aided in the establishment of a more effective seeding row.

To ensure a smaller lateral displacement of soil, the straight stubble blade angular speed should be reduced, but reducing blade rotary speed reduces the lateral displacement of straw. Considering the operation effect of the plain straight stubble-crushing blades, the better rotary speed is determined as 540 rpm, and the average unilateral lateral displacement of the soil is 40.43 mm, and the average unilateral lateral displacement of the straw is 102.46 mm. Therefore, the optimal sowing width of 80.86 mm can be formed, in theory.

3.3. *Simulations of Soil and Straw Dynamics Velocities*

3.3.1. Longitudinal Velocities

According to Table 2, the rotary speed of blades has a substantial ($p < 0.05$) influence on the X-direction velocity of soil particles. When the stubble-crushing blades speed increased from 270 to 540 rpm, the average longitudinal velocity of soil particles increased by 20% and the average longitudinal velocity of straw increased by 118.8%; when the rotary speed increased from 540 to 720 rpm, the average longitudinal velocity of the soil particles increased by 14.7% and the average longitudinal velocity of the straw increased by 108.9% (Figure 12); when the stubble-crushing blades rotary speed increased from 720 to 810 rpm, the average longitudinal velocity of soil particles increased by 16.8% and the average longitudinal velocity of straw increased by 38.8%, indicating that the rotary speed had an effect on the longitudinal velocities of the soil and the straw, with the effect on the longitudinal velocity of straw being more pronounced than that of the longitudinal velocity of soil. Increasing the rotary speed will boost the longitudinal throwing motions of the soil and the straw, which is also the primary throwing direction during stubble-crushing operations.

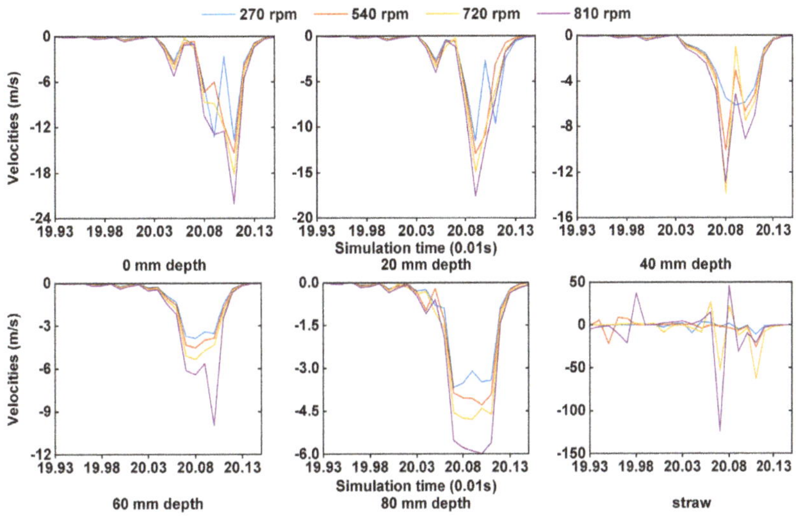

Figure 12. Longitudinal velocities of soil particles at different depths and surface straws at different rotary speeds.

The average longitudinal velocity of the soil particles fell by 19.7% as the soil depth increased from 0 to 20 mm, by 17% as the soil depth increased from 20 to 40 mm, by 23.7% as the soil depth increased from 40 to 60 mm, and by 8.3% as the soil depth increased from 60 to 80 mm. Therefore, the longitudinal velocity of the soil had a clear tendency to decrease with increasing soil depth, most likely because the interaction force between soil particles increased with the increasing soil depth and shallow soil impeded the vertical motion of deep soil, resulting in the weakening of the longitudinal motion of deep soil. Consequently, during stubble-crushing operations, shallow soil particles were mostly thrown at a greater distance, while deep soil particles were primarily thrown closer. To accomplish the separation motion of soil and straw, it is necessary to raise the blade rotary speed. This would, however, enhance the impact of blades on soil and straw, and the increased rotary speed would reflect the increased impact.

3.3.2. Lateral Velocities

The average lateral velocity of the soil particles increased by 60% and the average lateral velocity of the straw increased by 85.5% when the stubble-crushing blade rotary speed increased from 270 to 540 rpm (Figure 13). The average lateral velocity of the soil particles increased by 47.5% and the average lateral velocity of the straw increased by 79.2% when the rotary speed increased from 540 to 720 rpm. When the rotary speed was increased from 720 to 810 rpm, the average lateral velocity of the soil particles increased by 28.9% and the average lateral velocity of the straw increased by 74.7%. This shows that increasing the rotary speed of the stubble-crushing blades will enhance the lateral throwing motions of soil and straw.

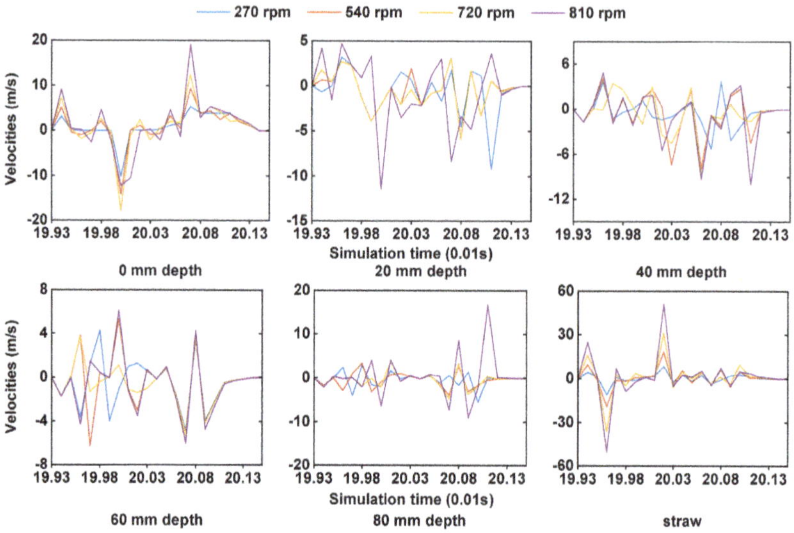

Figure 13. Lateral velocities particles at different depths and surface straws at different rotary speeds with or without airflow.

The average lateral velocity of the soil particles dropped by 29.3% as the soil depth increased from 0 to 20 mm, by 15.5% as the soil depth increased from 20 to 40 mm, by 43.23% as the soil depth increased from 40 to 60 mm, and by 67.8% as the soil depth climbed from 60 to 80 mm. Therefore, as the soil depth increased, the lateral velocity of the soil tended to decrease. This is likely because, as soil depth increases, the interaction force between the soil increases, which influences the trend of mutual motion between the soil particles and weakens the lateral motion of the soil. Therefore, the soil disturbed to the far sides during the stubble-crushing operation is mostly shallow soil, whereas the soil affecting the stubble-crushing operation region is often deep soil. In conclusion, if the lateral velocity of the soil is assured to be low (as a result of the weak contact between the stubble-crushing blades and the soil), the stubble-crushing blade rotary speed should be decreased, but this decreases the lateral velocity of straw. When the interaction between the stubble-crushing blades and the soil is poor, it is possible to create greater straw movement.

3.3.3. Vertical Velocities

According to the test results (Figure 14), the average vertical velocity of the soil particles increased by 15.5% and the average vertical velocity of the straw increased by 46.3% when the stubble-crushing blade speed was increased from 270 to 540 rpm. However, the average vertical velocity of the soil particles decreased by 75.5% and the average vertical velocity of the straw decreased by 9.5% when the rotary speed was increased from 540 to 720 rpm. The effect of the rotary speed on the vertical velocities of the soil and the straw was, thus, non-directional, and increasing stubble-crushing blades speed did not necessarily increase the vertical velocities of soil and straw.

The average vertical velocity of the soil particles decreased by 5.3% as the soil depth increased from 0 to 20 mm, by 79% as the soil depth increased from 20 to 40 mm, by 65.7% as the soil depth increased from 40 to 60 mm, and by 109.1% as the soil depth increased from 60 to 80 mm. Therefore, as the soil depth increased, the vertical velocity of the soil particles tended to decrease. This is likely because, as the soil depth increases, the interaction force between the soil increases, and the shallow soil will impede the vertical motion of the deep soil, resulting in the weakening of the deep soil vertical motion. Consequently, during stubble-crushing operations, the shallow soil is primarily thrown to a higher height, whereas the deep soil typically returns to the surface after being thrown to a lower height.

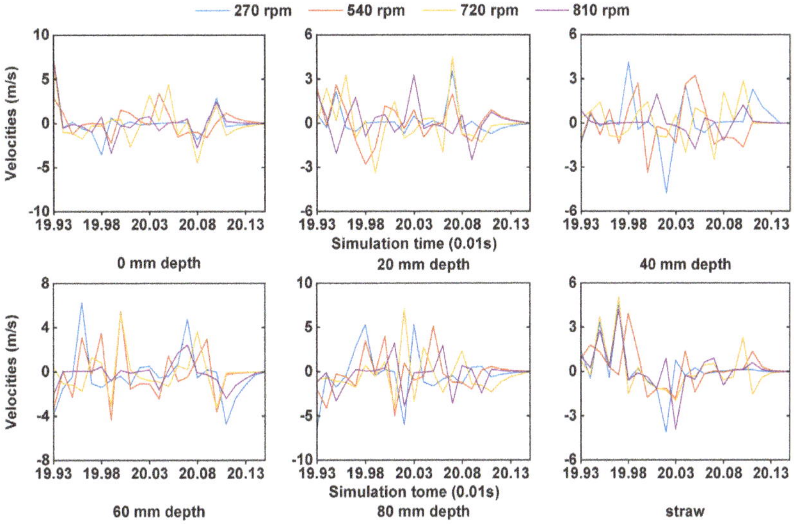

Figure 14. Vertical velocities particles at different depths and surface straws at different rotary speeds with or without air-flow.

4. Conclusions

In this study, a DEM-CFD model was developed to simulate the dynamic motion of soil and straw at different rotary speeds. The results of the field high-speed camera tests showed that the DEM-CFD coupling model established in this study could effectively and accurately capture the dynamic motions of soil and straw during a stubble-crushing operation. The simulation results showed that airflow substantially affected the soil and straw displacement and velocity. The airflow effect could accelerate the lateral and longitudinal throwing of the soil and the straw and expand the throwing range with little effect on their vertical position.

The longitudinal direction was the main throwing direction of soil and straw in the stubble-crushing operation. As the rotary speed increased, the lateral and longitudinal displacement and velocity of soil and straw increased evidently, which facilitated straw burial and soil layer exchange. The stubble-crushing blade rotary speed had no significant directional effect on the vertical motions of soil. After the stubble-crushing operation, the displacement of the shallow soil would fall and the displacement of the deep soil would rise, forming an exchange between the deep and shallow soil layers while achieving a certain degree of surface straw cover. The effect of the rotary speed on the lateral displacement and lateral velocity of the soil particles decreased with the increasing soil depth. After a stubble-crushing operation, the displacement of the shallow soil would fall and the displacement of the deep soil would rise, forming an exchange between the deep and shallow soil layers and realizing the change of soil layers. At the same time, most of the longitudinal soil was thrown further, which was shallow soil. In the field operation, the rotary speed should be increased appropriately to increase the lateral displacement of the straw, which in turn achieves the disturbance of the straw in the lateral direction. Taking into account the separation of the soil and straw and the formation of the desired seeding row, plain straight stubble-crushing blades with an angular speed of 540 rpm were able to form the optimal seeding row with a width of 80.86 mm. This also helps to achieve the separation of soil and straw and creates good seed bed conditions. These results have implications for the future design of improved components and their evaluation for strip tillage.

Author Contributions: Conceptualization, methodology, software, Y.Y. and J.W.; validation, formal analysis, investigation, resources, data curation, Y.Y. and X.Z.; writing—original draft preparation, Y.Y. and J.W.; writing—review and editing, Y.Y. and S.Z.; supervision, S.Z. All authors have read and agreed to the published version of the manuscript.

Funding: This research was funded by the National Key Research and Development Program of China (Grant No. 2020YFD1000903).

Institutional Review Board Statement: Not applicable.

Data Availability Statement: Not applicable.

Conflicts of Interest: The authors declare no conflict of interest.

References

1. Klik, A.; Rosner, J. Long-term experience with conservation tillage practices in Austria: Impacts on soil erosion processes. *Soil Tillage Res.* **2020**, *203*, 104669. [CrossRef]
2. Margenot, A.J.; Paul, B.K.; Sommer, R.R.; Pulleman, M.M.; Parikh, S.J.; Jackson, L.E.; Fonte, S.J. Can conservation agriculture improve phosphorus (P) availability in weathered soils? Effects of tillage and residue management on soil P status after 9 years in a Kenyan Oxisol. *Soil Tillage Res.* **2017**, *166*, 157–166. [CrossRef]
3. Wang, Z.; Chen, Q.; Liu, L.; Wen, X.; Liao, Y. Responses of soil fungi to 5-year conservation tillage treatments in the drylands of northern China. *Appl. Soil Ecol.* **2016**, *101*, 132–140. [CrossRef]
4. Kassam, A.; Friedrich, T.; Derpsch, R. Global spread of Conservation Agriculture. *Int. J. Environ. Stud.* **2019**, *76*, 29–51. [CrossRef]
5. Flower, K.C.; Ward, P.R.; Micin, S.F.; Cordingley, N. Crop rotation can be used to manipulate residue levels under no-tillage in a rainfed Mediterranean-type environment. *Soil Tillage Res.* **2021**, *212*, 105062. [CrossRef]
6. Jiang, Y.; Xie, H.; Chen, Z. Relationship between the amounts of surface corn stover mulch and soil mesofauna assemblage varies with the season in cultivated areas of northeastern China. *Soil Tillage Res.* **2021**, *213*, 105091. [CrossRef]
7. Celik, A.; Altikat, S.; Way, T.R. Strip tillage width effects on sunflower seed emergence and yield. *Soil Tillage Res.* **2013**, *131*, 20–27. [CrossRef]
8. Lowry, C.J.; Robertson, G.P.; Brainard, D.C. Strip-tillage decreases soil nitrogen availability and increases the potential for N losses in a cover cropped organic system. *Agric. Ecosyst. Environ.* **2021**, *319*, 107524. [CrossRef]
9. Licht, M.A.; Al-Kaisi, M. Strip-tillage effect on seedbed soil temperature and other soil physical properties. *Soil Tillage Res.* **2005**, *80*, 233–249. [CrossRef]
10. Lekavičienė, K.; Šarauskis, E.; Naujokienė, V.; Buragienė, S.; Kriaučiūnienė, Z. The effect of the strip tillage machine parameters on the traction force, diesel consumption and CO_2 emissions. *Soil Tillage Res.* **2019**, *192*, 95–102. [CrossRef]
11. Al-Kaisi, M.M.; Yin, X. Tillage and crop residue effects on soil carbon and carbon dioxide emission in corn–soybean rotations. *J. Environ. Qual.* **2005**, *34*, 437–445. [CrossRef] [PubMed]
12. Gathala, M.K.; Laing, A.M.; Tiwari, T.P.; Timsina, J.; Islam, S.; Bhattacharya, P.M.; Dhar, T.; Ghosh, A.; Sinha, A.K.; Chowdhury, A.K.; et al. Energy-efficient, sustainable crop production practices benefit smallholder farmers and the environment across three countries in the Eastern Gangetic Plains, South Asia. *J. Clean. Prod.* **2020**, *246*, 118982. [CrossRef]
13. Matin, M.A.; Hossain, M.I.; Gathala, M.K.; Timsina, J.; Krupnik, T.J. Optimal design and setting of rotary strip-tiller blades to intensify dry season cropping in Asian wet clay soil conditions. *Soil Tillage Res.* **2021**, *207*, 104854. [CrossRef] [PubMed]
14. Fang, H.M.; Ji, C.Y.; Farman, A.C.; Guo, J.; Zhang, Q.Y.; Chaudhry, A. Analysis of soil dynamic behavior during rotary tillage based on distinct element method. *Trans. Chin. Soc. Agric. Mach.* **2016**, *47*, 22–28.
15. Fang, H.M.; Ji, C.Y.; Ahmed, A.T.; Zhang, Q.Y.; Guo, J. Simulation analysis of straw movement in straw-soil-rotary blade system. *Trans. Chin. Soc. Agric. Mach.* **2016**, *47*, 60–67.
16. Zhao, H.; Li, H.; Ma, S.; He, J.; Wang, Q.; Lu, C.; Zheng, Z.; Zhang, C. The effect of various edge-curve types of plain-straight blades for strip tillage seeding on torque and soil disturbance using DEM. *Soil Tillage Res.* **2020**, *202*, 104674. [CrossRef]
17. Asl, J.H.; Singh, S. Optimization and evaluation of rotary tiller blades: Computer solution of mathematical relations. *Soil Tillage Res.* **2009**, *106*, 1–7. [CrossRef]
18. Barr, J.B.; Ucgul, M.; Desbiolles, J.M.A.; Fielke, J.M. Simulating the effect of rake angle on narrow opener performance with the discrete element method. *Biosyst. Eng.* **2018**, *171*, 1–15. [CrossRef]
19. Shmulevich, I.; Asaf, Z.; Rubinstein, D. Interaction between soil and a wide cutting blade using the discrete element method. *Soil Tillage Res.* **2007**, *97*, 37–50. [CrossRef]
20. Tekeste, M.Z.; Balvanz, L.R.; Hatfield, J.L.; Ghorbani, S. Discrete element modeling of cultivator sweep-to-soil interaction: Worn and hardened edges effects on soil-tool forces and soil flow. *J. Terramechanics* **2019**, *82*, 1–11. [CrossRef]
21. Zeng, Z.; Ma, X.; Chen, Y.; Qi, L. Modelling residue incorporation of selected chisel ploughing tools using the discrete element method (DEM). *Soil Tillage Res.* **2020**, *197*, 104505. [CrossRef]

22. Aikins, K.A.; Ucgul, M.; Barr, J.B.; Jensen, T.A.; Antille, D.L.; Desbiolles, J.M.A. Determination of discrete element model parameters for a cohesive soil and validation through narrow point opener performance analysis. *Soil Tillage Res.* **2021**, *213*, 105123. [CrossRef]
23. Zhao, M.Q.; Liu, Y.Q.; Hu, Y.W. An airflow field finite element analysis of the seed adsorption hole of the pneumatic seeder. *Appl. Mech. Mater.* **2012**, *117*, 1810–1815. [CrossRef]
24. Sakai, M.; Koshizuka, S. Large-scale discrete element modeling in pneumatic conveying. *Chem. Eng. Sci.* **2009**, *64*, 533–539. [CrossRef]
25. Aleshin, V.; Van Den Abeele, K. Preisach analysis of the Hertz–Mindlin system. *J. Mech. Phys. Solids.* **2009**, *57*, 657–672. [CrossRef]
26. de Billy, M.; Cohen Tenoudji, F. Resonance ultrasonic spectroscopy applied to normal and tangential contact modes investigation of a constrained metallic sphere. *Ultrasonics* **2021**, *117*, 106539. [CrossRef]
27. Wang, X.; Zhang, S.; Pan, H.; Zheng, Z.; Huang, Y.; Zhu, R. Effect of soil particle size on soil-subsoiler interactions using the discrete element method simulations. *Biosyst. Eng.* **2019**, *182*, 138–150. [CrossRef]
28. Hoseinian, S.H.; Hemmat, A.; Esehaghbeygi, A.; Shahgoli, G.; Baghbanan, A. Development of a dual sideway-share subsurface tillage implement: Part 1. Modeling tool interaction with soil using DEM. *Soil Tillage Res.* **2022**, *215*, 105201. [CrossRef]
29. Shaikh, S.A.; Li, Y.; Ma, Z.; Chandio, F.A.; Tunio, M.H.; Liang, Z.; Solangi, K.A. Discrete element method (DEM) simulation of single grouser shoe-soil interaction at varied moisture contents. *Comput. Electron. Agric.* **2021**, *191*, 106538. [CrossRef]
30. Su, Z.; Li, Y.; Dong, Y.; Tang, Z.; Liang, Z. Simulation of rice threshing performance with concentric and non-concentric threshing gaps. *Biosyst. Eng.* **2020**, *197*, 270–284. [CrossRef]
31. Johnson, K.L.; Kendall, K.; Roberts, A.D. Surface energy and the contact of elastic solids. *Proc. R. Soc. A.* **1971**, *324*, 301–313. [CrossRef]
32. Ajmal, M.; Roessler, T.; Richter, C.; Katterfeld, A. Calibration of cohesive DEM parameters under rapid flow conditions and low consolidation stresses. *Powder Technol.* **2020**, *374*, 22–32. [CrossRef]
33. Bahramian, A.; Olazar, M. Influence of restitution and friction coefficients on the velocity field of polydisperse TiO_2 agglomerates in a conical fluidized bed by the adhesive CFD-DEM simulation. *Powder Technol.* **2021**, *386*, 491–504. [CrossRef]
34. e Silva, B.B.; da Cunha, E.R.; de Carvalho, R.M.; Tavares, L.M. Modeling and simulation of green iron ore pellet classification in a single deck roller screen using the discrete element method. *Powder Technol.* **2018**, *332*, 359–370. [CrossRef]
35. Mudarisov, S.; Farkhutdinov, I.; Khamaletdinov, R.; Khasanov, E.; Mukhametdinov, A. Evaluation of the significance of the contact model particle parameters in the modelling of wet soils by the discrete element method. *Soil Tillage Res.* **2022**, *215*, 105228. [CrossRef]
36. Zhu, J.; Zou, M.; Liu, Y.; Gao, K.; Su, B.; Qi, Y. Measurement and calibration of DEM parameters of lunar soil simulant. *Acta Astronaut.* **2022**, *191*, 169–177. [CrossRef]
37. Ahn, S.-H.; Xiao, Y.; Wang, Z.; Luo, Y.; Fan, H. Unsteady prediction of cavitating flow around a three dimensional hydrofoil by using a modified RNG k-ε model. *Ocean Eng.* **2018**, *158*, 275–285. [CrossRef]
38. Dai, T.; Liu, S.; Liu, J.; Jiang, N.; Liu, W.; Chen, Q. Evaluation of fast fluid dynamics with different turbulence models for predicting outdoor airflow and pollutant dispersion. *Sustain. Cities Soc.* **2022**, *77*, 103583. [CrossRef]
39. Huo, J.; Wang, Z.; Luan, X.; Jing, M.; Hou, S.; Jiang, J.; Zhang, B. The CFD modeling of bund overtopping phenomena and prediction of dynamic pressure on the bund. *J. Loss Prevent Proc.* **2022**, *74*, 104653. [CrossRef]
40. Tiwary, B.; Kumar, R.; Singh, P.K. Thermofluidic characteristic of a nanofluid-cooled oblique fin heat sink: An experimental and numerical investigation. *Int. J. Therm. Sci.* **2022**, *171*, 107214. [CrossRef]
41. Zhao, S.; Liu, H.; Yang, C.; Yang, L.; Gao, L.; Yang, Y. Design and discrete element simulation of interactive layered subsoiler with maize straw returned to filed. *Trans. Chin. Soc. Agric. Mach.* **2021**, *52*, 75–87.
42. Wang, J.; Zhao, S.; Yang, Z.; Gao, L.; Yang, Y. Design and experiment of driving stubble cutter for corn strip with less tillage operation. *Trans. Chin. Soc. Agric. Mach.* **2021**, *52*, 51–61.
43. Saunders, C.; Ucgul, M.; Godwin, R.J. Discrete element method (DEM) simulation to improve performance of a mouldboard skimmer. *Soil Tillage Res.* **2021**, *205*, 104764. [CrossRef]
44. Verdoia, M.; Negro, F.; Kedhi, E.; Suryapranata, H.; Marcolongo, M.; De Luca, G. Benefits with drug-coated balloon as compared to a conventional revascularization strategy for the treatment of coronary and non-coronary arterial disease: A comprehensive meta-analysis of 45 randomized trials. *Vasc. Pharmacol.* **2021**, *138*, 106859. [CrossRef] [PubMed]
45. Zhou, S.; Buitrago, C.; Foong, A.; Lee, V.; Dawit, L.; Hiramoto, B.; Chang, P.; Schilperoort, H.; Lee, A.; de-Madaria, E.; et al. Comprehensive meta-analysis of randomized controlled trials of Lactated Ringer's versus Normal Saline for acute pancreatitis. *Pancreatology* **2021**, *21*, 1405–1410. [CrossRef] [PubMed]
46. Zhang, S.; Chen, X.; Jia, S.; Liang, A.; Zhang, X.; Yang, X.; Wei, S.; Sun, B.; Huang, D.; Zhou, G. The potential mechanism of long-term conservation tillage effects on maize yield in the black soil of Northeast China. *Soil Tillage Res.* **2015**, *154*, 84–90. [CrossRef]
47. Lee, K.S.; Park, S.H.; Park, W.Y.; Lee, C.S. Strip-tillage characteristics of rotary tiller blades for use in a dry land direct rice seeder. *Soil Tillage Res.* **2003**, *71*, 25–32. [CrossRef]
48. Matin, M.A.; Desbiolles, J.M.A.; Fielke, J.M. Strip-tillage using rotating straight blades: Effect of cutting edge geometry on furrow parameters. *Soil Tillage Res.* **2016**, *155*, 271–279. [CrossRef]

49. Fang, H.M.; Zhang, Q.Y.; Chandio, F.A.; Guo, J.; Sattar, A.; Arslan, C.; Ji, C.Y. Effect of straw length and rotavator kinematic parameter on soil and straw movement by a rotary blade. *Eng. Agric. Environ. Food.* **2016**, *9*, 235–241. [CrossRef]
50. Hendrick, J.G.; Gill, W.R. Rotary tiller design parameters part. III. Ratio of peripheral and forward velocities. *Trans. ASAE* **1971**, *14*, 679–683.
51. Matin, M.A.; Fielke, J.M.; Desbiolles, J.M.A. Furrow parameters in rotary strip-tillage: Effect of blade geometry and rotary speed. *Biosyst. Eng.* **2014**, *118*, 7–15. [CrossRef]

agriculture

Article

Robust Trajectory Tracking Control of an Autonomous Tractor-Trailer Considering Model Parameter Uncertainties and Disturbances

En Lu [1,2,*], Jialin Xue [1,2], Tiaotiao Chen [1,2] and Song Jiang [3]

1 Key Laboratory for Theory and Technology of Intelligent Agricultural Machinery and Equipment, Jiangsu University, No. 301 Xuefu Road, Zhenjiang 212013, China; 2222216049@ujs.edu.cn (J.X.)
2 School of Agricultural Engineering, Jiangsu University, No. 301 Xuefu Road, Zhenjiang 212013, China
3 School of Mechanical Engineering, Jiangsu University, No. 301 Xuefu Road, Zhenjiang 212013, China
* Correspondence: jsluen@163.com; Tel.: +86-187-9600-5357

Abstract: This paper discusses the robust trajectory tracking control of an autonomous tractor-trailer in agricultural applications. Firstly, considering the model parameter uncertainties and various disturbances, the kinematic and dynamic models of the autonomous tractor-trailer system are established. Moreover, the coordinate transformation is adopted to convert the trajectory tracking error into a new unconstrained error state space model. On this basis, the prescribed performance control (PPC) technique is designed to ensure the convergence speed and final tracking control accuracy of the tractor-trailer control system. Then, this paper designs a double closed-loop control structure. The posture control level adopts the model predictive control (MPC) method, and the dynamic level adopts the sliding mode control (SMC) method. At the same time, it is worth mentioning that the nonlinear disturbance observer (NDO) is designed to estimate all kinds of system disturbances and compensate for the tracking control system to improve the system's robustness. Finally, the proposed control strategy is validated through comparative simulations, demonstrating its effectiveness in achieving robust trajectory tracking of the autonomous tractor-trailer system.

Keywords: tractor-trailer; trajectory tracking; prescribed performance; model predictive control; robust sliding mode control

Citation: Lu, E.; Xue, J.; Chen, T.; Jiang, S. Robust Trajectory Tracking Control of an Autonomous Tractor-Trailer Considering Model Parameter Uncertainties and Disturbances. *Agriculture* **2023**, *13*, 869. https://doi.org/10.3390/agriculture13040869

Academic Editors: Muhammad Sultan, Redmond R. Shamshiri, Md Shamim Ahamed and Muhammad Farooq

Received: 20 March 2023
Revised: 7 April 2023
Accepted: 10 April 2023
Published: 14 April 2023

1. Introduction

With the gradual maturity and application of agricultural machineries' automatic driving technology based on the satellite navigation system, the automation level and working efficiency of agricultural machinery have been greatly improved [1,2]. As the main power machinery in agricultural production, research on autonomous control of tractors has received extensive attention. To further enhance agricultural productivity and autonomous operation, tractors are often connected to trailers through rigid shafts, forming tractor-trailer systems that enable cost-effective transportation in material collection, load carriage, crop harvesting, and so on [3]. Unfortunately, the nonlinear multi-body dynamics, coupling effects, and various disturbances of the tractor-trailer lead to the complexity and difficulty of modeling, and the uneven and harsh farmland working environment makes its autonomous operation control very challenging.

At present, there have been successful research studies aimed at addressing the motion control of tractor-trailers. Yuan et al. developed a trajectory tracking controller, based on the backstepping technique, aiming to drive the multi-steering tractor-trailer mobile robot's states towards their desired trajectories that result in convergence [4]. Yue et al. proposed an effective quintic polynomial-based trajectory planning approach combined with a robust tube-based MPC method for an underactuated tractor-trailer system during lane change maneuver [5]. Alipour et al. addressed the lateral and longitudinal slip of the wheel as a group of limited disturbances and designed a robust sliding mode trajectory tracking controller based on the established nonlinear dynamic model [6]. Kassaeiyan et al. designed a full state trajectory tracking controller that ensures the asymptotic stabilization of the output errors and enables tractor-trailer wheeled robots to follow the desired paths both in forward and backward [7]. Murillo et al. presented a novel nonlinear mathematical model of an articulated tractor-trailer system, which can be used to combine with receding horizon techniques to enhance the performance of path tracking tasks of articulated systems [8]. Shojaei et al. drew a neural adaptive PID tracking controller to guarantee that the tracking errors exponentially converge to an arbitrary small ultimate bound with a prespecified maximum overshoot and convergence rate [3].

The coordinated tracking control of the posture and dynamic of tractor-trailer systems is a critical control strategy that requires careful consideration of both kinematics and dynamics. Liu et al. proposed a composite control strategy that integrates a posture controller based on the MPC method and a dynamic controller based on the SMC method [9]. Zhang et al. researched the robust trajectory tracking control method of driverless vehicles and designed a hierarchical control framework based on conditional integral algorithm, which is composed of kinematic controller and dynamic controller [10]. Liao et al. proposed an integrated dynamic model that includes several critical factors, such as chassis kinematics, chassis dynamics, wheel-ground interaction, and wheel dynamics. Based on this model, they developed a model-based coordinated adaptive robust controller that features three-level designs for different parts of robot dynamics [11]. By effectively combining actor critical multilayer neural networks with adaptive robust controllers, Elhaki et al. proposed a unique intelligent prescribed performance output feedback multi-loop controller that improves robustness through identification of various nonlinear parameter uncertainties and external disturbances [12].

When towing trailers work in various complex environments, they will inevitably encounter the internal and external disturbances and model parameter changes, which will greatly reduce the performance of the controller and even lose the stability of the system. Besides the previously mentioned anti-disturbance control methods, another effective approach is to use a disturbance observer to identify and compensate for the observed disturbances, thereby suppressing their effects on the system. To address the challenge of accurately tracking specified paths with agricultural vehicles, Taghia et al. proposed a sliding mode controller with a NDO [13]. Aiming at the influence of trailer mass change on the stability of the tractor-trailer system, the trailer mass is estimated by the designed deep neural network (DNN) [14]. Han et al. proposed a novel estimation system for the hitch angle using the Kalman filter and deep-learning techniques [15]. Guevara et al. reported the use of active disturbance rejection control (ADRC) with a dual-stage NDO to improve the backward trajectory tracking performance of Generalized N-Trailers in no ideal conditions [16].

Regarding problems related to performance constraints of tracking, the transient and steady-state performances of the vehicle tracking control system have always been one of the issues that scholars focus on [17]; they directly determine the control effect. For this reason, Bechlioulis et al. proposed a prescribed performance control (PPC) method [18]. Its core idea is to manually set the performance envelope for the state (or error) of the control system, and describe the transient (such as convergence speed, up-regulation, down-regulation, etc.) and steady-state (such as control accuracy, etc.) performances of the control system through the convergence characteristics of the performance envelope function, so as to ensure that the prescribed tracking performance conditions are met while achieving the control objectives. To develop an open-loop error dynamic model based on the unconstrained filtered tracking error, a nonlinear transformation is used to convert the constrained errors to unconstrained ones based on the prescribed performance technique [19]. In the context of fixed configuration formation control of vehicles, Guo et al. proposed a finite-time vehicle formation control method that considers prescribed transient and steady-state performance constraints [20].

To solve all the above problems, a high-performance robust tracking controller for an autonomous tractor-trailer considering model uncertainties and disturbances is designed. The main contributions and innovations of this work are summarized as follows:

(1) The kinematic and dynamic models of the autonomous tractor-trailer system are established, taking into account model uncertainties and various disturbances. Moreover, considering the nonholonomic characteristics of the tractor-trailer, the nonlinear transformation is used to convert the trajectory tracking error into a new unconstrained error state space model. This greatly facilitates the design of a follow-up tracking controller.

(2) On this basis, this paper designs a double closed-loop control structure. The posture control level adopts the standard MPC method, and the dynamic level adopts the SMC method. At the same time, it is worth mentioning that the fast power reaching law with second-order sliding mode characteristics is selected to reduce the chattering of the traditional SMC method.

(3) Special application scenarios and complex working environments lead to significant model parameter changes and various disturbances in the tractor-trailer system. Therefore, the NDO is designed to estimate all kinds of system disturbances and compensate for the trajectory tracking control system of the tractor-trailer system to improve the system's robustness.

(4) Compared with many previous works [4–9,12,14–16], taking into account the convergence speed and final tracking control accuracy in the transient and steady-state performances, the PPC strategy is added to the front end of the double closed-loop controller to ensure the effective implementation of the robust trajectory tracking control of the tractor-trailer without any possible controller singularity.

This paper is organized as follows: Section 2 presents the kinematic and dynamic models of the tractor-trailer system, the tracking error model, and the prescribed performance function. Section 3 introduces the main results of this paper, including the design of the posture and dynamic controller. Results and analysis are provided in Section 4 to illustrate the effectiveness of the method proposed in this paper. Finally, Section 5 presents brief conclusions.

2. Problem Statement

2.1. Kinematic Model of Tractor-Trailer

The tractor-trailer is a typical complex multi-body system, and its plane motion diagram is shown in Figure 1. To enable facilitate modeling, several assumptions are as follows [9]: (1) the tractor-trailer only operates on the horizontal plane; (2) the tractor-trailer is composed of rigid components; and (3) the tractor-trailer does not slip laterally and longitudinally during movement.

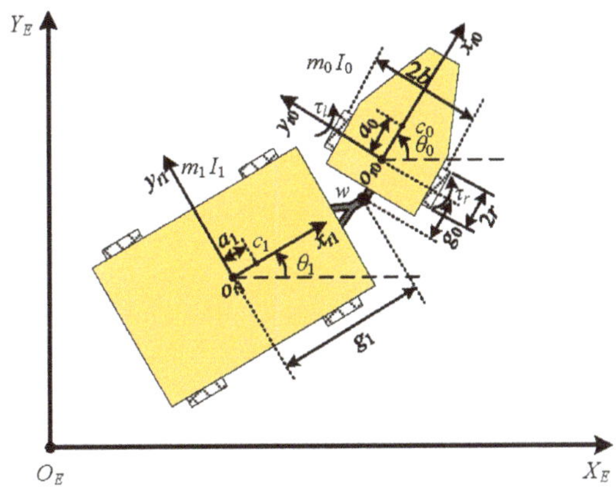

Figure 1. Plane motion diagram of the tractor-trailer. ($O_E X_E Y_E$ is the earth-fixed coordinate system.

$o_{t0}x_{t0}y_{t0}$ is the tractor-fixed coordinate system. $o_{t1}x_{t1}y_{t1}$ is the trailer-fixed coordinate system. θ_0 is the heading angle of the tractor. θ_1 is the heading angle of the tractor. τ_l and τ_r are the driving torque of the left and right wheels of the tractor. r is the radius of driving wheels. $2b$ is the distance between two active wheels of the tractor. a_0 is the distance of the centroid c_0 and the midpoint o_{t0} of tractor wheel axle. a_1 is the distance of the centroid c_1 and the center point o_{t1} of the trailer. g_0 is the distance between the hinge point w and the midpoint o_{t0} of tractor wheel axle. g_1 is distance between the hinge point w and the center point o_{t1} of the trailer. m_0 is the mass of the tractor. m_1 is the mass of the trailer without load. I_0 is the rotational inertia about vertical axis through c_0. I_1 is the rotational inertia about vertical axis through c_1 when the trailer is not loaded).

In the system of the autonomous tractor-trailer, in addition to describing the position and direction variables of the tractor and trailer, the relative heading angle information of the tractor and trailer must also be considered. Based on the previous assumption (3)—the wheels of the tractor and trailer do not slip laterally—the following nonholonomic constraints can be obtained:

$$\begin{cases} \dot{x}\sin\theta_0 - \dot{y}\cos\theta_0 - g_1\dot{\theta}_1\cos(\theta_0-\theta_1) = 0 \\ \dot{x}\sin\theta_1 - \dot{y}\cos\theta_1 = 0 \end{cases} \tag{1}$$

where x and y represent the position coordinates of the trailer, θ_1 represents the heading angle of the trailer, and θ_0 represents the heading angle of the tractor.

The posture vector of the tractor-trailer is defined as $q = [x, y, \theta_1, \theta_0]^T$, and Equation (1) can be rewritten into the following matrix form:

$$A(q)\dot{q} = \begin{bmatrix} \sin\theta_0 & -\cos\theta_0 & -g_1\cos(\theta_0-\theta_1) & 0 \\ \sin\theta_1 & -\cos\theta & 0 & 0 \end{bmatrix} \begin{bmatrix} \dot{x} \\ \dot{y} \\ \dot{\theta}_1 \\ \dot{\theta}_0 \end{bmatrix} = 0 \tag{2}$$

where $A(q) = \begin{bmatrix} \sin\theta_0 & -\cos\theta_0 & -g_1\cos(\theta_0-\theta_1) & 0 \\ \sin\theta_1 & -\cos\theta & 0 & 0 \end{bmatrix}$ and it is the constraint matrix of the system.

In addition, we introduce a full rank matrix $S(q) = [s_1(q), s_2(q)]^T$, which consists of a set of smooth and linearly independent vector fields, $s_i(q) \in \Re^n$, $i = 1, 2$, in the null space of $A(q)$, that is, $A(q)S(q) = 0$ [3]. We define the pseudo velocity vector $v(t) = [v(t), \omega(t)]^T$ as the input vector, and the system can be expressed as:

$$\dot{q} = \underbrace{\begin{bmatrix} \cos\theta_1 & 0 \\ \sin\theta_1 & 0 \\ (1/g_1)\tan(\theta_0-\theta_1) & -(g_0/g_1)\sec(\theta_0-\theta_1) \\ 0 & 1 \end{bmatrix}}_{S(q)} \underbrace{\begin{bmatrix} v(t) \\ \omega(t) \end{bmatrix}}_{v(t)} \tag{3}$$

where $v(t)$ represents the linear speed of o_t point on the trailer, $\omega(t)$ represents the angular speed of the tractor, and $S(q)$ is called the motion matrix.

2.2. Dynamic Model of Tractor-Trailer

The tractor-trailer system can be described by using the first Lagrange equation, which can be expressed as follows:

$$\frac{d}{dt}\left(\frac{\partial L}{\partial \dot{q}_k}\right) - \frac{\partial L}{\partial q_k} = f_k + A^T(q)\lambda \; k = 1,2,3,4 \tag{4}$$

where f_k represents the generalized force, λ represents the Lagrange multiplier, and k represents the number of generalized coordinates; L can be obtained by the following formula:

$$L = T(q, \dot{q}) - U(q) \tag{5}$$

where $T(q, \dot{q})$ represents the kinetic energy of the system and $U(q)$ represents the potential energy of the system.

According to assumption (1) in Section 2.1, the tractor-trailer moves only on the horizontal plane, so its gravitational potential energy will not change. According to assumption (2) in Section 2.1, the tractor-trailer is composed of rigid parts, so its elastic potential energy can be regarded as zero. Therefore, the body kinetic energy of the tractor and trailer can be considered to establish the dynamic model. Based on Equations (4) and (5), the dynamic model of the tractor-trailer system can be given by:

$$M_{a1}(t,q)\ddot{q}(t) + C_{a1}(t,q,\dot{q})\dot{q}(t) + D_{a1}(t,q)\dot{q}(t) = B_{a1}(q)\tau(t) + \tau_d(t) + A^T(q)\lambda \tag{6}$$

where τ_d represents the unknown nonlinearity vector caused by external disturbances, ground friction, and so on; M_{a1} represents the symmetric positive definite inertia matrix; C_{a1} represents the matrix of centripetal force and Coriolis force; D_{a1} represents the matrix of damping and viscous friction coefficient; B_{a1} represents the input transformation matrix; and $\tau = [\tau_l, \tau_r]^T$ represents the input torque vector. To simplify the design, an on-axle hitching tractor-trailer ($g_0 = 0$) is considered. Based on [3], the dynamic matrices are selected as follows:

$$M_{a1}(t,q) = \begin{bmatrix} m(t) & 0 & -A(t)\sin\theta_1 & -a_0 m_0 \sin\theta_0 \\ 0 & m(t) & A(t)\cos\theta_1 & a_0 m_0 \cos\theta_0 \\ -A(t)\sin\theta_1 & A(t)\cos\theta_1 & I_{\theta_1}(t) & a_0 d m_0 \cos(\theta_0 - \theta_1) \\ -a_0 m_0 \sin\theta_0 & a_0 m_0 \cos\theta_0 & a_0 d m_0 \cos(\theta_0 - \theta_1) & I_{\theta_0} \end{bmatrix}$$

$$C_{a1}(t,q,\dot{q}) = \begin{bmatrix} 0 & 0 & -A(t)\dot{\theta}_1 \cos\theta_1 & -a_0 m_0 \dot{\theta}_0 \cos\theta_0 \\ 0 & 0 & -A(t)\dot{\theta}_1 \sin\theta_1 & -a_0 m_0 \dot{\theta} \sin\theta_0 \\ 0 & 0 & 0 & -a_0 g_1 m_0 \dot{\theta}_0 \sin(\theta_0 - \theta_1) \\ 0 & 0 & a_0 g_1 m_0 \dot{\theta}_1 \sin(\theta_0 - \theta_1) & 0 \end{bmatrix}$$

$$D_{a1}(t,q) = diag[d_{11}(t), d_{22}(t), d_{33}(t), d_{44}(t)]$$

$$B_{a1}(q) = \frac{1}{r}\begin{bmatrix} \cos\theta_0 & \cos\theta_0 \\ \sin\theta_0 & \sin\theta_0 \\ g_1 \sin(\theta_0 - \theta_1) & g_1 \sin(\theta_0 - \theta_1) \\ b & -b \end{bmatrix}$$

where the model parameters can be defined as $m(t) = m_0 + m_1 + m_l(t)$, $A(t) = [a_1 + \delta_{a1}(t)][m_1 + m_l(t)] + g_1 m_0$, $I_{\theta 1}(t) = m_1 a_1{}^2 + m_0 g_1{}^2 + I_1 + I_l(t)$, $I_{\theta 0} = m_0 a_0{}^2 + I_0$, where the $m_l(t)$ represents the time-varying load on the trailer, $I_l(t)$ represents the rotational inertia on the trailer, and $\delta_{a1}(t)$ represents the unexpected change of the centroid of the trailer.

The derivative of Equation (3) with respect to time can be obtained $\ddot{q} = \dot{S}(q)v + S(q)\dot{v}$, and then it can be substituted into Equation (6) and multiply $S^T(q)$ on both sides of the equation to eliminate the nonholonomic constraints in the model [21]. Finally, the simplified equation is as follows:

$$M_{a2}(t,q)\dot{v}(t) + C_{a2}(t,q,v)v(t) + D_{a2}(t,q)v(t) = B_{a2}(q)\tau(t) + \tau_{ds}(t,q) \tag{7}$$

where $M_{a2}(t,q) = S^T(q)M_{a1}(t,q)S(q)$, $C_{a2}(t,q,v) = S^T(q)M_{a1}(t,q)\dot{S}(q) + S^T(q)C_{a1}(t,q,\dot{q})$ $S(q)$, $D_{a2}(t,q) = S^T(q)D_{a1}(t)S(q)$, $B_{a2}(q) = S^T(q)B_{a1}(q)$, $\tau_{ds}(t,q) = S^T(q)\tau_d(t)$.

2.3. Coordinate Transformation

In order to achieve the goal of trajectory tracking, it is necessary to construct a tracking error space based on the driving trajectory of the tractor-trailer and reference trajectory, and then derive the tracking error model. Let the state vector of the reference trajectory be $q_r = [x_r, y_r, \theta_{1r}, \theta_{0r}]^T$, the trajectory tracking error of the tractor-trailer can be expressed as $q_e = q - q_r = [q_{e1}, q_{e2}, q_{e3}, q_{e4}]^T = [x_e, y_e, \theta_{1e}, \theta_{0e}]^T$. We transform the posture errors between the driving trajectory of the tractor-trailer and the reference trajectory in the earth-fixed coordinate system ($O_E X_E Y_E$) are transformed into the trailer-fixed coordinate system ($o_{t1} x_{t1} y_{t1}$) using the following transformation [3]:

$$\begin{bmatrix} x_e \\ y_e \\ \theta_{1e} \\ \theta_{0e} \end{bmatrix} = \begin{bmatrix} \cos\theta_1 & \sin\theta_1 & 0 & 0 \\ -\sin\theta_1 & \cos\theta_1 & 0 & 0 \\ 0 & 0 & 1 & 0 \\ 0 & 0 & 0 & 1 \end{bmatrix} \begin{bmatrix} x - x_r \\ y - y_r \\ \theta_1 - \theta_{1r} \\ \theta_0 - \theta_{0r} \end{bmatrix} \tag{8}$$

The derivative of Equation (8) with respect to time is performed, and the Taylor series expansion is used at the reference trajectory point. The tracking error state space equation of the tractor-trailer after ignoring the higher-order term is as follows:

$$\dot{q}_e = A_e q_e + B_e \overline{v} \tag{9}$$

where $A = \begin{bmatrix} 0 & \dot{\theta}_{1r} & 0 & 0 \\ -\dot{\theta}_{1r} & 0 & v_r & 0 \\ 0 & 0 & -\frac{v_r \sec^2(\theta_{0r}-\theta_{1r})}{d_1} & \frac{v_r \sec^2(\theta_{0r}-\theta_{1r})}{d_1} \\ 0 & 0 & 0 & 0 \end{bmatrix}$, $B = \begin{bmatrix} 1 & 0 \\ 0 & 0 \\ \frac{tan(\theta_{0r}-\theta_{1r})}{d_1} & \frac{-d_0 \sec(\theta_{0r}-\theta_{1r})}{d_1} \\ 0 & 1 \end{bmatrix}$, $\overline{v} = \begin{bmatrix} v - v_r \\ \omega - \omega_r \end{bmatrix}$.

2.4. Prescribed Performance Function

In order to improve the transient performance and final tracking error of the control system, the prescribed performance function is introduced to set the performance envelope of the controlled system, so that the tracking error will always be within the prescribed boundary range [20]. To achieve this, a strictly positive, bounded, smooth and decreasing performance function of time $\rho_j(t) : \Re^+ \to \Re^+$ for each element of the tracking error vector $q_e(t) \in \Re^n$, i.e., $q_{ej}(t), j = 1, \ldots, 4$, to satisfy the following bounds:

$$-\underline{\lambda}_j \rho_j(t) < q_{ej}(t) < \overline{\lambda}_j \rho_j(t), \forall t > 0 \; j = 1, \ldots, 4 \tag{10}$$

where $\underline{\lambda}_j$ and $\overline{\lambda}_j$ are some positive constants, indicating overshoot suppression parameters. The following smooth continuous and monotonically decreasing prescribed performance function $\rho_j(t)$ is defined [22]:

$$\rho_j(t) = (\rho_{j0} - \rho_{j\infty}) e^{-l_j t} + \rho_{j\infty} \tag{11}$$

Moreover, the prescribed performance function $\rho_j(t)$ needs to meet the following conditions: (1) $\lim_{t\to 0} \rho_j(t) = \rho_{j0}$, $\lim_{t\to\infty} \rho_j(t) = \rho_{j\infty} > 0$; (2) $\rho_j(0) = (\rho_{j0} - \rho_{j\infty}) e^{-l_j t} + \rho_{j\infty} = \rho_{j0}$. Among them, ρ_{j0}, $\rho_{j\infty}$ and l_j are positive numbers. ρ_{j0} represents the prescribed initial value, $\rho_{j\infty}$ represents the prescribed maximum allowable steady-state error, and l_j represents the convergence rate of tracking error.

The following form of error transformation is introduced to convert the inequality constraint shown in Equation (10) into the form of equality constraint:

$$q_{ej}(t) = \rho_j(t) S(\zeta_j) \tag{12}$$

where ζ_j represents the new conversion error; then, the error conversion function $S(\zeta_j)$ meets the following conditions [23]: (1) $S(\zeta_j)$ is a smooth and strictly monotone increasing function; (2) $-\underline{\lambda}_j < S(\zeta_j) < \overline{\lambda}_j$; (3) $\lim\limits_{\zeta_j \to -\infty} S(\zeta_j) = -\underline{\lambda}_j$, $\lim\limits_{\zeta_j \to \infty} S(\zeta_j) = \overline{\lambda}_j$.

The selected error conversion function is as follows:

$$S(\zeta_j) = \frac{\overline{\lambda}_j e^{\zeta_j} - \underline{\lambda}_j e^{-\zeta_j}}{e^{\zeta_j} + e^{-\zeta_j}} \tag{13}$$

According to Equation (13), it can be further obtained:

$$\zeta_j = S^{-1}(\gamma_j) = \frac{1}{2} \ln\left(\frac{\underline{\lambda}_j + \gamma_j}{\overline{\lambda}_j - \gamma_j} \right) \tag{14}$$

where $\gamma_j = q_{ej}(t)/\rho_j(t)$. Therefore, the new conversion error ζ_j can be obtained by error equivalence transformation, and the tracking error $q_e(t) \in \Re^n$, i.e., $q_{ej}(t)$, $j = 1, \ldots, 4$ of the tractor-trailer is controlled within the predetermined boundary (Equation (10)) by designing the trajectory tracking controller. That is, the prescribed performance tracking control problem of the tractor-trailer system (Equation (9)) can be transformed into the stabilization problem of the equivalent error system (Equation (14)).

3. Main Results

For a robot with simple structure, it is easy to obtain the analytical solution of the kinematic and dynamic models and achieve accurate and stable trajectory tracking control. However, it is difficult or even impossible to obtain the accurate kinematic and dynamic models in practical application because of the complex mechanical structure of the tractor-trailer and various uncertain disturbances. The past control methods have many disadvantages, such as complex control laws and high requirements for model accuracy. To solve these problems, a robust trajectory tracking control algorithm with double closed-loop structure is proposed, as shown in Figure 2. The PPC method uses a prescribed performance function to constrain deviations from the desired trajectory within a specified range. Combined with the posture tracking error model, the MPC method is selected to construct the posture controller. Then, the robust dynamic controller is designed based on SMC and NDO. Through the methods proposed in this paper, the posture tracking and driving torque control of the tractor-trailer can be realized simultaneously.

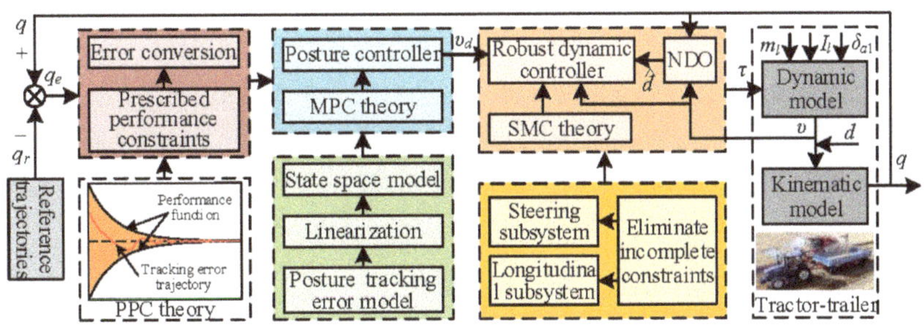

Figure 2. Schematic diagram of the proposed robust tracking control system.

3.1. Posture Controller

The posture tracking error model of the tractor-trailer is shown in Equation (9); it is a continuous-time system. The forward difference method is used to discretize the above linear system, which can be expressed as follows:

$$q_e(k+1) = A_{k,t} q_e(k) + B_{k,t} \overline{v}(k) \tag{15}$$

where $A_{k,t} = I + A_e T_s$, $B_{k,t} = B_e T_s$, I represents the identity matrix and T_s represents the sampling step size of the discretization process.

To design the MPC controller, Equation (15) can be converted into the following new state space form:

$$\begin{cases} \xi(k+1|t) = \overline{A}_{k,t}(k|t)\xi(k|t) + \overline{B}_{k,t}(k|t)\Delta U(k|t) \\ \eta(k|t) = \overline{C}_{k,t}\xi(k|t) \end{cases} \tag{16}$$

where $\xi(k+1|t) = \begin{bmatrix} q_e(k+1|t) \\ \overline{v}(k|t) \end{bmatrix}$, $\overline{A}_{k,t} = \begin{bmatrix} A_{k,t} & B_{k,t} \\ 0_{2\times4} & I_2 \end{bmatrix}$, $\overline{B}_{k,t} = \begin{bmatrix} B_{k,t} \\ I_2 \end{bmatrix}$, $\overline{C}_{ek} = [I_4, 0_{4\times2}]$, ΔU represents the increment of control input.

Define the prediction time domain as N_p and the control time domain as N_c, then the output of the system at the future time can be expressed in the following matrix form [9]:

$$Z(t) = \Psi_t \xi(t|t) + \Theta_t \Delta U(t) \tag{17}$$

where $Z(t) = \begin{bmatrix} \eta(t+1|t) \\ \eta(t+2|t) \\ \vdots \\ \eta(t+N_c|t) \\ \vdots \\ \eta(t+N_p|t) \end{bmatrix}$, $\Psi_t = \begin{bmatrix} \overline{C}_{t,t}\overline{A}_{t,t} \\ \overline{C}_{t,t}\overline{A}_{t,t}^2 \\ \vdots \\ \overline{C}_{t,t}\overline{A}_{t,t}^{N_c} \\ \vdots \\ \overline{C}_{t,t}\overline{A}_{t,t}^{N_p} \end{bmatrix}$, $\Delta U(t) = \begin{bmatrix} \Delta\overline{v}(t|t) \\ \Delta\overline{v}(t+1|t) \\ \vdots \\ \Delta\overline{v}(t+N_c|t) \end{bmatrix}$,

$$\Theta_t = \begin{bmatrix} \overline{C}_{t,t}\overline{B}_{t,t} & 0 & \cdots & 0 \\ \overline{C}_{t,t}\overline{A}_{t,t}\overline{B}_{t,t} & \overline{B}_{t,t} & \cdots & 0 \\ \vdots & \vdots & \ddots & \vdots \\ \overline{C}_{t,t}\overline{A}_{t,t}^{N_c-1}\overline{B}_{t,t} & \overline{C}_{t,t}\overline{A}_{t,t}^{N_c-2}\overline{B}_{t,t} & \cdots & \overline{C}_{t,t}\overline{B}_{t,t} \\ \overline{C}_{t,t}\overline{A}_{t,t}^{N_c}\overline{B}_{t,t} & \overline{C}_{t,t}\overline{A}_{t,t}^{N_c-1}\overline{B}_{t,t} & \cdots & \overline{C}_{t,t}\overline{A}_{t,t}\overline{B}_{t,t} \\ \vdots & \vdots & \ddots & \vdots \\ \overline{C}_{t,t}\overline{A}_{t,t}^{N_p-1}\overline{B}_{t,t} & \overline{C}_{t,t}\overline{A}_{t,t}^{N_p-2}\overline{B}_{t,t} & \cdots & \overline{C}_{t,t}\overline{A}_{t,t}^{N_p-N_c-1}\overline{B}_{t,t} \end{bmatrix}.$$

In order to avoid a situation where the optimal solution cannot be calculated, the following objective functions with soft constraints are designed:

$$\begin{gathered} J(k) = \sum_{i=1}^{N_p} \left\| \eta(k+i|t) - \eta_{ref}(k+i|t) \right\|_Q^2 + \sum_{i=0}^{N_c-1} \|\Delta U(k+i|t)\|_R^2 + \rho\varepsilon^2 \\ s.t. \begin{matrix} \overline{v}_{\min} \le \overline{v} \le \overline{v}_{\max} \\ \Delta\overline{v}_{\min} \le \Delta\overline{v} \le \Delta\overline{v}_{\max} \end{matrix} \end{gathered} \tag{18}$$

where Q and R are the weight matrices, ρ is the weight coefficient, and ε is the relaxation factor.

To obtain a standard quadratic programming model with inequality constraints, we transform the objective function using Equation (19). At the same time, considering the dynamic constraints and actuator constraints in the trajectory tracking control process of the tractor-trailer, the system control quantities and control increment constraints are set.

$$\begin{gathered} minJ(\xi(k), \overline{v}(k-1), \Delta U(k)) = \tfrac{1}{2}\Delta U^T(k)H_t(k)\Delta U + G_t{}^T(k)\Delta U(k) + \rho\varepsilon^2 \\ s.t. \begin{matrix} U_{\min} \le U \le U_{\max} \\ \Delta U_{\min} \le \Delta U \le \Delta U_{\max} \end{matrix} \end{gathered} \tag{19}$$

where $H_t = 2(\Theta_t{}^T Q \Theta_t + R_x)$, it is a Hessian matrix, and describes the quadratic term of the objective function; $G_t = 2\Theta_t{}^T Q \Psi \xi$, it describes the linear part of the objective function.

By solving Equation (19), a series of control input increments in the control horizon can be obtained as $\Delta U_t^* = \left[\Delta\overline{v}_t^*, \Delta\overline{v}_{t+1}^*, \cdots, \Delta\overline{v}_{t+N_c-1}^*\right]^T$. Subsequently, we set the first element of the control sequence as the actual control input increment of vehicle system, given by [5]:

$$\overline{v}(t) = \overline{v}(t-1) + \Delta\overline{v}_t^* \tag{20}$$

Then, the corresponding control input of the posture controller can be ultimately obtained by:

$$v_d(t) = v_r(t) + \overline{v}(t) = \begin{bmatrix} v_r + \overline{v} \\ \omega_r + \overline{\omega} \end{bmatrix} \tag{21}$$

3.2. Dynamic Controller

The goal of the inner loop dynamic controller is to track the desired speed signal v_d generated by the outer loop posture controller by controlling the driving torque on both sides of the tractor-trailer. The uncertainty of model parameters and internal and external disturbances will also affect the performance of the dynamic controller. In this section, the NDO is designed to estimate the total disturbance in the system to achieve disturbance compensation. At the same time, considering the advantages of SMC, such as strong robustness, simple design process, and low model dependency, the dynamic controller will be designed based on the SMC method.

3.2.1. Design of NDO

The dynamic model Equation (7) of the tractor-trailer can be rewritten as follows:

$$\dot{v}(t) = -{M_{a2}}^{-1}(q)[C_{a2}(q,v)v(t) + D_{a2}(q)v(t)] + {M_{a2}}^{-1}(q)B_{a2}(q)\tau(t) + d \tag{22}$$

where $d = {M_{a2}}^{-1}(q)\tau_{ds}(t,q)$ is the unknown disturbance term.

The basic design idea of the disturbance observer is to correct the estimated value by the difference between the actual output and the estimated output. Assuming that the time-varying unknown disturbance d is bounded and continuously differentiable, the observation error of the disturbance observer is defined as $\tilde{d} = d - \hat{d}$, and the linear disturbance observer is designed as [24]:

$$\begin{aligned} \dot{\hat{d}} &= L(v)\tilde{d} = L(v)(d - \hat{d}) \\ &= L(v)\{\dot{v}(t) + {M_{a2}}^{-1}(q)[C_{a2}(q,v)v(t) + D_{a2}(q)v(t)] - {M_{a2}}^{-1}(q)B_{a2}(q)\tau(t)\} - L(v)\hat{d} \end{aligned} \tag{23}$$

Generally, there is no differential prior knowledge of disturbance. Relative to the dynamic characteristics of the disturbance observer, we can assume that the change of disturbance is slow, that is, $\dot{d} = 0$. Then, the dynamic equation of observation error is as follows:

$$\dot{\tilde{d}} = \dot{d} - \dot{\hat{d}} = -L(v)\tilde{d} \tag{24}$$

where $L(v)$ represents observation error gain.

Further, the following NDO can be designed:

$$\begin{cases} \hat{d} = z + g(v) \\ \dot{z} = -L(v)z - L(v)\{g(v) - {M_{a2}}^{-1}(q)[C_{a2}(q,v)v(t) + D_{a2}(q)v(t)] + {M_{a2}}^{-1}(q)B_{a2}(q)\tau(t)\} \end{cases} \tag{25}$$

where $\hat{d}$ is the estimated value of the unknown disturbance d; z is the intermediate variable of the NDO; $g(v)$ is the nonlinear function to be designed; $L(v)$ is the gain coefficient of the NDO; and $L(v) = \partial g(v)/\partial t$. $L(v)$ is selected as a constant, the design function $g(v) = L(v - v_0)$, and v_0 is the initial value of the state variable.

According to Equations (23) and (24), the dynamic equation of observation error of NDO can be obtained as follows:

$$\begin{aligned}\dot{\tilde{d}} &= \dot{d} - \dot{\hat{d}} \\ &= L(v)[z + g(v)] - L(v)\{\dot{v}(t) - M_{a2}{}^{-1}(q)[C_{a2}(q,v)v(t) + D_{a2}(q)v(t)] - M_{a2}{}^{-1}(q)B_{a2}(q)\tau(t)\} \\ &= -L(v)\tilde{d}\end{aligned} \tag{26}$$

Through the above Equation (26), we can get $\tilde{d} = ce^{-L(v)t}$, where c is a constant. Therefore, if $L(v) > 0$, the observation error of the NDO can be converged, and the convergence rate can be determined by selecting the design parameter $L(v)$ [25].

3.2.2. Design of Sliding Mode Controller

According to the dynamic model shown in Equation (7), the tractor-trailer can be divided into two subsystems: longitudinal speed and steering angular speed, and the corresponding tracking error can be defined as follows:

$$\tilde{v} = v - v_d = \begin{bmatrix} \dot{v} - \dot{v}_d \\ \dot{\omega} - \dot{\omega}_d \end{bmatrix} \tag{27}$$

Select the sliding mode surface in the form of proportional-integral as follows [9]:

$$s = \begin{bmatrix} s_1 \\ s_2 \end{bmatrix} = \tilde{v} + \beta \int_0^t \tilde{v}(\mu) d\mu \tag{28}$$

where $\beta = \begin{bmatrix} \beta_1 & 0 \\ 0 & \beta_2 \end{bmatrix}$ is positive weight coefficient.

The derivative of the Equation (28) regarding time is shown as follows:

$$\dot{s} = \dot{\tilde{v}} + \beta\tilde{v} = \dot{v} - \dot{v}_d + \beta(v - v_d) \tag{29}$$

To reduce the chattering of the sliding mode surface, we choose a fast power reaching law with second-order sliding mode characteristics to ensure better dynamic characteristics in the reaching stage [26,27]; it is shown as follows:

$$slaw = \begin{bmatrix} -\kappa_{11}s_1 - \kappa_{12}|s_1|^{\alpha_1}\mathrm{sgn}(s_1) \\ -\kappa_{21}s_2 - \kappa_{22}|s_2|^{\alpha_2}\mathrm{sgn}(s_2) \end{bmatrix} = \begin{bmatrix} \dot{s}_1 \\ \dot{s}_2 \end{bmatrix} \tag{30}$$

where κ_{11}, κ_{12}, κ_{21} and κ_{22} are constants greater than zero, $\alpha_1 \in (0, 1)$, $\alpha_2 \in (0, 1)$.

According to Equations (7), (29), and (30), the driving torque control law of the tractor-trailer can be expressed as follows:

$$\begin{aligned}\tau &= B_{a2}{}^{-1}(q)M_{a2}(q)\left[\dot{v}(t) + M_{a2}{}^{-1}(q)C_{a2}(q,v)v(t) + M_{a2}{}^{-1}(q)D_{a2}(q)v(t) - M_{a2}{}^{-1}(q)\tau_{ds}(t,q)\right] \\ &= B_{a2}{}^{-1}(q)M_{a2}(q)\left[\dot{v}(t) - \dot{v}_d(t) + \dot{v}_d(t) + M_{a2}{}^{-1}(q)C_{a2}(q,v)v(t) + M_{a2}{}^{-1}(q)D_{a2}(q)v(t) - M_{a2}{}^{-1}(q)\tau_{ds}(t,q)\right] \\ &= B_{a2}{}^{-1}(q)M_{a2}(q)\left[\dot{v}_d(t) + M_{a2}{}^{-1}(q)C_{a2}(q,v)v(t) + M_{a2}{}^{-1}(q)D_{a2}(q)v(t) - M_{a2}{}^{-1}(q)\tau_{ds}(t,q) - \beta\tilde{v} - \kappa_1 s - \kappa_2|s|^{\alpha}\mathrm{sgn}(s)\right]\end{aligned} \tag{31}$$

where $d = M_{a2}{}^{-1}(q)\tau_{ds}(t,q)$ is the unknown disturbance term.

As mentioned earlier, $M_{a2}{}^{-1}(q)\tau_{ds}(t,q)$ in Equation (31) is abbreviated as d, and the NDO is designed to estimate it. Therefore, the final driving torque control law of the tractor-trailer can be expressed as follows:

$$\tau = B_{a2}{}^{-1}(q)M_{a2}(q)\left[\dot{v}_d(t) + M_{a2}{}^{-1}(q)C_{a2}(q,v)v(t) + M_{a2}{}^{-1}(q)D_{a2}(q)v(t) - \hat{d} - \beta\tilde{v} - \kappa_1 s - \kappa_2|s|^{\alpha}\mathrm{sgn}(s)\right] \tag{32}$$

In order to analyze the stability of the designed SMC, we select the Lyapunov function presented as:

$$V = \frac{1}{2}s^2 \tag{33}$$

The derivative of Equation (33) regarding time is shown as follows:

$$\begin{aligned} \dot{V} &= s\dot{s} \\ &= s\left[-\kappa_1 s - \kappa_2 |s|^{\alpha}\mathrm{sgn}(s)\right] \\ &= -\kappa_1 s^2 - \kappa_2 |s|^{1+\alpha} \\ &\leq 0 \end{aligned} \tag{34}$$

Therefore, according to the Lyapunov stability principle and LaSalle invariance theorem, it can be concluded that all signals are bounded and the velocity error can converge to zero with time.

4. Results and Analysis

4.1. Simulation Description

Assume that the trajectory of the trailer in the tractor-trailer meets the following equation:

$$\begin{cases} x(t) = 3t \\ y(t) = [2 + \cos(0.1t)][3 - \sin(0.2t)] \end{cases} \tag{35}$$

According to the geometric relationship, the tractor in the tractor-trailer should meet the following constraints [21]:

$$\begin{cases} \theta_1(t) = \arctan\frac{\dot{y}(t)}{\dot{x}(t)} \\ \theta_0(t) = \theta_1(t) + \arcsin\frac{g_1\dot{\theta}(t)}{v(t)} \\ x_0(t) = x(t) + g_1 \cos\theta_1(t) \\ y_0(t) = y(t) + g_1 \sin\theta_1(t) \end{cases} \tag{36}$$

According to Ref. [3], the model parameters of the tractor-trailer used in the simulation are defined as follows: r = 0.45 m, b = 0.8 m, a_0 = 0.45 m, a_1 = 0.25 m, g_0 = 0 m, g_1 = 3 m, m_0 = 700 kg, m_1 = 450 kg, I_0 = 280 kg·m^2, I_1 = 180 kg·m^2. The actuator input signals of the posture controller are saturated $|v| \leq 5$ m/s and $|\omega| \leq 4$ rad/s, respectively. The actuator input signals of the dynamic controller are also saturated $|\tau_l| \leq 200$ N·m and $|\tau_r| \leq 200$ N·m, respectively. The reference trajectory for the tractor-trailer begins at the initial posture $q = [0.5, 11, 0, 0]^T$.

Considering the existence of uneven ground, load interference, sensor measurement error, and other disturbances in the actual farmland environment, the following interference and noise vectors are selected to be added to the model of the tractor-trailer:

$$d = \begin{bmatrix} 0.35\sin(0.06t) + 0.15\cos(0.3t) + 0.01randn(1) \\ 0.25\sin(0.08t) - 0.1\cos(0.2t) + 0.01randn(1) \end{bmatrix} \tag{37}$$

As autonomous navigation technology has developed, there are more and more scenarios where the tractor-trailers need to work together with other vehicles, such as combine harvesters and tractor-trailers working together to unload grain. At this time, in addition to the influence of internal and external disturbances, the model parameters such as the mass, rotational inertia, and centroid of the tractor-trailer will change greatly. The following Equation (38) is used to simulate the change of trailer mass and rotational inertia in the trajectory tracking control process of the tractor-trailer:

$$\begin{cases} \dot{m}_l(t) = [2.4 + 1.2\sin(0.3t) + 0.4\cos(0.7t) + randn(1)] \times \left[H(t - T_s) - H(t - T_f)\right] \\ \dot{I}_l(t) = [0.7 + 0.4\sin(0.3t) + 0.12\cos(0.7t) + 0.7randn(1)] \times \left[H(t - T_s) - H(t - T_f)\right] \end{cases} \tag{38}$$

where T_s = 20 s and T_f = 80 s are the start and stop times of the crop loading. $H(t)$ is the Heaviside step function [3].

The sample time $T = 0.05$ s. The prescribed performance functions are used with parameter values of $\rho_{10} = 0.5$, $\rho_{1\infty} = 0.1$, $l_1 = 1$, $\underline{\lambda}_1 = -1$, $\overline{\lambda}_1 = 1$, $\rho_{20} = 1$, $\rho_{2\infty} = 0.1$, $l_2 = 1$, $\underline{\lambda}_2 = -1$, $\overline{\lambda}_2 = 1$, $\rho_{30} = 0.5$, $\rho_{3\infty} = 0.1$, $l_3 = 1.5$, $\underline{\lambda}_3 = -1$, $\overline{\lambda}_3 = 1$, $\rho_{40} = 0.5$, $\rho_{4\infty} = 0.1$, $l_4 = 1.5$, $\underline{\lambda}_4 = -1$, $\overline{\lambda}_4 = 1$. The parameters of posture controller based on the MPC method are selected as the predictive horizon $N_p = 15$, the control horizon $N_c = 5$, the weight matrix $Q = 100 \times I_{60}$, $R = I_{30}$ and $\rho = 10$. Moreover, the parameters of the dynamic controller based on the SMC method and NDO are selected as $\beta_1 = 5$, $\beta_2 = 5$, $\kappa_{11} = 3$, $\kappa_{12} = 0.1$, $\kappa_{21} = 3$, $\kappa_{22} = 0.1$, $\alpha_1 = 0.2$, $\alpha_2 = 0.2$, $L = 100 \times I_2$.

4.2. Simulation Results

4.2.1. NDO

Before the trajectory tracking control simulation of the tractor-trailer, we need to prove the effectiveness of the NDO designed in this paper. First of all, suppose that there is no change in the model parameters of the tractor-trailer during driving; the system disturbance is shown in Equation (37), and the disturbance observation results are shown in Figure 3. In Figure 3, the black solid line and red dashed line are the system disturbance curves added in the simulation, and the green dash-dotted line and blue dotted line are the observed value curves of the system disturbance. It shows that the proposed NDO can successfully realize the disturbance observation of the tractor-trailer system.

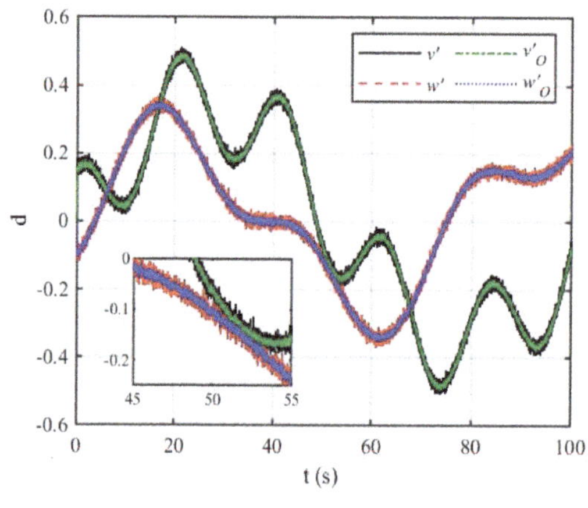

Figure 3. Disturbance observation curves of the tractor-trailer without considering the model parameter changes.

In some special application scenarios, for example, during the cooperative operation of the tractor-trailer and the combine harvester, the combine harvester will synchronously transport the grain to the trailer of the tractor-trailer. At this time, the mass, rotational inertia, and centroid of the trailer change significantly, and the simulated incremental change process is shown in Figure 4a, and the disturbance observation results are shown in Figure 4b. It can be seen from Figure 4b that the observation curve of the NDO in the angular velocity direction is consistent with the added disturbance curve, but the observation curve of the NDO in the velocity direction is inconsistent with the added disturbance. This is because the model parameter changes affect the size of the actual disturbances in the tractor-trailer. Therefore, the changes in model parameters can be incorporated into system disturbances, so that the actual disturbances of the tractor-trailer can be observed using the NDO designed in this paper. At the same time, it also reflects that the idea of designing the NDO to estimate the system disturbances and compensate them to the control system is correct.

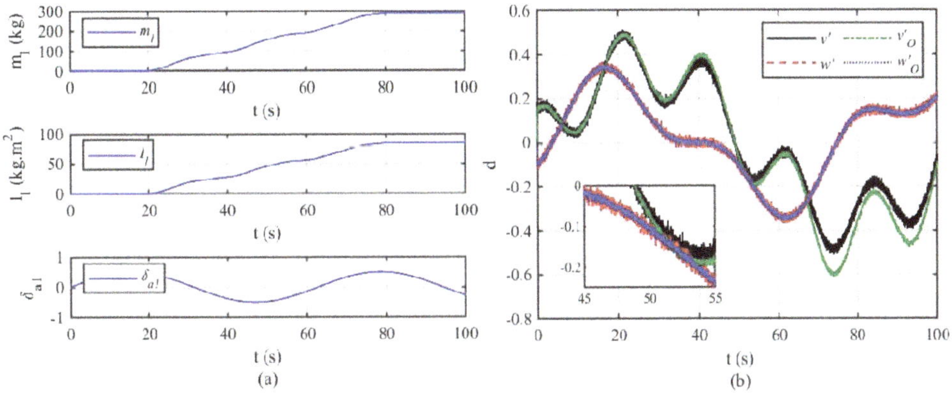

Figure 4. Disturbance observation curves of the tractor-trailer considering the model parameter changes: (**a**) the change curves of model parameters; (**b**) the observation curves of system disturbances.

4.2.2. Robust Tracking Control

In order to verify the effectiveness of the method proposed in this paper, it is compared with the control method in Refs. [5,9]. In the following simulation results, the legend Ref. represents the reference input, the legend MS represents the method proposed in Ref. [9], the legend RTMS represents the method proposed in Ref. [5], and the legend PMSO represents the method proposed in this paper.

According to the reference trajectories shown in Equations (35) and (36), the trajectory tracking control results of the tractor-trailer are shown in Figure 5. It can be seen from the position control curves in Figure 5a that the three control methods (MS-Ref. [9], RTMS-Ref. [5], PMSO-this paper) can successfully achieve high-precision trajectory tracking control, but the method proposed in this paper has the fastest response speed. In Figure 5b, the heading tracking control errors of the control method in Ref. [9] is the smallest at the beginning, but the response speed is the slowest. The Ref. [5] and the method proposed in this paper have faster response speed, but have larger tracking error at the beginning, and the tracking error of the method proposed in this paper is smaller.

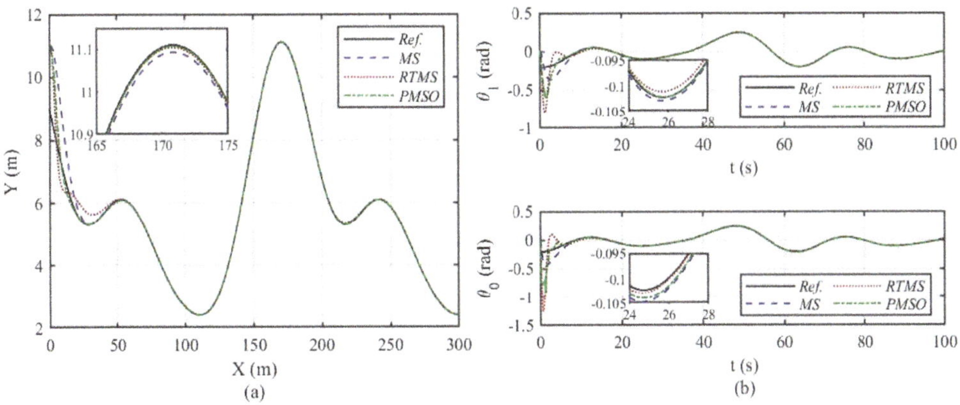

Figure 5. Trajectory tracking control results of the tractor-trailer: (**a**) position control curves; (**b**) heading control curves.

The tracking control errors of the tractor-trailer are shown in Figure 6. It can be seen from the position error curves in Figure 6a that the method proposed in this paper has a faster response speed and better tracking error stability than the methods in Refs. [5,9]. At the same time, they also prove the goal of the transient and steady-state performance constraint specified by the prescribed performance function. Similarly, the heading error curves in Figure 6b also prove this point.

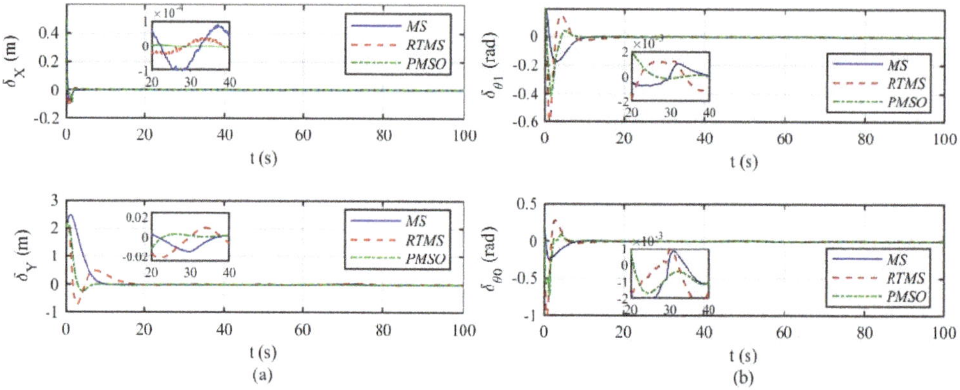

Figure 6. Tracking control errors of the tractor-trailer: (**a**) position error curves; (**b**) heading error curves.

In order to better illustrate the effectiveness of the method proposed in this paper, we have calculated the cumulative tracking control errors of the tractor-trailer, as shown in Figure 7. It can be seen from the cumulative posture error curves in Figure 7a that the cumulative posture errors rise rapidly at the beginning of control, and then reaches the stable tracking control stage. Furthermore, the subsequent cumulative posture errors basically do not increase, reflecting that the three control methods (MS-Ref. [9], RTMS-Ref. [5], PMSO-this paper) can achieve the stable and high-precision trajectory tracking control of the tractor-trailer. However, the cumulative posture errors of the proposed method in this paper are the smallest, which shows it has higher robustness and better performance. Then, we sum the position errors (x_e, y_e) and heading errors (θ_{1e}, θ_{0e}), respectively, to obtain Figure 7b. Similarly, the cumulative error curves of position and heading in Figure 7b also prove the effectiveness of the method proposed in this paper.

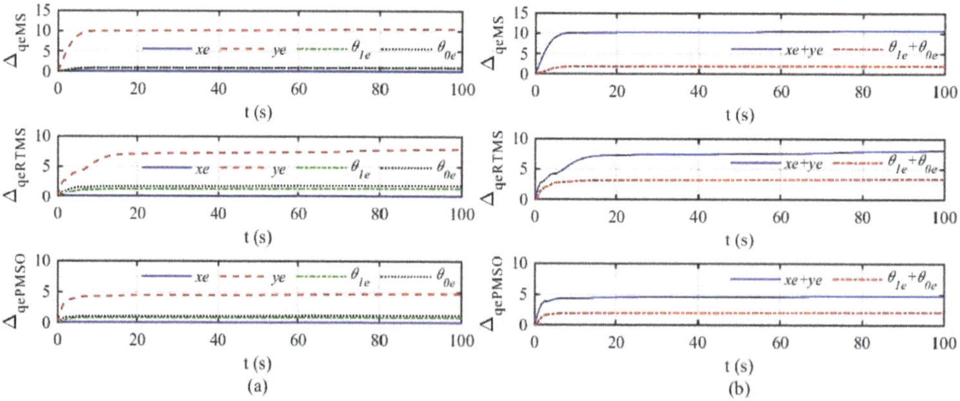

Figure 7. Cumulative tracking control errors of the tractor-trailer: (**a**) cumulative posture error curves; (**b**) cumulative error curves of position and heading.

The control quantities of the posture controller and dynamic controller of the tractor-trailer are shown in Figure 8. Figure 8a shows the control outputs of the posture controller. It can be seen from Figure 8a that the method proposed in this paper has changed greatly in the initial stage compared with the methods in Refs. [5,9], because the tractor-trailer has large initial trajectory tracking errors. In order to meet the prescribed transient and steady-state performances, the prescribed performance function used in this paper will restrict the tracking error, which leads to this phenomenon. However, when the trajectory tracking control reaches the stable stage, the method proposed in this paper shows better robustness and steady-state performance. Figure 8b shows the actual speed and angular speed of the tractor-trailer based on the three control methods (MS-Ref. [9], RTMS-Ref. [5], PMSO-this paper). Figure 8c shows the control outputs of the dynamic controller. It can be seen that compared with the traditional sliding mode reaching law in Ref. [9], the control output curves of the proposed fast power reaching law with second-order sliding mode characteristics are smoother and achieve the purpose of reducing chattering. Figure 8d shows the sliding mode surfaces of the dynamic controller.

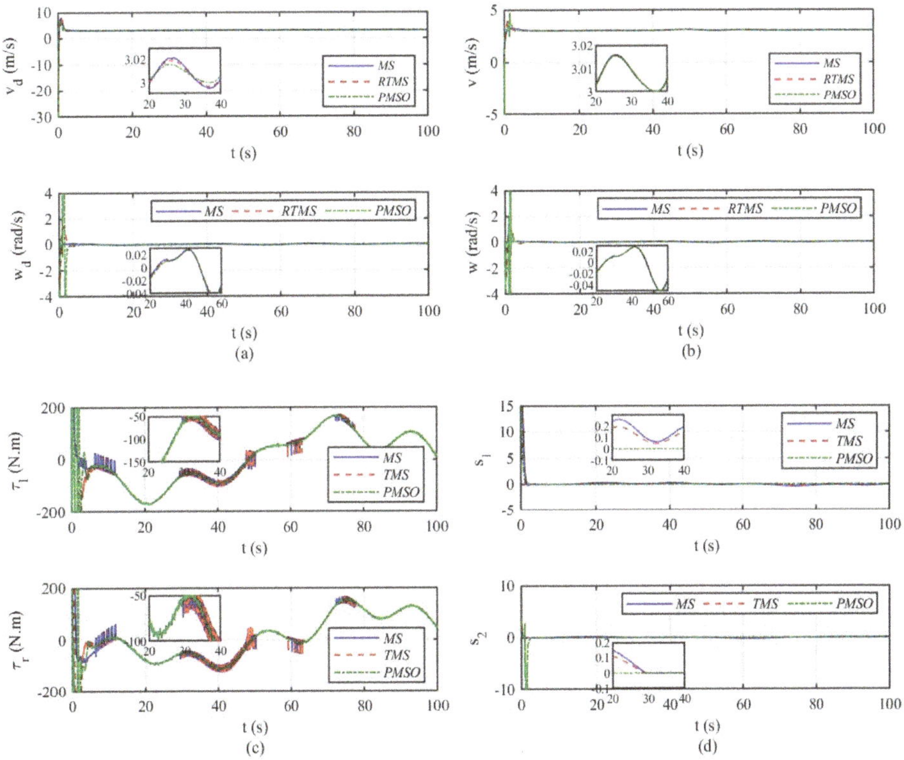

Figure 8. Control quantities of the posture controller and dynamic controller of the tractor-trailer: (**a**) control outputs of the posture controller; (**b**) actual speed and angular speed of the tractor-trailer; (**c**) control outputs of the dynamic controller; (**d**) sliding mode surfaces of the dynamic controller.

5. Conclusions

Aiming at the requirement of stable and high-precision tracking control of tractor-trailer vehicles in modern agriculture, this paper studies the robust trajectory tracking control of autonomous tractor-trailer systems. Based on the derived kinematic and dynamic model, the double closed-loop control structure is designed, the MPC method is used to construct the posture controller, and the SMC method and NDO strategy are used to construct the dynamic controller. Moreover, the convergence speed and final tracking control accuracy of the tractor-trailer control system can be guaranteed by an effective application of the PPC technique.

The disturbance observation results show the designed NDO can precisely estimate the system disturbances. The comparative simulation results show that under the proposed control method in this paper, even with system disturbances, the tractor-trailer can track the reference trajectory well. The tracking control error curves show that the method proposed in this paper has faster response speed and better tracking error stability than the other two control methods. At the same time, they also prove the goal of the transient and steady-state performance constraint achieved by the prescribed performance function. The cumulative tracking control error curves show that the cumulative errors of the proposed method are the smallest, and show higher robustness and better performance. In addition, the control output curves of the dynamic controller show that the proposed fast power reaching law with second-order sliding mode characteristics are smoother and achieve the purpose of reducing chattering. In future work, we plan to carry out application research under complex working conditions, and focus on controller saturation, time delay, control parameter optimization, and other related issues.

Author Contributions: Conceptualization, E.L. and J.X.; methodology, E.L.; software, T.C.; investigation, S.J.; writing—original draft preparation, J.X.; writing—review and editing, E.L. All authors have read and agreed to the published version of the manuscript.

Funding: This research was funded by the National Natural Science Foundation of China (Grant No. 52005220), the National Key Research and Development Program of China (Grant No. 2022YFD00150402), and A Project Funded by the Priority Academic Program Development of Jiangsu Higher Education Institutions (Grant No. PAPD-2018-87).

Data Availability Statement: Data available on request from the authors.

Conflicts of Interest: The authors declare no conflict of interest.

References

1. Bechar, A.; Vigneault, C. Agricultural robots for field operations: Concepts and components. *Biosyst. Eng.* **2016**, *149*, 94–111. [CrossRef]
2. Lu, E.; Ma, Z.; Li, Y.M.; Xu, L.Z.; Tang, Z. Adaptive backstepping control of tracked robot running trajectory based on real-time slip parameter estimation. *Int. J. Agric. Biol. Eng.* **2020**, *13*, 178–187. [CrossRef]
3. Shojaei, K. Intelligent coordinated control of an autonomous tractor-trailer and a combine harvester. *Eur. J. Control.* **2021**, *59*, 82–98. [CrossRef]
4. Yuan, J.; Sun, F.C.; Huang, Y.L. Trajectory generation and tracking control for double-steering tractor-trailer mobile robots with on-axle hitching. *IEEE Trans. Ind. Electron.* **2015**, *62*, 7665–7677. [CrossRef]
5. Yue, M.; Hou, X.Q.; Zhao, X.D.; Wu, X.M. Robust tube-based model predictive control for lane change maneuver of tractor-trailer vehicles based on a polynomial trajectory. *IEEE Trans. Syst. Man Cybern. Syst.* **2018**, *50*, 5180–5188. [CrossRef]
6. Alipour, K.; Robat, A.B.; Tarvirdizadeh, B. Dynamics modeling and sliding mode control of tractor-trailer wheeled mobile robots subject to wheels slip. *Mech. Mach. Theory* **2019**, *138*, 16–37. [CrossRef]
7. Kassaeiyan, P.; Alipour, K.; Tarvirdizadeh, B. A full-state trajectory tracking controller for tractor-trailer wheeled mobile robots. *Mech. Mach. Theory* **2020**, *150*, 103872. [CrossRef]
8. Murillo, M.; Sánchez, G.; Deniz, N.; Genzelis, L.; Giovanini, L. Improving path-tracking performance of an articulated tractor-trailer system using a non-linear kinematic model. *Comput. Electron. Agric.* **2022**, *196*, 106826. [CrossRef]
9. Liu, Z.Y.; Yue, M.; Guo, L.; Zhang, Y.S. Trajectory planning and robust tracking control for a class of active articulated tractor-trailer vehicle with on-axle structure. *Eur. J. Control.* **2020**, *54*, 87–98. [CrossRef]
10. Zhang, R.X.; Xiong, L.; Yu, Z.P.; Bai, M.F.; Fu, Z.Q. Robust trajectory tracking control of autonomous vehicles based on conditional integration method. *J. Mech. Eng.* **2018**, *54*, 129–139. [CrossRef]
11. Liao, J.F.; Chen, Z.; Yao, B. Model-based coordinated control of four-wheel independently driven skid steer mobile robot with wheel-ground interaction and wheel dynamics. *IEEE Trans. Ind. Inform.* **2019**, *15*, 1742–1752. [CrossRef]
12. Elhaki, O.; Shojaei, K. Output-feedback robust saturated actor-critic multi-layer neural network controller for multi-body electrically driven tractors with n-trailer guaranteeing prescribed output constraints. *Robot. Auton. Syst.* **2022**, *154*, 104106. [CrossRef]
13. Taghia, J.; Wang, X.; Lam, S.; Katupitiya, J. A sliding mode controller with a nonlinear disturbance observer for a farm vehicle operating in the presence of wheel slip. *Auton. Robot.* **2017**, *41*, 71–88. [CrossRef]

14. Korayem, A.H.; Khajepour, A.; Fidan, B. Trailer mass estimation using system model-based and machine learning approaches. *IEEE Trans. Veh. Technol.* **2020**, *69*, 12536–12546. [CrossRef]
15. Han, S.; Yoon, K.; Park, G.; Huh, K. Hybrid state observer design for estimating the hitch angles of tractor-multi unit trailer. *IEEE Trans. Intell. Veh.* **2022**, *8*, 1449–1458. [CrossRef]
16. Guevara, L.; Jorquera, F.; Walas, K.; Auat-Cheein, F. Robust control strategy for generalized N-trailer vehicles based on a dual-stage disturbance observer. *Control Eng. Pract.* **2023**, *131*, 105382. [CrossRef]
17. Guo, Y.K.; He, S.D.; Dai, S.L. Finite-time synchronization of nonlinear multi-agent systems with prescribed performance. In Proceedings of the 38th Chinese Control Conference, Guangzhou, China, 27–30 July 2019; pp. 5635–5640.
18. Bechlioulis, C.P.; Rovithakis, G.A. Robust adaptive control of feedback linearizable MIMO nonlinear systems with prescribed performance. *IEEE Trans. Autom. Control* **2008**, *53*, 2090–2099. [CrossRef]
19. Shojaei, K.; Abdolmaleki, M. Output feedback control of a tractor with N-trailer with a guaranteed performance. *Mech. Syst. Signal Process.* **2020**, *142*, 106746. [CrossRef]
20. Guo, G.; Zhang, Q.; Gao, Z.Y. Finite-time fixed configuration formation control of intelligent vehicles with prescribed transient and steady-state performance. *China J. Highw. Transp.* **2022**, *35*, 28–42.
21. Yue, M.; Wu, X.M.; Guo, L.; Gao, J.J. Quintic polynomial-based obstacle avoidance trajectory planning and tracking control framework for tractor-trailer system. *Int. J. Control. Autom. Syst.* **2019**, *17*, 2634–2646. [CrossRef]
22. Yin, Z.Y.; Suleman, A.; Luo, J.J.; Wei, C.S. Appointed-time prescribed performance attitude tracking control via double performance functions. *Aerosp. Sci. Technol.* **2019**, *93*, 105337. [CrossRef]
23. Chen, L.S.; Ning, X.M. Nonlinear PI cascade attitude control with prescribed performance for a quadrotor UAV. *J. Appl. Sci.* **2019**, *37*, 137–150.
24. Jin, Z.J.; Liu, S.; Zhang, L.H.; Lan, B.; Qin, H.R. Command-filtered backstepping control for stabilization of ship rolling based on nonlinear disturbance observer. *Shipbuild. China* **2019**, *60*, 121–130.
25. Li, C.X.; Meng, X.Y.; Wang, J. Design of aircraft trajectory tracking controller based on disturbance observer. *Syst. Eng. Electron.* **2022**, *44*, 2593–2600.
26. Li, P. *Research and Application of Traditional and Higher-Order Sliding Mode Control*; National University of Defense Technology: Changsha, China, 2011.
27. Lu, E.; Li, W.; Yang, X.F.; Liu, Y.F. Anti-disturbance speed control of low-speed high-torque PMSM based on second-order non-singular terminal sliding mode load observer. *ISA Trans.* **2019**, *88*, 142–152. [CrossRef] [PubMed]

agriculture

MDPI

Article

Advancement in Agriculture Approaches with Agrivoltaics Natural Cooling in Large Scale Solar PV Farms

Noor Fadzlinda Othman [1,2,*], Mohammad Effendy Ya'acob [2,3,*], Li Lu [4], Ahmad Hakiim Jamaluddin [5], Ahmad Suhaizi Mat Su [1], Hashim Hizam [4], Rosnah Shamsudin [3] and Juju Nakasha Jaafar [6]

1 Department of Agriculture Technology, Faculty of Agriculture, Universiti Putra Malaysia, Serdang 43400, Selangor, Malaysia; asuhaizi@upm.edu.my
2 Hybrid Agrivoltaic System Showcase (HAVs), Faculty of Engineering, Universiti Putra Malaysia, Serdang 43400, Selangor, Malaysia
3 Department of Process and Food Engineering, Faculty of Engineering, Universiti Putra Malaysia, Serdang 43400, Selangor, Malaysia; rosnahs@upm.edu.my
4 Department of Electrical and Electronic, Faculty of Engineering, Universiti Putra Malaysia, Serdang 43400, Selangor, Malaysia; hhizam@upm.edu.my
5 Department of Statistics, School of Mathematics and Statistics, University of New South Wales, Sydney 2052, Australia
6 Department of Crop Science, Faculty of Agriculture, Universiti Putra Malaysia, Serdang 43400, Selangor, Malaysia; jujunakasha@upm.edu.my
* Correspondence: gs51010@student.upm.edu.my (N.F.O.); m_effendy@upm.edu.my (M.E.Y.)

Abstract: The increasing concerns about the impact of large-scale solar photovoltaic farms on the environment and the energy crisis have raised many questions. This issue is mainly addressed by the integration of agriculture advancement in solar photovoltaic systems infrastructure facilities, commonly known as agrivoltaic. Through the use of these systems, the production of crops can be increased, and the efficiency of PV panels can be improved. Accordingly, adopting such synergistic paths forward can contribute toward building resilient energy-generation and food-production systems. The utilization of cooling techniques can provide a potential solution for the excessive heating of PV cells and lower cell temperatures. Effective cooling applied to PV cells significantly improves their electrical efficiency, as well as increasing their lifespan because of decreasing thermal stresses. This paper shares an overview of both active and passive cooling approaches in solar PV applications with an emphasis on newly developed agrivoltaic natural cooling systems. Actual data analysis at the 2 MWp Puchong agrivoltaic farm shows a significant value of 3% increase of the DC generation (on average) which is most beneficial to solar farm operators.

Keywords: agriculture advancement; large scale solar; natural cooling; agrivoltaic; sustainability

Citation: Othman, N.F.; Ya'acob, M.E.; Lu, L.; Jamaluddin, A.H.; Mat Su, A.S.; Hizam, H.; Shamsudin, R.; Jaafar, J.N. Advancement in Agriculture Approaches with Agrivoltaics Natural Cooling in Large Scale Solar PV Farms. *Agriculture* **2023**, *13*, 854. https://doi.org/10.3390/agriculture13040854

Academic Editors: Muhammad Sultan, Redmond R. Shamshiri, Md Shamim Ahamed and Muhammad Farooq

Received: 15 March 2023
Revised: 30 March 2023
Accepted: 31 March 2023
Published: 12 April 2023

1. Introduction

Across the globe, the amount of electricity produced by the large-scale solar (LSS) photovoltaic (PV) installations has shown an exponential growth in recent decades as concern has grown toward clean renewable energy for mitigating the energy crisis and environmental issues [1,2]. As an example, many LSS PV farms have been operated in Malaysia. The Malaysian government expects to achieve 45% deduction of CO_2, mainly by LSS PV farms, by 2030. Meanwhile, 10% of national electricity demands will be also satisfied through this continuous effort [3]. However, Barron-Gafford et al. [1] illustrated that LSS PV installations would cause a "heat island" effect. In other words, local surrounding temperatures over the LSS PV plant would increase. In some cases, therefore, the PV "heat island" effect has sparked public concerns and has indirectly led to resistance to the development of LSS PV farms.

As is known, PV cells generate electricity as well as heat. Up to 80% of the incident solar radiation can be absorbed by PV cells [4,5]. However, only a small portion of the absorbed incident solar energy is converted into electricity via PV photonic effect. Much of the remaining energy is dissipated as heat above ambient temperature, depending upon the conversion efficiencies of the PV cell technology utilized as shown in Figure 1. The elevated temperatures can be caused by heat accumulating on the surface of the PV cells [5]. Accordingly, the operating temperature of the PV cells also linearly increases, resulting in one of the most important factors that can influence the PV cells' performance: irreversible degradation and shortening of the cells' lifetime [6,7]. It is clear that the use of cooling techniques on the PV is of great importance. Currently, numerous cooling technologies for regulating the thermal issue of PV systems have been investigated in many studies [5].

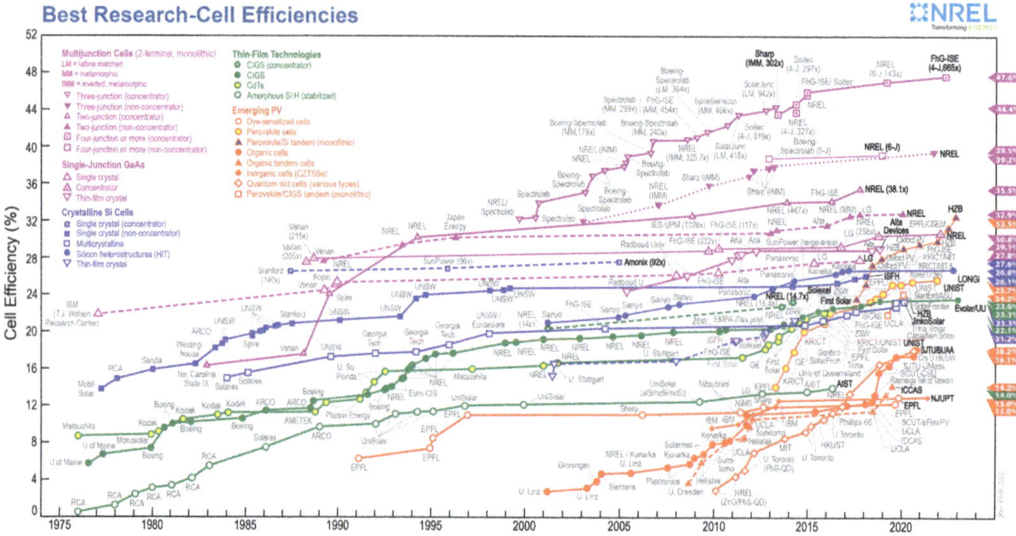

Figure 1. The cell efficiencies of various types of solar PV cells [8].

1.1. Six Common Cooling Approaches for Solar PV Cells

References [9,10] suggest that the temperature of the PV cells without cooling can increase up to 80 °C under warm and hot climate conditions. Depending on the PV cell technology used, every 1 °C increase in temperature of a PV module results in 0.4–0.5% reduction on the output power [9,11]. The utilization of cooling techniques can provide a potential solution for the excessive heating of PV cells and for lowering cell temperatures. Therefore, PV systems not only consist of inverters, as well as other electrical and mechanical devices, but also solar cell cooling [2,12]. Effective cooling of PV cells significantly improves their electrical efficiency, as well as increasing the lifespan of the PV cells because of the decreased thermal stresses. Approaches to cooling PV cells can be mainly classified as active and passive. Typically, the type of cooling (active or passive) approach and the materials adopted in cooling are selected in accordance with local weather conditions [13,14]. Figure 2 shows some active and passive cooling methods for PV cells.

1.1.1. Active Cooling

Active cooling is a process that removes the heat from the system by using external coolant devices such as pump water, forced air, or fans to cool the panels. One of the drawbacks of active cooling is that a part of generated electrical energy is used by the external coolant system. However, the total output of the PV system with active cooling is higher than that with passive cooling and is also more effective for cooling heat transfer rates [14]. Some studies regarding active cooling methods, such as air cooling, water cooling, and Thermoelectric cooling for PV cells are shown in Table 1.

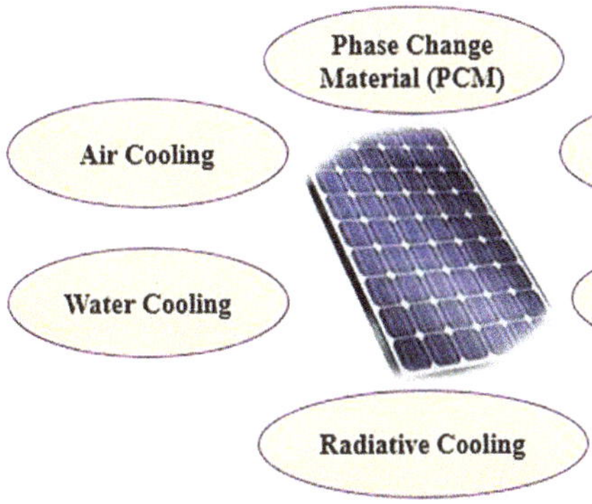

Figure 2. Six common cooling approaches for PV cells.

Air is regarded as a common cooling medium where air cooling systems are typically used in various devices for reducing the temperature and thermal management. Air cooling of PV cells is performed by using fans or other means to create forced convection airflow and then to decrease the temperature of PV cells. Despite the fact that using air as a coolant is less efficient than using liquids, air cooling offers some benefits, such as minimum material usage and cheap operating costs [15–17].

Water is the most frequently employed fluid in liquid-based cooling of PV systems for PV cells. The process of water cooling is performed by either spraying the water on the surface of the PV modules directly or passing the water behind the panel from inside the installed pipe [14]. Thermoelectric (TE) cooling technology is used to capture and convert excess heat from the PV cells directly into electricity. TE modules offer outstanding features of being lightweight, maintenance free, strongly reliable, noiseless in operation, and no complex parts. Thermoelectric generators (TEG) are one type of TE module that generates electrical power from the temperature gradient [5].

Table 1. A review of some studies on the active cooling methods for PV cells.

Air Cooling Approach		
Authors	**Study Aim**	**Result**
Teo et al. [18]	To compare the performance of the PV module with and without air cooling.	With air cooling, the operating temperature of PV module could be kept at 38 °C and electrical efficiency maintained around 12.5%, whereas these two values can rise to around 68 °C and 8.6%, respectively, without air cooling.
Sajjad et al. [19]	To improve the performance efficiency of PV modules by using forced air cooling.	Compared to PV modules without cooling, forced air cooling achieves 6% and 7.2% power ratio and higher electrical efficiency, respectively.
Water cooling approach		
Authors	**Study Aim**	**Result**
Krauter [20]	To investigate the impact of utilizing water flow as a coolant on the performance of the PV cells.	The temperature of cells was reduced to 22 °C by the water cooling approach, thus the output of the cells was improved by 10.3%.
Mah et al. [21]	To improve the performance of a crystalline silicon PV system via water cooling in a tropical region.	The power output of crystalline silicon PV system was increased by 15%, and each panel produced 0.0178 kW·h with the water cooling under the 1150 solar irradiance. In addition, water cooling also contributed to the uniform temperature distribution between the front and the back surfaces of the panels.

Table 1. *Cont.*

Thermoelectric cooling approach		
Authors	**Study Aim**	**Result**
Sark [22]	To determine the efficiency of PV panel with TE converter by using a numerical model.	TE converter resulted in 8–23% enhancement of PV module's electrical efficiency.
Benghanem et al. [23]	To present the performance of PV cells by using TE module as the cooling system.	The temperature of PV cells dropped from 83 °C to 65 °C with TE modules, while the efficiency of PV cells decreased by 0.5% along with per °C rise in temperature.

1.1.2. Passive Cooling

Passive cooling is a natural approach that provides air or liquid circulation to reduce the heat of the system, which is highly suitable for PV cooling projects. Unlike the aforementioned active cooling, passive cooling does not need to use any external power source for driving the cooling system, thus this benefit contributes to a simpler structure and lowers maintenance costs [24]. Some studies of passive cooling methods such as phase change material (PCM), heat pipe cooling, and radiative cooling are demonstrated in Table 2.

PCM is a useful passive cooling approach in the thermal management of PV cells due to its great capacity for heat storage with prolonged heat availability. PCM absorbs extra heat from PV cells through its latent heat, then keeps PV cells at the accepted temperature for a certain period. The melting temperature of PCM for thermal management of PV cells is recommended to be 25 °C in the summer. Additionally, the PCM's melting temperature should be lower than the PV cells' temperature for effective thermal management of the cells [25]. Heat pipes as coolant devices for PV cells are used due to their high thermal conductivity, uncertain heat flux, and ability to create uniform temperatures. Such devices are typically composed of a sealed pipe with high thermal conductivity material at both condenser and evaporator. Heat pipes can lower the temperature and then enhance the electrical efficiency of PV cells by transferring heat from PV cells to water or air [12,14]. Radiative cooling is a passive cooling method based on using an atmospheric window with a transparency in the wavelength range between 8 μm and 14 μm. In other words, radiative cooling is only achievable when the entrance heat flux caused by conduction, convection or radiation to the infrared spectral layer (with the thickness between 8 μm and 14 μm) is smaller than the output heat flux from the earth's body. It is valuable to note that the spectral alteration of the emissivity of modules' areas for thermal radiation and absorption determines the rate of radiative cooling [14,26].

Table 2. A review of some studies on the passive cooling methods for PV cells.

PCM Approach		
Authors	**Study Aim**	**Result**
Hasan et al. [27]	To study the performance of cooling the PV cell by using the paraffin-based PCM with melting temperature (38 °C–43 °C).	PCM cooling dropped 10.5 °C in PV temperature on average at peak time and contributed to increasing 5.9% in PV output power on annual basis.
Wongwuttanasatian et al. [28]	To investigate the performance of PV system by using palm wax as a low-cost PCM.	The temperature of PV system was decreased about 6.1 °C along with a 5.3% electrical efficiency increase via PCM cooling, which compared to the PV system without the cooling.

Table 2. *Cont.*

Authors	Study Aim	Result
Heat pipe cooling approach		
Authors	**Study Aim**	**Result**
Habeeb et al. [29]	To carry out the performance of cooling PV panels using thermosyphon heat pipe at Baghdad climate.	Compared to the traditional panel, the module temperature was colder at a rate of 15–35%, and its efficiency was enhanced by 11–14%.
Alizadeh et al. [30]	To investigate the thermal performance for PV cooling by using pulsating heat pipe.	Under 1000 Wm^{-2} solar irradiation, the generated electricity was increased to 18% with the heat pipe cooling.
Radiative cooling approach		
Authors	**Study Aim**	**Result**
Nishioka et al. [31]	To use a high radiative coating to improve the performance of CPVs.	The temperature of the solar cells reduced around 10 °C, and their efficiency increased about 0.5%. In addition, the uniform temperature distribution in the cells was improved.
Zhu et al. [32]	To utilize a radiative cooling strategy that consists of a sky-access photonic thermal emitter to avoid the high operating temperature of PV cells without affecting their absorption coefficient.	The temperature of this design was forecast to drop at 17.6 °C and thus enhanced the electrical efficiency up to 7.9% under 800 Wm^{-2} solar intensity.

Even though the aforementioned cooling methods have remarkable effects on improving the performance of PV cells in terms of temperature reduction and electrical efficiency, the overall investment cost of cooling systems is a considerable concern when dealing with LSS PV farms [26]. Additionally, employing gravel as ground cover for PV installations is a business-as-usual approach. The ground-mounted PV installations with gravel ground cover also create a "heat island" effect. In other words, temperatures around PV solar arrays increase. Replacing the gravel with vegetation by strategic planning, therefore, can help to counter the heat feedback loop. As such, agrivoltaic technology holds promising implications for the food-energy-water nexus [33].

1.2. Agrivoltaic Approach

The co-location of PV and agriculture, commonly known as an agrivoltaic system, offers a win-win solution through many benefits, such as reducing water loss, increasing crop production, and improving the conversion efficiency of PV panels. Accordingly, adopting such synergistic paths forward can contribute toward building resilient energy-generation and food-production systems [33].

In the USA, Barron-Gafford et al. [34] performed an agrivoltaic system by planting chiltepin peppers, jalapeños, and cherry tomatoes under PV arrays. The system was created to capture the effects of this approach on physical and biological features during the average three-month summer growing season. Compared to the traditional planting area (control) of the food production, the total productions of chiltepin peppers and cherry tomatoes in the agrivoltaic system were three and two times greater, respectively. With regard to water savings, as shown in Figure 3(i), soil moisture remained around 15% and 5% higher for irrigating every two days and irrigating every day before the next watering, respectively, in the agrivoltaic system. With regard to the improved renewable energy production, as shown in Figure 3(ii), PV panels in the agrivoltaic system were approximately 9 °C cooler during daytime hours. On balance, therefore, the agrivoltaic approach provides mutual benefits in drylands in terms of the food-energy-water nexus.

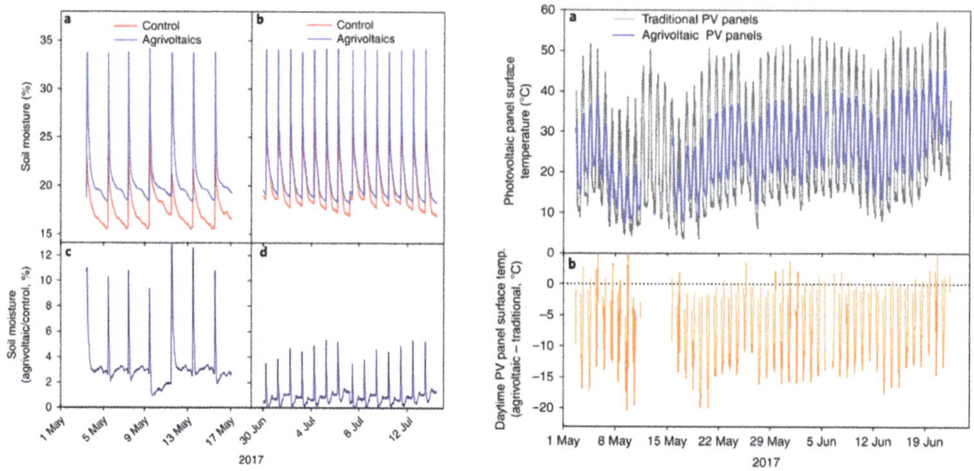

Figure 3. (i) Impacts of agrivoltaic over control installations on soil moisture: (**a**,**b**)—the comparison between control plots and agrivoltaics on soil moisture; (**c**,**d**)—differences between soil moisture in the agrivoltaic and in the control settings where positive values mean additional moisture in the agrivoltaic setting. (**ii**) impacts of agrivoltaic over traditional ground-mounted installations on the surface temperature of PV panels: (**a**)—the comparison between the surface temperature of traditional PV panels and the surface temperature of agrivoltaic PV panels; (**b**)—differences in PV panel temperature between the agrivoltaic and traditional settings where negative values mean the degree to which PV panels in the agrivoltaic were cooler. Reproduced with permission [34]. Copyright 2018, Springer Nature.

In Europe, agriculture technology company Sun'Agri from France showed that the trees shaded by the agrivoltaic system in the Durance Valley decreased the ambient temperature from 2 °C to 4 °C and also contributed to 63% reduction of water stress on the crops [35]. In China, the capacity of a 640 MW solar park was installed, while goji berries were planted under the solar panels. The results showed that the evaporation of land moisture for this solar park effectively reduced by 30–40%, and 85% vegetation coverage significantly improved the climate in this region. More interestingly, the ecosystem has also changed accordingly in this region. For instance, the number of small wild animals, such as hares, pheasants, and sparrows, has significantly increased [36]. In Singapore, Teng et al. [37] investigated the impact of agrivoltaic system on the surrounding rooftop microclimate by using ENVI-met simulation. Compared to the results without crops, on sunny days under the agrivoltaic approach, PV temperatures were on average reduced by 2.83 °C, and PV efficiency performance was improved by 1.13–1.42%. On cloudy days under the agrivoltaic approach, PV temperatures were also on average lower by 0.71 °C, and PV efficiency performance was enhanced by 0.28–0.35%.

An illustration of the practical implementation of agrivoltaic projects is shown in Figure 4 with the geographical details distributed worldwide. Herein, this paper aims to perform the tropical field validation for energy performance via agrivoltaic natural cooling approach in the LSS PV farm in Malaysia. The structure of this paper is as follows: Section 2 contains the details of field setup, data logging, and experimental approach. Section 3 shares some field analysis on the environmental parameters, namely ambient temperature, wind speed, and relative humidity for both weather stations. The main findings on herbal natural cooling in large scale solar PV farms is described based on Fisher ANOVA on energy at different plots and with some statistical justifications to support the results. Section 4 concludes the study with a significant DC energy increase via agrivoltaic approach as per the Welch two sample *t*-test between energy production at agrivoltaic and non-agrivoltaic plots.

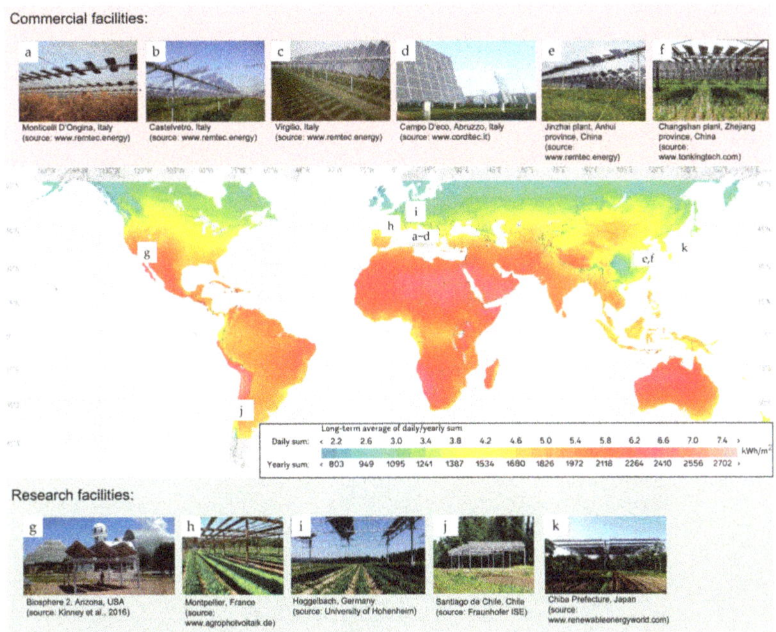

Figure 4. Geographical distribution of agrivoltaic projects for industrial and research facilities worldwide [38]. a; Monticelli D'Ongina Italy (www.remtec.energy (accessed on 22 March 2023)), b; Castelvelo Italy (www.remtec.energy (accessed on 22 March 2023)), c; Virgilio Italy (www.remtec.energy (accessed on 22 March 2023)), d; Campo D'eco Abruzzo Italy (www.corditec.it (accessed on 22 March 2023)), e; Jinzhai plant Anhui Province China (www.remtec.energy (accessed on 22 March 2023)), f; Changshan plant Zhejiang Province China (www.tonkingtech.com (accessed on 22 March 2023)), g; Biosphere 2, Arizona, USA [39], h; Montpellier France (www.agrophotvoltaik.de (accessed on 22 March 2023)), k; Chiba Prefecture Japan (www.renewableenergyworld.com (accessed on 22 March 2023)). Copyright 2019, Springer.

2. Methodology

2.1. Site Setup

The LSS PV field setup was located at UPM Agri Solar Power Plant in Puchong, Selangor, with 2 MWp generating capacity. It consisted of 8064 monocrystalline PV modules within five acres area, including 84 strings segregated into 12 plots. Each PV plot was divided by seven strings and could be further separated into four sections. As shown in Figures 5 and 6, plots were selected for the analysis of this research. The plots (rectangles in round yellow dots) were planted with Misai Kucing. Plot 7 is designated as the reference plot, and the condition is maintained as per normal Solarfarm structures without any Misai Kucing crops planted underneath. Figure 6 presents the experimental facilities of this research. Figure 7 shows the data logging platform via Sunny Explorer software from SMA Solar Technology AG for electrical output.

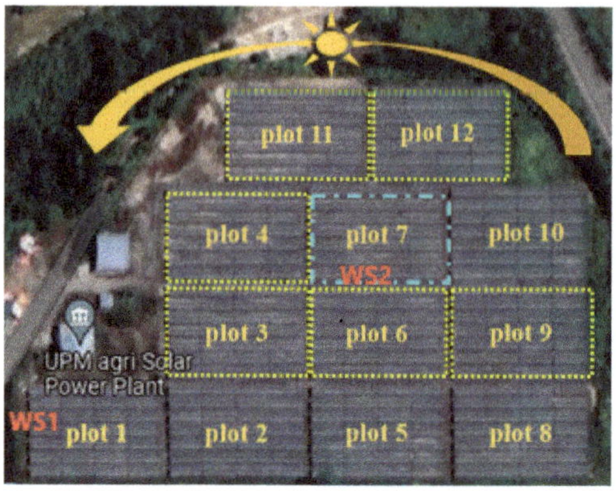

Figure 5. Google map layout of LSS PV farm in Puchong, Selangor (five acres). WS is the location for Weather Station.

Figure 6. (**a**) LSS PV farm setup facing south with 5° slanting angle; (**b**) Misai Kucing plants were planted beneath the PV panels as agrivoltaic setup; (**c**) the weather station for LSS PV farm.

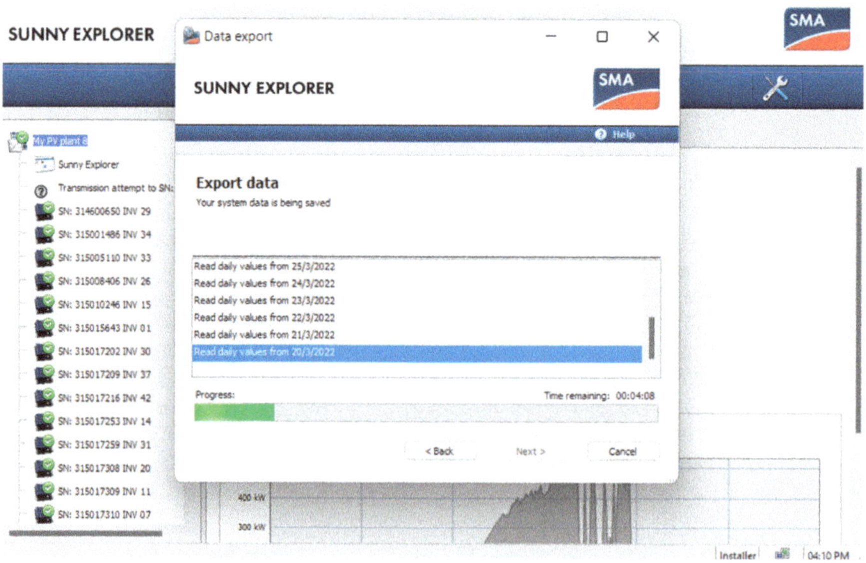

Figure 7. Data logging platform using Sunny Explorer for electrical output.

2.2. Experiment Process

The experiment was conducted from February 2022 to March 2022 for two months with continuous monitoring of all plots. The data logging process occurred from 7:00 a.m. to 7:00 p.m. each day (only during the presence of sunlight). Figure 8 shows PV modules' construction at Puchong Solarfarm with mounting structures.

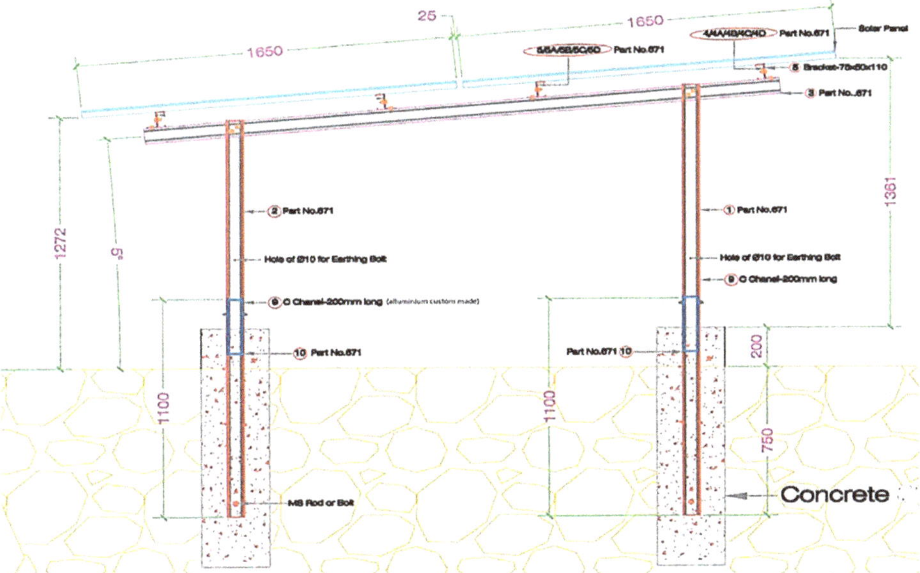

Figure 8. PV modules mounting structure with height distance from ground level.

Some descriptive statistics to obtain a preliminary understanding of energy production under different settings (e.g., sun level and plot) were computed. To assess the relationship between sun levels and energy production, we conducted a Welch one-way analysis of variance (ANOVA) [40] and a further pairwise comparison (via Games-Howell test) [41]. Next, to assess the relationship between agrivoltaic status and energy production, we conducted a Fisher ANOVA [42] and further pairwise comparison with reference to the non-agrivoltaic plot (via Student's t pairwise comparison test with Dunnett's method adjustment) [43]. Finally, to further look into the comparison between agrivoltaic and non-agrivoltaic plots, we grouped all agrivoltaic plots into a single group and conducted a two-sample mean test.

3. Results and Discussion

Based on the collected energy data for Puchong Solarfarm, it is observed that only plots 3, 4, 6, 7, 9, 11 and 12 (as shown in Figure 5) have recorded continuous data without errors. Thus, those without faulty data were used for analysis. These plots are considered sufficient based on their location surrounding the reference plot. The typical results on environmental parameters are further analyzed based on the two-sample location, i.e., near Plot 1 (agrivoltaic:WS1-corner) and near Plot 7 (non-agrivoltaic: WS2-middle). Based on the 24 h data collection, the ambient temperature in the agrivoltaic area and at the middle of the farm area is nearly the same with the maximum value of 41.2 °C recorded at the middle area as shown in Figure 9. The wind profiling for both locations is almost the same throughout the sample day as shown in Figure 10. As for the relative humidity shown in Figure 11, the value recorded for agrivoltaic areas shows a much higher humidity level with an average difference of 5%.

Higher humidity near solar panels can affect the lifespan of solar panels in several ways. First, increased humidity can lead to corrosion of metal parts and electrical components in the panels. Second, high humidity can cause moisture to enter the solar cells, which can degrade their performance and efficiency over time. Additionally, high humidity can promote the growth of microorganisms, such as algae, which can accumulate on the surface of solar panels and reduce their efficiency. Finally, high humidity can also lead to the formation of condensation on the surface of the panels, which can damage the electrical components and reduce the efficiency of the system. To mitigate the effects of high humidity on solar panels, it is important to ensure that the panels are properly sealed and protected from moisture, which is a critical parameter during the PV manufacturing process [44–47]. In this study, the focus will be on temperature reduction via the natural cooling approach. Regular cleaning of the panels can also help to prevent the accumulation of algae and other debris on the surface of the panels, which can reduce their efficiency over time.

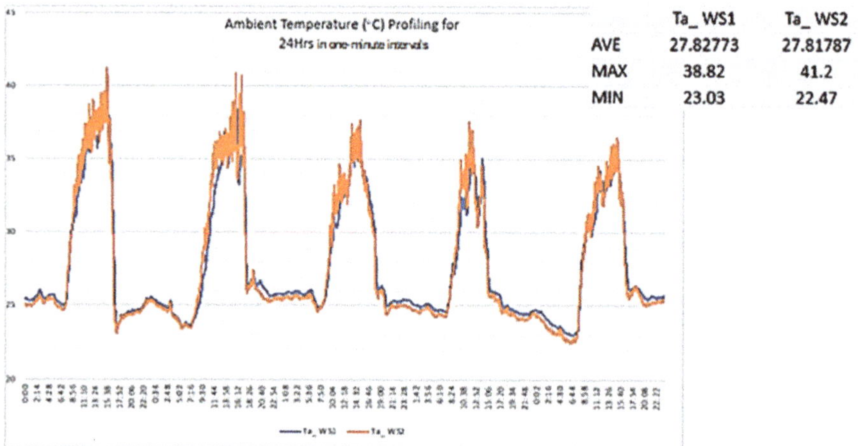

Figure 9. Ambient temperature profiling.

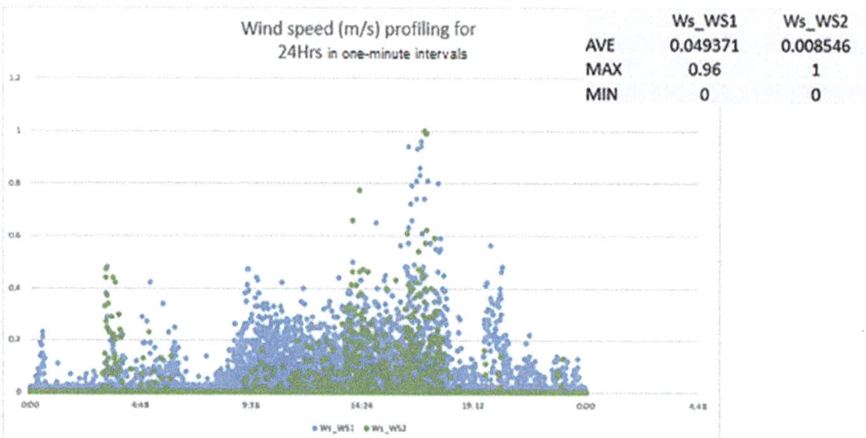

Figure 10. Wind speed profiling.

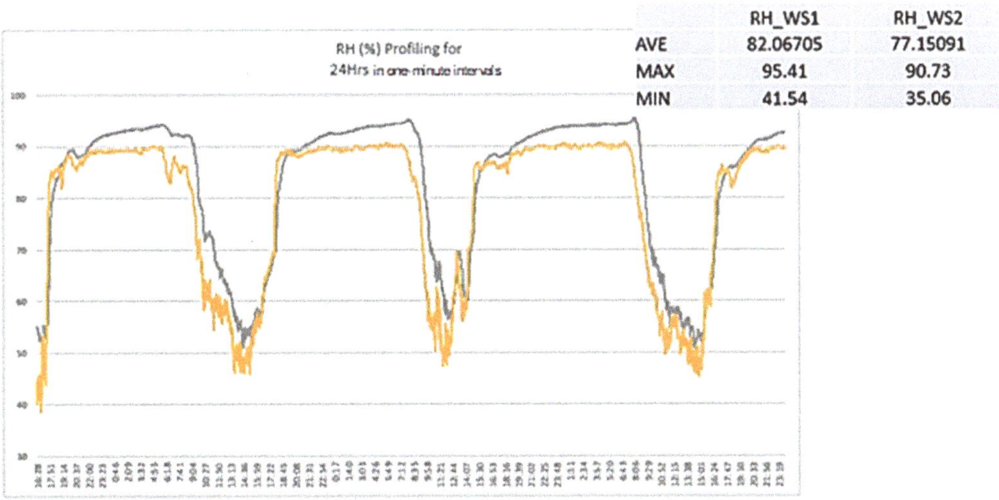

Figure 11. Relative humidity profiling.

A preliminary observation from descriptive statistics showed that Plot 7 (the non-agrivoltaic plot) had the lowest energy, and most energy was captured during peak sun (refer to Appendix A Table A1 for further detailed descriptive statistics). Some further analyses were conducted to examine the different energy level captured at different sun levels of the day as shown in Appendix A Figure A1. First, several main assumptions about the distribution of energy at different sun levels that were tested were the assumption of normality as well as constant variance of residuals (refer to Appendix B Figure A2). Since we have a large dataset, normality is assumed. Unfortunately, there is a fan-shape in the residuals vs. fitted plot which shows inconsistency in variance (heteroscedasticity) between sun levels. We further proved the presence of heteroscedasticity through the Bartlett test [48] of homogeneity of variances (Bartlett's K-squared = 8275.1, df = 4, p-value < 2.2×10^{-16}). Hence, Welch ANOVA was employed (refer to Appendix B Figure A3). There is a significant difference (F_{Welch} = 26,408.92; p-value = 0.00 < 0.05) in energy at different sun levels. The Games-Howell test showed that the energy is different between pairs of sun levels (all p-values < 0.05) (refer to Appendix B Table A2). That is, the energy level is different at different sun levels, and the highest was recorded during the peak sun, like that obtained in Othman et al. [49]. The sample comparative analysis of both agrivoltaic and non-agrivoltaic plots is shown in Figure 12 for two sample plots and Table 3 with details of the electrical outputs for 7 sample plots, which confirms the increase in DC electricity generation by means of natural plant cooling. Based on actual field data analysis, the location of each plot showed some varying values although they are installed at the same location. As an example, Plot 3 generated a higher power than Plot 9, with different fluctuation patterns throughout the day along with the sun movement where the location of PV module distribution does provide some impact in terms of DC generation. Shadow is not a factor because the location was selected and constructed to negate any shadow impact. With respect to Plot 7 as the non-agrivoltaic plot, this study has proven a significant increase in DC power generation.

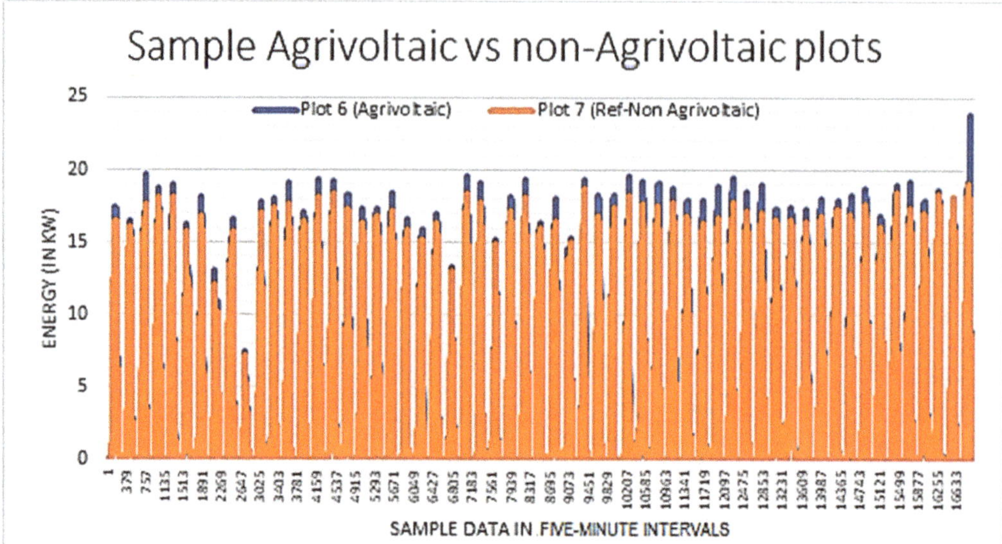

Figure 12. Sample agrivoltaic vs. non-agrivoltaic plots.

Table 3. The electrical output for the seven selected plots.

	Plot 3 (Agrivoltaic)	Plot 4 (Agrivoltaic)	Plot 6 (Agrivoltaic)	Plot 7 (Ref-Non Agrivoltaic)	Plot 9 (Agrivoltaic)	Plot 11 (Agrivoltaic)	Plot 12 (Agrivoltaic)
Sum (kW)	66,031.087	66,176.308	64,724.269	63,835.959	64,652.976	65,161.215	64,907.0955
Ave (kW)	3.886239009	3.894785946	3.809326644	3.757045436	3.805130716	3.835042964	3.82008684
Max (kW)	19.758	19.632	23.808	19.104	18.396	19.278	18.7275
Comparison	P3-P7	P4-P7	P6-P7	P9-P7	P11-P7	P12-P7	
kW	2195.128	2340.349	888.31	817.017	1325.256	1071.1365	
%	3.324385679	3.536536067	1.372452735	1.263695889	2.033811064	1.650261026	

The analysis of the effect of agrivoltaic plots on energy production begins with some assumptions required for the parametric statistical tests: i.e., normality and constant variance of residuals were fulfilled. Hence, Fisher ANOVA was used, and some reliable results are summarized in Figure 13.

There is strong evidence for the difference ($F_{Fisher} = 2.39$; p-value = 0.03 < 0.05) in energy from different plots. Student's t pairwise comparison test with Dunnett's method adjustment (a close approximation to the Dunnett adjustment) showed that the energy is different between Plot 7 and two other plots, i.e., Plot 3 and Plot 4, at 95% confidence level (refer to Table 4). The 95% confidence intervals of the mean difference between individual agrivoltaic plots (Plots 3, 4, 9 and 11) and non-agrivoltaic plot (Plot 7) further support the results, i.e., the intervals for the pairs do not include zero (refer to Figure 14).

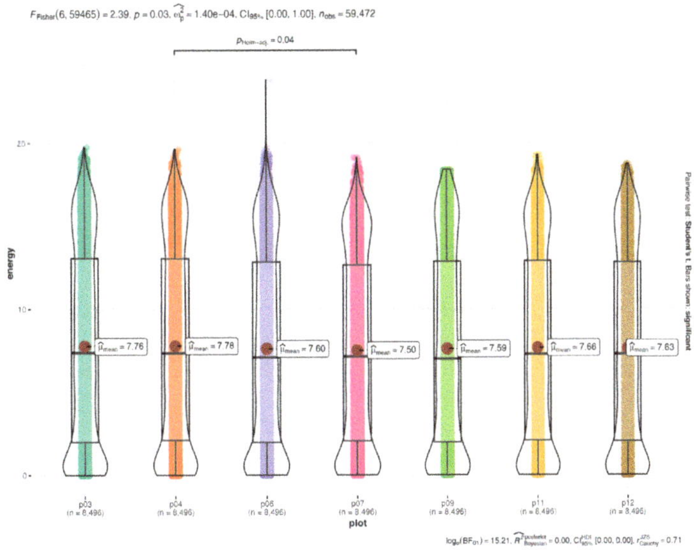

Figure 13. Fisher ANOVA on energy at different plots.

Table 4. Pairwise comparison test (with Dunnett's method adjustment) on energy from different plots (SE = 0.08792941; df = 59465).

Contrast	Estimate	*t*-Ratio	Confidence Interval	*p*-Value
p04–p07	0.2751	3.1292	(−0.0826, 0.5986)	0.0097
p03–p07	0.2559	2.9105	(−0.0591, 0.6221)	0.0193
p06–p07	0.1016	1.1552	(0.0536, 0.7348)	0.6825
p09–p07	0.0931	1.0591	(−0.0051, 0.6761)	0.7406
p11–p07	0.1563	1.7779	(−0.0476, 0.6336)	0.2992
p12–p07	0.1247	1.4185	(−0.6604, 0.0209)	0.5128

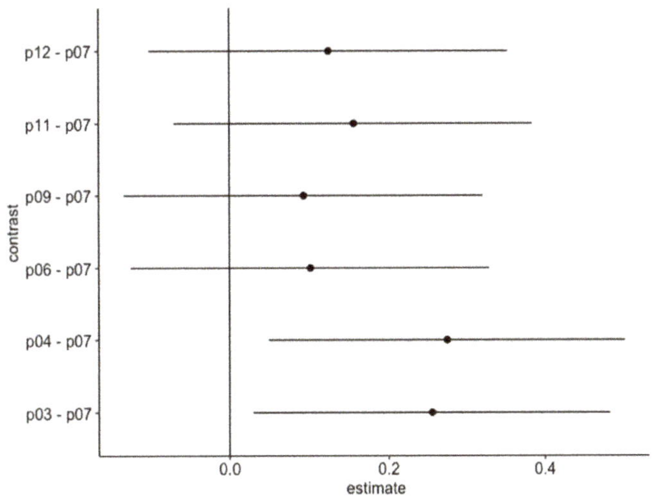

Figure 14. 95% confidence interval of mean difference between Agrivoltaic plots and non-Agrivoltaic plot.

Further, the two-sample mean comparison (Welch t = 2.5624, df = 11,743, p-value = 0.01041 < 0.05, 95% confidence interval = [0.0394, 0.2962]) shows strong evidence for the difference between a group of agrivoltaic plots and a non-agrivoltaic plot (refer Table 5). In summary, the highest average energy increase by plot and the overall plot energy average increase are 0.28 kW (3.73%) and 0.17 kW (2.24%), respectively, with the implementation of agrivoltaic.

Table 5. Welch two sample t-test between energy production at agrivoltaic and non-agrivoltaic plots.

t	df	Confidence Interval	p-Value
2.5624	11743	[0.0394, 0.2962]	0.0104

4. Conclusions

The agrivoltaic integration in large-scale solar PV farms adopts such synergistic paths forward which contribute toward building resilient energy-generation and food-production systems. This study provides the proof of concept where herbal plantation (in this case, Misai Kucing cultivation) as a means of agrivoltaic plot supports the operation of solar farms through natural cooling directly underneath the solar PV arrays. The significant DC energy increase of 3% (on average) via agrivoltaic cooling provides sufficient savings and surplus to the operators plus some means of secondary business with the fresh produce. Thus, it is greatly recommended that all large-scale solar farms, especially in Malaysia, should be transformed, not only by producing electricity for the grid, but also by integrating agriculture via agrivoltaic approach. Future recommendations on the impacts of PV module lifespan, economic perspective, and soil properties under agrivoltaic conditions are suggested.

Author Contributions: Conceptualization, N.F.O., M.E.Y. and H.H.; methodology, L.L., A.S.M.S. and H.H.; software, A.H.J.; validation, M.E.Y. and A.H.J.; formal analysis, A.H.J.; investigation, L.L.; resources, N.F.O.; data curation, A.H.J. and J.N.J.; writing—original draft, N.F.O. and L.L.; writing—review & editing, M.E.Y.; visualization, A.S.M.S.; supervision, R.S. and J.N.J.; project administration, N.F.O. and R.S.; funding acquisition, M.E.Y. All authors have read and agreed to the published version of the manuscript.

Funding: The authors delegate our thanks to the Ministry of Energy and Natural Resources (MENR) under the AAIBE Research Fund (Vote no. 6300921) and the Research Management Centre (RMC), Universiti Putra Malaysia for the approval of research funding under the Putra Grants Scheme (Vote no. 9709000).

Institutional Review Board Statement: Not Applicable.

Data Availability Statement: Not Applicable.

Conflicts of Interest: The authors declare no conflict of interest.

Appendix A. Descriptive Statistics

Table A1. Descriptive statistics of energy at different sun levels and plots.

Plot	Measures	Sun Level					
		Early Sun	Mild Sun (Evening)	Moderate Sun (Evening)	Moderate Sun (Morning)	Peak Sun	Overall
Plot 3							
	Mean (SD) Interval [min, max]	2 (2.41) [0, 10.89]	2.75 (2.8) [0, 12.29]	9.24 (5.13) [0, 19.74]	10.55 (3.87) [0.72, 18.85]	13.24 (4.07) [1.77, 19.76]	7.76 (5.83) [0, 19.76]
Plot 4							
	Mean (SD) Interval [min, max]	2.22 (2.5) [0, 11.53]	2.66 (2.75) [0, 12.12]	9.1 (5.13) [0, 19.63]	10.68 (3.84) [0.75, 18.74]	13.2 (4.07) [1.82, 19.26]	7.78 (5.8) [0, 19.63]

Table A1. *Cont.*

Plot	Measures	Sun Level					
		Early Sun	Mild Sun (Evening)	Moderate Sun (Evening)	Moderate Sun (Morning)	Peak Sun	Overall
Plot 6							
	Mean (SD) Interval [min, max]	1.94 (2.32) [0, 10.84]	2.75 (2.81) [0, 12.37]	9.16 (5.16) [0, 23.81]	10.24 (3.79) [0.7, 18.79]	12.96 (4.13) [1.71, 19.42]	7.6 (5.75) [0, 23.81]
Plot 7 (reference plot)							
	Mean (SD) Interval [min, max]	2.14 (2.41) [0, 11.23]	2.59 (2.69) [0, 11.71]	8.8 (4.92) [0, 18.48]	10.33 (3.68) [0.73, 18.41]	12.67 (3.83) [1.76, 19.1]	7.5 (5.56) [0, 19.1]
Plot 9							
	Mean (SD) Interval [min, max]	1.9 (2.27) [0, 10.78]	2.73 (2.8) [0, 12.59]	9.12 (5.17) [0, 18.34]	10.27 (3.85) [0.72, 18.24]	12.98 (4.22) [1.67, 18.4]	7.59 (5.78) [0, 18.4]
Plot 11							
	Mean (SD) Interval [min, max]	2.2 (2.48) [0, 11.27]	2.59 (2.7) [0, 11.84]	8.94 (5.03) [0, 19.26]	10.57 (3.78) [0.71, 18.6]	12.97 (3.97) [1.79, 19.28]	7.66 (5.7) [0, 19.28]
Plot 12							
	Mean (SD) Interval [min, max]	2.05 (2.37) [0, 10.74]	2.66 (2.73) [0, 11.78]	9.03 (5.01) [0, 18.72]	10.42 (3.75) [0.72, 18.27]	12.98 (3.92) [1.74, 18.73]	7.63 (5.69) [0, 18.73]
Overall							
	Mean (SD) Interval [min, max]	2.07 (2.4) [0, 11.53]	2.67 (2.75) [0, 12.59]	9.06 (5.08) [0, 23.81]	10.44 (3.79) [0.7, 18.85]	13 (4.04) [1.67, 19.76]	7.65 (5.73) [0, 23.81]

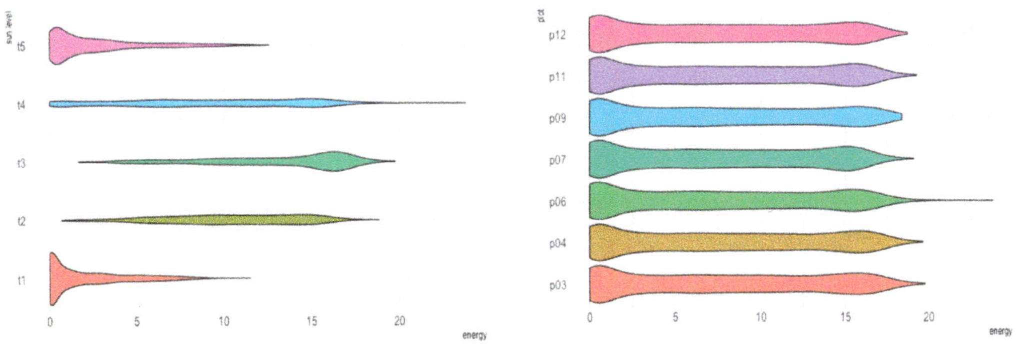

(**a**) (**b**)

Figure A1. (**a**) Distribution of energy at different sun level. (**b**) Distribution of energy at different plots.

Appendix B. Energy Level at Different Sun Levels

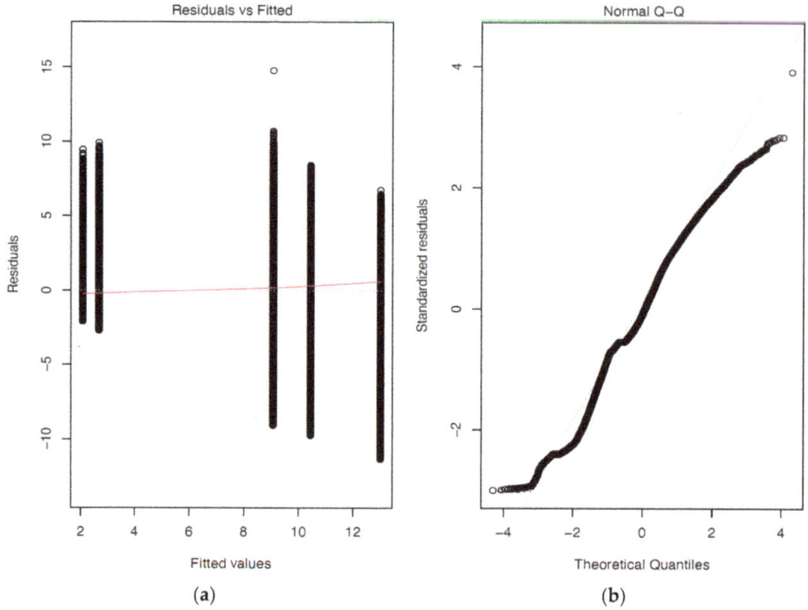

Figure A2. Equal variance and normality assessments on energy at different sun levels. (**a**) Residuals vs. fitted. (**b**) Normal Q-Q.

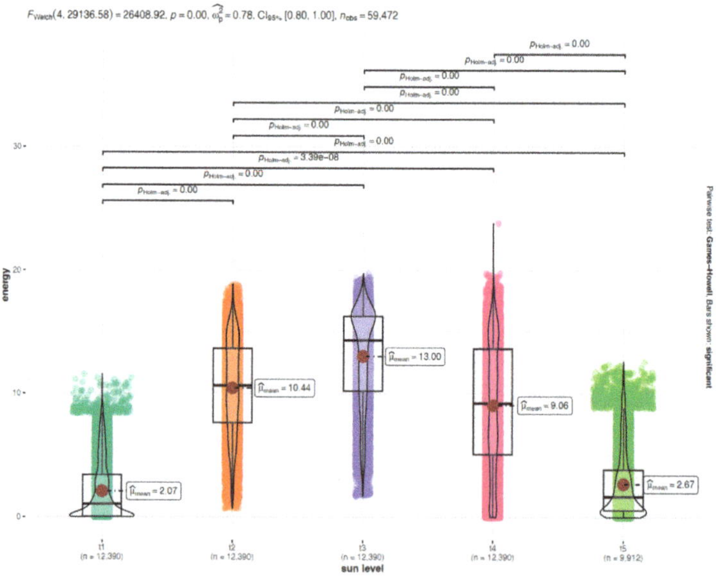

Figure A3. Welch ANOVA on energy at different sun levels.

Table A2. Pairwise comparison in energy at different sun levels using Games-Howell test.

Contrast Pair	Estimate	Confidence Interval	p-Value Adjusted
t1–t2	8.3706	[8.2607, 8.4806]	0.0000
t1–t3	10.9357	[10.8207, 11.0508]	0.0000
t1–t4	6.9907	[6.853, 7.1284]	0.0000
t1–t5	0.6094	[0.5137, 0.705]	0.0000
t2–t3	2.5651	[2.4293, 2.7008]	0.0000
t2–t4	−1.3800	[−1.5354, −1.2246]	0.0000
t2–t5	−7.7613	[−7.881, −7.6415]	0.0000
t3–t4	−3.9450	[−4.1041, −3.786]	0.0000
t3–t5	−10.3263	[−10.4507, −10.2019]	0.0000
t4–t5	−6.3813	[−6.5269, −6.2357]	0.0000

References

1. Barron-Gafford, G.A.; Minor, R.L.; Allen, N.A.; Cronin, A.D.; Brooks, A.E.; Pavao-Zuckerman, M.A. The photovoltaic heat island effect: Larger solar power plants increase local temperatures. *Sci. Rep.* **2016**, *6*, 35070. [CrossRef] [PubMed]
2. Lu, L.; Ya'acob, M.E.; Anuar, M.S.; Mohtar, M.N. Comprehensive review on the application of inorganic and organic photovoltaics as greenhouse shading materials. *Sustain. Energy Technol. Assess.* **2022**, *52*, 102077. [CrossRef]
3. Khan, R.; Go, Y.I. Assessment of Malaysia's Large-Scale Solar Projects: Power System Analysis for Solar PV Grid Integration. *Glob. Challenges* **2020**, *4*, 1900060. [CrossRef] [PubMed]
4. Gomaa, M.R.; Al-Dhaifallah, M.; Alahmer, A.; Rezk, H. Design, modeling, and experimental investigation of activewater cooling concentrating photovoltaic system. *Sustainability* **2020**, *12*, 5392. [CrossRef]
5. Makki, A.; Omer, S.; Sabir, H. Advancements in hybrid photovoltaic systems for enhanced solar cells performance. *Renew. Sustain. Energy Rev.* **2015**, *41*, 658–684. [CrossRef]
6. Wu, S.; Xiong, C. Passive cooling technology for photovoltaic panels for domestic houses. *Int. J. Low-Carbon Technol.* **2014**, *9*, 118–126. [CrossRef]
7. Sargunanathan, S.; Elango, A.; Mohideen, S.T. Performance enhancement of solar photovoltaic cells using effective cooling methods: A review. *Renew. Sustain. Energy Rev.* **2016**, *64*, 382–393. [CrossRef]
8. National Renewable Energy Laboratory. Best Research-Cell Efficiency Chart 2020. Available online: https://www.nrel.gov/pv/cell-efficiency.html (accessed on 1 October 2022).
9. Ahmad, E.Z.; Sopian, K.; Jarimi, H.; Fazlizan, A.; Elbreki, A.; Hamid, A.S.A.; Rostami, S.; Ibrahim, A. Recent advances in passive cooling methods for photovoltaic performance enhancement. *Int. J. Electr. Comput. Eng.* **2021**, *11*, 146–154. [CrossRef]
10. Kirpichnikova, I.M.; Sudhakar, K.; Makhsumov, I.B.; Martyanov, A.S.; Priya, S.S. Thermal model of a photovoltaic module with heat-protective film. *Case Stud. Therm. Eng.* **2022**, *30*, 101744. [CrossRef]
11. Ya'acob, M.E.; Hizam, H.; Bakri, M.A.; Radzi, M.A.M.; Khatib, T.; Rahim, A.H.M.A. Performance test conditions for direct temperature elements of multiple PV array configurations in Malaysia. *Energy Procedia* **2014**, *61*, 2387–2390. [CrossRef]
12. Dwivedi, P.; Sudhakar, K.; Soni, A.; Solomin, E.; Kirpichnikova, I. Advanced cooling techniques of P.V. modules: A state of art. *Case Stud. Therm. Eng.* **2020**, *21*, 100674. [CrossRef]
13. Elbreki, A.M.; Alghoul, M.A.; Sopian, K.; Hussein, T. Towards adopting passive heat dissipation approaches for temperature regulation of PV module as a sustainable solution. *Renew. Sustain. Energy Rev.* **2017**, *69*, 961–1017. [CrossRef]
14. Ghadikolaei, S.S.C. An enviroeconomic review of the solar PV cells cooling technology effect on the CO_2 emission reduction. *Sol. Energy* **2021**, *216*, 468–492. [CrossRef]
15. Siecker, J.; Kusakana, K.; Numbi, B.P. A review of solar photovoltaic systems cooling technologies. *Renew. Sustain. Energy Rev.* **2017**, *79*, 192–203. [CrossRef]
16. Maleki, A.; Haghighi, A.; El Haj Assad, M.; Mahariq, I.; Alhuyi Nazari, M. A review on the approaches employed for cooling PV cells. *Sol. Energy* **2020**, *209*, 170–185. [CrossRef]
17. Fikri, M.A.; Samykano, M.; Pandey, A.K.; Kadirgama, K.; Kumar, R.R.; Selvaraj, J.; Rahim, N.A.; Tyagi, V.V.; Sharma, K.; Saidur, R. Recent progresses and challenges in cooling techniques of concentrated photovoltaic thermal system: A review with special treatment on phase change materials (PCMs) based cooling. *Sol. Energy Mater. Sol. Cells* **2022**, *241*, 111739. [CrossRef]
18. Teo, H.G.; Lee, P.S.; Hawlader, M.N.A. An active cooling system for photovoltaic modules. *Appl. Energy* **2012**, *90*, 309–315. [CrossRef]
19. Sajjad, U.; Amer, M.; Muhammad, H.; Dahiya, A. Cost effective cooling of photovoltaic modules to improve efficiency. *Case Stud. Therm. Eng.* **2019**, *14*, 100420. [CrossRef]
20. Krauter, S. Increased electrical yield via water flow over the front of photovoltaic panels. *Sol. Energy Mater. Sol. Cells* **2004**, *82*, 131–137. [CrossRef]
21. Mah, C.Y.; Lim, B.H.; Wong, C.W.; Tan, M.H.; Chong, K.K.; Lai, A.C. Investigating the Performance Improvement of a Photovoltaic System in a Tropical Climate using Water Cooling Method. *Energy Procedia* **2019**, *159*, 78–83. [CrossRef]
22. Van Sark, W.G.J.H.M. Feasibility of photovoltaic—Thermoelectric hybrid modules. *Appl. Energy* **2011**, *88*, 2785–2790. [CrossRef]

23. Benghanem, M.; Al-Mashraqi, A.A.; Daffallah, K.O. Performance of solar cells using thermoelectric module in hot sites. *Renew. Energy* **2016**, *89*, 51–59. [CrossRef]
24. Ramkiran, B.; Sundarabalan, C.; Sudhakar, K. Sustainable passive cooling strategy for PV module: A comparative analysis. *Case Stud. Therm. Eng.* **2021**, *27*, 101317. [CrossRef]
25. Ali, H.M. Recent advancements in PV cooling and efficiency enhancement integrating phase change materials based systems—A comprehensive review. *Sol. Energy* **2020**, *197*, 163–198. [CrossRef]
26. Allouhi, A.; Rehman, S.; Buker, M.S.; Said, Z. Up-to-date literature review on Solar PV systems: Technology progress, market status and R&D. *J. Clean. Prod.* **2022**, *362*, 132339. [CrossRef]
27. Hasan, A.; Sarwar, J.; Alnoman, H.; Abdelbaqi, S. Yearly energy performance of a photovoltaic-phase change material (PV-PCM) system in hot climate. *Sol. Energy* **2017**, *146*, 417–429. [CrossRef]
28. Wongwuttanasatian, T.; Sarikarin, T.; Suksri, A. Performance enhancement of a photovoltaic module by passive cooling using phase change material in a finned container heat sink. *Sol. Energy* **2020**, *195*, 47–53. [CrossRef]
29. Habeeb, L.J.; Mutasher, D.G.; Ali, F.A.M.A. Cooling Photovoltaic Thermal Solar Panel by Using Heat Pipe at Baghdad Climate. *Int. J. Mech. Mechatronics Eng. IJMME-IJENS* **2017**, *17*, 6.
30. Alizadeh, H.; Ghasempour, R.; Shafii, M.B.; Ahmadi, M.H.; Yan, W.M.; Nazari, M.A. Numerical simulation of PV cooling by using single turn pulsating heat pipe. *Int. J. Heat Mass Transf.* **2018**, *127*, 203–208. [CrossRef]
31. Nishioka, K.; Ota, Y.; Tamura, K.; Araki, K. Heat reduction of concentrator photovoltaic module using high radiation coating. *Surf. Coatings Technol.* **2013**, *215*, 472–475. [CrossRef]
32. Zhu, L.; Raman, A.; Wang, K.X.; Anoma, M.A.; Fan, S. Radiative cooling of solar cells. *Optica* **2014**, *1*, 32. [CrossRef]
33. National Renewable Energy Laboratory. Benefits of Agrivoltaics Across the Food-Energy-Water Nexus 2019. Available online: https://www.nrel.gov/news/program/2019/benefits-of-agrivoltaics-across-the-food-energy-water-nexus.html (accessed on 8 January 2023).
34. Barron-Gafford, G.A.; Pavao-Zuckerman, M.A.; Minor, R.L.; Sutter, L.F.; Barnett-Moreno, I.; Blackett, D.T.; Thompson, M.; Dimond, K.; Gerlak, A.K.; Nabhan, G.P.; et al. Agrivoltaics provide mutual benefits across the food–energy–water nexus in drylands. *Nat. Sustain.* **2019**, *2*, 848–855. [CrossRef]
35. Sun'Agri. Arboriculture: La Pugère 2021. Available online: https://sunagri.fr/en/project/the-la-pugere-experimental-station/ (accessed on 15 August 2022).
36. Bellini, E. Giant Agrivoltaic Project in China. PV Mag 2020. Available online: https://www.pv-magazine.com/2020/09/03/giant-agrivoltaic-project-in-china/ (accessed on 26 August 2022).
37. Teng, J.W.C.; Soh, C.B.; Devihosur, S.C.; Tay, R.H.S.; Jusuf, S.K. Effects of Agrivoltaic Systems on the Surrounding Rooftop Microclimate. *Sustainability* **2022**, *14*, 7089. [CrossRef]
38. Weselek, A.; Ehmann, A.; Zikeli, S.; Lewandowski, I.; Schindele, S.; Högy, P. Agrophotovoltaic systems: Applications, challenges, and opportunities. A review. *Agron. Sustain. Dev.* **2019**, *39*, 35. [CrossRef]
39. Kinney, K.; Minor, R.; Barron-Gafford, G.; Lewandowski, I.; Schindele, S.; Högy, P. Testing predictions used to build an agrivoltaics installation on a small-scale educational model. 2016. Available online: http://digitalcommons.calpoly.edu/star/386/ (accessed on 14 March 2023).
40. Delacre, M.; Leys, C.; Mora, Y.L.; Lakens, D. Taking Parametric Assumptions Seriously: Arguments for the Use of Welch's F-test instead of the Classical F-test in One-Way ANOVA. *Int. Rev. Soc. Psychol.* **2019**, *32*, 13. [CrossRef]
41. Games, P.A.; Howell, J.F. Pairwise Multiple Comparison Procedures with Unequal N's and/or Variances: A Monte Carlo Study. *J. Educ. Stat.* **1976**, *1*, 113–125. [CrossRef]
42. Kaufmann, J.; Schering, A. Analysis of Variance ANOVA. *Wiley StatsRef Stat. Ref. Online* **2014**, 1–12. [CrossRef]
43. Dunnett, C.W. A Multiple Comparison Procedure for Comparing Several Treatments with a Control. *J. Am. Stat. Assoc.* **1955**, *50*, 1096–1121. [CrossRef]
44. Bich, N.N.; Hoa, L.V.; Phuong, L.T. The impact of environmental factors on the efficiency of solar panels. *Environ. Eng. Manag. J.* **2020**, *19*, 1439–1447.
45. Jafri, A.; Kwan, C.; Raga, R. Solar panels: Design, materials, and performance. In *Solar Energy Materials.*; Springer: Berlin/Heidelberg, Germany, 2020; pp. 1–24.
46. Xia, W.; Wang, J.; Li, Z.; Wang, J. An investigation of humidity effects on the power output of photovoltaic modules. *Int. J. Photoenergy* **2019**, 2763720.
47. Zhang, H.; Shen, H.; Wu, X.; Wang, Y. Effects of humidity on the performance of silicon solar cells. *J. Semicond.* **2020**, *41*, 022601.
48. Arsham, H.; Lovric, M. Bartlett's Test. In *International Encyclopedia of Statistical Science*; Springer: Berlin/Heidelberg, Germany, 2011; pp. 87–88.
49. Othman, N.F.; Yaacob, M.E.; Su, A.S.M.; Jaafar, J.N.; Hizam, H.; Shahidan, M.F.; Jamaluddin, A.H.; Chen, G.; Jalaludin, A. Modeling of stochastic temperature and heat stress directly underneath agrivoltaic conditions with orthosiphon Stamineus Crop Cultivation. *Agronomy* **2020**, *10*, 1472. [CrossRef]

agriculture

MDPI

Article

The Development of an LSTM Model to Predict Time Series Missing Data of Air Temperature inside Fattening Pig Houses

Jun-gyu Kim [1], Sang-yeon Lee [1] and In-bok Lee [2,3,*]

[1] Agriculture, Animal & Aquaculture Intelligence Research Center, Electronics and Telecommunications Research Institute, 218 Gajeong-ro, Yuseong-gu, Daejeon 34129, Republic of Korea

[2] Department of Rural Systems Engineering, Research Institute for Agriculture and Life Sciences, Global Smart Farm Convergence Major, College of Agriculture and Life Sciences, Seoul National University, 1 Gwanak-ro, Gwanak-gu, Seoul 08826, Republic of Korea

[3] Research Institute of Green Eco Engineering, Institute of Green Bio Science and Technology, Seoul National University, 1 Gwanak-ro, Gwanak-gu, Seoul 08826, Republic of Korea

* Correspondence: iblee@snu.ac.kr

Abstract: Because of the poor environment inside fattening pig houses due to high humidity, ammonia gas, and fine dust, it is hard to accumulate reliable long-term data using sensors. Therefore, it is necessary to conduct research for filling in the missing environmental data inside fattening pig houses. Thus, this research aimed to develop a model for predicting the missing data of the air temperature inside fattening pig houses using a long short-term memory (LSTM) model, which is one of the artificial neural networks (ANNs). Firstly, the internal and external environmental data of the fattening pig house were monitored to develop the LSTM models for data filling of the missing data and to validate the developed LSTM model. The LSTM model for data filling of the missing data was developed by learning the measured temperature inside the pig house. The LSTM model developed in this study was validated by comparing the air temperature data predicted by the LSTM model with the air temperature data measured in the fattening pig house. The LSTM model was accurate within a 3.5% error rate for the internal air temperature. Finally, the accuracy and applicability of the developed LSTM model were evaluated according to the order of learning data and the length of the missing data. In the future, for information and communication technologies (ICTs) and the convergence and application of smart farms, the LSTM models developed in this study may contribute to the accumulation of reliable long-term data at the fattening pig house.

Keywords: environmental monitoring; imputation; machine learning; pig house; recurrent neural network

Citation: Kim, J.-g.; Lee, S.-y.; Lee, I.-b. The Development of an LSTM Model to Predict Time Series Missing Data of Air Temperature inside Fattening Pig Houses. *Agriculture* **2023**, *13*, 795. https://doi.org/10.3390/agriculture13040795

Academic Editors: Muhammad Sultan, Redmond R. Shamshiri, Md Shamim Ahamed and Muhammad Farooq

Received: 21 February 2023
Revised: 24 March 2023
Accepted: 28 March 2023
Published: 30 March 2023

1. Introduction

The livestock industry is a major sector of agriculture and has been continuously growing in South Korea, reaching USD 18 billion in 2020 and constituting 40% of the total agricultural production. Additionally, the pig industry in South Korea accounts for 35% of the total livestock industry, ranking it as the largest livestock industry [1].

Traditionally, the internal environments of most livestock housings are controlled based on the measured data, such as internal air temperature and relative humidity. Here, air temperature has been generally used as a basis to regulate the exhaust fans and inlet windows in order to maintain the proper environment inside the livestock houses. Environmental control using only air temperature has a limitation since it cannot consider various factors such as humidity, ammonia concentration, and odor. In winter, the minimum ventilation for the management of the internal air temperature can cause excessive humidity, ammonia concentration, and odor. Recently, with the development of the smart farm based on information and communication technology (ICT), the algorithm for the environmental control of livestock houses has been changed from control based on an environmental factor to integrated control based on the big data of various environmental factors. Reliable long-term data on the internal environment should be accumulated to design a precise control algorithm for livestock houses.

The expansion and automation of livestock houses have induced active research and development (R&D) and the establishment of smart farms. The global demand of smart farms continues to increase from USD 2.81 billion in 2015 to USD 4.92 billion in 2020 [2]. The monitoring of environmental data, such as temperature, humidity, and gas concentration inside the livestock houses, is critical for obtaining basic data to design environment control algorithms for an automatic control system of smart farms. Therefore, the accumulation of reliable long-term data on the environment is important because it is the basic data for environmental control and the development of the smart farm.

The long-term accumulation of monitoring data can be hindered by the reduced durability of the monitoring devices installed at livestock houses for daily measurements due to the poor internal environment with high levels of humidity, gas, and fine particulate matter [3]. In addition, the frequent failure of sensors will add repair and replacement costs for the sensor. Therefore, it is necessary to conduct research for filling in the missing data on the environments inside the livestock houses.

Several studies across different research fields have employed various techniques to estimate the missing data, either by using statistical analysis such as linear regression, auto-regressive integrated moving average (ARIMA), and seasonal auto-regressive integrated moving average (SARIMA) [4–9]. In the field related to agriculture, several studies have been conducted focused on the imputation of meteorological data [6,7,9,10]. Xie et al. [11] used a hybrid deep learning-driven sequential concentration transport emission model (DL-CTEM) that predicts the emissions of ammonia, carbon dioxide, and hydrogen sulfide, which are major harmful gases in pig houses, and suggested optimal ventilation control using this. Additionally, since real-time thermal environment management is important, a discrete model was used to improve the accuracy of thermal environment prediction inside the laying hen house [12]. A discrete model was developed that incorporates time-period groups (TPGs), the group buffered rolling (GBR) mechanism, and TPG factors. In the case of control based on the internal environment data in livestock house environment control, it can be said that the continuity of measured data and the interpolation of the missing data are very important.

Recently, machine learning techniques have been actively applied for the imputation of the missing data, and studies for the imputation of time series data using machine learning techniques had been also conducted in the agricultural field [13–16]. Boomgard-Zagrodnik and Brown [13] developed the random forest imputation model for missing Mesonet temperature observation. The model predicted the growing degree days (GDDs) value within an average error rate of 1.4%. Moon, Lee, and Son [15] developed the two-dimensional convolutional neural network (CNN) model for imputation of missing tabular data of several greenhouses. Song, Gao, Zhao, and Zhao [16] used the recurrent neural network (RNN) and long short-term memory (LSTM) models to fill the missing data of stem moisture data of plants. However, few studies focused on predicting the missing data of the livestock houses. In particular, since there is a lot of dust and harmful gases in livestock houses, missing data may occur frequently. Moreover, the RNN model, which is one of the ANN models, is being applied more actively because it is very important to collect time series data in agricultural facilities that manage animals and plants [17–23]. Therefore, it is very likely that the RNN model will be used as a method to solve missing data in real time in livestock facilities.

Compared to other models dealing with time series data, the LSTM models do not require the nonlinear functions to be estimated, and it has demonstrated superior performance in a wide range of sequence modeling applications [24–27]. Also, if the model uses the same number of layers, LSTM models can have a more complex structure and more parameters than gated recurrent units (GRUs) [28], which are commonly used to predict data with higher accuracy [29]. In addition, LSTM models have already shown higher accuracy than GRU in other studies [30,31]. Accordingly, it can be seen that it is appropriate to use the LSTM model using long-term data of livestock house.

In this study, the RNN model was developed for expecting the missing data of air temperature inside the fattening pig house using the field-measured data of the internal air temperature. The internal and external environmental data of the fattening pig house were monitored during the field experiment. Based on the result of the field experiment, descriptive statistics were conducted to evaluate the environmental management of the fattening pig house. Some of the measured air temperature data inside the fattening pig house were assumed to be the missing data. The RNN model for data filling of the missing data was developed by learning the measured air temperature. The RNN model developed in this study was validated by comparing the air temperature data predicted by the RNN model with the measured data by the field experiment. Finally, the accuracy and applicability of the developed RNN model were evaluated according to the order of learning data and the length of the missing data.

2. Materials and Methods

The research flow chart for predicting missing data using the RNN model is as follows (Figure 1). First, a field experiment was conducted to monitor the internal and external environments of the fattening pig house. Through the field experiment, the air temperature, relative humidity, ventilation rate, ammonia concentration inside the fattening pig house, and the external weather were monitored in the fattening pig house. These monitoring data were used to develop the RNN models for data filling of the missing data and to validate the RNN model. Based on the result of the field experiment, descriptive statistics were conducted to evaluate the environment inside the fattening pig house. In this study, the air temperature inside the fattening pig house were initially assumed to be the missing data. The RNN model for data filling of the missing data was developed by learning the measured data in other period. The RNN model developed was validated by comparing the air temperature data predicted by the RNN model with the measured air temperature. The accuracy of the RNN model was further improved by considering the periodic parameter. Finally, the accuracy and applicability of the developed RNN model were evaluated according to the order of learning data and the length of the missing data.

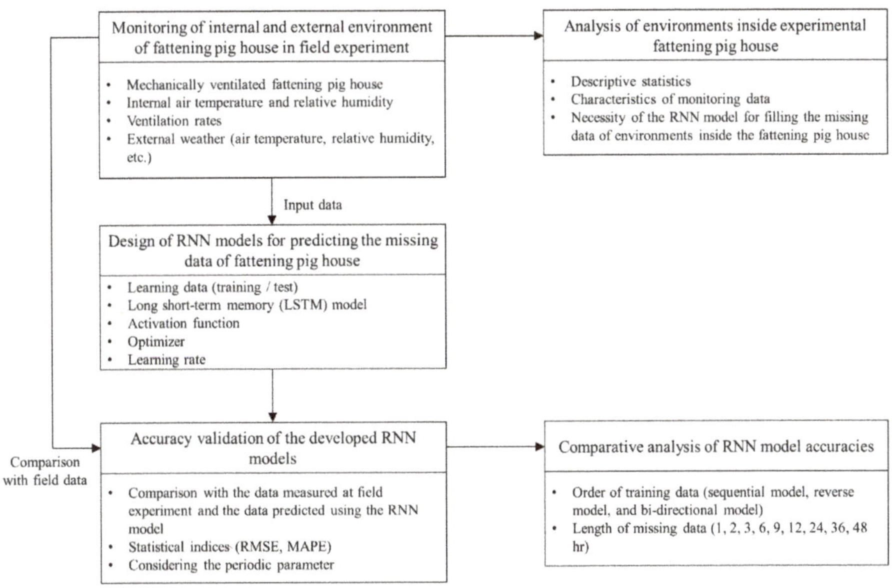

Figure 1. Flow chart of the experimental procedure of this study to develop the RNN model.

2.1. Experimental Facility (Fattening Pig House)

In this study, the experiment was conducted at the mechanically ventilated fattening pig house located at Incheon-myeon, Buyeo-gun, Chungcheongnam-do Province (126°90′ E, 36°21′ N). The experimental fattening pig house was shown in Figure 2. The experimental fattening pig house had a width of 42 m, a length of 145.1 m, and consisted of 24 pig rooms. The pig room where the environmental monitoring was conducted was strategically selected among the 24 pig rooms. The experimental pig room had a width of 13.3 m, a length of 18.8 m, a height of 2.6 m, and a pit depth of 1 m. The floor of the pig room was a concrete slatted floor. A total of 320 fattening pigs (about 70 kg) were reared within the experimental pig room, and the rearing density was 0.78 m^2 $animal^{-1}$. In the experimental fattening pig house, there were six sidewall slots (1.0 m × 0.4 m), twelve ceiling slots (0.6 m × 0.6 m), two 0.5 m diameter exhaust fans, and three 0.95 m diameter exhaust fans for mechanical ventilation at the sidewall. The exhaust fans were controlled following the three steps for operating the exhaust fans. The first operating condition was that two 0.5 m diameter fans were operated depending on the internal air temperature. When the air temperature inside the pig house increased, the ventilation fans for temperature control were operated. The 0.95 m diameter fan started to operate when the operation rate of the first fans was 100%. Then, all the remaining fans started operating as the third operating condition when the operation rate of the second fans was 100%. In addition, the inlet ducts installed in the longitudinal direction below the ceiling have not been used.

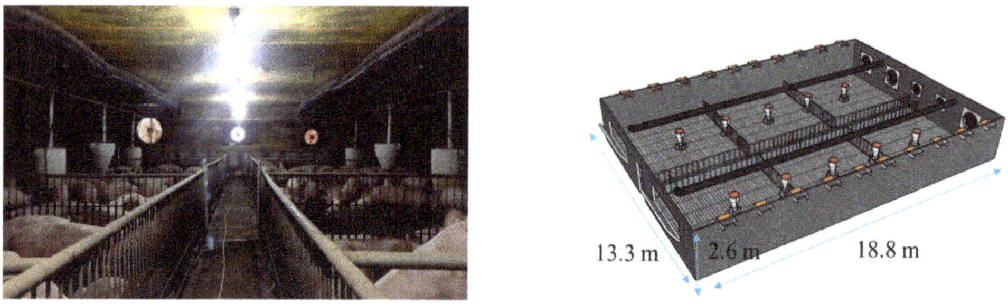

(**a**) (**b**)

Figure 2. Experimental fattening pig house. (**a**) Inside the fattening pig house. (**b**) A schematic of the fattening pig house.

2.2. Recurrent Neural Network

As computer performance develops, machine learning technology is widely used in various fields. In previous studies, machine learning models have been used to analyze animal behavior patterns [19,20,23,32,33], behavior before calving [31,34,35], and the voice of livestock [36] Some studies have also used machine learning to predict dependent variables according to the various environmental factors [17]. Among several machine learning technologies, artificial neural networks (ANNs) have been actively used as methods to accurately predict the dependent variables from independent variables. In this study, the RNN model, which is one of an ANN, was used to predict the missing data inside the fattening pig house. Through RNN, it is possible to use iterative learning through the memory of the ANN. The memory can store information from previous stages of learning and can provide a feedback function that considers information from previous stages as input data. The RNN structure is a form in which a path is added to the general ANN structure to re-insert the output value of the hidden layer at the previous time (t − 1) as the input value of the hidden layer at the next time (t). This structure repeats the process where the result at the current time (t) affects the next time (t + 1). A basic structure of the RNN model is shown in Figure 3a.

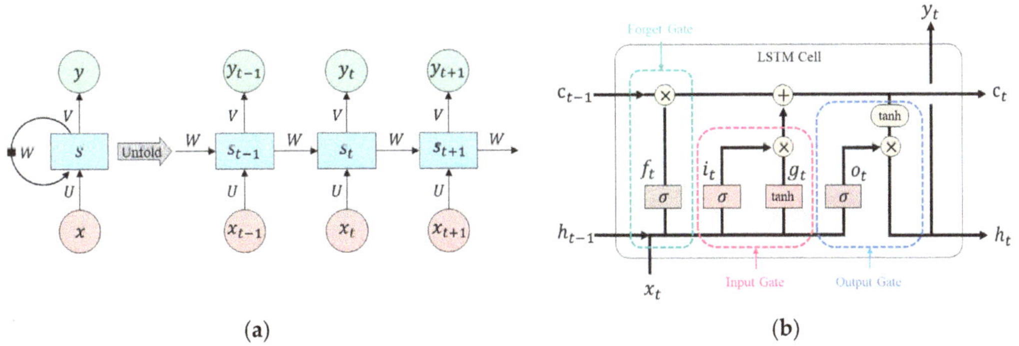

Figure 3. Basic architecture of the RNN and LSTM models [33]. (x: input data, y: output data, s: hidden data, U, V, W: weight, f_t: forget gate, i_t: input gate, o_t: output gate, h: hidden-state, c: cell-state). (**a**) RNN model. (**b**) LSTM model.

Meanwhile, the LSTM method was developed to solve the vanishing gradient problem of the general RNN algorithm [27]. A vanishing gradient is a gradient at a time step far away from time step (t) that has little effect on the learning process when learning data for a long time. While there are limitations to learning long-term dependencies using general RNN models, LSTMs can remember long previous sequences of data. The difference with the LSTM algorithm is a cell with multiple gates. LSTM accepts previous data with an additional operation, so the vanishing gradient problem does not occur. LSTM accomplishes this by using a set of gates (input, forget, and output) that control the flow of information into and out of the memory cell. The input gate can control how much new information is added to the memory cell, the forget gate controls which information is discarded from the memory cell, and the output gate controls how much information is output from the memory cell. A basic structure of the LSTM model is shown in Figure 3b.

2.3. Experimental Procedure

2.3.1. Monitoring the Environmental Factors inside the Fattening Pig House

For the data collection, the environmental factors inside and outside the experimental fattening pig house was monitored from 10 July 2021 to 31 July 2021. In this study, the monitoring data in summer was used as the learning data. The cooling system was not used at the experimental fattening pig house and the exhaust fans were maximally operated during daytime in summer. Otherwise, the internal air temperature of the fattening pig house significantly fluctuated with the ventilation, infiltration, heating, and so on. As shown in Figure 4, the six sensors (HTX 75 series, Dotech Inc., Ansan-si, Gyeonggi-do, Republic of Korea) for monitoring the air temperature and relative humidity were installed at the center of each pen at a height of 1.5 m in the fattening pig house. An ammonia sensor (Multirae-ir, RAEsystem Inc., San Jose, CA, USA) was also installed in front of the center exhaust fan. An electrometer (DW-6092, Newtech Inc., Seoul, Republic of Korea) was installed to measure the current flow when the exhaust fans were running. Then, the ventilation rates of the fattening pig house were calculated using the monitoring data of the electric current in real-time. The air temperature, relative humidity, ammonia concentration, and ventilation rate inside the fattening pig house were logged at one-second intervals. However, the data averaged for five minutes was used when the measured data were used to analyze the environments, develop the RNN model, and validate the developed RNN model. A portable weather station (Watchdog weather station 2900ET, Aurora, IL, USA) was installed on the roof of the management office near the pig house to observe the external weather conditions. External environmental data such as wind speed, wind direction, solar radiation, temperature, relative humidity, and rainfall were monitored at 1-second intervals, and the 5-minute average data was recorded on the equipment. However, the weather data were not used as data for developing the RNN model and were used only to analyze the environment inside the pig house during the model's development process. Descriptive statistics of the monitored environments inside and outside the fattening pig house during the monitoring period were calculated for the environmental analysis of the experimental fattening pig house. The temperature humidity index (THI) [37,38] was also calculated to analyze the heat stress on the pigs. Finally, the monitoring data of air temperature inside the fattening pig house was used to develop the RNN model in this study because the air temperature is one of the most important factors for the environmental control inside the fattening pig houses. The average data of internal air temperature measured at six points was used to develop the RNN model.

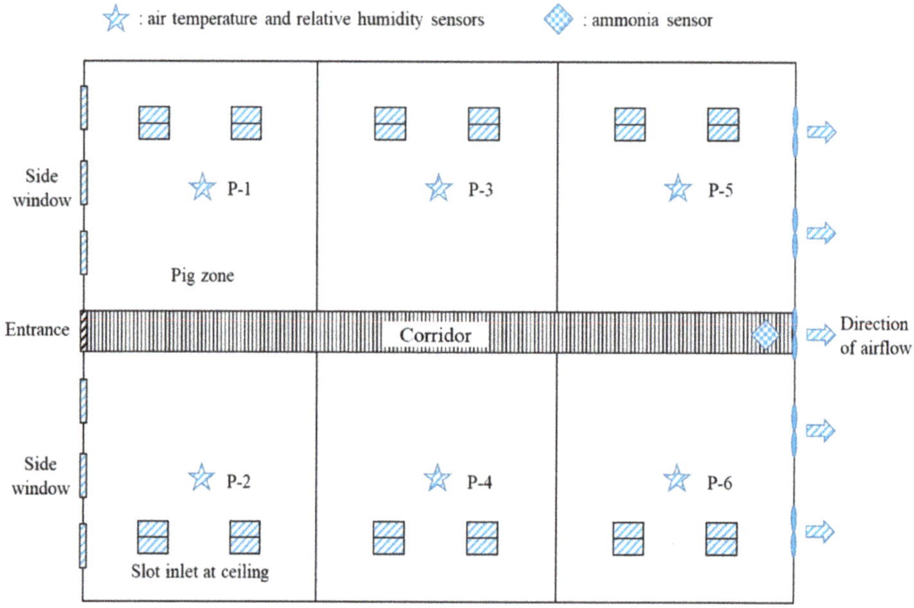

Figure 4. Sensor locations for monitoring the environmental factors inside the experimental fattening pig house.

2.3.2. Design of RNN Models for Expecting Missing Data

In this study, the RNN model was developed to predict the missing data of the air temperature from the measured data inside the fattening pig house. Since it was learned only with time series data of temperature data, it is expected that the developed model will be lightweight and highly applicable to the actual field. As previously mentioned, the measured air temperature from 10 July 2021 to 31 July 2021 was used for the RNN model's development, as shown in Figure 5. Detailed data information during the experimental period for the development and validation of the RNN model is presented in Table 1. The ratio between the lengths of training data and test data was 5:1 [16]. The data from 10 July 2021 to 19 July 2021 were used for learning in sequential order. The data house from 22 July 2021 to 31 July 2021 was used for learning in reverse order. The data house from 20 July 2021 to 21 July 2021 was used as the test data.

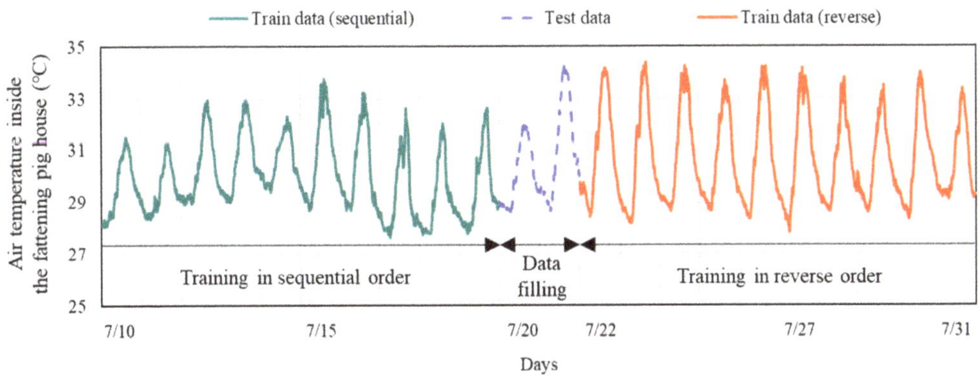

Figure 5. The train and test data set for developing the LSTM model.

Table 1. Date information periods during the experimental period.

Data Set	Monitoring Period	Days	Number of Data
Sequential data learning	10 July~19 July	10 days	2880
Validation	20 July~21 July	2 days	576
Reverse data learning	22 July~31 July	10 days	2880

Gradient vanishing problems can occur when using long-term data as training data for general RNN models. To solve this problem, a single-layer LSTM model suitable for long-term data learning was used. The learning rate was set to 0.01 and the tanh function with high accuracy of the RNN model was used as the activation function as the learning parameter. The commonly used AdamOptimizer was applied in this study, and the loss was trained to minimize the mean squared error. Since the range of learning variables can be different, the range of data according to several variables must be unified from 0 to 1. If the data range is not unified, the model diverges during the training process. Therefore, for successful learning, all training data were normalized in the range of 0 to 1 using the min–max scaler in Equation (1).

$$d_{scaled} = \frac{d - d_{min}}{(d_{max} - d_{min}) + 10^{-7}} \tag{1}$$

where d is learning data, d_{scaled} is scaled learning data, d_{max} is the maximum value of the variable, d_{min} is the minimum value of the variable, and 10^{-7} is the noise term for preventing zero division.

2.3.3. Validation of Accuracy of Developed Models

The developed RNN model was validated by comparing the temperature data predicted with the missing data assumption of the temperature data measured in the pig house. First, Model 1 was developed by learning only air temperature data inside the fattening pig house as a basic model. To improve the accuracy of the RNN model, Model 2 was developed by additionally considering a periodic parameter. The periodic parameter was set from 1 to 288 by dividing a day into five minutes to consider the periodic character of a day. The accuracies of the RNN models were compared by considering the periodic parameter (Table 2).

Table 2. Experimental conditions of learning data for developing the RNN model.

Conditions		Conditions		Number of Cases
Validation of RNN model accuracy	Learning data (independent variables)	Model 1	Monitored air temperature inside the fattening pig house	2
		Model 2	Monitored air temperature inside the fattening pig house and periodic parameter	
	Dependent variable	Missing data of air temperature		1
Total		-		2

To compare the predicted data and the measured data, statistical indices, such as the root-mean-square error (RMSE) and the mean absolute percentage error (MAPE), were calculated using Equations (2) and (3), respectively. The RMSE is commonly used to measure the difference between two pieces of data. However, there is no quantitative criterion for evaluating RMSE, and MAPE is a measure of prediction accuracy as a percentage of error. Therefore, this index was used to assess the accuracy of predicted data using the RNN model.

$$\text{RMSE} = \sqrt{\frac{\sum_{i=1}^{n}(R_i - C_i)^2}{n}} \quad (2)$$

$$\text{MAPE} = \frac{100}{n}\sum_{i=1}^{n}\left|\frac{R_i - C_i}{R_i}\right| \quad (3)$$

where RMSE is root-mean-square error (°C, %), MAPE is mean absolute percentage error (%), n is the total number of data according to time, R_i is the measured data at a specific time, and C_i is the predicted data at a specific time.

2.3.4. Comparative Evaluation of the RNN Model according to the Order of Learning Data and the Length of the Missing Data

A total of 27 cases of experimental conditions for learning data was used to develop the RNN model, as shown in Table 3. Previous studies have shown that when developing RNN models, training time-series data in reverse order can generally improve the accuracy of RNN models [39–41]. Therefore, in this study, the RNN model was developed by learning time series data in sequential and reverse orders. The bidirectional model was developed by combining the sequential model and reverse model for improving the accuracy of the RNN models. In general, the longer the length of the missing data, the lower the accuracy of the RNN model. Therefore, to identify this trend and analyze the accuracy of the RNN model, the accuracies of the RNN models were compared according to the length of the missing data at 1, 2, 3, 6, 9, 12, 24, 36, and 48 h.

The predicted results of the bidirectional model were calculated by adding the predicted results of the sequential model and reverse model considering the weight values, as shown in Equation (4). The weight values for the predicted results of the sequential model and reverse model have gradients, as shown in Figure 6. It was expected that the sequential model has high accuracy on the front of the missing data. Otherwise, it was expected that the reverse model has high accuracy on the back of the missing data. Finally, it was expected that the accuracy of the bidirectional model is improved by comparing it with the sequential and reverse models. Finally, the accuracies of the RNN models were developed and compared according to the sequential, reverse, and bidirectional models.

$$z_i = a_i x_i + b_i y_i \quad (4)$$

where z_i is the predictive value of the bidirectional model, x_i is the predictive value of the sequential model, y_i is the predictive value of the reverse model, a_i is the weight of x_i, b_i is the weight of y_i, and i is the sequence number of the missing data.

Table 3. Experimental conditions of learning data for developing the RNN model.

Conditions	Conditions		Number of Cases
Case study for enhancing the accuracy of the RNN model	Learning data (independent variables)	Monitored air temperature inside the fattening pig house and periodic parameter	1
	Dependent variable	Missing data of air temperature	1
	Order of learning data	Sequential, reverse, and bidirectional order	3
	The length of the missing data	1, 2, 3, 6, 9, 12, 24, 36, and 48 h	9
Total		-	27

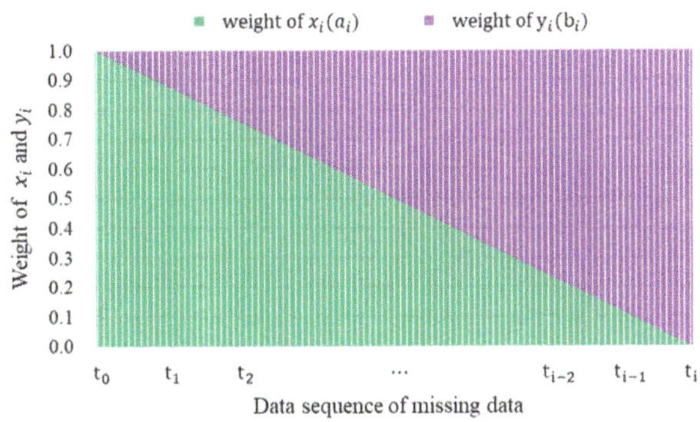

Figure 6. Weight of x_i and y_i according to the data sequence of the missing data.

3. Results and Discussions

3.1. Analysis of the Internal Environment of the Experimental Pig House

Environmental data inside and outside the experimental fattening pig house were monitored to analyze the environmental problems of the fattening pig house, to accumulate the learning data for the development of the RNN models, and to validate the developed RNN model. Descriptive statistical analysis was conducted to analyze the temperature and relative humidity characteristics inside the pig house. The box plots shown in Figure 7 describe the distributions of the air temperature and relative humidity inside the experimental fattening pig house. Figure 8 shows the measured environments inside and outside the fattening pig house during the experimental period. The average air temperature inside the fattening pig house was higher than the outside temperature due to solar radiation, insulation of the wall, heat generation of the pigs, etc. The internal air temperature was higher than 34 °C during daytime and the external air temperature was higher than 36 °C. It was expected that pigs had high-temperature stress because of the high air temperature inside the fattening pig house during the experimental period. The relative humidity inside the fattening pig house fluctuated from 50 to 90% due to the temperature difference between day and night. According to the measurement points, the analysis showed that relatively low air temperatures (28.7 °C and 29.0 °C) were measured near the slot opening at the sidewall (P-1 and P-2), whereas relatively high air temperatures (30.6 °C and 30.0 °C) were measured at P-5 and P-6, which were located near the exhaust fans. The airflow from the slot opening at the sidewall (inlets) to the exhaust fans (outlets) caused the distribution of the internal airflow.

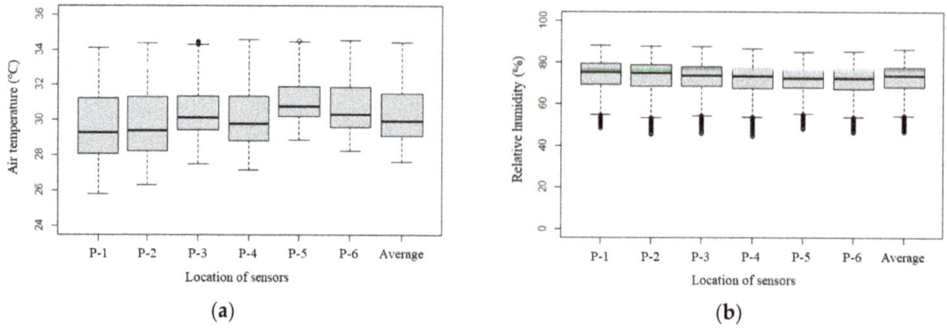

Figure 7. A box plot with the 1.5 IQR (Interquartile range) of the measured environments inside the fattening pig house. (**a**) Air temperature. (**b**) Relative humidity.

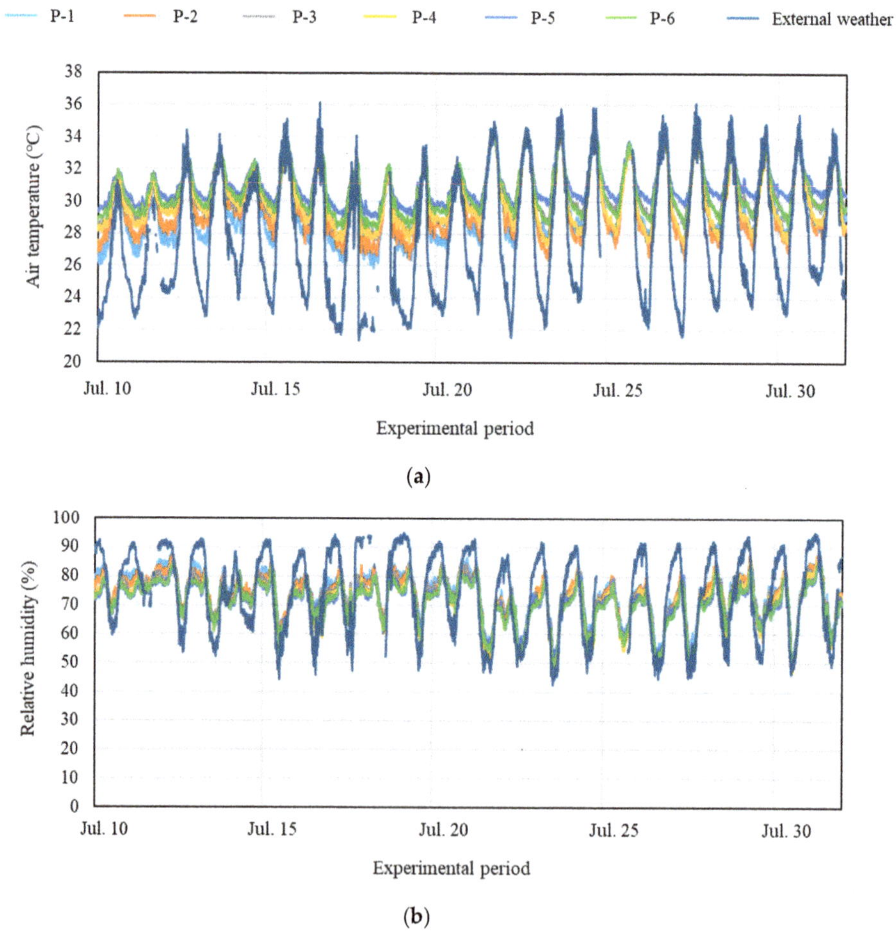

Figure 8. The results of measured air temperature and relative humidity inside the fattening pig house. (**a**) Air temperature. (**b**) Relative humidity.

Based on the results, it was expected that the pigs had high-temperature stress. The THI was additionally calculated by simultaneously considering the air temperature and relative humidity, as shown in Figure 9a. The ammonia concentration and ventilation rate inside the fattening pig house were also presented in Figure 9b,c. Accordingly, the measured THI ranged from 78 to 86 during the experimental period. This means that the internal environments were under the alert and danger sections for high-temperature stress on pigs. Although the ventilation rates were maximum in the daytime, the internal air temperature of the fattening pig house was higher than 34 °C. Therefore, it is recommended to install additional cooling systems to relieve the high-temperature stress in summer.

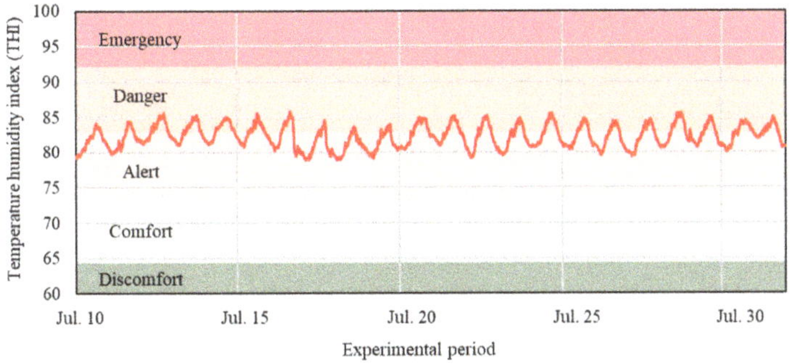

(a)

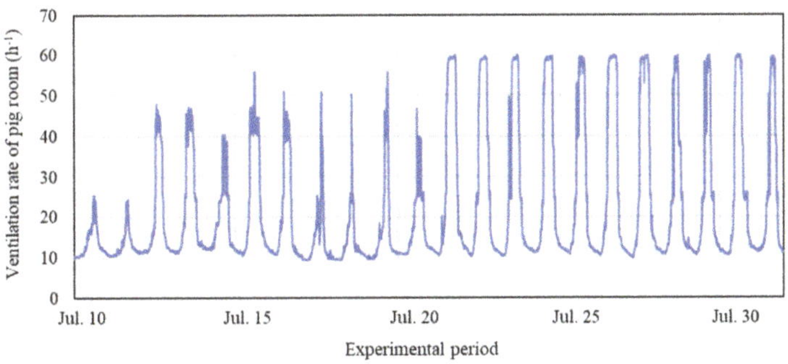

(b)

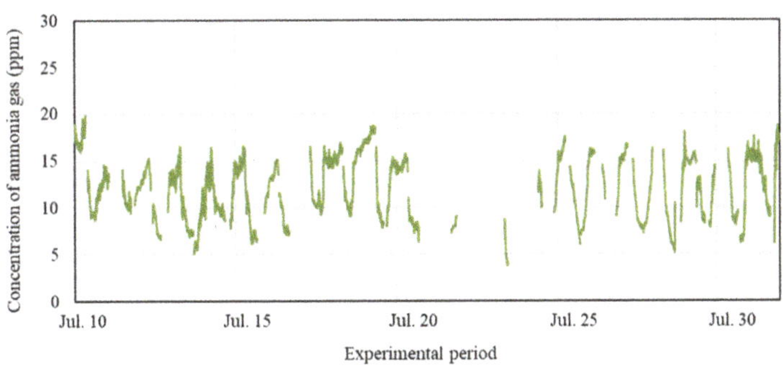

(c)

Figure 9. The evaluation results of the breeding environment inside the experimental pig house. (**a**) Temperature humidity index. (**b**) Ventilation rate of the pig room. (**c**) Concentration of ammonia gas inside the pig room.

In the case of ammonia concentration inside the fattening pig house, a relatively high range of 5 to 20 ppm was measured. The ammonia concentration fluctuated with the change in ventilation rates during the day and night. The ammonia concentration increased as the ventilation rate decreased at night. On the other hand, the concentration of ammonia gas was lowered due to the high ventilation rate in the daytime. As a result of the monitoring in the fattening pig house, it was identified that internal environments were poor with a high air temperature, relative humidity, and ammonia concentration. Due to the poor environment inside fattening pig houses, it is hard to accumulate reliable long-term data on the fattening pig house due to sensor failure. Therefore, it is necessary to conduct research for filling in the missing data on the environments inside livestock houses. In this study, the research for the data filling of the missing data as the basic research was conducted using air temperature data, which is one of the most important factors for the environmental control inside fattening pig houses.

3.2. Validation of the Accuracy of the Developed Models

The developed RNN model in this study was validated by comparing air temperature data predicted by the RNN model with the assumed missing data. The air temperature predicted by the RNN model was presented in Figure 10, according to the periodic parameter. The RMSE and MAPE as statistical indices were calculated for the quantitative comparison, as shown in Table 4. When the RNN model was developed by learning only internal air temperature, the value of the RMSE and MAPE were 1.92 °C and 4.70%, respectively. As shown in Figure 10a, the RNN model could not accurately predict the internal air temperature when only air temperature data was considered as the learning data. Therefore, additional processes, such as considering a multi-layer model and adding other parameters, were necessary to improve the accuracy of the RNN model for predicting the missing data of the air temperature inside the fattening pig house. However, the computing loads and learning time could increase for the development of the multi-layer RNN model. Therefore, the RNN model additionally considers the periodic parameter as the learning data. When the RNN model was developed by considering the periodic parameter, the value of the RMSE and MAPE were 1.41 °C and 3.55%, respectively. The accuracy of the RNN model was improved with a 1.15% decrease in the error rate. The tendency of air temperature predicted by the RNN model considering the periodic parameter was fitted with the tendency of air temperature measured in the field experiment. It was suitable to use the developed RNN model for predicting the air temperature inside the fattening pig house within a 3.5% error rate. The developed RNN model in this study could be applied to predict the internal air temperature when the monitoring sensors are not working. Furthermore, the RNN models were expected to be highly applicable because they could be continuously improved the RNN model through learning from the monitoring data.

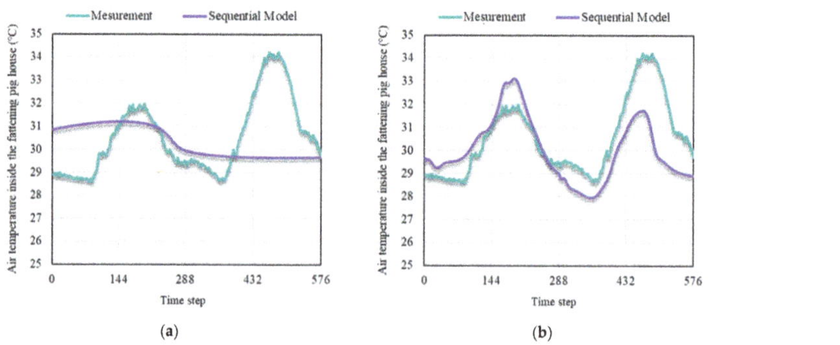

Figure 10. A comparative analysis of the predictive air temperature inside the fattening pig house according to the periodic parameter. (**a**) Learning without the periodic parameter. (**b**) Learning with the periodic parameter.

Table 4. Validation of the RNN model for predicting the missing data, according to learning parameters.

Statistical Indices	Learning without the Periodic Parameter	Learning with the Periodic Parameter
RMSE (°C)	1.92	1.41
MAPE (%)	4.70	3.55

3.3. A Comparative Evaluation of the RNN Models according to the Order of Learning Data

The RNN model was developed by learning time series data in sequential and reverse orders. The bidirectional model was additionally developed by combining the sequential model and reverse model to improve the accuracy of the RNN models. The air temperature predicted by the sequential, reverse, and bidirectional models were presented in Figure 11. A comparative analysis was also conducted using the statistical indices of the RMSE and MAPE, according to the order of learning data and the length of the missing data. The increase in the RNN model's accuracy was presented in Table 5, according to the order of learning data.

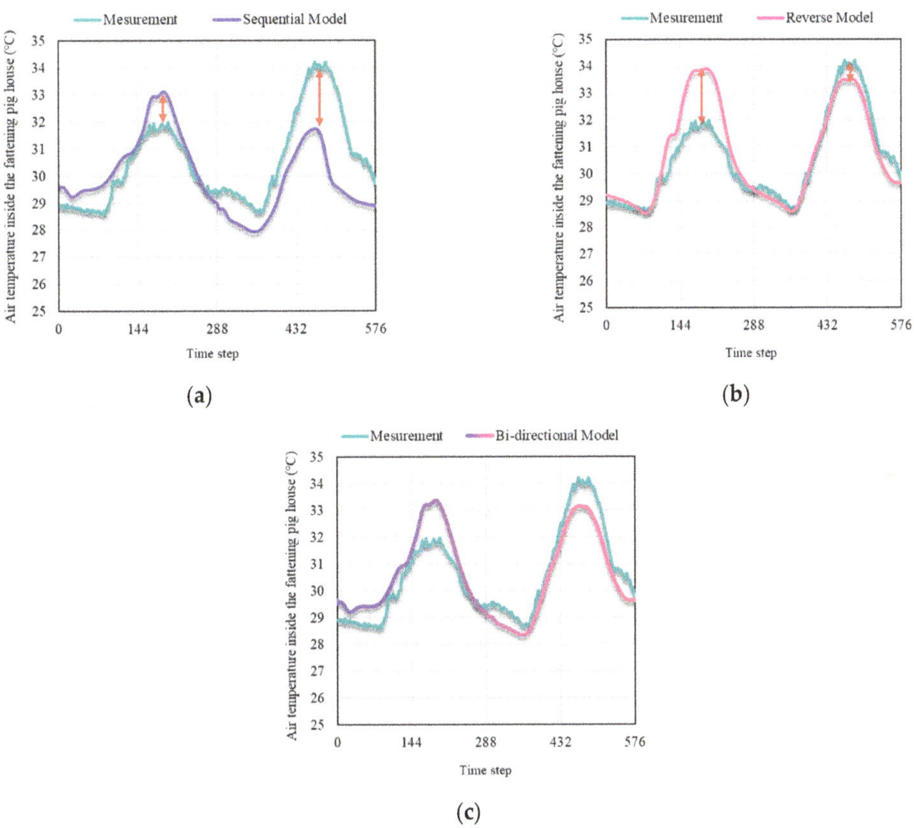

Figure 11. A comparative analysis of the predictive air temperature inside the fattening pig house according to the order of learning data. (**a**) Sequential model. (**b**) Reverse model. (**c**) Bidirectional model.

Table 5. An increase in the RNN model's accuracy according to the order of learning data.

MAPE (%)	The Length of the Missing Data								
	1 hr	2 hr	3 hr	6 hr	9 hr	12 hr	24 hr	36 hr	48 hr
Sequential model (A)	2.15	2.25	2.89	2.94	2.77	2.33	2.14	2.81	3.55
Bidirectional model (B)	0.24	0.77	0.43	0.93	1.06	1.12	1.48	2.60	2.64
Increase in accuracy (A-B)	1.91	1.48	2.46	2.01	1.71	1.21	0.66	0.21	0.91

As shown in Figure 11a, the sequential model was relatively more accurate at the 0–144 time step than at the 144–288 time step. On the contrary, the reverse model was relatively more accurate at the 144–288 time step than at the 0–144 time step, as shown in Figure 11b. As a result of the quantitative comparison (Table 6), the reverse model was more accurate than the sequential model. These results were in agreement with the results of previous studies [39–41]. The sequential model was able to accurately predict the missing data of the internal air temperature with an error rate of 2.14–3.55%, according to the length of the missing data. The reverse model was accurate with an error rate of 0.72–3.11%.

Table 6. The RNN model's accuracy, according to the order of learning data and the length of the missing data.

Sequential order	The Length of the Missing Data								
	1 hr	2 hr	3 hr	6 hr	9 hr	12 hr	24 hr	36 hr	48 hr
RMSE (°C)	0.63	0.67	0.84	0.85	0.82	0.73	0.86	0.87	1.41
MAPE (%)	2.15	2.25	2.89	2.94	2.77	2.33	2.14	2.81	3.55
Reverse order	**Sequence length for the LSTM model**								
	1 hr	2 hr	3 hr	6 hr	9 hr	12 hr	24 hr	36 hr	48 hr
RMSE (°C)	0.55	0.22	0.65	0.41	0.37	0.72	1.17	1.07	0.83
MAPE (%)	1.88	0.72	2.21	1.27	1.13	1.93	3.11	2.59	1.85
Bidirectional order	**Sequence length for the LSTM model**								
	1 hr	2 hr	3 hr	6 hr	9 hr	12 hr	24 hr	36 hr	48 hr
RMSE (°C)	0.08	0.25	0.14	0.30	0.35	0.36	0.51	0.87	0.99
MAPE (%)	0.24	0.77	0.43	0.93	1.06	1.12	1.48	2.60	2.64

The accuracy of the bidirectional model was improved and there was a decrease in the error rate of 0.21–2.46% compared with the accuracy of the sequential model. Especially, the shorter the length of the missing data, the higher the accuracy of the bidirectional model. When the length of the missing data was an hour, the values of the RMSE and MAPE for the sequential model were 0.63 °C and 2.15%, respectively. When the length of the missing data was an hour, the values of the RMSE and MAPE for the bidirectional model were 0.08 °C and 0.24%, respectively. When the length of the missing data was 12 h, the values of the RMSE and MAPE for the sequential model were 0.73 °C and 2.33%, respectively. When the length of the missing data was 12 h, the values of the RMSE and MAPE for the bidirectional model were 0.51 °C and 1.48%, respectively. When the length of the missing data was 24 h, the values of the RMSE and MAPE for the sequential model were 1.41 °C and 3.55%, respectively. When the length of the missing data was 24 h, the values of the RMSE and MAPE for the bidirectional model were 0.99 °C and 2.64%, respectively. The accuracies of the bidirectional models with the length of the missing data of 1, 12, and 24 h were improved, as error rates of 1.91, 1.21, and 0.91% were decreased by comparing with the accuracy of the sequential model, respectively.

The accuracy of the bidirectional model was improved by comparing it with the accuracy of the sequential model. Therefore, the missing data could be more accurately predicted using the bidirectional model when the air temperature data are accumulated after the missing data. Furthermore, the longer the length of the missing data, the higher the applicability of the bidirectional model because the accuracy of the RNN model was lower when the length of the missing data was longer.

3.4. A Comparative Evaluation of the RNN Models according to the Length of the Missing Data

In this study, the accuracies of the developed RNN models were evaluated according to the length of the missing data. Generally, the evaluation of the model should state that a larger number of missing data will result in a lower accuracy of the RNN model. The RMSE and MAPE values of the sequential, reverse, and bidirectional models, according to the length of the missing data, were presented in Figure 12. As the result of the comparative analysis, the tendencies of the sequential and reverse models were not clear, as shown in Figure 12a,b. However, the tendency of the bidirectional model was clear, as shown in Figure 12c. The bidirectional model developed in this study could accurately predict the missing data of air temperature inside the fattening pig house. When the length of the missing data was 6 h, the values of the RMSE and MAPE for the bidirectional model were 0.30 °C and 0.93%, respectively. When the length of the missing data was 24 h, the values of the RMSE and MAPE for the bidirectional model were 0.51 °C and 1.48%, respectively. When the length of the missing data was 48 h, the values of the RMSE and MAPE for the bidirectional model were 0.99 °C and 2.64%, respectively. Therefore, when the length of the missing data was less than 6 h, the bidirectional model could predict the missing data of the internal air temperature within a 1% error rate. When the length of the missing data was less than 24 and 48 h, the bidirectional model could predict the missing data of the internal air temperature within 1.5% and 3% error rates, respectively.

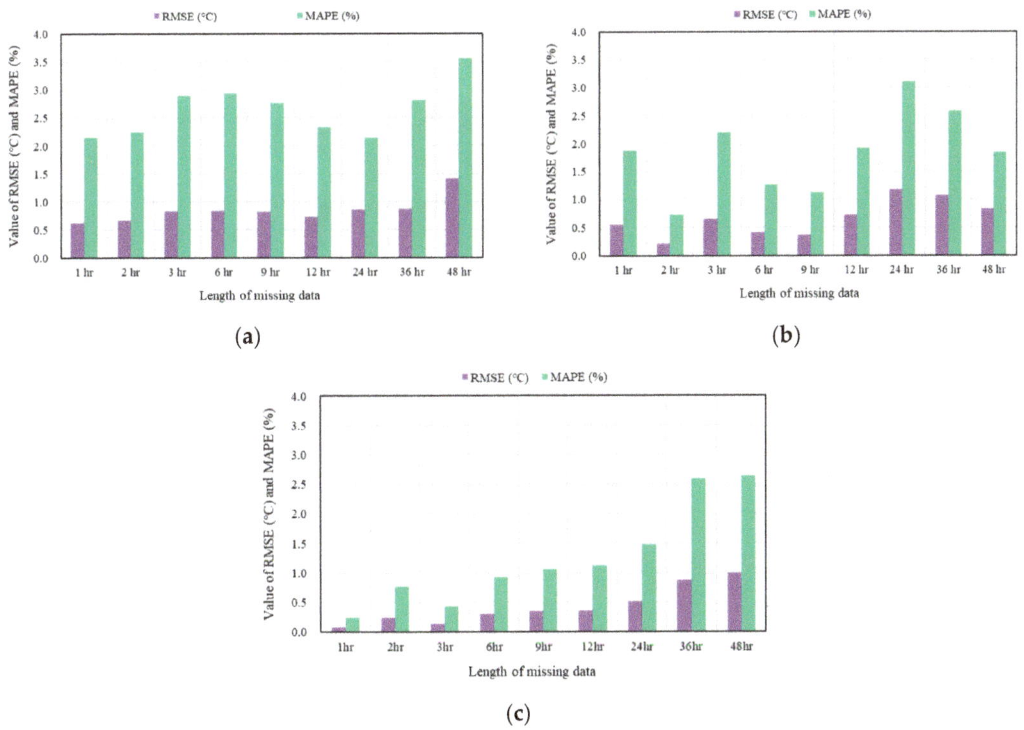

Figure 12. A comparative analysis of the model accuracies according to the length of the missing data. (**a**) Sequential model. (**b**) Reverse model. (**c**) Bidirectional model.

Finally, the RNN model developed in this study could be usefully applied to monitor the internal environments of the fattening pig houses, where it is difficult to accumulate reliable data for the long term due to the poor environments such as high air temperature, relative humidity, and ammonia concentration. RNN models have the advantage that the accuracy of RNN models can be continuously improved by learning from monitoring data. The RNN model could be improved to learn the data of several factors, such as external weather and ventilation rates, which directly impact the internal environments. Since the RNN model developed in this study was developed by learning only temperature data, it is highly applicable to livestock facilities where temperature-based control is used.

4. Conclusions

In this study, the developed RNN models could predict the missing data of air temperature inside the experimental fattening pig house. First, a field experiment was conducted to identify the environmental problems of the fattening pig house. During the field experiment, air temperature, relative humidity, ventilation rate, ammonia concentration inside the fattening pig house, and the external weather, were monitored. From these, RNN models for data filling of the missing data were developed and validated. Based on analyzed monitoring data, it was found that the pigs were expected to experience high-temperature stress. Although the ventilation rates were maximum in the daytime, the air temperature and THI inside the fattening pig house were higher than 34 °C and 85 during the experimental period, respectively. Therefore, it is necessary to conduct research for filling in the missing data on the environments inside livestock houses.

In this study, the research for the data filling of the missing data as the basic research was conducted using air temperature data, which is one of the most important factors for environmental control inside fattening pig houses.

The RNN model for predicting the missing data by learning air temperature data is one of the most important factors for environmental control inside fattening pig houses. Some of the measured air temperature data inside the fattening pig house were assumed to be the missing data. As a result of validating the accuracy of the RNN model considering the periodic parameter, the RNN model for predicting the air temperature inside the fattening pig house was accurate within a 3.5% error rate. Therefore, the RNN model was useful for predicting data to replace missing data when it occurs.

The accuracies of the RNN models were evaluated according to the order of learning data. Based on the result, the accuracy of the bidirectional model was improved by comparing it with the accuracy of the sequential model. Therefore, the missing data could be more accurately predicted using the bidirectional model when air temperature data are accumulated after the missing data. Furthermore, the accuracies of the RNN models were compared and analyzed according to the length of the missing data. The longer the length of the missing data, the lower the accuracy of the RNN model. When the length of the missing data was less than 26, 4, and 48 h, the bidirectional model could predict the missing data of the internal air temperature within 1%, 1.5%, and 3% error rates, respectively.

The main contribution of this study is the development of the RNN model for predicting the time series missing data of the air temperature inside fattening pig houses. The RNN models also have an advantage because the accuracies of the RNN models could be continuously improved by learning the measured data. Furthermore, the RNN models developed in this study have high applicability because the RNN model was developed by learning only air temperature data. In the future, it is possible to improve the accuracy of the RNN model by learning the data of several factors, such as external weather and ventilation rates, which directly impact the internal environments. Research predicting the data of ammonia concentration data is necessary because the ammonia sensor is more vulnerable to poor environments inside fattening pig houses.

Author Contributions: Conceptualization, J.-g.K. and I.-b.L.; data curation, S.-y.L.; methodology, J.-g.K. and S.-y.L.; software, J.-g.K.; supervision, I.-b.L.; visualization, J.-g.K.; writing—original draft, J.-g.K.; writing—review and editing, I.-b.L. All authors have read and agreed to the published version of the manuscript.

Funding: This project was funded by the Korea Institute of Planning and Evaluation for Technology in Food, Agriculture, and Forestry (IPET) through the Livestock Industrialization Technology Development Program, which was funded by the Ministry of Agriculture, Food, and Rural Affairs (MAFRA) (321085-5).

Institutional Review Board Statement: Not applicable.

Data Availability Statement: All data will be made available upon request to the correspondent author's email with appropriate justification.

Conflicts of Interest: The authors declare no conflict of interest.

References

1. Ministry of Agriculture Food and Rural Affairs. 2021. Available online: www.mafra.go.kr (accessed on 12 December 2022).
2. Korea Labor Institute. *Effects of Smart Farm Activation Policy on Employment*; KLI: Sejong, Republic of Korea, 2019; p. 55.
3. Seo, I.-H. Development of Wearable Device for Monitoring Working Environment in Pig House. *J. Korean Soc. Agric. Eng.* **2020**, *62*, 71–81.
4. Jinubala, V.; Lawrance, R. Analysis of Missing Data and Imputation on Agriculture Data With Predictive Mean Matching Method. *Int. J. Sci. Appl. Inf. Technol.* **2016**, *5*, 1–4.
5. Lokupitiya, R.S.; Lokupitiya, E.; Paustian, K. Comparison of missing value imputation methods for crop yield data. *Env. Off. J. Int. Env. Soc.* **2006**, *17*, 339–349. [CrossRef]
6. Zakaria, W.; Salleh, M.Z. Determination of the best single imputation algorithm for missing rainfall data treatment. *J. Qual. Meas. Anal.* **2016**, *12*, 79–87.
7. Staub, B.; Hasler, A.; Noetzli, J.; Delaloye, R. Gap-filling algorithm for ground surface temperature data measured in permafrost and periglacial environments. *Permafr. Periglac. Process.* **2017**, *28*, 275–285. [CrossRef]
8. Rubin, L.H.; Witkiewitz, K.; Andre, J.S.; Reilly, S. Methods for handling missing data in the behavioral neurosciences: Don't throw the baby rat out with the bath water. *J. Undergrad. Neurosci. Educ.* **2007**, *5*, A71.
9. Ferrari, G.T.; Ozaki, V. Missing data imputation of climate datasets: Implications to modeling extreme drought events. *Rev. Bras. Meteorol.* **2014**, *29*, 21–28. [CrossRef]
10. Afrifa-Yamoah, E.; Mueller, U.A.; Taylor, S.; Fisher, A. Missing data imputation of high-resolution temporal climate time series data. *Meteorol. Appl.* **2020**, *27*, e1873. [CrossRef]
11. Xie, Q.; Ni, J.-Q.; Li, E.; Bao, J.; Zheng, P. Sequential air pollution emission estimation using a hybrid deep learning model and health-related ventilation control in a pig building. *J. Clean. Prod.* **2022**, *371*, 133714. [CrossRef]
12. Wang, Y.; Zheng, W.; Li, B. A modified discrete grey model with improved prediction performance for indoor air temperatures in laying hen houses. *Biosyst. Eng.* **2022**, *223*, 138–148. [CrossRef]
13. Boomgard-Zagrodnik, J.P.; Brown, D.J. Machine learning imputation of missing Mesonet temperature observations. *Comput. Electron. Agric.* **2022**, *192*, 106580. [CrossRef]
14. Hamzah, F.B.; Mohd Hamzah, F.; Mohd Razali, S.F.; Jaafar, O.; Abdul Jamil, N. Imputation methods for recovering streamflow observation: A methodological review. *Cogent Environ. Sci.* **2020**, *6*, 1745133. [CrossRef]
15. Moon, T.; Lee, J.W.; Son, J.E. Accurate Imputation of Greenhouse Environment Data for Data Integrity Utilizing Two-Dimensional Convolutional Neural Networks. *Sensors* **2021**, *21*, 2187. [CrossRef]
16. Song, W.; Gao, C.; Zhao, Y.; Zhao, Y. A time series data filling method based on LSTM—Taking the stem moisture as an example. *Sensors* **2020**, *20*, 5045. [CrossRef] [PubMed]
17. Demmers, T.G.; Cao, Y.; Gauss, S.; Lowe, J.C.; Parsons, D.J.; Wathes, C.M. Neural predictive control of broiler chicken and pig growth. *Biosyst. Eng.* **2018**, *173*, 134–142. [CrossRef]
18. Lee, S.-Y.; Lee, I.-B.; Yeo, U.-H.; Kim, J.-G.; Kim, R.-W. Machine Learning Approach to Predict Air Temperature and Relative Humidity inside Mechanically and Naturally Ventilated Duck Houses: Application of Recurrent Neural Network. *Agriculture* **2022**, *12*, 318. [CrossRef]
19. Li, H.; Cryer, S.; Acharya, L.; Raymond, J. Video and image classification using atomisation spray image patterns and deep learning. *Biosyst. Eng.* **2020**, *200*, 13–22. [CrossRef]
20. Liu, D.; Oczak, M.; Maschat, K.; Baumgartner, J.; Pletzer, B.; He, D.; Norton, T. A computer vision-based method for spatial-temporal action recognition of tail-biting behaviour in group-housed pigs. *Biosyst. Eng.* **2020**, *195*, 27–41. [CrossRef]
21. Moon, T.; Choi, H.Y.; Jung, D.H.; Chang, S.H.; Son, J.E. Prediction of CO2 Concentration via Long Short-Term Memory Using Environmental Factors in Greenhouses. *Hortic. Sci. Technol.* **2020**, *38*, 201–209.
22. Wang, L.; Zhang, T.; Wang, X.; Jin, X.; Xu, J.; Yu, J.; Zhang, H.; Zhao, Z. An approach of improved Multivariate Timing-Random Deep Belief Net modelling for algal bloom prediction. *Biosyst. Eng.* **2019**, *177*, 130–138. [CrossRef]

23. Wu, D.; Wu, Q.; Yin, X.; Jiang, B.; Wang, H.; He, D.; Song, H. Lameness detection of dairy cows based on the YOLOv3 deep learning algorithm and a relative step size characteristic vector. *Biosyst. Eng.* **2020**, *189*, 150–163. [CrossRef]
24. Sherstinsky, A. Fundamentals of recurrent neural network (RNN) and long short-term memory (LSTM) network. *Phys. D Nonlinear Phenom.* **2020**, *404*, 132306. [CrossRef]
25. Pham, V.; Bluche, T.; Kermorvant, C.; Louradour, J. Dropout improves recurrent neural networks for handwriting recognition. In Proceedings of the 2014 14th International Conference on Frontiers in Handwriting Recognition, Hersonissos, Greece, 1–4 September 2014; pp. 285–290.
26. Khaki, S.; Wang, L.; Archontoulis, S.V. A cnn-rnn framework for crop yield prediction. *Front. Plant Sci.* **2020**, *10*, 1750. [CrossRef] [PubMed]
27. Schmidhuber, J.; Hochreiter, S. Long short-term memory. *Neural Comput.* **1997**, *9*, 1735–1780.
28. Cho, K.; Van Merriënboer, B.; Gulcehre, C.; Bahdanau, D.; Bougares, F.; Schwenk, H.; Bengio, Y. Learning phrase representations using RNN encoder-decoder for statistical machine translation. *arXiv* **2014**, arXiv:1406.1078.
29. Chung, J.; Gulcehre, C.; Cho, K.; Bengio, Y. Empirical evaluation of gated recurrent neural networks on sequence modeling. *arXiv* **2014**, arXiv:1412.3555.
30. Moon, T.; Ahn, T.I.; Son, J.E. Forecasting root-zone electrical conductivity of nutrient solutions in closed-loop soilless cultures via a recurrent neural network using environmental and cultivation information. *Front. Plant Sci.* **2018**, *9*, 859. [CrossRef]
31. Borchers, M.; Chang, Y.; Proudfoot, K.; Wadsworth, B.; Stone, A.; Bewley, J. Machine-learning-based calving prediction from activity, lying, and ruminating behaviors in dairy cattle. *J. Dairy Sci.* **2017**, *100*, 5664–5674. [CrossRef]
32. Peng, Y.; Kondo, N.; Fujiura, T.; Suzuki, T.; Yoshioka, H.; Itoyama, E. Classification of multiple cattle behavior patterns using a recurrent neural network with long short-term memory and inertial measurement units. *Comput. Electron. Agric.* **2019**, *157*, 247–253. [CrossRef]
33. Zhao, K.; Jin, X.; Ji, J.; Wang, J.; Ma, H.; Zhu, X. Individual identification of Holstein dairy cows based on detecting and matching feature points in body images. *Biosyst. Eng.* **2019**, *181*, 128–139. [CrossRef]
34. Keceli, A.S.; Catal, C.; Kaya, A.; Tekinerdogan, B. Development of a recurrent neural networks-based calving prediction model using activity and behavioral data. *Comput. Electron. Agric.* **2020**, *170*, 105285. [CrossRef]
35. Peng, Y.; Kondo, N.; Fujiura, T.; Suzuki, T.; Ouma, S.; Yoshioka, H.; Itoyama, E. Dam behavior patterns in Japanese black beef cattle prior to calving: Automated detection using LSTM-RNN. *Comput. Electron. Agric.* **2020**, *169*, 105178. [CrossRef]
36. Milone, D.H.; Galli, J.R.; Cangiano, C.A.; Rufiner, H.L.; Laca, E.A. Automatic recognition of ingestive sounds of cattle based on hidden Markov models. *Comput. Electron. Agric.* **2012**, *87*, 51–55. [CrossRef]
37. NRC. *Effect of Environment on Nutrient Requirements of Domestic Animals*; National Research Council: Washington, DC, USA, 1981.
38. St-Pierre, N.; Cobanov, B.; Schnitkey, G. Economic losses from heat stress by US livestock industries. *J. Dairy Sci.* **2003**, *86*, E52–E77. [CrossRef]
39. Srivastava, N.; Mansimov, E.; Salakhudinov, R. Unsupervised learning of video representations using lstms. In Proceedings of the International conference on machine learning, Lille, France, 6–11 July 2015; pp. 843–852.
40. Sutskever, I.; Vinyals, O.; Le, Q.V. Sequence to sequence learning with neural networks. *Adv. Neural Inf. Process. Syst.* **2014**, *27*, 3104–3112.
41. Vinyals, O.; Bengio, S.; Kudlur, M. Order matters: Sequence to sequence for sets. *arXiv* **2015**, arXiv:1511.06391.

agriculture

Article

Experimental and Numerical Simulations of a Solar Air Heater for Maximal Value Addition to Agricultural Products

Zuhair Qamar [1,*], Anjum Munir [2,*], Timothy Langrish [3], Abdul Ghafoor [1] and Muhammad Tahir [4]

1 Department of Farm Machinery & Power, University of Agriculture Faisalabad, Punjab 38000, Pakistan
2 Department of Energy Systems Engineering, University of Agriculture Faisalabad, Punjab 38000, Pakistan
3 School of Chemical and Biomolecular Engineering, The University of Sydney, Camperdown, NSW 2006, Australia
4 Department of Agronomy, University of Agriculture Faisalabad, Punjab 38000, Pakistan
* Correspondence: dr.zqamar@gmail.com (Z.Q.); anjum.munir@uaf.edu.pk (A.M.)

Abstract: Agriculture is the backbone of Pakistan's economy. Currently, the agricultural sector is facing many challenges, especially post-harvest losses, which result in lower yield and profitability. These losses may be reduced by developing indigenous post-harvest processing technologies, such as drying out of agricultural products to enhancement of their sustainability and reduce transportation costs. The country has the advantage of an abundant amount of solar insulation, which can be effectively utilized to operate post-harvest machinery, particularly solar heaters and dryers. Currently, conventional solar heaters face challenges due to lower efficiencies. Therefore, in this study, a solar air heater (SAH), having a size 1220 × 610 × 65 mm, was designed and developed to be connected to a milk powder spray drying system for converting raw milk to powder. Computational fluid dynamics (CFD) were used to anticipate air flow and temperature distribution across the SAH to evaluate optimal performance parameters. An air mass flow rate of 0.01 kgs^{-1} was required, with the CFD predicting an outlet temperature of 82 °C compared with the experimental observation of 73 °C. The pressure drop across the SAH was recorded to be 0.0000434 bar at this flow rate, which is negligible, with the SAH operating near ambient pressure. The overall heat loss coefficient for convection was 2.27 $W{\cdot}m^{-2}{\cdot}K^{-1}$. The energy losses from the SAH were 37% and the useful energy was 63% of the total energy provided to the SAH. The breakeven point of SAH at a minimum of 4 h of daily usage was 3700 h or 2.5 years. The solar air heater used as a preheater for a spray dryer can save 30 $PKR{\cdot}kWh^{-1}$ of energy.

Keywords: solar air heater (SAH); computational fluid dynamics; performance optimization; milk powder; agricultural drying; post-harvest losses

Citation: Qamar, Z.; Munir, A.; Langrish, T.; Ghafoor, A.; Tahir, M. Experimental and Numerical Simulations of a Solar Air Heater for Maximal Value Addition to Agricultural Products. *Agriculture* **2023**, *13*, 387. https://doi.org/10.3390/agriculture13020387

Academic Editor: John M. Fielke

Received: 19 December 2022
Revised: 29 January 2023
Accepted: 1 February 2023
Published: 6 February 2023

1. Introduction

Important concerns that require a linkage strategy include preventing climate change and ensuring a sustainable food supply for an expanding global population. The burden on the agricultural industry and food production has significantly increased because of the growing global population, restricted access to fossil fuels, and higher pricing. Renewable technology in the agricultural sector improves dependability and efficiency by reducing carbon emissions due to the use of fossil-fuel. The technologies adopted in agriculture may include distributed electricity generation, agricultural cultivation greenhouses, space cooling and heating application, the desalination of saltwater, water pumping and irrigation systems, the drying of agricultural products, solar powered agricultural machinery, and farm robots [1].

Pakistan is an agricultural country, with about 67 % of the country's population being directly or indirectly linked to agriculture [2]. The agriculture sector of Pakistan is facing many challenges that are significantly affecting the farming community. The country is facing a severe energy crisis that is having consequential severe effects on the agriculture sector. Therefore, renewable energy-operated technologies for the processing of agricultural commodities are urgently needed.

Among the largest milk-producing countries, Pakistan ranks as the fifth in the world. Milk production obtained from buffalos and cows was 55 million tons in 2019–2020 [3], compared with the maximum breed potential of 110 million tons. Several factors currently limit the opportunities in the dairy industry, i.e., low returns, restrictions for credit, the low genetic potential of mammals, a lack of proper marketing systems, improper animal care, and a lack of research [4]. A very large amount of milk is wasted due to the absence of better processing facilities. Furthermore, milk powder is imported into the country because of the above-mentioned problems. This problem can be solved by introducing better milk processing facilities.

Milk processing units operated by renewable energy are growing in developing countries that are suffering from an energy crisis. Pakistan is fortunate to have different sources of renewable energy, including solar energy and biomass, so further development of solar thermal energy in Pakistan would be a valuable and viable option. Pakistan is an equatorial country with a very high solar irradiance that ranges from 5 to 7 $kWh{\cdot}m^{-2}{\cdot}d^{-1}$ [5], which is available for use in solar heating.

The Efficiencies of the Flat plate collector (FPC) which is applied to transform solar power in thermal energy and the evaluation of the thermal performance greatly depends on how radiation is modelled and assessed. Monte Carlo radiation modelling was used by the authors [6]. This approach makes it possible to analyze spectral and diffuse surfaces as well as complex geometries. Due to the statistical character of this model, a disadvantage is that the surface heat transfer may vary between iterations [7]. The FLUENT program package also contains the surface-to-surface (S2S), discrete ordinates (DO), and discrete transfer radiation models (DTRM). For the investigation of heat transfer inside solar collectors, any one of the three models may be applied. An advantage of the S2S sub model is that it is effective for simulating enclosed radiation heat transfer, while it requires shorter computational times compared with DO and DTRM [8]. Solar radiation, used as an energy source, has been applied as a boundary condition for longwave radiation in the S2S sub model to calculate the heat transfer in previous work [9].

The solar load model provided in ANSYS 2020 R1 FLUENT contains a solar ray tracing method and a surface-to-surface radiation model (S2S). The solar ray tracing technique allows the evaluation of solar energy absorption, while the S2S radiation model assesses the internal distribution of energy. The surfaces exposed to the rays of the sun and consequently absorbing solar thermal energy are the sources. This method of modeling solar load has its limitations. However, the combination of these models should be able to give a realistic representation of an actual case. This S2S model may be used to predict the temperature distribution and the FLUENT model may be used for predicting the airflow patterns.

SAHs are simple, inexpensive, and are widely used in agro-based industries for low-range temperature applications [10]. These solar air heaters absorb radiation and subsequently transmit that energy to fluid flows through collector. SAHs are affordable and are widely used collectors, being used in a variety of applications for solar energy, space heating, wood seasoning, and agricultural drying. Through reviewing previous studies, it was observed that all elements of a SAH, for example, a damping tray, glazing, isolation, extensive surfaces, and the tilting angle, greatly affect the system's thermal efficiency.Deficiencies in the design and production of SAHs may lead to poor performance [11,12]. Jongpluempiti and Pannucharoenwong [13] described the construction of a solar air heater coupled with a spray dryer, and experiments were performed for various inclination angles of the air heater. The results indicated maximum efficiency is given for an angle of 15°, which was equal to the latitude of the site. Aghbashlo and Mobli [14] experimented with powder drying by using a portable solar dryer. The operational cost was negligible compared with that using conventional fuel (wood). The efficiency of the batch dryer was 54% (with solar heating), and it was found that solar drying was cost-effective, since the cost was one-third of that of wood-fired heating. Fara [15] estimated the performance parameter of an FPC using simulation techniques. Optimal efficiency can be attained by changing the collector area, the storage medium, and the mass flow rate. Tolga and Ural [16] determined that an SAH used in the textile sector has a better performance efficiency than a conventional solar air collector. Nikolic and Lukic [17] investigated the theoretical and experimented performance analysis of FPSC (Flat plate solar collector) which showed a significant increase of 48% in thermal efficiency over a simple solar collector.

Different approaches for enhancing the thermal efficiency of SAHs have been attempted, such as maximizing the size of air-heater modules, using large surfaces of varying shapes and sizes, using sensible or latency storage media, using solar radiation concentrators, and incorporating photovoltaic components in the heaters. In addition, there are several advantages of using SAHs. A conventional solar air heater may use different surface methods in the absorber plate or plates, such as roughened surfaces, extended surfaces, corrugated surfaces, and perforated surfaces to enhance thermal efficiency [18]. As Figure 1 shows, the collector includes an absorber plate and translucent selective cover. The sensible heat capacity of air is less than that of water and thus requires fans with greater flow rates. Black-painted aluminum, GI, and steel are preferred materials for absorber plates, where the plates have high aspect ratios. The cross-sectional area of the SAH duct decreases because of the increase in the aspect ratio. The air velocity and heat transfer rate are also increased by a high aspect ratio [19].

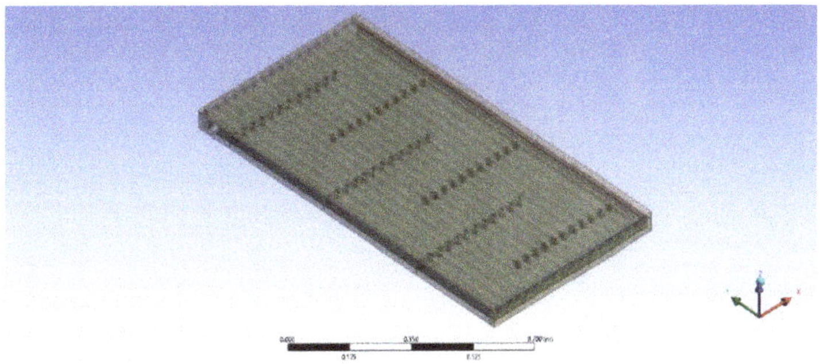

Figure 1. Geometric diagram of the solar air heater (SAH).

A transparent cover (glass, polycarbonate sheet) placed above the absorber plate on the top side, and insulation is inserted on the bottom and sides to decrease thermal heat losses. A solar air heater is a type of non-concentrating collector, so it is usually placed in a stationary position with the optimum tilt angle equal to the latitude, north-facing southern hemisphere's and vice versa. In this work, the solar air heater was designed and fabricated at the University of Sydney, Australia with locally and easily available materials.

Gill and Singh [20] created three different types of SAH of single-pass with different covers. The differences affected the heat losses for both convective and radiative heat transfer. The temperature was increased up to 35 °C in the packed bed at an air-flow rate of 0.013 kg $s^{-1}{\cdot}m^{-2}$. The increase in temperature was 12 °C for the first bed with a single glass cover, and it was 18 °C for the second bed with a double cover of glass at an airflow rate of 0.025 $kg{\cdot}s^{-1}{\cdot}m^{-2}$. A higher efficiency for the SAH was observed with a packed bed (72% at an airflow rate of 0.025 $kg{\cdot}s^{-1}{\cdot}m^{-2}$) while it was just 45% for the bed with two covers, and 30% for the bed with only one cover.

The aim of this work was to design and develop an SAH to couple with a milk powder spray drying system for converting raw milk to powder. We aimed to use computational fluid dynamics (CFD) to predict the air flow pattern and temperature distribution across the solar air heater to evaluate the best performance parameters. The structure of this paper is as follows. First, we describe the approach to the theoretical modelling technique used to guide the development of the solar air heater, before describing the experimental equipment (the SAH) that was built and discussing the performance evaluation and further development of the SAH. This is the first time that the performance parameters, i.e., heat transfer rate, air velocity, and baffle arrangements of SAH were optimized using a CFD approach.

2. Materials and Methods

The ANSYS 2020 FLUENT software was used to predict the airflow patterns and the S2S model was used for predicting the temperature distribution. A positioning vector was used by the Solar Ray Tracing method to specify the location and amount (power) of the solar energy input, and for characterizing its solar load. A solar calculator was used to state the sun's direction and magnitude. The solar calculator required coordinates or the location of the model, for calculating the vector giving the solar time, the date of the day, and the magnitudes of illumination parameters. The two illumination factors consisted of two irradiation terms, namely diffuse and direct solar irradiation. Transmissivity and absorptivity were used to define three directly visible bands, diffuse hemispherical, and direct IR, while smearing the solar load in cloudy conditions. While using solar ray tracing, the glazing materials were also involved. Reflectivity and transmissivity were specified in the boundary conditions of the wall. The ground reflectivity was set to a value of 0.2.

The collectors for the solar air heaters had a similar design to solar water heater collectors. The solar air heater absorber plate was made of corrugated metal.

It uses a solar-heated heat exchanger with baffles to create a serpentine pathway for the gas. The baffles increase the gas velocity and hence the heat-transfer coefficients from the absorber material to the gas, and the baffles also increase the heat-transfer area between the gas and the absorber surface. The materials were chosen for ease of construction in developing countries, such as Pakistan, so these materials are important due to their availability and low cost, as well as their reasonable heat-transfer performance. The solar air heater was fabricated locally from easily available materials. A schematic diagram of the solar air heaters has been shown in Figure 1. Marine plywood, a medium temperature tolerance wood material, was used in the solar air heater together with Tasmanian oak to make it cheap, portable, and reduce thermal losses. The collector area was 1220 mm × 610 mm with a corrugated iron sheet of 1182 mm × 572 mm as the absorber plate. For maximum absorbance of solar energy, the absorber plate was painted matte black. The optimal air gap was found to be 4–5 cm, so the air gap in the SAH was assumed to be 4 cm, because the polycarbonate sheet was flexible. Plywood having dimensions of 1220 mm × 610 mm × 6 mm has been used as a backing sheet, and fiberglass insulation of 30 mm thickness was used between the corrugated iron and the plywood, to reduce heat losses. Transparent cellular polycarbonate sheets of dimensions 1200 mm × 610 mm × 8 mm were used instead of brittle glass. These polycarbonate sheets provided good thermal insulation and have high strength, as well as being lighter in weight than glass. The airflow passage between the absorber plate and the glazing (polycarbonate sheet) was 30 mm. The cellular polycarbonate sheet had a light transmission of about 80%. The solar heat gain coefficient was 0.80, and the heat transfer coefficient for heat loss (U) was 2.9 Wm^{-2} K^{-1}. The polycarbonate sheet retains good light transmission properties for over ten years. The polycarbonate sheets were more durable and tolerated higher temperatures than polythene plastic film glass having a higher transmissivity (90%). Polythene sheets have been frequently used in SAHs, but they become more brittle over time than polycarbonate. Bakari and Minja [21] explained the characteristics of the top covers for the solar air heaters. A good top cover should have low reflection, absorption, and high transmission of light. Glass has 90% solar irradiance transmitting ability while plastic has some advantages over glass such as being less likely to break, having a lighter weight, and being cheaper. Plastic cannot withstand the high temperatures when the collector is idle, and UV light affects the transmissivity as well. Polystyrene foam insulation was used at the sides and the back of the SAH to reduce heat losses.

Kaushal and Goel [22] explained that the conventional SAHs (flat plate) have poor thermal performances as heat transfer coefficients between the absorber plates and the flowing medium (air) are typically low. The heat-transfer coefficient can be enhanced with artificial roughness which can create turbulent flow which has a positive impact on heat transfer. Based on a literature review, it was found in different research that the Reynolds numbers for solar air heaters range from 2000 to 16,000 [23,24].

2.1. Thermal Analysis of the Solar Air Heater

The efficiency of the solar air heater (SAH) has been assessed using Equation (1). The following parameters were calculated using Equations (2)–(10): the energy gain (Q_u), the mass flow rate, the equivalent diameter, the Reynolds number, the tilt angle, the friction factor for laminar or turbulent flow, the pressure drops, and the overall heat transfer coefficient. The thermal performance evaluation parameters of the SAH are presented in Table 1.

Table 1. Thermal properties of the solar air heater, [25].

Parameter	Equation	Equation No.
Efficiency of SAH	$\eta = \frac{Q_u}{I_r\ A_c}$	(1)
Energy Gain	$Q_u = \dot{m}C_P(T_{out} - T_{in})$	(2)
Mass flow rate	$\dot{m} = \rho.A_C.V$	(3)
Equivalent Diameter	$de = \frac{2LH}{L+H}$	(4)
Reynolds Number	$Re = \frac{V.\rho.2LH}{\mu(L+H)}$	(5)
Tilt angle	$\beta = L\ \pm 15^{\circ}$	(6)
Friction factor	$f = \frac{24}{Re}$ (for laminar)	(7)
	$f = 0.079\ Re^{-0.25}$(for turbulent)	(8)
Pressure Drop	$\Delta P = \frac{2f\ (\rho.V^2)\ L}{de}$	(9)
Overall heat transfer coefficient	$h = \frac{Q_u}{A_C(T_P\_T_o)}$	(10)

2.2. Heat Losses

The total energy flow rate in the SAH is the sum of the useful energy flow rate Q_u, the stored energy flow rate Q_s, which is neglected because no storage material is in use and the system is considered to be close to steady state, $Q_{loss,th}$ is the thermal loss rate, and $Q_{loss,opt}$ is the optical loss and U is the overall heat loss coefficient, which can be determined using Equations (11) to (16), respectively. The optical losses are caused by the cover transmission losses and the absorption ability of the absorber, the so-called transmission absorption product $(\tau\alpha)_e$. The optical losses are calculated when the SAH is exposed to solar light. The heat temperature T_{heater} is assumed here to be the average value of the inlet and outlet temperatures. The heat loss parameters of the solar air heater are presented in Table 2.

Table 2. Heat loss parameters of the solar air heater, [26].

Parameter	Equation	Equation No.
Total energy flow	$G = Q_u + Q_s + Q_{loss,th} + Q_{loss,\ opt}$	(11)
Heat loss	$Q_{loss} = C_p.m.(T_{in} - T_{out})$	(12)
Overall heat loss coefficient (W/m^2 K)	$U_L = \frac{Q_{loss}}{A_C(T_P - T_a)}$	(13)
Thermal heat loss	$Q_{loss,th} = U_L A_C(T_{heater} - T_a)$	(14)
Optical heat loss	$Q_{loss,opt} = G(1 - (\tau\alpha)_e)$	(15)
Efficient transmission and absorption of the product	$(\tau\alpha)_e = \frac{Q_u + U_L A_C\ (T_P - T_a)}{I_t.A_c}$	(16)

2.3. Economic Evaluation

Management of machinery includes two important elements of finance, which are the ownership or capital cost (fixed) and the operating cost, which occurs during the use of the machinery [27]. Different factors affect the fixed cost, including depreciation, interest, and taxes. Various methods may be used to determine depreciation, including the straight-line method, which is one of the most straightforward ones and has been used here [28]. Salvage value, which is 10–15% of the capital cost, has been assigned to the machine [27]. Here, the salvage value from the first installed cost, and difference was divided by useful life (years) for the machinery [29].

$$D_s = \frac{(P-S)}{L} \tag{17}$$

where D_s = depreciation amount per year by the straight-line method in rupees, P = purchase value (capital cost) in rupees, S = salvage value in rupees, and L = useful life in years.

The cost of machinery also includes the interest cost. Interest is also considered an opportunity, and the interest on the machinery may be calculated by the equation developed by Jacobs and William [27].

$$I = \frac{(P+S)}{2 \times i} \tag{18}$$

where

I = annual interest amount in rupees, and

i = interest rate as a fraction, and P and S have been defined above.

3. Results

The structure of this section is to report and discuss the theoretical performance of the SAH in Section 3.1, before discussing the actual heat losses and leakages from the experimental unit in Section 3.2. The complete technical performance of the SAH and its further development follow in Section 3.3, building on the important preliminary theoretical work in Section 3.1 and practical foundations in Section 3.2. To complete a techno-economic assessment of the SAH, the economics of the SAH are presented and discussed in Section 3.4.

3.1. Solar Air Heater: Theoretical Performance

This section discusses the results of the computational and theoretical performance of a solar heater containing the corrugated absorber plate with a baffled SAH, with respect to varying mass flow rates, solar intensities, and geometries. This section indicates the optimal results obtained at appropriate mass flow rates for a solar-assisted spray-drying process. The airflow analysis and temperature distribution of the SAH was conducted using the FLUENT software [30,31]. The key advantage of FLUENT is predicting the detailed air flow patterns in the whole equipment. Baffles were used because they gave higher air velocities and higher heat-transfer coefficients compared with arrangements without baffles in the solar air heater. The entire structure above absorber plate with baffles was meshed in triangular elements and rectangular meshing was used on the baffles. Triangular meshing was chosen rather than a hexagonal mesh because it is finer and gives a better prediction of results in the complicated regions shown in Figure 2. In computational analysis, the solar irradiance applied on the absorber plate of the SAH was 900 W·m^{-2}. The velocity flow rates for the air are also shown in Figure 3, which indicates that the velocity of the inlet and outlet air flows was the same and very little pressure drop was recorded in the equipment. Here, the color variation in the predicted temperature distributions for the solar air heater showed that the air temperature increased when the air moved towards the outlet in Figure 4. The temperature distribution in the solar air heater, its maximum plate temperature was computed to be 102 °C based on the boundary conditions, and the outlet temperature was predicted to be 82 °C for a mass flow rate of 0.01 kg·s^{-1}. The baffles were equally distributed over the plate which kept the air velocity high, giving more heat transfer than without the baffles. After analyzing the situation at different flow rates (0.005 kg·s^{-1}, 0.01 kg·s^{-1}), 0.01 kg·s^{-1}, a mass flow rate was chosen, which was also suitable for the spray drying system. A greater radiation intensity increases the outlet temperature which helps to increase the collector efficiency. The variation in the collector efficiency at different mass flow rates was connected with the changing outlet temperatures and geometries. The computational method predicts the outlet temperature which can be used to assess the efficiency of the system [32,33]. The theoretical efficiency of the solar air heater was calculated from the experimental data to be 69%; while using the computational method, the solar air heater efficiency was predicted to be 73%. To obtain the computational results, the input parameters were the inlet temperature (35 °C) and solar irradiance (950 W·m^{-2}). The experimental value for the outlet temperature was 82 °C and the actual thermal energy gain was calculated to be 473 W. These experimental results are close to the computational predictions, and the experiments included a 38 °C inlet temperature and 327 W of thermal gain. The difference in the rate of thermal gain is due to the different solar irradiance values of 700 W·m^{-2} and 950 W·m^{-2} for the actual and the computational experiments, respectively. The details of the actual laboratory experiment are summarized later in this paper.

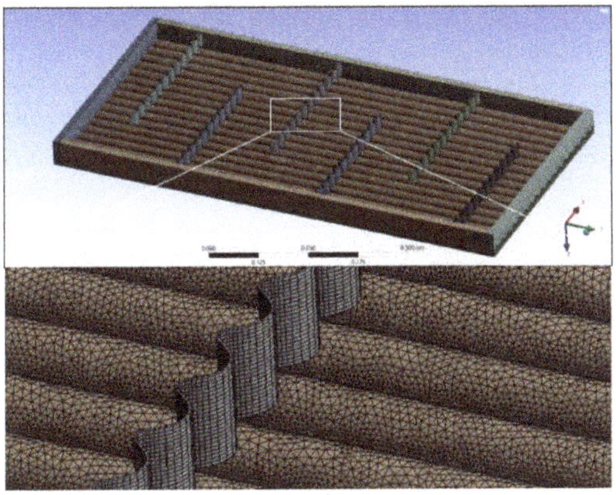

Figure 2. Surface mesh for the solar air heater.

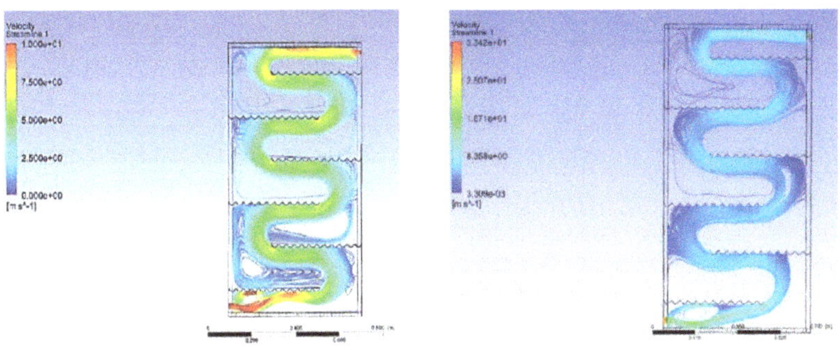

At mass flow rate 0.01 kg·s^{-1} **At mass flow rate 0.005 kg·s^{-1}**

Figure 3. Performance prediction of an air-flow distribution pattern using CFD for the SAH (e^{x} = 10^{x}).

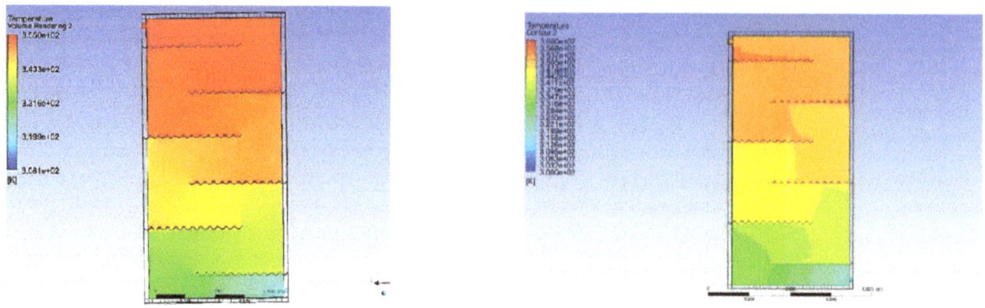

At mass flow rate 0.01 kg·s^{-1} At mass flow rate 0.005 kg·s^{-1}

Figure 4. Performance prediction of air temperature distribution pattern using CFD for the SAH ($e^x = 10^x$).

The CFD simulation with the baffles predicts a maximum collector efficiency of about 73%, and the temperature increase in the solar air heater with baffles results in better performance, due to the increased rate of heat transfer from the absorber plate to the flowing medium (air) [26,34]. The computational method predicts that this design of solar air heater has a slightly higher efficiency than the actual observation.

The fine meshing resulted in almost 84 K cells for the finest mesh. The pattern of the baffles on the corrugated SAH increases the heat transfer area as well as the heat transfer rate.

3.2. *Heat Losses from the Solar Air Heater*

Having assessed the theoretical performance of the solar air heater using CFD, the experimental aspects of its evaluation are now reported and discussed. The first aspect reported in this section, is the area of heat losses. An experiment was performed over one hour to achieve steady-state conditions, as shown in Figure 5 for the experiments in the laboratory (with no solar energy input) and Figure 6 for the experiments in the courtyard (with solar energy input). The ambient, inlet, outlet, and plate temperatures measured in the laboratory with no energy input were 26 °C, 59 °C, 48 °C, and 57 °C, respectively. From Equations (11)–(16), the overall heat loss coefficient for convection was then estimated as 4.9 W·m^{-2}·K^{-1} using Equation (13). The radiation and convective losses in the SAH were calculated in the courtyard where solar energy was used as the input energy. The thermal losses were then estimated as 154 W using Equation (14), and the optical losses were estimated as 46 W using Equation (15). The total losses were estimated to be 200 W using Equation (11). The useful energy flow rate provided to the air when the experiment was performed in the courtyard was 334 W using Equation (2). The losses were 37% and useful energy was 63% of the total energy provided by the SAH. The effective transmittance absorbance product $(\tau\alpha)_e$ of the SAH was 91% (Equation (16)) for this system with a double polycarbonate sheet at the top cover and a corrugated iron sheet with baffles at a spacing between the baffles of 17 cm.

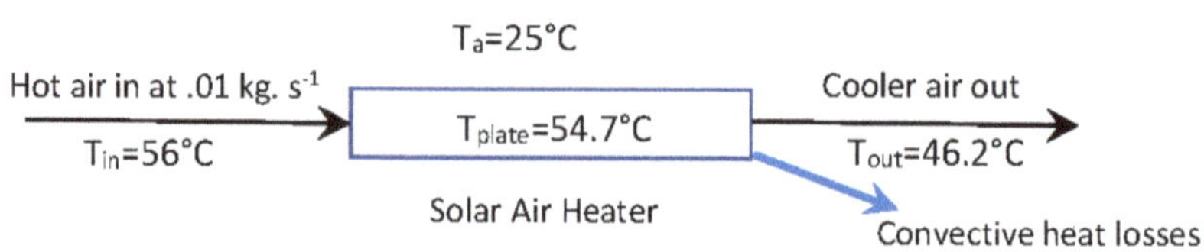

Figure 5. Schematic diagram of mass and energy flows in the experiments for the heat loss measurements in the laboratory.

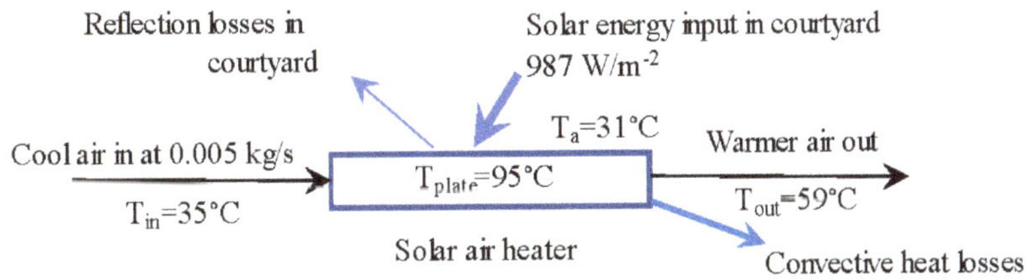

Figure 6. Schematic diagram of mass and energy flows in the experiments for the heat loss measurements in the courtyard.

In Figure 5, the temperature measurements in the laboratory (with no solar energy input) are given in Table 3 for a one-hour experiment to achieve steady-state conditions and a heat input from an electric heater to the air entering the SAH. The ambient, inlet, outlet, and plate temperatures measured in the laboratory with no energy input were 26 °C, 58 °C, 47 °C, and 54 °C, respectively. The overall heat loss coefficient for convection was then estimated as 4.9 $W{\cdot}m^{-2}{\cdot}K^{-1}$. In this experiment, no polystyrene and double poly-carbonate cover sheet was used. The spray dryer was attached for air circulation with a mass flow rate of 0.01 $kg{\cdot}s^{-1}$. The aspirator gave an air flow of 28.5 $m^3{\cdot}h^{-1}$. In addition, the laboratory test checked for any significant air leakage from the solar air heater, and no significant sources of leakage were found. The above two experiments show the results from a solar air heater in which the convective heat losses were calculated. The experiment was performed in a laboratory of the School of Chemical and Biomolecular Engineering, The University of Sydney, Australia.

Table 3. Temperature measurements in the laboratory for heat loss calculations.

	Inlet Temperature (°C)	Outlet Temperature (°C)	Ambient Temperature (°C)	Plate Temperature (°C)	Mass Flow Rate (kg/s)
Experiment 1	58	47	26	54	0.01
Experiment 2	56	46.2	25	54.7	0.01

Having assessed the heat losses, the overall performance evaluation and the further development of the design will be reported and discussed in the next section, Section 3.3.

3.3. Performance Evaluation and SAH Devlopment

The performance evaluation of the solar air heaters without baffles was investigated. The experiment was performed in the courtyard using a blower, having an air flow rate of 0.005 $kg{\cdot}s^{-1}$, in an open loop mode as shown in Figure 6. The experiment was performed during peak sunlight hours, and the average solar irradiance was recorded to be 987 $W{\cdot}m^{-2}$. The very large distinction between outlet air temperature and the plate temperature is shown in Figure 6.

The thermal losses of the solar air heater were high, since there was only one polycarbonate sheet, and there were no polystyrene sheets on the back and sides in this experiment. Due to excessive heat losses, the thermal efficiency of the system was low, and low amounts of useful energy were produced.

Figure 7 shows a graphical representation of the temperature of the air inlet, the air outlet, and the plate temperature over a time period of one hour. The solar irradiance is shown on the right axis and has a gradually declining trend. However, the use of baffles increases the air velocity inside the heating chamber and enhances the heat transfer rate. In addition, significantly reduced thermal losses were found when good insulation was provided. The analysis showed that the plate temperature attained a steady state condition after 25 min, as shown in Figure 7.

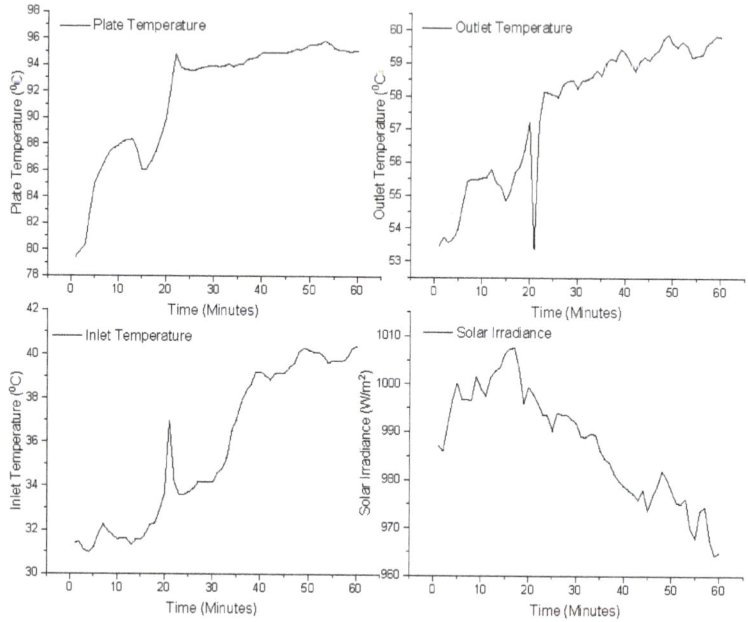

Figure 7. Graphical representation of different points temperatures and solar irradiance for parametric study of the solar air heater.

A performance evaluation of the solar air heaters with baffles and insulation was also investigated and is presented in Figure 8.

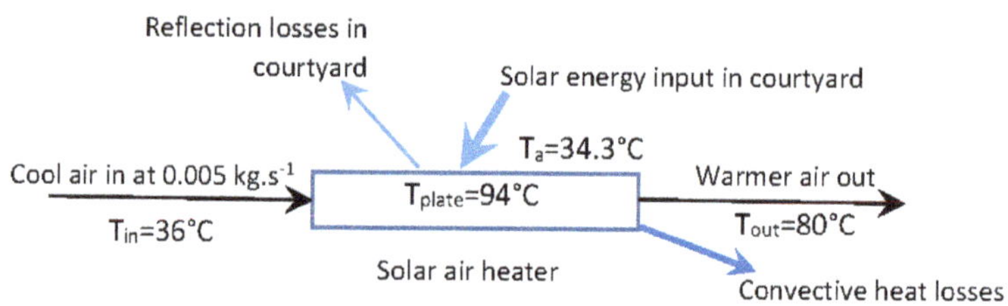

Figure 8. Schematic diagram of mass and energy flows in the experiments for performance evaluation and the heat loss measurements in the courtyard.

The experiment was performed in the courtyard using a blower, having an air flow rate of 0.005 kg·s^{-1} in open loop mode, as shown in Figures 8 and 9. The experimental setup was installed twice in the courtyard to get good results after insulating the sides and back of the solar air heater with polystyrene. The solar irradiance was measured to be, on average, at 740 W·m^{-2} and 840 W·m^{-2} on two days of experiments. The difference between the outlet air temperature and the plate temperature was reduced, compared with the previous experiment. Figure 8 represents the different point temperatures and solar irradiance for the experiments over a period of one hour at steady state. A comparison of the results has shown that there was an increase in the outlet air temperature and a reduction in the difference between plate temperature and the outlet air temperature. These results indicate that the heat transfer rate for a corrugated, black-painted, iron sheet with the working fluid (air) increased due to greater air velocities with the provision of baffle plates. In this experiment, the thermal heat losses from the solar air heaters were reduced from 4.9 to 2.27 W·m^{-2}·K^{-1}, because the back and the sides of the SAH were provided with appropriate insulation material (polystyrene sheet; thermal conductivity 0.033 W·m^{-1}·K^{-1}).

Due to the reduction in the heat losses, the thermal efficiency of the system was increased, and the system produced more useful energy. The mass flow rate in this experiment was low, which had a direct effect on the solar air heater efficiency. The use of obstacles or baffles enhanced the heat transfer rate, and the heat losses were significantly reduced with good insulation. The analysis showed, in both experiments, that the experimentally measured temperatures were effectively at steady state conditions during the whole trial.

Figure 9. Experimental setup for evaluating performance parameters of SAH.

In Figure 10, the solar irradiance value dropped to 250 $W \cdot m^{-2}$ due to some clouds moving across the SAH, and the outlet and plate temperature also showed some variation at that specific time. The radiative and convective losses in the SAH were calculated in the courtyard where solar energy was used as input energy. The thermal losses were then estimated as 154 W, and the optical losses were estimated as 46 W. The total losses were estimated to be 200 W. When the experiments were performed, the useful energy flow rate provided to the SAH outlet air in the courtyard experiment was 334 W. The losses were calculated to be 37%, and the useful energy was estimated to be 63% of the total energy provided to the SAH. The effective transmittance absorbance product $(\tau\alpha)_e$ of the SAH was calculated to be 81%. The system had a double polycarbonate sheet at the top cover and a corrugated iron sheet with baffles at a spacing between baffles of 17 cm. According to the findings of El-Said [35], the overall thermal performance of his air heater was approximately 77% at an air mass flow rate of 0.03 $kg \cdot s^{-1}$, an opening diameter of 3 mm, and a baffle angle of 7°. The CFD-predicted pressure drops across this current solar air heater varied between 0.0001505 and 0.000017 bar. These results are significant for the application of this enhanced solar air heater and demonstrate that the corrugated perforated baffles in the air channel are an effective design feature that improves the collector efficiency. In the above section, the results for the actual pressure drop were very similar, recorded to be 0.0000434 bar for a system operating near ambient pressure (1.013 bar), so the pressure drop was negligible.

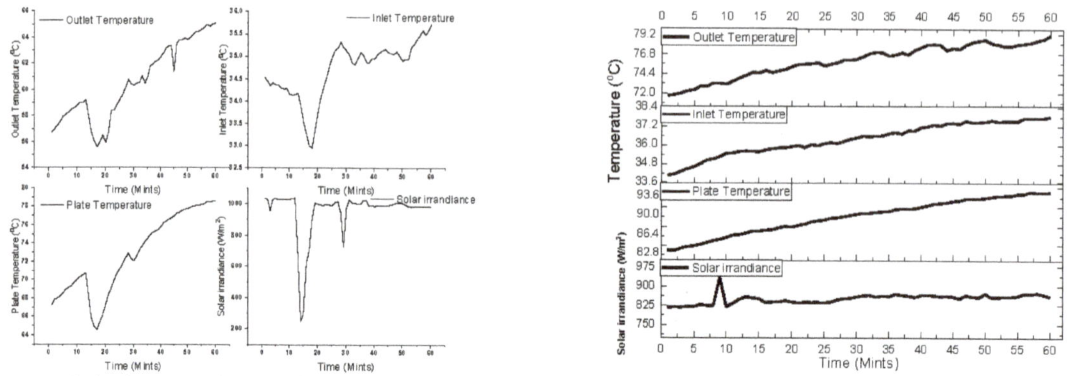

(**a**) Without insulation of polystyrene (**b**) With insulation of polystyrene

Figure 10. Graphical representation of different points temperatures and solar irradiance for parametric study of the solar air heater at mass flow rate of 0.005 kg·s^{-1}.

The final experiment was conducted in the courtyard of the School of Chemical and Biomolecular Engineering the University of Sydney, Australia (33.8905° S, 151.1929° E). The solar air heater was used as a pre-heater providing an outlet air temperature of 73 °C with a humid enthalpy of 164 kJ·kg^{-1}. The mass flow rate was 0.01 kg·s^{-1}

The ambient temperature measured was 31 °C. The energy flow rate into the electrical heater was 0.327 kW in the experiment using the solar air heater. Figure 11 shows the inlet temperature, the outlet temperature, the plate temperature, and the solar irradiance for the final experiment using a the solar-assisted spray drying system. The inlet temperature shows a steady value, while the plate and outlet temperatures dropped in the initial 5 min then reached relatively steady state conditions. The average outlet temperature under steady state conditions was measured to be 73 °C. A greater difference between the inlet and outlet air temperatures indicates the good efficiency of the solar air heater. The lower difference between the plate temperature and the outlet temperature indicated a good heat transfer rate. On the right *y*-axis, the solar irradiance was seen to rise, and fall. According to international standards, for a solar based experiment to be considered valid, the solar irradiance value cannot be increased or decreased by more than 100 W·m^{-2} over a ten minute time interval, and a solar irradiance value lower than 450 W·m^{-2} or greater than 1100 W·m^{-2} also render the experiment invalid. In addition, the experiment for performance evaluation must be conducted between 10.00 to 14.00 solar time for the experiment to be considered valid [36]. These validity requirements were all met in these experiments. The Graphical representation of the below experiment for one hour is shown in Figure 12.

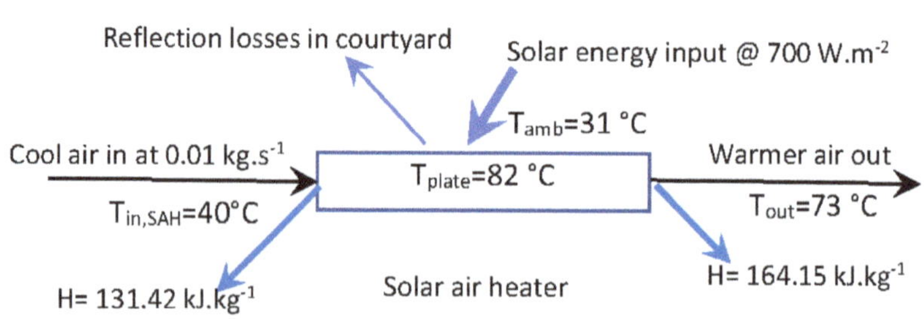

Figure 11. Schematic diagram of mass and energy flows in the experiments for performance evaluation and the heat loss measurements in the courtyard at mass flow rate of 0.01 kg·s^{-1}.

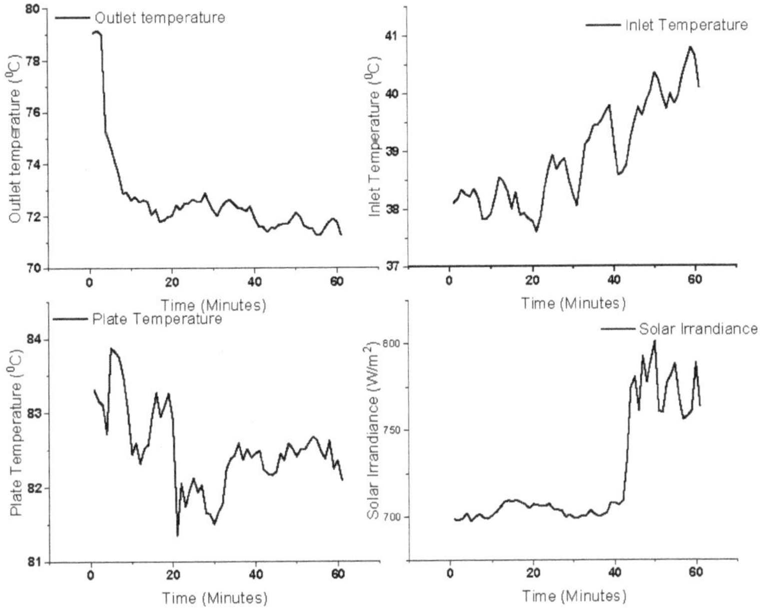

Figure 12. Graphical representation of the different temperatures and solar irradiance as function of time during the final experiment.

3.4. Economic Analysis

With the technical performance of the SAH being confirmed to be satisfactory, the economic assessment will now be given, to complete a techno-economic assessment of the SAH. The following assumptions were made during the economic cost analysis of the solar-assisted spray drying unit.

The life span of the solar-assisted spray dryer = 15 years

Daily usage = 4 h

Total utilized hours over the equipment life span = 21,900 h

The initial cost of complete solar air heater = P = 13,000 PKR

Salvage value, *S* (10% of the initial cost of solar air heater) = 1300 PKR

Fixed cost: The straight-line method below was used for calculating depreciation cost (D_s).

$D_s = \frac{13{,}000-1300}{15}$ = 780 PKR per year = 0.09 PKR per hour

Interest cost = 0.08 × $\frac{13{,}000+1300}{2}$ = 572 PKR per year = 0.065 PKR per hour

Total fixed cost = $D_s + I + P$ = 0.09 + 0.065 + 13,000 = 0.155 PKR per hour

Variable cost:

Repair and maintenance = 2.5% of SAH = 0.025 × 13,000 PKR = 325 PKR/year = 0.23 PKR/h

Electricity cost per unit = 30 PKR·kWh^{-1}

Total cost for forced circulation of air = 0.2 kW × 30 PKR·kWh^{-1} = 6 PKR·h^{-1}

The total variable cost of the spray drying system = 0.23 + 6 = 6.23 PKR·h^{-1}

The total cost of SAH = Total Fixed cost + Total variable cost = 0.155 + 6.23 = 6.38 PKR·h^{-1}

3.5. Breakeven Analysis

The economic viability of any organization often starts with a breakeven analysis. In this analysis, the focus is on finding the operation level where the organization has no profit at all. This analysis is a very important step when launching a new product for any industry. Through this analysis, a reference may be created for further operations as this helps in indicating those levels of operation where cost and revenue become equal. If a breakeven analysis is conducted for this design of solar air heater, which works for about 4 h a day, then in 3700 h, or in other words in 2.5 years, the breakeven point is achievable. About 30 PKR/kWh of energy is saved by this solar air heater, which works as a preheater for the spray dryer. The breakeven analysis of the solar air heater is shown in Figure 13.

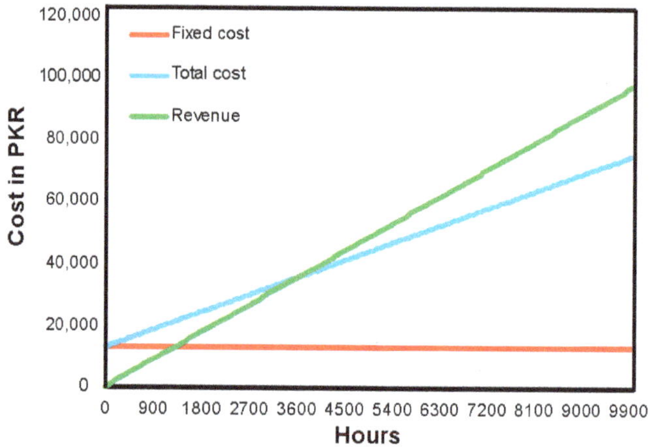

Figure 13. Breakeven analysis of the solar air heater.

4. Discussion

The radiative and convective losses in the SAH were calculated in the courtyard where solar energy was used as input energy. The thermal losses were then estimated as 154 W, and the optical losses were estimated as 46 W. The total losses were estimated to be 200 W. When the experiments were performed, the useful energy flow rate provided to the air in the courtyard was 334 W. The losses were calculated to be 37% and the useful energy was determined to be 63% of the total energy provided to the SAH. The effective transmittance absorbance product $(\tau\alpha)_e$ of the SAH was calculated to be 81%. The system included a double polycarbonate sheet as the top cover and a corrugated iron sheet with baffles at a spacing between baffles of 17 cm. This thermal performance may be compared with the SAH of El-Said [35] where this device had an efficiency of approximately 77% at an air mass flow rate of 0.03 kg·s^{-1}, an opening diameter of 3 mm, and a baffle angle of 7°. The experimental pressure drops across this SAH varied between 0.0001505 and 0.000017 bar.

In the above paper, the results showed a thermal efficiency of 63% for the single-pass roughened-surface SAH while the thermal efficiency was 77% for the double-pass SAH. The pressure drop across the SAH here was found to be 0.0000434 bar, and the agreement between the CFD predicted pressure drop and that measured experimentally was satisfactory. The solar-assisted spray dryer operated at atmospheric pressure (except for the compressed air to the atomizer at 5 bar), so the SAH pressure drop is negligible. These CFD numerical results are significant for the application of this enhanced SAH and demonstrate that the corrugated, perforated baffles in the air channel are an effective factor in improving collector efficiency.

5. Conclusions

This study is focused on design optimization and parametric studies of a solar air heater which has been conducted through an experimental study and a numerical prediction of the SAH performance. The Following conclusions can be extracted from this work.

There is very good agreement between the experimental results and the CFD predictions for the increase in the SAH outlet temperature. Despite some minor differences, the

CFD simulation tool has predicted a similar SAH performance to that which was observed, so the CFD approach may be applied in the future for more solar air heater applications. The effect of variable solar irradiance and mass flow rate values on the thermal efficiency of a solar air heater has also been investigated. The result shows a direct relationship between the solar irradiance, the mass flow rate, and the thermal efficiency of the system. The experimental study showed that 0.327 kW was produced and thermal heat losses from the solar air heater were reduced from 4.9 to 2.27 $W \cdot m^{-2} \cdot K^{-1}$ because the back and sides of SAH were appropriately insulated (polystyrene sheet; thermal conductivity 0.033 $W \cdot m^{-1} \cdot K^{-1}$). The pressure drop across the SAH was recorded to be 0.0000434 bar; therefore, the pressure drop is negligible.

Author Contributions: Conceptualization, Z.Q. and A.M.; Methodology, Z.Q. and T.L.; Software, Z.Q.; Validation, Z.Q. and A.M.; Writing—original draft, Z.Q. and A.M.; Writing—review and editing, Z.Q., T.L., A.G., and M.T.; Project administration, A.M. and T.L.; Funding acquisition, Z.Q. and A.M. All authors have read and agreed to the published version of the manuscript.

Funding: This research was funded by the Higher Education Commission of Pakistan.

Institutional Review Board Statement: This study is a mandatory part of my PhD studies and all other authors are the member of my PhD supervisory com-mittee. Hence there is no conflict of interest and lack of ethics in publishing this manuscript.

Data Availability Statement: The research profile and the data of my supervisory committee (Co authors) have been given in the links below. http://www.uaf.edu.pk/EmployeeDetail.aspx?userid=711; https://www.sydney.edu.au/engineering/about/our-people/academic-staff/timothy-langrish.html; http://www.uaf.edu.pk/EmployeeDetail.aspx?userid=387; http://uaf.edu.pk/EmployeeDetail.aspx?userid=158; For more research details, google scholar and researchgate sites of the individuals can be seen.

Acknowledgments: The authors acknowledge Financial and Technical support from the Higher Education Commission of Pakistan (HEC), University of Agriculture, Faisalabad Pakistan, and the School of Chemical and Biomolecular Engineering for this research, The University of Sydney is also acknowledged for support and guidance for this research.

Conflicts of Interest: The authors declare no conflict of interest.

Nomenclature

$\dot{m}$ = mass flow rate of air ($kg \cdot s^{-1}$)
C_p = specific heat capacity of air ($kJ \cdot kg^{-1} \cdot K^{-1}$))
T_{out} = outlet air temperature (°C)
T_{in} = inlet air temperature (°C)
I_r = solar irradiance ($W \cdot m^{-2}$)
A_c = area of the collector (m^2)
ρ = density of the fluid ($kg \cdot m^{-3}$)
V = velocity of air ($m \cdot s^{-1}$)
L = length of SAH (m)
H = height of the SAH (m)
V = velocity of the air in SAH ($m \cdot s^{-1}$)
μ = dynamic viscosity ($kg \cdot s^{-1} \cdot m^{-1}$)
β = collectors tilt angle (degrees)
L = latitude of the specific site (degrees)
Q_u = useful energy flow rate (W),
Q_s = stored energy flow rate (W),
$Q_{loss,th}$ = thermal loss rate (W),
$Q_{loss,opt}$ = optical loss (W),
$(\tau\alpha)_e$ = transmission absorption product
FPSC = Flat plate solar collector
SAH = Solar air heater

References

1. Gorjian, S.; Fakhraei, O.; Gorjian, A.; Sharafkhani, A.; Aziznejad, A. Sustainable Food and Agriculture: Employment of Renewable Energy Technologies. *Curr. Robot. Rep.* **2022**, *3*, 153–163. [CrossRef]
2. Baig, M.B.; Khan, N. Rural Development in Pakistan: From Vision to Action. In *The Rural Citizen: Governance, Culture and Well-Being in the 21st Century*; University of Plymouth: Plymouth, UK, 2006.
3. Sattar, A. Pide Knowledge Brief. Available online: https://econpapers.repec.org/paper/pidkbrief/ (accessed on 10 December 2022).
4. Sattar, A.; Ahmad, U. *What is Holding Back Milk Production Potential in Pakistan?* Research Square: Durham, NC, USA, 2022.
5. Ghafoor, A.; Rehman, T.U.; Munir, A.; Ahmad, M.; Iqbal, M. Current status and overview of renewable energy potential in Pakistan for continuous energy sustainability. *Renew. Sustain. Energy Rev.* **2016**, *60*, 1332–1342. [CrossRef]
6. Sultana, T.; Morrison, G.L.; Taylor, R.A.; Rosengarten, G. Numerical and experimental study of a solar micro concentrating collector. *Sol. Energy* **2015**, *112*, 20–29. [CrossRef]
7. Howell, J.R.; Mengüç, M.P.; Daun, K.; Siegel, R. *Thermal Radiation Heat Transfer*; CRC Press: Boca Raton, FL, USA, 2020.
8. Filipović, P.; Dović, D.; Ranilović, B.; Horvat, I. Numerical and experimental approach for evaluation of thermal performances of a polymer solar collector. *Renew. Sustain. Energy Rev.* **2019**, *112*, 127–139. [CrossRef]
9. Reichl, C.; Kramer, K.; Thoma, C.; Benovsky, P.; Lemée, T. Comparison of modelled heat transfer and fluid dynamics of a flat plate solar air heating collector towards experimental data. *Sol. Energy* **2015**, *120*, 450–463. [CrossRef]
10. Close, D.J. Solar air heaters for low and moderate temperature applications. *Sol. Energy* **2003**, *7*, 117–124. [CrossRef]
11. Said, Z.; Ghodbane, M.; Sundar, L.S.; Tiwari, A.K.; Sheikholeslami, M.; Boumeddane, B. Heat transfer, entropy generation, economic and environmental analyses of linear fresnel reflector using novel rGO-Co_3O_4 hybrid nanofluids. *Renew. Energy* **2021**, *165*, 420–437. [CrossRef]
12. Bellos, E.; Said, Z.; Tzivanidis, C. The use of nanofluids in solar concentrating technologies: A comprehensive review. *J. Clean. Prod.* **2018**, *196*, 84–99. [CrossRef]
13. Jongpluempiti, J.; Pannucharoenwong, N.; Benjapiyaporn, C.; Vengsungnle, P. Design and construction of the flat plate solar air heater for spray dryer. *Energy Procedia* **2017**, *138*, 288–293. [CrossRef]
14. Aghbashlo, M.; Mobli, H.; Rafiee, S.; Madadlou, A. Energy and exergy analyses of the spray drying process of fish oil microencapsulation. *Biosyst. Eng.* **2012**, *111*, 229–241. [CrossRef]
15. Luminosu, I.; Fara, L. Determination of the optimal operation mode of a flat solar collector by exergetic analysis and numerical simulation. *Energy* **2005**, *30*, 731–747. [CrossRef]
16. Ural, T. Experimental performance assessment of a new flat-plate solar air collector having textile fabric as absorber using energy and exergy analyses. *Energy* **2019**, *188*, 116116. [CrossRef]
17. Nikolić, N.; Lukić, N. Theoretical and experimental investigation of the thermal performance of a double exposure flat-plate solar collector. *Sol. Energy* **2015**, *119*, 100–113. [CrossRef]
18. Yadav, A.S.; Thapak, M.K. Artificially roughened solar air heater: Experimental investigations. *Renew. Sustain. Energy Rev.* **2014**, *36*, 370–411. [CrossRef]
19. Priyam, A.; Chand, P. Thermal performance of wavy finned absorber solar air heater. *Int. J. Heat Technol.* **2018**, *36*, 1393–1403. [CrossRef]
20. Gill, R.S.; Singh, S.; Singh, P.P. Low cost solar air heater. *Energy Convers. Manag.* **2012**, *57*, 131–142. [CrossRef]
21. Bakari, R.; Minja, R.J.A.; Njau, K.N. Effect of Glass Thickness on Performance of Flat Plate Solar Collectors for Fruits Drying. *J. Energy* **2014**, *2014*, 1–8. [CrossRef]
22. Kaushal, M.; Goel, V. Effective Efficiency of Solar Air Heaters of Different Types of Roughness Geometries over Absorber Plate. *Walailak J. Sci. Technol.* **2013**, *11*, 729–738.
23. Gill, R.S.; Hans, V.S.; Singh, S. Investigations on thermo-hydraulic performance of broken arc rib in a rectangular duct of solar air heater. *Int. Commun. Heat Mass Transf.* **2017**, *88*, 20–27. [CrossRef]
24. Singh, H.; Alam, T.; Siddiqui, M.I.H.; Ali, M.A.; Sagar, D. Experimental investigation of heat transfer augmentation due to obstacles mounted in solar air heater duct. *Exp. Heat Transf.* **2022**, 1–20. [CrossRef]
25. Saxena, A.; Varun; El-Sebaii, A.A. A thermodynamic review of solar air heaters. *Renew. Sustain. Energy Rev.* **2015**, *43*, 863–890. [CrossRef]
26. Kumar, D.; Kumar, S. Simulation Analysis of Overall Heat Loss Coefficient of Parabolic trough Solar Collector at Computed Optimal Air Gap. *Energy Procedia* **2017**, *109*, 86–93. [CrossRef]
27. Jacobs, C.O.; Williams, R.H. *Agricultural Power and Machinery*; National Book Foundation: Lahore, Pakistan, 1987.
28. Kepner, R.A. *Principles of Farm Machinery*, 3rd ed.; CBS Publishers & Distributors 4596/1-A Darya Ganj: New Delhi, India, 2000.
29. Yasin, M.; Farooq, M.A.; Tanveer, T.; Aslam, M. *Harvesting and Threshing Techniques in Pakistan*; Paper Prepared for RNAM/ESCAP; Workshop on Agriculture Machinery Testing Technology: Beijing, China, 1986.
30. Yadav, A.S.; Bhagoria, J. Heat transfer and fluid flow analysis of solar air heater: A review of CFD approach. *Renew. Sustain. Energy Rev.* **2013**, *23*, 60–79. [CrossRef]
31. Tagliafico, L.A.; Scarpa, F.; De Rosa, M. Dynamic thermal models and CFD analysis for flat-plate thermal solar collectors—A review. *Renew. Sustain. Energy Rev.* **2014**, *30*, 526–537. [CrossRef]

32. Yang, M.; Yang, X.; Li, X.; Wang, Z.; Wang, P. Design and optimization of a solar air heater with offset strip fin absorber plate. *Appl. Energy* **2014**, *113*, 1349–1362. [CrossRef]
33. Kessentini, H.; Bouden, C. Numerical and experimental study of an integrated solar collector with CPC reflectors. *Renew. Energy* **2013**, *57*, 577–586. [CrossRef]
34. Murali, G.; Sundari, A.T.M.; Raviteja, S.; Chanukyachakravarthi, S.; Tejpraneeth, M. Experimental study of thermal performance of solar aluminium cane air heater with and without fins. *Mater. Today Proc.* **2020**, *21*, 223–230. [CrossRef]
35. El-Said, E.M.S. Numerical investigations of fluid flow and heat transfer characteristics in solar air collector with curved perforated baffles. *Eng. Rep.* **2020**, *2*, e12142. [CrossRef]
36. Funk, P.A. Evaluating the international standard procedure for testing solar cookers and reporting performance. *Sol. Energy* **2000**, *68*, 1–7. [CrossRef]

agriculture

Article

Attention-Based Fine-Grained Lightweight Architecture for Fuji Apple Maturity Classification in an Open-World Orchard Environment

Li Zhang [1], Qun Hao [1,2,3] and Jie Cao [1,2,*]

1 Key Laboratory of Biomimetic Robots and Systems, School of Optics and Photonics, Beijing Institute of Technology, Beijing 100081, China
2 Yangtze Delta Region Academy, Beijing Institute of Technology, Jiaxing 314003, China
3 School of Opto-Electronic Engineering, Changchun University of Science and Technology, Changchun 130013, China
* Correspondence: caojie@bit.edu.cn

Abstract: Fuji apples are one of the most important and popular economic crops worldwide in the fruit industry. Nowadays, there is a huge imbalance between the urgent demand of precise automated sorting models of fruit ripeness grades due to the increasing consumption levels and the limitations of most existing methods. In this regard, this paper presents a novel CNN-based fine-grained lightweight architecture for the task of Fuji apple maturity classification (FGAL-MC). Our proposed FGAL-MC architecture has three advantages compared with related previous research works. Firstly, we established a novel Fuji apple maturity dataset. We investigated the Fuji apple's different growth stages using image samples that were captured in open-world orchard environments, which have the benefit of being able to guide the related methods to be more suitable for the practical working environment. Secondly, because maturity grades are difficult to discriminate due to the issues of subtle expression differences, as well as the various challenging disadvantages for the unstructured surroundings, we designed our network as a fine-grained classification architecture by introducing an attention mechanism to learn class-specific regions and discrimination. Thirdly, because the number of parameters of an architecture determines the time-cost and hardware configuration to some extent, we designed our proposed architecture as a lightweight structure, which is able to be applied or promoted for actual agriculture field operations. Finally, comprehensive qualitative and quantitative experiments demonstrated that our presented method can achieve competitive results in terms of accuracy, precision, recall, F1-score, and time-cost. In addition, extensive experiments indicated our proposed method also has outstanding performance in terms of generalization ability.

Keywords: fine-grained; attention mechanism; maturity classification; open-world

Citation: Zhang, L.; Hao, Q.; Cao, J. Attention-Based Fine-Grained Lightweight Architecture for Fuji Apple Maturity Classification in an Open-World Orchard Environment. *Agriculture* **2023**, *13*, 228. https://doi.org/10.3390/agriculture13020228

Academic Editors: Muhammad Sultan, Redmond R. Shamshiri, Md Shamim Ahamed and Muhammad Farooq

Received: 10 November 2022
Revised: 11 January 2023
Accepted: 12 January 2023
Published: 17 January 2023

1. Introduction

The Fuji apple is one of the most popular kinds of fruits worldwide [1]. Different ripeness levels have great differences in taste and usage [2]. For instance, at an early stage of maturity, apples can be used for cider vinegar or cider brewing, whereas apples are ready to be eaten or used for bread and pies at a later stage of ripeness. Moreover, there is a direct relationship between maturity level and the quality or price sold on the market [3]. In addition, different maturity levels have different storage condition requirements, such as temperature, humidity, duration, and so on. Due to this, agricultural managers may formulate a specific harvesting strategy or configure different quantitative environmental factors according to their current maturity stage, such as sunlight, soil, nutrients, etc. Therefore, an accurate and effective distinction of the maturity level is crucial to its management [4–7]. When the Fuji apples gradually mature, the content of internal chlorophyll decreases while the epidermal red pigment increases, and as the pulp begins to soften, the starch begins to convert into sugar and the acidity decreases. Therefore, the indicators of apple ripeness can be determined by the fruit's color, firmness, starch content, soluble solid content, and sugar content. Basically, the task of apple maturity classification can be divided into destructive and non-destructive methods. The destructive method of grading the ripeness of fruit is performed by quantitatively testing the content of various components in the fruit by chemical means [8,9], which usually causes certain damage to the fruit and makes it impossible to subsequently sell [10]. Computer vision-based apple maturity classification methods are very meaningful and hopeful for tasks in the practical agricultural working system [11]. In recent years, fruit maturity grades sorting or classification research has been widely carried out, such as citrus fruits maturity [12], tomato [13,14], strawberry [15,16], mango [17], and specifically for apples [2,8,9]. With the rapid development of apple harvesting robots in orchard environments, there is urgent demand for visual systems with diverse functions to meet the different custom requirements. Unlike operating in the indoor environment, where many uncontrollable influencing factors may be mitigated compared with the wild, open-world scenarios. In an outdoor environment, fruits are easily occluded by nearby branches, leaves, or even weeds [18]. Moreover, algorithms or models are also easily affected by different sunlight conditions and the background. Despite all these challenges, distinguishing different maturity levels of fruits effectively is one of the most important tasks to achieve intelligent orchard management. Moreover, orchard managers may arrange certain harvesting tasks or the numbers of employees for different maturity periods accordingly. In this study, we presented a novel way to classify different maturity grades of Fuji apples. Our proposed method has many significant differences compared with most related published work. The overall flowchart of our proposed method is shown in Figure 1.

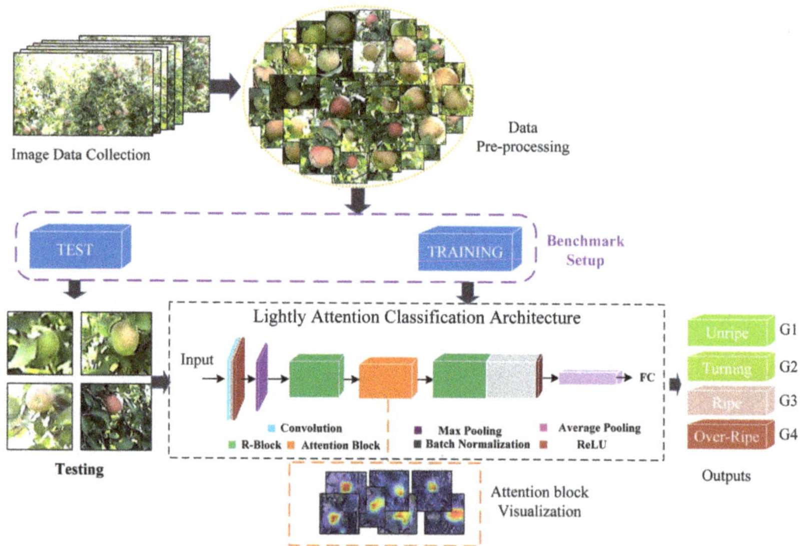

Figure 1. The overall flowchart of our Fuji apple maturity classification task.

Firstly, we captured image samples according to the different growth stages of Fuji apples. After annotation, cropping, labeling, etc., we established a dataset for the Fuji apple maturity classification task. Moreover, we proposed an attention-based fine-grained lightweight architecture for Fuji apple maturity classification (AFGL-MC). Our proposed AFGL-MC is able to learn class-specific regions and discrimination by improved attentional mechanisms, which helped the model to achieve high accuracy outputs. Finally, the results of comprehensive and extensive experiments demonstrated that our presented architecture provided a potential method for practical application and promotion. The main contributions of our proposed method can be summarized as follows:

- We captured images from outdoor orchard environments during various Fuji apple growth periods, and a Fuji apple maturity benchmark, which contains certain practical influencing factors, was established. Learning-based networks that are trained with diverse scenario data are more suitable to be applied and generalized for practical orchard working environments.
- We proposed a novel AFGL-MC architecture for Fuji apple maturity classification. To the best of our knowledge, it is the first time Fuji apple maturity classification has been applied as a fine-grained task. In addition, to reduce the probability of confusion caused by high similarity in different categories, an improved attentional mechanism was introduced to enhance the discriminative ability of our model. Moreover, we designed the structure as lightweight as possible to facilitate model promotion and practical use.
- Finally, comprehensive and extensive experiments were conducted to demonstrate that our proposed method not only has good performance for the task of Fuji apple maturity classification, but also has excellent performance in other fruit categories and quality classification tasks.

The rest of this paper is organized as follows. In Section 2, we introduce background information and related work. Section 3 shows how we collected and established the Fuji apple maturity dataset. Then, we present our proposed AFGL-MC architecture in Section 4, and describe our experiments and analysis in Section 5. Finally, in Section 6 we discuss the conclusions and plans for future work.

2. Related Work

2.1. Deep-Learning-Based Methods

Since color changes in the appearance of the fruit can be used as a key indicator for the stage of maturity [19], potential non-destructive automatic ways of assessing apple maturity are provided by computer vision techniques [20]. Ref. [21] proposed a method with two layers of a feed-forward back-propagation artificial neural network (ANN) to classify the maturity category of apples and count the number of ripe and unripe apples. The images used to evaluate this method were from online websites. Most of these images were captured under ideal conditions, having almost no background influence. Similarly, Ref. [22] graded Fuji apples into three stages, namely, immature, overripe, and mature according to the changes in apple appearance color. Furthermore, the ANN-based method was utilized to classify Fuji fruits. The methods based on machine vision need to extract features manually, then segment the fruit from the background before judging the level of standards, and such kinds of operations are quite complex. To improve this weakness, Ref. [23] exploited CNN-based architectures to achieve ripe, overripe, and unripe categorizations. In particular, Ref. [24] took Manalagi and Rome Beauty apples as research objects and presented two layers of CNN-based methods to classify the captured apple images into ripe, half-ripe, and raw categories. Such CNN-based methods are more robust in performance and limit the steps of the workflow. Ref. [16] proposed a strawberry ripeness detection system through a camouflage-based data augmentation technique to simulate the strawberry harvesting in natural environment conditions and achieved promising outputs.

All the above research works have inspired the vigorous development of the orchard management industry. However, when applied in working fields, the following shortcomings need to be addressed. First of all, all computer vision-based methods seldom consider the influence of the fruit background. The fruits are entirely in the middle of the image, with a pure different background color, and hardly occluded by leaves or branches. Such kinds of images are too idealistic to be suitable for actual use cases. Secondly, there is no set standard for the classification of maturity levels, so they are difficult to promote or apply practically. Finally, the robust performance of these methods is not strong enough in some regards, although such CNN-based methods reduce some manual operations and lead to a more intelligent workflow, the designs of these models exploit the CNN to extract features, or try to use some complex architecture, that hardly translates to real-world environments. For all these reasons, the model is difficult to generalize and apply practically.

2.2. Fine-Grained Visual Categorization (FGVC)

In recent years, CNN-based methods ahve achieved great improvements in many fields [25–27]. FGVC refers to a more detailed classification and has become a hot topic in recent years [28–32]. Some existing FGVC methods had used category-level information intuitively and completed FGVC tasks by combining visual information and text descriptions [33]. Generally, such kinds of methods require high requirements for data collection and annotation, which leads to high costs. Some are using the transfer learning method by adding domain adaptation or instance-level-weighted mechanisms, as [34]. Although these methods have taken adaptability into account, they are not suitable in practice due to the complicated training steps involved in its application. Other classical algorithms have focused on the informative parts of an image by introducing an attentional mechanism [35]. Focusing on both the representations and regions used to distinguish categories, attentional mechanisms are widely used in FGVC tasks [36,37]. Therefore, FGVC-based classification tasks for fruits and vegetables have made great progress and have brought valuable and meaningful thoughts to researchers. However, most of the aforementioned methods were applied in laboratory settings; rare investigation on the portability of these methods is one of the most serious limitations.

3. Data Acquisition

First and foremost, we captured images from an open-world orchard environment in Bologna (11°21′ E, 44°30′ N), Italy, with a digital camera (Cannon EOS Kiss X5). More than sixteen weeks of images were captured from 5 July to 15 November once per week, and the distances between the camera and the Fuji trees were around one to one and a half meters. In the first four weeks, the fruits were quite small and totally immature, therefore the images from the last twelve weeks, from which we captured 9852 images, were kept as our original image dataset and processed with a resolution of 1280 by 720 pixels. Some samples are shown in Figure 2.

Figure 2. Collected samples from a variety of growth periods. From item (1) to (12), it represents the process of changing the appearance of the Fuji fruit from immature to totally ripe.

As samples shown in Figure 2, when Fuji apples are in the early immature stages, their appearance is almost entirely turquoise. Gradually, the green part is lightly reduced, yellows, and slowly turns red during the progression from the immature to the mature stage. Moreover, images collected in the outdoor environment are easily affected by natural sunlight conditions. For instance, when images were captured in the case of strong sunlight, the color of the fruit's appearance is prone to a certain degree of whitening. For such kinds of images, traditional image pre-processing methods may meet great challenges. Therefore, a more robust method is urgently needed for real surroundings. Due to the crucial factor of establishing an accurate and appropriate dataset for the maturity model learning and prediction, we labeled the images according to the period of image collection and the appearance of color changes. Depending on the corresponding relationship between the changes in fruit epidermis and ripeness, the fruit ripeness grade was divided into four categories according to the United States Standards for Grades of Apples [38]. Ref. [38] indicated as a mature apple becomes overripe it will show varying degrees of firmness, depending upon the stage of the ripening process, hence "Hard", "Firm", "Firm ripe", and "Ripe" were the four terms used for describing the different stages. Due to the appearance of Fuji apples being highly related to its attributes, we invited experts to label these images. According to the captured image date and the appearance of each fruit, the experts classified these images into unripe, turning, ripe, and overripe. These four categories were respectively marked as grade one (G1), grade two (G2), grade three (G3), and grade four (G4) for short. Partial samples from our proposed benchmark are shown in Figure 3.

Figure 3. The partial samples of our proposed Fuji apple maturity benchmark.

As the samples show, when the fruits transition from G1 to G2, to G3, and to G4, the appearance color of the fruits change from green (G1) to light pink in some areas of the fruits (G2). By G3, most of the fruit has turned to pink, and finally most of the fruit's appearance presents a relatively dark red area (G4). This process takes an apple from unripe to fully ripe. Moreover, we collected the images totally from an outdoor orchard environment under natural sunlight and surroundings. Therefore, Fuji apples have a variety of poses and degrees of illumination, such as over-light, normal and insufficient. Ideally, if we completely ignore or eliminate the background factor when designing the technical algorithms for fruit classification, detection, counting, or depth estimation, the methods indeed have certain limitations for practical use. Therefore, we took unfavorable background factors into account and established our benchmark for the Fuji apple's maturity classification. The samples of Fuji apple objects from our proposed benchmark are affected by varying degrees of occlusion due to leaves, branches, or weeds, as shown in the image samples. In fact, the unfavorable influential factors of complex orchard background and diverse severity of occlusion posed a greater challenge to the robustness of the performance in the design of these methods. In summary, more serious challenges to models or methods have to be faced in realistic open-world environment conditions. Nevertheless, we hope to train and evaluate the model based on the challenging data obtained from the outdoor environments, which are likely to guide our model closer to practical application. Finally, 1740 images were obtained and the number of each grade was confirmed to be the same. We then split the images of each grade according to the ratio of the training set and the test set to 4:1. Compared with the other public datasets used for classification tasks, such as ImageNet [39] and MS COCO [40], our dataset has a smaller set of samples with lower resolution. In fact, such kinds of tiny-scale datasets are more prevalent in practical fields due to the high time and monetary costs for data collection in practical use cases. Training with such a tiny-scale dataset means that the network model with deep CNN layers is extremely prone to the phenomenon of overfitting, such as VGG16 [41] or AlexNet [42]. On the contrary, networks with very few CNN layers may be able to mitigate the issue of overfitting effectively. However, the accuracy of the prediction results is hard confirm, especially in an open-world environment, such as the influences of complex background information, sunlight conditions, and occlusions, etc. Therefore, designing a CNN-based architecture suited for this tiny-scale dataset meets even more challenges compared with large-scale datasets.

4. AFGL-MC Architecture

In recent years, CNN-based computer vision classification methods have achieved encouraging performance in many fields, for example, VGG16/19, InceptionNet [43,44], ResNet [45], AlexNet, etc. These methods require a large number of training epochs and a relatively large-scale dataset with diverse image samples. Generally, the model has deeper layers, and hence the time cost is higher for the final prediction and training. Therefore, a model has with a deep architecture is difficult to combine with agricultural robots and promote in practical fields. However, while improvements such as simply reducing or cutting the number of layers from the architecture, can overcome the high time cost and the requirements for large-scale and diverse samples to a certain extent, the model is still prone to encounter problems related to insufficient information learned from the samples of the dataset and the accuracy of the final prediction cannot be guaranteed. Furthermore, as mentioned previously, the acquisition and setup of a large-scale and diverse dataset is a time-consuming and labor-intensive task generally, hence the models that rely on such large-scale datasets may have limitations in their ability to be generalized. The ability of generalization is one of the most important indicators in evaluating whether the model has the potential to be applied in practice in an outdoor orchard environment. To strengthen its generalization, we consider designing a lightweight network model by introducing attentional mechanisms to lead the model to focus on areas with important features rather than them all. In trying to use the limited number of parameters effectively, it is possible to ensure the model has a lightweight architecture and excellent outputs at the same time. Based on the above considerations, we designed the model as shown in Figure 4.

Compared with the classical network models, such as VGG, InceptionNet, ResNet, and AlexNet, as well as in the case of MobileNet-Tiny, AFGL-MC has the notable characteristic of being lightweight. Specifically, the inputs to AFGL-MC are resized RGB color images with both the values of width and height being 64 pixels, and the output is a one-hot encoder that presents the classification results in four kinds of categories. We also detail the AFGL-MC architecture in the following section.

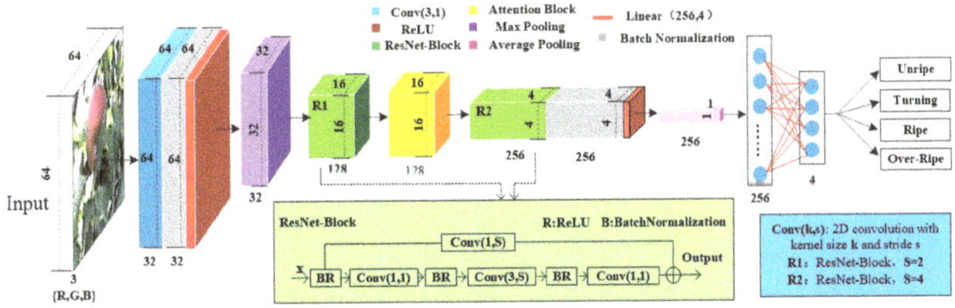

Figure 4. The architecture of our proposed AFGL-MC.

4.1. Layer Unit

The layer unit of our proposed architecture includes the layer of convolution, batch normalization, activation function, pooling, and fully connected layer. The convolution (CONV) layer plays a vital role in extracting image features. Each convolution kernel has the attributes of size, stride, and depth. Generally, the larger the kernel size, the better the quality of the bandpass filters learned [46]. However, for our Fuji apple maturity classification task, the distinction between different levels is so close, and the larger kernel size may lead the model to focus on the large shape or structure of the image, but not on the attributes of local characteristics. Based on such considerations, we chose 1 or 3 as our CONV kernel size. Therefore, the convolution in our architecture only has one or three of these two kinds of kernel sizes. Another benefit is that the number of parameters is reduced as much as possible, which allows it to be lightweight. Regarding the attributes of stride and depth, we designed the model according to the principle that the feature size is gradually reduced. Meanwhile, depth is gradually increased. The high degree of correlation and coupling between layers in the network is one of the factors that causes difficulties with deep neural network convergence and achieving good performance [47]. We exploited the batch normalization (BN) layer after each CONV layer to improve the internal covariate shift phenomenon. We chose ReLU as the activation function [48], shown as equation:

$$f(x) = max(0, x) \tag{1}$$

such that when $x \geq 0$, the output is a linear function. Otherwise, the output value is zero. Additionally, the ReLU has faster convergence than sigmoid and tanh when the outputs are generated using a linear function. We applied the fully connected (FC) layer [49,50] to connect every neuron from the previous layer to the last layer, adn finally the flattened matrix goes through a fully connected layer to classify the images. The FC is presented as equation:

$$u_l = \omega^l x^{(l-1)} + b^l \tag{2}$$

where u_l is the FC layer and $x^{(l-1)}$ is the output of the previous layer. In addition, ω^l and b^l are the weight and bias coefficients, respectively. The output vector u_l goes through softmax activation to obtain the classification likelihood for each potential category. The max-pooling and average-pooling layers provide a typical down-sampling operation that reduces the in-plane dimensionality of the feature maps. The advantages of a pooling [50] layer are to introduce a translation invariance for small shifts and distortions and to decrease the number of subsequent learnable parameters. For the shallow layer of the architecture, we applied max-pooling to filter out features with large amounts noise information, whereas an average pooling layer that is arranged before a fully connected layer based on the consideration of the features from deeper layers could help the final classification results in general.

4.2. Block Unit

The block unit of our presented architecture includes both ResNet [44] and attention blocks. Benefiting from the shortcut structure, the residual network is quite popular as a unit for the backbone in feature extraction, as applied to a large amount of CNN-based network models. The operation expression is shown as equation:

$$x_l = H_l(x_{l-1}) + x_{l-1} \tag{3}$$

here, we denote the output of the ℓth layer as x_l, and $H_l(\cdot)$ represents a nonlinear transformation. The final feature map is the element-wise addition of $H_l(x_{l-1})$ and x_{l-1}. According to its characteristics, we applied the idea of the ResNet block as our attention and the backbone module. To design our attention model, we kept the dimensions of the input and output the same. The improved module is presented as equation:

$$x_l = H_l(x_{l-1}) + Conv_l(x_{l-1}) \tag{4}$$

here, $Conv_l(\cdot)$ represents a convolution operation, and the element-wise addition of $H_l(x_{l-1})$ and $Conv_l(x_{l-1})$ is the output of the attention module. For the backbone network, we have optimized and reduced the module as much as possible, through means of the size of the input samples or in the number of layers of the network model. Intuitively, our model has several stack layers, thus we strove to reduce the values for height and width of the output features from residual blocks by improving its structure. Next, we discuss the attention block [51]. It is hard to guarantee accurate indicators for the final outputs of a CNN-based network with very shallow layers. We proposed an attentional mechanism to strengthen our proposed AFGL-MC to obtain important information from learned samples. Our improved attention structure is shown in Figure 5.

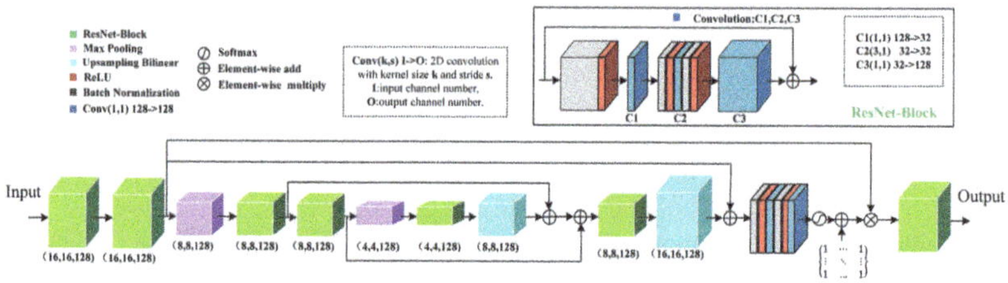

Figure 5. The architecture of our proposed attention module.

Specifically, our proposed attention module was inspired from a residual attention network [51] as multi-scale, and the attention module H is presented as equation:

$$H_{i,c}(x) = (1 + M_{i,c}(x)) * F_{i,c}(x) \tag{5}$$

where i values represent spatial positions and c is the index of the channel. $M(x)$ arranges the values from $[0, 1]$. By element-wise addition of a constant matrix and $M(x)$, we can get the equation $H_{i,c}(x) = F_{i,c}(x)$, when $M(x)$ is zero. As for our proposed attention block, the $M(x)$ includes three branch features that are concatenated to improve the utilization of the extracted features. Moreover, instead of the form of stacked attention blocks, we applied only one attention block by element-wise addition with different feature scales. In addition, we added residual modules at the head and tail to strengthen the functionality of our attention block.

4.3. Loss Function

To mitigate large changes in parameters that may be caused by one or a few noisy samples, we calculated the loss value over the whole batch. During the training processing, we smoothed the loss value by calculating a certain number of batch sizes to mitigate large changes in parameters that may be caused by one or a few noisy samples. The loss value L is presented as equation:

$$L = \frac{1}{N}\sum_{i}^{N} L_i \tag{6}$$

where N is the number of samples in a batch size for one training epoch. L_i is the loss value for each sample, and is shown as equation:

$$L_i = \sum_{c=1}^{M} y_{ic} log(p_{ic}) \tag{7}$$

where M is the number of categories. The probability that sample i belongs to category c is presented as p_{ic}, and the sign function y_{ic} is denoted as equation:

$$y_{ic} = \begin{cases} 1 & i = C \\ 0 & otherwise \end{cases} \tag{8}$$

when the class label i is equal to the true label C, then $y_{ic} = 1$. Otherwise, $y_{ic} = 0$.

4.4. Training and Evaluation

The main hardware configurations of our experiments were conducted on one Intel i5-7500 CPU @ 3.20 GHz processor with 8GB of memory. The GPU is a NVIDIA GeForce GTX 1080 with 11 GB of memory. The operating system is Ubuntu 18.04. We use Python as a programming language and pytorch as our framework. Moreover, due to the learning rate (LR) settings playing a crucial role during the network training, we set the LR as dynamic with the training epochs, correspondingly. In general, a larger LR is suitable for helping the network adjust more quickly and effectively in the beginning stage. As the number of epochs increases, the value of the parameter gets close to the optimal value, hence, if we maintain a large LR throughout the process, it is prone to lead the network to skip its optimal value. Based on the above considerations, we designed a dynamic LR that can be updated as the number of epochs increases with flexibility (Table 1).

Table 1. The relationship between the LR and the number of epochs.

Epochs	LR
$[1 \sim N \times 30\%)$	0.1
$[N \times 30\% \sim N \times 60\%)$	0.01
$[N \times 60\% \sim N \times 90\%)$	0.001
$[N \times 90\% \sim N)$	0.0001

Here, N is the number of total epochs, which was set to 200 in this experiment. The AFGL-MC was trained with a mini-batch size of 64 and tuned with 200 epochs. All the hyper-parameters were optimized with 60 update epochs at learning rates of 10^{-1}, 60 to 120, and 120 to 180, whereas the last 20 update epochs were optimized at lower learning rates of 10^{-2}, 10^{-3}, and 10^{-4}. The terms of precision, recall, F_1-score, and accuracy were taken as the evaluation indexes to verify our proposed network extensively and quantitatively. The corresponding formulas are:

$$Precision = \frac{TP}{TP + FP} \tag{9}$$

$$Recall = \frac{TP}{TP + FN} \tag{10}$$

$$F_1 = \frac{2TP}{2TP + FP + FN} \tag{11}$$

$$Accuracy = \frac{TP + TN}{TP + TN + FN + FP} \tag{12}$$

Specifically, the *TP*, *TN*, *FP*, and *FN* represent the relationship between the observed value and the predicted value, as shown in Table 2.

Table 2. Instructions for the elements of TP, TN, FP and FN.

Label Name	Predicted Positive	Predicted Negative
True Positive	TP	FN
True Negative	FP	TN

5. Experiments and Results Analysis

5.1. Fuji Apple Ripeness Results

We trained the AFGL-MC from scratch, setting the number of training epochs and batch sizes to 200 and 64, respectively. For our specific Fuji apple maturity classification task, we applied the test dataset to verify the accuracy for each ripeness grade after every epoch. These evaluation results are shown in Figure 6.

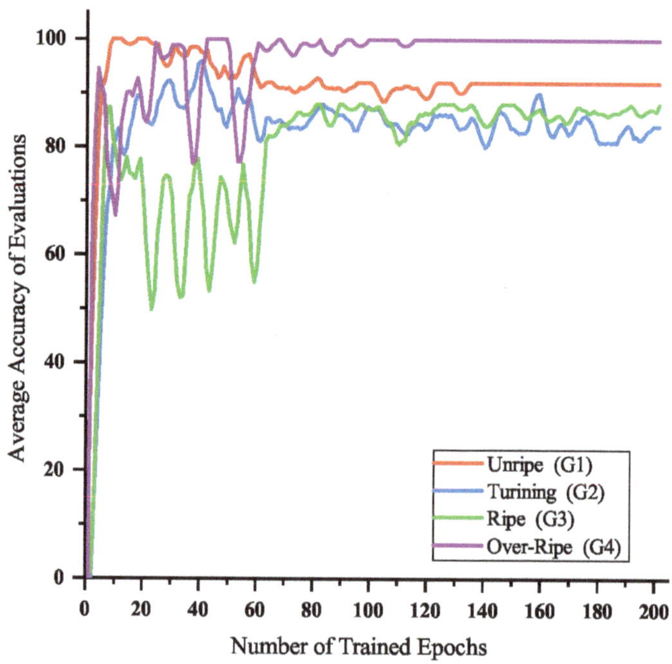

Figure 6. The relationship between the accuracy of each maturity grade and the different training epochs.

As the curves show, when the epochs progress from 1 to 60, each maturity category varies in accuracy severely, which is due to the set of the large learning rate values and the adjustment of the initialization parameters. In this range, G3 and G4 tend to have opposite results. Such a phenomenon is reflected in the practical situation, where fruits at the G3 and G4 levels have extremely similar features in terms of appearance. With the number of iterations increasing, the accuracy of G4 and G1 tends to be stable at in the early ranges compared with the other two grades, while G2 and G3 gradually smooth out and the accuracy tends to a more stable value. This phenomenon may be caused by the obvious appearance feature characteristics in G1 (totally green) and G4 (totally red). Whereas for a fruit in G2 or G3, the changing stage includes some influence factors from appearances that may cause a lower accuracy and a slower change to stable values. Moreover, we visualized the feature map and focused on the part that was predicted by our proposed attention module, as shown in Figure 7.

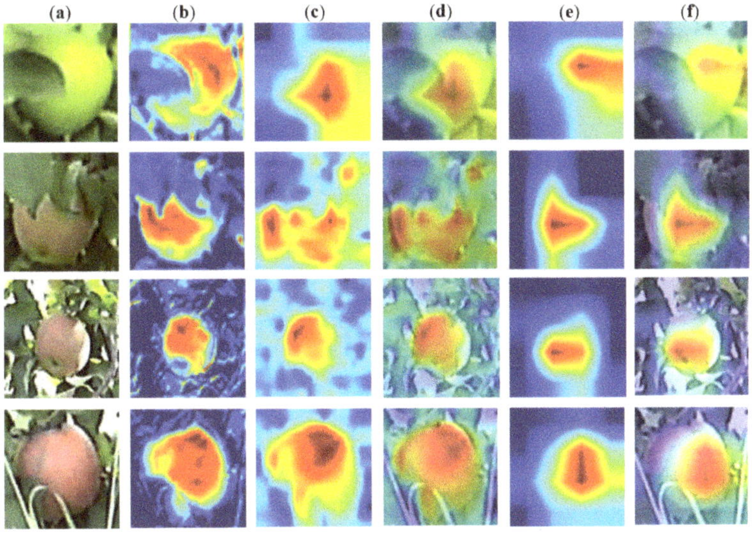

Figure 7. Visualizations of the feature map, highlighting the class-specific discriminative regions for the predicted examples from our proposed network. (**a**) Original four grades' images from our test dataset. (**b**) Visualization feature map by our trained attention block. (**c**,**d**) Localization of class-discriminative regions by the CAM technique in our attention layer, overlapped with the original image. (**e**,**f**) Grad_CAM visualizations of our attention block and overlapped with the original image. Note that in (**c**–**f**), red regions correspond to high class scores.

These samples are from our test dataset, which comprise four maturity grades, as well as images of the Fuji apples occluded by leaves, weeds, or partial appearance whitening due to over-sunlight. These unfavorable factors present great challenges for our fine-grained classification tasks. From the visualization feature map, the features predicted by our attention module can distinguish the background from the fruit accurately, even from the whitening part. As the highlighted region map shown by CAM [52] and Grad-CAM [53], our network evidently could discriminate specific classes for fine-grained maturity classification.

5.2. Model Comparison

In order to verify our proposed network comprehensively, we compared our proposed AFGL-MC with ResNet-18 [45], DenseNet-121 [54], MobileNet-Tiny [55], AlexNet [42], and VGG16 [41] on the task of Fuji apple ripeness classification. We trained all these network models independently on our tiny-scale dataset with 200 training epochs and confirmed the other conditions were as close as possible. We found AlexNet and VGG16 were hard to converge on such a limited number of training samples and number of iterations, therefore we only compared the AFGL-MC with ResNet-18, DenseNet-121, and MobileNet-Tiny. We saved the parameters for each trained epoch and predicted the images from the test dataset, and the results of which are shown in Figure 8.

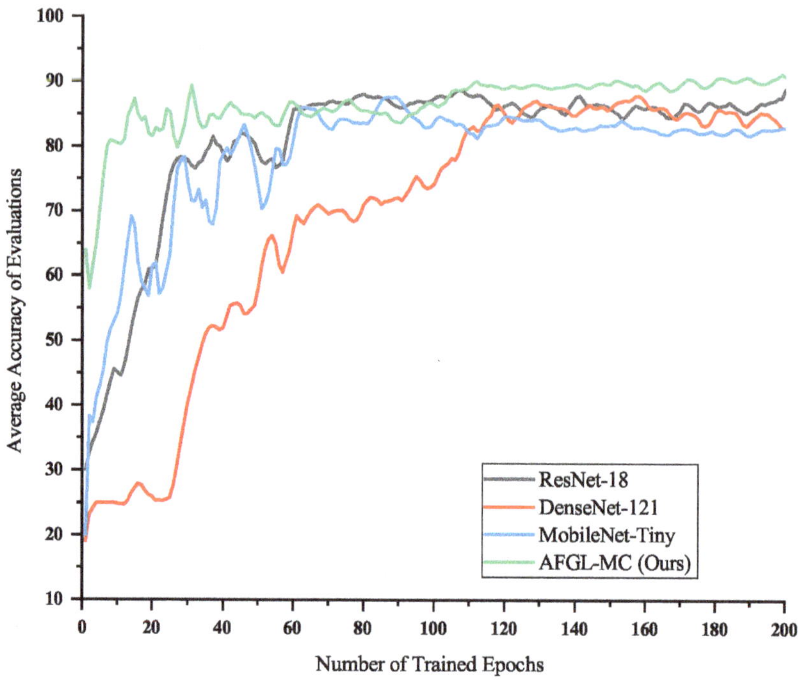

Figure 8. The evaluation results between accuracy and the saved trained models.

As can be seen in the curves shown in Figure 8, our proposed AFGL-MC has the fastest speed and tends to stabilize, followed by ResNet-18 and MobileNet-Tiny, respectively, with the slowest being DenseNet-121. Moreover, our proposed AFGL-MC achieved the best accuracy after 200 epochs of training compared with the others. For a more in-depth analysis, we took precision and recall as evaluation indicators to verify the performance of trained models for each maturity grade. To confirm the fairness of the experiments, we uniformly saved the parameters of the 200th epoch as the final evaluated pre-trained model. The results for each maturity level are shown in Table 3.

Table 3. The evaluation results of four comparison models on each maturity grade.

Level	Indicator	ResNet-18	Dense-121	Mobile-Tiny	Ours
G1	Precision	0.960	0.880	0.80	0.92
G1	Recall	0.960	0.846	0.833	1.0
G1	F_1	0.960	0.863	0.816	0.958
G2	Precision	0.840	0.760	0.80	0.840
G2	Recall	0.840	0.731	0.833	0.875
G2	F_1	0.840	0.745	0.816	0.875
G3	Precision	0.76	0.80	0.76	0.88
G3	Recall	0.86	0.870	0.731	0.846
G3	F_1	0.809	0.833	0.745	0.863
G4	Precision	1.0	0.88	0.88	1.0
G4	Recall	0.893	0.88	0.786	0.926
G4	F_1	0.943	0.88	0.830	0.962
Average	–	0.89	0.83	0.83	0.91

From the table, the results obtained by our proposed model at each ripeness level are generally good, and most of the indicators have achieved optimal values, indicating the effectiveness of our proposed model. In addition, our network achieved notable prediction results for G1 and G4 levels, with F1 scoring at 0.958 and 0.962, respectively. Although G2 and G3 have many appearance influences, as were explored previously, our proposed network obtained comparatively superior F1 values of 0.88 and 0.863, respectively. Moreover, we took the confusion matrix to more intuitively analyze the performance of each network at different maturity levels, as is shown in Figure 9.

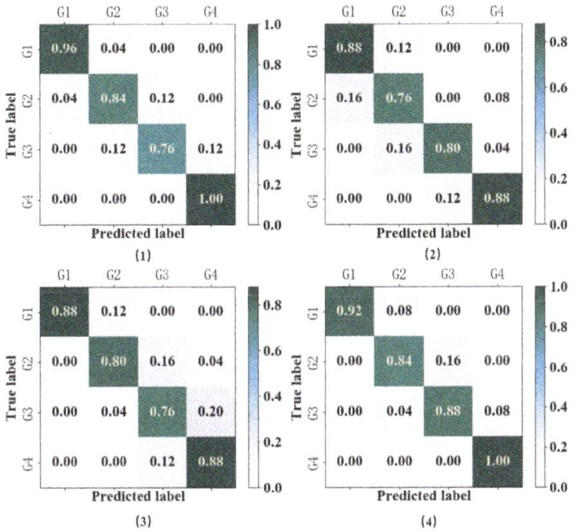

Figure 9. The confusion matrix results of comparison model. (1), (2), (3) and (4) are the results of the comparison model ResNet-18, Dense-121, Mobile-Tiny and our proposed AFGL-MC on each maturity grade, respectively.

From the confusion matrix, it can be seen that the model we proposed and the results obtained by ResNet-18, DenseNet-121, and MobileNet-Tiny on G1 and G4 have obvious advantages compared to the other two levels. However, our proposed network obtained comparable results with the other networks on G2 and G3. It may have benefited from the increased attention mechanism of our proposed network, which may lead our model to pay more attention to the fruit regions, while mitigating the influence of occlusion, sunlight conditions, etc. To analyze the performance of our proposed model under different occlusion and illumination conditions, we set up a experiment for the four previously compared models under normal light without occlusion, normal light with occlusion, and non-integrated light without occlusion. The final results are shown in Table 4.

Table 4. The accuracy results of the four comparison models under different Occlusion and illumination conditions.

Level	ResNet-18	Dense-121	Mobile-Tiny	Ours
No Occlusion	0.927	0.85	0.842	0.930
Occlusion	0.710	0.653	0.620	0.72
Over-Illumination	0.807	0.760	0.759	0.782

From the compared results, we found that all the compared models achieved better accuracy results under natural illumination and no occlusion conditions. When fruits were occluded by objects, the accuracy of the output results were largely decreased, which is probably due to the fact that occlusion factors, in a serious way, will affect the judgment of the fruit's appearance. For the occlusion condition, our proposed model obtained better outputs than the other three compared models, which may be thanks to the attention mechanism that focuses on the fruit region. As for the illumination changes, they are less affected than in the case of occlusions, and our proposed model handles this situation well.

5.3. Generalization Evaluation

In order to apply the AFGL-MC more prevalently, we took the Fruits Quality (FQ) dataset [56] and the Fruits and Vegetables Classification (FV) dataset [57] to verify the generalization performance of our proposed model. The FQ dataset is a public dataset that provides six fruit categories. Each category has a good, bad, and mixed class. The benchmark includes categories for apples, bananas, guava, lemon, orange, and pomegranate. We randomly chose a bad folder from each category as the benchmark. A selection of samples are shown in Figure 10.

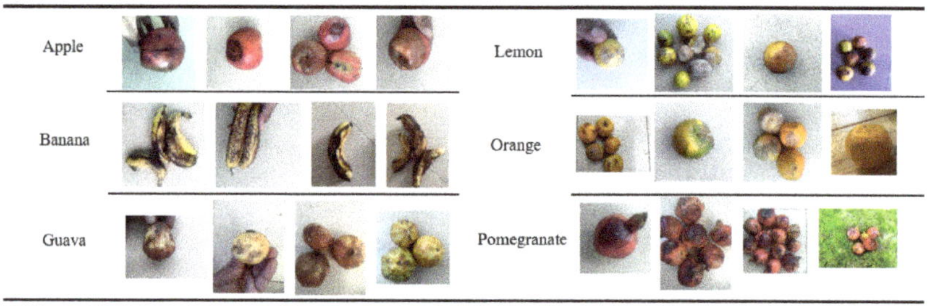

Figure 10. Examples of the FQ dataset on six categories.

In detail, all the samples from the FQ benchmark are image data with width and heights equal to 256 pixels, including the influences of occlusion, complex background, or varied illumination light. The number of samples in each category is around 1000. For example, there are 1104 and 419 samples as the training and test datasets for the apple category. We trained ResNet-18, DenseNet-121, MobileNet-Tiny, and our proposed network on this public dataset. The relationship between the training times and the average accuracy for the final prediction results is shown in Figure 11.

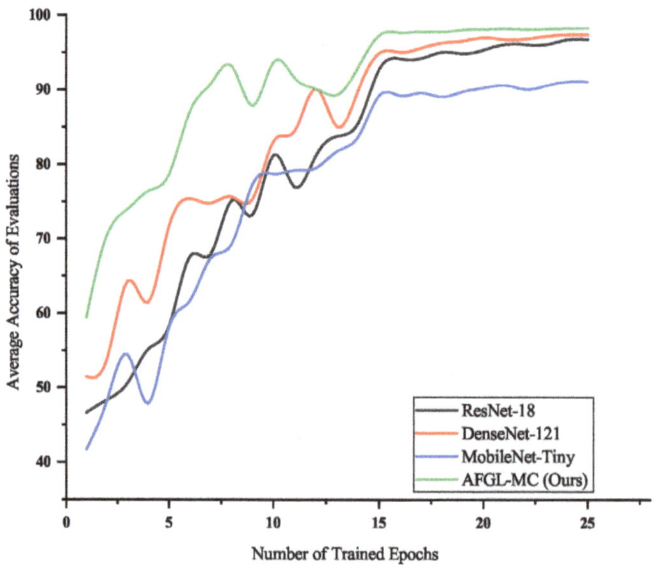

Figure 11. The evaluation results on FQ benchmark.

From the curves, the final prediction accuracy of ResNet-18, DenseNet-121, MobileNet-Tiny, and FGAL-MC are 0.97, 0.91, 0.96, and 0.98, respectively. The results evidently show that our proposed network has better average accuracy results at each epoch compared with the other three, which may reflect that the proposed method has good generalization ability in terms of this dataset. Similarly, we visualized the focused part using a proposed attention module with Grad-CAM highlight method, as shown in Figure 12.

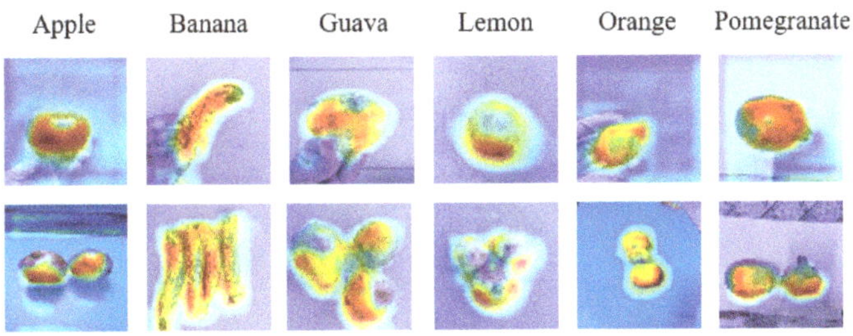

Figure 12. Highlights the class-specific discriminative regions for the predicted examples from our proposed network.

From the discriminative regions, it can be found that our attention model can effectively obtain regions with clear boundaries between objects and the background, which would be helpful for the classification task, thus improving the accuracy of the model. The FV dataset contains 36 categories, including fruits and vegetables. Here we randomly chose 5 fruit classes for our experiment dataset. A selection of samples are shown in Figure 13.

Figure 13. Examples of the FV dataset on six categories.

As for the FV dataset, each class only has 100 training samples and 20 testing samples, which is a relatively tiny dataset. More importantly, the sizes of samples are significantly different. We take the watermelon category as an example. The largest sample is 4416 by 3312 pixels and 9.2 MB in size, whereas the smallest sample is 267 by 260 pixels and

11.28 KB in size. The network is challenged in terms of training or evaluating under such disadvantage factors. Still, we trained ResNet-18, DenseNet-121, MobileNet-Tiny, and our proposed network on this public dataset. The relationship between training times and the average accuracy of the final prediction results are shown in Figure 14.

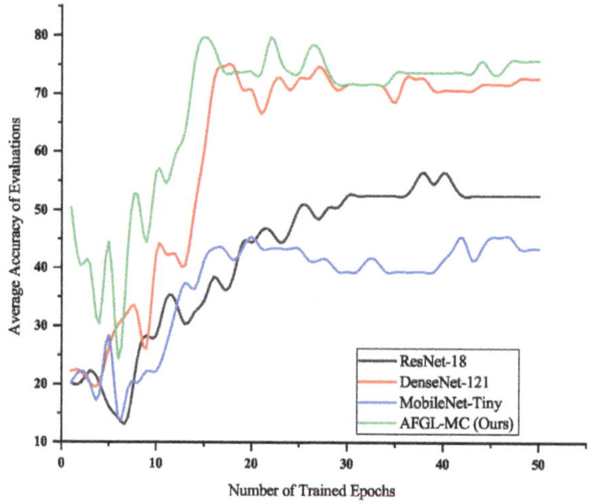

Figure 14. The evaluation results on FV benchmark.

From the curves, we can see the final values for prediction accuracy of ResNet-18, DenseNet-121, MobileNet-Tiny, and our proposed FGAL-MC are 0.53, 0.48,0.43, and 0.74, respectively. Therefore, the average accuracy of the networks trained on the FV dataset is greatly decreased compared with the FD dataset. This implies that the size of the samples and other related properties play a decisive role in a network's final prediction results. However, on such kinds of tiny-scale datasets, our proposed network has a comparatively higher accuracy over other classical network models with a limited number of training epochs. Finally, we visualized the focused part using the Grad-CAM highlight method, as shown in Figure 15.

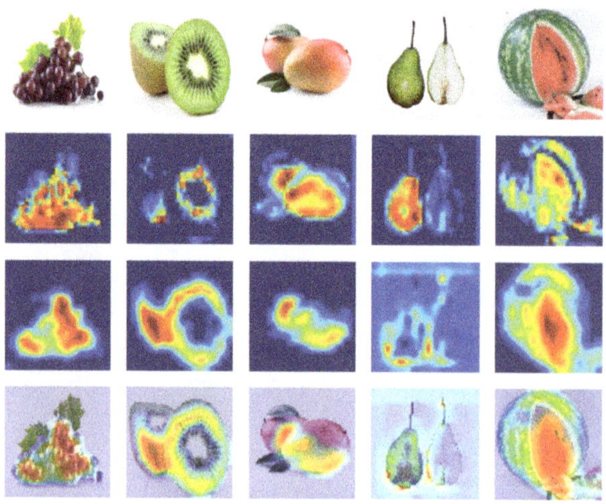

Figure 15. Visualizations of the feature map and class-specific discriminative regions for the predicted examples from the attention module of our proposed network.

Some feature maps lose the class-specific regions, e.g., kiwi and pear, whereas for mango and watermelon, the feature map did not cover all the regions of the objects. Therefore, compared to other datasets, the class-specific discriminative regions were not adequate in terms of the outputs of visualization from our attention module using the Gard-GAM method.

5.4. Parameters and Time Cost Evaluation

A model with the attributes of having lightweight parameters and low time cost is beneficial if it is to be conducted or promoted in practice. We took the number of parameters, the final pre-trained model size, and the time cost in terms of frames per second (FPS) as indicators, and the evaluation results are shown in Table 5.

Table 5. Comparison performance in parameters, memory and FPS.

Name	Number of Parameters (×105)	Memory (MB)	FPS (FQ)	FPS (FV)	FPS (Fuji Apple Maturity)
ResNet-18	111.79	44.8	472	27	386
DenseNet121	69.56	28.4	628	29	527
MobileNet-Tiny	5.40	2.3	857	30	836
AFGL-MC (Ours)	3.88	1.6	874	30	838

From the comparison results, our proposed AFGL-MC performs the best in terms of the number of parameters and memory. ResNet-18, DenseNet121, and MobileNet-Tiny are 20.7, 28.8, and 1.8 times that of AFGL-MC, respectively. The larger the number of parameters, the greater the number of parameters needed to be updated during the training process. The duration of each epoch is also increased. In terms of memory, which is proportional to the number of parameters, our proposed AFGL-MC achieved the best results. In terms of time cost, which is highly related to the resolution of the image being tested, we evaluated the performance of each comparison networks on the FQ, FV, and our proposed FM benchmark, respectively. The final results show that our proposed network achieved the best FPS values on FQ, FP, and our proposed FM dataset.

6. Conclusions

In this paper, we focused on the maturity classification task of Fuji apples. In order to achieve practical applications, we captured images from an open-world Fuji apple orchard and established a novel Fuji apple maturity benchmark that contains practical conditions. Then, we presented an attention-based fine-grained lightweight architecture for the Fuji apple maturity classification task, AFGL-MC. Our proposed AFGL-MC is advantageous in two ways. Firstly, the proposed AFGL-MC with an attention module guides the network to learn class-specific discrimination and regions for classification. Secondly, it has a slim structure, which has good performance in training and final prediction. Comprehensive experiments have shown that our proposed method has significant advantages in accuracy for maturity classification tasks, including many kinds of adverse factors from the orchard environment. Moreover, our proposed method is substantially superior to comparative models in terms of quantity of parameters and time cost. In addition, extensive experiments have shown that our proposed method has significant advantages in generalization ability, and the results implied our proposed method may provide a potential rethinking for more effective use of computer vision in the agriculture industry.

7. Future Work

In future work, we will attempt to deploy our model on a harvesting robot. Then, we will go one step further to focus on the fruit ripeness classification task under occlusion conditions, and try to realize the classification of fruit maturity under different occlusion conditions by complementing the multitude of information of nearby frames in the video.

Author Contributions: Conceptualization, L.Z., Q.H. and J.C.; methodology, L.Z.; software, L.Z.; validation, Q.H. and J.C.; formal analysis, L.Z.; investigation, J.C.; writing—original draft preparation, L.Z.; writing—review and editing, L.Z.; visualization, J.C.; supervision, Q.H.; project administration, J.C.; funding acquisition, J.C. and Q.H. All authors have read and agreed to the published version of the manuscript.

Funding: This work was supported by the Beijing Nature Science Foundation of China (No. 4222017). Funding of Science And Technology Entry program under grant (KJFGS-QTZCHT-2022-008). National Natural Science Foundation of China (62275022).

Institutional Review Board Statement: Not applicable.

Data Availability Statement: Data underlying the results presented in this paper are not publicly available at this time but may be obtained from the authors upon reasonable request.

Conflicts of Interest: The authors declare no conflict of interest.

References

1. Bonomelli, C.; Mogollón, R.; Tonetto de Freitas, S.; Zoffoli, J.P.; Contreras, C. Nutritional relationships in bitter pit-affected fruit and the feasibility of vis-nir models to determine calcium concentration in 'fuji'apples. *Agronomy* **2020**, *10*, 1476. [CrossRef]
2. Menesatti, P.; Zanella, A.; D'Andrea, S.; Costa, C.; Paglia, G.; Pallottino, F. Supervised multivariate analysis of hyper-spectral NIR images to evaluate the starch index of apples. *Food Bioprocess Technol.* **2009**, *2*, 308–314. [CrossRef]
3. Zhang, B.; Zhang, M.; Shen, M.; Li, H.; Zhang, Z.; Zhang, H.; Zhou, Z.; Ren, X.; Ding, Y.; Xing, L.; et al. Quality monitoring method for apples of different maturity under long-term cold storage. *Infrared Phys. Technol.* **2021**, *112*, 103580. [CrossRef]
4. Muscato, G.; Prestifilippo, M.; Abbate, N.; Rizzuto, I. A prototype of an orange picking robot: Past history, the new robot and experimental results. *Ind. Robot. Int. J.* **2005**, *32*, 128–138. [CrossRef]
5. Baeten, J.; Donné, K.; Boedrij, S.; Beckers, W.; Claesen, E. Autonomous fruit picking machine: A robotic apple harvester. In Proceedings of the Field and Service Robotics, Chamonix, France, 9–12 July 2007; Springer: Berlin/Heidelberg, Germay, 2008; pp. 531–539.
6. Tu, S.; Xue, Y.; Zheng, C.; Qi, Y.; Wan, H.; Mao, L. Detection of passion fruits and maturity classification using Red-Green-Blue Depth images. *Biosyst. Eng.* **2018**, *175*, 156–167. [CrossRef]
7. Faisal, M.; Albogamy, F.; Elgibreen, H.; Algabri, M.; Alqershi, F.A. Deep learning and computer vision for estimating date fruits type, maturity level, and weight. *IEEE Access* **2020**, *8*, 206770–206782. [CrossRef]
8. Pathange, L.P.; Mallikarjunan, P.; Marini, R.P.; O'Keefe, S.; Vaughan, D. Non-destructive evaluation of apple maturity using an electronic nose system. *J. Food Eng.* **2006**, *77*, 1018–1023. [CrossRef]
9. Chagné, D.; Lin-Wang, K.; Espley, R.V.; Volz, R.K.; How, N.M.; Rouse, S.; Brendolise, C.; Carlisle, C.M.; Kumar, S.; De Silva, N.; et al. An ancient duplication of apple MYB transcription factors is responsible for novel red fruit-flesh phenotypes. *Plant Physiol.* **2013**, *161*, 225–239. [CrossRef]
10. Lunadei, L.; Galleguillos, P.; Diezma, B.; Lleó, L.; Ruiz-Garcia, L. A multispectral vision system to evaluate enzymatic browning in fresh-cut apple slices. *Postharvest Biol. Technol.* **2011**, *60*, 225–234. [CrossRef]
11. Gao, F.; Fang, W.; Sun, X.; Wu, Z.; Zhao, G.; Li, G.; Li, R.; Fu, L.; Zhang, Q. A novel apple fruit detection and counting methodology based on deep learning and trunk tracking in modern orchard. *Comput. Electron. Agric.* **2022**, *197*, 107000. [CrossRef]
12. Chen, S.; Xiong, J.; Jiao, J.; Xie, Z.; Huo, Z.; Hu, W. Citrus fruits maturity detection in natural environments based on convolutional neural networks and visual saliency map. *Precis. Agric.* **2022**, *23*, 1515–1531. [CrossRef]
13. Huang, Y.P.; Wang, T.H.; Basanta, H. Using fuzzy mask R-CNN model to automatically identify tomato ripeness. *IEEE Access* **2020**, *8*, 207672–207682. [CrossRef]
14. Al-Mashhadani, Z.; Chandrasekaran, B. Autonomous Ripeness Detection Using Image Processing for an Agricultural Robotic System. In Proceedings of the 2020 11th IEEE Annual Ubiquitous Computing, Electronics & Mobile Communication Conference (UEMCON), New York, NY, USA, 28–31 October 2020; pp. 743–748.
15. Liming, X.; Yanchao, Z. Automated strawberry grading system based on image processing. *Comput. Electron. Agric.* **2010**, *71*, S32–S39. [CrossRef]
16. Sadak, F. Strawberry Ripeness Assessment Via Camouflage-Based Data Augmentation for Automated Strawberry Picking Robot. *Düzce Üniversitesi Bilim Ve Teknol. Derg.* **2022**, *10*, 1589–1602. [CrossRef]
17. Razak, T.R.B.; Othman, M.B.; bin Abu Bakar, M.N.; bt Ahmad, K.A.; Mansor, A.R. Mango grading by using fuzzy image analysis. In Proceedings of the International Conference on Agricultural, Environment and Biological Sciences (ICAEBS'2012), Phuket, Thailand, 26–27 May 2012.
18. Jia, W.; Zhang, Z.; Shao, W.; Ji, Z.; Hou, S. RS-Net: Robust segmentation of green overlapped apples. *Precis. Agric.* **2022**, *23*, 492–513. [CrossRef]
19. Bramlage, W.; Autio, W. Determining apple maturity. *Pa. Fruit News* **1990**, *70*, 78–82.
20. Hossain, M.S.; Al-Hammadi, M.; Muhammad, G. Automatic fruit classification using deep learning for industrial applications. *IEEE Trans. Ind. Inform.* **2018**, *15*, 1027–1034. [CrossRef]

21. Lal, S.; Behera, S.K.; Sethy, P.K.; Rath, A.K. Identification and counting of mature apple fruit based on BP feed forward neural network. In Proceedings of the 2017 Third International Conference on Sensing, Signal Processing and Security (ICSSS), Chennai, India, 4–5 May 2017; pp. 361–368.
22. Hamza, R.; Chtourou, M. Apple ripeness estimation using artificial neural network. In Proceedings of the 2018 International Conference on High Performance Computing & Simulation (HPCS), Orleans, France, 16–20 July 2018; pp. 229–234.
23. Xiao, B.; Nguyen, M.; Yan, W.Q. Apple ripeness identification using deep learning. In Proceedings of the International Symposium on Geometry and Vision, Auckland, New Zealand, 28–29 January 2021; pp. 53–67.
24. Gunawan, K.C.; Lie, Z.S. Apple Ripeness Level Detection Based On Skin Color Features With Convolutional Neural Network Classification Method. In Proceedings of the 2021 7th International Conference on Electrical, Electronics and Information Engineering (ICEEIE), Malang, Indonesia, 2 October 2021; pp. 1–6.
25. Mavridou, E.; Vrochidou, E.; Papakostas, G.A.; Pachidis, T.; Kaburlasos, V.G. Machine vision systems in precision agriculture for crop farming. *J. Imaging* **2019**, *5*, 89. [CrossRef]
26. Zhao, S.; Peng, Y.; Liu, J.; Wu, S. Tomato leaf disease diagnosis based on improved convolution neural network by attention module. *Agriculture* **2021**, *11*, 651. [CrossRef]
27. Lu, J.; Tan, L.; Jiang, H. Review on convolutional neural network (CNN) applied to plant leaf disease classification. *Agriculture* **2021**, *11*, 707. [CrossRef]
28. Fu, J.; Zheng, H.; Mei, T. Look closer to see better: Recurrent attention convolutional neural network for fine-grained image recognition. In Proceedings of the IEEE Conference on Computer Vision and Pattern Recognition, Honolulu, HI, USA, 21–26 July 2017; pp. 4438–4446.
29. Dubey, A.; Gupta, O.; Guo, P.; Raskar, R.; Farrell, R.; Naik, N. Pairwise confusion for fine-grained visual classification. In Proceedings of the European Conference on Computer Vision (ECCV), Munich, Germany, 8–14 September 2018; pp. 70–86.
30. Sun, M.; Yuan, Y.; Zhou, F.; Ding, E. Multi-attention multi-class constraint for fine-grained image recognition. In Proceedings of the European Conference on Computer Vision (ECCV), Munich, Germany, 8–14 September 2018; pp. 805–821.
31. Wang, Y.; Morariu, V.I.; Davis, L.S. Learning a discriminative filter bank within a cnn for fine-grained recognition. In Proceedings of the IEEE Conference on Computer Vision and Pattern Recognition, Salt Lake City, UT, USA, 18–22 June 2018; pp. 4148–4157.
32. Yang, Z.; Luo, T.; Wang, D.; Hu, Z.; Gao, J.; Wang, L. Learning to navigate for fine-grained classification. In Proceedings of the European Conference on Computer Vision (ECCV), Munich, Germany, 8–14 September 2018; pp. 420–435.
33. He, X.; Peng, Y. Fine-grained image classification via combining vision and language. In Proceedings of the IEEE Conference on Computer Vision and Pattern Recognition, Honolulu, HI, USA, 21–26 July 2017; pp. 5994–6002.
34. Gebru, T.; Hoffman, J.; Fei-Fei, L. Fine-grained recognition in the wild: A multi-task domain adaptation approach. In Proceedings of the IEEE International Conference on Computer Vision, Venice, Italy, 22–29 October 2017; pp. 1349–1358.
35. Xiao, T.; Xu, Y.; Yang, K.; Zhang, J.; Peng, Y.; Zhang, Z. The application of two-level attention models in deep convolutional neural network for fine-grained image classification. In Proceedings of the IEEE Conference on Computer Vision and Pattern Recognition, Boston, MA, USA, 7–12 June 2015; pp. 842–850.
36. Liu, X.; Milanova, M. Visual attention in deep learning: A review. *Int. Rob. Auto J.* **2018**, *4*, 154–155.
37. Luo, Y.; Jiang, M.; Zhao, Q. Visual attention in multi-label image classification. In Proceedings of the IEEE/CVF Conference on Computer Vision and Pattern Recognition Workshops, Long Beach, CA, USA, 15–20 June 2019.
38. Usda, U.; Ams, A. United States Standards for Grades of Apples. 2002. Available online: https://www.ams.usda.gov/sites/default/files/media/Apple_Standards.pdf (accessed on 15 January 2023).
39. Deng, J.; Dong, W.; Socher, R.; Li, L.J.; Li, K.; Fei-Fei, L. Imagenet: A large-scale hierarchical image database. In Proceedings of the 2009 IEEE Conference on Computer Vision and Pattern Recognition, Miami, FL, USA, 20–25 June 2009; pp. 248–255.
40. Lin, T.Y.; Maire, M.; Belongie, S.; Hays, J.; Perona, P.; Ramanan, D.; Dollár, P.; Zitnick, C.L. Microsoft coco: Common objects in context. In Proceedings of the European Conference on Computer Vision, Zurich, Switzerland, 6–12 September 2014; pp. 740–755.
41. Simonyan, K.; Zisserman, A. Very Deep Convolutional Networks for Large-Scale Image Recognition. *arXiv* **2015**, arXiv:1409.1556.
42. Krizhevsky, A.; Sutskever, I.; Hinton, G.E. Imagenet classification with deep convolutional neural networks. *Commun. ACM* **2017**, *60*, 84–90. [CrossRef]
43. Szegedy, C.; Vanhoucke, V.; Ioffe, S.; Shlens, J.; Wojna, Z. Rethinking the inception architecture for computer vision. In Proceedings of the IEEE Conference on Computer Vision and Pattern Recognition, Las Vegas, NV, USA, 27–30 June 2016; pp. 2818–2826.
44. Szegedy, C.; Ioffe, S.; Vanhoucke, V.; Alemi, A.A. Inception-v4, inception-resnet and the impact of residual connections on learning. In Proceedings of the Thirty-First AAAI Conference on Artificial Intelligence, San Francisco, CA, USA, 4–9 February 2017.
45. He, K.; Zhang, X.; Ren, S.; Sun, J. Deep residual learning for image recognition. In Proceedings of the IEEE Conference on Computer Vision and Pattern Recognition, Las Vegas, NV, USA, 27–30 June 2016; pp. 770–778.
46. Glorot, X.; Bordes, A.; Bengio, Y. Deep sparse rectifier neural networks. In Proceedings of the Fourteenth International Conference on Artificial Intelligence and Statistics. JMLR Workshop and Conference Proceedings, Fort Lauderdale, FL, USA, 11–13 April 2011; pp. 315–323.
47. Ioffe, S.; Szegedy, C. Batch normalization: Accelerating deep network training by reducing internal covariate shift. In Proceedings of the International Conference on Machine Learning. PMLR, Lile, France, 6–11 July 2015; pp. 448–456.
48. Hara, K.; Saito, D.; Shouno, H. Analysis of function of rectified linear unit used in deep learning. In Proceedings of the 2015 International Joint Conference on Neural Networks (IJCNN), Killarney, Ireland, 12–17 July 2015; pp. 1–8.

49. Hahnloser, R.H.; Sarpeshkar, R.; Mahowald, M.A.; Douglas, R.J.; Seung, H.S. Digital selection and analogue amplification coexist in a cortex-inspired silicon circuit. *Nature* **2000**, *405*, 947–951. [CrossRef]
50. Yamashita, R.; Nishio, M.; Do, R.K.G.; Togashi, K. Convolutional neural networks: An overview and application in radiology. *Insights Imaging* **2018**, *9*, 611–629. [CrossRef] [PubMed]
51. Wang, F.; Jiang, M.; Qian, C.; Yang, S.; Li, C.; Zhang, H.; Wang, X.; Tang, X. Residual attention network for image classification. In Proceedings of the IEEE Conference on Computer Vision and Pattern Recognition, Honolulu, HI, USA, 21–26 July 2017; pp. 3156–3164.
52. Yu, Y.; Xu, L.; Jia, W.; Zhu, W.; Fu, Y.; Lu, Q. CAM: A fine-grained vehicle model recognition method based on visual attention model. *Image Vis. Comput.* **2020**, *104*, 104027. [CrossRef]
53. Selvaraju, R.R.; Cogswell, M.; Das, A.; Vedantam, R.; Parikh, D.; Batra, D. Grad-cam: Visual explanations from deep networks via gradient-based localization. In Proceedings of the IEEE Conference on Computer Vision and Pattern Recognition, Honolulu, HI, USA, 21–26 July 2017; pp. 618–626.
54. Huang, G.; Liu, Z.; Van Der Maaten, L.; Weinberger, K.Q. Densely connected convolutional networks. In Proceedings of the IEEE Conference on Computer Vision and Pattern Recognition, Honolulu, HI, USA, 21–26 July 2017; pp. 4700–4708.
55. Sanjay, N.S.; Ahmadinia, A. MobileNet-Tiny: A deep neural network-based real-time object detection for rasberry Pi. In Proceedings of the 2019 18th IEEE International Conference On Machine Learning And Applications (ICMLA), Boca Raton, FL, USA, 16–19 December 2019; pp. 647–652.
56. Meshram, V.; Patil, K. FruitNet: Indian fruits image dataset with quality for machine learning applications. *Data Brief* **2022**, *40*, 107686. [CrossRef]
57. Oltean, M. *Fruits 360 Dataset: A Dataset of Images Containing Fruits and Vegetables;* Kaggle: San Francisco, CA, USA, 2019.

agriculture

MDPI

Review

Recent Advances in Carbon and Activated Carbon Nanostructured Aerogels Prepared from Agricultural Wastes for Wastewater Treatment Applications

Syaifullah Muhammad [1,2,*], Esam Bashir Yahya [3,4], H. P. S. Abdul Khalil [4,5,*], M. Marwan [1,2] and Yonss M. Albadn [5]

1. Chemical Engineering Department, Universitas Syiah Kuala, Banda Aceh 23111, Indonesia
2. ARC-PUIPT Nilam Aceh, Universitas Syiah Kuala, Banda Aceh 23111, Indonesia
3. Bioprocess Technology Division, School of Industrial Technology, Universiti Sains Malaysia, Penang 11800, Malaysia
4. Green Biopolymer, Coatings and Packaging Cluster, School of Industrial Technology, Universiti Sains Malaysia, Penang 11800, Malaysia
5. Bioresource Technology Division, School of Industrial Technology, Universiti Sains Malaysia, Penang 11800, Malaysia

* Correspondence: syaiful.muhammad@unsyiah.ac.id (S.M.); akhalilhps@gmail.com (H.P.S.A.K.)

Abstract: The global demand for clean water has become increasingly important in the past decade as a result of the growing world population, civilization, and the increase in sources of contaminations. Aerogels are an exceptional form of porous materials with extraordinary unique properties. The aerogel has been fabricated from different inorganic and organic materials and incorporated with a variety of novel compounds for specific applications and to enhance its performance in the desired application. Activated carbon is well known for its water-pollutant adsorption, it has been prepared from several organic materials including agricultural wastes and used to treat water from organic dyes, heavy metals, oils, and toxic chemicals. However, as a powder form, activated carbon must be incorporated either into a filter or undergo a post-treatment step to remove the adsorbent from treated water. This review highlighted the development of agricultural waste-based carbon and activated carbon loaded nano-structured aerogels. A review of the types of aerogels and the properties based on the precursor materials was conducted to extensively discuss the potential use of agricultural waste-based carbon and activated carbon loaded nano-structured aerogels in wastewater treatment applications. We also discussed the challenges and future prospects of carbon and activated carbon nano-structured aerogels for wastewater treatment applications.

Keywords: aerogels; porous materials; agricultural activated carbon; adsorption; wastewater treatment

Citation: Muhammad, S.; Yahya, E.B.; Abdul Khalil, H.P.S.; Marwan, M.; Albadn, Y.M. Recent Advances in Carbon and Activated Carbon Nanostructured Aerogels Prepared from Agricultural Wastes for Wastewater Treatment Applications. *Agriculture* **2023**, *13*, 208. https://doi.org/10.3390/agriculture13010208

Academic Editor: Bin Gao

Received: 9 November 2022
Revised: 4 January 2023
Accepted: 10 January 2023
Published: 14 January 2023

1. Introduction

Aerogels are three-dimensional (3D) nano-porous structures of non-fluid colloidal inter-connected polymeric or non-polymeric networks [1]. Aerogels exhibit several unique properties, such as ultra-low density, high porosity and extremely high surface area, which make them suitable in water treatment applications [2,3]. They are normally fabricated from silica through the conventional sol–gel method followed by a drying method such as ambient pressure drying, supercritical drying or freeze drying [4]. More recently, other forms of aerogels have been developed, including carbon aerogels, biopolymeric aerogels, synthetic polymers aerogels and metal aerogels, etc. The fabrication approaches have been also widely developed to become more facile, eco-friendly and unexpensive. Over the past few years, many applications have been identified for aerogels that suit their unique properties, including: thermal and non-thermal insulation [5], absorption [6], packaging [7], supercapacitors [8], catalysts [9], energy storage [10], filtration [11], conduction [12], sensing [13], and the cleaning and adsorption of wastes [14]. Hybrid aerogels were the key to enhance the properties and performance of the pure aerogel, as integration of new substances into a pure aerogel both significantly enhance and gave the material new properties for use in wider applications [15]. The past ten years have witnessed accelerated and widespread use of different aerogels in several applications, including water treatment applications. Figure 1 shows the accelerated grow of scientific publications related to aerogels in general and water treatment applications.

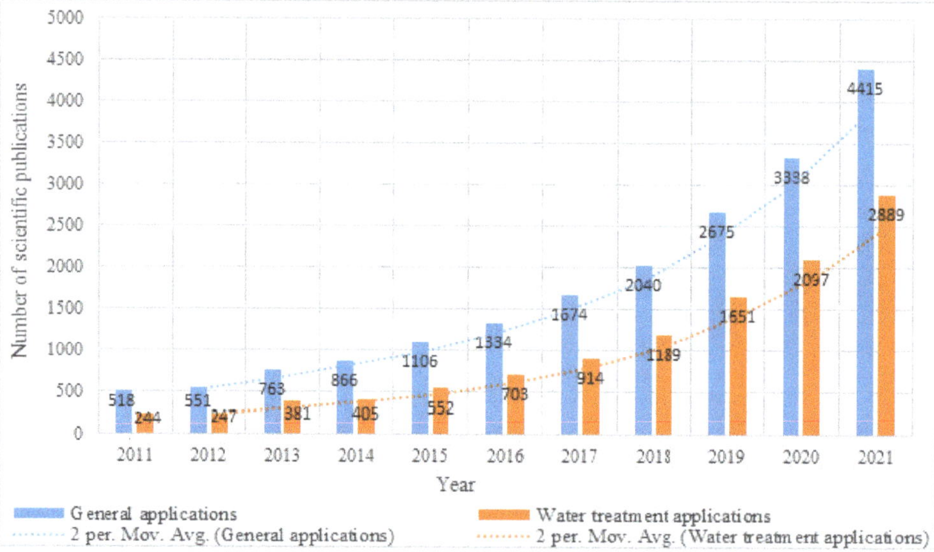

Figure 1. Illustration of the accelerated growth of scientific publications in the past decade related to aerogel and its general and wastewater treatment applications (The search done on 1 November 2022 through Science Direct database).

Activated carbon is a unique form of adsorbent material that has been extensively used for the treatment and purification of different aqueous solutions [16]. However, activated carbon is also linked with some drawbacks, including its separation from the water after it performs the adsorption, in addition to its limited dispersion in water [16–18]. Several scientists have worked to overcome this issue, and facilitate its integration into the aerogels to incorporate the magnificent properties of the two materials [19,20]. Carbon aerogels are a unique class of aerogels that have extensive potential applications in water treatment applications due to the combined effect of both materials. Carbon and activated carbon aerogels have been recently used in several wastewater treatment applications including the removal of organic and inorganic dyes, the elimination of heavy metals, pesticides, herbicides, and oil/water separation [21–23]. So far, there are no or only a limited number of review articles discussing the use of carbon and activated carbon nanostructured aerogels in wastewater treatment applications. Several review articles either discuss general carbon aerogels [24], single precursor aerogels [25], or the adsorption of specific materials from water [26]. Other works address aerogels and their applications in water treatment without specifying carbon and activated carbon aerogels [27]. The aim of this review was to present an introduction to aerogels and classify their types, including organic, inorganic and composite aerogels. We also discussed the preparation of agricultural waste-based carbon and activated carbon nano-structured aerogels in water treatment applications. Agricultural waste-based activated carbon is also discussed as a key player in functional materials, and we presented its main properties and preparations. Finally, we discussed the most recent research on the utilization of carbon and activated carbon nano-structured aerogels in wastewater treatment applications, including organic dyes adsorption, heavy metals removal, oil/water separation, water deionization, and the removal of toxic chemicals.

2. Development of Agricultural-Based Nano-Structured Aerogels

Aerogels are open-celled and mesoporous forms of materials composed of interconnected nanostructured networks exhibiting a porosity of more than 50% [28]. Aerogels have steadily developed since S. Kistler fabricated aerogels for the first time from silica gel in 1931 [29]. Abdul Khalil et al. [30] presented the chronological development of aerogel materials from inorganic silica aerogel until the current forms of nano-structured aerogels. The term nano is applied to materials that have at least one dimension in the nano-meter range (1 to 100 nm). Nano-structured aerogels are said to be materials with pores of less than 50 nm in diameter [31]. Having such a character, nano-structured aerogels possess an ultra-light weight and extremely high surface area, making them attractive materials for several advanced functionalities.

2.1. Classification and Properties of Nano-Structured Aerogels

The history of aerogel materials commenced with the fabrication of a silica aerogel by Kistler in 1931, who replaced the liquid in "Hydrogel" with gas without inducing any shrinkage [32]. Owing to the unique properties that suit particular applications, these materials have been widely utilized in several applications, including acting as: carriers for the delivery of different drugs [3]; electrodes in batteries [33]; filters for wastewater treatment [34]; and catalyst supports in fuel cells [35]. Several methods have been used to classify aerogels since there is no standard IUPAC classification yet for them [36]. The most convenient classification for aerogels is based on the precursor/s and/or the additives, which can be classified into three main groups, including inorganic, organic and composites [1]. Inorganic aerogels are the initial form of the materials prepared from metal alkoxides and/or metal salts to form metal oxide aerogels, chalcogenide aerogels and metallic aerogels. Organic aerogels are derived from several form of carbon, including biopolymers, phenol formaldehyde resin, etc to form biopolymeric aerogels, carbon nanotubes, graphene aerogels and other polymeric aerogels [37]. However, the third form of aerogels is formed by combining both inorganic and organic precursors to form composite aerogels (Table 1).

Table 1. Classification of aerogels based on the precursor material/s.

Type of Aerogel	Sub-Type	Example	Ref.
Inorganic based aerogels	(1) Inorganic oxide aerogels (2) Metallic aerogels (3) Chalcogenide aerogels (4) Silica aerogels	Gold/iron oxide aerogel Vanadium nitride nanosheets aerogel Chalcogenide aerogel $CuSb_2S_4$ Transparent silica aerogel fibers	[38] [39] [40] [41]
Organic based aerogels	(5) Carbon aerogels (1) Biomass aerogels (2) Synthetic organic aerogels (3) Biopolymeric aerogels	Shaddock peel-based carbon aerogel Marine biomass-derived aerogel Polylactide aerogel Nanocellulose/Chitosan aerogel	[42] [43] [44] [45]
Composites aerogels	(4) Mixed metal oxide aerogels (1) Mixed organic aerogels (2) Mixed inorganic aerogels (3) Organic/inorganic aerogels (4) Other composite aerogels	Magnetic iron oxide/graphene aerogel Alginate-chitosan aerogel Mesoporous Fe-silica aerogel Nanocellulose/silica aerogel Nanocellulose/poly ethylenimine aerogel	[46] [47] [48] [49] [50]

Aerogels have been developed from almost every material, which determines the properties of that aerogel. However, the lack of novel properties in a single material limits the multifunctionality of many pure aerogels. Therefore, composite aerogels provide a solution for many potential applications to enhance, introduce and develop novel materials for many new applications. Nanoparticles of silica were impregnated with a network of polymers and loaded with model drugs [51]. This composite aerogel possesses an excellent drug release properties, which can be a potential drug carrier for many medical applications, such as wound dressing [51]. A silica–gelatin aerogel hybrid was fabricated and incorporated with the anticancer drug methotrexate [52]. The incorporation of the drug to the composite aerogel provided an excellent control for drug release. Other composite aerogels have been also used in the field of adsorption, separation and filtration, such as the removal of antibiotics from polluted water using nanocellulose/graphene oxide hybrid aerogel [53], and the same composite aerogel has been used for water purification [54]. Gonçalves and co-workers developed a composite aerogel based on alginate by incorporation of an alginate biopolymer with some drugs in the form of microparticles for mucosal administration. The authors reported that drug release from the alginate-based hybrid was faster than in previous polymer hybrids such as alginate/pectin [55].

The silica aerogel is the first prepared inorganic aerogel characterized with extraordinary properties, which opened a new pathway to many potential applications. The high cross-linked structure, high porosity and ultra-low density of silica-based aerogels raised a lot of interest in many applications [56]. They are mainly composed of air (more than 95%) and only the rest are silica, therefore, it exhibits poor thermal conductivity, and a high surface area [57,58], and being a novel material with unusual properties, has a promising and bright future in many important fields. Several precursors have been used for the preparation of silica aerogels including; Na_2SiO_3, $Si(OR)_4$, MTMS, TEOS, and TMOS [59]. The production expenses of silica aerogels minimized their usage at the time they were discovered. However, in recent years, silica aerogels have been prepared from cheaper precursors, such as bamboo leaf [60], fly ash [61], oil shale ash [62], water glass [63], agricultural wastes, such as wheat husk [64] and bagasse ash [65] instead of with expensive organoalkoxysilanes [64]. Moreover, instead of using expensive, health risky supercritical drying in the large-scale production of aerogels, which is expensive may pose a risk to health, numerous researches have used alternative methods such as ambient pressure [66], freeze drying [67] and microwave drying [68]. The fabrication of flexible, smokeless, super thermal insulators have been always a novel objective for humankind. However, Kim, Y.-G., et al. [69] developed a silica-based aerogel which exhibited an ultra-low thermal conductivity compared to previous fabrications. Similarly, an unexpansive silica aerogel was developed from rice husk and incorporated into cement for thermal insulation purposes [70]. The authors observed an excellent thermal insulation in their aerogels, which may potentially be used as green materials for building applications.

2.2. *Fabrication of Agricultral Carbon Nano-Structured Aerogels*

Several polymeric materials can easily form gels when they are suspended in water, such as carrageenan, gelatine, and starch, etc., but others, such as cellulose and synthetic polymers require a cross-linker for this purpose [30]. However, aerogels can be simply prepared from both types of materials, with or without the formation of a wet-gel of the material. The shape of particles can be preservative and fixed during the drying phase either by the formation of wet-gel or by freezing of the suspension. The fabrication of an aerogel can be initiated by dissolving or dispersing the precursor material/s in distilled water or any other liquid solvent. The homogenization of precursor materials are required for a non-solvable material to achieve homogeneous suspension; the solution then requires some time for the aging process to form a wet-gel [71]. The precursor particles are linked together during the aging process, forming a viscous semi-solid material known as wet-gel. This material is formed by the network of precursor/s surrounded by the solvent. The next step is to remove the solvent without shrinking or disturbing the structure of that network. In order to achieve this, the wet-gel is frozen to keep the structure integrity intact, and then the frozen material is either freeze-dried or supercritical-dried. Freezing of the wet-gel will result in removing the liquid from the system, leaving the precursor network, which appears as a porous 3D material known as aerogel [72]. Li et al. [73] fabricated cocoon-based carbon aerogel by pyrolyzing the initially prepared aerogel at 800 °C for two hours. The authors were able to achieve a nano-porous structure with excellent catalytic activity for their carbon aerogel (Figure 2). The porosity, pore size and volume of the aerogel can be determined by the type and concentration of the precursor material/s. The fabrication technique and the conditions of preparation also highly affect the physical, chemical, and mechanical properties of the aerogels.

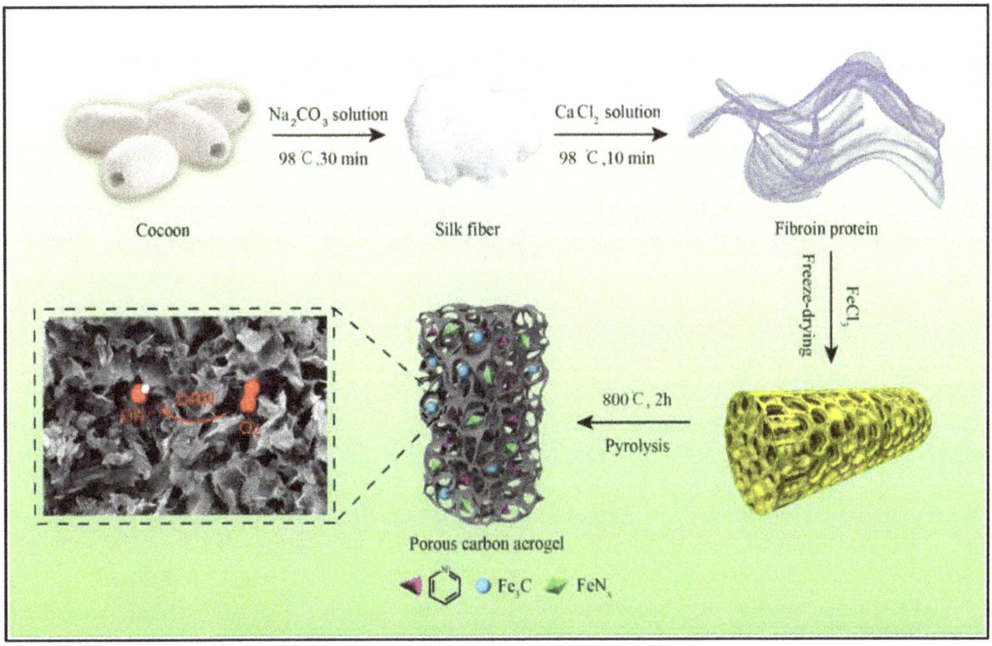

Figure 2. Schematic representation of agricultural carbon aerogel fabrication using cocoon as precursor material. Adapted from ref. [73].

The preparation techniques of aerogels are divided into two major classes; conventional preparation techniques, which include those techniques that do not involve computer aid and follow fully manual lab-based routes. Advanced preparation techniques (rapid prototyping techniques), include those techniques that involve the aid of a computer in the fabrication process, as described below [74].

Conventional preparation techniques of aerogels such as supercritical CO_2 drying, freeze-drying and thermal-induced phase separation, etc., do not involve any computer aid and are fully controlled by technicians. Since they are fully manmade materials, these techniques have the drawback of preparing accurate shapes for particular applications. However, due to their cost effectiveness and simplicity, many of these approaches are still in use even now. These techniques are associated with several issues, including time consumption and manpower requirements; thus, the past two decades have witnessed the development in the computer-aided design of aerogels in term of composition ratios and the final shape of the material. Such fabrication techniques are referred to as rapid prototyping techniques, due to the ease of preparation compared with traditional techniques. Moreover, using the computer in mixing the precursor material also helps in determining the optimal combination of each hybrid, in addition to the control of the physical, morphological and mechanical properties of the aerogels [75]. Table 2 highlights the main differences between traditional and advanced bioaerogel fabrication techniques.

Table 2. Comparison between conventional and advanced bioaerogel fabrication techniques.

Functionality	Conventional Techniques	Rapid Prototyping Techniques
Time consuming and computer-aid	Consume longer time, without computer aid	Rapid fabrication with computer-aid
Easiness and Manpower requirements	Require more manpower	Minimize manpower required
Aerogel homogeneity	Difficult to obtain homogeneous structures	Easier to obtain homogeneous structures
Accurate controllable properties and aerogel shape	Relatively unspecific	Highly specific
Aerogel porosity and pore shape	Random and irregular pores shape	Highly regular and interconnected pores
Aerogel toxicity	Depending on the technique (more toxic)	Less toxic in some techniques
Production costs	More expensive, consume more materials	Consume minimum materials, generally low cost of production

3. Agricultural Activated Carbon and Nano-Structured Aerogels

Activated carbon is a carbonaceous organic material produced by pyrolysis process and mostly used as a remarkable adsorbent material [76]. Nanostructured aerogels have been also utilized in wastewater treatment applications. Powdered activated carbon has been immobilized into several types of filters to overcome the drawbacks associated with the post-treatment stage. Activated carbon has been synthesized from a variety of plant-based precursors including rice husk, cotton shell, straw, corn stalk, palm shell, grass, and other plant-based biomass materials. Generally, any plant-based biomass can be transformed into carbonaceous material, which is then activated by chemical and/or physical activator to result in activated carbon of the particular material. Conditions of high temperatures and a limited or absent amount of oxygen is used during the pyrolysis process, such conditions play an essential role in the properties of the resulted activated carbon [77]. Figure 3 presents the fabrication process of activated carbon from different biomass materials.

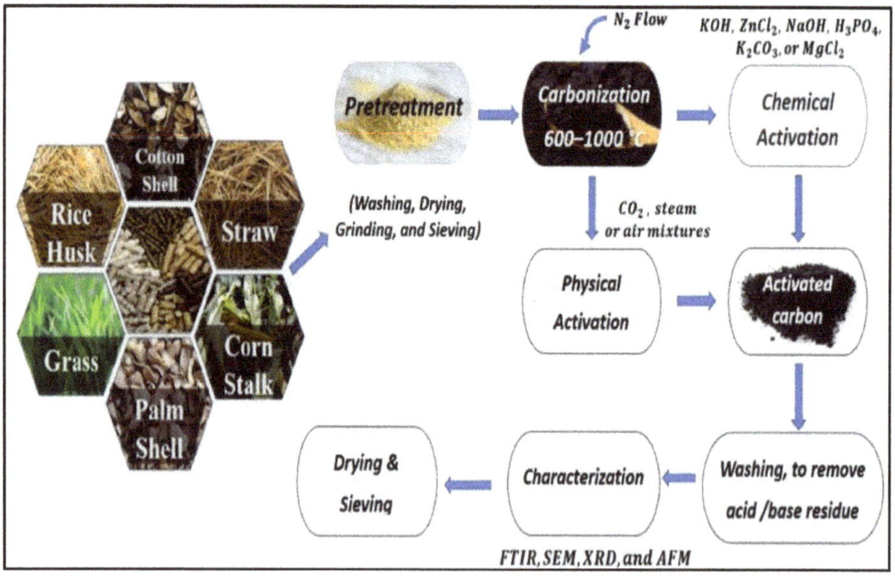

Figure 3. Fabrication routes of agricultural-based-activated carbon from different precursors. Adapted from [78].

The activation of carbonaceous material can be done by several physical activators such as carbon dioxide, air, steam, or a combination of two or more of these agents, which is more preferable to the chemical ones due to their non-toxicity and eco-friendliness. However, chemical activators, particularly KOH-based activators were found to be more effective than physical activation as reported in [79]. The authors reported that such activation resulted in higher porosity, larger pore volume and increased specific surface areas. Huang and co-workers confirmed the results of Song and used KOH activation of their activated carbon at 800 °C and reported significant increases in specific surface areas, which led to better adsorption performance [80]. Several researches have reported the ability of biomass to perform adsorption of many pollutants such as heavy metals, pesticide and organic dyes, etc., [81,82]. Thus, the activation is only enhancing the adsorption performance of the precursor material by altering its physical, chemical and morphological properties [83].

3.1. Activated Carbon Aerogel

Activated carbon aerogel is a unique class of aerogels with 3D porous networks and extremely high specific surface areas resulting from the double porosity of activated carbon particles and the whole aerogel itself. Therefore, it has extensive potential as an adsorbent material in water treatment and purification. Several studies have reported high performance of activated carbon aerogel in the adsorption of dyes [84], heavy metals and other organic pollutants [24]. Gan and co-workers reported that biopolymers can be a sustainable precursor for the preparation of activated carbon aerogels [24]. Owing to their cost-effectiveness, sustainability and easy scale up, biopolymer-based activated carbon aerogels have great potential as an advanced functional material. Unlike toxic precursors such as formaldehyde, furfural and resorcinol and the expensive process of conventional aerogel fabrication that hamper its large-scale production, biopolymer-activated carbon aerogel is characterized by its promising adsorption capacity but the hydrophilicity of many biopolymers remains challenging and thus, chemical modification is mostly required. Yang and co-workers compared the electrochemical performances of both the commercial activated carbon and their prepared activated carbon aerogel and reported that the activated carbon aerogel had significantly better performance due to the higher specific surface area [85]. Activated carbons only have micropores, which limitate their adsorption; unlike activated carbon aerogel that has both the porosity of the activated carbon particles and that of the aerogel.

3.2. Applications of Activated Carbon Aerogels

The desire of scientists all over the world to design materials with controlled characteristics has significantly grown, with rapid and accelerated advances in materials science and technology. Since its development in 1989, carbon aerogels have been developed from several carbon precursors, such as plant biomass, graphene and carbon nano-tubes, etc., and investigated for multiple applications. Despite the unique advantages of carbon aerogels, impurities in some natural carbon as well as the difficulty of structural control are considered major issues in the application of high-quality materials. The properties of carbon and activated carbon aerogels have led to their employment in a wide range of applications, including water treatment filters, energy storage materials, sound insulators, chemical adsorbents, catalysts, thermal insulators, and catalytic supports (Figure 4).

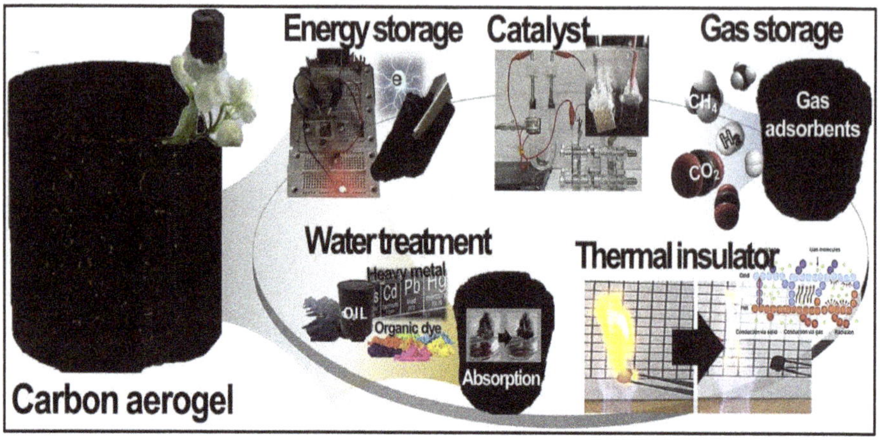

Figure 4. Illustration of carbon and activated carbon aerogel fields with applications. Adapted from [86].

4. Agricultural Activated Carbon Nano-Structured Aerogels for Wastewater Treatment

Water quality is a general term used to describe the physical (such as turbidity, color, suspended solids, etc.), chemical (such as organic contaminant, inorganic contaminant, heavy metals, etc.), and biological (such as bacteria, plankton, algae, fungi, etc.) parameters that may be present in water [87,88]. The consideration of water quality in most cases corresponds to the origin and the purpose of water. In this section we discuss the utilization of activated carbon loaded nano-structured aerogels in the adsorption of organic and inorganic dyes, the removal of heavy metals from water, removal of harmful toxins and toxic materials, in addition to the separation of pharmaceutical compounds.

4.1. Organic Dye Adsorption

In recent years, great developments have been made in all industrial sectors, which has resulted in the excessive use of different chemical compounds that have polluted surface water bodies [89]. As part of the chemical compounds, organic and inorganic dyes that pollute the water have caused a serious threat to humans, animals and aquatic ecosystems due to their toxic nature. Several studies have linked artificial food and non-food dyes with many major health problems including cancer, hyperactivity, hives and asthma, in addition to behavioral changes such as irritability and depression in children and adults [90,91]. Activated carbon is known for its high adsorption ability, it has been widely incorporated with several forms of materials including aerogels [16]. In one study, Yu and co-workers developed an eco-friendly approach for the fabrication of nano-structured aerogel from sodium carboxymethyl cellulose using sol–gel processing and freeze-drying [84]. The authors pyrolyze their obtained aerogel and activated it via a KOH activation. The aerogel exhibited an extremely high surface area as a result of its nano-porosity and connected 3D nanostructures. The same authors reported 249.6 and 245.3 mg/g adsorption capacity of their optimum aerogel for the two tested organic dyes, which recommends this material for further water treatment applications. In different work, Wang and co-workers compared unattached activated carbon prepared from trichosanthes kirilowii maxim shell, nickel alginate-graphene oxide-based aerogels and nickel alginate-activated carbon aerogels for the adsorption of methylene blue dye [21]. The authors reported that the two aerogels loaded with the activated carbon were found to be more effective in the adsorption and to control the dispersion state. The adsorption of methylene blue dye on the nickel alginate-graphene oxide-based aerogels was 505.050 mg/g, compared with nickel alginate-activated carbon aerogel was 465.12 mg/g. The production of activated carbon is relatively expensive, and many researchers have developed approaches to reduce the cost of production. Wang and co-workers developed nano-structured carbon aerogels from agarose biomass [92]. The authors further enhanced their aerogels by introducing zeolitic imidazolate framework-8 to increase the surface area (up to 516 m^2/g) and improve its adsorption capacity for different organic pollutants. In a different study, Li et al. [93] prepared carbon aerogels using an easy, cost-effective and eco-friendly fabrication approach. The authors used glucose as a precursor material, which was activated by potassium hydroxide, which built interconnected coral-like micro-structure during the sol–gel and activation processes. These structures made the aerogel more efficient, enhanced its porosity, and increased its surface area. The authors reported superior adsorption capacity of their preparation toward wide range of organic pollutant including phenols, antibiotics, and dyes; the adsorption of these compounds varied, depending on the material, from 194.07 to 1030.05 mg/g. In the same manner, Huang et al. [94] developed modified cellulose carbon aerogel using wet ball-milling and TEMPO mediated oxidation approaches followed by pyrolysis (Figure 5). The authors reported that the cellulosic fibres effectively turned into plane or wrinkle structures due to treatment conditions. These graphite-like structures made the aerogel exhibit high specific surface areas of more than 2825 m^2/g, which resulted in maximum adsorption capacities to organic dyes of 644 mg/g and 1078 mg/g for alizarin reds and methylene blue, respectively. The adsorption mechanism of dyes for the prepared aerogel was found to be via pore-filling, hydrophobic partition, p/π-π interactions of electron donor–acceptor and H-bonding. For the methylene blue as a cationic dye, the adsorption was reinforced by electrostatic attraction, compared with the anionic alizarin reds that exhibited weakened electrostatic repulsion due to the high salt level [95].

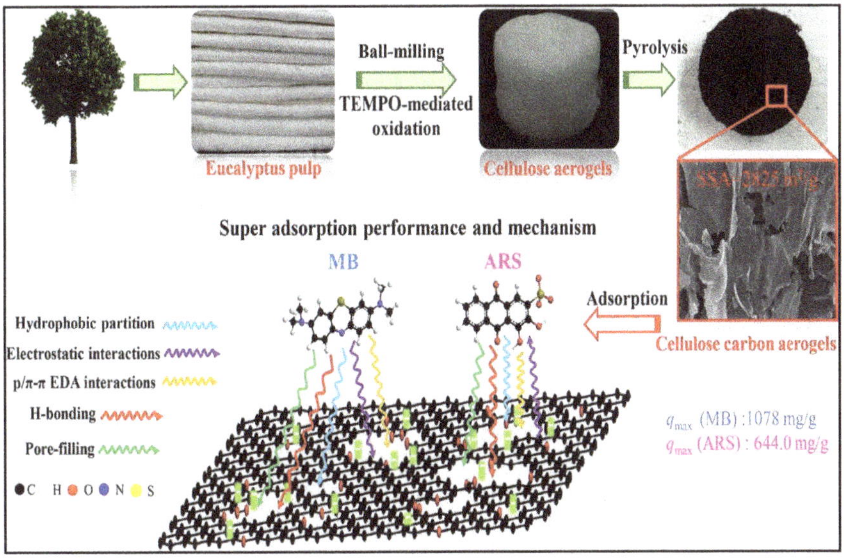

Figure 5. Illustration of cellulose carbon aerogel for enhanced organic dyes adsorption from an aqueous solution. Adapted from Huang et al. [94].

4.2. Deionization of Water

Several studies have warned of the global depletion of freshwater resources, which will result in difficulties in accessing clean and fresh water [96,97]. Remarkable achievements in desalination technology have been recently made, but this technology has yet to meet the social and global demand for water capacitive deionization. Aerogels have been utilized in water deionization as a revolutionary solution derived from renewable and green precursors. Many aerogels have been made from biomass- derived carbon and investigated as active materials for capacitive deionization electrodes [98]. Zhang et al. [99]

developed nanostructured activated carbon aerogel via ambient pressure-drying for capacitive deionization (Figure 6a). The authors reported that their aerogel was able to adsorb salts from water at a capacity of 10.34 mg/g. In a recent study, Liu et al. [100] developed a cost-effective and one-step process to develop a leather wast- based carbon aerogel and use it for water deionization (Figure 6c). The low-cost and eco-friendly aerogel was mainly fabricated from leather wastes, which is rich in N, O and S doping. The authors used one-step activated pyrolysis to reduce the construction costs; the aerogels had a significantly enhanced adsorption capacity to salts due to the extremely high specific surface area (2523 m^2/g), which is highly favorable to salt ions absorption and storage. The same authors were able to achieve a maximum salt adsorption of up to 20.92 mg/g with a high adsorption rate compared to conventional active carbon.

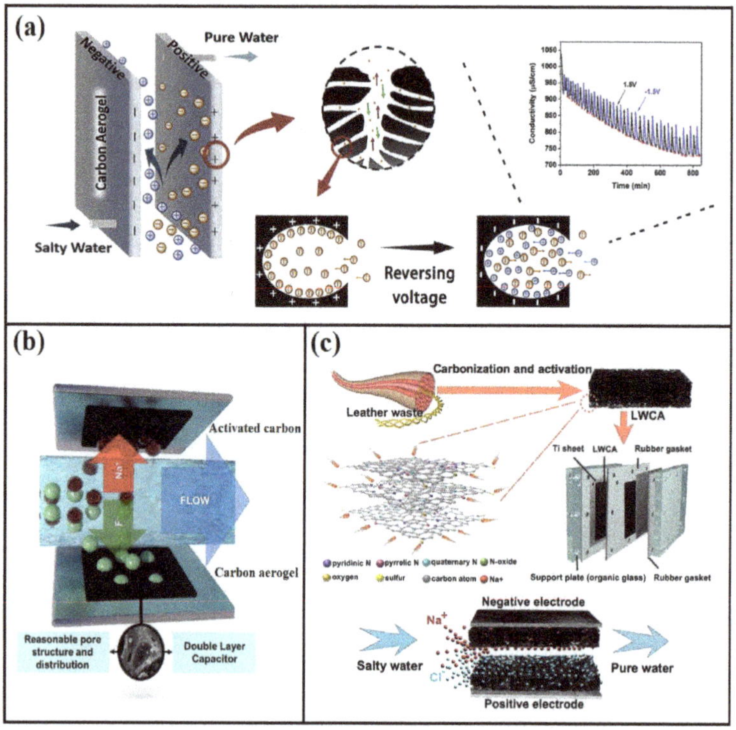

Figure 6. Schematic illustration of carbon aerogels for water deionization; (**a**) nanostructured carbon aerogel for capacitive deionization, adapted from Zhang et al. [99]. (**b**) Schematic diagram of carbon aerogel's electrode and its defluorination, adapted from Zhang et al. [101]. (**c**) leather waste-based carbon aerogel using one-step pyrolysis for enhanced capacitive water deionization. Adapted from Liu et al. [100].

4.3. Removal of Heavy Metals

Heavy metals are those metals that exhibit high density which can cause advert health effect even at very tiny amounts. There are about 23 known heavy metals that are of concern, including arsenic, cadmium, chromium, copper, iron and lead, etc. In recent years and with industrial development, most surface water bodies have been polluted with various amounts of heavy metals, which require smart and sustainable solutions. Despite the excellent adsorption performance of activated carbon, the requirements of the post-treatment stage increase the process costs in the case of using activated carbon particles. Thus, it has been incorporated with filtering membranes and aerogels to overcome the issues related to the post-treatment stage. In the research of Chen and co-workers, the authors fabricated nano-aerogels from cotton-derived porous carbon oxide and investigated its ability in heavy metal elimination as well as the removal of organic pollutants [102]. The aerogel was synthesized from natural cotton waste, the authors reported extremely heigh surface area of 1160 m^2/g, which induced superior sorption capacities for heavy metal ions including strontium (II) (33.3 mg/g), lead (II) (111.1 mg/g) copper (II) (71.4 mg/g) and cadmium (II) (40.2 mg/g). In a different study, the citrate sol-gel method was used to fabricate an Fe3C/carbon aerogel for arsenic removal from water [103]. The authors reported an extremely high surface area and average pore size of 290 m^2/g and 2.7 nm, respectively. The carbon aerogel had a maximum adsorption capacity for arsenic of 56.2 mg/g at pH 7.0. Such excellent performance makes it a potentially attractive material for the removal of hazardous substances from water. Li et al. [104] fabricated biochar-loaded aerogels for heavy metals elimination. Owing to the biochar loading and high surface area of the aerogel, the aerogel had excellent adsorption capacity with a maximum removal of 205.07, 137.89 and 105.56 mg/g for Pb(II), Zn(II) and Cd(II) respectively. Cao and co-workers developed a high porose carbon aerogel with a huge specific surface area using the sol-gel approach and atmospheric drying, and used it for Co (II) in water [105]. The authors reported the promising potential of their aerogel in copper removal in addition to its potential use in water deionization. Wang and co-workers fabricated nano-structured aerogel from carbon quantum dot combined with graphene and investigated its ability to reduce chromium (VI) from an aqueous solution [106]. The authors reported that their aerogel possessed superior photocatalyst activity and was further enhanced by the combination of graphene aerogel with carbon quantum dot, which immobilized the carbon dots and made the liquid phase reaction reusable. This novel aerogel was able to reduce up to 91% of Cr(VI) from the aqueous solution within only 40 min by photocatalytic reaction under the UV–Vis light irradiation (Figure 7).

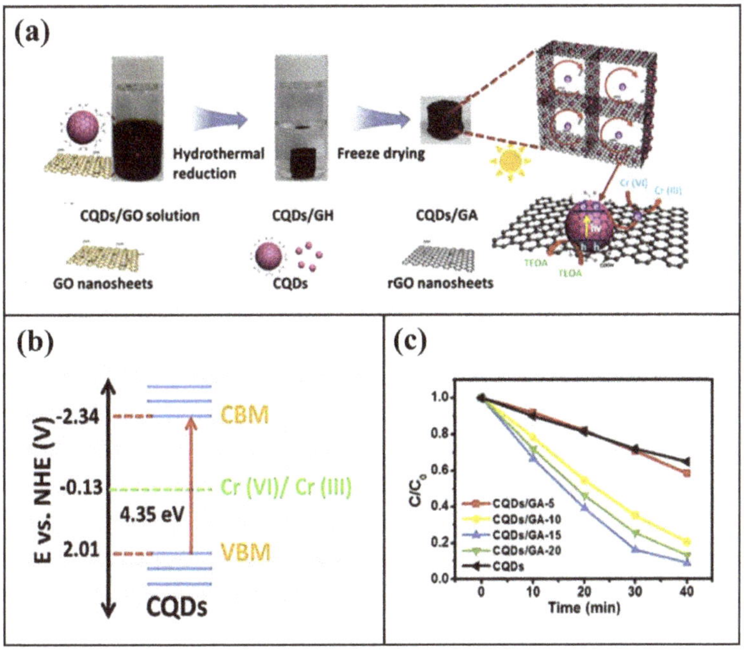

Figure 7. Carbon quantum dot graphene nano-aerogel: (**a**) the fabrication process; (**b**) band structure of carbon quantum dot; and (**c**) photocatalytic performance of chromium (VI) reduction during UV–Vis light irradiation. Adapted with permission from Wang et al. [106] and Lee et al. [86].

In a different study, a magnetic carbon aerogel was synthesized using sodium alginate as a precursor material for carbon and gelatine as a cross-linker as well as a secondary carbon source [107]. The authors loaded their carbon aerogel with Fe_3O_4 nanoparticles as magnetic components to facilitate the separation and removal of aerogel after heavy metal adsorption. The carbon aerogel had an extremely high surface area of 145 m^2/g with a variety of surface functional groups, which resulted in outstanding adsorption performance (143.88 mg/g) to Cd(II) from aqueous solutions [107]. Refer to Table 3 for the summary of using carbon and activated carbon nano-structured aerogels for heavy metals removal.

Table 3. Characteristics and adsorptive capacities of carbon and activated carbon nano-structured aerogels for heavy metal removal.

Precursor Material/s	Preparation Technique	Type of Heavy Metal	Adsorption Capacity	Ref.
Versatile cellulose-based carbon aerogel	Carbonization and freeze-drying approach	Cr(VI) and Pb(II)	68 and 240 mg/g	[108]
Modified carbon aerogel	Sol–gel approach	Cd(II)	15.53 mg/g	[109]
Luminescent carbon quantum dots/nanocellulose aerogel	Freeze-drying under a vacuum atmosphere	Cr3+	104.5 mg/g	[110]
Commercial carbon aerogel	sol–gel polymerization of metal alkoxides	Pb(II), Hg(II) and Cd(II)	34.72, 34.96 and 15.53 respectively	[111]
Straw biochar-loaded N-doped carbon aerogel	Hydrothermal carbonization process	Pb(II) and Cd(II)	205.07 and 105.56	[104]
Amyloid fibrils-based carbon aerogels	Hydrothermal carbonization process	Au(III)	650.08 mg/g	[22]

4.4. Oil/Water Separation

Nano-structured aerogels have been also utilized in oil/water separation due to their facile, rapid, selectivity, reusability, and recyclability advantages. Different forms of aerogels possess different adsorption capacities; Qu and co-workers displayed the ability of their prepared N-doped graphene framework compared with several previous works [112]. The authors reported that their aerogel was able to adsorb oil up to 200–600 times its weight, compared with previous works; conventional graphene aerogel (10–37 times) [113], carbon nanofiber aerogels (51–139 times) [114] and carbon-nanotube aerogel (80–180 times) [115]. Bi and co-workers developed an easy approach for developing cost effective nano-structured carbon micro-belt aerogels using waste papers as a precursor material [116]. The authors reported superior properties for their aerogel including hydrophobicity, low density and a high specific surface area, which was able to adsorb organic liquids including pump oil; (up to 188 g/g), and chloroform (up to 151 g/g). Using such an aerogel opens many doors for largescale production of advanced adsorption materials, and the same authors stated that their aerogel can be regenerated and reused many times without any significant decrease in sorption performance by distillation or squeezing [116]. A mesoporous silica aerogel membrane was recently fabricated by Wang and co-workers for oil adsorption applications [117]. As an inorganic material, the aerogel exhibited high hydrophobicity, and was able to adsorb up to 99.9% of surfactant water-in-oil. A carbon nanotube aerogel was also used for oil adsorption due to its oil absorption capacity and excellent mechanical properties [118,119]. In one study done by Gui and co-workers, the authors reported an oil absorption capacity of up to 180 times for a carbon nanotube aerogel [115]. Despite the excellent performance of this aerogel, it is still associated with high production costs, which may not be applicable in at large-scale levels. To solve this issue, several studies have used biopolymers as a sustainable precursor material for aerogels. Cervin and co-workers developed a hydrophobic cellulose-based aerogel and reported encouraging performance in the separation of oil from water [120]. Eom et al. [121] fabricated an octylamine reduced graphene oxide aerogel able to effectively separate oil from water (Figure 8). The authors also modified Mg_2(dobpdc) by using monoamines of several alkyl chain lengths, which led to long-term stabilization of the mixture and facile fine-tuning of its wettability. The same authors investigated the potential of their aerogel in absorbing other organic solvents and reported the promising potential of their materials (Figure 8c).

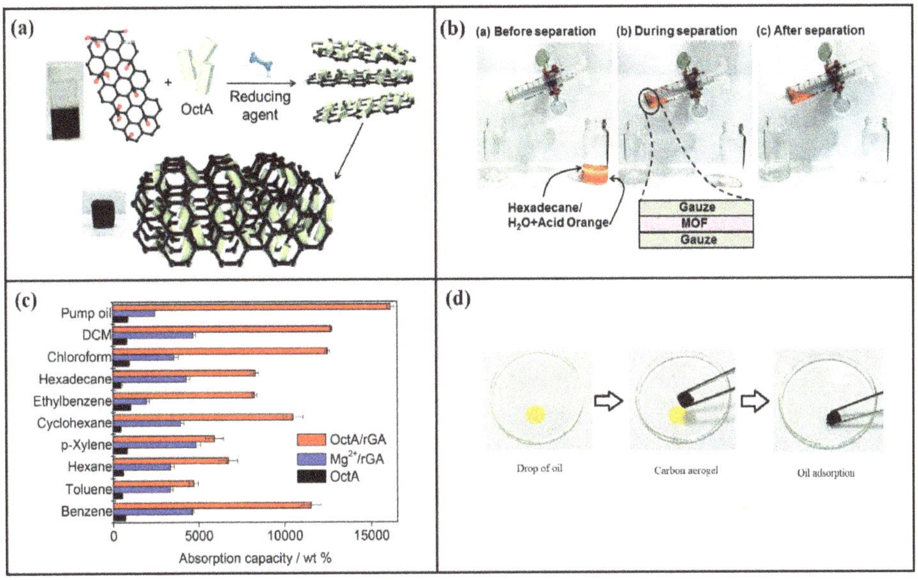

Figure 8. Illustration of octylamine reduced graphene oxide aerogel fabrication for oil/water separation: (**a**) the fabrication process; (**b**) organic solvent separation experiment; (**c**) absorption of different organic solvents and dyes; and (**d**) absorption performance of the aerogel. Reproduced with permission from Eom et al. [121].

4.5. Removal of Toxic Chemicals

A huge amount of toxic chemicals are deposited in surface and ground water everyday as a result of industrial and agricultural processes. Although these chemicals have brought significant benefits and aid to our industries and agricultural production, the excessive use, and misuse, of such chemicals has polluted the environment including the water and soil, which becomes a serious threat to our lives and to the entire ecosystem [122]. In a recent study, a 3D graphene-based nano-aerogel able to detect and remove six different pesticides was developed through a chemical reduction process [123]. The aerogel was able to detect these toxic chemicals even at a tiny amount of 0.12 to 0.58 μg/L depending on the type. Such innovations could have great potential for environmental screening and other monitoring applications. In a different study, a carbon nanotube-based aerogels was prepared and investigated for the removal of herbicides from water [124]. The aerogel was able to absorb both chipton and alachlor herbicides at a high efficiency rate of 227.3 mg/g. The addition of metal–organic framework nanoparticles to the aerogel improved the adsorption performance of both compounds. Li and co-workers used a facile and eco-friendly approach to develop carbon aerogels from glucose [93]. The authors reported that the sol–gel process built the carbon interconnected and coral-like microstructure, which then developed microporosity and mesoporosity after the potassium hydroxide activation process. The authors reported extremely a high specific surface area of 2413 m^2/g, which resulted in a high adsorption capacity ranging from 194.07 to 1030.05 mg/g for phenols, antibiotics, and even dyes. In a different study, Ahamad et al. [125] developed an N/S doped carbon aerogel by using sugarcane bagasse as a source of cellulose for the elimination of bisphenol-A toxic compound from an aqueous solution (Figure 9). The authors were able to fabricate the aerogel with a high yield and high purity and reported a maximum removal of the bisphenol-A of 98 to 99% at natural pH and initial bisphenol-A of 100 ppm. The same authors also investigated the adsorption kinetics and isotherms and reported that it followed pseudo second order and Langmuir for the kinetics and isotherms, respectively.

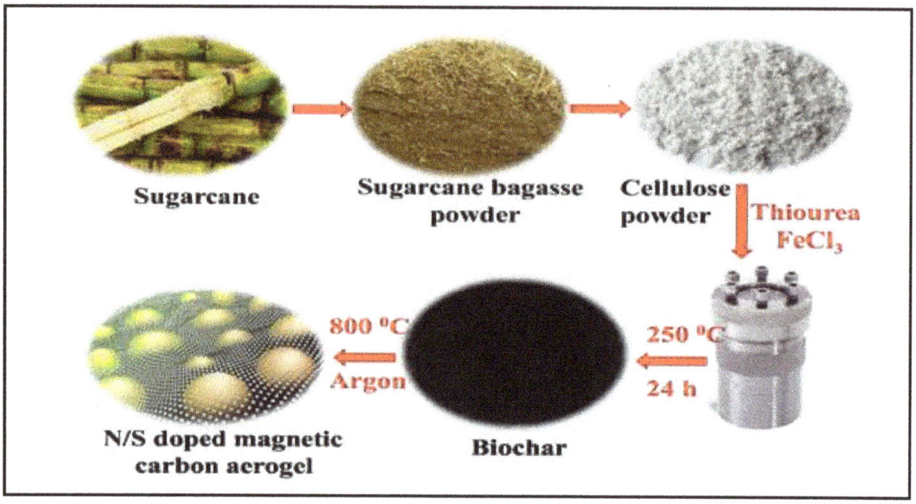

Figure 9. The fabrication of N/S doped carbon aerogel by using sugarcane bagasse for bisphenol-A absorption. Adapted from Ahamad et al. [125].

4.6. Other Applications

Water treatment is a vital process for our daily life use including drinking, cleaning and agricultural use. The presence of a certain amount of one material could make the water undrinkable or even unusable for specific purposes [126]. Dassanayake and co-workers developed an activated carbon-loaded aerogel from chitin and KOH activation [127]. The aerogel exhibited high CO_2 adsorption at different temperature. Owing to the KOH activation, the authors reported about a 37-fold increase in the aerogel's surface area and a more than 95-fold increase in the micropore volume. This activated carbon aerogel suggests further applications for these materials. In a different study, Aylaz et al. [128] developed carbon aerogel from waste paper sources for the effective adsorption of hygromycin B, gentamicin, and vancomycin antibiotics from water. The authors reported that their aerogel had a porosity of more than 90.80% and a surface area of 795.15 mm^2, thus it had an adsorption capacity of 104.1, 107.5 and 81.3 mg/g for hygromycin B, vancomycin and gentamicin, respectively. The authors claim that their study represents the first study in terms of antibiotic adsorption based on carbon aerogels obtained from waste paper. Such waste utilization to develop functional materials will help to ensure the world's sustainable development. Fluoride contamination in groundwater has been taking a heavy toll on human life, with only about 2.5 billion people having access to safe and consumable water [129]. To solve this issue, Zhang et al. [101] prepared a carbon aerogel for excellent fluoride removal; the authors were able to achieve high removal performance of up to 24.44 mg/g. Ling et al. [130] developed a novel carbon aerogel to carry molybdenum trioxide (MoO_3) for the removal of gaseous elemental mercury (Hg0) using an impregnation approach. The huge specific surface area and high porosity of the carbon aerogel, the authors stated that molybdenum trioxide (the active compound) was well distributed in the aerogel scaffold, which led to enhanced Hg0 removal performance. The maximum Hg0 removal capacity was found to be 74%, which reported for the carbon aerogel at 300 °C. Higher temperatures (500–700 °C) was found to reduce the adsorption performance till around 60% [130]. Zhang et al. recently developed carbon aerogel based microbial fuel cell able to generate electricity from wastewater by the oxidation of its organic substrates [131].

The authors used anodic exoelectrogenic bacteria for the oxidation at a neutral pH and were able to achieve 1.7 times higher maximum power density (2300 mW m^{-2}) than conventional used Pt/C air cathodes. The carbon aerogel is a cost-efficient catalyst and can be effectively used for harvesting electrical energy from different organic polluted wastewater.

5. Challenges and Future Prospective

Despite the various encouraging advantages of carbon and activated carbon nanostructured aerogels, they are still in their initial research and evaluation stages, and not yet ready for commercialization and industrial applications. Sam et al. [132] stated that the reasons for the delay in the commercialization of carbon aerogels are the long and time-consuming preparation processes, which mostly involve a sol-gel polymerization stage. The same authors mentioned that using high capillary tension such as supercritical CO_2 drying or freeze-drying techniques during the preparation raises the preparation cost and make the large-scale industry of these materials challenging. Thus, the operational costs of preparing carbon and activated carbon nanostructured aerogels need to be reduced by minimizing the processing steps or using alternative approaches that could also minimize the required time and speed of the preparation process. The use of agricultural waste materials, including grass clippings, bamboo fibres, cocoa shells, rice husk waste, wood chips, sawdust waste, and horse manure as sources of activated carbon and aerogel preparations can significantly lower the production costs of the adsorption materials. It is expected that the global market size and development of carbon aerogel will grow in the next few years, due to the high demand for such materials. The precursor materials of carbon and activated carbon nanostructured aerogels can be made from low-cost and sustainable material such as biomasses and plant waste. Although these aerogels may exhibit lower performance in term of water treatment applications compared to those nano aerogels prepared from graphene and carbon nanotubes [133], green modification may enhance their performance to reach and even exceed the performance of unsustainable materials. The control or design of carbon and activated carbon nanostructured aerogels porose structures is still a challenge, and several factors have been found to affect these characteristics but it also either reduce or affect other characters [134], which deserves extensive explorations. Using advance techniques such as 3D printing of carbon and activated carbon nanostructured aerogels may solve this issue and lead to the fabrication of materials with the desired architecture [135]. The use of such advanced techniques could also solve other issues such as the long preparation time and high preparation costs. The large-scale production of carbon and activated carbon nanostructured aerogels should benefit from these techniques to overcome such issues.

Author Contributions: Conceptualization, S.M., H.P.S.A.K. and E.B.Y.; validation, Y.M.A. investigation, E.B.Y.; resources, S.M. and H.P.S.A.K.; data curation, Y.M.A. and M.M.; writing—original draft preparation, E.B.Y.; writing—review and editing, E.B.Y. and H.P.S.A.K.; visualization, M.M. and H.P.S.A.K.; supervision, H.P.S.A.K.; project administration, S.M. and H.P.S.A.K.; funding acquisition, Y.M.A. All authors have read and agreed to the published version of the manuscript.

Funding: This work was supported by International Research Grant, No: 304/PTEKIND/ 6501194.A158 and by LPPM-Universitas Syiah Kuala, grant number 345/UN11.2.1/PT.01.03PNBP/2021.

Institutional Review Board Statement: Not applicable.

Data Availability Statement: Not applicable.

Acknowledgments: The authors would like to thank the collaboration between Universiti Sains Malaysia, Penang 11800, Malaysia and Universitas Syiah Kuala, Banda Aceh 23111, Indonesiathat made this work possible. Also, thanks to Lembaga Penelitian dan Pengabdian Masyarakat (LPPM) Universitas Syiah Kuala for supporting this work and providing the required resources.

Conflicts of Interest: The authors declare no conflict of interest.

References

1. Alwin, S.; Sahaya Shajan, X. Aerogels: Promising nanostructured materials for energy conversion and storage applications. *Mater. Renew. Sustain. Energy* **2020**, *9*, 7. [CrossRef]
2. Iskandar, M.; Yahya, E.B.; Abdul Khalil, H.P.S.; Rahman, A.; Ismail, M. Recent Progress in Modification Strategies of Nanocellulose-Based Aerogels for Oil Absorption Application. *Polymers* **2022**, *14*, 849. [CrossRef] [PubMed]
3. Yahya, E.B.; Jummaat, F.; Amirul, A.; Adnan, A.; Olaiya, N.; Abdullah, C.; Rizal, S.; Mohamad Haafiz, M.; Abdul Khalil, H.P.S. A review on revolutionary natural biopolymer-based aerogels for antibacterial delivery. *Antibiotics* **2020**, *9*, 648. [CrossRef] [PubMed]
4. Abdul Khalil, H.P.S.; Jummaat, F.; Yahya, E.B.; Olaiya, N.; Adnan, A.; Abdat, M.; NAM, N.; Halim, A.S.; Kumar, U.S.U.; Bairwan, R. A review on micro-to nanocellulose biopolymer scaffold forming for tissue engineering applications. *Polymers* **2020**, *12*, 2043. [CrossRef]
5. Ren, S.; Li, X.; Fan, J.; Liang, J. Preparation and characterization of organic-inorganic hybrid ZrOC/PF aerogel used as high-temperature insulator. *Ceram. Int.* **2020**, *46*, 6326–6332. [CrossRef]
6. Talebi, Z.; Soltani, P.; Habibi, N.; Latifi, F. Silica aerogel/polyester blankets for efficient sound absorption in buildings. *Constr. Build. Mater.* **2019**, *220*, 76–89. [CrossRef]
7. de Oliveira, J.P.; Bruni, G.P.; el Halal, S.L.M.; Bertoldi, F.C.; Dias, A.R.G.; da Rosa Zavareze, E. Cellulose nanocrystals from rice and oat husks and their application in aerogels for food packaging. *Int. J. Biol. Macromol.* **2019**, *124*, 175–184. [CrossRef]
8. Wang, C.; Guan, Z.; Wei, M.; Yu, S.; Sun, R.; Wong, C.-P. Synthesis and characterization of 3D CoMoO 4/rGO aerogel for supercapacitor electrodes. In Proceedings of the 2018 19th International Conference on Electronic Packaging Technology (ICEPT), Shanghai, China, 8–11 August 2018; pp. 1297–1300. [CrossRef]
9. Arfaoui, J.; Ghorbel, A.; Petitto, C.; Delahay, G. Novel V_2O_5-CeO_2-TiO_2-SO_4^{2-} nanostructured aerogel catalyst for the low temperature selective catalytic reduction of NO by NH_3 in excess O_2. *Appl. Catal. B Environ.* **2018**, *224*, 264–275. [CrossRef]
10. Xia, W.; Qu, C.; Liang, Z.; Zhao, B.; Dai, S.; Qiu, B.; Jiao, Y.; Zhang, Q.; Huang, X.; Guo, W. High-performance energy storage and conversion materials derived from a single metal–organic framework/graphene aerogel composite. *Nano Lett.* **2017**, *17*, 2788–2795. [CrossRef]
11. Dai, J.; Tian, Q.; Sun, Q.; Wei, W.; Zhuang, J.; Liu, M.; Cao, Z.; Xie, W.; Fan, M. TiO_2-alginate composite aerogels as novel oil/water separation and wastewater remediation filters. *Compos. Part B Eng.* **2019**, *160*, 480–487. [CrossRef]
12. An, F.; Li, X.; Min, P.; Li, H.; Dai, Z.; Yu, Z.-Z. Highly anisotropic graphene/boron nitride hybrid aerogels with long-range ordered architecture and moderate density for highly thermally conductive composites. *Carbon* **2018**, *126*, 119–127. [CrossRef]
13. Dolai, S.; Bhunia, S.K.; Jelinek, R. Carbon-dot-aerogel sensor for aromatic volatile organic compounds. *Sens. Actuators B Chem.* **2017**, *241*, 607–613. [CrossRef]
14. Zhang, H.; Lyu, S.; Zhou, X.; Gu, H.; Ma, C.; Wang, C.; Ding, T.; Shao, Q.; Liu, H.; Guo, Z. Super light 3D hierarchical nanocellulose aerogel foam with superior oil adsorption. *J. Colloid Interface Sci.* **2019**, *536*, 245–251. [CrossRef] [PubMed]
15. Yahya, E.B.; Abdul Khalil, H.P.S.; Ahmad, M.I.; Rizal, S.; Muhammad, S. Cleaner approach of preparing antibacterial bioaerogel scaffolds using oil palm waste nanocellulose. *Ind. Crops Prod.* **2023**, *191*, 115897. [CrossRef]
16. Mariana, M.; Abdul Khalil, H.P.S.; Mistar, E.; Yahya, E.B.; Alfatah, T.; Danish, M.; Amayreh, M. Recent advances in activated carbon modification techniques for enhanced heavy metal adsorption. *J. Water Process Eng.* **2021**, *43*, 102221. [CrossRef]
17. Gao, Y.; Yue, Q.; Gao, B.; Li, A. Insight into activated carbon from different kinds of chemical activating agents: A review. *Sci. Total Environ.* **2020**, *746*, 141094. [CrossRef]
18. Al-Amshawee, S.; Yunus, M.Y.B.M.; Azoddein, A.A.M.; Hassell, D.G.; Dakhil, I.H.; Hasan, H.A. Electrodialysis desalination for water and wastewater: A review. *Chem. Eng. J.* **2020**, *380*, 122231. [CrossRef]
19. Muhammad, S.; Abdul Khalil, H.P.S.; Abd Hamid, S.; Albadn, Y.M.; Suriani, A.; Kamaruzzaman, S.; Mohamed, A.; Allaq, A.A.; Yahya, E.B. Insights into Agricultural-Waste-Based Nano-Activated Carbon Fabrication and Modifications for Wastewater Treatment Application. *Agriculture* **2022**, *12*, 1737. [CrossRef]
20. Kim, H.G.; Kim, Y.-S.; Kwac, L.K.; Shin, H.K. Characterization of activated carbon paper electrodes prepared by rice husk-isolated cellulose fibers for supercapacitor applications. *Molecules* **2020**, *25*, 3951. [CrossRef]
21. Wang, Y.; Pan, J.; Li, Y.; Zhang, P.; Li, M.; Zheng, H.; Zhang, X.; Li, H.; Du, Q. Methylene blue adsorption by activated carbon, nickel alginate/activated carbon aerogel, and nickel alginate/graphene oxide aerogel: A comparison study. *J. Mater. Res. Technol.* **2020**, *9*, 12443–12460. [CrossRef]
22. Peydayesh, M.; Vogt, J.; Chen, X.; Zhou, J.; Donat, F.; Bagnani, M.; Müller, C.R.; Mezzenga, R. Amyloid-based carbon aerogels for water purification. *Chem. Eng. J.* **2022**, *449*, 137703. [CrossRef]
23. Li, H.; Li, J.; Thomas, A.; Liao, Y. Ultra-high surface area nitrogen-doped carbon aerogels derived from a schiff-base porous organic polymer aerogel for CO2 storage and supercapacitors. *Adv. Funct. Mater.* **2019**, *29*, 1904785. [CrossRef]
24. Gan, G.; Li, X.; Fan, S.; Wang, L.; Qin, M.; Yin, Z.; Chen, G. Carbon Aerogels for Environmental Clean-Up. *Eur. J. Inorg. Chem.* **2019**, *2019*, 3126–3141. [CrossRef]
25. Li, W.; Chen, Z.; Yu, H.; Li, J.; Liu, S. Wood-Derived Carbon Materials and Light-Emitting Materials. *Adv. Mater.* **2021**, *33*, 2000596. [CrossRef] [PubMed]
26. Sabzehmeidani, M.M.; Mahnaee, S.; Ghaedi, M.; Heidari, H.; Roy, V.A. Carbon based materials: A review of adsorbents for inorganic and organic compounds. *Mater. Adv.* **2021**, *2*, 598–627. [CrossRef]

27. Hasanpour, M.; Hatami, M. Application of three dimensional porous aerogels as adsorbent for removal of heavy metal ions from water/wastewater: A review study. *Adv. Colloid Interface Sci.* **2020**, *284*, 102247. [CrossRef] [PubMed]
28. Mariana, M.; Abdul Khalil, H.P.S.; Yahya, E.B.; Olaiya, N.; Alfatah, T.; Suriani, A.; Mohamed, A. Recent trends and future prospects of nanostructured aerogels in water treatment applications. *J. Water Process Eng.* **2022**, *45*, 102481. [CrossRef]
29. Ziegler, C.; Wolf, A.; Liu, W.; Herrmann, A.K.; Gaponik, N.; Eychmüller, A. Modern inorganic aerogels. *Angew. Chem. Int. Ed.* **2017**, *56*, 13200–13221. [CrossRef]
30. Abdul Khalil, H.P.S.; Adnan, A.; Yahya, E.B.; Olaiya, N.; Safrida, S.; Hossain, M.; Balakrishnan, V.; Gopakumar, D.A.; Abdullah, C.; Oyekanmi, A. A review on plant cellulose nanofibre-based aerogels for biomedical applications. *Polymers* **2020**, *12*, 1759. [CrossRef]
31. Wang, H.-L.; Hsu, C.-Y.; Wu, K.C.; Lin, Y.-F.; Tsai, D.-H. Functional nanostructured materials: Aerosol, aerogel, and de novo synthesis to emerging energy and environmental applications. *Adv. Powder Technol.* **2020**, *31*, 104–120. [CrossRef]
32. Yahya, E.B.; Alzalouk, M.M.; Alfallous, K.A.; Abogmaza, A.F. Antibacterial cellulose-based aerogels for wound healing application: A review. *Biomed. Res. Ther.* **2020**, *7*, 4032–4040. [CrossRef]
33. Dong, X.; Zheng, X.; Deng, Y.; Wang, L.; Hong, H.; Ju, Z. SiO_2/N-doped graphene aerogel composite anode for lithium-ion batteries. *J. Mater. Sci.* **2020**, *55*, 13023–13035. [CrossRef]
34. Yu, Y.; Shi, X.; Liu, L.; Yao, J. Highly compressible and durable superhydrophobic cellulose aerogels for oil/water emulsion separation with high flux. *J. Mater. Sci.* **2021**, *56*, 2763–2776. [CrossRef]
35. Öner, E.; Öztürk, A.; Yurtcan, A.B. Utilization of the graphene aerogel as PEM fuel cell catalyst support: Effect of polypyrrole (PPy) and polydimethylsiloxane (PDMS) addition. *Int. J. Hydrogen Energy* **2020**, *45*, 34818–34836.
36. Salimian, S.; Zadhoush, A.; Naeimirad, M.; Kotek, R.; Ramakrishna, S. A review on aerogel: 3D nanoporous structured fillers in polymer-based nanocomposites. *Polym. Compos.* **2018**, *39*, 3383–3408. [CrossRef]
37. Yahya, E.B.; Amirul, A.; Abdul Khalil, H.P.S.; Olaiya, N.G.; Iqbal, M.O.; Jummaat, F.; AK, A.S.; Adnan, A. Insights into the role of biopolymer aerogel scaffolds in tissue engineering and regenerative medicine. *Polymers* **2021**, *13*, 1612. [CrossRef]
38. Wang, C.-T. Photocatalytic activity of nanoparticle gold/iron oxide aerogels for azo dye degradation. *J. Non-Cryst. Solids* **2007**, *353*, 1126–1133. [CrossRef]
39. Zhang, H.; Han, W.; Xu, K.; Zhang, Y.; Lu, Y.; Nie, Z.; Du, Y.; Zhu, J.; Huang, W. Metallic sandwiched-aerogel hybrids enabling flexible and stretchable intelligent sensor. *Nano Lett.* **2020**, *20*, 3449–3458. [CrossRef]
40. Ahmed, E.; Rothenberger, A. Enhancement in CO_2 adsorption capacity and selectivity in the chalcogenide aerogel $CuSb_2S_4$ by post-synthetic modification with LiCl. *Microporous Mesoporous Mater.* **2016**, *220*, 247–252. [CrossRef]
41. Du, Y.; Zhang, X.; Wang, J.; Liu, Z.; Zhang, K.; Ji, X.; You, Y.; Zhang, X. Reaction-spun transparent silica aerogel fibers. *ACS Nano* **2020**, *14*, 11919–11928. [CrossRef]
42. Gu, W.; Sheng, J.; Huang, Q.; Wang, G.; Chen, J.; Ji, G. Environmentally friendly and multifunctional shaddock peel-based carbon aerogel for thermal-insulation and microwave absorption. *Nano-Micro Lett.* **2021**, *13*, 102.
43. Yang, L.; Li, N.; Guo, C.; He, J.; Wang, S.; Qiao, L.; Li, F.; Yu, L.; Wang, M.; Xu, X. Marine biomass-derived composite aerogels for efficient and durable solar-driven interfacial evaporation and desalination. *Chem. Eng. J.* **2021**, *417*, 128051. [CrossRef]
44. Chen, P.; Bai, D.; Tang, H.; Liu, H.; Wang, J.; Gao, G.; Li, L. Polylactide aerogel with excellent comprehensive performances imparted by stereocomplex crystallization for efficient oil-water separation. *Polymer* **2022**, *255*, 125128. [CrossRef]
45. Rizal, S.; Yahya, E.B.; Abdul Khalil, H.P.S.; Abdullah, C.; Marwan, M.; Ikramullah, I.; Muksin, U. Preparation and Characterization of Nanocellulose/Chitosan Aerogel Scaffolds Using Chemical-Free Approach. *Gels* **2021**, *7*, 246. [CrossRef] [PubMed]
46. Dang, N.H.; Tu, T.H.; Linh, V.N.P.; Thy, L.T.M.; Nam, H.M.; Phong, M.T.; Hieu, N.H. Preparation of magnetic iron oxide/graphene aerogel nanocomposites for removal of bisphenol A from water. *Synth. Met.* **2019**, *255*, 116106.
47. Batista, M.; Gonçalves, V.S.; Gaspar, F.; Nogueira, I.; Matias, A.A.; Gurikov, P. Novel alginate-chitosan aerogel fibres for potential wound healing applications. *Int. J. Biol. Macromol.* **2020**, *156*, 773–782. [CrossRef]
48. Tang, R.; Hong, W.; Srinivasakannan, C.; Liu, X.; Wang, X.; Duan, X. A novel mesoporous Fe-silica aerogel composite with phenomenal adsorption capacity for malachite green. *Sep. Purif. Technol.* **2022**, *281*, 119950. [CrossRef]
49. Zuo, K.; Wu, J.; Chen, S.; Ji, X.; Wu, W. Superamphiphobic nanocellulose aerogels loaded with silica nanoparticles. *Cellulose* **2019**, *26*, 9661–9671. [CrossRef]
50. Mo, L.; Pang, H.; Tan, Y.; Zhang, S.; Li, J. 3D multi-wall perforated nanocellulose-based polyethylenimine aerogels for ultrahigh efficient and reversible removal of Cu (II) ions from water. *Chem. Eng. J.* **2019**, *378*, 122157.
51. Follmann, H.D.; Oliveira, O.N.; Lazarin-Bidóia, D.; Nakamura, C.V.; Huang, X.; Asefa, T.; Silva, R. Multifunctional hybrid aerogels: Hyperbranched polymer-trapped mesoporous silica nanoparticles for sustained and prolonged drug release. *Nanoscale* **2018**, *10*, 1704–1715. [CrossRef]
52. Nagy, G.; Király, G.; Veres, P.; Lázár, I.; Fábián, I.; Bánfalvi, G.; Juhász, I.; Kalmár, J. Controlled release of methotrexate from functionalized silica-gelatin aerogel microparticles applied against tumor cell growth. *Int. J. Pharm.* **2019**, *558*, 396–403. [CrossRef] [PubMed]
53. Yao, Q.; Fan, B.; Xiong, Y.; Jin, C.; Sun, Q.; Sheng, C. 3D assembly based on 2D structure of cellulose nanofibril/graphene oxide hybrid aerogel for adsorptive removal of antibiotics in water. *Sci. Rep.* **2017**, *7*, 45914. [CrossRef] [PubMed]
54. Wei, J.; Gui, S.-H.; Wu, J.-H.; Xu, D.-D.; Sun, Y.; Dong, X.-Y.; Dai, Y.-Y.; Li, Y.-F. Nanocellulose-Graphene Oxide Hybrid Aerogel to Water Purification. *Appl. Environ. Biotechnol.* **2019**, *4*, 11–17. [CrossRef]

55. Gonçalves, V.S.; Gurikov, P.; Poejo, J.; Matias, A.A.; Heinrich, S.; Duarte, C.M.; Smirnova, I. Alginate-based hybrid aerogel microparticles for mucosal drug delivery. *Eur. J. Pharm. Biopharm.* **2016**, *107*, 160–170. [CrossRef] [PubMed]
56. Iswar, S.; Malfait, W.J.; Balog, S.; Winnefeld, F.; Lattuada, M.; Koebel, M.M. Effect of aging on silica aerogel properties. *Microporous Mesoporous Mater.* **2017**, *241*, 293–302. [CrossRef]
57. Chandradass, J.; Kang, S.; Bae, D.-S. Synthesis of silica aerogel blanket by ambient drying method using water glass based precursor and glass wool modified by alumina sol. *J. Non-Cryst. Solids* **2008**, *354*, 4115–4119. [CrossRef]
58. Lee, K.-J.; Choe, Y.-J.; Kim, Y.H.; Lee, J.K.; Hwang, H.-J. Fabrication of silica aerogel composite blankets from an aqueous silica aerogel slurry. *Ceram. Int.* **2018**, *44*, 2204–2208. [CrossRef]
59. Gurav, J.L.; Jung, I.-K.; Park, H.-H.; Kang, E.S.; Nadargi, D.Y. Silica aerogel: Synthesis and applications. *J. Nanomater.* **2010**, *2010*, 409310. [CrossRef]
60. Kow, K.-W.; Yusoff, R.; Aziz, A.A.; Abdullah, E. From bamboo leaf to aerogel: Preparation of water glass as a precursor. *J. Non-Cryst. Solids* **2014**, *386*, 76–84. [CrossRef]
61. Shi, F.; Liu, J.-X.; Song, K.; Wang, Z.-Y. Cost-effective synthesis of silica aerogels from fly ash via ambient pressure drying. *J. Non-Cryst. Solids* **2010**, *356*, 2241–2246. [CrossRef]
62. Gao, G.-M.; Liu, D.-R.; Zou, H.-F.; Zou, L.-C.; Gan, S.-C. Preparation of silica aerogel from oil shale ash by fluidized bed drying. *Powder Technol.* **2010**, *197*, 283–287. [CrossRef]
63. He, P.; Gao, X.-D.; Li, X.-M.; Jiang, Z.-W.; Yang, Z.-H.; Wang, C.-L.; Gu, Z.-Y. Highly transparent silica aerogel thick films with hierarchical porosity from water glass via ambient pressure drying. *Mater. Chem. Phys.* **2014**, *147*, 65–74. [CrossRef]
64. Liu, S.-W.; Wei, Q.; Cui, S.-P.; Nie, Z.-R.; Du, M.-H.; Li, Q.-Y. Hydrophobic silica aerogel derived from wheat husk ash by ambient pressure drying. *J. Sol-Gel Sci. Technol.* **2016**, *78*, 60–67. [CrossRef]
65. Nazriati, N.; Setyawan, H.; Affandi, S.; Yuwana, M.; Winardi, S. Using bagasse ash as a silica source when preparing silica aerogels via ambient pressure drying. *J. Non-Cryst. Solids* **2014**, *400*, 6–11. [CrossRef]
66. Shi, F.; Wang, L.; Liu, J. Synthesis and characterization of silica aerogels by a novel fast ambient pressure drying process. *Mater. Lett.* **2006**, *60*, 3718–3722. [CrossRef]
67. Hyun, S.; Kim, T.; Kim, G.; Park, H.-H. Synthesis of low-k porous silica films via freeze drying. *J. Mater. Sci. Lett.* **2000**, *19*, 1863–1866. [CrossRef]
68. Nocentini, K.; Achard, P.; Biwole, P.; Stipetic, M. Hygro-thermal properties of silica aerogel blankets dried using microwave heating for building thermal insulation. *Energy Build.* **2018**, *158*, 14–22. [CrossRef]
69. Kim, Y.-G.; Kim, H.S.; Jo, S.M.; Kim, S.Y.; Yang, B.; Cho, J.; Lee, S.; Cha, J.E. Thermally insulating, fire-retardant, smokeless and flexible polyvinylidene fluoride nanofibers filled with silica aerogels. *Chem. Eng. J.* **2018**, *351*, 473–481. [CrossRef]
70. Abbas, N.; Khalid, H.R.; Ban, G.; Kim, H.T.; Lee, H.-K. Silica aerogel derived from rice husk: An aggregate replacer for lightweight and thermally insulating cement-based composites. *Constr. Build. Mater.* **2019**, *195*, 312–322. [CrossRef]
71. Zheng, Q.; Tian, Y.; Ye, F.; Zhou, Y.; Zhao, G. Fabrication and application of starch-based aerogel: Technical strategies. *Trends Food Sci. Technol.* **2020**, *99*, 608–620. [CrossRef]
72. Abdul Khalil, H.P.S.; Yahya, E.B.; Jummaat, F.; Adnan, A.; Olaiya, N.; Rizal, S.; Abdullah, C.; Pasquini, D.; Thomas, S. Biopolymers based Aerogels: A review on revolutionary solutions for smart therapeutics delivery. *Prog. Mater. Sci.* **2022**, *131*, 101014.
73. Li, C.; Sun, F.; Lin, Y. Refining cocoon to prepare (N, S, and Fe) ternary-doped porous carbon aerogel as efficient catalyst for the oxygen reduction reaction in alkaline medium. *J. Power Sources* **2018**, *384*, 48–57. [CrossRef]
74. Smirnova, I.; Gurikov, P. Aerogel production: Current status, research directions, and future opportunities. *J. Supercrit. Fluids* **2018**, *134*, 228–233. [CrossRef]
75. Chia, H.N.; Wu, B.M. Recent advances in 3D printing of biomaterials. *J. Biol. Eng.* **2015**, *9*, 4. [CrossRef] [PubMed]
76. Zhou, Y.; Li, W.; Qi, W.; Chen, S.; Tan, Q.; Wei, Z.; Gong, L.; Chen, J.; Zhou, W. The comprehensive evaluation model and optimization selection of activated carbon in the O3-BAC treatment process. *J. Water Process Eng.* **2021**, *40*, 101931. [CrossRef]
77. Tan, X.-F.; Liu, S.-B.; Liu, Y.-G.; Gu, Y.-L.; Zeng, G.-M.; Hu, X.-J.; Wang, X.; Liu, S.-H.; Jiang, L.-H. Biochar as potential sustainable precursors for activated carbon production: Multiple applications in environmental protection and energy storage. *Bioresour. Technol.* **2017**, *227*, 359–372. [CrossRef]
78. Naji, S.Z.; Tye, C.T. A review of the synthesis of activated carbon for biodiesel production: Precursor, preparation, and modification. *Energy Convers. Manag. X* **2022**, *13*, 100152. [CrossRef]
79. Song, M.; Jin, B.; Xiao, R.; Yang, L.; Wu, Y.; Zhong, Z.; Huang, Y. The comparison of two activation techniques to prepare activated carbon from corn cob. *Biomass Bioenergy* **2013**, *48*, 250–256. [CrossRef]
80. Huang, F.-C.; Lee, C.-K.; Han, Y.-L.; Chao, W.-C.; Chao, H.-P. Preparation of activated carbon using micro-nano carbon spheres through chemical activation. *J. Taiwan Inst. Chem. Eng.* **2014**, *45*, 2805–2812. [CrossRef]
81. Alkherraz, A.M.; Ali, A.K.; Elsherif, K.M. Removal of Pb (II), Zn (II), Cu (II) and Cd (II) from aqueous solutions by adsorption onto olive branches activated carbon: Equilibrium and thermodynamic studies. *Chem. Int.* **2020**, *6*, 11–20.
82. Guo, J.; Song, Y.; Ji, X.; Ji, L.; Cai, L.; Wang, Y.; Zhang, H.; Song, W. Preparation and characterization of nanoporous activated carbon derived from prawn shell and its application for removal of heavy metal ions. *Materials* **2019**, *12*, 241. [CrossRef] [PubMed]
83. Abdullah, N.; Othman, F.; Yusof, N.; Matsuura, T.; Lau, W.; Jaafar, J.; Ismail, A.; Salleh, W.; Aziz, F. Preparation of nanocomposite activated carbon nanofiber/manganese oxide and its adsorptive performance toward leads (II) from aqueous solution. *J. Water Process Eng.* **2020**, *37*, 101430. [CrossRef]

84. Yu, M.; Li, J.; Wang, L. KOH-activated carbon aerogels derived from sodium carboxymethyl cellulose for high-performance supercapacitors and dye adsorption. *Chem. Eng. J.* **2017**, *310*, 300–306. [CrossRef]
85. Yang, I.; Kwon, D.; Kim, M.-S.; Jung, J.C. A comparative study of activated carbon aerogel and commercial activated carbons as electrode materials for organic electric double-layer capacitors. *Carbon* **2018**, *132*, 503–511. [CrossRef]
86. Lee, J.-H.; Park, S.-J. Recent advances in preparations and applications of carbon aerogels: A review. *Carbon* **2020**, *163*, 1–18.
87. Palansooriya, K.N.; Yang, Y.; Tsang, Y.F.; Sarkar, B.; Hou, D.; Cao, X.; Meers, E.; Rinklebe, J.; Kim, K.-H.; Ok, Y.S. Occurrence of contaminants in drinking water sources and the potential of biochar for water quality improvement: A review. *Crit. Rev. Environ. Sci. Technol.* **2020**, *50*, 549–611. [CrossRef]
88. Yahya, E.; Abdulsamad, M.A. In-vitro Antibacterial Activity of Carbopol-Essential Oils hydrogels. *J. Appl. Sci. Process Eng.* **2020**, *7*, 564–571. [CrossRef]
89. Chkirida, S.; Zari, N.; Bouhfid, R. Insight into the bionanocomposite applications on wastewater decontamination. *J. Water Process Eng.* **2021**, *43*, 102198. [CrossRef]
90. Mittal, J. Permissible synthetic food dyes in India. *Resonance* **2020**, *25*, 567–577. [CrossRef]
91. Mota, I.G.C.; Neves, R.A.M.D.; Nascimento, S.S.D.C.; Maciel, B.L.L.; Morais, A.H.D.A.; Passos, T.S. Artificial dyes: Health risks and the need for revision of international regulations. *Food Rev. Int.* **2021**, 1–16. [CrossRef]
92. Wang, C.; Kim, J.; Tang, J.; Na, J.; Kang, Y.M.; Kim, M.; Lim, H.; Bando, Y.; Li, J.; Yamauchi, Y. Large-scale synthesis of MOF-derived superporous carbon aerogels with extraordinary adsorption capacity for organic solvents. *Angew. Chem.* **2020**, *132*, 2082–2086. [CrossRef]
93. Li, K.; Zhou, M.; Liang, L.; Jiang, L.; Wang, W. Ultrahigh-surface-area activated carbon aerogels derived from glucose for high-performance organic pollutants adsorption. *J. Colloid Interface Sci.* **2019**, *546*, 333–343. [CrossRef] [PubMed]
94. Huang, P.; Zhang, P.; Min, L.; Tang, J.; Sun, H. Synthesis of cellulose carbon aerogel via combined technology of wet ball-milling and TEMPO-mediated oxidation and its supersorption performance to ionic dyes. *Bioresour. Technol.* **2020**, *315*, 123815. [CrossRef]
95. Lai, K.C.; Lee, L.Y.; Hiew, B.Y.Z.; Thangalazhy-Gopakumar, S.; Gan, S. Environmental application of three-dimensional graphene materials as adsorbents for dyes and heavy metals: Review on ice-templating method and adsorption mechanisms. *J. Environ. Sci.* **2019**, *79*, 174–199. [CrossRef]
96. Ahmed, M.; Wiese, D.N. Short-term trends in Africa's freshwater resources: Rates and drivers. *Sci. Total Environ.* **2019**, *695*, 133843. [CrossRef]
97. Pradinaud, C.; Northey, S.; Amor, B.; Bare, J.; Benini, L.; Berger, M.; Boulay, A.-M.; Junqua, G.; Lathuillière, M.J.; Margni, M. Defining freshwater as a natural resource: A framework linking water use to the area of protection natural resources. *Int. J. Life Cycle Assess.* **2019**, *24*, 960–974.
98. Elisadiki, J.; Kibona, T.E.; Machunda, R.L.; Saleem, M.W.; Kim, W.-S.; Jande, Y.A. Biomass-based carbon electrode materials for capacitive deionization: A review. *Biomass Convers. Biorefinery* **2020**, *10*, 1327–1356. [CrossRef]
99. Zhang, C.; Wang, X.; Wang, H.; Wu, X.; Shen, J. A positive-negative alternate adsorption effect for capacitive deionization in nano-porous carbon aerogel electrodes to enhance desalination capacity. *Desalination* **2019**, *458*, 45–53. [CrossRef]
100. Liu, Y.; Zhang, X.; Gu, X.; Wu, N.; Zhang, R.; Shen, Y.; Zheng, B.; Wu, J.; Zhang, W.; Li, S. One-step turning leather wastes into heteroatom doped carbon aerogel for performance enhanced capacitive deionization. *Microporous Mesoporous Mater.* **2020**, *303*, 110303. [CrossRef]
101. Zhang, X.; Li, Y.; Yang, Z.; Yang, P.; Wang, J.; Shi, M.; Yu, F.; Ma, J. Industrially-prepared carbon aerogel for excellent fluoride removal by membrane capacitive deionization from brackish groundwaters. *Sep. Purif. Technol.* **2022**, *297*, 121510. [CrossRef]
102. Chen, H.; Wang, X.; Li, J.; Wang, X. Cotton derived carbonaceous aerogels for the efficient removal of organic pollutants and heavy metal ions. *J. Mater. Chem. A* **2015**, *3*, 6073–6081. [CrossRef]
103. Shen, G.; Xu, Y.; Liu, B. Preparation and adsorption properties of magnetic mesoporous Fe_3C/carbon aerogel for arsenic removal from water. *Desalination Water Treat.* **2016**, *57*, 24467–24475. [CrossRef]
104. Li, J.; Xia, C.; Cheng, R.; Lan, J.; Chen, F.; Li, X.; Li, S.; Chen, J.; Zeng, T.; Hou, H. Passivation of multiple heavy metals in lead–zinc tailings facilitated by straw biochar-loaded N-doped carbon aerogel nanoparticles: Mechanisms and microbial community evolution. *Sci. Total Environ.* **2022**, *803*, 149866. [CrossRef]
105. Cao, Z.; Zhang, C.; Yang, Z.; Qin, Q.; Zhang, Z.; Wang, X.; Shen, J. Preparation of carbon aerogel electrode for electrosorption of copper ions in aqueous solution. *Materials* **2019**, *12*, 1864. [CrossRef] [PubMed]
106. Wang, R.; Lu, K.-Q.; Zhang, F.; Tang, Z.-R.; Xu, Y.-J. 3D carbon quantum dots/graphene aerogel as a metal-free catalyst for enhanced photosensitization efficiency. *Appl. Catal. B Environ.* **2018**, *233*, 11–18. [CrossRef]
107. Li, Y.; Zhou, M.; Waterhouse, G.I.; Sun, J.; Shi, W.; Ai, S. Efficient removal of cadmium ions from water by adsorption on a magnetic carbon aerogel. *Environ. Sci. Pollut. Res.* **2021**, *28*, 5149–5157. [CrossRef]
108. Alatalo, S.-M.; Pileidis, F.; Makila, E.; Sevilla, M.; Repo, E.; Salonen, J.; Sillanpaa, M.; Titirici, M.-M. Versatile cellulose-based carbon aerogel for the removal of both cationic and anionic metal contaminants from water. *ACS Appl. Mater. Interfaces* **2015**, *7*, 25875–25883. [CrossRef]
109. Goel, J.; Kadirvelu, K.; Rajagopal, C.; Garg, V. Cadmium (II) uptake from aqueous solution by adsorption onto carbon aerogel using a response surface methodological approach. *Ind. Eng. Chem. Res.* **2006**, *45*, 6531–6537. [CrossRef]
110. Song, Z.; Chen, X.; Gong, X.; Gao, X.; Dai, Q.; Nguyen, T.T.; Guo, M. Luminescent carbon quantum dots/nanofibrillated cellulose composite aerogel for monitoring adsorption of heavy metal ions in water. *Opt. Mater.* **2020**, *100*, 109642. [CrossRef]

111. Kadirvelu, K.; Goel, J.; Rajagopal, C. Sorption of lead, mercury and cadmium ions in multi-component system using carbon aerogel as adsorbent. *J. Hazard. Mater.* **2008**, *153*, 502–507. [CrossRef]
112. Zhao, Y.; Hu, C.; Hu, Y.; Cheng, H.; Shi, G.; Qu, L. A versatile, ultralight, nitrogen-doped graphene framework. *Angew. Chem. Int. Ed.* **2012**, *51*, 11371–11375. [CrossRef] [PubMed]
113. Niu, Z.; Chen, J.; Hng, H.H.; Ma, J.; Chen, X. A leavening strategy to prepare reduced graphene oxide foams. *Adv. Mater.* **2012**, *24*, 4144–4150. [CrossRef] [PubMed]
114. Wu, Z.-Y.; Li, C.; Liang, H.-W.; Zhang, Y.-N.; Wang, X.; Chen, J.-F.; Yu, S.-H. Carbon nanofiber aerogels for emergent cleanup of oil spillage and chemical leakage under harsh conditions. *Sci. Rep.* **2014**, *4*, 4079. [CrossRef] [PubMed]
115. Gui, X.; Wei, J.; Wang, K.; Cao, A.; Zhu, H.; Jia, Y.; Shu, Q.; Wu, D. Carbon nanotube sponges. *Adv. Mater.* **2010**, *22*, 617–621. [CrossRef] [PubMed]
116. Bi, H.; Huang, X.; Wu, X.; Cao, X.; Tan, C.; Yin, Z.; Lu, X.; Sun, L.; Zhang, H. Carbon microbelt aerogel prepared by waste paper: An efficient and recyclable sorbent for oils and organic solvents. *Small* **2014**, *10*, 3544–3550. [CrossRef] [PubMed]
117. Wang, J.; Wang, H. Ultra-hydrophobic and mesoporous silica aerogel membranes for efficient separation of surfactant-stabilized water-in-oil emulsion separation. *Sep. Purif. Technol.* **2019**, *212*, 597–604. [CrossRef]
118. Zhao, D.; Yu, L.; Liu, D. Ultralight graphene/carbon nanotubes aerogels with compressibility and oil absorption properties. *Materials* **2018**, *11*, 641. [CrossRef] [PubMed]
119. Wang, C.; Yang, S.; Ma, Q.; Jia, X.; Ma, P.-C. Preparation of carbon nanotubes/graphene hybrid aerogel and its application for the adsorption of organic compounds. *Carbon* **2017**, *118*, 765–771. [CrossRef]
120. Cervin, N.T.; Aulin, C.; Larsson, P.T.; Wågberg, L. Ultra porous nanocellulose aerogels as separation medium for mixtures of oil/water liquids. *Cellulose* **2012**, *19*, 401–410. [CrossRef]
121. Eom, S.; Kang, D.W.; Kang, M.; Choe, J.H.; Kim, H.; Kim, D.W.; Hong, C.S. Fine-tuning of wettability in a single metal–organic framework via postcoordination modification and its reduced graphene oxide aerogel for oil–water separation. *Chem. Sci.* **2019**, *10*, 2663–2669. [CrossRef]
122. Gould, F.; Brown, Z.S.; Kuzma, J. Wicked evolution: Can we address the sociobiological dilemma of pesticide resistance? *Science* **2018**, *360*, 728–732. [CrossRef] [PubMed]
123. Sun, P.; Gao, Y.; Xu, C.; Lian, Y. Determination of six organophosphorus pesticides in water samples by three-dimensional graphene aerogel-based solid-phase extraction combined with gas chromatography/mass spectrometry. *RSC Adv.* **2018**, *8*, 10277–10283. [CrossRef] [PubMed]
124. Liang, W.; Wang, B.; Cheng, J.; Xiao, D.; Xie, Z.; Zhao, J. 3D, eco-friendly metal-organic frameworks@ carbon nanotube aerogels composite materials for removal of pesticides in water. *J. Hazard. Mater.* **2021**, *401*, 123718. [CrossRef]
125. Ahamad, T.; Naushad, M.; Alhabarah, A.N.; Alshehri, S.M. N/S doped highly porous magnetic carbon aerogel derived from sugarcane bagasse cellulose for the removal of bisphenol-A. *Int. J. Biol. Macromol.* **2019**, *132*, 1031–1038. [CrossRef] [PubMed]
126. Show, S.; Chakraborty, P.; Karmakar, B.; Halder, G. Sorptive and microbial riddance of micro-pollutant ibuprofen from contaminated water: A state of the art review. *Sci. Total Environ.* **2021**, *786*, 147327. [CrossRef]
127. Dassanayake, R.S.; Gunathilake, C.; Abidi, N.; Jaroniec, M. Activated carbon derived from chitin aerogels: Preparation and CO_2 adsorption. *Cellulose* **2018**, *25*, 1911–1920. [CrossRef]
128. Aylaz, G.L.N.; Okan, M.; Duman, M.; Aydin, H.M. Study on cost-efficient carbon aerogel to remove antibiotics from water resources. *ACS Omega* **2020**, *5*, 16635–16644. [CrossRef]
129. Mukherjee, S.; Halder, G. A review on the sorptive elimination of fluoride from contaminated wastewater. *J. Environ. Chem. Eng.* **2018**, *6*, 1257–1270. [CrossRef]
130. Ling, Y.; Man, X.; Zhang, W.; Wang, D.; Wu, J.; Liu, Q.; Gu, M.; Lin, Y.; He, P.; Jia, T. Molybdenum trioxide impregnated carbon aerogel for gaseous elemental mercury removal. *Korean J. Chem. Eng.* **2020**, *37*, 641–651. [CrossRef]
131. Zhang, X.; He, W.; Zhang, R.; Wang, Q.; Liang, P.; Huang, X.; Logan, B.E.; Fellinger, T.P. High-performance carbon aerogel air cathodes for microbial fuel cells. *ChemSusChem* **2016**, *9*, 2788–2795. [CrossRef]
132. Sam, D.K.; Sam, E.K.; Durairaj, A.; Lv, X.; Zhou, Z.; Liu, J. Synthesis of biomass-based carbon aerogels in energy and sustainability. *Carbohydr. Res.* **2020**, *491*, 107986. [CrossRef] [PubMed]
133. Hong, P.N.; Minh, D.N.; Van Hung, N.; Minh, P.N.; Khoi, P.H. Carbon nanotube and graphene aerogels—The world's 3D lightest materials for environment applications: A review. *Int. J. Mater. Sci. Appl* **2017**, *6*, 277.
134. Xu, X.; Wang, R.; Nie, P.; Cheng, Y.; Lu, X.; Shi, L.; Sun, J. Copper nanowire-based aerogel with tunable pore structure and its application as flexible pressure sensor. *ACS Appl. Mater. Interfaces* **2017**, *9*, 14273–14280. [CrossRef] [PubMed]
135. Mensing, J.P.; Lomas, T.; Tuantranont, A. 2D and 3D printing for graphene based supercapacitors and batteries: A review. *Sustain. Mater. Technol.* **2020**, *25*, e00190. [CrossRef]

agriculture

MDPI

Article

Design and Simulations of a Self-Assembling Autonomous Vertical Farm for Urban Farming

Bhanu Watawana *,† and Mats Isaksson †

Department of Mechanical Engineering and Product Design Engineering, Swinburne University of Technology, John St, Hawthorn, VIC 3122, Australia
* Correspondence: bwatawana@swin.edu.au
† These authors contributed equally to this work.

Abstract: Vertical farming has been proposed as a solution for diminishing arable land as it significantly reduces the footprint of farming. Most contemporary solutions use a low level of automation; however, automation of vertical farming is currently attracting attention from researchers. This paper introduces a conceptual design for an autonomous vertical farm where the main novelty is the self-assembling feature. The proposed system is designed to be installed and used by a non-specialist. The system is designed for cost minimisation, using one set of resources moved by a robot arm to service the plants. These resources include sensors, a depth camera, and the required farming tools. The farm has the capacity of self-powering, greenhouse conversion, data sharing and learning, and several other features. The paper provides the conceptual design in addition to an analysis of the dimensioning of the robot arm, time studies for operation, and an analysis of the self-powering ability.

Keywords: backyard farming; self-assembly; vertical expansion; factory in a box; automated farming; vertical farming; sustainable farm

Citation: Watawana, B.; Isaksson, M. Design and Simulations of a Self-Assembling Autonomous Vertical Farm for Urban Farming. *Agriculture* **2023**, *13*, 112. https://doi.org/10.3390/agriculture13010112

Academic Editors: Muhammad Sultan, Redmond R. Shamshiri, Md Shamim Ahamed, Muhammad Farooq and Luis Manuel Navas Gracia

Received: 20 October 2022
Revised: 7 December 2022
Accepted: 23 December 2022
Published: 30 December 2022

1. Introduction

The diminishing availability of farmable land per capita is a global issue. The global population is projected to reach 9.8 billion by 2050 [1]. According to population growth statistics from the United Nations Food and Agriculture Organisation (FAO) [2], by 2050, the arable land per capita is anticipated to diminish to 33% of the availability in 1970. A joint study conducted by the University of Melbourne and Deakin University [3] concluded that due to shortage of arable land and water and an increasing population, Melbourne's current foodbowl capacity of 41% of Melbourne's total food requirement will be reduced to 18% by 2050. The loss of farmable land is mainly due to climatic changes, urbanisation, desertification, and unavailability of fresh water [4]. The increasing consumer demand for freshness in food products is hampered by transport logistics as farming lands are often situated in distant rural areas. An increasing number of people view the use of artificial fertilisers and pesticides to be a health issue, changing consumer demand focus towards organically grown food. The home garden concept promoted in developing countries [5] caters to the quality requirements but not the quantity. Additionally, the majority of people in developed countries are unlikely to have the time or interest in managing their own farms. The introduction of a low-cost autonomous platform for robotic farming could support the increase in household farming. As a result of the high demand for fresh greens in the urban environment, solutions, such as portable farming inside containers [6], personal food computers [7], and household hydroponic farms [8], are currently being introduced.

Vertical farming is a concept aiming to increase the amount of arable land by 'growing upwards' [9]. The concept can be explained by a single tall glasshouse with multiple layers of racks for growing crops. This is an extension of a hydroponic greenhouse, addressing the lack of arable land and soil, and aiming to reduce the use of herbicides and pesticides [10].

Vertical farming can be implemented in an urban environment, providing the urban community access to fresh food. However, the sustainability of vertical farming is in question as the cost of production is considerably high [11]. One of the main costs is the requirement for artificial lighting. Lighting within a greenhouse makes up 30–50% of the operation costs [12]. Another drawback is that the plants in the system are typically interconnected, sharing water, nutrients and air. Hence, the risk of losing the entire yield due to a disease or malfunction of the system is considerable [13,14].

Automation of vertical farming can provide automated or semi-automated operations within the vertical farm, including automated monitoring and maintenance of the system. Vertical farming automation can be analysed in five stages, basic growth automation, conveyor automation, adaptive automation, system automation, and full automation [15]. Adaptive automation, system automation, and full automation are still being researched [15].

Ease of use and installation are crucial features for a household farming solution. The self-assembling vertical farm solution presented in this paper can be considered as the first step towards a 'Robot-based self-assembling mobile factory', a novel concept, where the factory can assemble itself, automate the process with the same resources used for assembly, and when required, dismantle and pack itself for transportation. Concepts such as 'factory-in-a-box' [16] and 'mobile manufacturing system' [17,18] have been discussed in academia since 2007. Self-assembly is defined as the autonomous organisation of components into a pattern or structure without human intervention [19]. Self-assembly in a multi-robot system allows robots to physically connect to each other to create a distinctive collective robot morphology. Autonomous self-assembly is still uncommon, and most of the current multi-robot systems have pre-programmed morphology to conduct specific tasks [20]. Multi-robot systems or collective robot systems can be categorised into two specific groups, second-order robotics and swarm robotics. A robotic system that can be physically connected to another robot system is considered a second-order system. For a second-order system, the two main classes are self-configurable robots and self-assembling robots [21,22]. Self-configurable robots are composed of modules with little or no independent mobility and only a few sensors, where the modules can connect to each other to form a complex mechanism [23]. A robotic system with the ability to move and operate independently and connect to one another to perform a task that needs collaborative action can be identified as a self-assembling robotic system. The main limitation of a self-assembling robotic system is the lack of full autonomy at the level of robotic modules [21].

Self-assembly, in the context of the mechanism presented in this paper, refers to a mechanism able to assemble itself to conduct specific operations. Examples of such mechanisms include tower cranes, a robot on a track where the robot lays its own track, a multi-arm robot where some arms can be used to modify the other arms. A climbing tower crane uses a mechanism to increase the height of the crane using a prefabricated module structure known as a mast. The climbing frame jacks the slewing structure from the topmost mast and inserts a mast replacing the climbing frame to increase the height [24]. The expansion mechanism of a tower crane is straightforward; however, in contrast to the mechanism proposed in this paper, a tower crane is not designed to access its own base structure.

This paper introduces a conceptual design of a self-assembling, low-cost, fully automatic vertical farm with a small footprint, suitable for the decreasing backyards in capital cities. In order to reduce cost, the design utilises a single set of sensors and tools moved by a robot arm. The authors believe that the concept presented herein may be the first complete solution based on the 'factory-in-a-box' concept. In addition, the proposed mechanism for self-assembly could potentially be used in other applications. This paper is organised as follows. Section 2 begins with a discussion on the requirements for a self-assembling vertical farm, followed by a detailed discussion of the proposed design to meet those requirements. Section 3 details the technical solutions, including dimensioning, a workspace analysis, and time studies for operation. Finally, Section 4 provides conclusions and ideas for future work.

2. A Self-assembling Vertical Farm

2.1. Requirements

The conceptual design presented in this paper targets autonomous vertical farming of vegetables. A prestudy led to the following objectives:

- Autonomous assembly, operation, and disassembly;
- Small footprint to suit an urban environment such as a small backyard;
- Centralised resources for cost reduction;
- Ability to be used both indoors and outdoors;
- Low power consumption;
- Ability to use solar energy;
- Integrated computer vision for phenotyping;
- Ability of greenhouse conversion.

2.2. Self-Assembly

As illustrated in Figure 1a, the proposed system would be delivered in two modules. The base module includes a base, the first stem, a robot arm, and growing platforms (leaves), while the resource module comprises stem expansion modules (SEMs), a power distribution system, water storage, and a battery pack. Both modules are wheeled and designed to be docked.

As illustrated in Figure 1b, when assembled, the farm has a tree-like structure with the base platform acting as the roots, the vertical guideway of the robot arm being the stem, and the growing platforms being the leaves.

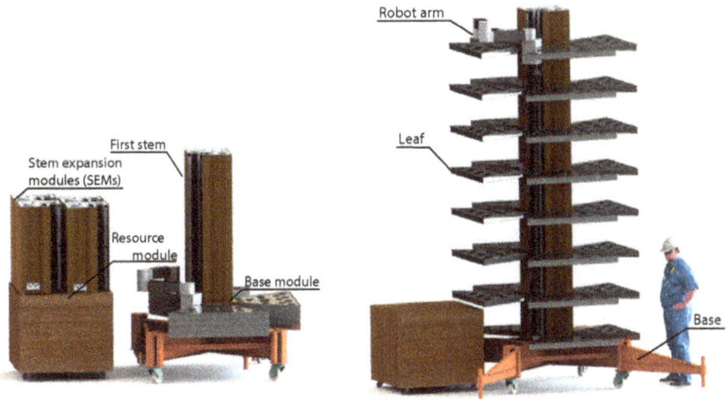

(a) (b)

Figure 1. A self-assembling autonomous vertical farm. Figure (**a**) includes a base module (right) and a resource module (left): (**a**) before assembly; (**b**) after assembly.

2.3. Self-Assembling Mechanism

The proposed system has the ability to assemble and disassemble itself with minimal user interactions as described below:

1 **Base expansion and levelling (Figure 2a,b)**
The first stage of the installation involves moving the wheeled base module to the desired location and expanding the base. The base includes four stabilising legs, which are manually extended. Once extended, the support legs in the corners of the base and stabilising legs are adjusted to level the base. Once the base is fully extended, installed; and levelled, the weight of the system rests on eight support legs. The support legs include optional bolt-down brackets for outdoor use.

2 **Resource module docking (Figure 2c,d)**
The resource module requires manual docking to the expanded base. Once the resource module is moved into the correct position, locks on two sides of the module

are engaged. After docking, the wiring harness and the water tube from the base module are connected to the resource module. Docking the station allows the robot arm on the base module to identify the exact position of the expansion modules.

3 **Power and water supply**
Single-phase power is connected to the power inlet on the resource module. A power connection is not required if using the optional solar panel solution, as the initial system setup would be conducted using the battery. A water supply is connected to the resource module, and the system will fill the tank automatically. The water tank also contributes to stabilise the system by lowering the centre of gravity.

4 **Initiating the system**
After booting the system, initial parameters must be provided, including the number of SEMs and the number of leaves. This is performed using the mobile app or web-based user interface by connecting to WiFi.

5 **Stem assembly (Figure 2e–m)**
The robot arm will reach the first SEM and secure it to the arm using the locking mechanism. Initially, the robot arm moves vertically to clear the resource module. Once cleared, the robot arm will position the SEM on top of the base stem section.
Once the robot arm releases the lock that secures the SEM to the arm, a spring-loaded latching lock will activate and lock the SEM to the stem. Additional SEMs are installed following the same procedure.

6 **Leaf assembly (Figure 2n–t)**
A leaf is composed of a water catchment tray and a growing tray mounted on top of the water tray. The growing tray is delivered with pre-installed coconut fibre medium (coir) pellets. Coir is a low-cost growing medium, lighter than soil and can be compressed into pellets [25]. The coir pellets are replaced after a growing cycle is completed. The leaves are assembled by installing the uppermost layer first and then working downwards. Within a layer, the robot mounts the two leaves farthest from the robot vertical drive first. Thereafter, the remaining two leaves are installed before continuing with the next lower layer. The assembly order is designed to minimise interference with obstacles during the assembly. The vertical distance between leaves can be configured depending on the available light, plant height. and other relevant variables.

7 **Self-disassembling**
During a potential relocation, the robot would conduct a disassembly procedure where it would drain the leaves, remove leaves one by one and stack them, and finally remove the SEMs and stack them.

(**a**) Shipped base (**b**) Extending support legs (**c**) Docking modules (**d**) Docking completed

Figure 2. *Cont.*

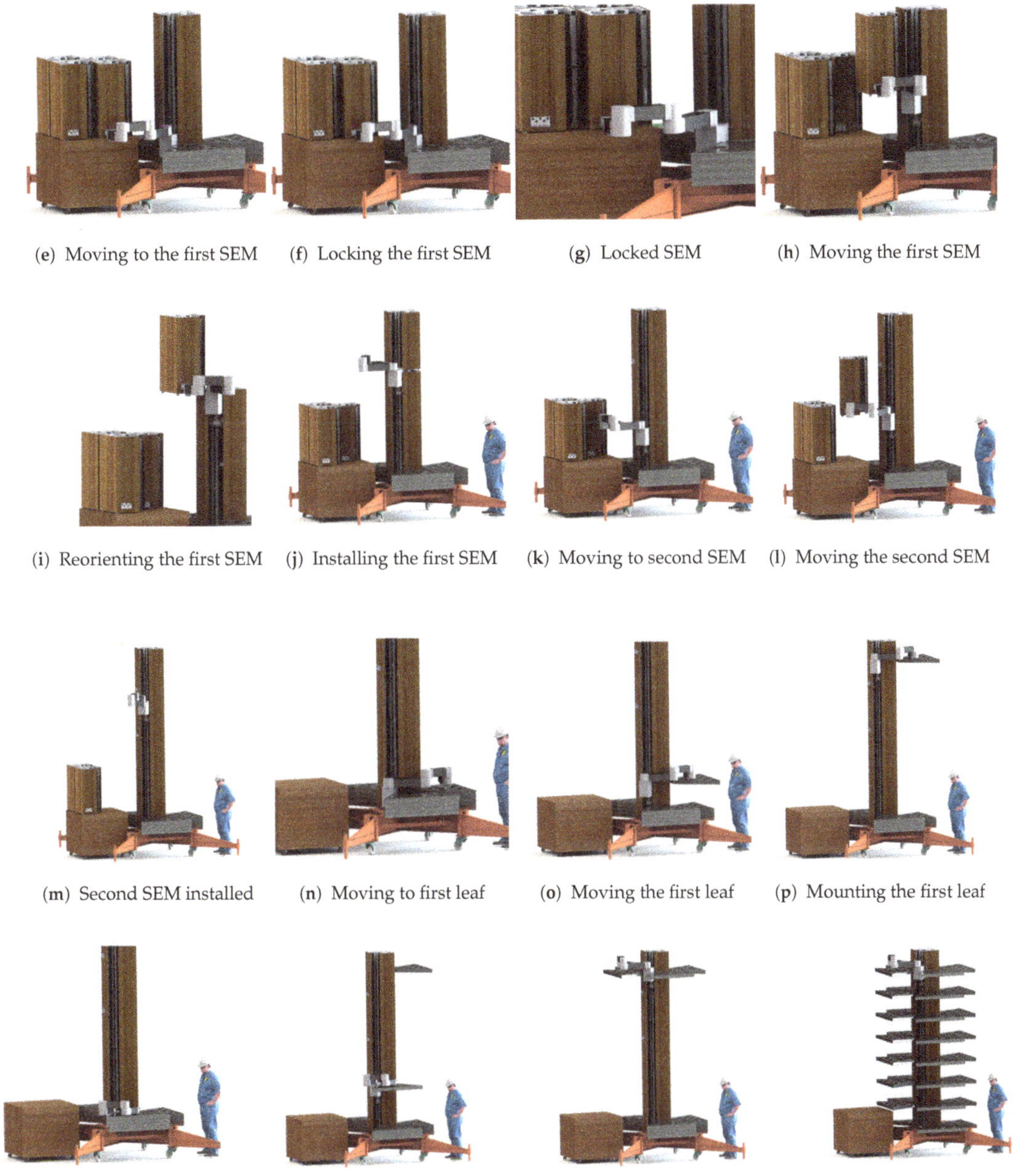

(**e**) Moving to the first SEM (**f**) Locking the first SEM (**g**) Locked SEM (**h**) Moving the first SEM

(**i**) Reorienting the first SEM (**j**) Installing the first SEM (**k**) Moving to second SEM (**l**) Moving the second SEM

(**m**) Second SEM installed (**n**) Moving to first leaf (**o**) Moving the first leaf (**p**) Mounting the first leaf

(**q**) Moving to second leaf (**r**) Moving the second leaf (**s**) First layer completed (**t**) All leaves assembled

Figure 2. Stages of the self-assembly process.

2.4. Functionality

The descriptions below provide details of the system functionality:

- **Vertical motion of the robot arm**
  The structure is assembled and serviced by a robot arm that can position axis-symmetric tools in three Cartesian directions. The self-assembling feature is enabled by imple-

menting the vertical motion with a rack-and-pinion drive with bearing guides, The rack and bearing guides are part of both the base module and all SEMs, meaning that each added SEM increases the robot's vertical workspace.

- **Water and power ducting**
  The water and power supply duct design considers a spring-loaded winding duct mechanism that runs through the stem. This mechanism connects the water supply and the power supply from the resource module to the moving robotic arm.
- **Water supply**
  A water supply may be permanently connected to the resource module, or the built-in stock tank located inside the resource module may be manually filled. The high-pressure diaphragm pump inside the resource module delivers a measured amount of water to the tools mounted on the robot arm. The previously described greenhouse conversion is advised for optimum water usage.
- **Growing**
  The robot arm would continually service each leaf of the structure to perform seeding, watering, monitoring, fertilising, weeding, and other required operations for growing the plants. The growing trays or leaves are delivered with pre-installed coir pellets for growing plants. Once hydrated, the coir pellets expand, creating a rich soil-like growing base. The coir pellets are replaced manually before each growing cycle.
- **Harvesting**
  The vision system provides real-time information about the plants and continuous estimations on when the plants are ready to be harvested. Once a leaf is ready for harvest, the system would drain the bottom tray to remove excess water. Thereafter, the tray will be locked in and moved down onto the resource module for manual harvesting. After harvesting, the coir must be replaced before the leaf is remounted for the next growing cycle.
- **Growth optimisation**
  The solution is designed to monitor and adjust parameters continually. These parameters include plant density, water supply, sunlight exposure, plant tray positions, and potential fertiliser use. The objective is to optimise plant growth and allow the system to learn optimum growth parameters for different plant varieties and allocate resources accordingly. There may also be a possibility of modifying the growing parameters to delay the harvesting time if that would be required.
- **Self-powering capability**
  Section 3.4 shows the calculation of the power requirement and results from a solar panel compatibility analysis. The system is designed with the ability of harnessing energy from solar panels. The only modification required for this feature is to replace the top growing trays (leaves) with solar panels, as shown in Figure 3a. This does not affect the operations of the system except for reducing the growing area by one layer of leaves.
- **Greenhouse conversion**
  As shown in Figure 3b, the system may be converted into a greenhouse. Similar to the solar panel attachment, an umbrella-like structure is mounted on top of the structure, acting as a frame for a transparent membrane covering the entire tree.
- **IoT enabled system**
  The system is intended to share the optimum growth parameters between all installed systems. This information would be collected in a central database where any connected system may access optimum parameters to start the growing process. Hence, the set of optimal growing parameters would continually evolve.
- **Indoor-ready design**
  The system may be used indoors by providing the growing platforms (leaves) with light emitting diodes (LEDs). The LED panels may be installed underneath each leaf to provide the required light for the leaf below. In an indoor system, the uppermost leaves would not be used for growing plants.

- **Advanced agricultural research platform**
  The system monitors and stores growth data, which can be collected from multiple geographical locations with various environmental conditions. The controlled environment would help researchers understand and quantify parameters, such as the environmental impact on agriculture.

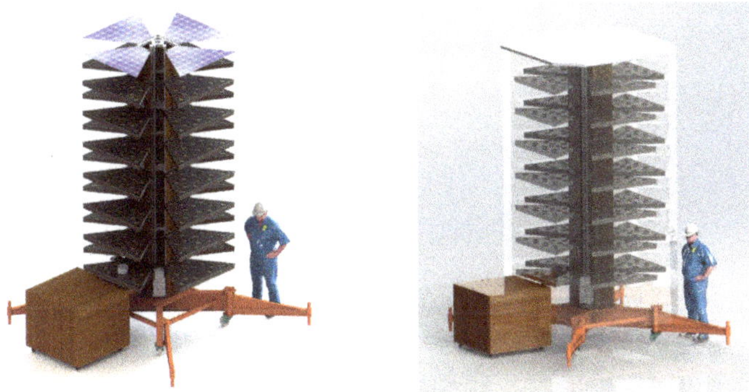

(**a**) (**b**)

Figure 3. Optional features: (**a**) inclusion of solar panels; (**b**) greenhouse conversion.

3. Technical Analysis

The mechanism has four degrees of freedom. Due to their low cost and ease of control, stepper motors were selected for both the prismatic joint and the three revolute joints of the robot arm. Worm and wheel gearboxes were selected for all joints to increase the detent torque and avoid using an electronic brake. This is particularly important for the prismatic joint as the robot should abort vertical motion in case of a power failure to prevent damaging the system.

3.1. Kinematics

As shown in Figure 4, a coordinate system is introduced with the origin in the centre of the stem at the lowest level the robot arm can travel. The axis of the first revolute joint of the robot arm is mounted at a distance of l_0 from the origin. This offset is introduced to increase the working angles of Joint 1 without interference between link L_1 and the stem. The last link in the chain is required in order to reach the stem with the orientation needed to mount the SEM.

The joint coordinate of the prismatic joint is denoted by α, while the joint coordinates of the three revolute joints are denoted by θ_1, θ_2, and θ_3 as specified in Figure 4. The coordinates of the TCP are denoted by x, y, z, and θ according to Figure 4. The forward kinematics can be expressed as:

$$\begin{cases} x = l_0 + l_1\, C(\theta_1) + l_2\, C(\theta_1+\theta_2) + l_3\, C(\theta_1+\theta_2+\theta_3) \\ y = l_1\, S(\theta_1) + l_2\, S(\theta_1+\theta_2) + l_3\, S(\theta_1+\theta_2+\theta_3) \\ z = \alpha \\ \theta = \theta_1+\theta_2+\theta_3 \end{cases} \tag{1}$$

where C and S denote the cosine and sine functions, respectively.

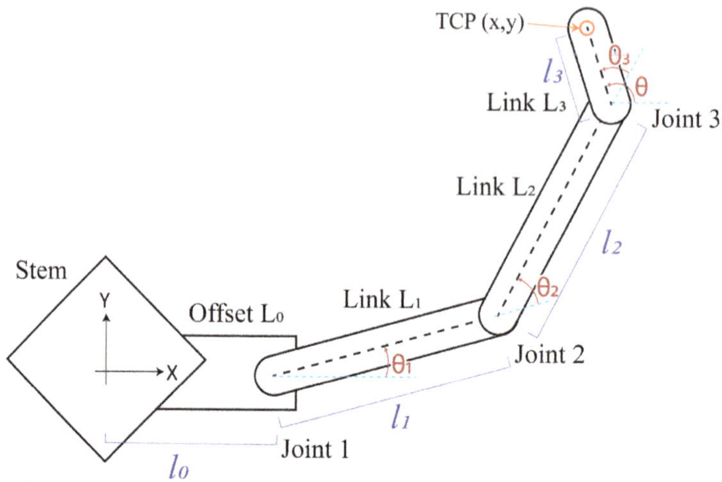

Figure 4. Kinematic parameters of the robot arm.

The joint angles of the manipulator are obtained from the TCP coordinates by first determining the coordinates of Joint 3 (x_3,y_3) according to

$$\begin{cases} x_3 = x - l_3 \cos\theta \\ y_3 = y - l_3 \sin\theta \end{cases} \tag{2}$$

and thereafter utilising the theorem of cosines to calculate

$$\begin{cases} \theta_1 = \gamma - \beta & y \leqslant 0 \\ \theta_1 = \gamma + \beta & y > 0 \end{cases} \tag{3}$$

where

$$\gamma = \text{atan2}\left(\frac{-y_3}{\sqrt{(x_3+l_0)^2 + y_3^2}}, \frac{-(x_3+l_0)}{\sqrt{(x_3+l_0)^2 + y_3^2}}\right) \tag{4}$$

$$\beta = \text{acos}\left(\frac{-(x_3+l_0)^2 - y_3^2 - l_1^2 + l_2^2}{2l_1\sqrt{(x_3+l_0)^2 + y_3^2}}\right) \tag{5}$$

The solution for $y \leqslant 0$ in Equation (3) corresponds to a left-arm solution, while the solution for $y > 0$ corresponds to a right-arm solution. Once a value of θ_1 has been found, θ_2 and θ_3 are calculated according to

$$\begin{cases} \theta_2 = \text{atan2}\left(\frac{y_3 - l_1 S(\theta_1)}{l_2}, \frac{(x_3+l_0) - l_1 C(\theta_1)}{l_2}\right) - \theta_1 \\ \theta_3 = \theta - \theta_1 - \theta_2 \end{cases} \tag{6}$$

In order to avoid interference between link L_1 and the stem, a joint limit of $-135^\circ < \theta_1 < 135^\circ$ was introduced. Adjacent manipulator links operate in different planes and cannot interfere. However, in order to reduce the overall height of the robot arm, link L_1 and link L_3 operate in the same plane. To avoid interference between these links, a joint limit of $-165^\circ < \theta_2 < 165^\circ$ was derived using the collision check in Solidworks.

3.2. *Link Length Selection and Workspace Analysis*

The length l_0 was obtained using Solidworks collision check, identifying the minimum length required to avoid collision when link L_1 is parallel to the stem wall, as shown in Figure 5a. The length of link L_3 is the minimum length required to reach the SEM mounting orientation, as illustrated in Figure 5b.

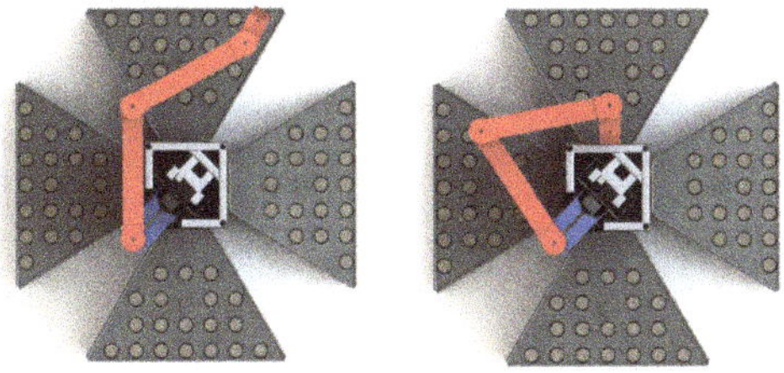

(**a**) (**b**)

Figure 5. Manipulator orientations (top view): (**a**) reaching the furthest point; (**b**) SEM mounting orientation.

In order to reach positions close to the stem, the lengths l_1 and l_2 were selected to be equal. As the mechanism is designed for axis-symmetric tools, this length was determined so that all positions on the leaves are reachable with at least one orientation. Workspace plots of the type shown in Figure 6 were calculated with increasing values of $l_1 = l_2$. The plot was obtained by evaluating all xy-positions in a grid using a step size of 0.1 m. Workspace positions that could be reached with at least one orientation θ are coloured in green.

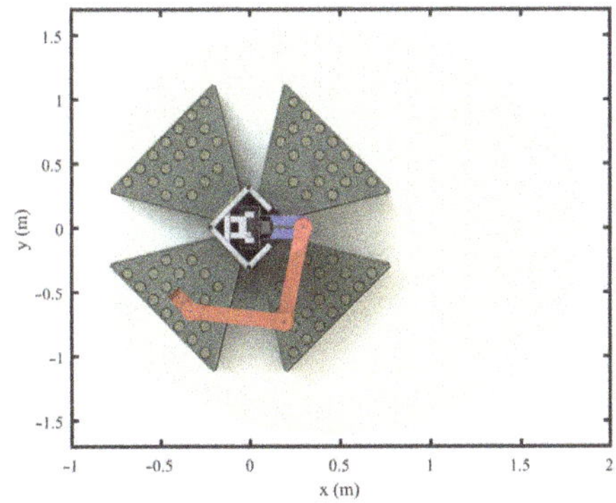

Figure 6. The horizontal workspace of the robot arm using the selected values of l_0, l_1, l_2, and l_3. The green area can be reached with at least one tool orientation.

The obtained values for the link lengths are l_0 = 500 mm, l_1 = l_2 = 600 mm, and l_3 = 120 mm.

3.3. *Growth Area Analysis*

The main advantage of vertical farming is the large arable area relative to the footprint. The footprint of the proposed structure is 9 m^2. This is the total area covered by the expanded legs without considering the optional bolted solution. The average arable area per leaf layer is 3.6 m^2. The base stem may include up to four leaf layers, whereas an SEM may include up to two leaf layers. Hence, with only two SEMs installed, the structure may include up to eight layers providing a total of 29 m^2 arable area, three times the footprint.

The number of SEMs per system depends on the use of optional bolted solution, indoor or outdoor use, ground support, and installation location.

3.4. Power Requirement Analysis

In order to evaluate the potential of utilising solar panels, a power requirement analysis was performed. The analysis was based on the assumptions of an outdoor application without artificial lighting. In order to obtain some margin for the calculations, continuous operation of the components was assumed. It was estimated that components operate at 80% of the nominal power ratings and that the robot arm is perfectly horizontal, meaning there is no work against gravity during horizontal motions. The loss due to friction was not considered in the calculations. The mechanism was designed to use NEMA 17 stepper motors with a worm gear mechanism for revolute joints. The maximum continuous power consumption for a NEMA 17 stepper motor is 5 W [26]. A NEMA 23 stepper motor was selected for the prismatic joint. The maximum power consumption of the NEMA 23 motor is 108 W [27]. A high-pressure diaphragm pump with a power rating of 98 W is used to pump water from the tank to the tool. The control system power consumption was estimated assuming the use of a Raspberry Pi for controlling. As detailed in Table 1, the maximum power consumption for the system is 266 W.

Table 1. Estimated power use of the system.

	Rated (W)	Loss (%)	Actual (W)
Manipulator	15	20%	18
Vertical drive	108	20%	130
Water pump	96		96
Control system	18	20%	22
Total maximum power consumption			266 W

In an outdoor configuration, the optional solar panels and the battery bank can be used to self-power the system. Considering a 15% efficient photovoltaic (PV) solar panel with a maximum power point tracker (MPPT) regulator, the average power output per square meter is 150 W [28]. Hence, at a minimum, the system requires 1.8 m^2 of solar panels. A single layer of leaves covers 3.6 m^2 and would produce twice the required power for operating the system. Additional power generated may be used as a backup power supply or may be connected to the power grid through a grid connect inverter.

3.5. Time Study

A time study was conducted to identify the time required for the robot to service the tree and to determine achievable values of the Cartesian acceleration and maximum Cartesian speed used in the linear interpolation.

The simulation was based on a robot with eight layers of leaves, similar to Figure 5. The robot arm starts at the bottom of the tree and initially moves vertically to the uppermost layer. The TCP thereafter follows the path outlined in Figure 7 before moving to the next lower layer and repeating this process. The end configuration of the robot arm is the same as the start configuration. The acceleration and maximum speed used in the interpolation of the vertical motion were 0.2 ms^{-2} and 0.5 ms^{-1}, respectively. These values are achievable with the proposed NEMA 23 stepper motor and transmission. For each leaf position, a reachable orientation was first determined, as close as possible to the orientation in the previous point. Starting from the previous orientation, θ-angles were evaluated with an increment of $\pm$10 degrees, and the first valid solution was used as the grid point orientation. The TCP follows a linear path between the points, starting and stopping at zero speed. Between each point, a delay of one second was introduced, corresponding to a seeding operation; however, this time would be highly dependent on the actual operation. As shown in Table 2, the horizontal Cartesian interpolation between positions was performed using varying acceleration and maximum speed.

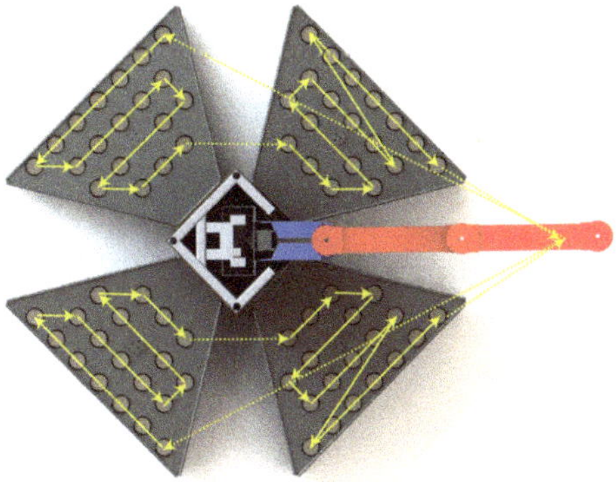

Figure 7. Robot path for servicing one layer of leaves.

Table 2. Time studies for different Cartesian accelerations and maximum speeds.

Simulation	1	2	3	4
Max Cartesian speed (ms^{-1})	1.0	2.0	1.0	2.0
Cartesian acceleration (ms^{-2})	1.0	1.0	2.0	2.0
Max joint speed ($°s^{-1}$) [a]	8.2	8.0	14.9	14.3
Max joint acceleration ($°s^{-2}$)	3.4	3.4	5.8	6.4
Total time (s)	1451.0	1449.0	1286.0	1106.0

[a] Calculated values.

3.6. Estimated Material Cost for a Prototype

Inverse kinematics were utilised to calculate the resulting joint speed and joint accelerations to verify if the selected Cartesian acceleration and maximum speed were achievable with the selected motors.

Table 2 includes the resulting execution time, maximum joint speed, and maximum joint acceleration for different Cartesian accelerations and maximum speeds when servicing a tree with eight layers of leaves according to the path in Figure 7.

The robot arm was designed to utilise NEMA 17 stepper motors with 1:15 gearing. Assuming continuous operation at 50% nominal speed, the maximum angular velocity is 9 $°s^{-1}$ and acceleration/deceleration is 6 $°s^{-2}$. The maximum angular velocity is calculated considering the maximum step frequency of the motor, gearbox speed reduction, and safety factors for joint rotation. Based on the simulation results, the selected Cartesian acceleration and maximum speed were 2 ms^{-2} and 1 ms^{-1}, respectively.

The simulation results indicate that the effect of maximum velocity on the total time is negligible. This is expected as the majority of motion segments are too short for the robot to reach maximum speed. However, the effect of the used acceleration is considerable and selecting motors with a high acceleration is an important design consideration. As shown in Table 2, with the selected values of acceleration and maximum speed, the robot takes nearly 21 min to service a structure of eight layers.

Table 3 provides an estimation of the material cost for a prototype. At the time of publication, the cost for a standard configuration is estimated about AUD 10,700. It should be noted that the unit price for a one-off prototype is significantly higher compared to the cost at mass production. As a comparison, a container farm is priced in between AUD 20,000 and AUD 25,000 [29].

Table 3. Cost estimation for prototyping the self-assembling robotic farm.

Segment	Rate (AUD)	Qty	Cost (AUD)
Base	1000	1	1000
Base stem*	1500	1	1500
Robot arm including tools	1000	1	1000
Control system	600	1	600
Resource module	1000	1	1000
SEM (per module)	800	2	1600
growing platform (per leaf)	200	20	4000
LED grow lights #	1000	20	20,000
Solar panels and battery #	4500	1	4500
Total of the standard (non optional) components			10,700

* The base stem includes a vertical drive, # Optional.

4. Conclusions

This paper introduced a conceptual architecture for a self-assembling vertical farm with a small footprint, designed for urban use. It is designed for autonomous farming by non-specialist users. The main novelty is the self-assembling mechanism which would be useful in any application where a temporary vertical structure must be assembled and serviced by a robot arm.

The computer-aided design was verified by mathematical modelling, confirming the ability of the robotic arm to reach all plants with at least one tool orientation. It was also verified that all operations of the robot can be carried out without mechanical interference. A power usage estimation confirms the viability of the solar-powered self-sustaining system. The proposed robotic farm has the potential to be a cost-effective solution compared to similar farming solutions. An array of the proposed mechanisms might be useful in commercial-scale agriculture. As the architecture is designed for both outdoor and indoor usage, an array of the proposed mechanisms may also be operated in a factory setting.

Although the presented design is only conceptual, effort has been made to theoretically evaluate the practical feasibility of various components.

Prospects for Further Research

Planned future research includes building a functional prototype for real-world evaluation. Further work should also include detailed studies on cultivation-related factors, such as cycle time, soil requirements, plant height, plant weight, and plant width and density which are essential to fully evaluate the proposed concept. The system can be further developed to tilt the leaf structures for sun tracking to increase light utilisation, adding a robotic platform for self-positioning and automatic docking functionality.

Author Contributions: Conceptualization and methodology, B.W and M.I.; software, B.W; validation, B.W and M.I.; formal analysis, B.W and M.I.; data curation, B.W; writing—original draft preparation, B.W; writing—review and editing, M.I.; visualization, B.W.; supervision, M.I.; project administration, M.I. All authors have read and agreed to the published version of the manuscript.

Funding: This research received no external funding.

Institutional Review Board Statement: Not applicable.

Informed Consent Statement: Not applicable.

Data Availability Statement: Not applicable.

Conflicts of Interest: The authors declare no conflict of interest.

References

1. Cohen, J.E.; Lloyd, C.B.; Lutz, W.P. World population in 2050: Assessing the projections/Discussions. In *Onference Series-Federal Reserve Bank of Boston*; Federal Reserve Bank of Boston: Boston, MA, USA, 2001; pp. 83–129.
2. Food and Agriculture Organization. Arable Land (Hectares Per Person). 2018. Available online: https://data.worldbank.org/indicator/AG.LND.ARBL.HA.PC (accessed on 10 November 2021).
3. Carey, R.; Larsen, K.; Sheridan, J.; Candy, S. *Planning a Resilient City Foodbowl Melbourne's Food Future A Foodprint Melbourne Report*; Technical Report, Victorian Eco-Innovation Lab, The University of Melbourne, Melbourne, Australia, 2016.
4. Fedoroff, N.V. Food in a future of 10 billion. *Agric. Food Secur.* **2015**, *4*, 11. [CrossRef]
5. Galhena, D.; Freed, R.; Maredia, K.M. Home gardens: A promising approach to enhance household food security and wellbeing. *Agric. Food Secur.* **2013**, *2*, 8. [CrossRef]
6. Grant, Z. *Commercial Microgreen Production*; University of Illinois Extension: Urbana, IL, USA, 2018.
7. Castelló Ferrer, E.; Rye, J.; Brander, G.; Savas, T.; Chambers, D.; England, H.; Harper, C. Personal food computer: A new device for controlled-environment agriculture. In *Proceedings of the Advances in Intelligent Systems and Computing*; Springer: Berlin/Heidelberg, Germany, 2019; Volume 881, pp. 1077–1096. [CrossRef]
8. AeroGarden. AeroGarden Official Store—Shop and Save on AeroGardens, Seed Kits, Grow Bulbs & More'. 2021. Available online: https://www.aerogarden.com/ (accessed on 19 January 2022)
9. Beacham, A.M.; Vickers, L.H.; Monaghan, J.M. Vertical farming: A summary of approaches to growing skywards. *J. Hortic. Sci. Biotechnol.* **2019**, *94*, 277–283. [CrossRef]
10. Benke, K.; Tomkins, B. Future food-production systems: Vertical farming and controlled-environment agriculture. *Sustain. Sci. Pract. Policy* **2017**, *13*, 13–26. [CrossRef]
11. Banerjee, C.; Adenaeuer, L. Up, Up and Away! The Economics of Vertical Farming. *J. Agric. Stud.* **2014**, *2*, 40. [CrossRef]
12. Nelson, J.A.; Bugbee, B. Economic Analysis of Greenhouse Lighting: Light Emitting Diodes vs. High Intensity Discharge Fixtures. *PLoS ONE* **2014**, *9*, e99010. [CrossRef] [PubMed]
13. Kalantari, F.; Tahir, O.M.; Joni, R.A.; Fatemi, E. Opportunities and challenges in sustainability of vertical farming: A review. *J. Landsc. Ecol.* **2018**, *11*, 35–60. [CrossRef]
14. Perez, V.M. Study of the Sustainbility Issues of Food Production Using Vertical Farm Methods in an Urban Environment within the State of Indiana. Ph.D. Thesis, Purdue University, West Lafayette, IN, USA, 2014.
15. van Delden, S.H.; SharathKumar, M.; Butturini, M.; Graamans, L.J.A.; Heuvelink, E.; Kacira, M.; Kaiser, E.; Klamer, R.S.; Klerkx, L.; Kootstra, G.; et al. Current status and future challenges in implementing and upscaling vertical farming systems. *Nat. Food* **2021**, *2*, 944–956. [CrossRef]
16. Jackson, M.; Zaman, A. Factory-In-a-Box-Mobile Production Capacity on Demand. Technical Report, In Proceedings of the 2006 IJME-INTERTECH Conference, Kean University, Union, NJ, USA, 19–21 October 2006.
17. Stillström, C.; Jackson, M. The concept of mobile manufacturing. *J. Manuf. Syst.* **2007**, *26*, 188–193. . 2008.03.002. [CrossRef]
18. Benama, Y.; Alix, T.; Perry, N. Framework definition for the design of a mobile manufacturing system. In *Advances on Mechanics, Design Engineering and Manufacturing*; Springer: Berlin/Heidelberg, Germany, 2017. [CrossRef]
19. Wei, H.; Li, N.; Liu, M.; Tan, J. A novel autonomous self-assembly distributed swarm flying robot. *Chin. J. Aeronaut.* **2013**, *26*, 791–800. [CrossRef]
20. Mathews, N.; Christensen, A.L.; Stranieri, A.; Scheidler, A.; Dorigo, M. Supervised morphogenesis: Exploiting morphological flexibility of self-assembling multirobot systems through cooperation with aerial robots. *Robot. Auton. Syst.* **2019**, *112*, 154–167. [CrossRef]
21. Trianni, V. *Evolutionary Swarm Robotics*; Springer: New York, NY, USA, 2008; Volume 108, pp. 171–172. [CrossRef]
22. Wei, H.; Chen, Y.; Tan, J.; Wang, T. Sambot: A self-assembly modular robot system. *IEEE/ASME Trans. Mechatronics* **2011**, *16*, 745–757. [CrossRef]
23. Rayguru, M.M.; Mohan, R.E.; Parween, R.; Yi, L.; Le, A.V.; Roy, S. An Output Feedback Based Robust Saturated Controller Design for Pavement Sweeping Self-Reconfigurable Robot. *IEEE/ASME Trans. Mechatronics* **2021**, *26*, 1236–1247. [CrossRef]
24. Sohn, H.W.; Hong, W.K.; Lee, D.; Lim, C.Y.; Wang, X.; Kim, S. Optimum Tower Crane Selection and Supporting Design Management. *Int. J. Adv. Robot. Syst.* **2014**, *11*, 1. [CrossRef]
25. Xiong, J.; Tian, Y.; Wang, J.; Liu, W.; Chen, Q. Comparison of Coconut Coir, Rockwool, and Peat Cultivations for Tomato Production: Nutrient Balance, Plant Growth and Fruit Quality. *Front. Plant Sci.* **2017**, *8*, 1327. [CrossRef] [PubMed]
26. Pololu-Stepper Motor: Unipolar/Bipolar, 200 Steps/Rev, 42×48mm, 4V, 1.2 A/Phase. 2021. Available online: https://www.pololu.com/product/1200 (accessed on 8 May 2020)
27. Maker Store PTY LTD. *Nema 23 Stepper Motor Datasheet*; Maker Store PTY Ltd.: Carrum Downs, VIC, Australia, 2021.

28. Dhimish, M.; Holmes, V.; Mather, P.; Sibley, M. Novel hot spot mitigation technique to enhance photovoltaic solar panels output power performance. *Sol. Energy Mater. Sol. Cells* **2018**, *179*, 72–79. [CrossRef]
29. Modular Farms Australia | Purpose Built Container Farms. Available online: https://www.modularfarms.com.au (accessed on 20 March 2022).

agriculture

Article

Feature Extraction on the Difference of Plant Stem Structure Based on Ultrasound Energy

Danju Lv *, Jiali Zi, Xin Huang, Mingyuan Gao, Rui Xi, Wei Li and Ziqian Wang

College of big data and intelligent engineering, Southwest Forestry University, Kunming 650224, China
* Correspondence: lvdanjv@swfu.edu.cn

Abstract: Plant growth is closely related to the structure of its stem. The ultrasonic echo signal of the plant stem carries much information on the stem structure, providing an effective means for analyzing stem structure characteristics. In this paper, we proposed to extract energy features of ultrasonic echo signals to study the structure of the plant stem. Firstly, it is found that there are obvious different ultrasonic energy changes in different kinds of plant stems whether in the time domain or the frequency domain. Then, we proposed a feature extraction method, density energy feature, to better depict the interspecific differences of the plant stems. In order to evaluate the extracted 24-dimensional features of the ultrasound, the information gain method and correlation evaluation method were adopted to compute their contributions. The results showed that the mean density, an improved feature, was the most significant contributing feature in the four living plant stems. Finally, the top three features in the feature contribution were selected, and each two of them composed as 2-D feature maps, which have significant differentiation of the stem species, especially for grass and wood stems. The above research shows that the ultrasonic energy features of plant stems can provide a new perspective for the study of distinguishing the structural differences among plant stem species.

Keywords: online non-destructive detection; structure of plant stems; ultrasonic feature; feature extraction; feature contribution

Citation: Lv, D.; Zi, J.; Huang, X.; Gao, M.; Xi, R.; Li, W.; Wang, Z. Feature Extraction on the Difference of Plant Stem Structure Based on Ultrasound Energy. *Agriculture* **2023**, *13*, 52. https://doi.org/10.3390/agriculture13010052

Academic Editors: Muhammad Sultan, Redmond R. Shamshiri, Md Shamim Ahamed and Muhammad Farooq

Received: 16 November 2022
Revised: 13 December 2022
Accepted: 20 December 2022
Published: 24 December 2022

1. Introduction

The study of variability in the internal structure of plant stems is one aspect of research for exploring the physiological functions of plant stems [1]. The marked interspecific differences in the structure of plant stems lead to variations in the distribution and proportion of the composition of the stem. The plant stem body is organized, from the outside to the inside, by bark, formative layer, xylem, and pith. The differences in stem cells lead to different stem characteristics. Due to many lignified cells, woody stems have a firm and tall stem texture. Herbaceous stems have a less developed or undeveloped xylem, so the stem is juicy and soft in texture. At the same time, as the growth of the internal structure of the stem is a dynamic and organic change process [2], the search for non-destructive online stem-condition detection has received increasing attention from researchers [3,4]. In recent years, researchers have initiated a large number of studies on dynamic online plant-stem growth detection and the proposal of new detection methods [5,6]: the γ-ray method [7], nuclear-magnetic-resonance method [8–10], X-ray computed tomography [11], resistance method [12], time-domain reflection method TDR [13], frequency-domain capacitance method [14,15] and ultrasonic detection [16,17] methods.

Because of ultrasound's accuracy and nondestructive characteristics in detecting the internal structure and properties of substances, it is widely used in medical and industrial testing. In biomedicine, ultrasound on different tissues and organs of the human body with different reflection and absorption characteristics can effectively detect the structural properties of tissues and organs and their dynamic characteristics. At present, it can effectively detect the thickness of the fat layer [18,19], masses [20,21], the shape and flow rate of human blood vessels [22], and the water content when water accumulation occurs in organs [23].

Ultrasonic testing has been widely used in inspecting wood since it was found that the velocity of low-frequency waves in the longitudinal direction of wood can effectively measure wood strength property indexes such as the smooth grain flexural modulus and the flexural strength [24]. Researchers have found that the ultrasonic velocity decreases with the increasing moisture content of wood [25,26]. In detecting wood by ultrasound, its axial ultrasound signal has the most negligible energy attenuation and the fastest ultrasound echo velocity, while the radial attenuation is the largest and the ultrasound echo velocity is the most minor [27,28]. Recent studies have pointed out that the ultrasonic attenuation characteristics of wood are closely related to the structure of the plant stem tissue [26]: because the plant stem bears the transport of water and organic matter in the axial direction, forming a direct connection or perforated connection between the axial cell tissues, so the ultrasonic impedance is small, while complete cell walls separate the radial cell tissues, so the ultrasonic impedance is large [28,29]. Therefore, it is important to extract features, which reflect the attenuation of the ultrasound energy in plant stems, to discover the performance of living plant stems. However, there is little research on this aspect.

At present, the study of plant stem structure difference by ultrasound still uses ultrasonic velocity. The freezing condition of plant stems was first studied by Guillaume in 2014 using the ultrasound echo signals [30], which detected the process of freezing and thawing of sampled plant stems by the velocity of the ultrasound echo signal. However, researchers still only adopted the features of the ultrasound echo velocity and energy decay in the time domain to study the freeze–thaw phenomenon of plant stems. Since the experiments were performed with intercepted stem samples, it was impossible to study the stress response of living stems after freezing. At the same time, because the ultrasound signals in the experiments were not ultrasonic radio frequency (RF) signals, more feature extraction could not be achieved to study the differences in living plant stems on ultrasonic. Lv [31] used the ultrasound velocity to track the moisture changes of living stem plants. They adopted velocity to analyze the change of plant stems, which assumed that plant stem structure is uniform. However, the truth is that because the plant stem is non-uniform and anisotropic, it is difficult for the velocity to reflect such internal structural differences. Therefore, it is necessary to extract new features that can reflect the plant stem structure other than velocity.

It is possible to explore the sensitivity of the ultrasound propagation to the density and structure of plant stems by ultrasound RF [32]. In this paper, we focused on extracting the ultrasound energy features from the ultrasound RF in the time and frequency domain to discover the difference of plant stem structure. What is more, the evaluation of the feature contribution was ranked by information gain and correlation.

2. Methods

2.1. Common Energy Features of Ultrasound in Plant Stems

2.1.1. Mean and Variance of Ultrasound Echo Signals

The change in the ultrasound echo signal's mean value reflects the plant stem conditions. As the ultrasound propagates through the plant stem, its energy is absorbed, attenuated, and scattered by the material in the stem, resulting in an exponential attenuation of the echo signal envelope. Different plant stems have different attenuation exponents, which ultimately lead to variations in the mean and variance of the ultrasound echoes. In this regard, the expression for calculating the mean value of the ultrasonic echo is presented in Equation (1).

$$\mu_x = E\{x\} = \lim_{T\to\infty} \frac{1}{T}\int_0^T x(t)dt \approx \frac{1}{N}\sum_{n=1}^{N} x[n] \tag{1}$$

where $x(t)$ is the ultrasonic echo continuous signal, and $x[n]$ is the ultrasound echo signal sequence values. T is the observation time of the ultrasound echoes. $n,\ n = 1,2,\ldots,N$ is the sampling point for the ultrasound echoes.

The ultrasonic echo signal variance reflects the fluctuating characteristics of the signal above and below the average ultrasonic energy, reflecting the changing energy characteristics of the ultrasonic stem signal. The mathematical expression for calculating the variance of an ultrasonic echo sampling signal is

$$\sigma_x^2 = E\{|x(n) - \mu_x|^2\} \tag{2}$$

2.1.2. Spectral Features

The propagation of the ultrasound waves is closely related to the density and structure of the material. Hence, when the structure of the plant stem changes, it causes changes in the spectrum of the ultrasonic echo signal.

The FFT is used to get the spectrum $X(k)$ of the ultrasound echo sequence $x(n)$. The energy in the frequency domain is expressed in Equation (3).

$$G_x(k) = \frac{1}{N}|X(k)|^2 \tag{3}$$

where $X(k)$ is the FFT sequence value of the ultrasound echo, and k represents the kth spectral line. The relationship between spectral line k and each harmonic frequency of the signal is

$$f = \frac{k}{NT_s} \tag{4}$$

where f is the sampling frequency, which is 5–7 times the center frequency of the ultrasonic probe, and T_s is the sampling interval.

2.2. Improvement of Ultrasound Features

The traditional ultrasonic feature is the ultrasonic velocity computed by the envelope of the ultrasonic RF signal, which is the signal retained after removing the high-frequency components in the FR ultrasonic. It is easy to identify the difference of the structure by the different velocity in the uniform medium. However, because the plant stem is non-uniform and anisotropic, the traditional feature velocity has limitations in reflecting structural differences. Therefore, we tried to extract new features based on the ultrasonic RF signal, carrying the inhomogeneous and anisotropic structural characters.

When ultrasonic energy is used as features, the processing of the detection depth is a key problem. Different plant diameters at height can cause various interference with ultrasound detection. This interference is manifested in the following ways: the internal structure of the same plant is similar, and its difference is mainly due to the different detection positions, forming different detection plant stem diameters, which cause different ultrasound energy parameters, such as mean and variance. While, if the plant diameters are the same, the different parameters of the ultrasound would show the difference of the plant stem structure. In live plant stem detection, at the radial direction, the stem diameter at the detection site is the detection depth of the ultrasound. As the structures of the plant stems at breast height can vary significantly depending on the species and growing period, this causes the difference in the depth of the ultrasound detection and the attenuation of the ultrasound. Suppose the depth is determined and is considered constant during the detection period, such as in a day. In this case, changes in the ultrasonic parameters can be caused by changes in the plant structure, such as water deficiency or lack of water. At the same time, to reduce the interference caused by the different detection depths, this paper proposes to use the thoracic diameter parameter to obtain the feature density of the ultrasound energy to suppress or eliminate the interference of the thoracic diameter on the premise of obtaining the standard ultrasound features.

$$den_f = \frac{f_{eature}}{d} \tag{5}$$

where den_f is the density of the ultrasonic feature, f_{eature} is the ultrasonic feature, and d is the diameter at breast height at the plant detection.

Based on the above method, the improved features in this paper are extracted, which are energy density, mean density, variance density, and first to third resonance peak amplitude density.

The equations of all the extracted features are shown in Table 1.

Table 1. Equations of the extracted features.

No.	Feature Name	Equation
1	Energy	$E = \sum_{i=1}^{N} \lvert x_i \rvert, \ i = 1 \cdots N$
2	Energy density	$den_E = \frac{E}{d}$
3	Mean value	$M = mean(x_i), \ i = 1 \cdots N$
4	Mean density	$den_M = \frac{M}{d}$
5	Variance	$v_{ar} = var(x_i), i = 1 \ldots .N$
6	peak-to-peak value	$peak = \lvert max(x) - \min(x) \rvert$
7	Spectrum DC	$S_{DC} = Z_0, Z = abs(fft(x_i)), i = 1, \ldots, N; \ Z = Z_0, \ldots, Z_{N-1}$
8	Spectral DC density	$dens_{DC} = \frac{Z_0}{d}$
9	1st formant value	$formant_1 = max(Z)$
10	1st formant density	$den_{formant_1} = \frac{formant_1}{d}$
11	1st formant frequency	$Index_{formant_1} = index(max(Z))$
12	2nd formant value	$formant_2 = second_{max(Z)}$
13	2nd formant density	$den_{formant2} = \frac{formant_2}{d}$
14	2nd formant frequency	$Index_{formant_2} = index(second_{max(Z)})$
15	3rd formant value	$formant_3 = third_{max(Z)}$
16	3rd formant frequency	$Index_{formant_3} = index(third_{max(Z)})$
17	removed DC energy	$E_{-DC} = E - mean(x), x = x_1, \ldots, x_N$
18	removed DC variance	$Var_{E_{DC}} = var(E_{-DC})$
19	removed DC 1st formant value	$formant_1 = max(Z_{-DC}), Z_{-DC} = Z_1, \ldots, Z_{N-1}$
20	removed DC 1st formant frequency	$Index_{formant_1} = index(max(Z_{-DC}))$
21	removed DC 2nd formant value	$formant_2 = second_{max(Z_{-DC})}$
22	De-rectified 2nd formant frequency	$Index_{formant_2} = index\left(second_{max(Z_{-DC})}\right)$
23	removed DC 3rd formant value	$formant_3 = third_{max(Z_{-DC})}$
24	removed DC 3rd formant frequency	$Index_{formant_3} = index\left(third_{max(Z_{-DC})}\right)$

2.3. *The Evaluttion of the Contribution of the Feature*

In order to achieve a multi-perspective study of the energy features of the ultrasound stem differences, it is necessary to evaluate the most discriminating features from the multidimensional ultrasound energy features. When multiple parametric features are extracted from the time and frequency domains, compelling ultrasound features or feature sets are first selected so that they can effectively reflect the structural differences in the plant stems. Tracking the structural properties within the stem body is also a crucial issue in feature selection. Therefore, this paper uses the information gain and correlation evaluation method to rank the ultrasound energy features of the plant stems by their contribution.

2.3.1. Information Gain

Information gain as shown in Equation (6) indicates the extent to which the information of a known feature vector A reduces the information uncertainty of the partitioned dataset D [33].

$$\mathrm{Gain}(D, A) = H(D) - H(D|A) \tag{6}$$

$H(D)$ is the information entropy of the dataset D. $H(D|A)$ is the conditional entropy of D given feature A. A is the feature vector, and D is the dataset. $H(D)$ is defined as shown in Equation (7).

$$H(D) = -\sum_{j-1}^{M} \frac{|C_j|}{|D|} log_2 \frac{|C_j|}{|D|} \tag{7}$$

where M is the number of categories C in dataset D. $|C_j|$ is the number of samples of category j. $|D|$ is the total number of samples of dataset D.

For a given dataset D, different features have different information gains. The greater the information gain, the more information the feature contains that helps in classification and the stronger the classification ability. The method of feature contribution is based on information gain. For a given dataset D, the information gain of each feature A is calculated, and the feature with the highest information gain is selected.

2.3.2. Evaluation of Correlation

Correlation is mainly used to assess the contribution of features to classification by measuring the correlation between feature vector A and category vector C [34].

The correlation coefficient r is calculated as in Equation (8).

$$r = \frac{\sum AC - \frac{\sum A \sum C}{|D|}}{\sqrt{(\sum A^2 - \frac{(\sum A)^2}{|D|})(\sum C^2 - \frac{(\sum C)^2}{|D|})}} \tag{8}$$

where C is the category vector, and where the dataset D is related to the sample category vector C as in Equation (9).

$$\sum_{j=1}^{M} |C_j| = |D| \tag{9}$$

where C_j is a vector of M categories, $j = 1, 2, \ldots, M$.

The correlation coefficient indicates the degree of linear correlation between the two variables, feature vector A and category C. When $0 < |r| < 1$, it indicates a certain degree of linear correlation between the two variables. Moreover, the closer $|r|$ is to 1, the closer the linear relationship between the two variables. The closer $|r|$ is to 0, the weaker the linear correlation between the two variables.

In this paper, the multiple stem ultrasound energy features were extracted based on time-domain acoustic and frequency-domain amplitude-frequency signals, combined into a multidimensional set of plant stem ultrasound energy features. The optimal ultrasound energy features or feature sets were selected to represent different stem structure differences using the information gain attribute evaluation and correlation evaluation methods.

3. Results

3.1. Simulation Experiment

The ultrasonic plant stem detection system is the same as the paper [31] shown in Figure 1. The system includes a non-metallic ultrasonic probe, ultrasonic pulse signal generator and receiver (CTS-8077PR). It also includes a Nextkit data acquisition instrument and computer. The acquisition function of Nextkit is developed on labView.

(1) CTS-8077PR parameters: pulse repetition frequency PRF is set to 1 kHz to get 1000 ultrasonic RF echo signals in each second, the pulse width is 1000 ns, and the conversion voltage is 200 v.

(2) Ultrasonic probe parameters: probe diameter is 30 mm, and the primary frequency is 1 MHz, which is suitable for ultrasonic testing of nonmetals.

(3) Data acquisition equipment parameters: sampling depth is 2000 points to record each ultrasonic RF echo signal, and the sampling frequency is 10 MHz to get more details of higher frequency of the signal.

Radial ultrasonic testing was conducted on four healthy plant stems: basho, palm, magnolia denudata, and potted sunflowers shown in Figure 2.

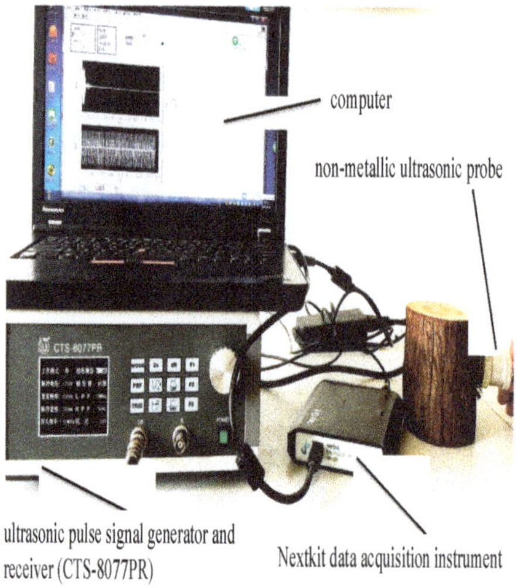

Figure 1. The ultrasonic plant stem detection system.

Figure 2. Testing samples. (**a**) basho; (**b**) palm; (**c**) sunflower; (**d**) magnolia denudata.

3.2. *Different Ultrasonic Manifestations of Stem and Body Structure of Different Tree Species*

The structural characteristics of the stem bodies of these four types of plants are shown in Table 2. The testing took place on 9 May 2019 from 5:00 p.m. to 5:30 p.m.

Table 2. Attributes of plant stems.

Varieties of Trees	Stem Texture	Growth Characteristics	Plant Category	Vascular Properties
Basho	Herbaceous stems	Perennial	Monocotyledon	Vascular bundles dispersed within the stem, no formative layer within bundles
Sunflower	Herbaceous stems	Annual	Dicotyledon	Vascular bundles forming a circular shape, without a forming layer within the bundle.
Magnolia denudata	Woody stem	Deciduous plant	Dicotyledon	Vascular bundles forming a circular shape, with a formative layer within the bundle.
palm	Woody stem	Casuarina	Monocotyledon	Vascular bundles scattered within stem, with forming layer within bundle.

The plant stem body ultrasound signal acquisition was completed using the experimental platform in the literature [31]. Assay parameters are shown in Table 3.

Table 3. Parameters of test data.

Samples	Basho 1	Sunflower 1	Sunflower 2	Palm 1	Palm 2	Magnolia Denudata 1	Magnolia Denudata 2
Height of the test point from the ground (cm)	48.00	20.00	20.00	57.00	79.00	75.00	80.00
Test circumference (cm)	20.00	5.00	7.00	49.00	50.00	46.30	63.20
Samples of ultrasound pulse echoes (pcs)	57	33	42	61	53	60	48

3.2.1. Ultrasonic Signals of Plant Stems in Time and Frequency Domain

Ultrasound Signal in Time Domain

The ultrasound signals of the four samples in the time domain are shown in Figure 3.

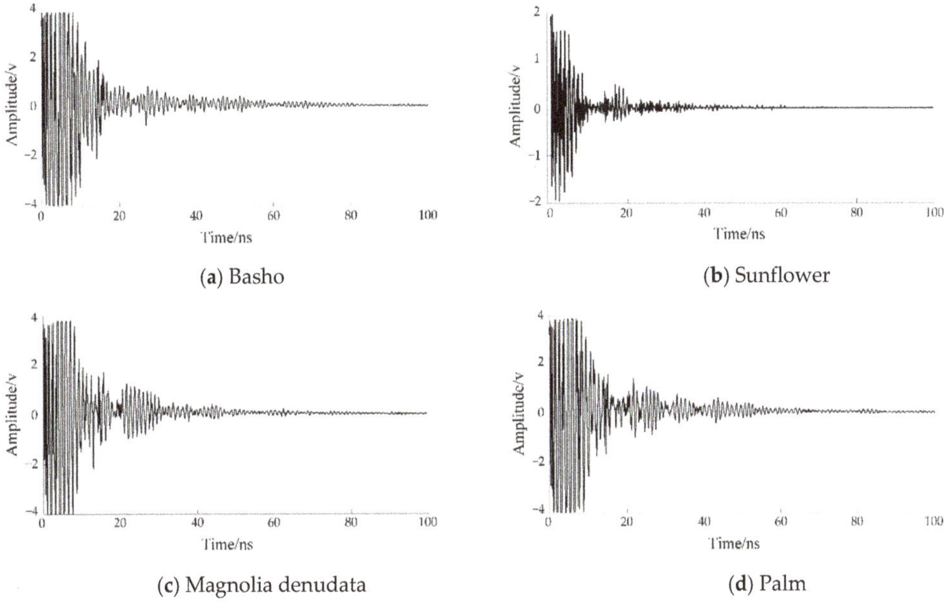

Figure 3. Ultrasonic wave of plant stems in time domain.

The ultrasonic echo signals from the stem bodies of the four plant species in Figure 3 showed an obvious difference in the envelope of their signals. The peaks of the signals of the sunflowers ranged from −2 to 2 v, while the peaks of the other three species ranged from −4 to 4 v. Moreover, waveforms of these four plants varied in the time domain. It showed that the signals displayed fluctuation at about 20 ns for the stems of magnolia denudata and palm, which was caused by the bark of these two kinds of stems. Because the texture structure of the bark has some extensive differences with the xylem of these stems, the waveform energy of ultrasonic RF signal has obvious increase, rather than constant attenuation. While this phenomenon did not appear in the stems of sunflower and basho, because these two kinds of stems have no obvious barks. However, it is a challenge to describe the difference in the time domain.

Frequency Domain Ultrasound Echo Amplitude and Frequency Signals

In order to further analyze the energy distribution characteristics of the ultrasonic wave of the four plant species stems, the experiments used the Fourier transform to obtain the amplitude-frequency signals. In the experiment, the sampling frequency was 10 MHz, and the spectrum could be analyzed from 0 to 5 MHz. The experimental results are shown in Figure 4.

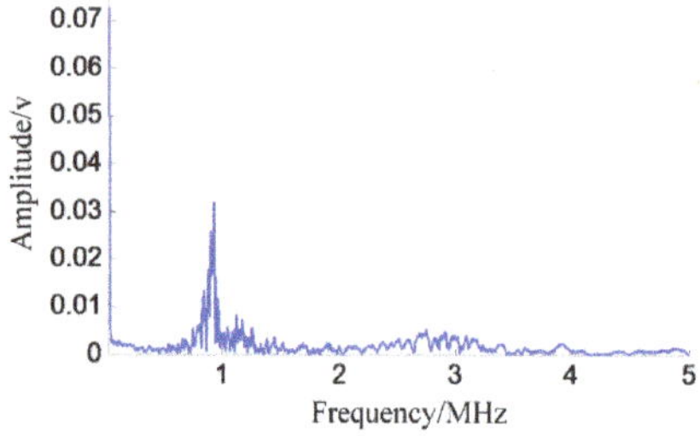

(a) Basho

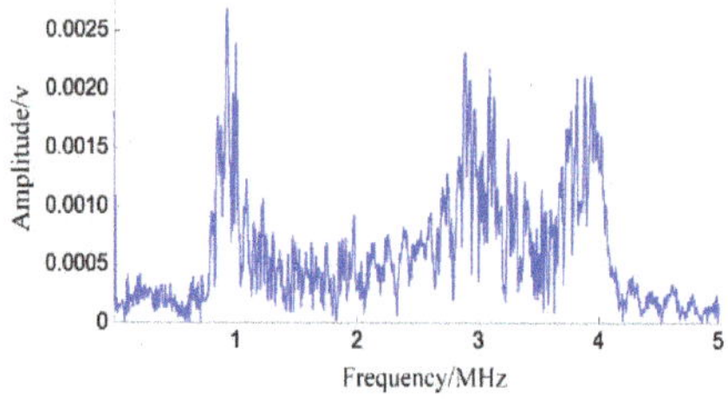

(b) Sunflower

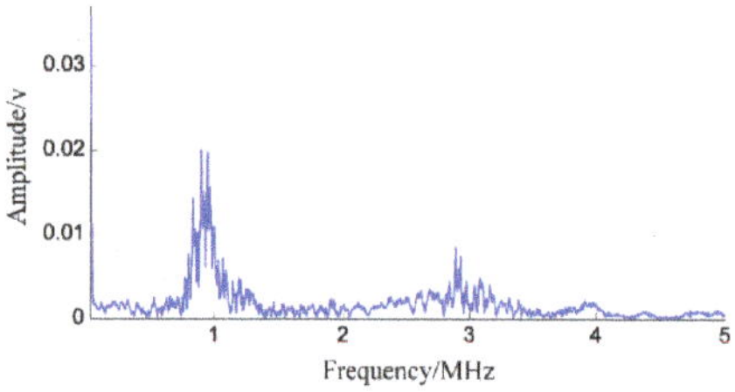

(c) Magnolia denudata

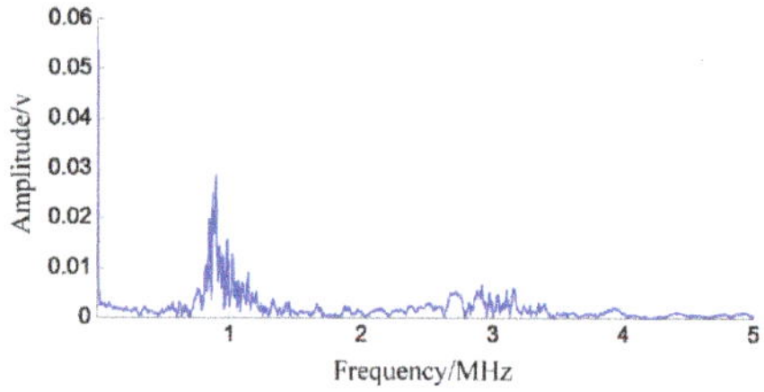

(d) Palm

Figure 4. Ultrasound amplitude-frequency signals of plant stems.

Figure 4 shows that the energy distribution of the four plants was prominent at 1 MHz and 3 MHz, but their spectral envelopes had apparent differences.

Figure 5 segments the spectrum to show the resonance peaks are mainly distributed near 1 MHz and 3 MHz.

The results showed that the stems of the monocotyledon (basho and palm) showed a transparent envelope of single peaks around the first resonance frequency (1 MHz). In contrast, the dicotyledon (sunflower and magnolia) showed a clear multi-peak feature. Comparing the energy peaks around 1 MHz and 3 MHz, the peak energy distribution of the monocotyledonous stems around 3 MHz was less than 4 to 8 times that around 1 MHz. In contrast, the distribution of the peak energy at the two frequencies for the dicotyledon sunflower and magnolia was closer, between 1 and 2 times, as shown in Table 4.

Table 4. Energy distribution of resonance peak in ultrasonic spectrum of plant stems.

Name of Tree	Envelope around 1 M.	Envelope around 3 M.	P1: Peak Value around 1 M (mv).	P3: Peak Value around 3 M (mv).	Times of P1/P3
Basho	Single peak	Multi-peak	32.00	4.20	8.00
Sunflower	Multi-peak	Twin peaks	2.60	2.30	1.00
Magnolia denudata	Multi-peak	Single peak	20.10	8.70	2.00
Palm	Single peak	Multi-peak	28.80	6.80	4.00

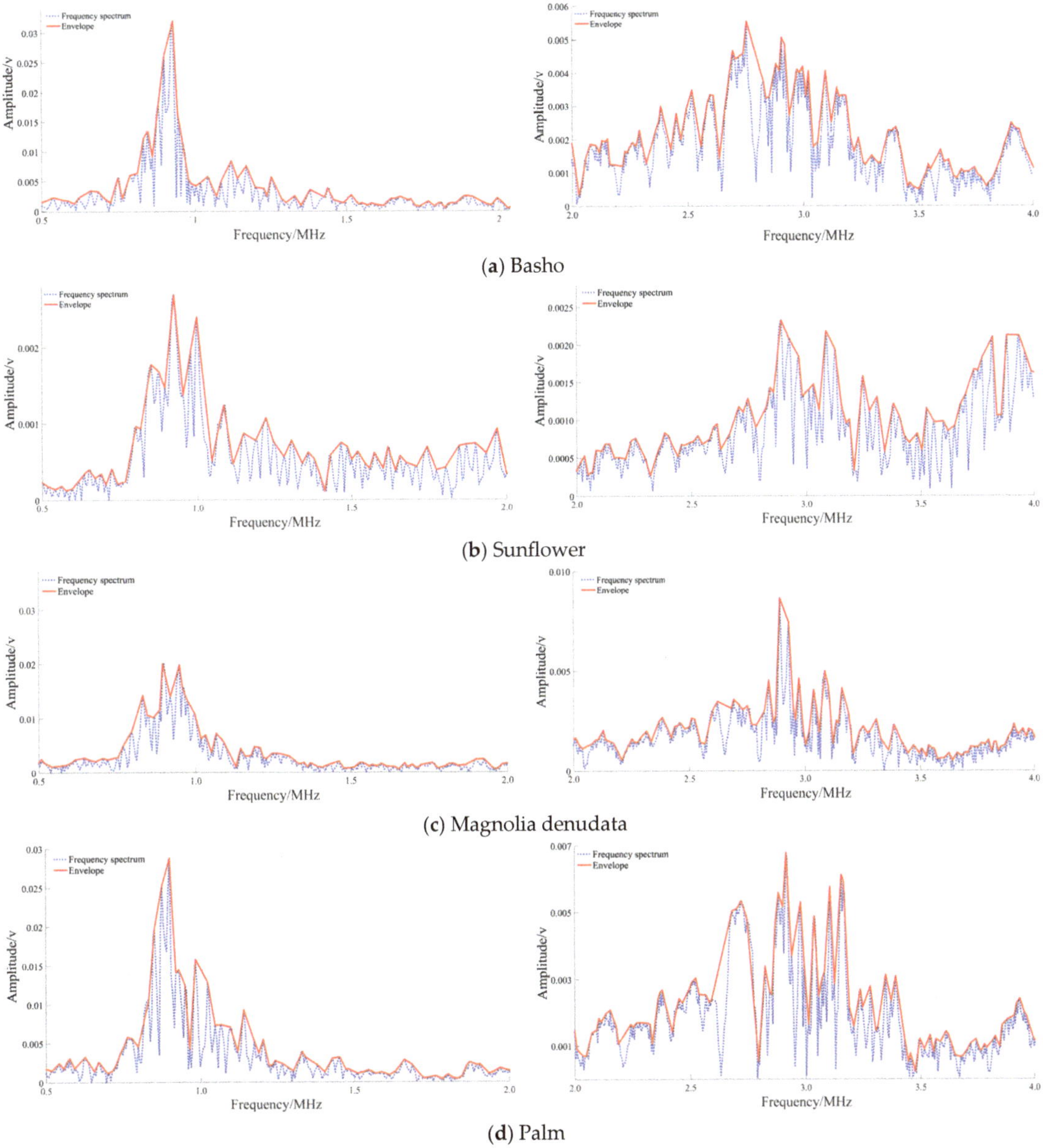

Figure 5. Segments of ultrasound amplitude-frequency of plant stems.

3.2.2. Ultrasonic Energy Features and Feature Contribution of Plant Stems

(1) Energy features in the time domain

The energy features of the radial ultrasound of the stems of the above four plant species were extracted in the time domain, as shown in Table 5. The main features are conventional energy features as well as improved energy features. The conventional energy features mainly contained total energy, mean value, variance, and peak-to-peak values. The improved energy features were mainly the energy density and the mean density.

Table 5. Ultrasound energy features of plant stems in time domain.

Name of Sample	Circumference (cm)	Energy (v)	Energy Density (v/cm·10^{-4})	Average Value (v)	Average Density (v/cm)	Variance	Peak-to-Peak Value (v)	Removed DC Energy (v)	Removed DC Variance
Basho 1	20.00	31.57	5.44	0.11	0.16	1.43	7.86	31.54	1.43
Sunflower 1	5.00	21.64	1.75	0.01	0.43	0.16	5.41	21.63	0.16
Sunflower 2	7.00	21.90	3.61	0.03	0.31	0.16	5.56	21.88	0.16
Magnolia denudata 1	63.50	31.76	0.74	0.05	0.05	1.51	7.84	31.76	1.51
Magnolia denudata 2	46.50	32.23	1.50	0.07	0.07	1.68	7.82	32.22	1.68
Palm 1	49.50	31.22	1.97	0.10	0.06	1.32	7.83	31.19	1.32
Palm 2	50.50	31.31	1.89	0.10	0.06	1.34	7.83	31.29	1.34

(2) Energy features in the time domain

The frequency-amplitude signals based on the Fourier transform were used to extract the radial ultrasound frequency energy features in the frequency domain, as shown in Table 6. The main features were the mean value, each resonance peak's energy value, and frequency location. At the same time, the mean value density and resonance amplitude density parameters were computed.

Table 6. Ultrasound features of plant stems in frequency domain.

Name of Sample	Basho 1	Sunflower 1	Sunflower 2	Magnolia Denudata 1	Magnolia Denudata 2	Palm 1	Palm 2
Average (dB)	125.65	8.78	27.08	28.73	29.95	70.37	109.92
Average density (dB/cm)	0.63	0.18	0.39	0.04	0.05	0.15	0.22
1st formant value (dB)	254.02	24.86	28.94	243.92	246.71	269.51	215.13
1st formant value density (dB/cm)	1.27	0.50	0.41	0.36	0.39	0.58	0.43
sampling position of 1st formant	184.00	186.00	191.00	184.00	185.00	183.00	183.00
1st formant frequency (MHz)	0.92	0.93	0.95	0.92	0.92	0.92	0.91
2nd formant value (dB)	83.52	25.82	34.43	83.52	84.99	97.19	75.24
2nd formant density (dB/cm)	0.42	0.52	0.49	0.11	0.13	0.21	0.15
sampling position of 2nd formant	579.00	583.00	585.00	585.00	584.00	595.00	574.00
2nd formant frequency (MHz)	2.90	2.91	2.92	2.92	2.92	2.97	2.87
3rd formant value (dB)	33.97	49.72	43.38	38.53	36.71	32.88	35.18
3rd formant density (dB/cm)	0.17	0.99	0.62	0.08	0.06	0.07	0.07
sampling position of 3rd formant	826.00	782.00	777.00	845.00	845.00	801.00	822.00
3rd formant frequency (MHz)	4.13	3.91	3.88	4.23	4.23	4.00	4.11
Circumference	20.00	5.00	7.00	63.50	46.50	49.50	50.50

(3) Feature combination

The 24-dimension ultrasound energy features were extracted from the time and frequency domains, as shown in Table 7.

Table 7. Features of plant stems and feature numbers.

Feature No.	Feature Name	Feature No.	Feature Name
1	Energy	13	2nd formant density
2	Energy density	14	2nd formant frequency
3	Mean value	15	3rd formant value
4	Mean density	16	3rd formant frequency
5	Variance	17	removed DC energy
6	peak-to-peak value	18	removed DC variance
7	Spectrum DC	19	removed DC 1st formant value
8	Spectral DC density	20	removed DC 1st formant frequency
9	1st formant value	21	removed DC 2nd formant value
10	1st formant density	22	De-rectified 2nd formant frequency
11	1st formant frequency	23	removed DC 3rd formant value
12	2nd formant value	24	removed DC 3rd formant frequency

The ultrasound energy features were selected based on information gain attributes and correlation evaluation methods. The ultrasound energy features were selected to reflect the differences in the stem structure of the different plant species. The number of samples used in the final samples is shown in Table 8.

Table 8. Sample number of radial features of stems of four species.

Detecting Tree Species	Basho	Sunflower	Palm	Magnolia Denudata	Total Number of Samples (pcs)
Ultrasound samples (pcs)	52	62	35	40	189

(4) Results of feature contributions

The information gain and correlation evaluation methods were used to compute the 24-dimensional feature to identify the most distinguishing energy features. The evaluated features are shown in Table 9.

Table 9. Contribution of ultrasound features of four plant stems.

Feature Sorting	Feature No.	Information Gain	Feature No.	Correlation Values	Feature No.	Mean Value of Contributions
1	4	1.71	18	0.71	4	1.20
2	3	1.55	5	0.71	3	1.11
3	7	1.29	9	0.70	7	0.96
4	5	1.14	19	0.70	5	0.92
5	18	1.14	4	0.69	18	0.92
6	17	1.12	1	0.69	17	0.90
7	1	1.10	17	0.69	1	0.89
8	2	1.10	6	0.69	9	0.85
9	8	1.07	12	0.68	19	0.85
10	13	1.00	21	0.68	6	0.84
11	9	1.00	13	0.67	12	0.84
12	12	1.00	3	0.67	21	0.84
13	19	1.00	7	0.64	13	0.83
14	21	1.00	20	0.41	2	0.66
15	6	1.00	11	0.41	8	0.63
16	20	0.83	23	0.34	11	0.62
17	11	0.83	15	0.34	20	0.62
18	10	0.45	10	0.29	10	0.37
19	22	0.35	2	0.23	15	0.33
20	14	0.35	24	0.23	23	0.33
21	23	0.32	16	0.23	14	0.25
22	15	0.32	8	0.20	22	0.25
23	16	0.13	14	0.16	16	0.18
24	24	0.13	22	0.16	24	0.18

The final contribution is the mean contribution that is the average of the information gain and correlation.

The results showed that the top five feature contributions by the information gain method were the mean density, mean value, spectral DC variance, and removed DC variance. The top five feature contributions by the correlation evaluation method were the removed DC variance, variance, 1st formant value, removed DC 1st formant valued, and mean density. The extracted features were ranked by the mean contribution of the two methods. The mean density was the most compelling feature for structural differences, followed by variance and 1st formant value.

The three features (the mean density, variance, and 1st formant value) were analyzed in 2-dimensional feature maps to explore plant species differentiation, and the experimental results are shown in Figure 6.

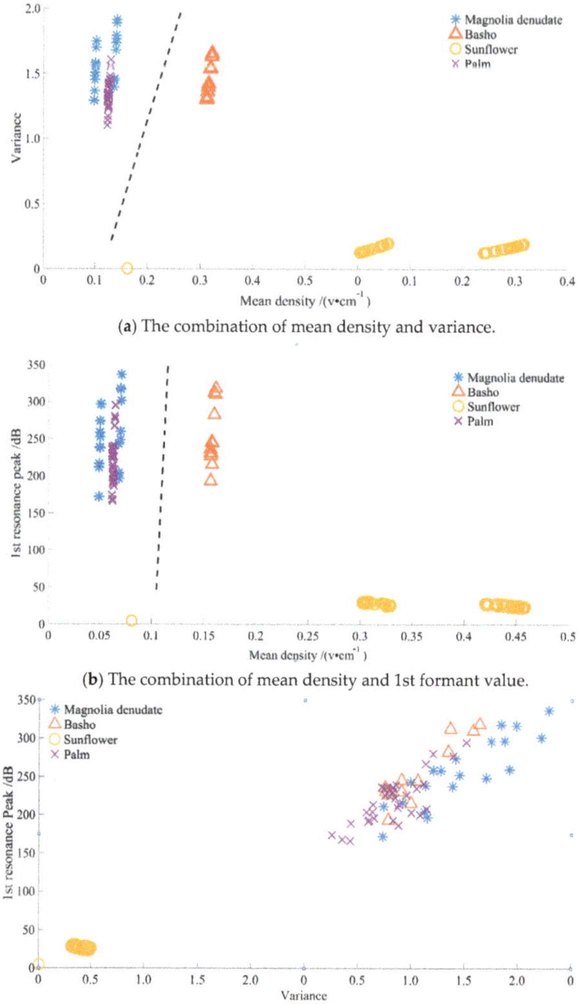

(**a**) The combination of mean density and variance.

(**b**) The combination of mean density and 1st formant value.

(**c**) The combination of variance and 1st formant value.

Figure 6. Differential identification of plant stems based on feature combination.

Figure 6 shows that the combination of features with the mean density and variance and features with the mean density and 1st formant value, as shown in Figure 6a,b, were linearly distinguishable between herbaceous and woody plant stem bodies with good differentiation. In contrast, the combination of features using variance and the 1st formant value, as shown in Figure 6c, was linearly indistinguishable between the herbaceous and woody classes. However, the difference between sunflowers and the other three stem types was still evident.

4. Discussion and Conclusions

In this paper, the ultrasonic energy signals of the differences in the stems of basho, sunflower, palm, and magnolia denudata were analyzed in the time and frequency domains, which provide a basis for the feature extraction from an ultrasonic energy perspective.

In order to reduce the interference caused by different detection depths, this paper proposed improved ultrasound energy features, that is, the ultrasound energy feature density, which constitutes energy density, mean density, variance density, and resonance peak density features.

The 24-dimensional energy features were extracted from the time and frequency domains. They were evaluated by the information gain and correlation method. Based on the top three energy features in feature contribution, the 2-dimensional feature maps were formed to discover the structure differences of the plant stems. The feature map of the variance and the first formant had good differentiation between herbaceous and woody plant stems. Our future study will test more stem plants to verify the universality of these extracted features.

Based on these extracted features, the feature selection, the filtered or wrapper method, is our future study. The classifier is intended to select random forest (RF), support vector machine (SVM) and Linear discriminant analysis (LDA).

Author Contributions: Conceptualization, D.L.; methodology, D.L., J.Z. and X.H.; software, M.G.; validation, R.X.; formal analysis, X.H.; resources, W.L.; data curation, Z.W.; writing—original draft preparation, D.L., J.Z. and R.X.; writing—review and editing, D.L., J.Z. and M.G. All authors have read and agreed to the published version of the manuscript.

Funding: This research was funded by the National Natural Science Foundation of China (Grant No. 31860332), Kunming Forestry Information Engineering Technology Research Center (Grant No. 2015FIA02), Major Special Projects in Yunnan Province (key technology research and application demonstration of blockchain serving key industries, Grant No. 202002AD080002), Major scientific and technological projects in Yunnan Province (Grant No. 202002AA10007) and Education Foundation of Yunnan Province (Grant No. 2022J0495).

Institutional Review Board Statement: Not applicable.

Informed Consent Statement: Not applicable.

Data Availability Statement: Not applicable.

Conflicts of Interest: The authors declare that there is no conflict of interest.

References

1. Li, R.; Jiang, Z.; Zhang, S.; Cai, J. A review of new research progress on the vulnerability of xylem embolism of woody plants. *Chin. J. Plant Ecol.* **2015**, *39*, 838–848.
2. Sun, L.; Tang, B.; Gao, R. Response and Adaptability of Anatomical Structure of Plant Stem to the Arid Environment. *For. Explor. Des.* **2016**, *2*, 43–46.
3. Zhao, Y.; Gao, C.; Zhang, X.; Xu, Q. Non-destructive measurement of plant stem water content based on standing wave ratio. *Trans. Chin. Soc. Agric. Mach.* **2016**, *47*, 310–316.
4. Ji, H. The Sap Flow Measurement Based on the Heat Ratio Method (HRM) and Environment Factor Correction Research. Master's Thesis, Northeast Forestry University, Harbin, China, 2012.
5. Hultine, K.R.; Nagler, P.L.; Morino, K.; Bush, S.E.; Burtch, K.G.; Dennison, P.E.; Glenn, E.P.; Ehleringer, J.R. Sap Flux-Scaled Transpiration by *Tamarisk* (*Tamarix* Spp.) before, during and after Episodic Defoliation by the Saltcedar Leaf Beetle (*Diorhabda Carinulata*). *Agric. For. Meteorol.* **2010**, *150*, 1467–1475. [CrossRef]
6. Choat, B.; Jansen, S.; Brodribb, T.J.; Cochard, H.; Delzon, S.; Bhaskar, R.; Bucci, S.J.; Feild, T.S.; Gleason, S.M.; Hacke, U.G.; et al. Global Convergence in the Vulnerability of Forests to Drought. *Nature* **2012**, *491*, 752–755. [CrossRef] [PubMed]
7. Edwards, W.R.N.; Jarvis, P.G. A Method for Measuring Radial Differences in Water Content of Intact Tree Stems by Attenuation of Gamma Radiation. *Plant Cell Environ.* **1983**, *6*, 255–260. [CrossRef]
8. Byrne, G.F.; Fenn, M.D.; Burgar, M.I. Nuclear Magnetic Resonance Studies of Water in Tree Sections. *Agric. For. Meteorol.* **1986**, *38*, 307–317. [CrossRef]
9. Borisjuk, L.; Rolletschek, H.; Neuberger, T. Surveying the Plant's World by Magnetic Resonance Imaging. *Plant J.* **2012**, *70*, 129–146. [CrossRef] [PubMed]
10. Windt, C.W.; Blümler, P. A Portable NMR Sensor to Measure Dynamic Changes in the Amount of Water in Living Stems or Fruit and Its Potential to Measure Sap Flow. *Tree Physiol.* **2015**, *35*, 366–375. [CrossRef]
11. Raschi, A.; Tognetti, R.; Ridder, H.-W.; Béres, C. Water in the Stems of Sessile Oak (*Quercus Petraea*) Assessed by Computer Tomography with Concurrent Measurements of Sap Velocity and Ultrasound Emission. *Plant Cell Environ.* **1995**, *18*, 545–554. [CrossRef]
12. Song, Z.; Lin, Y.; Gu, G. The influence of wood cell activity on moisture content measurement by electric resistance. *J. Northeast For. Univ.* **1994**, *22*, 113–116.

13. Sparks, J.P.; Campbell, G.S.; Black, A.R. Water Content, Hydraulic Conductivity, and Ice Formation in Winter Stems of Pinus Contorta: A TDR Case Study. *Oecologia* **2001**, *127*, 468–475. [CrossRef]
14. Hao, G.-Y.; Wheeler, J.K.; Holbrook, N.M.; Goldstein, G. Investigating Xylem Embolism Formation, Refilling and Water Storage in Tree Trunks Using Frequency Domain Reflectometry. *J. Exp. Bot.* **2013**, *64*, 2321–2332. [CrossRef] [PubMed]
15. Zhou, H.; Sun, Y.; Tyree, M.T.; Sheng, W.; Cheng, Q.; Xue, X.; Schumann, H.; Schulze Lammers, P. An Improved Sensor for Precision Detection of in Situ Stem Water Content Using a Frequency Domain Fringing Capacitor. *New Phytol.* **2015**, *206*, 471–481. [CrossRef] [PubMed]
16. Arciniegas, A.; Brancheriau, L.; Lasaygues, P. Tomography in Standing Trees: Revisiting the Determination of Acoustic Wave Velocity. *Ann. For. Sci.* **2015**, *72*, 685–691. [CrossRef]
17. Ou, M.; Hu, T.; Hu, M.; Yang, S.; Jia, W.; Wang, M.; Jiang, L.; Wang, X.; Dong, X. Experiment of Canopy Leaf Area Density Estimation Method Based on Ultrasonic Echo Signal. *Agriculture* **2022**, *12*, 1569. [CrossRef]
18. Gómez, M.; Aguado, F.; Menéndez, J.M.; Revilla, M.; Villa, L.F.; Cortés, J.; Rico, H. Influence of Soft Tissue (Fat and Fat-Free Mass) on Ultrasound Bone Velocity: An In Vivo Study. *Investig. Radiol.* **1997**, *32*, 609–612. [CrossRef] [PubMed]
19. Abdul Halim, M.H.; Buniyamin, N.; Mohamad, Z. Improving Intramuscular Fat Measurement by Considering the Thickness of Protective Layer in Ultrasonic Transducer. In Proceedings of the 2014 2nd International Conference on Electrical, Electronics and System Engineering (ICEESE), Kuala Lumpur, Malaysia, 9–10 December 2014; pp. 101–107.
20. Nagaoka, R.; Iwasaki, R.; Arakawa, M.; Kobayashi, K.; Yoshizawa, S.; Umemura, S.; Saijo, Y. Basic Study of Intrinsic Elastography: Relationship between Tissue Stiffness and Propagation Velocity of Deformation Induced by Pulsatile Flow. *Jpn. J. Appl. Phys.* **2015**, *54*, 07HF08. [CrossRef]
21. Bolla, D.; In-Albon, S.; Papadia, A.; Di Naro, E.; Gasparri, M.L.; Mueller, M.M.; Raio, L. Doppler Ultrasound Flow Evaluation of the Uterine Arteries Significantly Correlates with Tumor Size in Cervical Cancer Patients. *Ann. Surg. Oncol.* **2015**, *22*, 959–963. [CrossRef]
22. Aaslid, R.; Markwalder, T.M.; Nornes, H. Noninvasive Transcranial Doppler Ultrasound Recording of Flow Velocity in Basal Cerebral Arteries. *J. Neurosurg.* **1982**, *57*, 769–774. [CrossRef]
23. Jambrik, Z.; Gargani, L.; Adamicza, A.; Kaszaki, J.; Varga, A.; Forster, T.; Boros, M.; Picano, E. B-Lines Quantify the Lung Water Content: A Lung Ultrasound versus Lung Gravimetry Study in Acute Lung Injury. *Ultrasound Med. Biol.* **2010**, *36*, 2004–2010. [CrossRef] [PubMed]
24. Sandoz, J.L. Moisture Content and Temperature Effect on Ultrasound Timber Grading. *Wood Sci. Technol.* **1993**, *27*, 373–380. [CrossRef]
25. de Oliveira, F.G.R.; Candian, M.; Lucchette, F.F.; Luis Salgon, J.; Sales, A. A Technical Note on the Relationship between Ultrasonic Velocity and Moisture Content of Brazilian Hardwood (*Goupia Glabra*). *Build. Environ.* **2005**, *40*, 297–300. [CrossRef]
26. Yang, H.; Yu, L.; Wang, L. Effect of Moisture Content on the Ultrasonic Acoustic Properties of Wood. *J. For. Res.* **2015**, *26*, 753–757. [CrossRef]
27. Dikrallah, A.; Kabouchi, B.; Hakam, A.; Brancheriau, L.; Bailleres, H.; Famiri, A.; Ziani, M. Study of Acoustic Wave Propagation through the Cross Section of Green Wood. *Comptes Rendus Mécanique* **2010**, *338*, 107–112. [CrossRef]
28. Tomppo, L. Novel Applications of Electrical Impedance and Ultrasound Methods for Wood Quality Assessment. Ph.D. Thesis, Itä-Suomen Yliopisto, Itä-Suomi, Suomi, 2013.
29. Hasegawa, M.; Takata, M.; Matsumura, J.; Oda, K. Effect of Wood Properties on within-Tree Variation in Ultrasonic Wave Velocity in Softwood. *Ultrasonics* **2011**, *51*, 296–302. [CrossRef]
30. Guillaume, C.; Katline, C.-V.; Benoit, L.; Thierry, A.; Stefan, M. Changes in Ultrasound Velocity and Attenuation Indicate Freezing of Xylem Sap. *Agric. For. Meteorol.* **2014**, *185*, 20–25. [CrossRef]
31. Lv, D.; Shi, X.; Dong, Y.; Wang, Y.; Wang, X.; Wang, C. Non-destructive measurement of plant stem water content based on Ultrasonic Radio frequency. *Trans. Chin. Soc. Agric. Mach.* **2017**, *48*, 195–201.
32. Arciniegas, A.; Prieto, F.; Brancheriau, L.; Lasaygues, P. Literature Review of Acoustic and Ultrasonic Tomography in Standing Trees. *Trees* **2014**, *28*, 1559–1567. [CrossRef]
33. Mao, Y.; Cao, H.; Ping, P.; Li, X. Feature selection based on maximum conditional and joint mutual information. *J. Comput. Appl.* **2019**, *39*, 734–741.
34. Zhan, L. Research on Evaluation and Selection of Feature in Pattern Recognition. Master's Thesis, Tianjin University of Science and Technology, Tianjin, China, 2012.

agriculture

MDPI

Article

Development and Application of a Crossed Multi-Arch Greenhouse in Tropical China

Jian Liu [1], Xuyong Wu [1], Fangyuan Sun [1] and Baolong Wang [1,2,*]

[1] Key Laboratory for Quality Regulation of Tropical Horticultural Crops of Hainan Province, School of Horticulture, Hainan University, Haikou 570208, China
[2] Sanya Nanfan Research Institute, Hainan University, Sanya 572025, China
* Correspondence: wangbaolong@hainanu.edu.cn

Abstract: Deep analysis and demonstration of the developed crossed multi-arch greenhouse were conducted from the perspectives of conceptual design, architectural and structural design, functional design, loading parameters, and structural internal forces. The results show that the crossed multi-arch greenhouse combines the ventilation area between the floor-standing round-arch greenhouse and the unsuitable operation area under the arch bars into one to form a multi-span crossed arch structure with good ventilation and heat dissipation, land savings, and fine mechanical behaviors. The main arch structure uses 32.4% less steel and 25% less foundation volume than the control greenhouse under the same load, which can save about CNY 10,184.00/667m^2 of investment according to the current cost level. In the meantime, ventilation simulation analysis of the developed crossed multi-arch greenhouse was carried out using the software Design Builder. A comparison shows that, under the condition of no wind and breeze (1 m/s) in summer, the setting of the ventilation channel has obvious advantages for the heat dissipation of the greenhouse, and the average temperature is about 2 °C lower than that of the greenhouse without a ventilation channel; under the breeze condition, the temperature in the crossed multi-arch greenhouse is more evenly distributed than that of an ordinary round-arch greenhouse with ventilation channels. Considering the greenhouse function, building cost, using effect, and other evaluation factors, the crossed multi-arch greenhouse can meet the production environment requirements of tropical coastal areas (rain protection, sunshade, and ventilation), with obvious structural advantages, good typhoon resistance, and low construction costs, which is a preferable choice of greenhouse type.

Keywords: vegetable greenhouses; multi-arch greenhouses; arch cross type; resistance to the typhoon

Citation: Liu, J.; Wu, X.; Sun, F.; Wang, B. Development and Application of a Crossed Multi-Arch Greenhouse in Tropical China. *Agriculture* **2022**, *12*, 2164. https://doi.org/10.3390/agriculture12122164

Academic Editors: Muhammad Sultan, Redmond R. Shamshiri, Md Shamim Ahamed and Muhammad Farooq

Received: 31 August 2022
Accepted: 11 December 2022
Published: 15 December 2022

1. Introduction

Tropical areas have frequent typhoons and rainstorms with high temperatures, high humidity, and long durations of heat. The summer and autumn seasons are not suitable for open-air planting, but greenhouse facilities can provide a rain-proof, protected cultivation environment for crops, which becomes the main production method for summer and autumn off-season production in tropical areas [1,2]. The round plastic arch greenhouse has a simple structure, low cost, and wide application in China, which is one of the main greenhouse structures applied in the Hainan region. Ventilation and rain are the main considerations for Hainan greenhouses. However, due to the roll film window setting, round plastic arch greenhouses cannot be closed in time when a sudden rainfall occurs, and the crops inside the greenhouse can be damaged as a result [3]; conversely, a closed roof causes poor ventilation and heat accumulation problems, and many greenhouses in Hainan have been abandoned due to this reason, resulting in the adverse social impact of field desolation. Moreover, many greenhouses could be blown down, as they are not strong enough to resist typhoon weather, causing large amounts of property losses [4–9]. Therefore, greenhouse facilities in the tropics require special attention to rain, ventilation, and typhoon resistance. However, the dilemma in the process of solving these problems is that a better wind resistance requires a higher cost of the greenhouse, making it hard to recover the investment cost, whereas simple, lower-cost greenhouses do not have the ability to resist strong wind. Some greenhouses even have to sacrifice their functions of protecting the crops (such as not being rain-proof, no shade, poor ventilation, etc.) to reduce the building cost.

Driven by this market demand, the crossed multi-arch greenhouse was developed. This greenhouse type can fully absorb the advantages of various greenhouse structures, with function as the major consideration, i.e., suitable for local climate conditions [10] and meeting the requirements of storm prevention, ventilation, cooling, shading, and other functions, while developing reasonable structural forms, improving the wind resistance of the structure, and using less steel for the main structure, thus reducing the building cost without sacrificing function.

2. Development Idea

2.1. Evolution of the Round Arc Greenhouse Structure

The arch bar of the floor-type round-arch greenhouse structure (Figure 1a) has a good mechanical performance and an easy-to-install structure [11]. However, there are also problems such as the difficulties in using and operating the area below shoulder height in the greenhouse or between greenhouses. Thus, an upright side wall type single arch greenhouse structure (Figure 1b) was developed during real production. This structure solves the problem of shoulder space use. To further improve the land use and make full use of structural materials, a multi-span round-arch greenhouse structure (Figure 1c) was developed [12–15]. This structure is spacious and has a high land use rate, which is being gradually developed into a major structural form of the current multi-span plastic greenhouse [16].

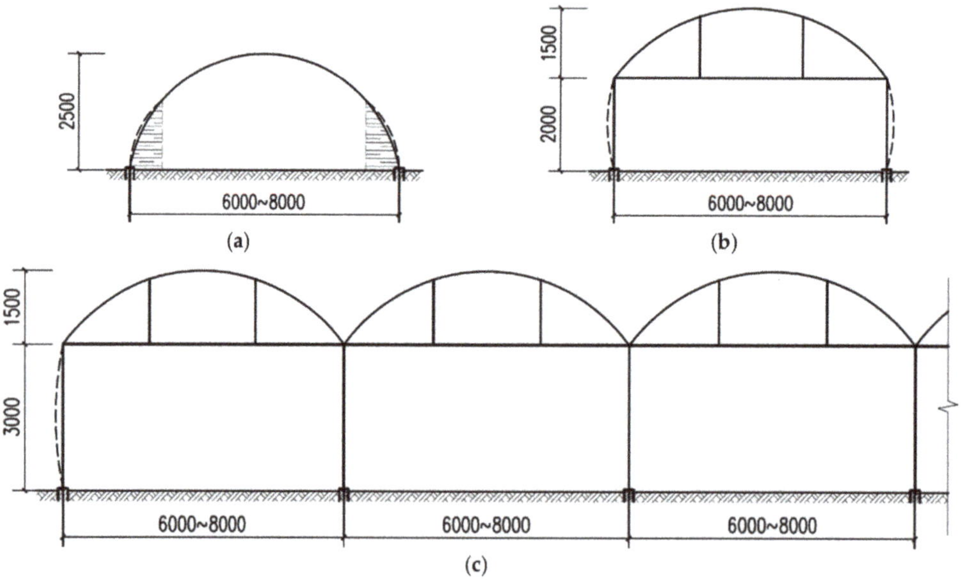

Figure 1. Evolution of the round-arch greenhouse structure (unit: mm). (**a**) Floor-type round-arch greenhouse structure; (**b**) Upright side wall type single arch greenhouse structure; (**c**) Multi-span round-arch greenhouse structure.

When the multi-span round arch plastic greenhouse is used in tropical areas such as Hainan, it has ventilation and rain protection problems. For example, when a sudden rainstorm occurs in summer and autumn, the roll film ventilation mechanism at the top of the greenhouse cannot be closed quickly, and the crops inside the greenhouse can thus be damaged by a rainstorm. Many users stopped using the roll film ventilation mechanism and covered the roof with roll film to ensure the rainproof performance of the greenhouse, the ventilation instead being solved by setting ventilation channels between the arches; thus, a multi-arch split structure model was formed [17]. This structure was first used in Sanya, Dongfang, Ledong, and other places in Hainan for cantaloupe greenhouses, and it quickly became the mainstream cantaloupe greenhouse type (Figure 2a) due to its simple structure (no need to consider the typhoon load; the column can be directly inserted into the ground without setting the foundation) and low cost because it is only used in winter and spring (no typhoon and fewer rainstorms). Inspired by this, structural designers made some improvements (Figure 2b) to the multi-arch split structure and used it for summer and autumn vegetable cultivation in Hainan. After years of practice, the structure has achieved good results. As it is mainly used for off-season production in summer and autumn, it is necessary to consider the resistance against typhoons. An independent foundation needs to be set up at the column, larger and stronger materials should be used, and the cost should be at least twice that of cantaloupe greenhouses.

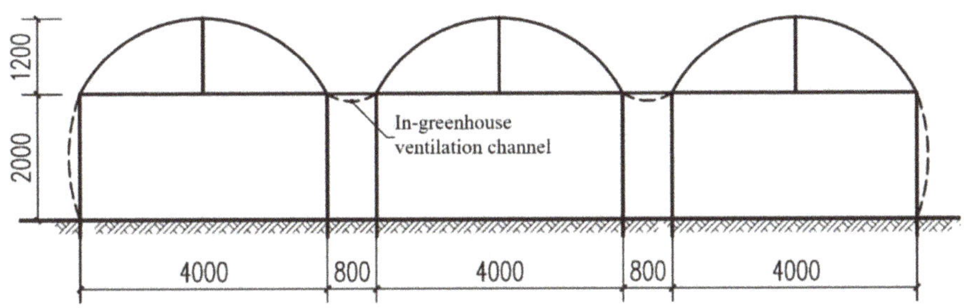

(a)

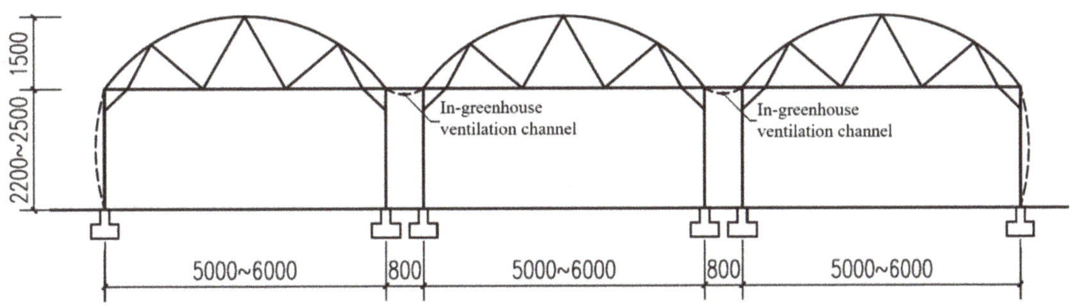

(b)

Figure 2. Multi-span round arch plastic greenhouse with an in-greenhouse ventilation channel (unit: mm). (**a**) Structure of the cantaloupe greenhouse; (**b**) Structure of the Hainan year-round vegetable production greenhouse.

2.2. Structural Concept Design

The arch bar of the floor-type round-arch greenhouse (Figure 3a) is directly connected to the foundation. The overall structure is a round arch that is evenly stressed with good stability. However, there is a large vacant space at the in-greenhouse ventilation channel and the inoperable area below the shoulder arc, which is a waste of land. A multi-span greenhouse structure (with a ventilation channel) (Figure 3b) has turned the arch bar below the shoulder into a vertical column and saved the in-greenhouse ventilation channel, which has a relatively good heat dissipation performance and a higher amount of available land to use. However, this structure has also changed its stress form and requires more column materials, which is not economical. Thus, is it possible to have a structure that retains the good stress resistance of a round arch as well as the function of the in-greenhouse ventilation channel whilst achieving a high amount of available land? After considerable analyses and calculations, and by combining the advantages of floor-type single-arch greenhouses and multi-span greenhouses (with a ventilation channel), this paper developed a crossed multi-arch greenhouse (Figure 3c; multi-arch greenhouse for short). The multi-arch greenhouse not only has the stress advantage of being a floor-type round-arch greenhouse but also retains the heat dissipation advantage of the in-greenhouse ventilation channel. Meanwhile, the stressed arches form a truss structure, which has better stress performance than a floor-type single greenhouse.

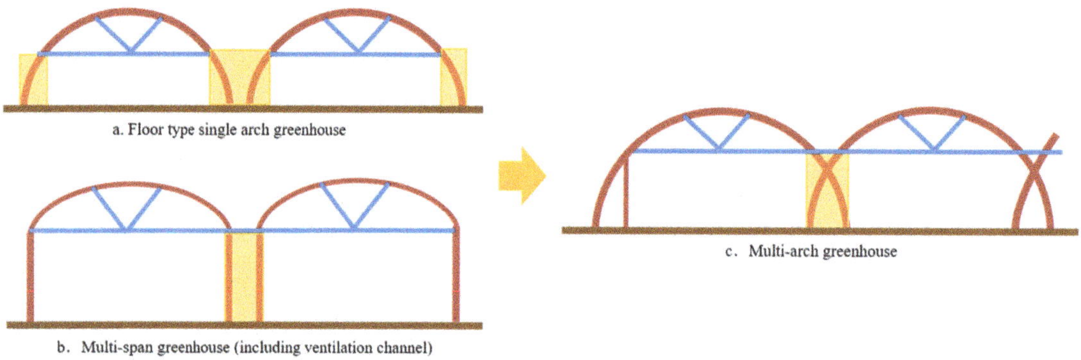

Figure 3. Development idea of a crossed multi-arch greenhouse.

2.3. *Architectural and Structural Design*

The scale of the structure is mainly based on the convenience of farming management and construction. The minimum height of the internal space of the greenhouse is 2.2 m, the span is 6m, and the width of the in-greenhouse ventilation channel is 0.8 m. The crossed arch land part is located in the ventilation channel, so the farming operation is not affected. An open space of 6 m × 2.2 m is formed inside each span, which can meet the operation of small- and medium-sized machinery. The size of the opening direction can be 2~4 m according to the actual load, although 3 m is recommended for the machinery to pass through.

In terms of the arrangement of structural components, the arches are crossed to form a multi-arch structure in the span direction, and the tie beam and reinforcement bar are set between the greenhouses. Compared with an independent arch structure, the crossed arch structure is evenly stressed, and the horizontal wind load on either side can be evenly dispersed to each span of the arch bar, which is more conducive to improving the overall stability of the structure. At the same time, the crossed arches, together with the tie beams and supports, form a geometrically invariant system (as shown in Figure 4), greatly improving the wind resistance of the greenhouse structure.

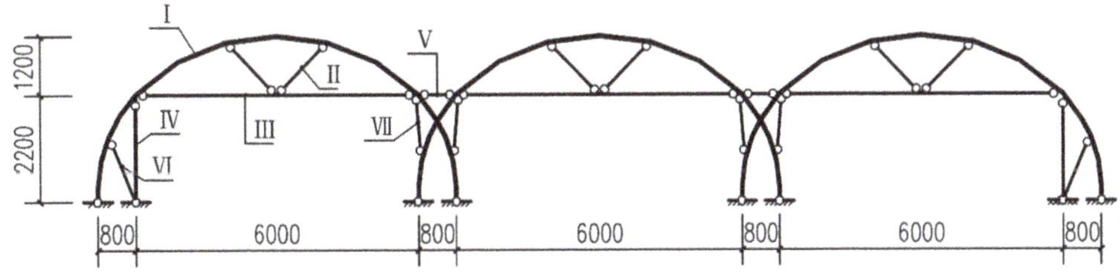

Figure 4. Systematic diagram of the multi-arch greenhouse structure (unit: mm). Notes: I—arch bar, II—web member, III—tie beam, IV—side wind-resisting column, V—in-greenhouse beam, VI—side reinforcement bar (adopted at a high wing load), VII—reinforcement bar; the icon of "o" is the hinged point or hinged support.

2.4. Functional Design

Summer and autumn vegetable greenhouse facilities in the tropical area of Hainan need to overcome some unfavorable meteorological factors such as high summer temperatures and frequent typhoons and rainstorms in order to meet the basic production conditions of ventilation, rain protection, pest control, and suitability for cultivation. Large-scale production applications should also achieve low building costs, less operating energy consumption, land saving, etc.

The top of the developed multi-arch greenhouse is covered with roof film (the thickness of the film is determined according to the actual loading conditions), and the vents between the arches and the surrounding vents are covered with an insect-proof screen. The inner curtain is installed at the internal shoulder-height beams. The entrance and exit are located at the hill wall, with a net width of no less than 2 m, which can meet the needs of small- and medium-sized farm machinery. The above basic dimensions and practices can achieve the basic functions of rain protection, ventilation, insect protection, internal shading, and suitability for cultivation (Figure 5). The multi-arch structure has combined the vacant area below the shoulder of the single-arch greenhouse (operating height restriction) and the ventilation and drainage space between the greenhouses to form a 0.8 m-wide ventilation belt. This belt also has the function of drainage, which improves the land use rate compared with a single-arch greenhouse. Only the top of the planting area of the greenhouse is covered with impermeable plastic film for rain protection; the surrounding and ventilation channels are covered with an insect-proof screen, allowing wind to blow from any direction. In addition, the height of the vent was increased from 1.6 m to 2.2 m, which has reduced the space for heat storage. During the growing stage, no extra energy is required for environmental control because the natural ventilation can remove excess heat from the greenhouse, thus realizing a low-energy-consumption operation.

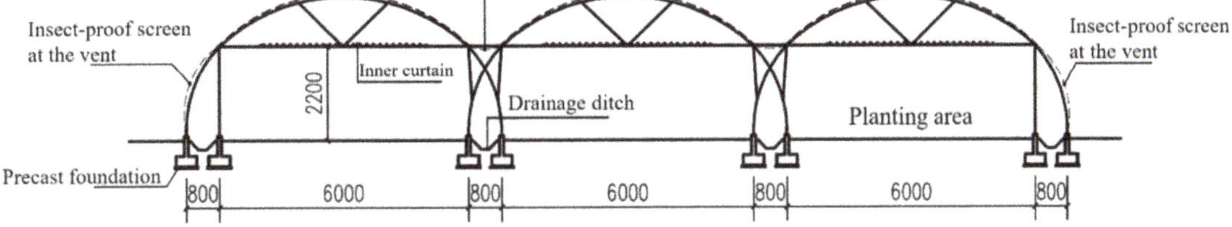

Figure 5. Diagrammatic cross-section of the multi-arch greenhouse structure (unit: mm).

Through continuous development and improvement, the structure of the crossed multi-arch greenhouse (Figure 6) has been improved, and relevant achievements have been made [18]. It is already used for vegetable cultivation in Hainan (Figure 7). This type of greenhouse is suitable for year-round vegetable cultivation in typhoon-resistant areas along the tropical coast, and by increasing the proportion of plastic film coverage and decreasing the proportion of insect-proof screen coverage or adding a film-rolling device, it can also be applied in winter production without heating in areas such as middle and low latitudes, especially in provinces with typhoons such as eastern China, where the advantages of their application are obvious.

Figure 6. Design sketch of a multi-arch greenhouse.

Figure 7. Production application example diagram of a multi-arch greenhouse.

3. Structural Calculation and Material Specifications

With the structural composition determined, different load parameters were selected for different regions and different design wind resistance classes to calculate the components that can meet the design load requirements [13]. This paper takes Haikou as an example. The structural calculation was conducted according to the parameters specified in GB/T 51183-2016 Structural Load Practice for Agricultural Greenhouses. The design life is 10 years.

3.1. Load Parameters

1. The permanent load, taking the self-weight of the frame, is 0.025 kN/m^2. For light steel greenhouses with wind resistance requirements, the self-weights of the structure and the equipment are both favorable loads, and only the self-weight of the frame structure is taken into account during the calculation;
2. The variable load, including the wind load and other variable loads. Multi-arch greenhouses are developed for year-round vegetable production in Hainan, which takes leafy vegetables as the production object, without considering the lifting weight of the crop.
    (1) Other variable loads: 0.15 kN/m^2;
    (2) Wind load: basic wind load (W0), 0.79 kN/m^2.

3.2. Calculation of the Load Standard Value and Design Value

The most unfavorable combination is constant load + wind load. Wind load is the dominant load, which can be calculated by the formula for the variable load control effect design value of S_d: $S_d = \gamma_G S_{Gk} + \gamma_{Q1} S_{Q1k} + \sum_{i=2}^{n} \gamma_{Qi} \psi_{ci} S_{Qik}$ [19], where S_{Gk} is the constant load standard value, S_{Q1k} is the wind load standard value, and S_{Qik} is the live load standard value (n = 2); take the constant load partial coefficient γ_G as 1, the wind load partial coefficient γ_{Q1} as 1, the live load partial coefficient γ_{Qi} as 1.2, the combined value coefficient of live load ψ_{ci} as 0.7, and the structural importance coefficient γ_0 as 0.9; the ground roughness is Category B, the height of the greenhouse is 3.4 m, and the wind pressure height variation coefficient μ_z is taken as 0.76 (the ground roughness is Category B); calculate with the open structure, and the wind load shape factor μ_s acting on the roof is taken as −0.75 [20]. The bay (opening) L is 4.0 m.

Constant load standard value S_{Gk} = 0.025 kN/m^2, wind load standard value $S_{Q1k} = W_0 \mu_Z \mu_S = 0.79 \times 0.76 \times (-0.75) = -0.4503$ kN/m^2, live load standard value S_{Qik} = 0.15 kN/m^2.

The design values for each load acting on the arch are calculated as follows:

(1) Constant load design value: $\gamma_0 \gamma_G S_{Gk} L = 0.9 \times 1 \times 0.025 \times 4 = 0.09$ kN/m;
(2) Wind load design value: $\gamma_0 \gamma_{Q1} S_{Q1K} L = 0.9 \times 1 \times (-0.4503) \times 4 = -1.6211$ kN/m;
(3) Live load design value: $\gamma_0 \gamma_{Qi} \psi_{ci} S_{Qik} L = 0.9 \times 1.2 \times 0.7 \times 0.15 \times 4 = 0.4536$ kN/m.

The wind load at the insect-proof screen should also be multiplied by the wind resistance factor for the wind load. Take 0.75 for the wind resistance factor of a 32-mesh insect-proof screen and 0.81 for a 40-mesh insect-proof screen [21].

3.3. Structural Calculation

Arch frame part: Arch bar: φ42 × 1.5 mm; web member: φ25 × 1.5 mm; tie beam: φ42 × 1.5 mm; side wind-resisting column: φ42 × 1.5 mm; in-greenhouse beam: φ25 × 1.5 mm; reinforcement bar: φ25 × 1.5 mm. The bars are connected with U-bolts and clamps, bolts, and cross springs, and all steel pipes and connections are treated by hot dip galvanization.

Components other than the arch frame include the roof longitudinal tie bar, the tie bar on the internal horizontal pull bar, inter-column supports, greenhouse head (octagonal) horizontal supports, greenhouse head roof supports, the secondary arch bar, the gable auxiliary column, etc. The specifications of the components are: roof longitudinal tie bar: φ42 × 1.5 mm; tie bar on internal horizontal pull bar: φ32 × 1.5 mm; inter-column support: φ42 × 1.5 mm; greenhouse head (octagonal) horizontal support: φ32 × 1.5 mm; greenhouse head roof support: φ25 × 1.5 mm; secondary arch bar: φ32 × 1.5 mm; gable auxiliary column: φ32 × 1.5 mm.

PKPM (2010 edition) was used for the structural calculation. The stress ratios are shown in Figure 8, and the arch bars were hinged to the foundation. The calculation results show that the selected components were all qualified in specifications (the stress ratios were all below 0.8). It can be seen from the graphical results that the internal force is distributed evenly inside the components. The multi-arch form makes each component form a space system, and the whole arch presents a truss structure with an excellent stress performance. The connection between the column and the foundation is fixed at the hinge support, and in this paper, for example, the connection point is at a typical location (the second one on the left) for basic analysis and calculation. According to the calculation, the basic reaction force (standard combination) corresponding to the maximum foundation base area when the above connection point is under the effect of load is: M = 0 kN.m, N = −5.9 kN, V = 0.62 kN, and the calculated foundation base size required is 0.6 m × 0.6 m, with a burial depth of 0.7 m.

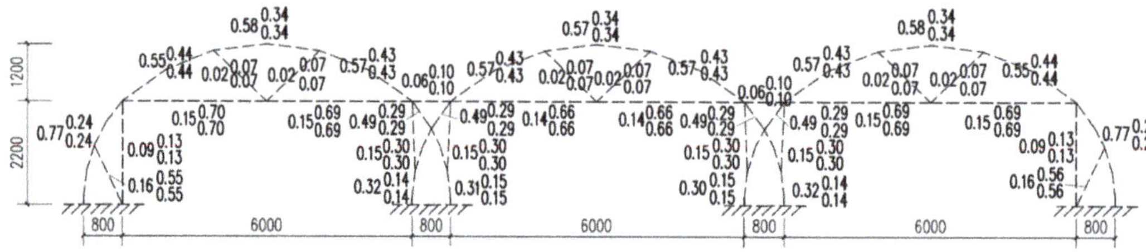

Figure 8. Stress ratios of structural calculation (unit: mm). Notes: The number on the left of each component is the strength stress ratio, that on the upper right is the in-plane stability stress ratio, and that on the lower right is the out-of-plane stability stress ratio.

3.4. Comparison of Structural Properties

In order to compare the structural properties between the multi-arch greenhouse and the multi-span round-arch greenhouse (with a ventilation channel), structural calculations of the multi-span round-arch greenhouse (with a ventilation channel) were conducted based on the load values of the previously mentioned multi-arch greenhouse. The column is rigidly connected to the foundation, and the stress ratios are shown in Figure 9. The results of the calculations show that the specifications of the selected components were all qualified (most of the stress ratios were under 0.8, and the stress ratio of the side column was under 0.9). The connection between the column and the foundation is fixed at the hinge support, and in this paper, for example, the connection point is at a typical location (the second one on the left) for basic analysis and calculation. According to the calculation, the basic reaction force (standard combination) corresponding to the maximum foundation base area when the above connection point is under the effect of load is: M = −2.35 kN.m, N = 1.52 kN, V = −2.34 kN, and the calculated foundation base size required is 0.8 m × 0.8 m, with a burial depth of 0.7 m.

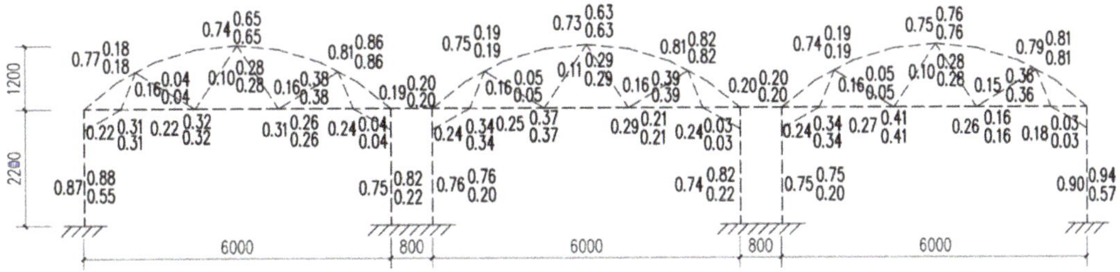

Figure 9. Stress ratios of structural calculation (unit: mm). Notes: the number on the left of each component is the strength stress ratio, that on the upper right is the in-plane stability stress ratio, and that on the lower right is the out-of-plane stability stress ratio.

Specifications of the main components: column: □80 × 60 × 2 mm; arc bar: φ50 × 1.5 mm; web member: φ25 × 1.5 mm; tie beam: φ42 × 1.5 mm; in-greenhouse beam: φ25 × 1.5 mm. The bars are connected with U-bolts and clamps, bolts, and cross springs, and all steel pipes and connections are treated by hot dip galvanization.

The materials of the multi-arch greenhouse and the multi-span round-arch greenhouse (with a ventilation channel) were calculated (Tables 1 and 2). The results show that the consumption of the steel pipe material for the main steel frame of the multi-arch greenhouse is 96.94 kg, while the steel pipe material consumption of the multi-span round-arch greenhouse (with a ventilation channel) is 143.45 kg, which is 46.51 kg higher than that of the multi-arch greenhouse; hence, the multi-arch greenhouse can save about 32.4% of the consumption of steel pipe materials. When calculated with a coverage area of $3 \times 6 \times 4 = 72$ m^2, 0.646 kg of material per unit area is saved; when calculated using the greenhouse steel structure processing and the installation completed price of CNY 12.80/kg, CNY 8.27 per square meter is saved (CNY 5515/667 m^2).

Table 1. Components of the multi-arch greenhouse.

No.	Item	Specification	Length (m)	Quantity	Unit Weight (kg)	Total Weights (kg)
1	Arch bar	φ42 × 1.5 mm	11.2	3	16.78	50.33
2	Web member	φ25 × 1.5 mm	1.5	6	1.30	7.82
3	Tie beam	φ42 × 1.5 mm	6	3	8.99	26.96
4	Side wind-resisting column	φ42 × 1.5 mm	2.1	2	1.82	3.65
5	In-greenhouse beam	φ25 × 1.5 mm	0.8	2	0.70	1.39
6	Side reinforcement bar	φ25 × 1.5 mm	1.3	2	1.13	2.26
7	Reinforcement bar	φ25 × 1.5 mm	1.3	4	1.13	4.52
8			Total			96.94

Table 2. Components of the multi-span round-arch greenhouse (with a ventilation channel).

No.	Item	Specification	Length (m)	Quantity	Unit Weight (kg)	Total Weights (kg)
1	Column	□80 × 60 × 2 mm	2.2	6	9.39	56.36
2	Arch bar	φ50 × 1.5 mm	6.62	3	11.88	35.63
3	Web member	φ25 × 1.5 mm	7.2	3	6.26	18.77
4	Tie beam	φ42 × 1.5 mm	6	3	8.99	26.96
5	In-greenhouse beam	φ25 × 1.5 mm	0.8	2	0.70	1.39
6	Reinforcement bar	φ25 × 1.5 mm	0.83	6	0.72	4.33
7			Total			143.45

Note: The above material table only presents the steel structure materials of the previously mentioned calculation model of a three-span main arch structure. The load of the structural calculation is the same.

The foundation area of the multi-arch greenhouse is 0.6 m × 0.6 m, with a burial depth of 0.7 m, and the foundation area of the multi-span round-arch greenhouse (with a ventilation channel) is 0.8 m × 0.8 m, with a burial depth of 0.7 m. According to the estimation of similar projects, the overall unit price of the foundation of the greenhouse is about CNY 750/m^3 (with excavation, backfill, concrete, formwork, etc.), the cost of the foundation of the multi-arch greenhouse structure (eight independent bases) is 0.6 × 0.6 × 0.7 × 8 × 750 = CNY 1512, and the cost of the foundation of the multi-span round-arch greenhouse structure (six independent bases, with a ventilation channel) is 0.8 × 0.8 × 0.7 × 6 × 750 = CNY 2016. When calculated using a foundation bearing area of 3 × 6 × 4 = 72m^2, the average investment of the multi-arch greenhouse foundation is 1512 ÷ 72 = CNY 21/m^2, and the average investment of the multi-span round-arch greenhouse foundation is 2016.00 ÷ 72 = CNY 28/m^2. The multi-arch greenhouse can save CNY 7/m^2 or CNY 4669/667m^2.

Therefore, a total amount of CNY 10,184/667m^2 can be saved in the main steel frame and foundation of a multi-arch greenhouse compared with a multi-span round-arch greenhouse (with a ventilation channel).

4. Ventilation Simulation

4.1. Greenhouse Modeling

Natural ventilation is a simple and economical way of cooling. For open greenhouse structures in the tropics, the main purpose of ventilation is to remove excess heat and water vapor from the greenhouse to avoid high summer temperatures [22,23]. During the architectural and structural design phase of the greenhouse, the greenhouse vertical face is covered with an insect-proof screen to maximize the use of natural ventilation, but the roof is covered with plastic film for rain protection, which will inevitably cause heat accumulation, and the ventilation area of each greenhouse type and the distribution of ventilation areas have a direct effect on the cooling effect of natural ventilation. At the present stage, CFD simulations [24–30] were carried out using the software Design Builder to build a 3D model to test the spatial and temporal distribution of temperature in the greenhouse under given climatic conditions. The results were compared with the widely used multi-span round-arch greenhouses with and without ventilation channels in the industry.

Model parameter settings: the roofs of all three types of greenhouses were covered with 0.1 mm-thick transparent plastic film with 85% transmittance, and the heat transfer coefficient was 0.17 W/(m^2.K); the vertical faces were all covered with a 0.2 mm-thick insect-proof screen, and the heat transfer coefficient was 203 W/(m^2.K); the ground surface was covered with 13 cm-thick soil with a heat transfer coefficient of 3.188 W/(m^2.K) (Handbook, 2007) which are shown in Figure 10.

Model a. Crossed multi-arch greenhouse: 6 m span, 0.8 m in-greenhouse ventilation channel (arch cross spacing), 4 m opening, single unit size 62 m × 36 m = 2232 m^2 (nine spans and nine openings), 3.4 m vector height, 2.2 m shoulder height.

Model b. Multi-span round-arch greenhouse with an in-greenhouse ventilation channel: the greenhouse models were all set to the same dimensions for the convenience of comparison, with a span of 6 m, an in-greenhouse ventilation channel of 0.8 m, an opening of 4 m, a single unit size of 60.4 × 36 m = 2174.4 m^2 (nine spans and nine openings), a vector height of 3.4 m, and a shoulder height of 2.2 m.

Model c. Multi-span round-arch greenhouse without an in-greenhouse ventilation channel: without the ventilation channel, the opening span height was the same as that of the above two models: span 6 m, opening 4 m, single unit size 54 × 36 m = 1944 m^2 (nine spans and nine openings), vector height 3.4 m, and shoulder height 2.2 m.

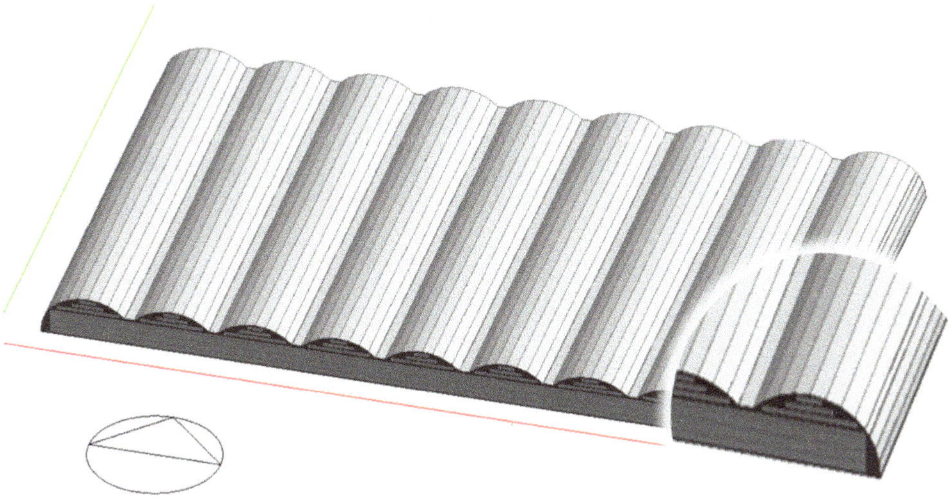

(**a**)

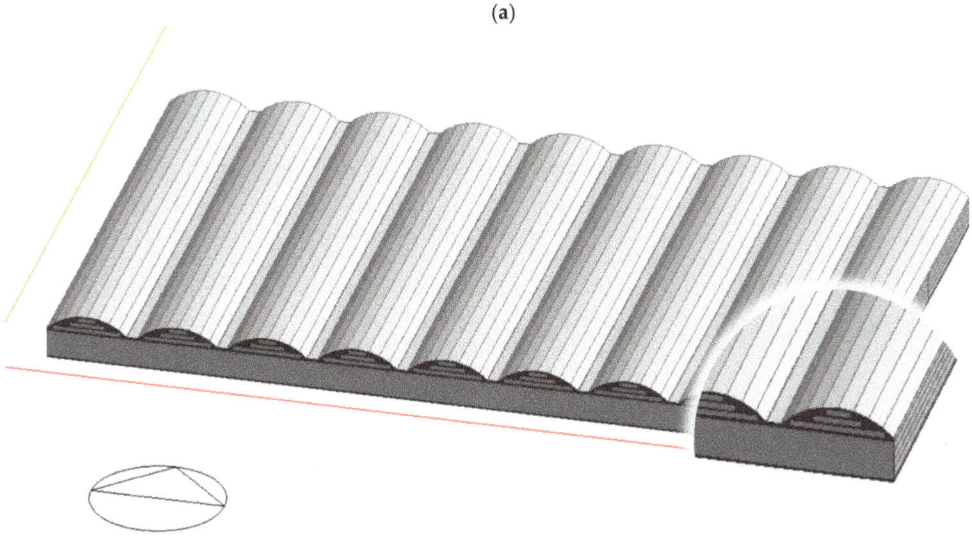

(**b**)

Figure 10. *Cont*.

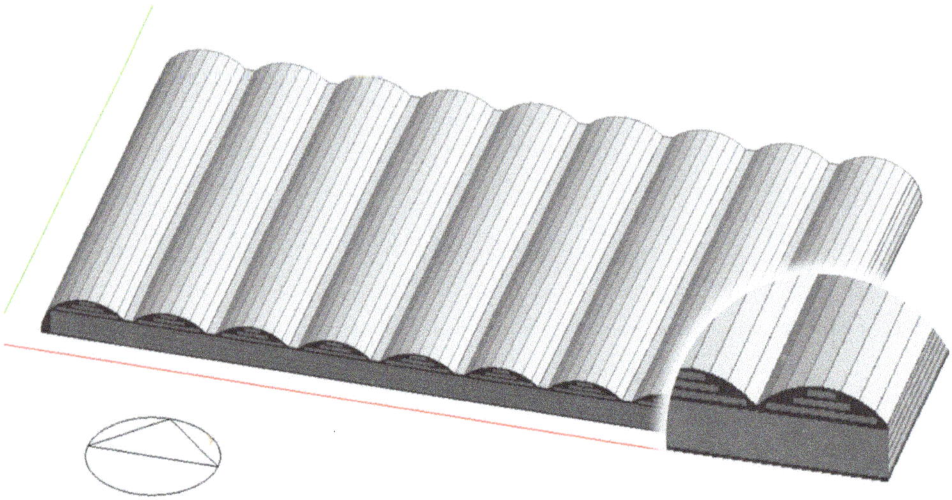

(c)

Figure 10. Simulation models of three types of greenhouses. (**a**) Crossed multi-arch greenhouse; (**b**) Multi-span round-arch greenhouse with an in-greenhouse ventilation channel; (**c**) Multi-span round-arch greenhouse without an in-greenhouse ventilation channel.

4.2. Simulation Environment Setting and Simulation Results Analysis

This paper simulates the natural environment of a greenhouse in Haikou, Hainan (11°33′ E, 20°05′ N), and relevant environmental parameters were determined based on the natural environment. The greenhouse was oriented in a north–south direction (this type of open greenhouse can make full use of the wind coming from any direction, but there is no special requirement for the orientation of the greenhouse in practical production applications). The monsoon winds alternate in the Haikou area, with southeasterly winds prevailing in summer (May to October), which are formed by the northward expansion of warm tropical marine air masses, resulting in hot and humid summers, and northeasterly winds prevailing in winter (November to April), which are formed by the southward expansion of cold polar continental air masses, resulting in dry winters with little rain. The average wind speed in the Haikou area is 3.1 m/s throughout the year and 2.8 m/s in summer [31]. The purpose of the simulation is to analyze the cooling effect of the greenhouse model using thermal and wind pressure ventilation under the natural ventilation state. The wind speed, wind direction, solar radiation, outdoor air temperature, and surface temperature use the software's built-in time-by-time weather data, and the inlet air temperature is set as the outdoor air temperature. When judged by experience, the highest temperature in a day in the greenhouse often occurs around midday when solar radiation is at its maximum, and the earth is uniformly heated at midday, usually with no wind or low wind speed. For this reason, the simulation selected a wind speed of 0 m/s and 1 m/s (the wind direction is along the long axis of the greenhouse) on a certain day in summer as the most unfavorable conditions to compare the ventilation and cooling effect of each model, and the simulation environment parameters are set, as shown in Table 3.

Table 3. Parameter setting of the simulation environment.

Location	Latitude and Longitude	Simulation Date	Outdoor Wind Speed (m/s)	Wind Direction (°)	Outdoor Temperature (°C)
Haikou, Hainan	East longitude 110°33′, Northern latitude 20°05′	8 July, 13:00	1	90	32.7
		12 July, 15:00	0	–	35.0

Note: The 0° wind comes from due north, the 90° wind comes from due east, the 180° wind comes from due south, and the 270° wind comes from due west.

Figures 11–13 show the simulated temperature clouds inside the greenhouse at 15:00 on 12 July, when the wind speed was 0. The inlet temperature of the simulated environment was 35 °C. The cloud map takes the midline of each row of arch bars to make temperature slices. The slices are located 2 m away from the front and rear rows of the arch bars.

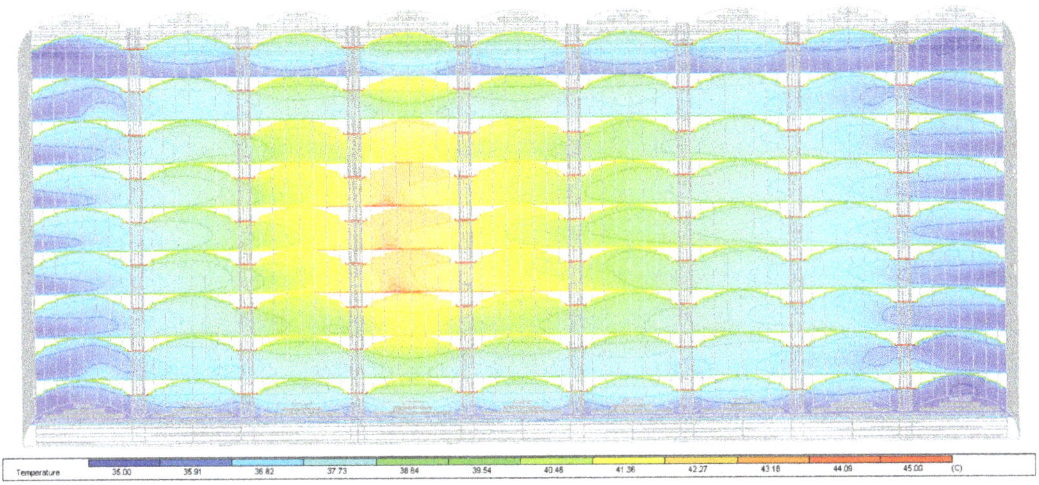

Figure 11. Temperature cloud map of the crossed multi-arch greenhouse (outdoor wind speed 0 m/s, outdoor temperature 35 °C).

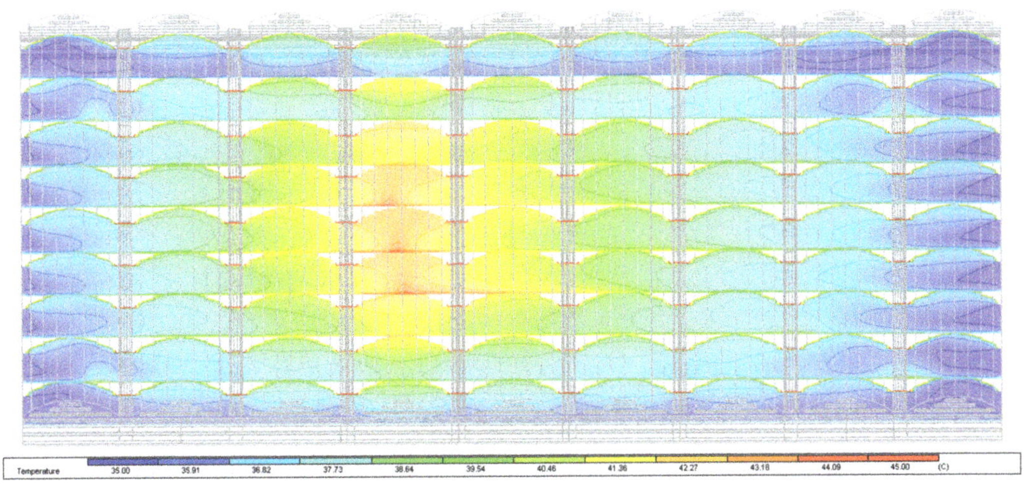

Figure 12. Temperature cloud map of the multi-span round-arch greenhouse (with an in-greenhouse ventilation channel) (outdoor wind speed 0 m/s, outdoor temperature 35 °C).

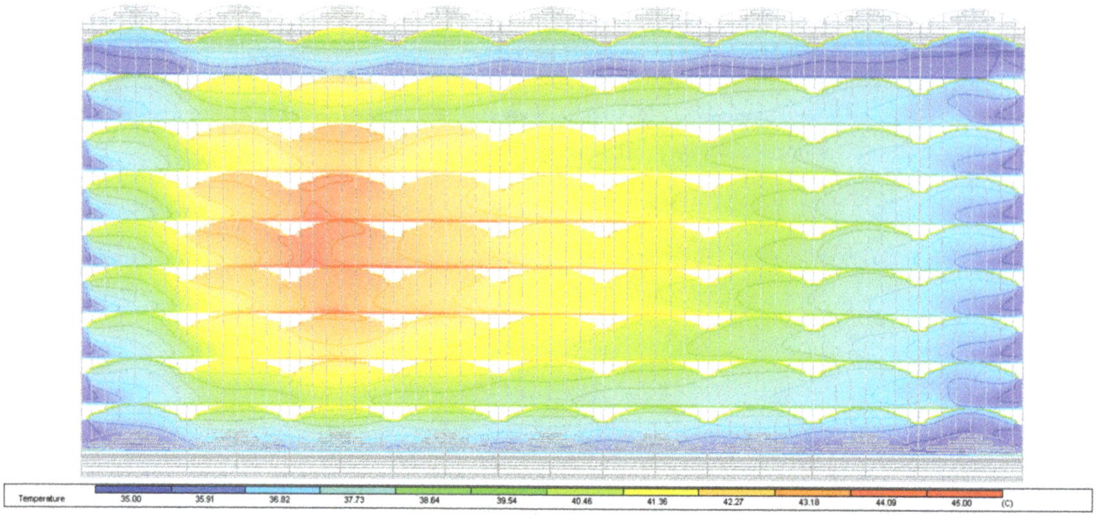

Figure 13. Temperature cloud map of the multi-span round-arch greenhouse (without an in-greenhouse ventilation channel) (outdoor wind speed 0 m/s, outdoor temperature 35 °C).

According to Figures 11–13, in the conditions of no wind (wind speed 0 m/s) and an external temperature of 35 °C, the internal ventilation of the greenhouse is only the heat pressure effect, and the heat is gathered in the middle of the greenhouse. The crossed multi-arch greenhouse and the multi-span round-arch greenhouse with a ventilation channel showed a good heat dissipation effect, with the edge temperature similar to the external temperature; the internal temperature at most points is around 36~39 °C, the highest temperature at the center is about 42 °C, and the average temperature difference is about 2~3 °C; as for the multi-span round-arch greenhouse without a ventilation channel, there is an obvious heat accumulation in the greenhouse, and the range of the high-temperature area is larger than that of the multi-span round-arch greenhouse with a ventilation channel, with most of the internal temperature being 37~41 °C, the highest temperature being about 44 °C, and the average temperature difference being about 4~5 °C. In the condition of no wind, the ventilation effect of the multi-span round-arch greenhouse with a ventilation channel is significantly higher than that of the greenhouse without a ventilation channel because the ventilation channel and surrounding vents can form good thermal pressure ventilation, and the air flows well inside and outside such that the heat collected inside can be exchanged with cooler air from outside more quickly. When there is no wind, the heat distributed asymmetrically, which is mainly affected by the monsoon and the solar orientation, especially the southeast monsoon that prevails in Hainan in summer. Although the outdoor wind speed is 0 m/s at the CFD simulation time, there may be southeast wind before and after the model time.

Figures 14–16 show the simulated temperature clouds inside the greenhouse at 13:00 on 8 July, when the outdoor wind speed was 1 m/s (windy hours around noon) and the inlet temperature of the simulated environment was 32.7 °C. The cloud map takes the midline of each row of arch bars to make temperature slices. The slices are located 2 m away from the front and rear rows of arch bars.

As can be seen in Figures 14–16, during windy hours (wind speed 1 m/s), the greenhouses simultaneously ventilated through the heat and wind pressure, with the wind coming from the east side along the greenhouse's long axis direction. The long axis direction has a long heat exchange path and inadequate heat exchange, causing heat accumulation in the greenhouse. The temperature at the edge of the two greenhouse types with ventilation channels is similar to that outside, and the internal temperature is roughly distributed between 34 and 37 °C, locally about 38 °C, with a maximum average temperature difference of about 2 to 3 °C. In contrast to Figure 14, the high-temperature areas of the multi-span round-arch greenhouse without a ventilation channel are more widely distributed than the two greenhouse types with ventilation channels, and the internal temperatures are mostly distributed between 34 and 38 °C, locally about 39 °C, with a maximum average temperature difference of about 4 to 5 °C.

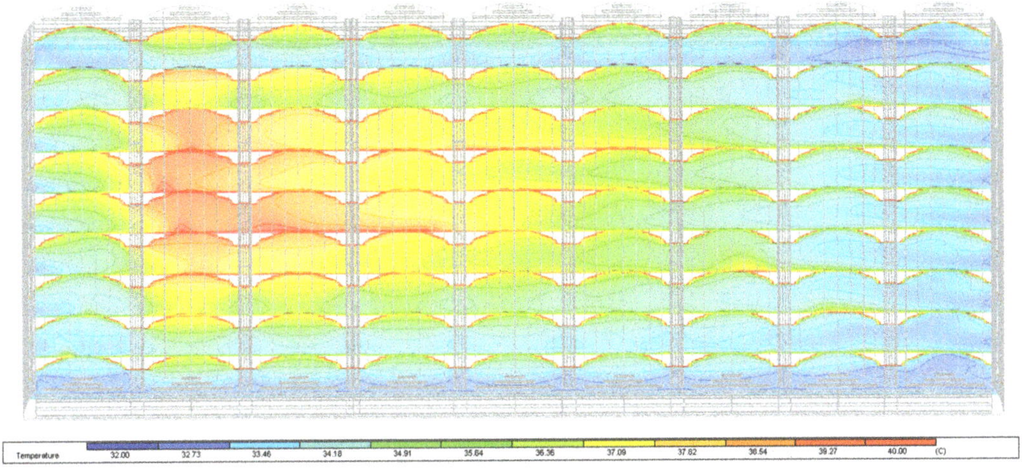

Figure 14. Temperature cloud map of the crossed multi-arch greenhouse (with an in-greenhouse ventilation channel) (outdoor wind speed 1 m/s, outdoor temperature 32.7 °C).

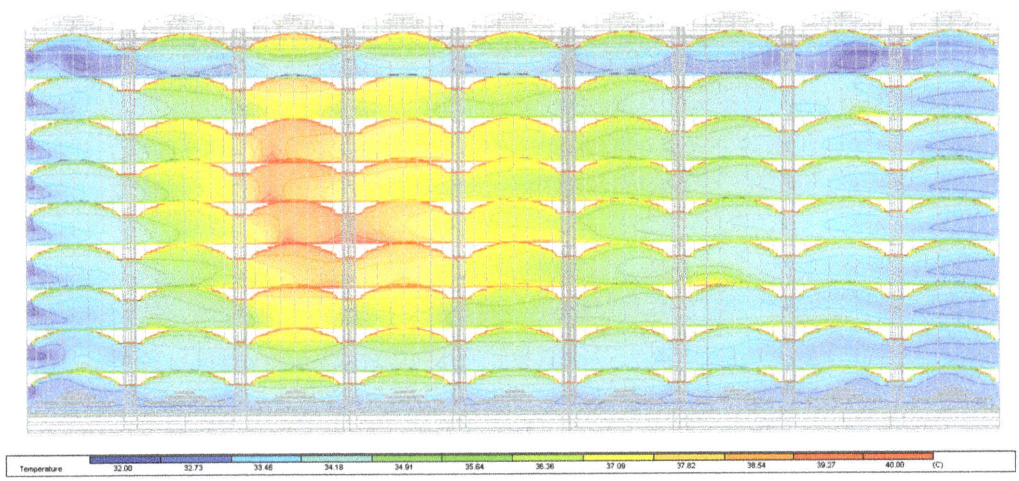

Figure 15. Temperature cloud map of the multi-span round-arch greenhouse (with an in-greenhouse ventilation channel) (outdoor wind speed 1 m/s, outdoor temperature 32.7 °C).

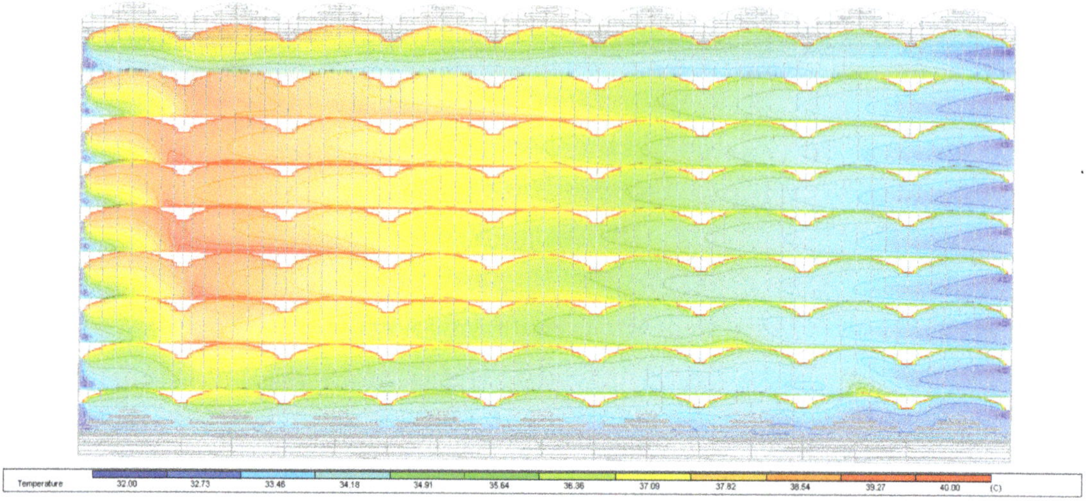

Figure 16. Temperature cloud map of the multi-span round-arch greenhouse (without an in-greenhouse ventilation channel) (outdoor wind speed 1m/s, outdoor temperature 32.7 °C).

Some studies [32] have shown that when the outdoor wind speed is higher than 2 m/s, the ventilation is mainly dominated by wind pressure, and the thermal pressure is negligible; when the outdoor wind speed is between 0.5 and 2 m/s, the ventilation process is dominated by wind pressure and supplemented by thermal pressure; when the outdoor wind speed is lower than 0.5 m/s, the role of thermal pressure ventilation cannot be neglected. From the temperature distribution in the greenhouse in the afternoon simulated in this study without wind and with a light breeze, it can be seen that the average temperature difference between the interior and exterior of the greenhouse is smaller when there is a ventilation channel, and the high-temperature area is significantly smaller than that of the greenhouse without a ventilation channel. It is concluded that a ventilation channel can effectively promote the flow of air inside and outside the greenhouse and lower the high temperature inside the greenhouse, and the ventilation and cooling capacity of the crossed multi-arch greenhouse and the multi-span round-arch greenhouse (with an in-greenhouse ventilation channel) are basically the same.

To further analyze the advantages and disadvantages of Model A and Model B, slice maps of the wind speed of the two types of greenhouses with a ventilation channel were formulated. Figures 17 and 18 show the wind speed cloud maps inside the greenhouse under the windless condition, and Figures 19 and 20 show the wind speed cloud maps inside the greenhouse under the windy condition (wind speed 1 m/s).

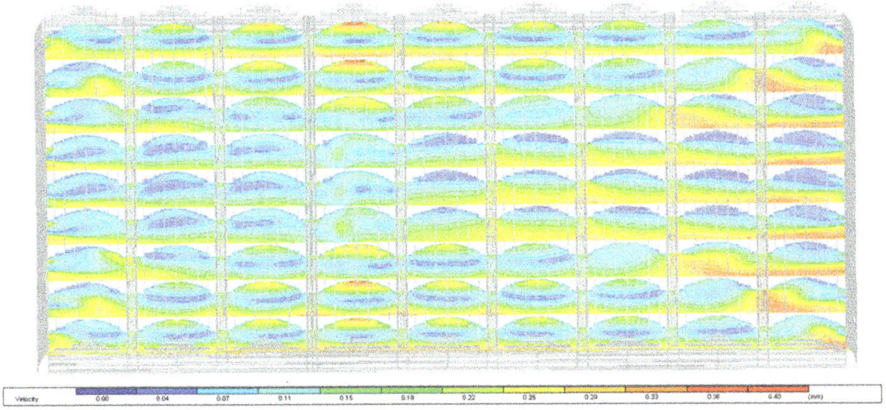

Figure 17. Wind speed cloud map of the crossed multi-arch greenhouse (outdoor wind speed 0 m/s).

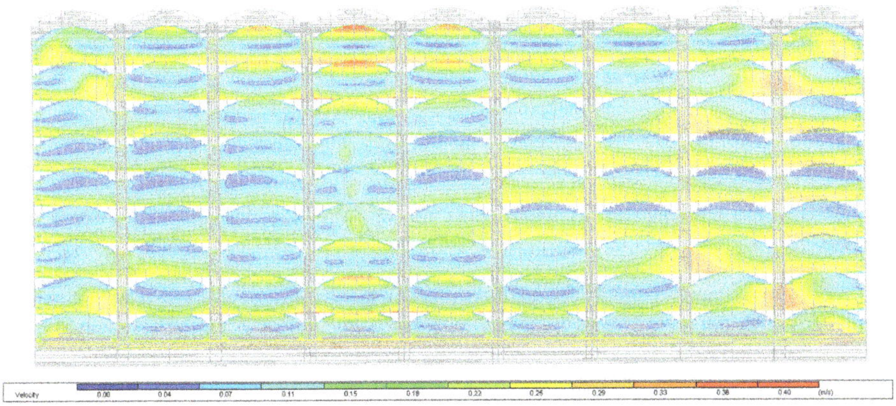

Figure 18. Wind speed cloud map of the multi-span round-arch greenhouse with an in-greenhouse ventilation channel (outdoor wind speed 0 m/s).

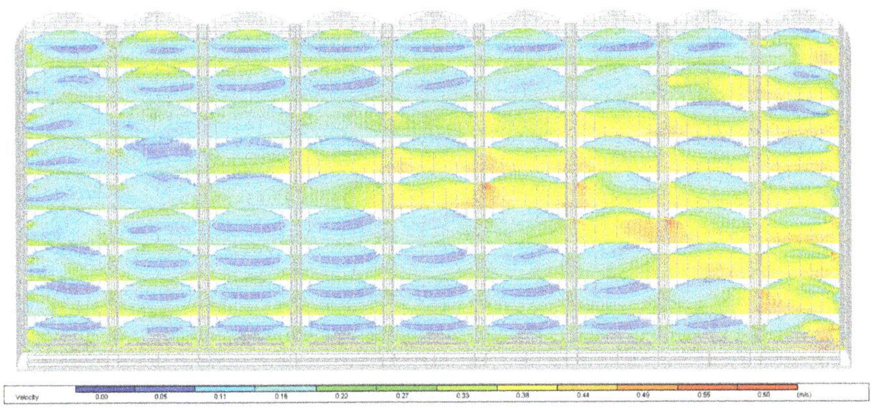

Figure 19. Wind speed cloud map of the crossed multi-arch greenhouse (outdoor wind speed 1 m/s).

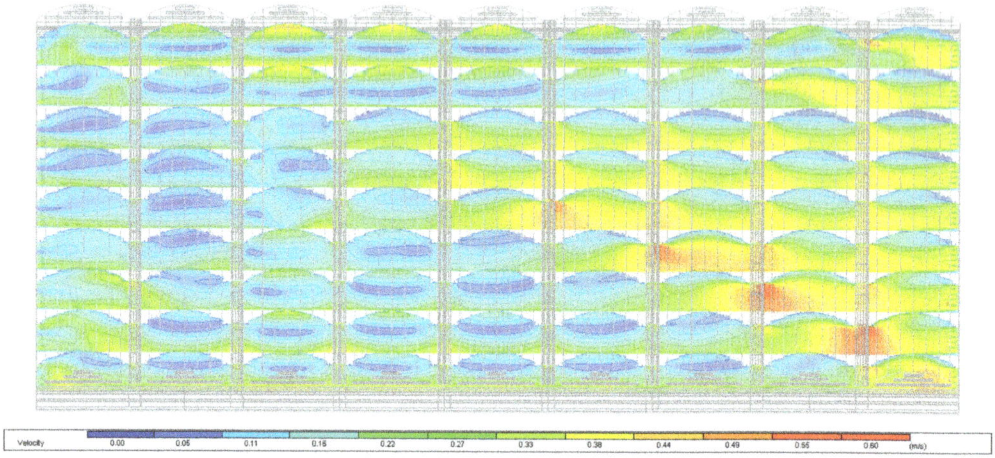

Figure 20. Wind speed cloud map of the multi-span round-arch greenhouse with an in-greenhouse ventilation channel (outdoor wind speed 1 m/s).

As shown, when the outdoor wind speed is 0 m/s, the maximum indoor wind speed of the two greenhouse types with a ventilation channel is 0.4 m/s; when the outdoor wind speed is 1 m/s, the maximum indoor wind speed of the two greenhouse types is 0.6 m/s. The speed of outside air entering the room decreased by about 40%, and there was no significant difference in the ventilation effect between the two types of sheds. The non-significant difference is because, in the same conditions, the wind speed near the ventilation opening of the round-arch greenhouse is larger than that of the multi-arch greenhouse, and the wind speed of the crossed multi-arch greenhouse is more evenly distributed. Some scholars [21,33–35] studied the ventilation of an insect-proof net from the perspective of the environment, obtained parameters such as the wind resistance coefficient, pressure drop coefficient, and flow coefficient of the screen net to describe the negative impact of insect-proof netting on natural ventilation, and fitted a series of formulas for quantifying the ventilation effect. However, due to the difference in the use environment, insect-control netting is used as the ventilation covering material in tropical China. In the same rainy and hot season in tropical China, the humid and hot environment speeds up the spread of diseases and pests. The selection of insect-control netting in production depends more on whether its pores can meet the plant-protection function [36]. Insect-proof netting and plastic film are the two most commonly used covering materials in tropical areas. Plastic film is used for rain protection, and insect-proof netting is used for insect control and ventilation. The ventilation effect of the greenhouse is affected by the size and distribution of the ventilation surface [37]. For greenhouses in tropical areas, the ventilation surface is the coverage of insect-control netting. On the premise of not affecting the internal rain protection function, increasing the proportion of insect protection net coverage can improve the ventilation and cooling effect in the shed.

5. Conclusions

Facility protection land cultivation can overcome summer and autumn rainstorms, high temperatures, frequent typhoons, and other unfavorable meteorological factors. A crossed multi-arch greenhouse is developed based on the existing production experience through extensive surveys, analogies, and analyses, which is a comprehensive application that combines various methods such as structural analyses and computations, environmental simulation analyses, and so on. As a new type of greenhouse structure suitable for planting in hot areas, a crossed multi-arch greenhouse is featured, with the following innovations and advantages:

(1) Structural innovation: the crossed multi-arch structure is similar to the integral truss structure, which can resist large wind loads with a smaller steel tube section. Under the same load condition, the structural force analysis and calculation of the crossed multi-arch greenhouse and the multi-span round-arch greenhouse (with a ventilation channel) show that the steel consumption of the main steel frame of the crossed multi-arch greenhouse is reduced by about 32.4%. At the same time, due to the uniform force inside the main steel frame, the connection between the steel frame and the foundation can be hinged in the form of column footings, so the bending moment generated by the wind load is not transferred to the foundation. Moreover, the size of the foundation is reduced compared to that of the solid jointed column footings, and the amount of material used for the foundation structure can be further reduced.

(2) Economic advantages: the crossed multi-arch greenhouse showed significant economic advantages, as it uses fewer materials. According to the statistics, a crossed multi-arch greenhouse can save 0.646 kg of materials per unit area compared with a multi-span round-arch greenhouse (with a ventilation channel). According to the material price and construction quota in the Hainan area, the cost of material processing and the installation of the greenhouse's steel structure is about CNY 12.8/kg, so it can save CNY 8.27 per square meter or CNY 5515/667 m^2 for steel investment; for the foundation part, the crossed multi-arch greenhouse can save CNY 7 per square meter or CNY 4669/667 m^2 compared with the multi-span round-arch greenhouse (with a ventilation channel). Thus, the crossed multi-arch greenhouse can save a total investment of CNY 10,184/667 m^2, which has obvious economic advantages.

(3) Ventilation and cooling: the temperature parameters inside different structural forms of greenhouses are simulated by simulation software under typical summer weather conditions. The results show that, under the conditions of no wind (wind speed 0 m/s) and breeze (wind speed 1 m/s), the crossed multi-arch greenhouse and the multi-span round-arch greenhouse with ventilation channels have a better heat-dissipation effect, and the average temperature difference between indoors and outdoors is about 2~3 °C. The greenhouse type without a ventilation channel has obvious heat accumulation, and the high-temperature area is larger than that of greenhouses with a ventilation channel. The average temperature difference between the indoor and outdoor space of this greenhouse type is about 4~5 °C.

In this paper, the research-and-development process of multi-arch greenhouses has been described from the perspectives of structure, the advantages of multi-arch greenhouses were compared from the perspective of cost, the ventilation simulation was used to prove that the multi-arch greenhouses have achieved the expected ventilation effect, and the ventilation performance was matched with the multi-span round arch plastic greenhouse with an in-greenhouse ventilation channel. At present, the multi-arch greenhouse structure has been built and applied to many planting bases in Hainan Province. It can meet the production function and has a good use effect.

6. Patents

One patent has been applied in China in this manuscript (Patent No. CN202021432481).

Author Contributions: Conceptualization, J.L.; methodology, B.W.; software, X.W.; validation, F.S.; formal analysis, B.W.; investigation, X.W.; resources, J.L.; data curation, J.L; writing—original draft preparation, F.S.; writing—review and editing, B.W.; visualization, F.S.; supervision, J.L.; project administration, J.L.; funding acquisition, J.L. All authors have read and agreed to the published version of the manuscript.

Funding: This research was funded by the Hainan Provincial Natural Science Foundation of China, grant number 322RC583.

Institutional Review Board Statement: Not applicable.

Data Availability Statement: All the data are presented in this article in the form of figures and tables.

Conflicts of Interest: The authors declare no conflict of interest.

References

1. Liu, J. Development and application of several typhoon-resistant vegetable planting greenhouses in Hainan. *Agric. Eng. Technol. (Greenh. Hortic.)* **2018**, *38*, 32–38. [CrossRef]
2. Zhuang, H.; Chen, L.; Du, C.; Gao, F.; Zheng, H.; Li, H. Efficient cultivation techniques of leafy vegetables under rain proof shelter in summer or fall in South China. *Guangdong Agric. Sci.* **2014**, *41*, 86–88. [CrossRef]
3. Zhou, C. Dr. Zhou's Field Investigation (34) Greenhouse in Taiwan ⑤ Continuous window opening method in plastic film greenhouse with round arch roof. *Agric. Eng. Technol. (Greenh. Hortic.)* **2014**, *6*, 26–28+30. [CrossRef]
4. Liu, J.; Huang, Y. Influence of Super Typhoon Sarika on facility agriculture in Hainan and suggestions. *Appl. Eng. Technol.* **2017**, *37*, 34–37. [CrossRef]
5. Yang, X.; Li, J.; Yang, M.; Huang, Z.; Cao, M.; Yang, G. Influence of Typhoon "Shanshen" on facility agriculture in south Hainan and reflections. *Agric. Eng. Technol. (Greenh. Hortic.)* **2013**, *1*, 20–25. [CrossRef]
6. Yang, X.; Li, J.; Yang, M.; Chen, M.; Liang, Z. Investigation and analysis of Typhoon Conson's damage on facility agriculture in Sanya, Hainan. *Agric. Eng. Technol. (Greenh. Hortic.)* **2011**, *2*, 20–21. [CrossRef]
7. Yang, X.; Li, J.; Yang, M.; Chen, M. Influence of Typhoon Ketsana on protected agriculture in south Hainan. *Chin. J. Trop. Agric.* **2009**, *29*, 68–69+74.
8. Yang, X.; Liu, J.; Cao, M.; Yang, G.; Zhang, X.; Huang, X. Structural optimization design of economical steel tube plastic greenhouse with typhoon resistance in tropical region. *J. Chin. Agric. Mech.* **2016**, *37*, 48–52. [CrossRef]
9. Wang, C.; Nan, B.; Wang, T.; Bai, Y.; Li, Y. Wind pressure acting on greenhouses: A review. *Int. J. Agric. Biol. Eng.* **2021**, *4*, 1–8. [CrossRef]
10. Ren, Y.; Wang, M.; Saeeda, I.A.; Chen, X.; Gao, W. Progress, problems and prospects for standardization of greenhouse-related technologies. *Int. J. Agric. Biol. Eng.* **2018**, *11*, 40–48. [CrossRef]
11. Wang, J. *Optimal Design and Ventilative Test of Mutual Insert Multi-Span Greenhouse*; Nanjing Agricultural University: Nanjing, China, 2007; Available online: https://webvpn.hainanu.edu.cn/https/77726476706e69737468656265737421fbf952d2243e635930068cb8/kcms/detail/detail.aspx?dbcode=CDFD&dbname=CDFD9908&filename=2008031525.nh&uniplatform=NZKPT&v=MQBqjRkXn37A932ISKhjE92f2XspsP72tVdPHJq1DqErtrHQaIEbUtzY5A3O4IjE (accessed on 14 July 2022).
12. Zhou, C. The current use of the main greenhouse type and performance (I). *Pract. Eng. Technol. Rural. Areas* **2000**, *1*, 8–9.
13. Zhou, C. The current use of the main greenhouse type and performance (II). *Pract. Eng. Technol. Rural. Areas* **2000**, *2*, 8–9.
14. Zhou, C. The current use of the main greenhouse type and performance (III). *Pract. Eng. Technol. Rural. Areas* **2000**, *3*, 8–9.
15. Zhou, C. The current use of the main greenhouse type and performance (IV). *Pract. Eng. Technol. Rural. Areas* **2000**, *4*, 8–9.
16. Zhou, C. *Handbook of Greenhouse Engineering Design*; China Agriculture Press: Beijing, China, 2007.
17. Zhang, Z.; Zhou, C. Zhou's Field Investigation (26) Investigation report of facility horticulture development status in Hainan Province. *Agric. Eng. Technol. (Greenh. Hortic.)* **2013**, 11–18. [CrossRef]
18. Liu, J.; Wang, B.; Chen, Y.; Zhu, G. A Fully Assembled Crossed Multi-Arch Greenhouse Structure. CN212696864U, 16 March 2021.
19. *GB/T 51183-2016*; Code for the Design Load of Horticultural Greenhouse Structures. China Planning Press: Beijing, China, 2016.
20. *GB 51022-2015*; Technical Specification for Steel Structure of Lightweight Buildings with Gabled Frames. China Construction Industry Press: Beijing, China, 2015.
21. Yan, D.; Xu, K.; Zhang, Q.; Li, X. Experimental study on wind load of insect control screen with different mesh sizes. *Agric. Eng. Technol.* **2020**, *40*, 57–63. [CrossRef]
22. Emekli, N.; Kendirli, B.; Kurunc, A. Structural analysis and functional characteristics of greenhouses in the Mediterranean region of Turkey. *Afr. J. Biotechnol.* **2010**, *9*, 3131–3139.
23. Wu, Y.; Li, H. Structure and design analysis of ventilation and cooling facilities in greenhouse in south china. *Mod. Agric. Equip.* **2013**, *12*, 60–62. Available online: https://kns.cnki.net/kcms/detail/detail.aspx?dbcode=CJFD&dbname=CJFDHIS2&filename=GDLJ201305020&uniplatform=NZKPT&v=77BL2gXUDMo6Ay7YkUufIXW3Bp3SfqV5Q7MHIMc9nQSt_NbHCVBbdzzf2m_cq9SE (accessed on 14 July 2022).
24. Hou, S. *Research on Microclimate Environment Prediction and Control of Facility Greenhouse Based on CFD Model*; Nanjing Agricultural University: Nanjing, China, 2016.
25. Floresvelazquez, J.; Montero, J.; Baeza, E.; Lopez, J. Mechanical and natural ventilation systems in a greenhouse designed using computational fluid dynamics. *Int. J. Agric. Biol. Eng.* **2014**, *7*, 1–16. [CrossRef]
26. Maraveas, C. Wind Pressure Coefficients on Greenhouse Structures. *Agriculture* **2020**, *10*, 149. [CrossRef]
27. Molina-Aiz, F.; Fatnassi, H.; Boulard, T.; Roy, J.; Valera, D. Comparison of finite element and finite volume methods for simulation of natural ventilation in greenhouses. *Comput. Electron. Agric.* **2010**, *72*, 69–86. [CrossRef]
28. Wu, F.; Zhang, L.; Xu, F.; Chen, J.; Chen, X. Application and Development of Computational Fluid Dynamics in Ventilation for Greenhouses. *J. Agric. Mech. Res.* **2008**, *6*, 1–5. [CrossRef]
29. Zhang, F.; Fang, H.; Yang, Q.; Cheng, R.; Zhang, Y.; Ke, X.; Lu, W.; Liu, H. Ventilation Simulation in a Large-scale Greenhouse Based on CFD. *Chin. J. Agrometeorol.* **2017**, *38*, 221–229. [CrossRef]

30. Lee, G.; Kim, J.; Kim, J.; Lee, S.; Ha, H.; Kang, T. A study of the triangular cross section type greenhouse using fluid-structure interaction. *J. Korean Soc. Vis.* **2019**, *17*, 17–24.
31. *GB/T 50178-93*; Standard of Climatic Regionalization for Architecture. China Planning Press: Beijing, China, 1993.
32. Fang, H.; Yang, Q.; Zhang, Y.; Cheng, R.; Zhang, F.; Lu, W. Simulation on ventilation flux of solar greenhouse based on the coupling between stack and wind effects. *Chin. J. Agrometeorol.* **2016**, *37*, 531–537. [CrossRef]
33. López-Martinez, A.; Granados-Ortiz, F.; Molina Aiz, F.; Lai, C.; Moreno-Teruel, M.; Valera-Martínez, D. Analysis of Turbulent Air Flow Characteristics Due to the Presence of a 13 × 30 Threads·cm^{-2} Insect Proof Screen on the Side Windows of a Mediterranean Greenhouse. *Agronomy* **2022**, *12*, 586. [CrossRef]
34. López-Martínez, A.; Molina-Aiz, F.; Valera-Martínez, D.; López-Martínez, J.; Peña-Fernández, A.; Espinoza-Ramos, K. Application of semi-empirical ventilation models in a mediterranean greenhouse with opposing thermal and wind effects. Use of non-constant cd (pressure drop coefficient through the vents) and cw (wind effect coefficient). *Agronomy* **2019**, *9*, 736. [CrossRef]
35. López, A.; Molina-Aiz, F.; Valera, D.; Peña, A. Wind tunnel analysis of the airflow through insect-proof screens and comparison of their effect when installed in a Mediterranean greenhouse. *Sensors* **2016**, *16*, 690. [CrossRef]
36. Dong, H.; Yan, Z.; Ji, Y.; Tan, X.; Shen, Y. Effects of Covering with Insect Proof Net on Growth and Anti-insect Efficiency of Brassica campestris ssp. chinensis Makino. *J. Chang. Veg.* **2013**, *20*, 63–65. [CrossRef]
37. Espinoza, K.; Lopez, A.; Valera, D.; Molina-Aiz, F.; Torres, J.; Pena, A. Effects of ventilator configuration on the flow pattern of a naturally-ventilated three-span Mediterranean greenhouse. *Biosyst. Eng.* **2017**, *164*, 13–30. [CrossRef]

agriculture

MDPI

Article

Effects of Uplift Resistance on Continuous-Pipe-Foundation of Single-Span Plastic Greenhouse by Steel Plate Pipe Connector

Myeonghwan Kim

Department of Agricultural and Rural Engineering, Chungbuk National University, Chungdea-ro 1, Seowon-Gu, Cheongju 26844, Chungbuk, Republic of Korea; mhkim3306@cbnu.ac.kr

Abstract: The single-span plastic greenhouses are affected by strong winds which generate uplift resistance causing the bending of members, damage to protective films, and damage to crops. This study performed a field test using the static axial tensile load method to present basic data to prevent damage to a single-span plastic greenhouse. Three representative areas were selected, and the effects of pipe connectors, rafter spacing, and embedding depth were tested. In the field test results, it was found to be greatly affected by the pipe connector. The pull-out resistance at the site fixed by welding instead of the pipe connector was measured as 4.5 times the sliding resistance standard value of the Rural Development Administration. In other sites, the measurement was below the standard value of the sliding resistance of the pipe connector. It was confirmed that the uplift resistance is determined by the sliding resistance of the pipe connector, the rafters, and the crossbar pipe. Therefore, it seems possible to increase the uplift resistance of a single-span plastic greenhouses continuous foundation through the reinforcement of the pipe connector. The field test results can be utilized as basic data for the reinforcement of the commercialization of single-span plastic greenhouses and new standards.

Keywords: uplift resistance; continuous pipe foundation; single-span plastic greenhouse; pipe connector

Citation: Kim, M. Effects of Uplift Resistance on Continuous-Pipe-Foundation of Single-Span Plastic Greenhouse by Steel Plate Pipe Connector. *Agriculture* **2022**, *12*, 1998. https://doi.org/10.3390/agriculture12121998

Academic Editors: Muhammad Sultan, Redmond R. Shamshiri, Md Shamim Ahamed and Muhammad Farooq

Received: 28 September 2022
Accepted: 21 November 2022
Published: 24 November 2022

1. Introduction

In Korea, from 1998 to 2015, the amount of damage to plastic houses due to wind resistance and heavy snowfall added up to KRW 76.7 billion annually, and the cumulative area of damage was 20.279 ha (about 40% of the total site) [1]. Single-span plastic greenhouses using galvanized steel pipes as rafter pipes accounted for 84.9% of the facility site [2].

The single-span plastic greenhouse industry for facility horticulture suffered economic losses in the early 2000s due to meteorological disasters such as heavy snow and strong winds. In order to reduce economic losses, MIFAFF and RDA (2014) released data such as snow depth and wind speed for July 2014 [3]. The main causes of damage to single-span plastic greenhouses were strong winds and heavy snowfall, and the damage caused by heavy rains weakens the ground supporting the structure, causing uplift or overturning damage when strong winds occur [4]. As of 2019, about 99% of all greenhouses were commercialized, and among them, single-span plastic greenhouses were predominantly used, accounting for 78.7% of the facility site [5].

The single-span plastic greenhouse is a lightweight structure that is very vulnerable to strong winds or heavy snow [6–8], with the damage caused by strong winds in 2019 accounting for 98% of the total damage, which is a serious level [9]. As of 2020, 91.8% of the damage to sites for single-span plastic greenhouses in private facilities was found to be caused by typhoons and strong winds [10]. However, the data on the uplift resistance of single-span plastic greenhouse foundations are insufficient [11,12]. In order to compensate for weak uplift resistance, a lot of effort is needed to build greenhouses with heavy foundations or that are made with concrete on site and backfilled with soil [13]. In addition, research on uplift resistance is steadily progressing, through indoor model tests on the type of foundation and depth of embedding [8,14–21].

The single-span plastic greenhouse presented in the anti-disaster design standards of the Ministry of Agriculture, Food and Rural Affairs (MAFRA) is intended to improve the uplift resistance or settlement. As for the design method, a continuous pipe foundation in the form of a purlin was installed at the midpoint (around 25 cm) of the embedded part of the rafter (40–50 cm) [3]. The connection between the rafter and the purlin was achieved using a pipe connector, forming a continuous pipe foundation that is embedded in the ground [4,11].

Yun et al. (2015) suggested a new pipe standard for the continuous pipe foundation because the pipe diameter and thickness had little effect on uplift resistance as a model test (Figure 1) [22]. In the greenhouse, lift is generated due to the pressure difference caused by the strong wind current, causing the foundation to be uplift-resistant and structural damage to the entire structure. The anti-disaster design standards of the Ministry of Agriculture, Food and Rural Affairs only formulate regulations for the sliding resistance, structural material and design strength (snow depth and wind speed) [3]. The pipe connector plays an important role in helping single-span plastic greenhouses withstand heavy snowfall and strong winds. Most farmers use single-span plastic greenhouses, so the performance of the pipe connector is important [23].

However, damage to a single-span plastic greenhouse may cause damage to the plastic used as a structural material. In these cases, plastic decomposes to microplastics through physical, chemical, and biological process [24]. The amount of plastic waste generated from plastic houses was 68,758 tons as of 2018, accounting for 21.6% of the total plastic waste [25]. However, the negative emissions caused by waste plastic is a source of pollution in rural and agricultural environments [26].

Figure 1. Case of foundation damage of single-span plastic greenhouse [27].

In order to minimize such secondary damage, it is necessary to prevent damage from strong winds in single-span plastic greenhouses, and for this, reinforcement measures for the pipe connector are essential.

Therefore, in this study, an in situ down-scaled model test was carried out on a continuous pipe foundation of the single-span plastic greenhouse. As a down-scaled model test object, the rafter and purlin are connected by a pipe connector of steel plate. In the field down-scaled model test, only one case was selected for each site, and the results were compared and evaluated by testing the installation direction and the presence or absence of the pipe connector differently.

2. Materials and Methods

2.1. Sample Preparation

The test sites where the characteristics of the embedded continuous pipe foundation of the single-span plastic greenhouse were identified are as follows: Cheongju, Chungcheongbuk-do (A), Jeonju, Jeollabuk-do (B), and Gyehwa, Jeollabuk-do (C). Site A is an agricultural high school practice site and is classified as SM according to the unified soil classification system (USCS). Site B is a clinical trial complex within the Rural Development Administration, which is classified as SC. Site C is an agricultural technology center located on reclaimed land and is classified as ML in the unified soil classification system. Table 1 shows the physical and mechanical properties of soil by the site. Characteristics are classified by USCS and USDA: SM is silty sand, loamy sand; SC is clayey sand, sandy loam, and ML is silt, sandy loam. G_s is the specific gravity and refers to the ratio of the weight of soil in the air to the weight of distilled water of the same volume. LL is the liquid limit, and it refers to the water content at the boundary where soil is mixed with water to change to a liquid state and a plastic state. PI is the water content at the boundary where the soil changes from a semi-solid state to a solid state. It is used to determine the physical condition of the soil. The γ is the unit weight and refers to the value obtained by dividing the weight of the soil mass by the corresponding volume. It is the weight ratio of soil per unit volume. The firmness and looseness of the ground can be determined. O.M.C is the water content corresponding to the maximum unit weight.

Table 1. Mechanics and physical properties of soils on site A, site B, and site C.

Site	G_s	LL (%)	PI	$\gamma_{d\ field}$ (kN/m^3)	$\gamma_{d\ A\ com.}$ (kN/m^3)	O.M.C	4.76 mm (%)	1.0 mm (%)	0.2 mm (%)	0.075 mm (%)	2μ (%)	USCS
A	2.62	N.P	N.P	13.56	19.17	11.4	83.4	77.2	49.4	35.5	7.5	**SM**
B	2.66	39.8	16.3	12.45	17.29	16.7	97.9	94.2	70.9	42.2	12.2	**SC**
C	2.61	N.P	N.P	10.53	15.84	18.4	100	99.8	99.0	65.1	5.5	**ML**

G_s: Specific gravity, LL: Liquid limit, PI: Plastic index, $\gamma_{d\ field}$: Field dry density(field); $\gamma_{d\ A\ com.}$: Dry unit weight of A compaction, O.M.C: Optimum water content; USCS: Unified soil classification system, N.P: Non plastic.

2.2. Field Test Condition

The continuous pipe foundation embedded in the ground of the single-span plastic greenhouse is composed of rafter and purlin. Rafter and purlin are embedded in the ground and are connected together with a pipe connector.

The pipe connector connects two members together by binding parts composed of a steel wire and a plate shape. In the field, cost and testing sites were limited, so only some parts of the rafter (column) and continuous pipe foundation were modeled. For the rafter and purlin used in the field test, galvanized steel pipes, as suggested in the anti-disaster design standards of the Rural Development Administration, were used. The rafter used in the test has a diameter of 31.8 mm, and the purlin has a diameter of 25.4 mm (Figure 2).

(a) (b) (c)

Figure 2. Construction materials for testing: (**a**) test pipe and pipe connector; (**b**) horizontal pipe connector; (**c**) vertical pipe connector.

In the anti-disaster design standards of the Rural Development Administration, a total of 19 types of single-span plastic greenhouses are classified. The specifications of the rafter and the purlin are the same, and the spacing of the rafter and the embedded depth of the continuous pipe foundation are presented in Table 2.

Table 2. Foundation types of single-span greenhouse [3].

Type	07-Single -Span 1~4	10-Single -Span 1~5	10-Single -Span 6~9	10-Single -Span 10~13	07-Single -Span 18	12-Single -Span 1
Rafter spacing (mm)	500	400	500	400	500	500
Embedded depth (mm)	250	250	250	200	250	250

The spacing of the rafter and the embedded depth of the continuous pipe foundation suggested in the field test were determined by referring to the anti-disaster design standards of the Rural Development Administration. Rafter spacing widely used by the Rural Development Administration at a standard of 500 mm, so the spacing was set to 500 mm, 600 mm, and 700 mm. The depths of the continuous pipe foundation were increased to 300 mm, 400 mm, and 500 mm by increasing the depth more than the standard because strong wind damage was occurring. As a test condition for each site, the rafter and continuous pipe foundation were welded in site A, and the pipe connector was connected in the rafter (vertical) direction in site B. In the site C, the pipe connector was connected in the crossbar (horizontal) direction.

2.3. Field Test Method

In order to classify the conditions of the field uplift resistance test, the first and English symbols indicate the site, the second number the embedded depth, and the third the rafter spacing (Table 3). The uplift resistance test method is based on JSF 1821 and ASTM D 3689-07 [28,29]. In this study, ASTM D 3689-07 standards were applied. This is a test method using the static axial tensile load of ASTM D 3689-07. The uplift resistance test was performed by increasing the load step by step by 5% to the estimated failure load. The step-by-step measurements were taken until the settlement gauge did not change and the load was maintained for at least 4 min to 15 min. The field test method using the axial tensile load control method is shown in Figure 3. The hydraulic load gauge used for the test was 50 kN, and the amount of settlement was measured up until 100 mm displacement. The area of the load plate was 0.08 m^2.

Table 3. Test condition identification number.

Site	Embedded Depth (mm)			Rafter Spacing (mm)		
	3	4	5	5	6	7
A, B, C	300	400	500	500	600	700

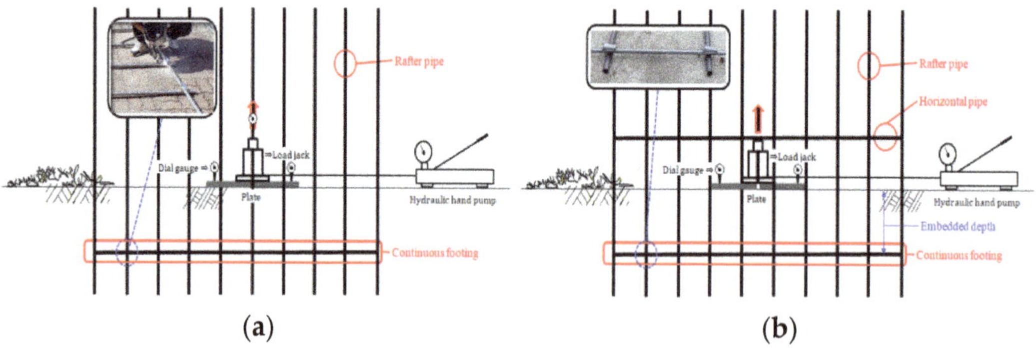

Figure 3. Field static axial tensile load test: (**a**) installed continuous foundation of site A and (**b**) installed continuous foundation of site B and site C.

The construction procedure of the field test is shown in Figures 4–6. Rafter and purlin were installed horizontally and vertically. The excavated soil was backfilled and compacted with a small compactor. The static axial tensile load test was conducted 24 h after construction.

(**a**) (**b**) (**c**)

Figure 4. Field test procedure of site A: (**a**) ground grading work; (**b**) embedded depth leveling; (**c**) uplift testing of continuous foundations.

(**a**) (**b**) (**c**)

Figure 5. Field test procedure of site B: (**a**) ground grading work; (**b**) embedded depth leveling; (**c**) uplift testing of continuous foundations.

(**a**) (**b**) (**c**)

Figure 6. Field test procedure of site C: (**a**) ground grading work; (**b**) embedded depth leveling; (**c**) uplift testing of continuous foundations.

2.4. Measured Uplift Resistance

There are four major methods for measuring the bearing capacity in load–displacement curves. The first depends on the shape of the curvature method [30], and the other three take into account the vertical displacement limit and the total settlement amount [31].

In this study, the load-settlement relationship was analyzed by the axial tensile load test. This method can use log–log plotting to obtain two straight lines with a clear difference between the elastic part and the plastic part, as shown in Figure 7b [32].

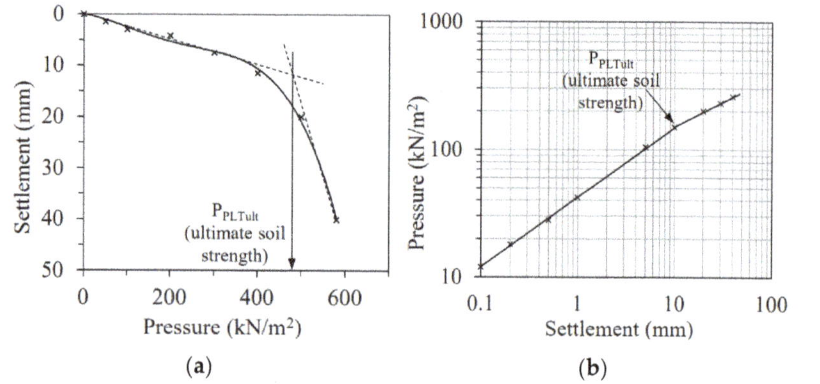

Figure 7. Load–settlement curve from PLT: (**a**) shape of curvature and (**b**) log-log plot (AbdelSalam et al., 2020).

3. Results

3.1. Results of Field Compaction on Site A, Site B, and Site C

Before the axial tensile load test, a field density test (Figure 8) was performed to determine the density of the original ground. The field density was compared with the density of the ground measured before the field axial tensile load test. Table 4 summarizes the field density results after 24 h of being embedded in the continuous pipe foundation in the test site. In site A, the average values were 12.84 kN/m^3, 13.62 kN/m^3, and 13.99 kN/m^3 by each measure station. In site B, the average values were 11.10 kN/m^3, 11.05 kN/m^3, and 12.70 kN/m^3 by each measure station. In site C, the average values were 12.96 kN/m^3, 17.53 kN/m^3, and 16.77 kN/m^3 by each measure station. The deviation of the field density was the smallest in site A, and it was found to be smaller than the field density measured before the test of 13.56 kN/m^3. Site B was found to be smaller than the measured field density of 12.45 kN/m^3 before the test. Site C showed the largest variation in field density by each measure station and was found to be larger than the field density of 10.53 kN/m^3 before the test. The differences in field density in site A, site B, and site C are thought to be due to the narrow width of the backfill excavation and the failure to compact the floor. In addition, site C is reclaimed and has high groundwater. Due to the high groundwater level, excess pore water pressure occurred during compaction, so the load was not evenly distributed during compaction, and the soil was not evenly spread out, so the field density was also different at these locations. It seems that it was not possible to allow sufficient time for the ground to be stabilized after installing the continuous pipe foundation.

Table 4. Field dry unit weight on site A, site B, and site C.

Site	Field Density (kN/m^3)								
	3–5	4–5	5–5	3–6	4–6	5–6	3–7	4–7	5–7
A	14.11	11.64	12.787	13.51	14.24	13.10	13.71	14.47	13.79
B	12.37	10.81	10.12	12.18	10.89	10.07	11.24	13.43	13.43
C	9.48	10.49	18.91	17.89	23.53	11.18	10.27	19.73	20.32

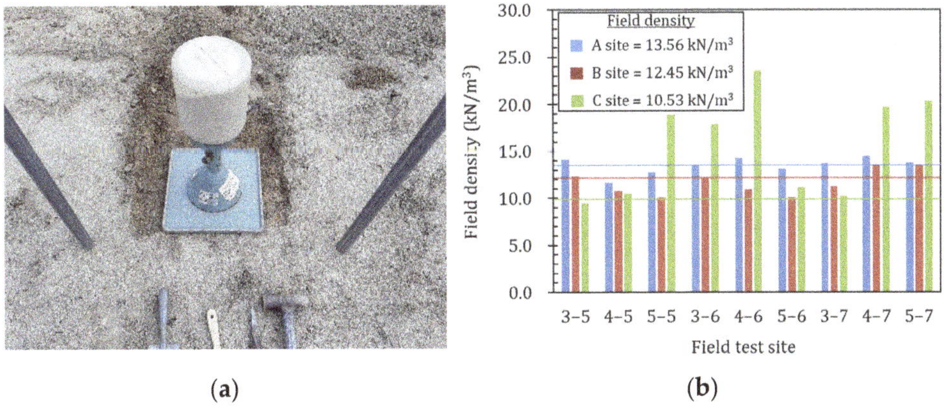

Figure 8. Field dry unit weight of soils. (**a**) field dry unit weight test; (**b**) results of field dry unit weight for site A, site B, and site C.

3.2. Results of Field Test of Uplift Resistance on Site A, Site B, and Site C

The results of the static axial tensile load test by site A, site B, and site C are shown in Figures 9–11. Processing the static axial tensile load test, when the uplift load increased, the rafters rose vertically and the load plate on the ground settlement. It was observed that the settlement of the load plate on the ground increased linearly as the load increased. As for the uplift loading stage, as described in the test method, the expected load was divided into five stages, and the degree of settlement was observed during uplift loading. In order to determine the yield point due to the destruction of the ground as the uplift load increases, an inflection point must occur in the load–settlement curve, but it increased linearly at all three sites.

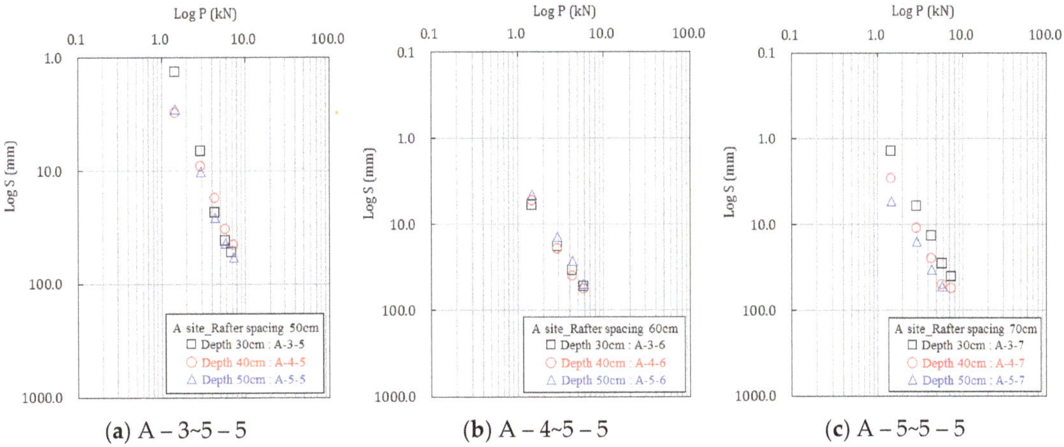

Figure 9. Results of field uplift resistance on site A.

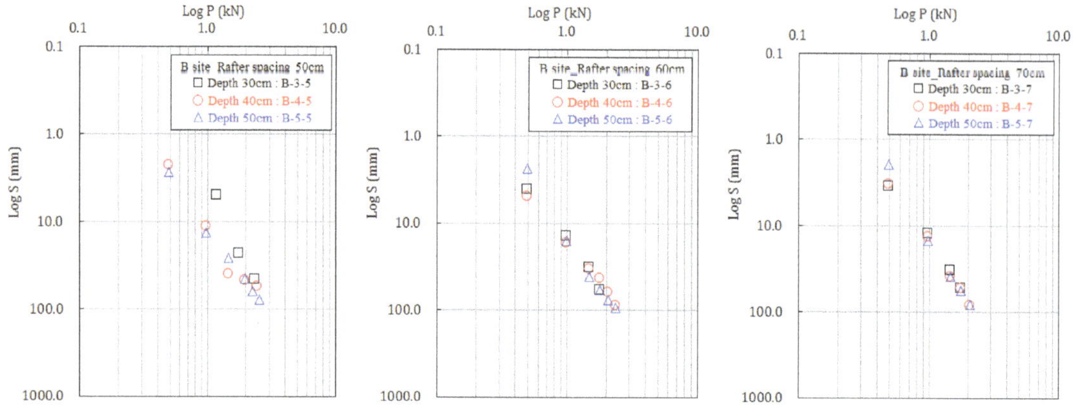

(a) B – 3~5 – 5 (b) B – 4~5 – 5 (c) B – 5~5 – 5

Figure 10. Results of field uplift resistance on site B.

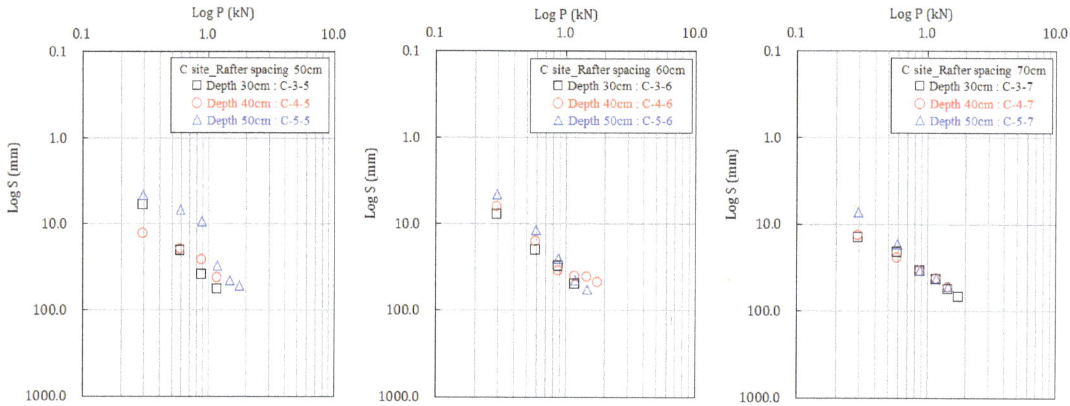

(a) C – 3~5 – 5 (b) C – 4~5 – 5 (c) C – 5~5 – 5

Figure 11. Results of field uplift resistance on site C.

Lee et al. (2014) reported that the maximum strength of the pipe connector was achieved upon the initial displacement in sliding resistance tests in the rafter and purlin directions and maintained a strength equivalent to 75% of the maximum strength up to 20.0 mm [33]. In addition, it was said that the influence of the test results on compaction management could not be ignored. In this study, after the settlement of up to 20.0 mm, the degree of settlement due to the increase in the load decreased. It maintains this strength of up to 20 mm, and it has been shown that there is little change in the degree of settlement due to an increase in load.

The reason why the inflection point was not found when assessing the relationship between the load and settlement is thought to be as follows: In site A, there is no purlin at the top of the ground. Therefore, the rafter was fixed and lifted directly. When the rafter was uplifted, the uplift load was transferred to the left and right rafters, and the ground surface was destroyed (Figure 12a). The destruction of the ground does not appear to cause an increase in the load and settlement. In site B and site C, bending deformation occurred due to the uplift load of the crosspiece (Figure 12b). After the bending deformation occurred, the fixed part of the pipe connector slid, and the bending deformation increased (Figure 12c).

(a) **(b)** **(c)**

Figure 12. Slipping of pipe connector: (**a**) uplift resistance test method of site A; (**b**) uplift resistance test method of site B; (**c**) uplift resistance test method of site C.

The sliding resistance limit of the pipe connector is 1360 N, which is suggested in the anti-disaster design standards for special agricultural and horticultural facilities [3]. The settlement amount increased significantly from the uplift load of 2.9 kN in site A, 1.8 kN in site B, and 0.9 kN in site C. As the uplift load increased, the pipe connector slipped at the connection between the rafter and the purlin pipe. As a result, settlement of the ground did not occur. For this reason, it is unlikely that an inflection point due to an increase in uplift load will occur.

Table 5 summarizes the uplift resistance test results by site A, site B, and site C. The ultimate stress and yield stress were in the order of site A > site B > site C. There was no difference in yield stress due to the effect of embedded depth. In addition, rafter spacing did not affect yield stress. This is believed to be caused by the sliding of the pipe connector. Faizi et al. (2014) stated that the maximum uplift resistance of an embedded pipe is affected by the embedded depth [34]. However, in this study, it was found that the depth of embedding and the rafter spacing had no effect on the uplift resistance due to the sliding of the pipe connector.

Table 5. Ultimate and yield stresses of embedded depth and rafter spacing on site A, site B, and site C.

Test Condition		Ultimate Stress (kN/m^2)			Yield Stress (kN/m^2)		
		A	B	C	A	B	C
Spacing 50 cm	Depth 30 cm	128.72	44.13	22.07	85.81	29.42	14.71
	Depth 40 cm	137.91	27.59	16.55	91.94	18.39	11.03
	Depth 50 cm	137.91	58.50	16.55	91.94	31.87	11.03
Average		134.85	43.41	18.391	89.90	26.560	12.26
Spacing 60 cm	Depth 30 cm	110.33	70.40	22.07	73.55	22.07	14.71
	Depth 40 cm	110.33	38.61	16.55	73.55	25.74	11.03
	Depth 50 cm	110.33	38.61	27.59	73.55	25.74	18.39
Average		110.33	49.21	22.07	73.55	24.52	14.71
Spacing 70 cm	Depth 30 cm	137.91	33.11	33.11	91.94	22.07	22.07
	Depth 40 cm	110.33	38.61	27.59	73.55	25.74	18.39
	Depth 50 cm	110.33	38.61	16.55	73.55	25.74	11.03
Average		119.52	36.78	25.75	79.68	24.52	17.16

3.3. Results of Uplift Resistance by Embedded Depth

The magnitudes of the uplift resistance for the embedded depth and the rafter spacing were compared (Figures 13 and 14). There was a difference in the size of the embedded depth due to the influence of the pipe connector by site A, site B, and site C, but there was no difference in the trend. The same results were also obtained for the distance between the rafters. The difference in yield stress could not be explained, regardless of the connecting direction of the rafter and purlin and whether they were installed or not. In the field test using the static axial tensile load control method, it is necessary to determine the point where the yield stress occurs due to the sudden subsidence of the ground, the same as in the plate load test method. However, the sliding resistance of the pipe connector reduces the increase in load and settling. Wu et al. (2019) stated that the pipe diameter is independent of the failure mechanism and the maximum uplift resistance has no effect on the outcome [35]. This means that the diameter of the pipe does not affect the test results. In addition, it was difficult to determine the effect on the rafter spacing and embedding depth in this test, so we tried to compare these values with other data. However, there was no field test study on pipe connectors, so comparisons could not be made.

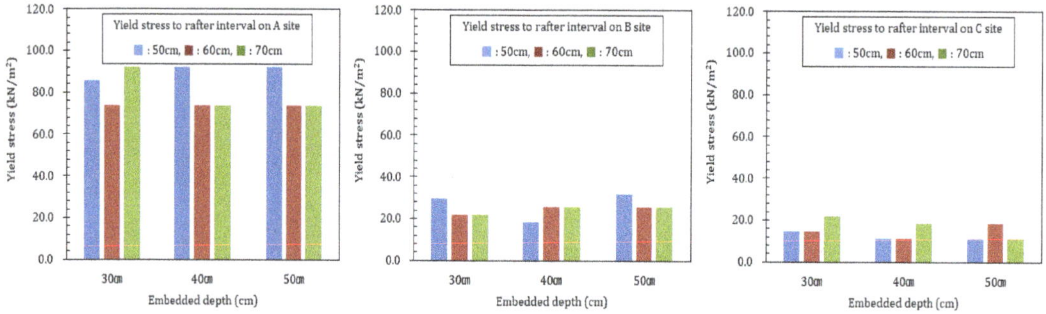

(a) (b) (c)

Figure 13. Yield stress to rafter spacing on site A, site B, and site C: (**a**) yield stress to rafter spacing on site A; (**b**) yield stress to rafter spacing on site B; (**c**) yield stress to rafter spacing on site C.

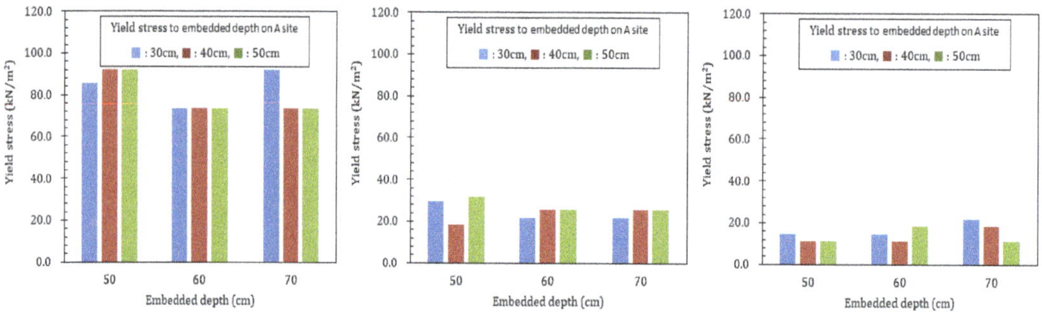

(a) (b) (c)

Figure 14. Yield stress to embedded depth on site A, site B, and site C: (**a**) yield stress to embedded depth on site A; (**b**) yield stress to embedded depth on site B; (**c**) yield stress to embedded depth on site C.

3.4. Discussion of Test Results for Site A, Site B, and Site C

The uplift resistance values of each test site, as determined using the field test, were compared. The uplift resistance was converted into a load and compared with the design standards of the Rural Development Administration. Considering the fact that the maximum uplift resistance is related to the sliding resistance of the pipe connector in the field test, the ultimate and yield stress was rearranged as a load, as shown in Table 6 and Figure 15. The uplift resistance by the test conditions showed a significant deviation in the order of A (1027.4) > B (999.5) > C (465.5). The uplift resistance showed a considerable variation in all three sites. Site A showed an average yield load of 6483.4 N, site B showed an average yield load of 1780.2 N, and site C showed an average yield load of 1176.8 N. It can be observed that the yield load does not satisfy the sliding resistance limit of the Rural Development Administration in site C. Site B also showed similar results to the slip resistance limit. The load-bearing capacity of a single-span plastic greenhouse varies greatly depending on the type of pipe connector. Experts assert that the pipe connector plays a central role in resistance to heavy snow and strong winds. However, the risk of damage to single-span plastic greenhouses is further increased by using pipe connectors with lower standard values than the standard pipe connectors suggested by the RDA [36].

Table 6. Yield load of slipping to pipe connector on site A, site B, and site C.

Test Condition		Ultimate Uplift Load (N)			Yield Uplift Load (N)		
		A	B	C	A	B	C
Interval 50 cm	Depth 30 cm	10,297.2	3530.4	1765.2	6864.8	2078.5	1176.8
	Depth 40 cm	11,032.8	2206.8	1323.6	7355.2	1299.3	882.4
	Depth 50 cm	11,032.8	4680.0	1323.6	7355.2	2251.6	882.4
Average		10,787.6	3472.4	1470.8	7191.7	1876.5	980.5
Interval 60 cm	Depth 30 cm	8826.0	5632.0	1765.2	5884.0	1559.2	1176.8
	Depth 40 cm	8826.0	3088.8	1323.6	5884.0	1818.5	882.4
	Depth 50 cm	8826.0	3088.8	2206.8	5884.0	1818.5	1471.2
Average		8826.0	3936.5	1765.2	5884.0	1732.1	1176.8
Interval 70 cm	Depth 30 cm	11,032.8	2648.4	2648.4	7355.2	1559.2	1765.6
	Depth 40 cm	8826.0	3088.8	2206.8	5884.0	1818.5	1471.2
	Depth 50 cm	8826.0	3088.8	1323.6	5884.0	1818.5	882.4
Average		9561.6	2942.0	2059.6	6374.4	1732.1	1373.1
Standard deviation. P		1027.4	999.5	465.5	684.9	267.9	310.3

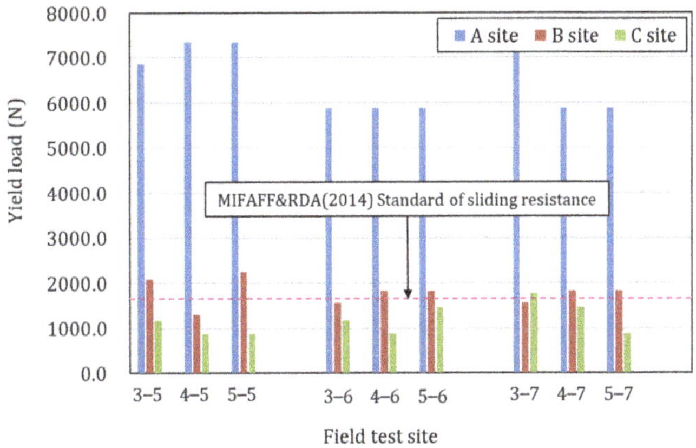

Figure 15. Comparison of field dry unit weight and MIFAFF and RDA standard.

Lee et al. (2014) reported that they did not meet the Rural Development Administration standard as a result of testing the same product as the aperture used in this study with different connecting directions [21]. In the field displacement control test results (Table 7) of Yun et al. (2015), it was reported that there are differences caused by the type of pipe connector [23]. In another recent study, we attempted to compare field uplift resistance results with soil box test results. However, they could not be compared because different test conditions were used with regard to the continuous pipe foundation.

Table 7. Ultimate uplift capacity according to pipe connector type [23].

No.	Ultimate Uplift Capacity (N)	
	Steel Wire	Steel Plate
1	198.0	1441.0
2	175.0	1763.0
Average	186.0	1602.0

In this study, it was found that the uplift resistance had no effect on the spacing and embedded depth of the rafter and purlin. The ultimate uplift resistance for the steel plate connector of Yun et al. (2015) was compared with the results of this study. The comparison results showed a significant difference of 507.99% increase at the site, 118.26% increase at the site B, and 8.52% increase at the site C. As shown in Figure 16, this can explain the cause of error between the results in the difference in field test conditions.

(a) (b)

Figure 16. Continuous pipe foundation for field test. (**a**) Steel wire, (**b**) steel plate [23].

However, it was found that the sliding resistance of the steel plate pipe connector affects the uplift resistance. As reported by previous researchers, the connecting method for each type of pipe connector is important, and improving the sliding resistance of the pipe connector can reduce damage from strong winds in single-span plastic greenhouses. Therefore, in order to reduce the damage caused by strong winds, it seems necessary to increase the sliding resistance limit of the rafter and purlin.

4. Conclusions

In this study, field tests of uplift resistance were performed on a continuous pipe foundation of a single-span plastic greenhouse according to the connection direction of the pipe connector and whether or not it was used. The ability of the pipe connector was evaluated by measuring the uplift resistance of the continuous pipe foundation using a static axial tensile load test.

The backfill conditions of the field ground were smaller in site A and site B than the field density of the undisturbed ground, but site C was highly compacted. However, the uplift resistance measured by the static axial tensile load test was found to be the smallest in site C. This seems to be because the time allowed for the stabilization of the backfill ground was insufficient, and it was difficult to reproduce the field test. It was found that the uplift resistance for site A, site B, and site C did not show an inflection point in the load–settlement curve due to the influence of the pipe connector. It can be observed that the uplift resistance due to the difference between the spacing of the rafter and the embedded depth of the continuous pipe foundation was not affected. Additionally, the sliding resistance of the pipe connector was dominantly acting on the uplift resistance. The reason is that the pipe connection of site A was welded, so the slip resistance value of the pipe connector was measured 4.5 times higher than the value suggested by the Rural Development Administration design standard. Therefore, it seems that the damage to the single-span plastic greenhouse caused by strong winds can be reduced by supplementing the pipe connector. Additionally, obviously, sustainable agriculture will be possible by establishing reinforcement measures for 99% of single-span greenhouses commercialized in Korea. It seems that related organizations can use these data as a sliding resistance standard of the pipe connector.

Funding: This research received no funding.

Institutional Review Board Statement: Not applicable.

Data Availability Statement: The data used to support the findings of this study are included within the article.

Conflicts of Interest: The authors declare that they have no conflict of interest.

References

1. RDA. *Development of Foundation Design Technology on Single Span Plastic Greenhouse According to Soil Property*; Rural Development Administration: Jeonju, Korea, 2021. (In Korean)
2. Ministry of Agriculture, Food and Rural Affairs (MAFRA). *Present Condition of Vegetable Grown in Facilities and Vegetable Production Performance*; Ministry of Agriculture, Food and Rural Affairs: Sejong City, Korea, 2018. (In Korean)
3. Ministry for Food, Agriculture, Forestry and Fisheries (MIFAFF); Rural Development Administration (RDA). *Designated Notice of Standards to Endure Disaster for Horticultural and Special Facilities*; MIFAFF & RDA: Gwacheon and Suwon, Korea, 2014. (In Korean)
4. Rural Development Administration (RDA). *Symposium for Reducing of Meteorological Disasters of Agricultural Facilities*; RDA: Suwon, Korea, 2007; pp. 160–161. (In Korean)
5. MAFRA, Facility Vegetable Greenhouse Status and Vegetable Production Output Report, Ministry of Agriculture, Food and Rural Affairs. 2021, pp. 105–108. Available online: 2020%20%EC%8B%9C%EC%84%A4%EC%B1%84%EC%86%8C%20%EC%98%A8%EC%8B%A4%ED%98%84%ED%99%A9%20%EB%B0%8F%20%EC%B1%84%EC%86%8C%EB%A5%98%20%EC%83%9D%EC%82%B0%EC%8B%A4%EC%A0%81.pdf (accessed on 15 September 2022). (In Korean).
6. Robertson, A.; Roux, P.; Gratraud, J.; Scarascia Mugnozza, G.; Castellano, S.; Dufresne de Virel, M.; Palier, P. Wind pressures on permeably and impermeably clad structures. *J. Wind. Eng. Ind. Aerodyn.* **2002**, *90*, 461–474. [CrossRef]
7. Lee, S.G.; Lee, J.W.; Lee, H.W. Development of One-Span Vinyl House for Reduction of Damage by Heavy Snow. *J. Bio-Environ. Control* **2005**, *14*, 138–144. (In Korean)
8. Lee, B.G.; Yun, S.W.; Choi, M.K.; Lee, S.Y.; Moon, S.D.; Yu, C.; Yoon, Y.C. Uplift Bearing Capacity of Spiral Steel Peg for the Single Span Greenhouse. *J. Bio-Environ. Control* **2014**, *23*, 109–115. (In Korean) [CrossRef]
9. Lee, W.-G.; Woo, J.-H.; Lee, H.-D.; Shin, K.-J. Analysis of Uplift Capacity on Single-span Greenhouse's Foundation According to Wind Load and Evaluation for Reinforcing Methods. *J. Archit. Inst. Korea* **2022**, *24*, 33–39. (In Korean)
10. MOIS. *Statistical Year Book*; Ministry of the Interior and Safety: Sejong City, Korea, 2021. (In Korean)
11. Ogawa, H.; Tsuge, I.; Sato, Y.; Hoshiba, S.; Yamashita, S. Experimental analysis on strength of pipe—House with Ground anchoring (II). *J. Soc. Agric. Struct.* **1990**, *20*, 262–269.
12. RDA; NAAS; Department of Agricultural Engineering. *Guide Book for Reduction of Meteorological Disaster of Agricultural Facilities*; Rural Development Administration National Institute of Agricultural Sciences: Wanju-gun, Korea, 2007. (In Korean)
13. Tanaka, H.; Fujikawa, M.; Matsuzaki, T.; Sumikawa, O.; Daikoku, M.; Inooku, K. Application of a screw type pile for the foundation of greenhouse. *Misc. Pub. NARO. West. Reg. Agric. Res. Cent.* **2005**, *3*, 21–37.
14. Kim, M.K.; Nam, S.W.; Son, J.E.; Yun, N.K. Analyses of actual state and structural safety of regionally characterized greenhouse in Korean. *Prot. Hortic. Plant Fact.* **1994**, *3*, 128–135. (In Korean)
15. Kim, M.K.; Nam, S.W. Experimental studies on the structural safety of pipe-houses. *Prot. Hortic. Plant Fact.* **1995**, *4*, 17–24. (In Korean)
16. Cho, Y.S.; Kwon, O.Y. Uplift Capacity of a Single Screw Anchor. *KSCE J. Civ. Eng.* **1997**, *17*, 275–283. (In Korean)
17. Lee, B.J.; Lee, J.K. A study on the critical embedment ratio of screw anchors. *J. Ind. Sci.* **1999**, *6*, 475–492. (In Korean)
18. Yoon, Y.C.; Suh, W.M.; Cho, J.H. The uplift capacity of plane and corrugated piles for pipe frame greenhouse. *Prot. Hortic. Plant Fact.* **2001**, *10*, 148–154. (In Korean)
19. Yoon, Y.C.; Lee, K.H.; Yu, C. A study on the uplift capacity improvement of pipe-framed greenhouse foundation using circular horizontal anchors. *KCID J.* **2003**, *10*, 55–61. (In Korean)
20. Kim, D.H.; Yoo, C.S. Pull-out capacity of screw anchor pile in sand using reduced-scale model tests. *J. KGS.* **2013**, *29*, 121–133. (In Korean)
21. Choi, M.K.; Yun, S.W.; Kim, H.N.; Lee, S.Y.; Kang, D.H.; Yoon., Y.C. Uplift capacity of spiral bar through the model experiment. *Prot. Hortic. Plant Fact.* **2015**, *24*, 202–209. (In Korean) [CrossRef]

22. Yun, S.W.; Choi, M.K.; Lee, S.Y.; Yoon, Y.C. Uplift capacity of meteorological disaster. *J. Agric. Life Sci.* **2015**, *49*, 303–310. (In Korean) [CrossRef]
23. Song, H.S.; Lim, S.Y.; Ryu, H.R. Slip-Resistance Test for Plastic Greenhouse Pipe Connectors According to Loading Rate. In Proceedings of the Korean Society of Agricultural Engineers Conference, Daejeon, Korea, 20 October 2016.
24. Ding, L.; Zhang, S.; Wang, X.; Yang, X.; Zhang, C.; Qi, Y.; Guo, X. The occurrence and distribution characteristics of microplastics in the agricultural soils of Shaanxi Province in north-western China. *Sci. Total Environ.* **2020**, *720*, 137525. [CrossRef]
25. Korean Statistics Information Service, Production of Agricultural Waste Vinyl 2004–2020, Last Modified 03-282020. 2020. Available online: https://kosis.kr/statHtml/statHtml.do?orgId=392&tblId=DT_AGRI01&vw_cd=MT_ZTITLE&list_id=T_20&seqNo=&lang_mode=ko&language=kor&obj_var_id=&itm_id=&conn_path=MT_ZTITLE (accessed on 15 September 2022). (In Korean).
26. Heo, J.; Knag, C.Y.; Kim, D.W.; Kim, N.Y.; Gi, Y. *Public Promotion Measures on the Adequate Disposal of Waste Agricultural Plastics and Collection Enhancement*; Korea Rural Economic Iistitute: Naju-si, Korea, 2002. (In Korean)
27. Lee, S.G. *Development of Small Scale Vinyl House Model for Reduction of Damage by Wind and Show*; National Research Foundation of Korea: Taejon-jikhalsi, Korea, 2004. (In Korean)
28. *JS F 1821*; Uplift Load Testing Method of Piles. Japanese Society of Soil Mechanics and Foundation Engineering. The Japanese Society of Soil Mechanics and Foundation Engineering. National Research Foundation of Korea: Taejon-jikhalsi, Korea, 1992; pp. 1821–1989.
29. *ASTM D 3689-07*; Standard Test Methods for Deep Foundations under Static Axial Tensile Load. ASTM International: West Conshohocken, PA, USA, 2010.
30. Butler, H.D.; Hoy, H.E. *User's Manual for the Texas Quick-Load Method for Foundation Load Testing*; FHWA-IP-77-8; FHWA: Washington, DC, USA, 1997.
31. FHWA. *Design and Construction of Driven Pile Foundations. Workshop Manual*; National Highway Institute: Arlington, VA, USA, 1997; Volumes I, FHWA-HI-97-013.
32. AbdelSalam, S.S.; Anwar, M.B.; Esmail, D.I. *LRFD for Shallow Foundations Using Plate Load Test Data*; Geo-Congress: Minneapolis, MN, USA, 2020; pp. 1–11.
33. Lee, S.H.; Shin, K.J. Slip Test of Pipe Connector for Greenhouse Frame. *J. Archit. Inst. Korea Struct. Constr.* **2014**, *30*, 9–17. (In Korean) [CrossRef]
34. Faizi, K.; Jahed Armaghani, D.; Momeni, E.; Nazir, R.; Mohamad, E.T. Uplift Resistance of Buried Pipelines Enhanced by geogrid. *Soil Mech. Found. Eng.* **2014**, *51*, 188–195. [CrossRef]
35. Wu, J.; Kouretzisa, G.; Suwala, L.; Ansaria, Y.; Scott, W.S. Shallow and deep failure mechanisms during uplift and lateral dragging of buried pipes in sand. *Can. Geotech. J.* **2019**, *27*, 1–17. [CrossRef]
36. Lee, S.I. The Farmers Newspaper. 2013. Available online: https://www.nongmin.com/plan/PLN/SRS/77249/view (accessed on 16 November 2022).

agriculture

Article

Is the Rural Population Caught in the Whirlwind of the Digital Divide?

Hayet Kerras [1,*], María Francisca Rosique Contreras [2], Susana Bautista [3] and María Dolores de-Miguel Gómez [1,*]

1 Departamento de Economía de la Empresa, Escuela Técnica Superior Ingeniería Agronómica (ETSIA), Universidad Politécnica de Cartagena, Paseo Alfonso XIII, 48, 30203 Cartagena, Spain
2 Departamento de Tecnologías de la Información y las Comunicaciones, Escuela Técnica Superior de Ingeniería de Telecomunicación (ETSIT), Universidad Politécnica de Cartagena, Edificio Cuartel de Antigones, Plaza del Hospital, 1, 30202 Cartagena, Spain
3 Escuela Politécnica Superior, Universidad Francisco de Vitoria, Ctra. Pozuelo-Majadahonda, Km 1.800, 28223 Pozuelo de Alarcón, Spain
* Correspondence: hayet.kerras@edu.upct.es (H.K.); md.miguel@upct.es (M.D.d.-M.G.); Tel.: +34-968325784 (M.D.d.-M.G.)

Abstract: Technology represents a benchmark ally for today's rural world and is a sine qua non-condition for achieving sustainable development. Indeed, today the arrival of digitization and information and communication tools makes life easier for the inhabitants of the rural world in general and for those who work in agriculture. However, not everyone has and knows how to use these technologies. There are very visible differences between the rural world and the urban one in the accessibility and use of technology, especially among vulnerable people (unemployed, elderly, women, etc.), causing a digital divide that reflects the great discrimination suffered by the rural world, full of stereotypes and very traditional role assignments. The objective of this study is to evaluate the differences in terms of access and use of technology. For this reason, the results of a survey carried out on the Spanish rural population have been analyzed with the structural equations tool "PLS-SEM". They show digital gaps, as well as a disturbance between the different gaps and the socioeconomic situation of users, which imposes the need to take immediate measures to reduce and fight against this type of inequality.

Keywords: sustainable development goals; agri-food security; equality; information and communication technologies; technology 4.0

Citation: Kerras, H.; Rosique Contreras, M.F.; Bautista, S.; de-Miguel Gómez, M.D. Is the Rural Population Caught in the Whirlwind of the Digital Divide? *Agriculture* **2022**, *12*, 1976. https://doi.org/10.3390/agriculture12121976

Academic Editors: Muhammad Sultan, Redmond R. Shamshiri, Md Shamim Ahamed and Muhammad Farooq

Received: 19 October 2022
Accepted: 17 November 2022
Published: 22 November 2022

1. Introduction

Today, Information and Communication Technologies (ICTs) have invaded all scenarios in the lives of human beings, from personal to professional [1]. The rural world has not been able to escape this boom in technology as it is closely related to agriculture, thanks to Artificial Intelligence (AI), which is the catalyst for the new revolution in the agricultural sector (agriculture 4.0) [2–4], called "AgriTech", which comprises the evolution of tractors with new functions such as the Global Positioning System (GPS), the appearance of autonomous robotics that today supports a large collection of agricultural data, the sensors that measure climatic factors: such as sunlight, wind speed, the degree of humidity, the drones that allow better remote control of crops, or identification of weeds, the smart irrigation systems the vertical farming, etc. [5–8]. This innovative method of vertical farming based on increased productivity, considering the limitations of agricultural plots, would make it possible to combat hunger (Achieve Sustainable Development Goal 2), poverty and fight against nutrition problems (Sustainable Development Goal 1, 3 and 12) and climate change (Sustainable Development Goal 13), achieving the sustainable development goals set by The United Nations Educational, Scientific and Cultural Organization (UNESCO) for the year 2030, which represents a fundamental element of social, economic, and environmental progress [7,9,10]. These initiatives affect agriculture as a whole and rural development specifically, either in terms of saving time, increasing efficiency, or terms of added value that they provide to farmers, allowing consumers to find products fresh and healthy to face food security problems [2,11–13].

According to the World Resources Institute, to feed the population sustainably in 2030: it will be necessary to produce 56% more calories; devote an additional 593 million hectares to agriculture. Farmers will have to produce more with fewer available resources, preserving the environment to meet the needs of a world population that is estimated to reach 10 billion people in 2050. Hence the need to transform the sector and bet on new management methods by introducing new technologies [14].

However, many people and many farm workers do not have the luck or the means to adapt to these rapid changes and, consequently, do not have the access and the necessary skills to enjoy the contribution of this technology, especially in the rural world where the main actors of agriculture tend to be concentrated [15–17].

This phenomenon, called the "Digital Divide," is separated into three categories: first, the ICT access gap, which refers to the difficulty that people have in accessing this resource due to socioeconomic differences since digitization requires very costly investments and infrastructure for less developed areas such as rural areas. The second is the gap in the use of ICTs, which refers to the lack of digital skills that prevent the use of technology [18,19]. In this sense, and to give an example, the International Telecommunication Union (ITU) indicates that there are 40 countries in which more than half of its inhabitants do not know how to attach a file to an email [20]. The third is the gap in quality of use, which refers to digital skills to manage ICTs and make good use of them.

The objective of this work is to analyze the correlation that exists between the access, and the different levels of use of ICTs in the Spanish rural world, relating them to the socioeconomic factors that most limit the population in rural areas, such as gender, age and professional status, and that lead us to ask ourselves the following questions: Are there gaps in access and use of ICTs among rural populations? Is this gap different between women and men? Does this gap have more influence on agricultural activity? To do this, data collection has been carried out through a survey aimed at the rural population, and the data obtained has been analyzed through structural equation modeling.

2. Theoretical Framework and Development of Hypotheses

In a world increasingly faced with the vagaries of the weather, working with precision following irregular changes in the weather requires high-precision tools that help make decisions. With the accelerated development of digitalization, the agricultural sector has had to adapt to the technological revolution, finding itself in full mutation thanks to modernization and innovation, which provides exponential growth of agricultural, livestock and agri-food products [21].

These imposed changes have shown once again the importance of having and knowing how to use all kinds of technologies, starting with the simplest and reaching the handling of big data or the algorithms of agricultural machinery, which is admitting that many countries and many areas (especially rural) succeed in this transition [21].

However, the Technology of Information and Communication, similarly to all innovations, has been imposed so rapidly in the last decade that it has not allowed the actors involved in agricultural activity and those responsible for the development of rural areas to have the necessary time to adopt it and adapt to it, creating a digital divide between various categories of people, especially within the rural population, mainly affecting women, the elderly and people who are unemployed or employed in precarious positions who do not usually have daily contact with ICTs [22,23]. This digital divide has been defined as the inability of certain groups to access and use ICTs due to the socioeconomic differences that exist between population groups or between countries since not all people have the sufficient economic level to buy electronic devices or pay access to these digital tools [24].

Traditionally, the existing literature on the subject distinguishes a first digital divide, referring to access to new technologies [25–29], and a second digital divide resulting from the use made of these new technologies [30–32]. To these two categories, a third digital divide is added, which refers to the quality of use of these ICTs [33–35].

Castaño [36] and Castaño et al. [37] specify that the first digital gender gap appears with access to technology and is quantitative in nature. The second is the use that is made of it and marks the degree of effective incorporation into it (of greater scope and of a qualitative nature). The third circumscribed to the use of the most advanced ICT services (also of a qualitative nature and of great importance for the evolution of the previous two).

To this end, Ferro et al. [38] identify three main approaches to understanding the digital divide: the access digital divide, the multi-dimensional digital divide, and the multi-perspective digital divide. The first type (digital access divide) observes the digital divide as a simple separation between "those who have" and "those who do not have" with attention to access to computer equipment or the Internet. The second type (multi-dimensional digital divide) perceives this phenomenon through a complex group of endogenous and exogenous factors involving specific groups of the population, such as stereotypes and traditional mentality, the assignment of roles by gender, the training gap, geographic isolation, etc. The third type (multi-perspective digital divide) reveals that no social group uses technologies in an inherently different way from others but recognizes that ICTs and the Internet are used to satisfy very specific objectives, often linked to their histories and social locations.

If this digital gap is significant in rural areas, it shows even more segregation between the urban population and the rural population [39–42]. In fact, Sevilla and Márquez [43] point out that despite Spain being a leading country in connectivity, there is a significant coverage gap in intermediate and high-quality networks (over 30 Mbps and over 100 Mbps), which exceeds 30 points (according to the values established by the Digital Economy and Society Index of the European Commission) between both zones. This is reflected in the degree of adoption of Internet access through the fixed network.

Indeed, in Spain, the rural environment represents 84% of the total area of the country but only comprises 16% of the total population [44], which makes the territories quite unpopulated. This depopulation means that the towns are generally disconnected, lacking a good Internet connection or with limited access, while others do not have the necessary ICT tools to connect, or if they do, they do not know how to use it [45–48]. That negatively affects families in rural environments who suffer from this digital divide, making it difficult for them to telecommute, access online education, or perform administrative tasks online.

Along these lines, Jimenez [49] indicates that the Spanish rural environment shows a development differential with respect to the urban environment, especially in those areas where the phenomenon of depopulation is stronger. This situation motivated the enactment of Law 45/2007 [50] for the sustainable development of rural areas, laying the foundations and measures for coordinated public action and comprehensive planning of rural areas. This is due, according to the author, to several factors, among them the depopulation and masculinization of the environment, the job insecurity suffered by this population, and the traditional mentality that limits the autonomy of women and their empowerment, due to the double shift that they must combine in this environment.

On the other hand, the Spanish General Union of Workers (UGT) [51] points out that the factors responsible for the differences in terms of ICT access and use between different population groups are summarized in low income, gender, advanced age, sparsely populated habitat, or low levels of training, which influences the integration of people in the information society.

Also, recognition and self-realization at work must be considered an essential element for access to ICTs, since, on many occasions, the assessment of work within the company entails greater responsibility and the use of more technology [51,52]. Starting from this context and considering the different factors that can affect digital divides, the following hypotheses have been considered in this study:

Hypothesis 1 (H_1). *The socioeconomic situation of rural users affects their access to ICT tools (First digital divide).*

This hypothesis has been nuanced by Korup and Sydlik [53], who associated the digital divide with the social context in which the ICT user finds himself and specified that this is linked to human capital (level, type of studies and job), the family context (income and family structure), and the social context (function of gender, age, cultural background, and place of residence).

Also, Mossberger et al. [54] and Helsper [55] point out that access to such technology is unequally distributed among individuals with different demographic characteristics, such as age, gender, socioeconomic level, ethnicity, and geographic location.

In fact, in our first hypothesis, three determining elements of the socioeconomic situation are considered, which are "age", "job position" and "well-being in their job position" since these variables determine whether users have been in direct contact with technology or if they have been in a situation of technological marginalization.

Several authors have commented on the close link that exists between the access and use of ICTs and the age of their users, including Tsai et al. [56] and Lee et al. [57] that indicate that age represents one of the main factors the digital divide, placing older people in the worst situations of access and use of ICTs. In this sense, Hernández [58] indicates that while 80% of young Spaniards between 16 and 24 years old have basic digital skills, only 35% of people between 55 and 74 years old have them.

Also, there is a very close link between the job position a person occupies and their degree of adoption of technologies [59–61]. In fact, being in continuous contact with technological tools and faced with the need to adapt to their daily use, in general, employees suffer less of a digital divide than unemployed people who do not have this opportunity to use ICTs frequently [61]. The unemployed are not the only ones who suffer from this digital divide; this is valid for operator positions in which ICTs are not usually used or little technology is used, given that in certain positions, the workers do not usually enjoy this training, and it is the which is why the well-being [62,63]. The importance perceived by the employee in the company is also considered. In this sense, the Infojobs report [64] indicates that 66% of the active population affirms that digital transformation and robotization have modified their job position and digital skills.

In addition to the above, UGT Communications [65] points out that the groups that suffer the most from technological inequalities in Spain are the unemployed and the inhabitants of the smallest population centers. The organization justifies it by the infrequent use that the unemployed make of ICT tools and the Internet and indicates that their digital skills are 15 percentage points lower than those used by employed people.

All this shows us that the socioeconomic conditions of users affect access to ICTs and their use and vice versa [66]. This is closely related to the economic means that the user has and that allow him to acquire ICT tools. This socioeconomic situation generally differs between urban and rural areas, where populations are isolated and have fewer job opportunities [67]. Esparza Chamba [66] indicates that the socioeconomic conditions in which ICTs are implemented can trigger new gaps in social inequality or widen existing ones. In addition, Ramírez and Sepulveda [68] and Gutierrez-Provecho et al. [69] assure that the digital divide is conditioned by the economic resources of its users. Therefore, the higher their job position, the more likely they are to have technological tools.

This hypothesis is closely related to the use of ICTs, since by having access to them, the skills of use can be developed, and this is what we are going to analyze in the following hypothesis.

Hypothesis 2 (H_2). *The access to technological tools (First digital divide) affects the achievement of basic abilities to use ICTs (Second digital divide).*

This hypothesis consists of determining if having access to ICTs, and basic skills of its use are automatically acquired or if they are two dissociable elements; in this study, two technological tools are mainly considered, which are the "laptop" and the "printer". The Internet element has not been considered in this case because it is a tool that does not always depend on the socioeconomic situation of the person and even more so in rural areas, since sometimes, even if the person is old enough to use the Internet or that they have a job that allows it because it is a necessary resource, there remains an element that depends on the coverage that does not always reach these areas [48,70,71]. Along these lines, Van Djik [72] indicates that the digital divide does not refer only to access to ICTs, but also to use since two people can have the same access to ICTs but not the same skills nor the same strategy in using them.

Chong [73] points out in this regard that when the digital divide is discussed as a problem of access to technology, it is forgotten that it is also necessary to use technology efficiently. Indeed, the use of technology inherently requires the development of skills that allow users to understand the processes by which information is sought and reached, ranging from turning on devices and connecting them to the Internet to the understanding of processes [74].

This led us to differentiate the use according to the different skills that a person may have of ICTs, considering first in this hypothesis the basic skills of turning on and off the computer, saving or modifying files, surfing the Internet, and writing a text email, and then to other more developed skills in the following hypotheses [75,76].

According to Ghobadi [77], the concept of "use" refers to the differential use of ICT applications daily. This could include both actual use of ICT and 'active versus passive use (Ghobadi [77] describes the active users of TICs as the ones who have a creative use of these tools, such as publishing a personal website, creating a weblog, posting a contribution on an online bulletin board, and newsgroup community; Contrary to the passive user, who does not usually have regular use of these technologies. This kind of use is largely linked to demographic characteristics of users and technical connections (e.g., social class, education, age, gender and ethnicity, the effectiveness of the connection, and the motivation to use ICT, material access, and having appropriate skills).). According to the author, the use is largely related to the demographic characteristics of the users, such as social class, age, gender, etc.

In this sense, Area [78] defines digital literacy for basic use in three dimensions; the first is instrumental and consists of having the skills to use hardware and software, such as "turning the computer on and off" or "use the Microsoft package", the second is cognitive and is summarized in the information search and analysis capabilities, such as "Internet surfing and looking for information", and the third is socio-communicational and deals with the abilities to express themselves and communicate through technology, such as "sending and receiving emails".

Other authors, such as Czerniewicz [79], indicate that there is a very strong correlation between the availability of ICTs and their use due to the frequency of access to them, but indicates that the high use of ICTs does not necessarily mean that there is a varied use of these ICTs. Moreover, this is what we are going to try to confirm in the following hypothesis:

Hypothesis 3 (H_3). *Access to technological tools affects the achievement of advanced skills in the use of ICTs (Second digital divide).*

In this hypothesis, the most advanced skills in the use of ICTs are considered, among them "computer configuration and hardware problem solving", "installation of a computer system", and "configuration of a computer program" [80,81]. The objective is whether the fact of making available to a user of a certain age and a certain job the necessary technological tools would allow him to have these advanced skills that define the second digital divide or if two people with different socioeconomic statuses but with exactly the same access, take very different advantage of this technology, as pointed out by the Organization for Economic Co-operation and Development (OECD) [82] and Pedró [83].

Jimenez et al. [84] categorize digital skills related to managing programs, installing devices, or using programming languages as more advanced and technical related to the management of specialized terminology (for example, plugins, cookies, etc.). In this sense, the Spanish Statistics National Institute (INE) [85] indicates that people who manage this task do not reach 40% of Internet users who have advanced digital skills (41.2% of men and 38.4% of women).

Calderón [86] identifies two types of competencies: the basic ones that include operational skills (management of technological devices and performance of basic tasks) and office automation skills (use of office automation software), mainly word processors and presentation programs, and advanced skills, linked to more complicated tasks, such as programming. Consequently, we formulate the following research hypothesis:

Hypothesis 4 (H_4). *Access to technological tools affects the achievement of operational/administrative skills in the use of ICTs (Second digital divide).*

This hypothesis is based on the existing correlation between access to ICT tools and the skills necessary to process or carry out daily tasks that are required in such a digitized world [87,88]. In this sense, the gap in the use of ICTs is defined as the lack of digital skills due to the lack of contact with ICTs or the lack of personal or professional training, which leaves a good part of the population outside the margin of certain services such as making medical appointments or carrying out administrative or financial procedures that improve the quality of life of the people who use them [24].

Also, Del Castillo [89] points out that the digital gap between active workers and unemployed people goes further and explores the data of the procedures with the Public Administrations through the Internet since, according to the INE [90], the unemployed have suffered more problems in carrying out these procedures than active workers. This has been justified by the lack of skills and knowledge since, according to this source, 83.3% of students have advanced ICT skills, compared to 50.5% of the employed and 32.2% of the unemployed.

Not accessing technology can prevent social insertion since it conditions the administrative management of many procedures (City Hall, Treasury, Social Security, hospitals, police, banks, etc.) and limits access to the facilities it offers the world of technology today, as is the case with online shopping [91–95]. Therefore, the two skills that have been evaluated at this point have been "management of an administrative file" and "management of a banking operation". To complete this work, the relationship of the first digital divide with the third is considered in the following hypothesis:

Hypothesis 5 (H_5). *Access to technological tools affects the achievement of professional skills in the use of ICTs (Third digital divide).*

In this hypothesis, a more professional aspect of the use of ICTs is considered, which is considered more specific to an activity or a task and requires specific training, as is the case of the skills of agricultural activity, and in this situation, two tasks have been selected for being the ones that use ICTs the most today: "remote control of crops" and "management of a geographic information system" [96–98].

The idea is to analyze whether, by having access to ICTs, farmers are more likely to develop their abilities to control the parameters of temperature, humidity, irrigation, and fertilization of crops remotely through the computer, which could facilitate the laborious work that is usually done in the field [99,100].

ICTs are focusing on this activity to create new functionalities that allow increasing, through better management, efficiency, and sustainability in the use of natural resources [98,101,102]. In recent years, agricultural yields have been increasing due to an increase in both the quantity and quality of marketed products [103–108].

Research aimed at reducing the agricultural digital divide also indicates that having access to ICT tools helps farmers in their daily work, but different strategies must be considered in the design of ICT tools so that they can use them correctly [109]. This strategy must be applied in the same way in the achievement of the different skills of use since the different use levels are related to each other, as we see in hypothesis 6:

Hypothesis 6 (H_6). *The basic ICT use skills are related to the other levels of useful skills.*

Hypothesis 6 considers the relationship between basic skills and other ICT use skills. It is about defining whether the degrees of advanced technological skills affect the degree of basic technological skills or not [110,111]. Indeed, developing increasingly high technological skills could improve the use made of technology in professional and personal life. For example, the fact of having to use technologies related to bank procedures or purchases over the Internet pushes citizens to learn basic skills since they would have to contact the bank by email to fix a problem or use an Excel form to find out the calculations for their accounts. The same happens with administrative procedures that sometimes require a more developed use of technology, as is the case of installing the digital signature on the computer, for example [69,112,113]. In the same way, the tasks of managing a geographic information system can be related to other basic tasks, such as surfing the Internet and searching for information [76,114]. In fact, Moeller et al. [115] indicate that learning ICTs allows greater inclusion in society since it is not only a powerful resource for learning but also an increasingly relevant tool for life.

Also, several authors, among them Van Deursen and Van Djik [116] and Hidalgo et al. [23], define the relationship between the different digital divides and indicate that people with more resources would obtain greater benefits from technology. This is due, according to them, to the high levels of resources and skills that allow the generation of higher levels of digital capital, which in turn favors an instrumental use of ICTs aimed at further increasing social, personal, economic, and political capital [117–122]. This theory is aligned with the hypothesis that access to technology plays the role of a moderating variable within the model that will be presented in the next section. Helsper [123] supports this assumption and indicates that higher levels of digital resources would correspond to a greater probability of avoiding potential adverse effects that may arise from the use of ICTs. In summary, Van Deursen et al. [118] indicate that different levels of resources correspond to different levels of skills that, in turn, generate different levels of involvement in technological activities and different levels of benefit from ICTs.

On the other hand, to check whether gender affects the different gaps, this variable that defines hypothesis 7 is analyzed:

Hypothesis 7 (H_7). *The gender variable affects the digital divide.*

If the previous bibliographic review has shown many inequalities that impede rural development, gender inequalities in this environment are even worse for women, who face a triple challenge: digital access, rurality, and gender. The digital divide depends on various sociodemographic characteristics, such as gender, which prevent women from enjoying the opportunities and benefits of digital transformation in the same way as men [124–129]. This is because women spend most of their time in unpaid activities and less in formal work, which means that they do not have the adequate financial capacity to access technological tools and are not faced with their daily use [105,130,131].

In this sense, Herrero [132] confirms that the digital gap between women and men that exists in society is more evident in rural areas since the use of these technologies is directly related to employment, which reduces the possibility of women due to the prioritization of their domestic activities, in which ICTs are not an indispensable resource. Indeed, in agriculture, women normally play a secondary role [133] in what could be called "the exploitation of the agrarian family". According to the Mundubat-CERES [134], only 32% of farm owners are women, although this percentage is gradually increasing.

These hypotheses lead us to the qualitative-quantitative study of analysis of the structural equations that are defined below in the methodology section.

3. Materials and Methods

To carry out this work and respond to the hypotheses raised, the qualitative and quantitative documentary method was used in two phases, which consisted of first analyzing the literature and the theoretical part related to the object of the study to acquire some basic knowledge and become familiar with the different concepts of the digital divide. To do this, articles, books, and publications related to digitization in rural and agricultural areas and generational and gender gaps have been reviewed. Secondly, the information and statistics related to the digital divide in Spain have been analyzed and compared.

For the study, as a primary source, outstanding data from the Spanish National Institute of Statistics (INE) were used, together with other statistical sources such as the Food and Agriculture Organization (FAO), the International Telecommunication Union (ITU), the World Bank and the Word Economic Forum. In addition, information was collected from different secondary sources for the consultation of scientific articles available in Internet databases and other works cited in the bibliography.

Once the data has been collected, the hypotheses related to the different gender gaps and the socioeconomic situation of the users have been defined. These hypotheses have constituted a basis that allowed us to carry out our second phase, which consisted of preparing a survey aimed at the population of the rural world to analyze the influence that the variables had between them. The questions have been prepared considering the socioeconomic factors that can create the digital gender gap in terms of access and use of ICTs.

Once the questionnaire has been configured in accordance with the requirements of the selected methodology, which consists of ensuring that it is reliable and capable of measuring without error the weight of the relationship between the variables and appropriate to the sample and the context, it was validated by experts from different areas related to this topic (Gender, agriculture, ICTs, sustainability and economic) and by the Ethics Commission of the University, this process was followed by a hybrid form. On the one hand, the surveys were carried out using in person in different Spanish rural areas, which required the displacement of the authors to different Spanish rural areas, and on the other hand, it was sent by email and disseminated through social networks. For this reason, different Spanish rural organizations, associations, and chambers of commerce have been contacted, which have kindly offered to send it to the contacts they had in their databases, share it on their web pages and social networks, and even do it in person during their events. This massive dissemination was intended to extend the survey in all Spanish communities and reach the target sample "Rural population" regardless of their socioeconomic status or economic activity; and the decision to do it in person was also motivated by the need to reach the elderly, people who live in isolated or unconnected areas, and people who do not have or do not know how to use technology in the same way as people who have access to ICTs and who can consult it remotely.

This survey, composed of 27 questions, was conducted in the period between 1 April 2022 and 25 July 2022, during which period 408 responses were obtained using the google form for the digital survey and the printed survey for the presential one, including 137 responses from men and 271 responses from women. This represents a sample greater than the minimum number of responses required (384) for the sample to be considered reliable and representative according to the "Sample Size Calculator" (www.surveysystem.com) (accessed on 6 April 2022).

The questionnaire was divided into 3 sections. The first section consisted of reflecting on the socioeconomic situation of the sample, considering the elements previously analyzed after reading the literature, such as gender, age, geographical area, and job position. The questions about the demographic and socioeconomic situation have been all closed.

The second section consisted of determining the gaps in terms of access and the questions related to these dichotomous variables and offered two options, Yes or No.

The last section, related to the use of ICTS, was made up of questions about the frequency of use of these tools, where the following options were proposed: daily, weekly, monthly, annually, and never.

The survey was sent and presented to the rural inhabitants regardless of their income level, gender, age, educational level, or job position. The only filter element that was considered was residing in a rural area. The objective was to obtain a wide and varied opinion of the rural population for maximum representativeness of the data. During the completion of this survey, the anonymity of the participants has been ensured, and the responses have been collected in accordance with the provisions of Article 40 of Law 3/2018 on data protection.

To analyze the data, the Structural Equations Model has been used through the PLS-SEM version 4 software. The selection of this tool is since it is the most appropriate to test the proposed theoretical model because it supports the simultaneous estimation of multiple relationships between the constructs [135–137]. In addition, the PLS-SEM allows the simultaneous evaluation of the reliability and validity of the measures of the theoretical constructs since it is designed mainly for causal-predictive analyses [138–143].

The constructs and items used in this study appear in Table 1 and have been determined after reviewing the literature.

Table 1. Constructs and scale ítems.

Construct/Indicator	Code	Scale Items	Source
	SSE		
Age	EEDAD	Age of rural people	
Job	EPTRAB	Professional activity	[34–45]
Importance at work	ETRABIMPL	You feel essential at work	
	AHI		
Access to technological tool 2	EDISHTEC2	Laptop availability	[34–45]
Access to technological tool 6	EDISHTEC6	Printer availability	
	UTN1		
Use of technology 1	EFRECMAN1	Turn on, turn off the computer	
Use of technology 2	EFRECMAN2	Use the Ms Office suite	
Use of technology 3	EFRECMAN3	Save and modify files	[46–51,72–78]
Use of technology 4	EFRECMAN4	Navigate in Internet	
Use of technology 5	EFRECMAN5	Write an e-mail	
	UTN2		
Use of technology 7	EFRECMAN7	Install a computer system	
Use of technology 8	EFRECMAN8	Set up the computer	[52–56,72–78]
Use of technology 17	EFRECMAN17	Set up a computer program	
	UTN3		
Use of technology 10	EFRECMAN10	Manage an administrative file	[17,54,58–63,72–78]
Use of technology 11	EFRECMAN11	Manage a banking operation	
	UTN4		
Use of technology 13	EFRECMAN13	Control a crop remotely	[48,64–78]
Use of technology 14	EFRECMAN14	Manage a geographic I.S.	

Nomenclature of Latent Variables (LV): SSE: Socioeconomic situation; AHI: Access to computer tools; UTN1: Basic technology use; UTN2: Advanced technology use; UTN3: Administrative/Operational technology use; UTN4: Professional technology use. Source: Own elaboration.

4. Results

To respond to the hypotheses raised above, and after collecting, cleaning, and coding the data, the different analyzes have been carried out, beginning by reflecting the socioeconomic composition of the selected sample, and separating the information by gender (Table 2).

Table 2. Sample Distribution.

	Male		Female		Total Sample	
	N	%	N	%	N	%
Age						
Between 18–25	10	7.30	11	4.06	21	5.15
Between 25–40	44	32.12	85	31.37	129	31.62
Between 40–55	60	43.76	105	38.75	165	40.44
Between 55–65	20	14.60	44	16.24	64	15.69
Over 65	3	2.19	26	9.59	29	7.10
Personal and family situation						
Married	73	53.28	138	50.92	211	51.71
Divorced	7	5.11	22	8.12	29	7.11
Single	55	40.15	103	38.01	158	38.73
Widower	2	1.46	8	2.95	10	2.45
Number of dependent family members						
None	66	48.18	125	46.13	191	46.81
Between 1–3	62	45.25	140	51.66	202	49.51
4 or more	9	6.57	6	2.21	15	3.68
Educational level						
Primary studies (School, college)	7	5.11	22	8.12	29	7.11
Secondary studies (Institute)	12	8.76	19	7.01	31	7.60
1st cycle vocational training	12	8.76	15	5.54	27	6.62
2nd cycle vocational training	21	15.33	36	13.28	57	13.97
Higher studies (University)	82	59.85	167	61.62	249	61.03
Training and specialization courses	2	1.46	8	2.95	10	2.45
No studies	1	0.73	4	1.48	5	1.23
Employment situation						
Salaried	77	56.20	157	57.93	234	57.35
Owner	33	24.09	46	16.97	79	19.36
Retired	2	1.46	23	8.49	25	6.13
Unpaid work outside the home	3	2.19	38	14.02	41	10.05
Unemployed	22	16.06	7	2.58	29	7.11
Total sample	137		271		408	

Source: Own elaboration.

Table 2 shows that the sample is very varied. In terms of age, the one that stands out is the one between 25–40 (Between 32.12% for men and 31.37% for women), and the one that predominates is the 40–55 years old (43.76% for men and 38.75% for women). Regarding the family situation, the married one stands out (53.28% for men and 50.92% for women). It is observed that in terms of dependents, in the case of men, the situation of not having dependents predominates, while in the case of women, 51.66% have responsibility for between 1 and 3 dependents. As far as educational levels are concerned, university studies stand out the most (59.85% for men and 61.62% for women). On the other hand, the employment situation shows that, in both cases, salaried workers prevail (56.20% for men and 57.93% for women).

To relate the variables to each other according to the hypotheses raised, we expose the structural model in a schematic way in the following figure (See Figure 1).

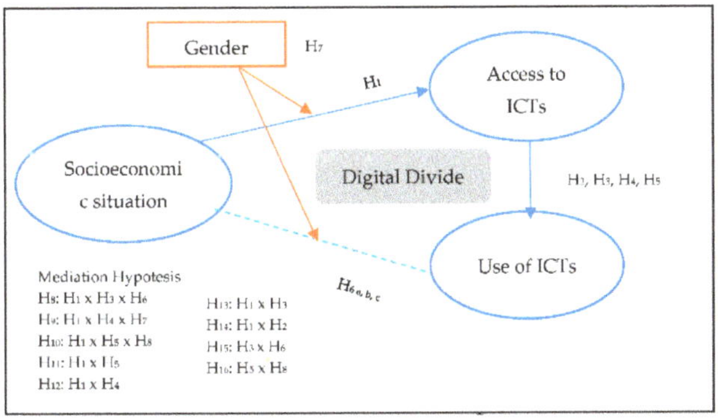

Figure 1. Research Model. **Source:** Own elaboration.

The analyzed data is reflected in Table 3, which represents the set of latent variables (LV which is reflected in the first column, with the different indicators that make it up. In addition, in Appendix A (Table A1), you can consult the table of the Covariances. Where:

- μ: average of the indicator with its values Min: minimum value reached by the indicator and Max: maximum value reached by the indicator; SD: standard deviation; SL: Standardized loadings.
- Convergent validity and reliability of latent variables are defined by: AVE: Average variance extracted; AC: Cronbach's Alpha; and CR: Composite reliability.

Table 3. Descriptive statistics of the model.

		μ	min	máx	SD	SL	Q^2	AVE	CA	CR
LV	EGEN	1.34	0	1	0.47					
SSE								0.56	0.58	0.78
	EEDAD	2.88	1	5	0.97	0.54				
	EPTRAB	3.75	1	8	2.45	0.84				
	ETRABIMPL	1.73	1	3	0.92	0.82				
AHI							0.11	0.65	0.47	0.79
	EDISHTEC2	1.15	1	2	0.36	0.86	0.12			
	EDISHTEC6	1.34	1	2	0.47	0.76	0.09			
UTN1							0.16	0.62	0.84	0.89
	EFRECMAN1	1.74	1	5	1.43	0.82	0.17			
	EFRECMAN2	2.06	1	5	1.56	0.84	0.16			
	EFRECMAN3	1.90	1	5	1.46	0.86	0.17			
	EFRECMAN4	1.40	1	5	1.10	0.62	0.15			
	EFRECMAN5	2.04	1	5	1.62	0.76	0.12			
UTN2							0.01	0.63	0.70	0.83
	EFRECMAN7	3.10	1	5	1.20	0.85	0.02			
	EFRECMAN8	3.25	1	5	1.21	0.84	0.02			
	EFRECMAN17	3.30	1	5	1.17	0.68	0.01			
UTN3							0.04	0.67	0.53	0.80
	EFRECMAN10	2.87	1	5	1.48	0.91	0.06			
	EFRECMAN11	3.25	1	5	1.64	0.71	0.02			
UTN4							0.002	0.68	0.52	0.81
	EFRECMAN13	3.74	1	5	0.86	0.81	0.002			
	EFRECMAN14	3.44	1	5	1.20	0.83	0.002			

Source: Own elaboration.

To determine the predictive capacity of the model, the Q^2 values obtained from the Stone-Geisser test have been included in the table applying blindfolding of both the constructs (VL) and the validated indicators, and all of them are positive [144].

Table 3 shows that of the three indicators that make up the socioeconomic situation construct (SSE), the variable with the most weight is "The job (EPTRAB)" and "The importance at work (ETRABIMPL). In terms of access to computer tools (AHI), the one with the most weight is "The laptop availability (EDISHTEC 2)". On the other hand, in the basic technology use (UTN1) it is observed that the one that has more weight is "Turn on, turn off the computer (EFRECMAN 1), "Use the Microsoft Office package (EFRECMAN2)" and "Save and modify files (EFRECMAN3)". In the advanced technology use (UTN2), "Install a computer system and configure the computer (EFRECMAN 7)" and "Solve hardware problems (EFRECMAN 8)" predominate in this construct. While in terms of administrative/operational technology use (UTN3), it is "Manage a computer administrative file (EFRECMAN 10)" the one that has more weight. As far as professional technology use is concerned (UTN4), "Managing a geographic information system (EFRECMAN 14)" is the most important, although the difference is very low with Controlling a crop remotely (EFRECMAN 13). This is reflected in Figure 2 below:

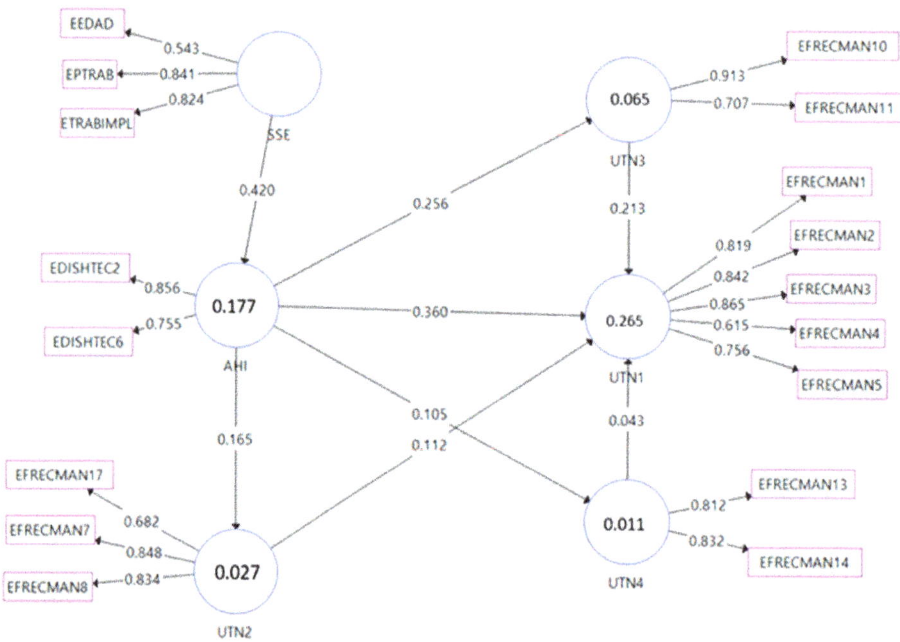

Figure 2. Structural model. Source: Own elaboration.

Graph 2 shows that the economic situation construct shows a great impact on access to ICTs. This, in turn has quite an effect on the first level of use of ICTs (UTN1) and secondarily on the second level of use (UTN2). Regarding the incidence of access to technologies on the third level of use (UTN3) and the fourth level of use (UTN4), it is less than in the two previous uses.

On the other hand, the graph highlights that, among the different uses of technologies, the one that has the greatest influence is the third level of use (UTN3) over the first level of use (UTN1).

To detail the degree of significance of all the constructs of the latent variables, the descriptive statistics are presented below in Table 4:

Table 4. Convergent validity and reliability.

		β	t	lo95%	hi95%	VIF	f^2
H_1	SSE → AHI	0.42	8.08 ***	0.33	0.50	1.00	0.21
H_2	AHI → UTN1	0.36	7.31 ***	0.28	0.44	1.08	0.17
H_3	AHI → UTN2	0.17	3.50 ***	0.08	0.23	1.00	0.03
H_4	AHI → UTN3	0.26	6.29 ***	0.19	0.32	1.00	0.07
H_5	AHI → UTN4	0.11	2.35 ***	0.02	0.17	1.00	0.01
H_{6a}	UTN2 → UTN1	0.11	2.57 ***	0.04	0.18	1.18	0.01
H_{6b}	UTN3 → UTN1	0.21	5.02 ***	0.14	0.28	1.17	0.05
H_{6c}	UTN4 → UTN1	0.04	0.98	−0.03	0.11	1.07	0.002

Note: β: Path coefficients; t = (β/SD); *p*-value: degree of significance *** $p < 0.01$; lo95% low confidence interval; hi95% high confidence interval: f^2: f squared. R^2: UTN1(B):0.265; UTN2(C): 0.027; UTN3 (D): 0.065; UTN4 (E): 0.011; AHI (F): 0.177. Estimated model fit: SRMR: 0.104; d_ULS:1.664; d_G: 0.367; Chi square: 896.868; NFI:0.570. Source: Own elaboration.

In Table 4, it is observed that all the paths have a high degree of significance except the fourth level of use (UTN4)-> the first level of use (UTN1) path, which is also reflected in the collinearity, where there is no interference between the constructs when acquiring a value below of 3, which validates the model. Regarding f^2, the Socioeconomic Situation (SSE) → Access to computer tools (AHI), and Access to computer tools (AHI) → the first level of use (UTN1) constructs contributes to a better fit of each path.

Also, it has been verified that the values of loads of the different indicators are greater in their own latent variable than in the rest.

Next, in Table 5, a discriminant validation table of the internal model is presented, where the values of the Fornell-Larcker test are shown, which consists in verifying that the correlations between latent variables are smaller than the square root of the Average Variance Extracted (AVE). It has been completed with the HeteroTrait-MonoTrait ratio test (HTMT) to show that the latent variables are sufficiently different from each other and do not measure the same concept twice [145].

Table 5. Discriminant validity.

		A	B	C	D	E	F
A	SSE	**0.81**	*0.80*	*0.69*	*0.29*	*0.48*	*0.21*
B	UTN1	0.42	**0.75**	*0.47*	*0.38*	*0.39*	*0.28*
C	UTN2	0.44	0.33	**0.78**	*0.32*	*0.50*	*0.19*
D	UTN3	0.17	0.24	0.25	**0.79**	*0.50*	*0.42*
E	UTN4	0.26	0.22	0.34	0.32	**0.82**	*0.16*
F	AHI	0.11	0.15	0.13	0.24	0.09	**0.82**

Source: Own elaboration.

The following Table 6 presents the results of mediation, which occurs when a third variable or construct intervenes between two related constructs.

Table 6. Mediation (Regardless of gender).

	Specific Indirect Effects	μ	SD	t	lo95%	hi95%
H_8	SSE → AHI → UTN2 → UTN1	0.01	0.004	1.88 ***	0.002	0.02
H_9	SSE → AHI → UTN3 →UTN1	0.02	0.01	3.42 ***	0.01	0.04
H_{10}	SSE → AHI → UTN4→ UTN1	0.002	0.002	0.86	−0.001	0.01
H_{11}	SSE → AHI → UTN4	0.05	0.02	2.10 ***	0.01	0.08
H_{12}	SSE → AHI → UTN3	0.11	0.02	4.42 ***	0.07	0.15
H_{13}	SSE → AHI → UTN2	0.07	0.02	2.88 ***	0.03	0.11
H_{14}	SSE → AHI → UTN1	0.16	0.03	4.08 ***	0.10	0.21
H_{15}	AHI → UTN2 → UTN1	0.02	0.01	2.12 ***	0.01	0.04
H_{16}	AHI → UTN4 → UTN1	0.004	0.01	0.90	−0.002	0.01

Note: *p*-value: degree of significance *** $p < 0.01$. Source: Own elaboration.

Table 6 shows that there is a mediation of the Socioeconomic Situation (SSE) to the first level of use (UTN1) through Access to Computer Tools (AHI) and the second level of use (UTN2). The same happens with the hypothesis. H_9, H_{11}, H_{12}, H_{13}, H_{14} and H_{15}. As for H_{10} and H_{16}, this mediation is not significant.

To determine whether gender, due to its dichotomous nature, presents a significant difference between the groups made up of the female gender and the male gender, the moderation study reflected in Table 7 has been carried out, which is also shown has added the difference in behavior between the rural population in general, and the population that is dedicated to agricultural activity [145,146].

Table 7. Moderation Statistics.

	Women		Men		Permutation Mean Differences (W–M)			
Path	β_W		β_M		$\beta_{(W-M)}$	IC [5–95%]	β	IC [5–95%]
	Rural	Agrarian	Rural	Agrarian	Rural		Agrarian	
SSE → AHI	0.50	0.32	0.22	(−0.00)	−0.007 ***	[−0.18–0.18]	0.006	[−0.44–0.47]
AHI → UTN1	0.48	0.31	0.21	0.05	−0.007 ***	[−0.18–0.17]	0.002	[−0.36–0.35]
AHI → UTN2	0.28	0.14	(−0.21)	(−0.29)	−0.002	[−0.17–0.17]	0.002 **	[−0.41–0.40]
AHI → UTN3	0.33	0.28	0.05	(−0.08)	−0.000 ***	[−0.14–0.15]	0.003 **	[−0.31–0.34]
AHI → UTN4	0.15	0.18	(−0.01)	0.02	−0.005 **	[−0.15–0.15]	−0.003	[−0.46–0.46]
UTN2 → UTN1	0.004	0.06	0.29	0.41	−0.001 ***	[−0.16–0.16]	0.004 **	[−0.27–0.29]
UTN3 → UTN1	0.19	0.18	0.24	0.17	0.002	[−0.15–0.15]	0.001	[−0.27–0.28]
UTN4 → UTN1	0.04	0.12	0.07	0.16	−0.003	[−0.16–0.16]	−0.002	[−0.30–0.31]
N	271	61	137	55	*** $p < 0.001$	** $p < 0.05$	* $p < 0.1$	
	μ(W–M)		*p*-value		σ		*p*-value	
MICON Analysis	Rural	Agrarian	Rural	Agrarian	Rural	Agrarian	Rural	Agrarian
SSE	0.349	0.012	0.001	0.513	0.004	0.007		0.375
AHI	0.100	−0.189	0.178	0.166	0.012	0.010	0.032	0.462
UTN1	0.042	−0.358	0.354	0.028	0.008	0.009	0.332	0.080
UTN2	0.321	0.324	0.001	0.044	0.005	0.002	0.160	0.344
UTN3	−0.009	−0.094	0.458	0.313	0.004	0.003	0.378	0.447
UTN4	0.094	0.051	0.187	0.410	0.009	0.005	0.080	0.415

Note: *p*-value: degree of significance *** $p < 0.01$; ** $p < 0.05$; * $p < 0.1$.

The results of Table 7 show a degree of significance in the difference in means between women and men in the rural world compared to the agrarian world. This appears in Socioeconomic Situation (SSE) → Access to computer tools (AHI), Access to computer tools (AHI) → Basic technology use (UTN1), Access to computer tools (AHI) → Administrative/operational technology use (UTN3), and Advanced technology use (UTN2) → Basic technology use (UTN1), all four in favor of men in the rural world. The different questions in the agrarian world in which they appear with a positive sign, although with a lower degree of significance: Access to computer tools (AHI) → Advanced technology use (UTN2), Access to computer tools (AHI) → Administrative/operational technology use (UTN3), Advanced technology use (UTN2) → Basic technology use (UTN1) in favor of women.

To determine the effect of moderation, the calculation of the measurement invariance of composite models (MICOM) has been applied in Table 7, which completes the study of moderation with the results of this analysis that reflect that there is a partial measurement invariance in the world rural and a complete measure invariance in the agricultural sector.

5. Discussion

The results obtained support the first six hypotheses (H_1: The socioeconomic situation of rural users affects the access to ICT tools, H_2: The access to technological tools affects the achievement of basic abilities to use ICTs, H_3: The access to technological tools affects the achievement of advanced skills in the use of ICTs, H_4: The access to technological tools affects the achievement of operational/administrative skills in the use of ICTs, H_5: The access to technological tools affects the achievement of professional skills in the use of ICTs, H_{6a}: the Advanced technology use affects the Basic technology use, H_{6b}: The Administrative technology use affects the basic technology use), except H_{6c} (The professional technology use affects the basic technology use), as it does not have a degree of significance. This is explained by the absence of a link between the professional skills specific to an agricultural activity with the other basic, advanced, or administrative skills. Also, it has been observed how the socioeconomic situation affects the first digital gap and, consequently, the second and third. This indicates that age, job position and the fact of feeling essential in their work represent important elements in situations of digital divides since not all generations have the same access to and use of ICTs. In this sense, it can be affirmed that many of the so-called digital natives (Digital native refers to young people exposed to Communication and Information Technologies since their birth according to Salas Delgado [147]) [148] are familiar with technology, but in addition to accessing it, it is necessary that they have digital literacy, since, as Pérez-Rodríguez and Delgado [149] put it, "having information does not automatically produce knowledge", since "transforming information into knowledge requires reasoning skills to organize, relate, analyze, synthesize and make inferences" and deductions of different levels of complexity; in short, understand it and integrate it into previous knowledge schemes".

Serrano and Martínez [150] and Moreno Gálvez [151] point out in this sense that the digital divide is not only related to exclusively technological aspects but also, therefore, a reflection of a combination of socioeconomic factors.

On the other hand, it has been seen how access to ICTs has affected the different degrees of use. This is explained, as indicated in the theoretical part, by the ease of learning that is achieved once the tool is made available. In this sense, Ndou [152] states that there is a positive correlation between access to ICTs and their use in rural areas.

Regarding the relationship between the different types of use, it has been observed that advanced technology use (UTN2) and administrative/operational technology use (UTN3) have a significant incidence on basic technology use (UTN1), which is not the case with the professional technology use (UTN4). This is because they are tasks closely related to the basic management of ICTs. Without knowing how to turn the computer on and off, you cannot, for example, install a computer program or manage a purchase or an administrative procedure.

Regarding the gender issue, it has been observed that this variable has a direct effect on the digital divide, sometimes in favor of men and other times in favor of women, depending on whether we are referring to the rural world or to agricultural activity, which confirms hypothesis H_7. Indeed, the results have shown that the differences in means have a significant impact on the hypothesis: H_1, H_2, H_4 and H_5 and with less significance on the hypothesis H_6. At the same time, they go in favor of women in the agricultural sector in the hypothesis: H_3, H_4, and H_6. This is justified by the greater involvement of women in activities in the agricultural sector, including rural tourism activities, which require a high level of involvement and the fact that, in other activities, women do not have this contact with the ICTs, due to the traditional mentality of rural areas, which conditions them to assume more family responsibilities, as indicated in Table 2. In reality, women continue to be the most affected by disproportionately assuming unpaid care, a burden that has increased due to the closure of schools, preventive isolation in homes, and the need for emotional support from other family members in the face of the uncertainty caused by the pandemic [153], an element also observed in Table 2, where 14.02% of women indicate working in unpaid activity, compared to only 2.19% of men. Sometimes, it is due to stereotypes that make women unwilling to use ICTs, which causes a reaction of rejection and a phobia towards technology. Indeed, different investigations show that men have more positive attitudes toward computers and more stereotyped attitudes regarding who can use them [154]. Additionally, women experience more computer-related anxiety than men and generally have lower levels of achievement around information technology [155]. Although the gender gap in physical access shows a decreasing trend [156], men use ICTs more than women due to higher prior exposure to technology and work-related requirements [157].

On the other hand, it is important to point out that the economic autonomy of rural women could influence their access and use of ICTs since economic resources greatly influence the acquisition of these tools and the training that can be done to manage them [158]. In this sense, the report of the Digital Economy and Society index indicates that only 37% of adults receive training on a regular basis, which generates a lack of basic digital skills [159].

In short, it has been seen that the digital divide in the rural world is due to a host of injustices of different categories, all of which are related, in one way or another, to a lack of training. In this regard, Bennett [160] indicates that, in rural areas, they have not received a formal education, and it is likely that they will not use a computer in their entire lives. This lack of use has a strong impact on the economic, social, health and community reality of the countryside because access to ICTs is, above all, the central tool for promoting personal, collective, and productive development.

6. Conclusions

According to Estefanía [161], living conditions have improved in the last century more than in the rest of human history. However, it is the period of history in which inequalities of all kinds are highest: economic, gender, educational, labor, generational, technological, digital, etc.

Indeed, the results of this work have shown that there are differences in the access and use of ICTs in rural Spain. This is due to several socioeconomic factors such as age, job position and the responsibility that each person has in their job.

On the other hand, it has been observed in this study that the gender variable affects the different digital gaps since men and women do not have the same abilities to use ICTs.

One of the consequences of these digital divides is the isolation of the rural population, which prevents them from enjoying the same training and employment opportunities as the urban population, in addition to the disadvantages in the telecommunications infrastructure that they suffer due to the distance from the main cities and the low population density.

This situation highlights how essential it is to have a public intervention to close the digital gaps in the rural world [162], especially among women and men, as well as the elderly and young people, since these gaps represent a real impediment to the rural environment to develop in the same way as the urban world, and this despite the natural resources it enjoys. In this sense, Otero et al. [163] argue that equalizing the access and use of ICTs would allow the improvement of food security and contribution to the achievement of a competitive, inclusive, and sustainable agriculture, which leads to food self-sufficiency among all the inhabitants of this environment. For this reason, it is essential to value the benefit that all the inhabitants of the rural world (even the elderly) can have from technological training.

Regarding the gender aspect, it is important to design and implement effective and useful public policies for the empowerment of women living in rural areas and the development of their digital potential, as well as their involvement in decision-making [163–167].

On the other hand, it is necessary to highlight the need to eliminate stereotypes and gender roles in the rural world, either in terms of education or employment, and to make visible the role of women in this environment [168].

This work has some limitations since it focuses only on the rural world. This leads us to consider the idea of continuing to investigate the digital gender gap between the urban and rural worlds as a future perspective, to see the existence of the gap between both areas. The intergenerational digital divide could also be analyzed in more detail, comparing the differences between both media, and examining their economic resources.

Author Contributions: Conceptualization, H.K. and M.D.d.-M.G.; methodology, M.D.d.-M.G.; software, M.D.d.-M.G.; validation, M.D.d.-M.G.; S.B.; M.F.R.C.; formal analysis, H.K. and M.D.d.-M.G.; investigation, H.K.; resources, H.K. and M.D.d.-M.G.; data curation, H.K. and M.D.d.-M.G.; writing—original draft preparation, H.K.; writing—review and editing, H.K.; S.B.; M.F.R.C.; and M.D.d.-M.G.; visualization, H.K. and M.D.d.-M.G.; supervision, M.D.d.-M.G. and S.B. All authors have read and agreed to the published version of the manuscript.

Funding: This research is funded by Greenworal Project: Grant Agreement nº: 2021-1-ES01-KA220-ADU-000033740 and by the AgriCamBio project: Grant PID2020-114576RB-I00 funded by MCIN/AEI/10.13039/501100011033.

Institutional Review Board Statement: Before the launch of the survey, it was sent to the Cartagena University Ethics Committee for approval to guarantee data protection and the anonymity of the people surveyed.

Acknowledgments: This article is a tribute to the rural woman, on the occasion of their world day. The authors thank all the people who have participated in this survey and essentially all the associations promoting the rural world that have contributed to the dissemination of the survey.

Conflicts of Interest: The authors declare no conflict of interest.

Appendix A

Table A1. Correlation of the observed variables.

		A	B	C	D	E	F	G	H	I	J	K	L	M	N	O	P	Q
A	EEDAD	1.000																
B	EPTRAB	0.117	1.000															
C	EDISHTEC2	0.334	0.248	1.000														
D	EDISHTEC6	0.126	0.303	0.308	1.000													
E	EFRECMAN1	0.205	0.233	0.345	0.263	1.000												
F	EFRECMAN2	0.209	0.189	0.292	0.210	0.645	1.000											
G	EFRECMAN3	0.234	0.193	0.299	0.247	0.651	0.756	1.000										
H	EFRECMAN4	0.300	0.152	0.342	0.307	0.374	0.312	0.362	1.000									
I	EFRECMAN5	0.093	0.222	0.177	0.265	0.500	0.545	0.577	0.380	1.000								
J	EFRECMAN7	0.180	0.155	0.204	0.012	0.152	0.204	0.218	0.078	0.109	1.000							
K	EFRECMAN8	0.219	0.060	0.166	0.064	0.136	0.166	0.187	0.198	0.159	0.568	1.000						
L	EFRECMAN10	0.113	0.175	0.232	0.176	0.252	0.279	0.300	0.203	0.293	0.266	0.265	1.000					
M	EFRECMAN11	0.043	0.130	0.134	0.091	0.134	0.198	0.122	0.114	0.222	0.151	0.173	0.357	1.000				
N	EFRECMAN13	0.208	0.069	0.056	0.069	0.081	0.059	0.095	0.099	0.095	0.118	0.126	0.082	0.051	1.000			
O	EFRECMAN14	0.141	0.035	0.093	0.060	0.101	0.124	0.063	0.060	0.040	0.123	0.135	0.066	0.032	0.352	1.000		
P	EFRECMAN17	0.199	0.067	0.155	-0.036	0.164	0.197	0.209	0.052	0.042	0.409	0.332	0.173	0.148	0.266	0.211	1.000	
Q	ETRABIMPL	0.119	0.703	0.223	0.260	0.183	0.171	0.167	0.088	0.160	0.152	0.101	0.145	0.175	0.076	0.018	0.123	1.000

References

1. Solórzano, F.; García, A. Fundamentos del aprendizaje en red desde el conectivismo y la teoría de la actividad. *Rev. Cuba. Educ. Super.* **2016**, *35*, 98–112.
2. Vasselle, L. Les Agritechs ou l'essor du Digital Dans l´agriculture. DigitalCorner. 2018. Available online: https://bit.ly/3SG61R8 (accessed on 30 August 2022).
3. Maffezzoli, F.A.; Ardolino, M.; Bacchetti, A. The Impact of the 4.0 Paradigm in the Italian Agricultural Sector: A Descriptive Survey. *Appl. Sci.* **2022**, *12*, 9215. [CrossRef]
4. Jellason, N.; Robinson, E.; Ogbaga, C. Agriculture 4.0: Is Sub-Saharan Africa Ready? *Appl. Sci.* **2021**, *11*, 5750. [CrossRef]
5. Fox, G.; Mooney, J.; Rosati, P.; Lynn, T. AgriTech Innovators: A Study of Initial Adoption and Continued Use of a Mobile Digital Platform by Family-Operated Farming Enterprises. *Agriculture* **2021**, *11*, 1283. [CrossRef]
6. Chaudhary, S.; Suri, P. Agri-tech: Experiential learning from the Agri-tech growth leaders. *Technol. Anal. Strateg. Manag.* **2022**, 1–14. [CrossRef]
7. Spanaki, K.; Sivarajah, U.; Fakhimi, M.; Despoudi, S.; Irani, Z. Disruptive technologies in agricultural operations: A systematic review of AI-driven AgriTech research. *Ann. Oper. Res.* **2020**, *308*, 491–524. [CrossRef]
8. Alcon, F.; Tapsuwan, S.; Martínez-Paz, J.M.; Brouwer, R.; de Miguel, M.D. Forecasting déficit irrigation adoption using a mixed stakeholder assessment methodology. *Technol. Forecast. Soc. Chang.* **2014**, *83*, 183–193. [CrossRef]
9. Sharif, A.M.; Irani, Z. Policy making for global food security in a volatile, uncertain, complex and ambiguous (VUCA) world. *Transform. Gov. People Process Policy* **2017**, *11*, 523–534. [CrossRef]
10. UN: United Nations. The Sustainable Development Goals Report 2016. United Nations Publications. 2017. Available online: https://bit.ly/3St3OJ5 (accessed on 2 September 2022).

11. Pérez, A.; Leyva, D.A.; Gómez, F.C. Desafíos y propuestas para lograr la seguridad alimentaria hacia el año 2050. *Rev. Mex. Cienc. Agrícolas* **2018**, *9*, 175–189. [CrossRef]
12. Zárate, A.H.; Miranda, G.A. Impacto del cambio climático en la seguridad alimentaria en zonas campesinas vulnerables de los Andes del Perú. *Rev. Mex. Cienc. Agrícolas* **2016**, *7*, 71–82. [CrossRef]
13. Turrent-Fernández, A.; Cortés-Flores, J.I.; Espinosa-Calderón, A.; Turrent-Thompson, C.; Mejía-Andrade, H. Cambio climático y algunas estrategias agrícolas para fortalecer la seguridad alimentaria de México. *Rev. Mex. Cienc. Agrícolas* **2016**, *7*, 1727. [CrossRef]
14. FAO. Promoting Sustainable Land Management Practices Is Critical to Improve People's Livelihoods. 2020. Available online: https://bit.ly/3SSfoxg (accessed on 2 September 2022).
15. OECD. Adoption of Technologies for Sustainable Farming Systems. *Wageningen Workshop Proceedings*. 2001. Available online: http://bit.ly/3TuTHmW (accessed on 11 November 2022).
16. Geodde, L.; Katz, J.; Ménard, A.; Revellat, J. *Agriculture ´s Connected Future: How Technology Can Yield New Growth*; McKinsey & Company: Washington, DC, USA, 2020; Available online: http://bit.ly/3UKYUYI (accessed on 11 November 2022).
17. Mehrabi, Z.; McDowell, M.J.; Ricciardi, V.; Levers, C.; Martinez, J.D.; Mehrabi, N.; Wittman, H.; Ramankutty, N.; Jarvis, A. The global divide in data-driven farming. *Nat. Sustain.* **2021**, *4*, 145–160. [CrossRef]
18. Lythreatis, S.; Sign, S.K.; El-Kassar, A.N. The digital divide: A review and future research agenda. *Technol. Forecast. Soc. Chang.* **2022**, *175*, 121359. [CrossRef]
19. Saeed, S.A.; Masters, R.M. Disparities in health care and the digital divide. *Curr. Psychiatry Rep.* **2021**, *23*, 1–6. [CrossRef]
20. Iberdrola. La Brecha Digital en el Mundo y por qué Provoca Desigualdad. 2022. Available online: https://bit.ly/3M4EF53 (accessed on 1 September 2022).
21. Cherif, E. Quand le Numérique Révolutionne l´Agriculture. *Agroligne*. 2019, 112, pp. 1–68. Available online: https://bit.ly/3qDe2KG (accessed on 5 September 2022).
22. Grishchenko, N. The gap not only closes: Resistance and reverse shifts in the digital divide in Russia. *Telecommun. Policy* **2020**, *44*, 102004. [CrossRef]
23. Hidalgo, A.; Gabaly, S.; Morales-Alonso, G.; Urueña, A. The digital divide in light of sustainable development: An approach through advanced machine learning techniques. *Technol. Forecast. Soc. Chang.* **2020**, *150*, 119754. [CrossRef]
24. Santander. *La Brecha Digital: Qué es y Cómo Podemos Reducirla*; Santander Universidades: Santander, Columbia, 2021; Available online: https://bit.ly/3RLa9OO (accessed on 2 September 2022).
25. Rogers, E.M. The digital divide. *Convergence* **2001**, *12*, 96–111. [CrossRef]
26. Wilson, K.R.; Wallin, J.S.; Reiser, C. Social Stratification and the Digital Divide. *Soc. Sci. Comput. Rev.* **2003**, *21*, 133–143. [CrossRef]
27. Stewart, C.M.; Gil-Egui, G.; Tian, Y.; Pileggi, M.I. Framing the digital divide: A comparison of US and EU policy approaches. *New Media Soc.* **2006**, *10*, 731–751. [CrossRef]
28. Livingstone, S.; Helsper, E. Gradations in digital inclusion: Children, young people and the digital divide. *New Media Soc.* **2007**, *8*, 671–696. [CrossRef]
29. Morales Martín, J.J. La tercera brecha digital: Estratificación social, ciencia e tecnología. In Proceedings of the VI Congreso Portugués de Sociología: Mundos Sociais: Saberes e Prácticas, Lisboa, Portugal, 25–28 June 2008; Universidad Nova de Lisboa, Facultad de Ciencia Sociales e Humanas: Lisboa, Portugal, 2008.
30. Attewell, P. The first and second digital divides. *Sociol. Educ.* **2001**, *74*, 252. [CrossRef]
31. Sassi, S. Cultural differentiation or social segregation? Four approaches to the digital divide. *New Media Soc.* **2005**, *10*, 684–700. [CrossRef]
32. Jamil, S. From digital divide to digital inclusion: Challenges for wide-ranging digitalization in Pakistan. *Telecommun. Policy* **2021**, *45*, 102206. [CrossRef]
33. Ragneda, M. *The Third Digital Divide: A Weberian Approach to Digital Inequalities*; Routledge: New York, NY, USA, 2017.
34. López-Montero, R.; Delgado-Baena, A.; Vela-Jiménez, R.; Slanes, A. Digital borders: The impact of the digital divide in the educational process of minor migrants living in marginalized urban areas. In Proceedings of the INTED2021 Conference, Online, 8–9 March 2021.
35. Morales Romo, N. El Reto de la Brecha Digital y Las Personas Mayores En El Medio Rural Español. El Caso De Castilla y León. Fonseca. *J. Commun.* **2016**, *13*, 165–185. [CrossRef]
36. Castaño, C. *La Segunda Brecha Digital*; Ediciones Cátedra: Madrid, Spain, 2008.
37. Castaño, C.; Martín Fernández, J.; Vásquez Cupeiro, S. La e-inclusión y el bienestar social—Una perspectiva de género. *Rev. Econ. Ind. Retos Y Opor. Para El Sist. Product. Español* **2008**, *367*, 139–152.
38. Ferro, E.; Helbig, N.C.; Gil-Garcia, J.R. The role of IT literacy in defining digital divide policy needs. *Gov. Inf. Q.* **2011**, *28*, 3–10. [CrossRef]
39. Lembani, R.; Gunter, A.; Breines, M.; Dalu, M.; Mwazvita, T.B. The same course, different access: The digital divide between urban and rural distance education students in South Africa. *J. Geogr. High. Educ.* **2020**, *44*, 70–84. [CrossRef]
40. Li, Y.; Ranieri, M. Educational and social correlates of the digital divide for rural and urban children: A study on primary school students in a provincial city of China. *Comput. Educ.* **2013**, *60*, 197–209. [CrossRef]
41. Hindman, D.B. The rural-urban digital divide. *J. Mass Commun. Q.* **2000**, *77*, 549–560. [CrossRef]

42. Ye, L.; Yang, H. From digital divide to social inclusion: A Tale of Mobile Platform Empowerment in Rural Areas. *Sustainability* **2020**, *12*, 2424. [CrossRef]
43. Sevilla, J.; Márquez, J.M. Brecha Entre Analógicos y Digitales. In *Colección Brechas Sociales*; Fundación "La CAIXA": Valencia, Spain, 2021; Available online: https://bit.ly/3RMciuD (accessed on 7 September 2022).
44. Agroinformación. Radiografía de la Realidad de la España Rural: Solo Representa el 16% de la Población, Pero Ocupa el 84% del Territorio. 2021. Available online: https://bit.ly/3e3Fej1 (accessed on 7 September 2022).
45. Pontones-Rosa, C.; Pérez-Morote, R.; Santos-Peñalver, J.F. Ict-Based Public Policies and Depopulation in Hollowed-Out-Spain: A survey analysis on the digital divide and citzen satisfaction. *Technol. Forecast. Soc. Chang.* **2021**, *169*, 1–14. [CrossRef]
46. Esteban-Navarro, M.A.; García-Madurga, M.A.; Morte-Nadal, T.; Nogales-Bocio, A.I. The rural digital divide in the fase of the COVID-19 Pandemic in Europe-Recommendations from a scoping review. *Informatics* **2020**, *7*, 54. [CrossRef]
47. Pérez-Morote, R.; Pontones-Rosa, C.; Núñez-Chicharro, M. Determinant factors of individuals´decision to emigrate in rural Spain: The role of ICT-based public policies. *Technol. Soc.* **2021**, *67*, 101777.
48. Ruiz-Martínez, I.; Esparcia, J. Internet Access in rural áreas: Brake or stimulate as Pst-Covid 19 opportunity. *Sustainability* **2020**, *12*, 9619. [CrossRef]
49. Jimenez, M. Coste de oportunidad de la brecha de género en el mundo rural. Women for a healthy economy. *Closinggap Inf.* **2022**, *12*. Available online: https://bit.ly/3qIe58c (accessed on 5 September 2022).
50. BOE. Ley 45/2007, de 13 de Diciembre, Para el Desarrollo Sostenible del Medio Rural. 2007. Available online: https://bit.ly/3SOMKNO (accessed on 19 September 2022).
51. UGT. La brecha digital en España: Estudio sobre la desigualdad postergada. 2015. Available online: https://bit.ly/3RM5YCv (accessed on 8 September 2022).
52. Olarte Encabo, S. Brecha digital, pobreza y exclusión social. *Rev. Andal. De Trab. Y Bienestar Soc.* **2017**, *138*, 285–313.
53. Korupp, S.; Szydlik, M. Causes and Trends of the Digital Divide. *Eur. Sociol. Rev.* **2005**, *21*, 409–422. [CrossRef]
54. Mossberger, K.; Tolbert, C.; Stansbury, M. *Virtual Inequality: Beyond the Digital Divide*; Georgetown University Press: Hoboken, NJ, USA, 2003.
55. Helsper, E. Gendered Internet use across generations and life stages. *Commun. Res.* **2010**, *37*, 352–374. [CrossRef]
56. Tsai, H.S.; Shillair, R.; Cotten, S.R.; Winstead, V.; Yost, E. Getting Grandma Online: Are Tablets the Answer for Increasing Digital Inclusion for Older Adults in the US? *Educ. Gerontol.* **2015**, *41*, 695–709. [CrossRef]
57. Lee, H.Y.; Kanthawala, S.; Choi, E.Y.; Kim, Y.S. Rural and non-rural digital divide persists in older adults: Internet access, usage, and attitudes toward technology. *Gerontechnology* **2021**, *20*, 1–9. [CrossRef]
58. Hernández, R. ¿Cómo Reducer la Brecha Digital Entre Trabajadores de Distintas Generaciones? *HRTRENDS by Infoempleo*. 2022. Available online: https://bit.ly/3CxJtwK (accessed on 7 September 2022).
59. Wong, Y.C.; Kwong, C.; Chu Fung, J.Y.; Ping Lee, V.W. Digital divide and social inclusion: Policy challenge for social development in Hong Kong and South Korea. *J. Asian Public Policy* **2010**, *3*, 37–52. [CrossRef]
60. Neagu, G.; Berigel, M.; Lendzhova, V. How digital inclusion increase opportunities for young people: Case of Neets from Bulgaria, Romania and Turkey. *Sustainability* **2020**, *13*, 7894. [CrossRef]
61. Vasilescu, M.D.; Serban, A.C.; Dimian, G.C.; Aceleanu, M.I.; Picatoste, X. Digital divide, skills and perceptions on digitalisation in the European Union—Towards a smart labour market. *PLoS ONE* **2020**, *15*, e0232032. [CrossRef] [PubMed]
62. Genz, S.; Schnabel, C. *Digging into the Digital Divide: Workers' Exposure to Digitalization and Its Consequences for Individual Employment*; IZA Institute of Labor Economics: Bonn, Germany, 2021; Available online: http://bit.ly/3Tz4A78 (accessed on 11 November 2022).
63. Van Djik, J. *The Digital Divide*; Polity Press: Cambridge, UK, 2020.
64. Infojobs. Brecha en la Digitalización: La Mitad de los Trabajadores Reconocía Antes de la Covid No estar Formándose en Competencias Sobre Nuevas Tecnología. 2022. Available online: https://bit.ly/3Ecjpsl (accessed on 8 September 2022).
65. UGT Comunicaciones. La Brecha Digital se Ceba Con los Desempleados y la España Vaciada. 2021. Available online: https://bit.ly/3yjpPlB (accessed on 10 September 2022).
66. Esparza Chamba, K.J. La Brecha Digital Aumenta las Desigualdades Socioeconómicas. 2021. Available online: https://bit.ly/3CxuGSM (accessed on 2 November 2022).
67. Kayumova, M. The role of ICT regulations in Agribusiness and rural development. In *Enabling the Business of Agriculture*; World Bank Group: Washington, DC, USA, 2018.
68. Ramírez, L.A.; Sepúlveda, J.J. Brecha digital e inclusión digital: Fenómenos socio-tecnológicos. *Rev. Univ. EIA* **2018**, *15*, 89–97.
69. Gutiérrez-Provecho, L.; López-Aguado, M.; García Llamas, J.L.; Quintanal Díaz, J. La brecha digital en población en riesgo de exclusión social. *Rev. Interuniv.* **2021**, *39*, 123–138. [CrossRef]
70. Díaz Chang, B.; Ayala, D. Red de alta velocidad que permite la cobertura de acceso a internet en parroquias rurales de América Latina. *J. Bus. Entrep. Stud.* **2020**, *4*, 19.
71. Domínguez Álvarez, J.L. Internet y nuevas tecnologías como punta de lanza para la revitalización de territorios rurales despoblados. La necesaria reconstrucción de la idea de servicio público. *Rev. Digit. De Derecho Adm.* **2021**, *26*, 91–124. [CrossRef]
72. Van Dijk, J.A.G.M. *The Network Society, Social Aspects of New Media*; Sage: London, UK; Thousand Oaks, CA, USA; New Delhi, India, 1999.
73. Chong, B. La agenda digital en México. In *Retos y Perspectivas*; Fontamara S.A.: Barcelona, Spain, 2014.

74. Díaz Navarrete, A.K. Propuesta de elementos para diseñar una estrategia de alfabetización tecnológica. In *En Vías del E-Gobierno*; Universidad Autónoma del Estado de México, Facultad de Ciencias Políticas y Sociales: Toluca, Mexico, 2019.
75. Aydin, M. Does the digital divide matter? Factors and conditions that promote ICT literacy. *Telemat. Inform.* **2021**, *58*, 101536. [CrossRef]
76. Song, Z.; Wang, C.; Bergmann, L. China´s prefectural digital divide: Spatial análisis and multivariate determinants of ICT diffusion. *Int. J. Inf. Manag.* **2020**, *52*, 102072. [CrossRef]
77. Ghobadi, S.; Ghobadi, Z. How Access Gaps Interact and Shape Digital Divide: A cognitive Investigation. *Behav. Inf. Technol.* **2013**, *34*, 330–340. [CrossRef]
78. Area, M. *La Competencia Digital e Informacional en la Escuela*; Universidad Internacional Menéndez y Pelayo: Santander, Spain, 2009.
79. Czerniewicz, L. *ICT Access-User-Use Relationships in Teaching/Learning in Higher Education in South Africa*; University of Cape Town: Cape Town, South Africa, 2012.
80. Perifanou, M.A.; Economides, A. Gender Digital Divide in Europe. *Int. J. Bus. Humanit. Technol.* **2020**, *10*, 7–14. [CrossRef]
81. Seung-Yoon, S.; Dongwook, K.; Soon, A.C. Digital Divide in Advanced Smart City Innovations. *Sustainability* **2021**, *13*, 4076.
82. OECD. Are the New Millennium Learners Making the Grade? In *Technology Use and Educational Performance in PISA*; OECD Publishing: Paris, France, 2010.
83. Pedró, F. Tecnología y escuela: Lo que funciona y por qué. In *XXVI Semana Monográfica de la Educación*; La Educación En La Sociedad Digital; Fundación Santillana: Madrid, Spain, 2011.
84. Jiménez, R.; Vega, L.; Vico, A. Habilidades en internet de mujeres estudiantes y su relación con la inclusión digital: Nuevas brechas digitales. *Educ. Knowl. Soc.* **2016**, *17*, 29–48. [CrossRef]
85. INE. *Encuesta Sobre Equipamiento y Uso de Tecnologías de Información y Comunicación en Los Hogares*; INE: Madrid, Spain, 2019.
86. Calderón, D. Jóvenes y Desigualdad Digital: Las Brechas de Acceso, Competencias y Uso. Centro Reina Sofía Sobre Adolescencia y Juventud. 2020. Available online: https://bit.ly/3yok3iH (accessed on 12 September 2022).
87. Oreku, G.S. Adopting the ICT Innovation to Administrative and Activity Procedures in a University. *Interdiscip. J. Educ. Res.* **2021**, *3*, 60–73. [CrossRef]
88. Abraham-Ibe, I.G. Information and Communication Technology (ICT) and improved method of Office Management/Administration. *Afr. Sch. J. Mgt. Sci. Entrep.* **2021**, *23*, 199–214.
89. Del Castillo, C.; La Brecha Digital que no se Cierra: Más de un 20% de Españoles no Busca en Internet ni Manda Emails. El Diario del 1 de Diciembre del 2020. Available online: https://bit.ly/3SQogDw (accessed on 13 September 2022).
90. INE. Encuesta Sobre Equipamiento y Uso de Tecnologías de Información y Comunicación en los Hogares Año. 2020. Available online: https://bit.ly/3Cb6T9M (accessed on 18 September 2022).
91. Vázquez, S.; Castaño, C. La brecha digital de género: Prácticas de e-inclusión y razones de la exclusión de las mujeres. *Asparkía* **2011**, *22*, 33–49.
92. Qureshi, S.; Davis, A. Overcoming the digital divide through electronic commerce: Harnessing opportunities in IT for development. In Proceedings of the 40th Hawaii International Conference on System Sciences–2007, Washington, DC, USA, 3–6 January 2007.
93. Liu, L.; Wu, F.; Tong, H.; Hao, C.; Xie, T. The Digital Divide and Active Aging in China. *Int. J. Environ. Res. Public Health* **2021**, *18*, 12675. [CrossRef]
94. Díaz Dávila, E.M. Brecha digital de género en la gestión administrativa. In *Trabajo de Fin de Grado*; Universidad Técnica del Norte, Facultad Ciencias Administrativas y Económicas: Ibarra, Ecuador, 2020.
95. Criado, I.; Barrero, D. Estudio de la Brecha Digital y la Demanda de Administración Electrónica. In Proceedings of the XI Congreso de la Asociación Española de Ciencia Política y de la Administración Sevilla, Análisis Empírico del Caso Español. Sevilla, Spain, 18–20 September 2013; AECPA: Sevilla, Spain.
96. Rware, H.; Kansiime, M.K.; Oyango, D.; Tambo, J.A.; Mloza Banda, C.; Phiri, N.A.; Chipabika, G.; Matimelo, M.; Kabuya Chaaba, D.; Davis, T.; et al. Is radio an effective method for delivering actionable information for responding to emerging pest threats? A case study of fall armyworm campaign in Zambia. *CABI Agric. Biosci.* **2021**, *2*, 32. [CrossRef]
97. Ceballos, F.; Kannan, S.; Kramer, B. *Assessing Feasibility and Effects of Personalized Remote Advisories Based on Smartphone Pictures: A Formative Evaluation in India*; CCAFS Working Paper No 406; CGIAR Research Program on Climate Change; Agriculture and Food Security (CCAFS): Wagenigen, The Netherlands, 2021.
98. Vitali, G.; Francia, M.; Golfarelli, M.; Canavari, M. Crop Management with the IoT: An interdisciplinary survey. *Agronomy* **2021**, *11*, 181. [CrossRef]
99. Beaumont, A.M. Cultivos Inteligentes: Un Programador Permite su Control a Distancia. Es Diario, 2022. Available online: https://bit.ly/3CbGuZm (accessed on 8 October 2022).
100. Rubio, I. La Tecnología que Controla los Cultivos a Partir de Imágenes Tomadas por Satélites. El país, 2017. Available online: https://bit.ly/3CFxc9I (accessed on 8 October 2022).
101. Giraldo Cerón, A.F. Tan cerca y tan lejos de la agricultura 4.0 en Colombia. *Rev. Univ. EAFIT* **2020**, *55*, 78–85.
102. Barrachina Aurell, A. Agricultura de Precisión, Nuevas Tecnologías y Sostenibilidad en la Producción Hortícola. Ph.D. Thesis, Ingeniería Agroambiental y del Paisaje, Barcelona, Spain, June 2020.
103. Ballantyne, P.; Maru, A.; Porcari, E.M. Information and Communication Technologies—Opportunities to Mobilize Agricultural Science for Development. *Crop Sci.* **2010**, *50*, S-63–S-69. [CrossRef]

104. Xin, J.; Zazueta, F.S.; Vergot, P.; Mao, X.; Kooram, N.; Yang, Y. Delivering knowledge and solutions at your fingertips: Strategy for mobile app development in agriculture. *Agric. Eng. Int. CIGR J. Special issue* **2015**, 317–325.
105. Rubio, M. Innovación Y Desarrollo Sostenible: El Papel De Las Tic En La Agricultura Del Medio Rural Remoto. *Rev. Delos* **2019**, *12*, 5245.
106. Cañadas, J.; Del Águila, I.M.; Palma, J. Un sistema web de ayuda a la decisión para control de plagas. In Proceedings of the II Simposio Nacional de Ingeniería Hortícola, Almeira, Spain, 10–12 February 2016; Automatización y TICs en Agricultura: Almería, Spain, 2016.
107. Fountas, S.; Sørensen, C.G.; Tsiropoulos, Z.; Cavalaris, C.; Vatsanidoud, A. Farm management information systems: Current situation and future perspectives. *Comput Electron. Agric.* **2015**, *115*, 40–50. [CrossRef]
108. Monali, T.; Payal, J.; Komal, R.; Prachi, S.; Dipak, B. Android Based Solution for Indian Agriculture Management A Design Paper. *Int. J. Curr. Trends Eng. Res.* **2016**, *2*, 17–19.
109. Bacco, M.; Barsocchi, P.; Ferro, E.; Gotta, A.; Ruggeri, M. The Digitisation of Agriculture: A Survey of Research Activities on Smart Farming. *Array* **2019**, *3*, 100009. [CrossRef]
110. López Baldominos, I.; Fernández Sanz, L.; Pospelova, V. Análisis de las competencias digitales básicas en Europa y en España. *Actas De Las Jenul.* **2020**, *5*, 77–84.
111. Garay Sanchez, M.D. Estilos de Aprendizaje y Habilidades Digitales de los Estudiantes de Ingenieria de Sistemas de una Universidad Privada de Lima Norte. Ph.D. Thesis, Académico de Maestro en Docencia Universitaria, Perú, Lima, 2021.
112. Menéndez Sebastián, E.; Ballina Díaz, J. Digital Citizenship: Fighting the digital divide. *Eur. Rev. Digit. Adm. Law* **2021**, *2*, 149–155.
113. Barrabés Escartín, M.; Kissová, M. La Evolución de la Notificación en las Administraciones Públicas: El Problema de la Brecha Digital. Ph.D. Thesis, Universidad de Zaragoza, Zaragoza, Spain, 2021.
114. King, S. GIS a bridge across the digital divide: Engaging participatory methods to build capacity in research communities. *Pract. Anthropol.* **2021**, *43*, 35–40. [CrossRef]
115. Moeller, S.; Joseph, A.; Lau, J.; Carbo, T. *Towards Media and Information Literacy Indicators*; UNESCO: Paris, France, 2011; Available online: https://goo.gl/04syAg (accessed on 8 October 2022).
116. Van Deursen, A.J.; Van Dijk, J.A. Measuring internet skills. *Int. J. Hum. Comput. Interact* **2010**, *26*, 891–916. [CrossRef]
117. Gómez, C. The Third Digital Divide and Bourdieu: Bidirectional Conversion of Economic, Cultural, and Social Capital to (and from) Digital Capital among Young People in Madrid. *New Media Soc. D* **2020**, *23*, 2534–2553. [CrossRef]
118. Van Deursen, A.J.A.M.; Helsper, E.; Eynon, R.; Van Dijk, J.A.G.M. The Compoundness and Sequentiality of Digital Inequality. *Int. J. Commun.* **2017**, *11*, 452–473.
119. Ragnedda, M.; Ruiu, M.L. Social Capital and the Three Levels of Digital Divide. In *Theorizing Digital Divides*; Ragnedda, M., Muschert, G.W., Eds.; Routledge: London, UK, 2017; pp. 27–40.
120. Helsper, E.J.; Van Deursen, A.J.A.M. The Third-Level Digital Divide: Who Benefits Most from Being Online? In *Communication and Information Technologies Annual 2015*; Robinson, L., Shelia Cotten, R., Schulz, J., Hale, T.M., Williams, A., Eds.; Howard House: Felixstowe, UK, 2015; Volume 10, pp. 29–52.
121. Ragnedda, M. Conceptualizing digital capital. *Telemat. Inform.* **2018**, *35*, 2366–2375. [CrossRef]
122. De Marco, S. El Comercio Electrónico En España (2019): Un Ejemplo De Tercera Brecha Digital. *Rev. Int. Sociol. RIS* **2022**, *80*, e206. [CrossRef]
123. Helsper, E.J. *The Digital Disconnect*; Sage Publications: London, UK, 2021.
124. Sey, A.; Hafkin, N. *Taking Stock: Data and Evidence on Gender Equality in Digital Access, Skills, and Leadership*; United Nations University Institute on Computing and Society/International Telecommunications Union: Macau, China, 2019; pp. 1–341.
125. Gauld, R.; Goldfinch, S.; Horsburgh, S. Do they want it? Do they use it? The "Demand-Side" of e-government in Australia and New Zealand. *Gov. Inf. Q.* **2010**, *27*, 177–186. [CrossRef]
126. Reddick, C. Citizen interaction and e-government. Evidence for the managerial, consultative, and participatory models. *Transform. Gov. People Process Policy* **2011**, *5*, 167–184.
127. Reddick, C.G.; Turner, M. Channel choice and public service delivery in Canada: Comparing e-government to traditional service delivery. *Gov. Inf. Q.* **2012**, *29*, 1–11. [CrossRef]
128. Al Rababah, B.; Abu-Shanab, E. E-government and gender digital divide: The case of Jordan. *Int. J. Electron. Bus. Manag.* **2010**, *8*, 1–8.
129. Choi, Y.-T.; Park, S. Understanding gender inequality in central egovernment: A Korean case study. *Gov. Inf. Q.* **2013**, *30*, 300–309. [CrossRef]
130. Aguirre, R. Procesos de empobrecimiento y desigualdades de género. In *Desafíos Para la Medición*; Universidad de la Republica de Uruguay: Montevideo, Uruguay, 2014.
131. Pivoto, D.; Waquil, P.D.; Talamini, E.; Finocchio, C.P.S.; Dalla Corte, V.F.; De Vargas Mores, G. Scientific development of smart farming technologies and their application in Brazil. *Inf. Process. Agric.* **2018**, *5*, 21–32. [CrossRef]
132. Herrero Pulgar, S. La brecha digital de género en el ámbito rural (Frómista, Palencia). In *Trabajo fin de grado (Profesional en Educación Social)*; Escuela Universitaria de Educación de Palencia: Madrid, Spain; Universidad de Valladolid: Valladolid, Spain, 2012; 46p.
133. Contzen, S.; Forney, J. Family farming and gendering division of labour on the move: A typology of farming-family configurations. *Agric. Hum. Values* **2017**, *34*, 27–40. [CrossRef]

134. Mundubat-CERES. Derechos Económicos, Sociales y Culturales de las Mujeres en el Medio Rural: Una Aproximación Crítica. 2020. Available online: https://bit.ly/3rK13Hy (accessed on 12 September 2022).
135. Sarstedt, M.; Ringle, C.M.; Hair, J.F. *Partial Least Squares Structural Equation*; Springer Nature: Cham, Switzerland, 2021.
136. Hair, J.F.; Ringle, C.M.; Sarstedt, M. PLS-SEM: Indeed a silver bullet. *J. Mark. Theory Pract.* **2011**, *19*, 139–151. [CrossRef]
137. Hair, J.F.; Hult, G.T.; Ringle, C.M.; Sarstedt, M.; Danks, N.P.; Ray, S. *Partial Least Squares Structural Equation Modeling (PLS-SEM) Using R. Business*; Springer: Berlin/Heidelberg, Germany, 2021; ISBN 978-3-030-80518-0.
138. Ruiz, M.A.; Pardo, A.; San Martín, R. Modelos de ecuaciones estructurales. *Pap. Del Psicólogo* **2010**, *31*, 34–45.
139. Escobedo Portillo, M.T.; Hernández Gómez, J.A.; Estebané Ortega, V.; Martínez Moreno, G. Modelos de Ecuaciones Estructurales: Características, Fases, Construcción, Aplicación y Resultados. *Cienc. Trab.* **2016**, *55*, 16–22. [CrossRef]
140. Cepeda-Carrion, G.; Cegarra-Navarro, J.-G.; Cillo, V. Tips to use partial least squares structural equation modelling (PLS-SEM) in knowledge management. *J. Knowl. Manag.* **2019**, *23*, 67–89. [CrossRef]
141. Leguina, A. A primer on partial least squares structural equation modeling (PLS-SEM). *Int. J. Res. Method Educ.* **2015**, *38*, 220–221. [CrossRef]
142. Shiau, W.L.; Sartstedt, M.; Hair, J.F. Internet research using partial least squares structural equation modeling (PLS-SEM). *Internet Res.* **2019**, *29*, 398–406. [CrossRef]
143. Hair, J.; Alamer, A. Partial Least Squares Structural Equation Modeling (PLS-SEM) in second language and education research: Guidelines using an applied example. *Res. Methods Appl. Liguistics* **2022**, *1*, 100027. [CrossRef]
144. Casado, G.; Fernández-Gámez, M.A.; Aldeanueva, I.; Ruiz-Palomo, D. Competitive Intelligence and Sustainable Competitive Advantage in the Hotel Industry. *Sustainability* **2019**, *11*, 159.
145. Ruiz-Palomo, D.; León-Gómez, A.; García-Lopera, F. Disentangling organizational commitment in hospitality industry: The roles of empowerment, enrichment, satisfaction and gender. *Int. J. Hosp. Manag.* **2020**, *90*, 102637. [CrossRef]
146. Ruiz-Palomo, D.; Fernández-Gámez, M.A.; León-Gómez, A. Analyzing the Effect of Financial Constraints on Technological and Management Innovation in SMEs: A Gender Perspective. *J. SAGE* **2022**, *12*, 21582440221079925. [CrossRef]
147. Salas Delgado, M.V. *Convergence between Digital Natives and Digital Immigrants*; Universidad Católica de Santa María (UCSM): Santa María, CA, USA, 2020; Available online: http://bit.ly/3G9bUmU (accessed on 12 November 2022).
148. Bennett, S.; Maton, K.; Kervin, L. The "digital natives" debate: A critical review of the evidence. *Br. J. Educ. Technol.* **2008**, *5*, 775–786. [CrossRef]
149. Pérez-Rodríguez, M.A.; Delgado, A. De la competencia digital y audiovisual a la competencia mediática: Dimensiones e indicadores. *Comunicar* **2012**, *39*, 25–34. [CrossRef]
150. Serrano, A.; Martínez, E. *La Brecha Digital: Mitos y Realidades*; Departamento Editorial Universitaria de la Universidad Autónoma de Baja California: Nueva, Mexico, 2003.
151. Moreno Gálvez, F.J. Ciudadanía y apropiación tecnológica: Herramientas para el análisis de las prácticas comunicativas de los sujetos en torno a las nuevas tecnologías. In Proceedings of the En Move.Net, Actas del I Congreso Internacional Move.Net Sobre Movimientos Sociales y Tic, Sevilla, Spain, 4–7 February 2015; pp. 251–264.
152. Ndou, A.S. Relationship between access to ICT and the use of electronic library resources by scholars and postgraduate students in a rural-based South African university. *S. Afr. J. Libr. Inf. Sci.* **2021**, *87*, 42–50. [CrossRef]
153. Cediel, N.M.; Sánchez, M.J.; Sánchez, K.D.; Castro, P.M. Pandemia de la COVID-19: Un obstáculo para el logro de la equidad de género y el cierre de la brecha entre lo urbano y lo rural en Colombia. *Rev. De La Univ. De La Salle* **2021**, *84*, 122–145. [CrossRef]
154. Huffman, A.; Whetten, J.; Huffman, W. Using technology in higher education: The influence of gender roles on technology self-efficacy. *Comput. Hum. Behav.* **2013**, *29*, 1779–1786. [CrossRef]
155. Dixon, L.; Correa, T.; Straubhaar, J.; Covarrubias, L.; Graber, D.; Spence, J.; Rojas, V. Gendered space: The Digital divide between male and female users in internet public access sites. *J. Comput. Mediat. Commun.* **2014**, *19*, 991–1009. [CrossRef]
156. Poushter, J. Sma Rtphone Ownership and Internet Usage Continues to Climb in Emerging Economies. 2016. Available online: http://s1.pulso.cl/wp-content/uploads/2016/02/2258581.pdf (accessed on 1 October 2022).
157. Zillien, N.; Hargittai, E. Digital distinction Status-specific types of internet usage. *Soc. Sci. Q.* **2009**, *90*, 274–291. [CrossRef]
158. Martínez-Cantos, J.L.; Castaño, C. La brecha digital de género y la escasez de mujeres en el ámbito TIC. *Panor. Soc.* **2017**, *25*, 49–65.
159. ITUSER. La Comisión Propone que 2023 sea el Año Europeo de las Capacidades. 2022. Available online: https://bit.ly/3Tclmth (accessed on 17 October 2022).
160. Bennett, N. La red rusa de información rural. In *Secreto a Voces Radio, NTICs e Interactividad (Editado por Bruce Girard)*; Organización de las Naciones Unidas para la Agricultura y la Alimentación: Roma, Italy, 2004.
161. Estefanía, J. *La Cara Oculta de la Prosperidad: Economía Para Todos*; Taurus: Madrid, Spain, 2003.
162. Jung, M.C.; Park, S.; Lee, J.Y. Information Network Villages A community-focused digital divide reduction policy in rural Korea. *Aust. J. Telecommun. Digit. Econ.* **2014**, *2*, 98–112. [CrossRef]
163. Otero, M.; Cabrol, M.; Polastri, R.; Monden, C. *Desigualdad Digital de Género en América Latina y el Caribe*; Instituto Interamericano de Cooperación para la Agricultura: San José, Costa Rica, 2022.
164. Lekhanya, L.M. The Digitalisation of Rural Entrepreneurship. In *Entrepreneurship—Trends and Challenges*; Brito, S.M., Ed.; Books on Demand GmbH: Norderstedt, Germany, 2018; pp. 36–63.

165. Agu, M.N. Application of ICT in Agricultural Sector: Women's Perspective Monica. *Int. J. Soft Comput. Eng.* **2013**, *2*, 58–60.
166. Díez-Gutiérrez, E.J.; Rodríguez-Rejas, M.J. Retos y propuestas para repoblar la España vaciada: Por una España Repoblada. *Cuad. Desarro. Rural* **2021**, *18*, 1–19.
167. Villela, F.; Contreras, D.S. La brecha digital como una nueva capa de vulnerabilidad que afecta el acceso a la educación en México. *Acad. Y Virtualidad* **2021**, *14*, 169–187. [CrossRef]
168. Ramirez, F.; Hernandez, L.; Gutierrez, I.G.R.; Padilla, D. *La Perspectiva de Género en los Procesos de Desarrollo Comunitario y Sostenible*; Centro Agronómico Tropical de Investigación y Enseñanza: Turrialba, Costa Rica, 2012.

agriculture

Article

Vibration Analysis of a Roller Bearing Condition Used in a Tangential Threshing Drum of a Combine Harvester for the Smooth and Continuous Performance of Agricultural Crop Harvesting

Shankar Bhandari * **and Eglė Jotautienė**

Department of Agricultural Engineering and Safety, Agriculture Academy, Vytautas Magnus University, LT-53362 Akademija, Lithuania
* Correspondence: shankar.bhandari@vdu.lt; Tel.: +370-6355-4571

Abstract: Testing the reliability of the threshing unit is difficult and thus often neglected before the harvesting season, which can result in breakdown maintenance during peak harvesting time in difficult-to-access areas for sensor mounting. In this paper, the vibration analysis of the threshing condition of the combine harvester was performed by introducing the bracket for inaccessible locations. The Adash A4900 Vibrio M analyzer (Adash spol. s.r.o., Ostrava, Czech Republic) was used for a vibration signal measurement and the DDS Adash software was used for signal processing. The demodulated fast Fourier transform (FFT) root mean square (RMS) (500 Hz–16 kHz) method was used to evaluate the bearing condition and DDS Adash Fault Source Identification Tool (FASIT) technology was used to evaluate other mechanical conditions such as the looseness, misalignment, and unbalance of the threshing unit of the Massey Ferguson series of combine harvesters. Modal and random vibration analyses were performed on the bracket and components and compared to prevent the resonance phenomenon using the Ansys Software (Ansys, Inc., Canonsburg, PA, USA). The constrained modal analysis of the threshing drum was performed to observe the deformation. Decent results were obtained from the method used. The method was again validated by the tangential threshing test bench and successfully determined the bearing fault condition. The method used is an uncomplicated and effective way of performing the bearing analysis of the tangential unit of the combine harvester.

Keywords: combine harvester; tangential threshing; vibration analysis; FFT demodulation; random vibration analysis; bearing fault; machine condition

Citation: Bhandari, S.; Jotautienė, E. Vibration Analysis of a Roller Bearing Condition Used in a Tangential Threshing Drum of a Combine Harvester for the Smooth and Continuous Performance of Agricultural Crop Harvesting. *Agriculture* **2022**, *12*, 1969. https://doi.org/10.3390/agriculture12111969

Academic Editors: Muhammad Sultan, Redmond R. Shamshiri, Md Shamim Ahamed and Muhammad Farooq

Received: 21 October 2022
Accepted: 16 November 2022
Published: 21 November 2022

1. Introduction

Threshing is one of the most important parameters to complete the harvesting process which is time-sensitive [1]. The grain damage and grain loss are consistent significant parameters of the threshing mechanism which can be monitored by vibration analysis for optimal performances [2]. A discharge beater is one of the essential tools constituting the threshing mechanism which is adjoined to the outlet and selectively rotates in response to the feeding rate. It is arranged to transfer the crop remains from the threshing drum and conveyor arrangement and is responsible for the net production of grains outlets [3]. The partial load is the major parameter that determines the unbalanced vibration on a tangential threshing unit which disturbs the working accuracy of the machine and the comfort of the operator [4].

The mode of the cylinder used, the loads on the threshing units, and the frequency range of measurement play vital roles in the vibration analysis of the threshing unit of a combine harvester [5]. Ji Jangtoa studied the interaction between the threshing unit and the plant materials, concluding that the amplitude of the vibration signal experienced the maximum elevation of amplitude on the horizontal radial axis (feeding direction)—more than that of the vertical axis during the loading threshing condition [6]. The sensor mounted on the vertical radial axis encountered maximum vibration amplitude and the sensor mounted at 45° with the horizontal radial axis is capable of displaying the vibration from the vertical and horizontal directions combined under no-load conditions [7].

Zhong analyzed the comparative study of the idling state of the no-load and loading states of the tangential threshing and shaft vibration by solving the modalities of a tangential threshing cylinder. It was observed that the rotation speed frequency would not induce the occurrence of the resonance phenomenon and the vibration amplitude increases 22–25 times with a loading condition compared to the no-load condition [8]. Investigations were made into the swirl features and instability of a combine harvester's thresher under loading conditions as it is subjected to stalk winding. The stalks looped around the thresher significantly impacts the stability and swirling properties of the thresher–stalk system [9]. Xiang made a comparative study of machine learning, ensemble learning, and deep learning with his approach to the fast Fourier transform (FFT) with the decimation-in-time (DIT) and XGBoost algorithm to quickly and accurately identify the fault type of bearing [10].

Zhou detected the bearing defect by collectively using signal collective processing techniques including self-adaptive noise cancellation (SANC), kurtogram, and envelope analysis [11]. To enable a more effective calculation in impulse signal analysis, online bearing vibration detection and analysis were carried out using the enhanced fast Fourier transform algorithm, which is based on a straightforward arithmetic operation [12]. The selection of the suitable demodulation frequency band is seen as an important and demanding stage in bearing fault diagnostics. It decides whether envelope analysis may be used to derive fault information from the demodulated signal [13].

If bearing defects exist, the measured vibration signal would be amplitude modulated at its characteristic defect frequency. The modulating wave is a pass vibration signal corresponding to local defects. Wang applied an envelope demodulation method based on the Hilbert transform to extract the characteristics of defect frequencies. He obtained the FFT after the Hilbert transform for the envelope spectrum [14].

Wide vibration analysis techniques and their applications are very popular in automobile industries [15–18], however, these have few implementations concerning the threshing unit of combine harvesters. Most vibration analyses are conducted in the cabin area of the harvester and focus on driver safety and comfort [19,20], simulated vibration analysis [21], and under laboratory conditions [22]. However, the vibration analysis techniques of the threshing unit's health monitoring are mostly under laboratory conditions. The vibration analysis performed for the threshing unit of a combine harvester includes the complexity of instrumentation due to its complex structure, as the components need to be dismantled to obtain the raw vibration signal for processing. To avoid such complexities, in this paper, the manufactured bracket is mounted on the bearing housing to obtain the vibration signal. The modal analysis is performed for bracket and connected components (housing, bolt, threshing shaft, disk, and hub) for comparative analysis to avoid resonance. The demodulated fast Fourier transform (FFT) (500 Hz–16 kHz) method was used to evaluate the bearing condition and DDS Adash Fault Source Identification Tool (FASIT) technology was used to evaluate other mechanical conditions such as the looseness, misalignment, and unbalance of the tangential threshing unit of the Massey Ferguson series of combine harvesters.

2. Materials and Methods

The Massey Ferguson 7374 s ACTIVA with 1300 engine hours (EH) and the new Massey Ferguson 7370 Beta were investigated for vibration analysis as shown in Figure 1.

The threshing drum bearings consist of 6310 (SKF, Gothenburg, Sweden) for both series of Massey Ferguson. The vibration measurement was conducted three times for each location. The detail of the technical condition is shown in Table 1.

(a)

(b)

Figure 1. (**a**) Massey Ferguson 7370 Beta; and (**b**) Massey Ferguson 7347 s Activa.

Table 1. Threshing unit technical parameters of the Activa and Beta version of Massey Ferguson.

	Massey Ferguson 7374 s (1300 EH) ACTIVA	Massey Ferguson 7370 (New) BETA
Engine power	306 KW	365 KW
Threshing system	Tangential threshing	Tangential threshing
Threshing drum	1340 mm × 600 mm	1600 mm × 600 mm
Threshing bearing	6310 SKF	6310 SKF
No. of drums	6	6
No. of rasp bar	12	12

The measurement was taken at 45° with the horizontal radial axis of the threshing shaft. The measurement was taken in the bracket and the housing on the new Beta version, but only in housing for the Activa version of Massey Ferguson as the bracket mounting was not viable for the Activa version due to the compact housing design. The variable working measuring speed was the same for both combine harvesters. The measuring speeds were 900, 1100, 1800, and 2500 engine RPMs, corresponding to 220, 290, 410, and 470 threshing drum RPMs, respectively. These engine speeds, which are the real operating speeds used during the harvesting period on the farm from low- to high-speed range, were obtained from the operators themselves, whilst the corresponding speeds at a threshing unit were obtained from the inbuilt sensor mounted in the Massey Ferguson combine harvesters. The major parameters of the threshing unit are shown in Table 2. The measuring position of the accelerometer on the bearing housing is shown in Figure 2.

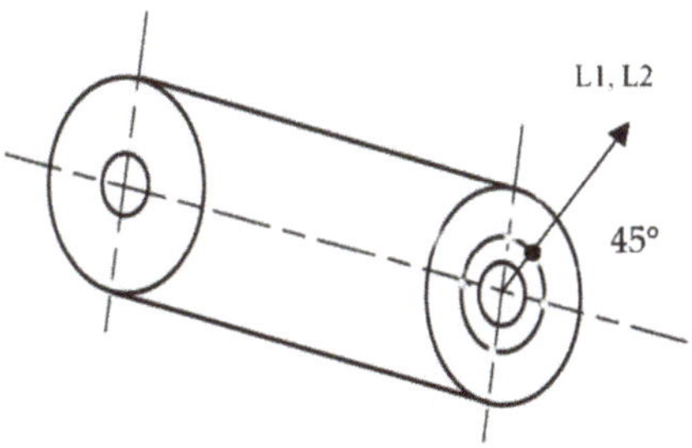

(a)

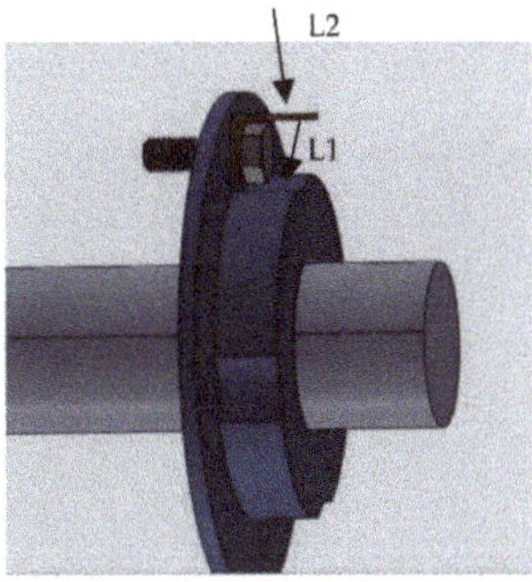

(b)

Figure 2. Sensor mounting position: L1 when the accelerometer is placed directly on a bearing seat; and L2 when the accelerometer is placed on a manufactured bracket: (**a**) The angle of the accelerometer mounted, and (**b**) The position of the accelerometer mounted.

Table 2. 6310 SKF bearing details.

Bore diameter	50 mm
Outer diameter	110 mm
Width	27 mm
Static load rating	65 kN
Dynamic load rating	38 kN

For the vibration measurement and data analysis, the Adash A4900 Vibrio M vibration accelerometer with a magnetic base and the Adash A4900 Vibrio M vibration analyzer were both used. This instrument complies with the ISO 10816-3 standard [23] and the requirements for mechanical vibration testing listed in ISO 10816-3, 2009 ISO10816-3:2009.

The Adash A4900Vibrio M, an accelerometer of type AC 150 with a magnetic base and a vibration spectrum analyzer with a sensitivity of 100 mV/g and an accuracy range of 2.5%, was used in the vibration instrumentation [24]. The DDS software was used by the Adash vibration analyzer to show the machine's overall vibration severity, bearing condition, speed, looseness, imbalance, misalignment, and other unidentified source severity.

For the bracket and housing bearing, modal analysis was carried out using the ANSYS software to locate and eliminate the typical natural frequencies that could lead to resonance. This is required to undertake the adequate design, material selection, and simulation of that material to prevent the resonance problem before taking the true vibration measurement of the bearing from the made–added materials. Before measuring the vibration from the added bracket, the choice of material was examined and recommended from the standpoint of material qualities.

The rotating cylinder drum and the concaves make up the majority of the threshing apparatus, which separates the grain from the stalks. The rasp bar cylinders have slotted plates that are attached to the cylinder rings so that one plate's slot direction is the opposite of another plate's slot direction. The drum with the rasp bar on its surface will impact the plant entering in the tangential direction to remove the seeds from the stalks and husks. The bearing, housing, bolt, and bracket system components were subjected to a modal analysis in Ansys. To avoid resonance in the system, the findings were compared to the fundamental and natural frequencies of the individual components.

To validate the measuring positions, using the vibration method and FASIT tool technology, all the processes were again verified on the threshing stand test bench. The tangential threshing drum bearings test bench was constructed and investigated for vibration analysis, as shown in Figure 3. The threshing unit was procured from a NIVA SK-5 combine harvester and modified. The main units of the tangential threshing units consist of the bearing, bushing, top cover, shaft, disks, rasp bar, and bottom cover, as shown in Figure 4.

The threshing drum bearings consist of KBC 6313 BS: 65 × 140 × 33 (KBC Industrial Co., Ltd., Zhejiang, China) with 3620 MH (motor hours). The vibration measurement was conducted three times for each location. The detailed dimensions of the stand and bearing are shown in Table 3.

Figure 3. Tangential threshing unit bearing test bench.

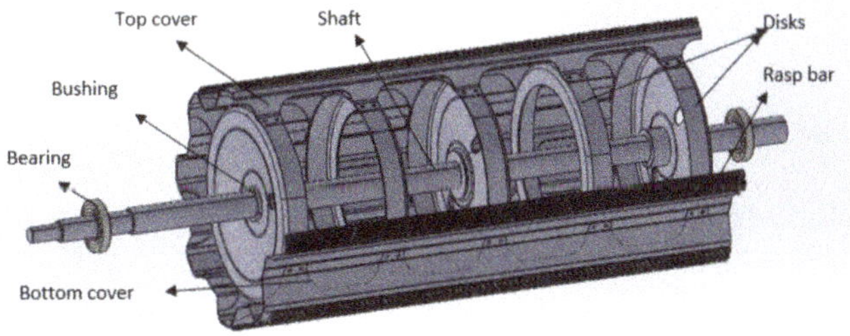

Figure 4. Computer-aided design of the main units of the tangential threshing units of the test bench.

Table 3. Tangential threshing unit test bench parameters.

Segment	Description	Dimension
Platform	Length Width Height	3700 mm 2060 mm 1540 mm
Threshing unit	Length Drum diameter	1940 mm 597.88 mm
	No. of rasp bar	8
	Fillet parameter of rasp bar (symmetric)	3 mm
	Rasp bar length	1176 mm
	Width	45 mm
	Distance between rasp bar	178 mm
	Length	65 mm
Bearing (KBC 6313 BS)	Bore diameter Outer diameter Raceway width	65 mm 140 mm 33 mm

The threshing drum bearing vibration measurement was taken from a sensor mounted directly on the bearing housing. The variables measuring the threshing speed were the same for both the driving and non-driving sides of the threshing. The measuring speeds were 210, 350, and 420 threshing drum RPMs, respectively.

The threshing cylinder rotation frequency was changed by a voltage frequency converter DeltaVFD-C2000 SERIES and a cylinder gear variator. The test bench was driven by a 30 kW electric motor. The measurement was taken at 45° with the horizontal radial axis of the threshing. The vibration measuring device Adash A4900 Vibrio M device with an AC150 piezoelectric sensor was used, as shown in Figure 5.

Figure 5. a. Adash A4900 Vibrio M device; and b. AC150 piezoelectric sensor.

3. Results

3.1. Modal Analysis

The 3D CAD modal of components (bearing, bracket, housing, bolt, hub, threshing shaft, and threshing disk), as shown in Figure 6, are designed in the solid works software and exported to the Ansys software for modal analysis.

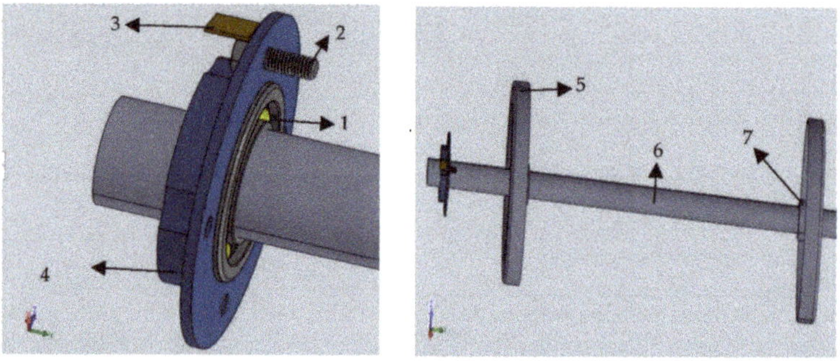

Figure 6. CAD modal of components: 1. Bearing, 2. Bolt, 3. Bracket, 4. Housing, 5. Threshing disk, 6. Threshing shaft, 7. Hub.

The modal analysis was performed for each component individually to obtain the eigenvalues and eigenvectors of the components for the first six different modes. The bracket is set as fixed onto the contact region with the bolt for modal analysis. Furthermore, the bearing, housing, bolt, and bracket used in the system were subjected to modal analysis. The dimension of each component is presented in Appendix A (Appendices A.1–A.9). The material was set as S235 steel for the bracket, gray cast iron for the housing, and stainless steel for other components. The material properties of the stainless steel and gray cast iron were obtained from Ansys material library, whereas the material properties of S235 are shown in Table 4. Since the eigenvalue of the bracket does not coincide with the components (Table 5) and the fundamental frequencies of the bearing, the bracket is processed for manufacturing.

Table 4. Material properties of S235.

Material Properties	S235
Density ρ	7850 kg/m^3
Unit weight γ	78.5 KN/m^3
Young's modulus	210,000 MPa
Shear modulus G	81,000 Mpa
Yield strength f_y	235 Mpa
Ultimate strength f_u	360 Mpa
Poisson's ratio in elastic range ν	0.3
Coefficient of linear thermal expansion α	12×10^{-6} °K^{-1}

Table 5. Eigenvalues of the bracket and components.

	Mode 1 (Hz)	Mode 2 (Hz)	Mode 3 (Hz)	Mode 4 (Hz)	Mode 5 (Hz)	Mode 6 (Hz)
Bracket	1918.5	3598.2	6566.8	8610.8	16,586	21,780
Housing	5027.4	5078.4	5086	5153.3	5203.4	5748.7
Bolt	43,893	44,422	47,372	85,087	94,161	94,580
Threshing shaft	130.02	131.64	355.46	359.73	689.14	687.11
Disk	779.03	948.93	953.32	1253.7	1379.8	1586.9
Hub	165.13	166.81	253.97	256.03	281.54	471.13
Threshing drum assembly	70.415	76.563	77.855	158.19	185.61	187.27

The rotational constraint of 470 RPM was imposed based on the established threshing cylinder finite element model, and the natural frequencies of the sixth-order constraint mode and the related modal shapes are presented in Figure 7. The first-order modal shape has the maximum deformation, measuring 0.123 m, occurring at the outer covers. The second-order modal shape was general bending; the maximum deformation, measuring 0.14 m, occurred at the intersection of the outer cover, inner cover, drum, and shaft. The third-order modal shape was general bending; the maximum deformation, measuring 0.14 m, occurred at the intersection of the outer cover, inner cover, drum, and shaft. The fourth-order modal shape has the maximum deformation, measuring 0.119 m, occurring at the outer covers. The fifth-order modal shape was general bending; the largest maximum deformation, measuring 0.168 m, occurred at the shaft. The sixth-order modal shape was general bending; the largest maximum deformation, measuring 0.168 m, occurred at the shaft.

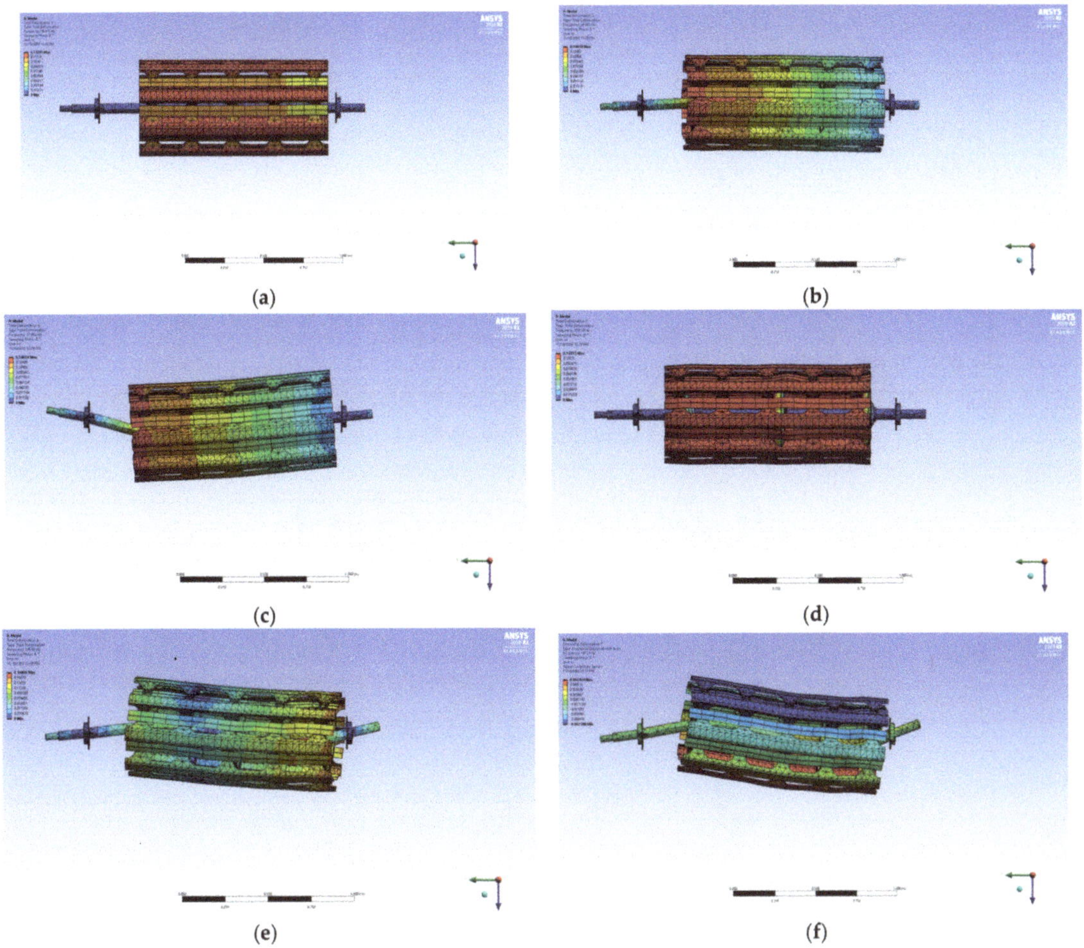

Figure 7. Modal analysis of the threshing cylinder. (**a**) First-order modal shape (70.41 Hz); (**b**) Second-order modal shape (76.56 Hz); (**c**) Third-order modal shape (77.85 Hz); (**d**) Fourth-order modal shape (158.9 Hz); (**e**) Fifth-order modal shape (185.61 Hz); and (**f**) Sixth-order modal shape (187.27 Hz).

3.2. Vibration Analysis

To determine the bearing vibration differences with and without a bracket, the easily accessible bearing housing of the combine harvester was selected. The measurements were performed by comparing the data. The main objective of these measurements was to develop a database to determine the differences between a measurement with the manufactured bracket and without the bracket. These results are presented in Figure 8. Also, the comparison of the measurement of Beta and Activa variants using a bracket is shown in Figure 9 where demodulated RMS value is slightly higher for the Beta version due to higher loads and dimensions compared to the Activa version.

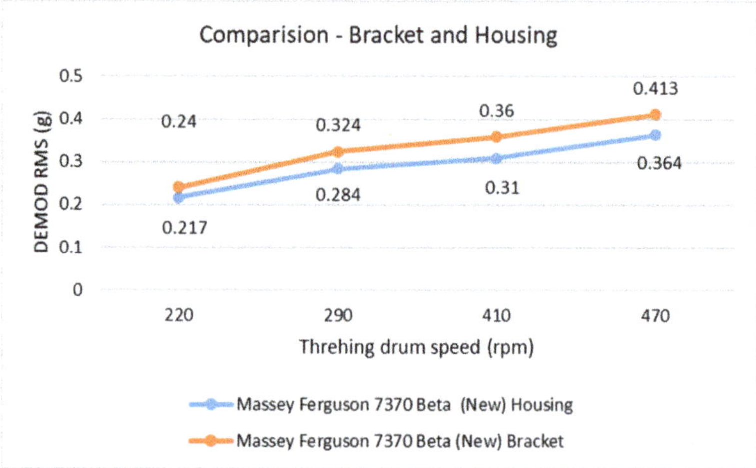

Figure 8. Comparison of the demodulated RMS value when the sensor is mounted on a bracket with respect to the housing of the Beta version in different threshing rotating speeds.

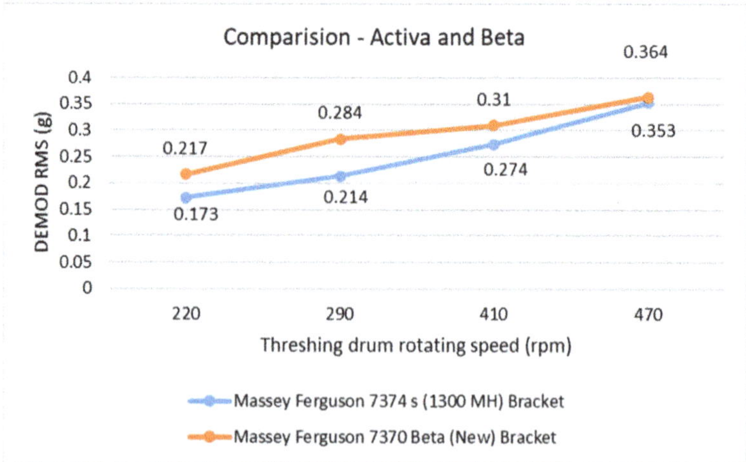

Figure 9. Comparison of the demodulated RMS value when the sensor is mounted on a bracket of the Beta and Activa versions at different threshing rotating speeds.

As the demodulated RMS is responsible for the evaluation of the bearing conditions, the difference between when the accelerometer is mounted in housing compared to on the bracket is analyzed, which is up to 16.13% increased for the data measured on the bracket compared to those on the housing, as shown in Figure 8. The metal bracket can therefore be utilized to help with the vibration measurements of the threshing drum bearings if it is properly fitted in specific spots. The demodulated RMS value increases with respect to the speed for both conditions, which satisfies the proper vibration measurement.

The demodulated RMS (g) values are compared between the two versions of the Massey Ferguson (Figure 10) as the constructions are with some changes in shaft and drum dimensions, as shown in Tables 1 and 2. The recorded value increases with the speed and is higher for the new Massey Ferguson 7374 s (1300 EH), which satisfies the condition as the dimensions of the threshing units such as the drum and shaft are larger compared to the Activa version. The comparison of the vibration measurement is collectively analyzed in the Activa and Beta versions with and without bracket with the vibration analysis parameters, as shown in Figure 5. The difference in the vibration measurement was compared in the Beta (New) version and Activa version with the vibration parameters such as the spectrum (0-P, g), spectrum (RMS, g), demodulated RMS (g), and velocity RMS (g). The recorded vibration data show a higher magnitude with the bracket measurement compared to the housing measurement in the Beta version. As the threshing unit of both combine harvesters was similar in construction with a similar bearing, the recorded data were compared with both harvesters when measured on a bracket. The maximum amplitude was obtained for the new Beta version as the dimension is higher which results in a larger radial load and high vibration magnitude.

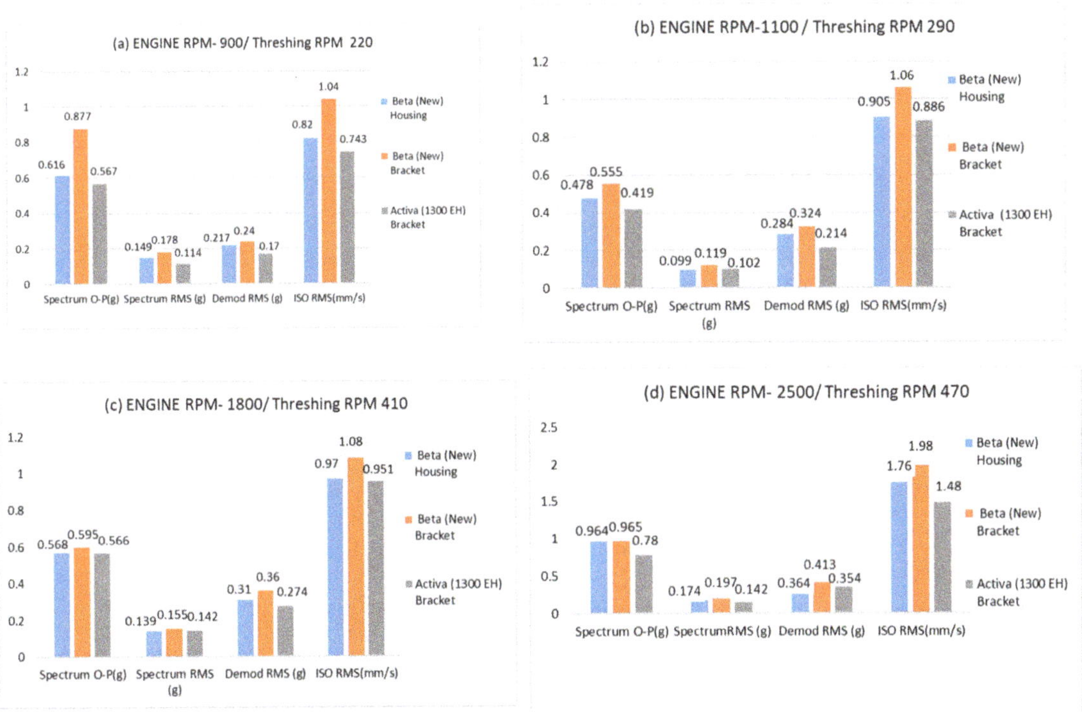

Figure 10. Comparison of the vibration parameters values when the sensor is mounted on bracket and the housing of Beta and Activa versions at different threshing rotating speeds: (**a**) Threshing drum rotating at 220 RPM; (**b**) Threshing drum rotating at 290 RPM; (**c**) Threshing drum rotating at 410 RPM; and (**d**) Threshing drum rotating at 470 RPM.

To study the bearing used in both combine harvesters, the demodulated FFT was considered for both variants of combine harvester when the measurement was taken with the bracket and housing for the Beta version and just with the bracket for the Activa version. The fundamental frequency was obtained from the DDS Adash bearing library and verified from the SKF catalog for bearing 6310 (SKF, Gothenburg, Sweden), as shown in Table 3. Multiplying the rotational speed of the threshing drum with the fundamental frequencies of the bearing gives fault frequencies for different operating speeds.

Peak amplitude is under the noise threshold for envelope analysis. The demodulated FFT spectrum with an amplitude RMS(g) and a threshing speed of 470 RPM for bearing failure frequencies FTF (2.98 Hz, 0.024 g), BPFO (23.9 Hz, 0.007 g), BSF (31 Hz, 0.006 g), and BPFI (38.8 Hz, 0.003 Hz) were measured (Figure 11). The higher RMS values should be as near to zero as possible when being assessed.

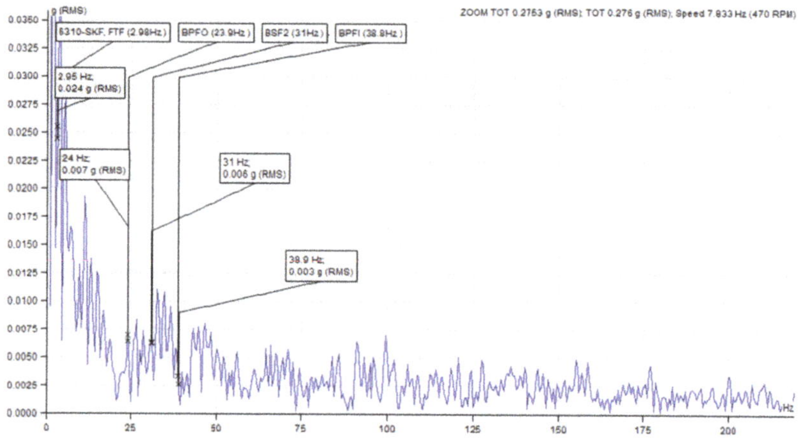

Figure 11. The fast Fourier transform (FFT) (500 Hz–16 kHz) demodulation of the 6310 SKF of the non-driving end of the threshing drum of Massey Ferguson 7374 s (1300 EH) Activa, a sensor mounted on a bracket.

Similarly, the demodulated FFT spectrum RMS analysis was performed for the Beta version when the sensor was mounted on the housing and on the bracket, respectively. The demodulated FFT spectrum with an amplitude RMS (g) with a threshing speed of 470 RPM for bearing failure frequencies FTF (2.98 Hz, 0.004 g), BPFO (23.9 Hz, 0.003 g), BSF (31 Hz, 0.003 g), and BPFI (38.8 Hz, 0.004 Hz) were measured when the data were recorded from the accelerometer mounting on the housing (Figure 12). The demodulated FFT spectrum with an amplitude RMS (g) with a threshing speed of 470 RPM for bearing failure frequencies FTF (2.98 Hz, 0.037 g), BPFO (23.9 Hz, 0.005 g), BSF (31 Hz, 0.008 g), and BPFI (38.8 Hz, 0.008 Hz) were measured when the data were recorded from the accelerometer mounting on the bracket (Figure 13).

To observe the higher amplitude peaks in the spectrum and to identify any sidebands and harmonics resembling the bearing and other parts, the overall spectrum (g) is considered for both combine harvesters, as shown in Figures 14 and 15. No higher amplitude peaks, harmonics, and side bands are present in the analyses. The bearing is considered safe to operate. The magnitudes of the measured RMS value indicate no bearing faults in the Massey Ferguson 7370 Beta (New). The bearing condition is in good condition to operate. The RMS bearing static amplitude is 0.353 g, which indicates a lower number than the alarm limit specified on ISO 10800: Noise and Vibration standard alarm set on the DDS software.

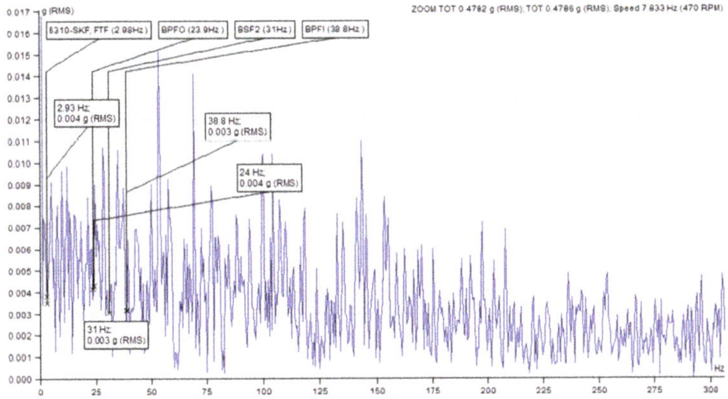

Figure 12. The fast Fourier transform (FFT) (500 Hz–16 kHz) demodulation of the 6310 SKF of the non-driving end of the threshing drum of Massey Ferguson 7370 Beta (New), a sensor mounted on housing.

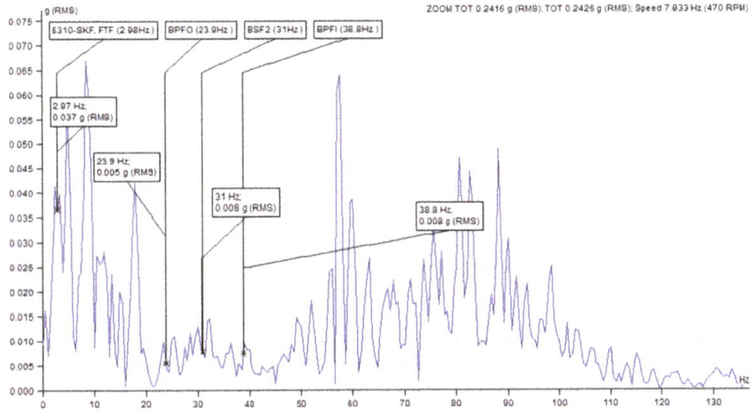

Figure 13. The fast Fourier transform (FFT) (500 Hz–16 kHz) demodulation of the 6310 SKF of the non-driving end of the threshing drum of Massey Ferguson 7370 Beta (New), a sensor mounted on a bracket.

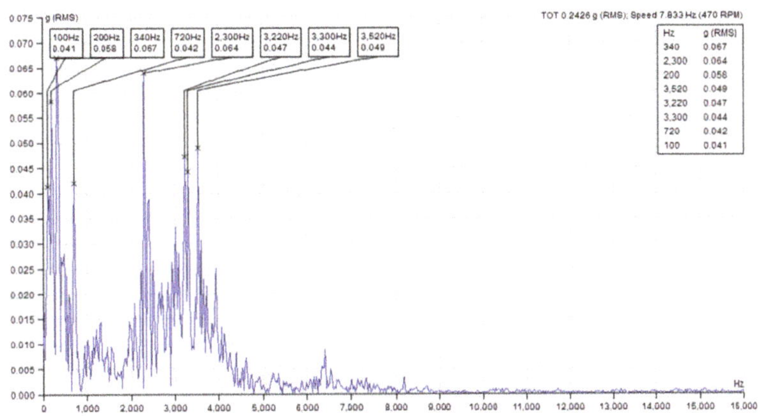

Figure 14. Top 8 peaks from the overall spectrum (10 Hz–16 kHz) of Massey Ferguson Activa.

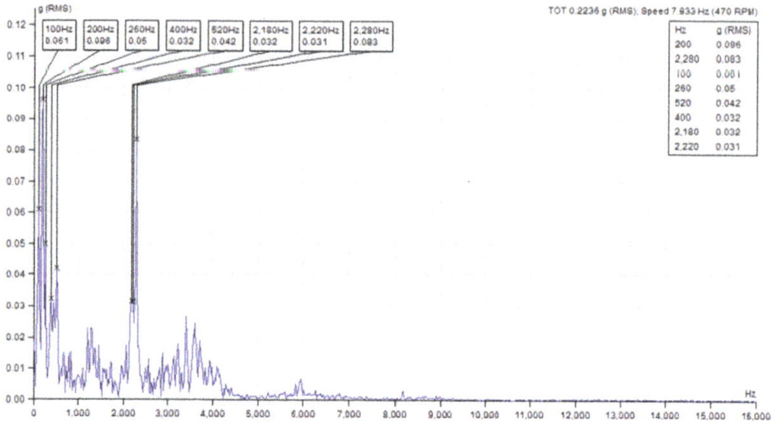

Figure 15. Top 8 peaks from overall spectrum (10 Hz–16 kHz) of Massey Ferguson Beta.

The FASIT tool provided by the Adash DDS helps identify the other mechanical condition of the threshing unit. The results show 0.5% looseness, 1% unbalance, 1% misalignment, and 5% other mechanical faults, as shown in Figure 16.

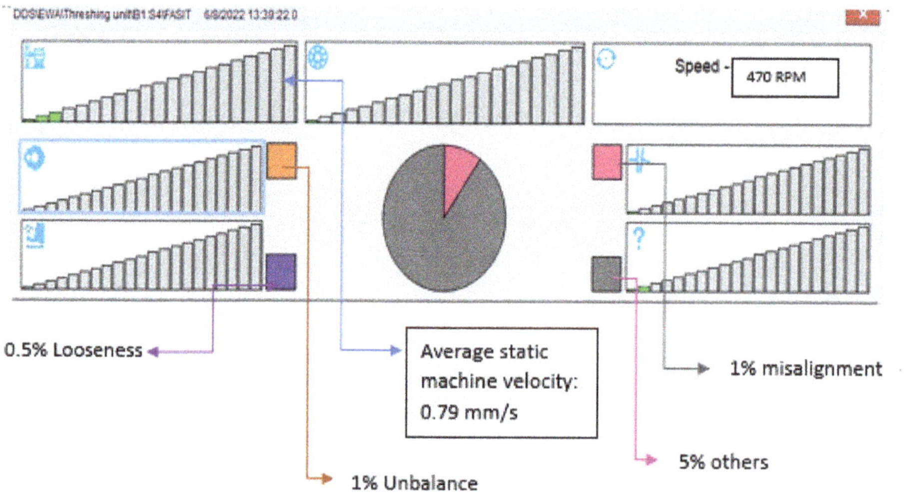

Figure 16. The Adash FASIT tool display for correctable mechanical conditions (looseness, misalignment, bearing, unbalance, and speed).

The validation of the method was again performed on the tangential threshing test bench. To study the bearing used in the stand demodulated FFT is considered when the measurement is from an accelerometer-mounted housing at 45° from the horizontal radial axis. When the sensor was directly mounted on the housing, 30% unbalance, 0% misalignment, and 40% looseness were observed from the FASIT tool display. The unbalance and looseness issues were fixed for the tangential threshing stand and the vibration analysis was performed. The fundamental frequency was obtained from the DDS Adash bearing library, as shown in Table 6.

Table 6. Fundamental frequencies of the 6310 SKF bearing.

Fundamental train frequency (FTF)	0.381 Hz
Ball spin frequency (BSF)	1.98 Hz
Ball passing frequency of outer race (BPFO)	3.05 Hz
Ball passing frequency of inner race (BPFI)	4.95 HZ

The demodulated FFT spectrum, with an amplitude RMS (g) with a threshing speed of 420 RPM (maximum operating speed), and when the sensor was directly mounted onto the housing for the bearing fundamental train frequency (FTF) (2.69 Hz, 0.005 g), ball passing frequency of outer race (BPFO) (21.5 Hz, 0.009 g), ball spin frequency (BSF) (28.5 Hz, 0.004 g), and ball passing frequency of inner race (BPFI) (34.5 Hz, 0.01 Hz), was measured and is shown in Figure 17. The RMS higher values should be as near to zero as possible when being assessed. Since the BPFI amplitude is close to the second digit after decibel and higher compared to the other amplitudes, it is further diagnosed with the digital camera with 12-megapixel-wide and ultrawide lenses (Figure 18). It is observed that the bearing has initiated the abrasive marking with contamination. The main causes include filthy hands, contaminated workspaces, dirty instruments, and foreign objects in lubricants.

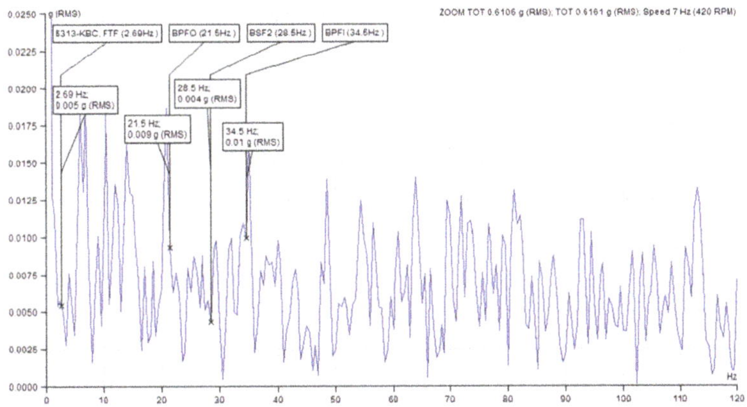

Figure 17. The fast Fourier transform (FFT) (500 Hz–16 kHz) demodulation 6313 KBC of the non-driving end of the tangential threshing drum rotating at 420 rpm with the sensor mounted on the housing.

Figure 18. Image of bearing surface inner ring.

Therefore, the fast Fourier transform (FFT) (500 Hz–16 kHz) demodulated RMS vibration analysis technique successfully determined the bearing initial fault condition and is suitable for the easy and effective analysis of the tangential threshing bearing condition. The FASIT tool successfully determined the other machine condition.

3.3. Random Vibration Analysis

The reaction of structures exposed to random vibration loads must be examined by random vibration analysis. Compared to the scenario of a deterministic time history loading, random vibration loads were not always quantifiable with the confidence of the size and duration. The outcomes are also statistical in nature because the input loads are described using statistical values.

The design was assembled in Solid works software and imported to the Ansys software. The shaft extruded and was cut to 50 mm (millimeters) to reduce the computational time. The inner race-shaft and roller-bearing races have a friction coefficient of 0.21, the bearing races-housing and bracket and housing have a friction coefficient of 0.4, whilst the bolts-bracket has a friction coefficient of 0.23.

The FFT (500–16,000 Hz) data range between 400 Hz and 800 lines. The fundamental frequencies of the bearing corresponding to the FFT frequency of measurement (Figure 16) were converted into power spectral density (PSD), as shown in Table 7. The frequency domain analysis with PSD estimation based on FFT [25] is given by:

$$\text{PSD} = \text{Amplitude}^2/(\Delta f \times Wf) \tag{1}$$

where Wf is the correction value (3.671441636) and Δf is the frequency resolution obtained by:

$$\Delta f = \frac{\text{Frequency range}}{\text{No. of Lines}} \tag{2}$$

Table 7. Conversion from FFT to PSD.

Fundamental Bearing Frequencies at 420 rpm of 6313 KBC	FFT Amplitude Demodulated RMS (g)	PSD Amplitude (g^2/Hz)
Fundamental train frequency (FTF) (2.69 Hz.)	0.005	3.405×10^{-5}
Ball spin frequency (BSF) (21.5 Hz.)	0.009	1.1031×10^{-4}
Ball passing frequency of outer race (BPFO) (25.5 Hz.)	0.004	2.179×10^{-5}
Ball passing frequency of inner race (BPFI) (34.5 Hz.)	0.01	1.3619×10^{-4}

The corresponding PSD frequency and amplitude are applied to all the connections and PSD deformation and response PSD due to the adjacent PSD input being observed on the bracket in the vertical radial direction. The first sigma scale factor showed 3.211×10^{-8} mm. minimum deformation and 1.23×10^{-3} mm. maximum deformation and the third sigma scale factor showed 9.6329×10^{-8} mm. minimum deformation and 3.7184×10^{-3} mm. maximum deformation on the x axis (vertical radial direction). Again, the random vibration was applied to the system, as shown in Table 8. These values are random in nature and obtain the PSD response in frequency range from 10 to 2000 Hz. The first sigma scale factor showed 1.73×10^{-9} mm. minimum deformation and 2.288×10^{-7} mm. maximum deformation and the third sigma scale factor showed 5.16×10^{-9} mm. minimum deformation and 6.57×10^{-3} mm. maximum deformation on the x axis (vertical radial direction).

The first sigma represents 68.26% deformation, and the third sigma represents 99.73% deformation. Since the deformations on the brackets are negligibly very small, the deformation on the other parts can be calculated and analyzed using a bracket. Furthermore, the PSD response is taken on the vertex of the bracket, as shown in Figure 19 in the x direction.

Table 8. Random Input PSD (10–2000 Hz.).

Frequency (Hz)	G Acceleration (G^2/Hz)
10	0.1
25.5	0.5
100	0.5
1000	0.5
2000	0.2

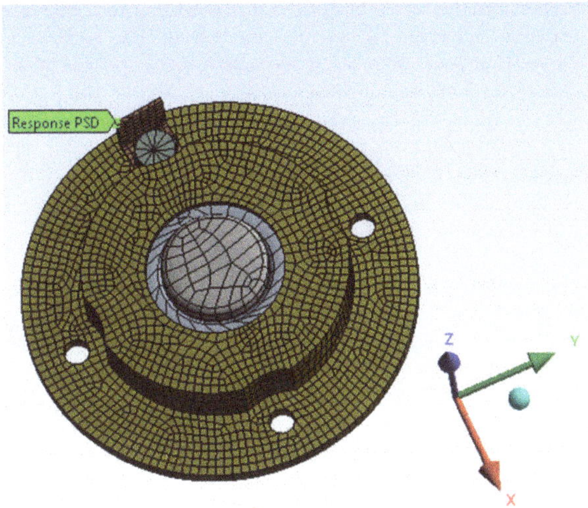

Figure 19. PSD response vertex.

The PSD response (RPSD) provides the spectral response of a structure subjected to random excitation and the RPSD plot gives the information as to where the average power is distributed as a function of frequency. This gives the RMS value of the selected frequency range over the entire available frequency range as well as information about the peak g acceleration responses that occur at the resonant frequency on the assembly (Figure 20).

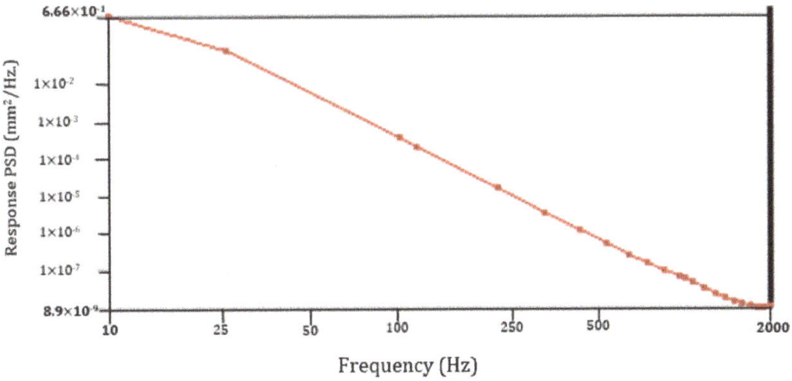

Figure 20. PSD response graph displaying peak g acceleration responses.

Therefore, the bracket deformation is comparatively negligible for the PSD corresponding to the test result and random vibration PSD inputs. Thus, the bracket can provide the deformation information of other associated components during vibration analysis when the sensor is mounted onto its surface in the radial direction.

4. Discussion

The most suitable vibration signals were selected to assess the threshing drum technical characteristics of a combine and to spot flaws when the combine is in function under nonloading conditions. The best method for determining the bearing quality condition is demodulated FFT spectrum analysis, which offers the necessary vibration information. The FASIT tool technology provided by Adash DDS is effective and can easily identify machine conditions such as misalignment, looseness, unbalance, and other mechanical conditions to account for before every harvesting season.

The measurement was taken by introducing the simple bracket into the bearing housing to locations where sensor mounting is difficult, as the vibration analysis is conducted for the Massey Ferguson Beta and Activa variants. The comparison was performed to identify the difference in measurement taken when the sensor was mounted on the bracket and housing. The modal analysis was performed on Ansys to identify the eigenvalues and eigenvectors to prevent the resonance phenomenon. The static demodulated RMS recording was found to be 16.23% more on the bracket compared to the housing which satisfies the evaluation of the bearing condition. The magnitudes of the measured RMS value indicated no bearing faults in the Massey Ferguson 7370 Beta (New) and Ferguson 7374 s (1300 EH) Activa. The RMS bearing static amplitude was 0.353 g and 0.364 g for the Beta and Activa variants, respectively, which indicates a lower number than the alarm limit specified on ISO 10800: Noise and Vibration standard alarm set on the DDS software. The FASIT tool provided by the Adash DDS helps identify the other mechanical conditions of the threshing unit. The results show 0.5% looseness, 1% unbalance, 1% misalignment, and 5% other mechanical faults. This indicates no severe mechanical machine condition problems and that it is safe to operate for the harvesting season. This method is again validated from the tangential threshing stand and successfully determines the bearing fault condition. Additionally, a PSD response plot (RPSD) is obtained to study the peak g acceleration responses that occur at the resonant frequency on the assembly. Therefore, the fast Fourier transform (FFT) (500 Hz-16 kHz) determined the bearing condition, and the FASIT technology can determine other mechanical conditions such as looseness, misalignment, and the unbalance of the tangential threshing unit.

Author Contributions: Conceptualization, S.B. and E.J.; methodology, S.B.; software, S.B.; validation, E.J. and S.B.; formal analysis, E.J.; investigation, S.B.; resources, E.J.; data curation, S.B.; writing—original draft preparation, S.B.; writing—review and editing, S.B.; visualization, S.B.; supervision, E.J.; project administration, E.J. All authors have read and agreed to the published version of the manuscript.

Funding: This research received no external funding.

Data Availability Statement: Not applicable.

Acknowledgments: We acknowledge the East West Agro (EWA) group, Kaunas, Lithuania, for providing the Massey Ferguson series of combine harvesters for investigation and the Agriculture Academy of Vytautas Magnus University.

Conflicts of Interest: The authors declare no conflict of interest.

Appendix A

Appendix A.1

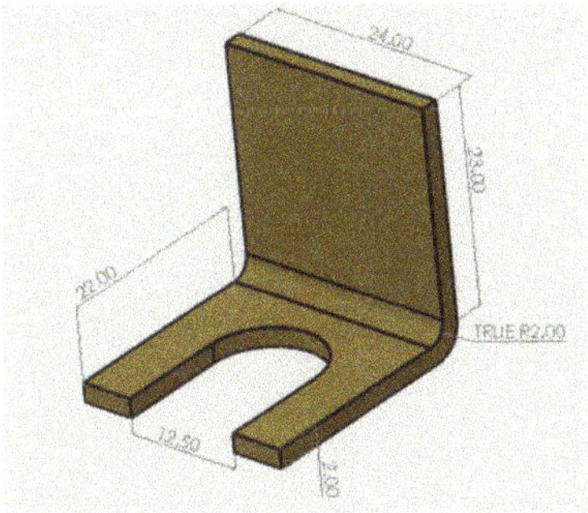

Figure A1. Manufactured Bracket (Dimensions in mm).

Appendix A.2

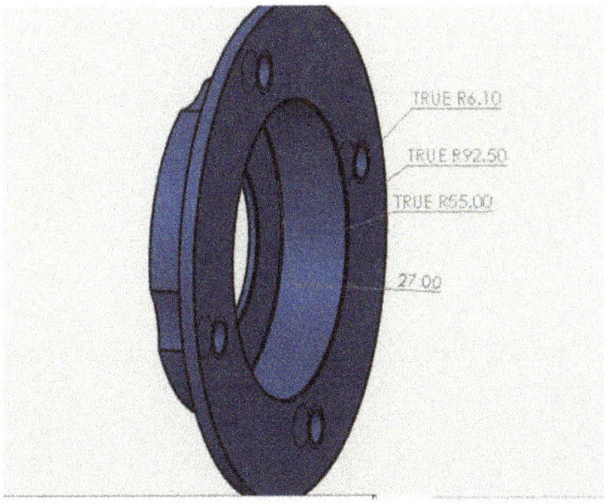

Figure A2. Bearing housing (Dimensions in mm).

Appendix A.3

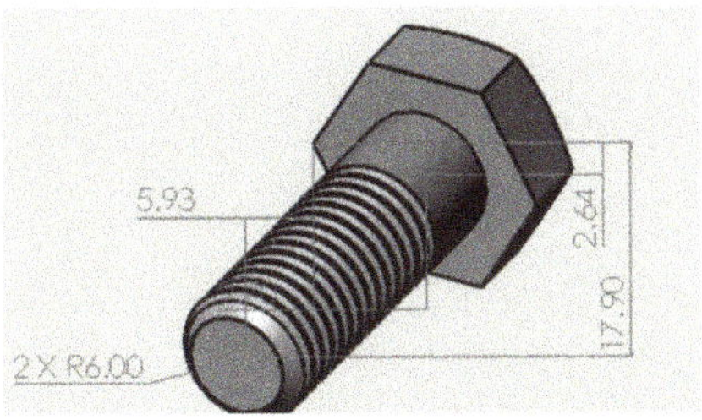

Figure A3. Bolt M12 (Dimensions in mm).

Appendix A.4

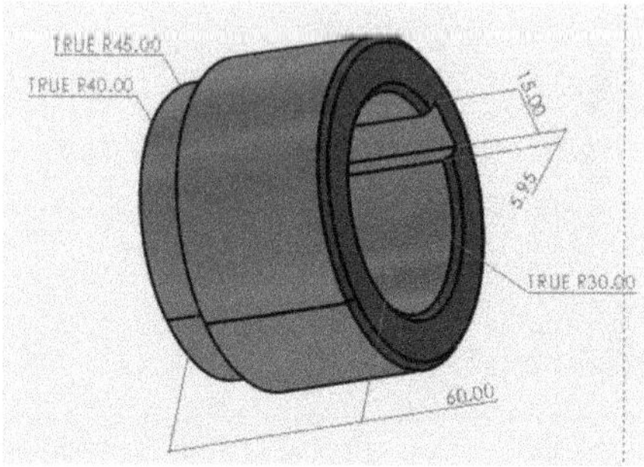

Figure A4. Hub (Dimensions in mm).

Appendix A.5

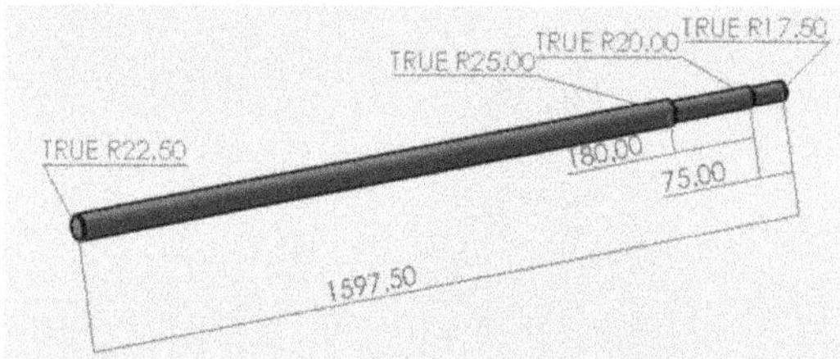

Figure A5. Threshing shaft (Dimensions in mm).

Appendix A.6

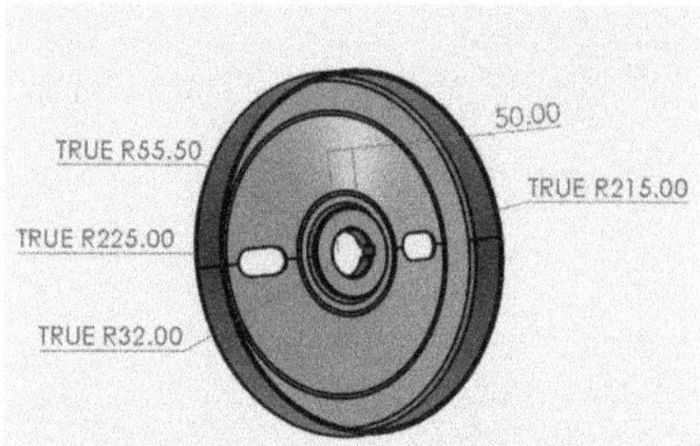

Figure A6. Threshing disk (Dimensions in mm).

Appendix A.7

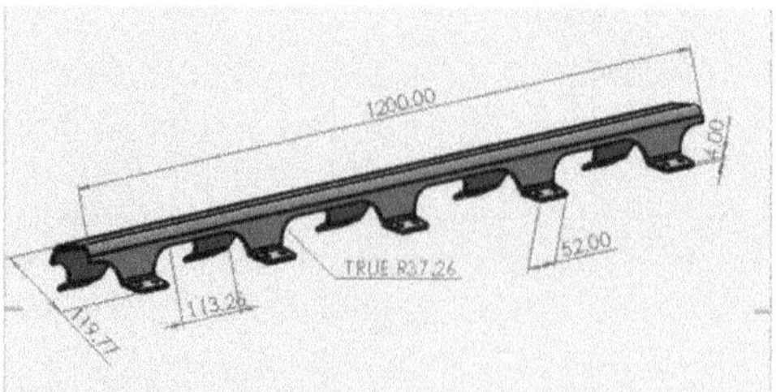

Figure A7. Top cover (Dimensions in mm).

Appendix A.8

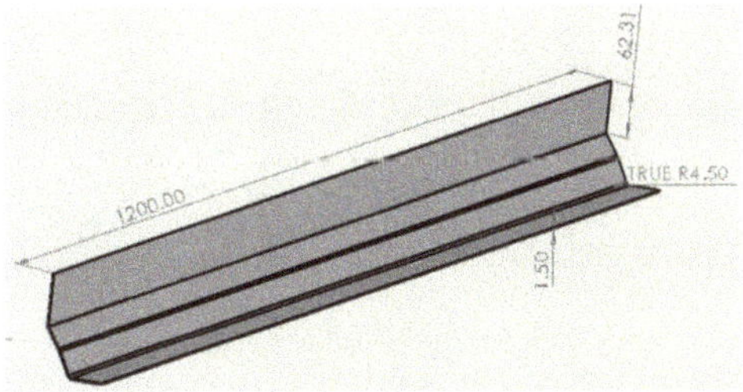

Figure A8. Bottom cover (Dimensions in mm).

Appendix A.9

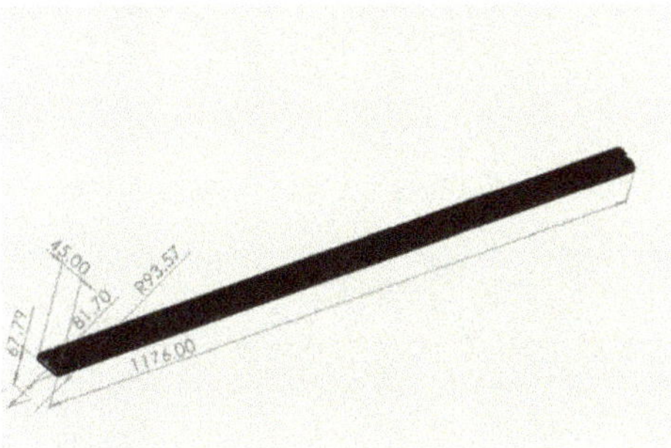

Figure A9. Rasp bar (Dimensions in mm).

References

1. Špokas, L.; Adamčuk, V.; Bulgakov, V.; Nozdrovick, L. The experimental research of combine harvesters. *Res. Agric. Eng.* **2016**, *62*, 106–112. [CrossRef]
2. Zhao, Z.; Li, Y.; Chen, J.; Xu, J. Grain separation loss monitoring system in combine harvester. *Comput. Electron. Agric.* **2011**, *76*, 183–188. [CrossRef]
3. Hollaz, B.; Alegra, P. Beater and Feed Mechanism for a Combine Harvester. U.S. Patent US 8,118,652 B2, 21 February 2012.
4. Yao, Y.; Song, Z.; Du, Y.; Zhao, X.; Mao, E.; Liu, F. Analysis of vibration characteristics and its major influenced factors of header for corn combine harvesting machine. *Nongye Gongcheng Xuebao/Trans. Chin. Soc. Agric. Eng.* **2017**, *33*, 40–49.
5. Zhong, T.; YaoMing, L.; Rui, W.; JingSong, C. Experimental study of multi-source dynamic load of combine harvester in rice harvesting. *Int. Agric. Eng. J.* **2016**, *25*, 46–52.
6. Jiangtao, J.; Jingpeng, H.; Shengsheng, W.; Ruihong, Z.; Jing, P. Vibration and Impact Detection of Axial- Flow Threshing Unit Under Dynamic Threshing Conditions. *Inmateh Agric. Eng.* **2020**, *60*, 183–192. [CrossRef]
7. Shamsha, S.M.; Sinha, J.K. Rotor Unbalance Estimation with Reduced Number of Sensors. *Machines* **2016**, *4*, 19. [CrossRef]
8. Tang, Z.; Zhang, H.; Zhou, Y. Unbalanced Vibration Identification of Tangential Threshing Cylinder Induced by Rice Threshing Process. *Shock. Vib.* **2018**, *2018*, 4708730. [CrossRef]
9. Wang, G.; Wang, X.; Yuan, H. Dynamic analysis for the thresher of a combined harvester subjected to stalk winding. *AIP Adv.* **2022**, *12*, 095320. [CrossRef]
10. Xiang, C.; Ren, Z.; Shi, P.; Zhao, H. Data-Driven Fault Diagnosis for Rolling Bearing Based on DIT-FFT and XGBoost. *Complexity* **2021**, *2021*, 4941966. [CrossRef]
11. Zhou, L.; Duan, F.; Corsar, M.; Elasha, F.; Mba, D. A study on helicopter main gearbox planetary bearing fault diagnosis. *Appl. Acoust.* **2019**, *147*, 4–14. [CrossRef]
12. Lin, H.C.; Ye, Y.C.; Huang, B.J.; Su, J.L. Bearing vibration detection and analysis using enhanced fast Fourier transform algorithm. *Adv. Mech. Eng.* **2016**, *8*, 1687814016675080. [CrossRef]
13. Smith, W.A.; Fan, Z.; Peng, Z.; Li, H.; Randall, R.B. Optimised Spectral Kurtosis for bearing diagnostics under electromagnetic interference. *Mech. Syst. Signal Process.* **2016**, *75*, 371–394. [CrossRef]
14. Wang, H.Q.; Hou, W.; Tang, G.; Yuan, H.F.; Zhao, Q.L.; Cao, X. Fault detection enhancement in rolling element bearings via peak-based multiscale decomposition and envelope demodulation. *Math. Probl. Eng.* **2014**, *2014*, 329458. [CrossRef]

15. Senthil Kumar, M.; Naiju, C.D.; Chethan Kumar, S.J.; Kurian, J. Vibration analysis and improvement of a vehicle chassis structure. *Appl. Mech. Mater.* **2013**, *372*, 528–532. [CrossRef]
16. Liu, Z.; Yuan, S.; Xiao, S.; Du, S.; Zhang, Y.; Lu, C. Full Vehicle Vibration and Noise Analysis Based on Substructure Power Flow. *Shock. Vib.* **2017**, *2017*, 8725346. [CrossRef]
17. Yu, W. Analysis and Optimization of Low-Speed Road Noise in Electric Vehicles. *Wirel. Commun. Mob. Comput.* **2021**, *2021*, 5537704. [CrossRef]
18. Qi, F.; Lei, Y.; Deng, P.; Huang, Q. Car body vertical vibration analysis under track medium wave irregularity and the influence factors of ballast bed. *J. Low Freq. Noise Vib. Act. Control.* **2019**, *38*, 1160–1177. [CrossRef]
19. Jahanbakhshi, A.; Ghamari, B.; Heidarbeigi, K. Vibrations analysis of combine harvester seat in time and frequency domain. *J. Mech. Eng. Sci.* **2020**, *14*, 6251–6258. [CrossRef]
20. Xu, L.H.; Chai, X.Y.; Gao, Z.P.; Li, Y.M.; Wang, Y.D. Experimental study on driver seat vibration characteristics of crawler-type combine harvester. *Int. J. Agric. Biol. Eng.* **2019**, *12*, 90–97. [CrossRef]
21. Chen, S.; Zhou, Y.; Tang, Z.; Lu, S. Modal vibration response of rice combine harvester frame under multi-source excitation. *Biosyst. Eng.* **2020**, *194*, 177–195. [CrossRef]
22. Wang, J.; Xu, C.; Xu, Y.; Qi, X.; Liu, Z.; Tang, H. Vibration analysis and parameter optimization of the longitudinal axial flow threshing cylinder. *Symmetry* **2021**, *13*, 571. [CrossRef]
23. *ISO 10816-3:2019*; Mechanical Vibration. Evaluation of Machine Vibration by Measurements on Non-Rotating Parts. ISO Standard: Geneva, Switzerland, 2009.
24. User 's Guide Adash 4900. Available online: https://adash.com/documents/A4900/Adash-A4900-Vibrio-manual.pdf (accessed on 1 March 2022).
25. Qinghua, H.; Li, J.; Ye, F.; Carpinteri, A.; Lacidogna, G. A new frequency domain method for random fatigue life estimation in a wide-band stationary Gaussian random process. *Fatigue Fract. Eng. Mater. Struct.* **2019**, *42*, 97–113.

agriculture

Article

Lightweight Corn Seed Disease Identification Method Based on Improved ShuffleNetV2

Lu Lu, Wei Liu *, Wenbo Yang, Manyu Zhao and Tinghao Jiang

School of Information and Electronic Engineering, Shandong Technology and Business University, Yantai 264005, China
* Correspondence: liuwei@sdtbu.edu.cn

Abstract: Assessing the quality of agricultural products is an essential step to reduce food waste. The problems of overly complex models, difficult to deploy to mobile devices, and slow real-time detection in the application of deep learning in agricultural product quality assessment requiring solutions. This paper proposes a lightweight method based on ShuffleNetV2 to identify phenotypic diseases in corn seeds and conduct experiments on a corn seed dataset. Firstly, Cycle-Consistent Adversarial Networks are used to solve the problem of unbalanced datasets, while the Efficient Channel Attention module is added to enhance network performance. After this, a 7×7 depthwise convolution is used to increase the effective receptive field of the network. The repetitions of basic units in ShuffleNetV2 are also reduced to lighten the network structure. Finally, experimental results indicate that the number of model parameters are 0.913 M, the computational volume is 44.75 MFLOPs and 88.5 MMAdd, and the recognition accuracy is 96.28%. The inference speed of about 9.71 ms for each image was tested on a mobile portable laptop with only a single CPU, which provides a reference for mobile deployment.

Keywords: image classification; lightweight neural networks; zea mays; ECA attention; CycleGAN

Citation: Lu, L.; Liu, W.; Yang, W.; Zhao, M.; Jiang, T. Lightweight Corn Seed Disease Identification Method Based on Improved ShuffleNetV2. *Agriculture* **2022**, *12*, 1929. https://doi.org/10.3390/agriculture12111929

Academic Editors: Muhammad Sultan, Redmond R. Shamshiri, Md Shamim Ahamed and Muhammad Farooq

Received: 14 October 2022
Accepted: 10 November 2022
Published: 17 November 2022

1. Introduction

Evaluating the quality agricultural products is an issue to which countries have always attached great importance. Recent years have seen the introduction of the concept of precision agriculture, with stricter requirements for the quality assessment of agricultural products, which is important to guarantee accurate identification and effective control of seed pests and diseases, in addition to playing a vital role in grain storage and distribution management, helping to ensure seed quality, avoid food waste, and ensure food security [1,2].

Fusarium graminearum, F.cepacia, F.proliferatum, and F.subglutinans are common causes of root, stalk, and cob rot in maize [3]. Diseased seeds are an important source of primary infestation resulting in plant diseases, leading to the long-distance spread of plant diseases and reducing the germination rate of seeds [4]. Infected seeds are not conducive to storage and can cause other seeds in the storage to become moldy, thus causing huge food losses and further leading to declines in seed quality or rendering these seeds altogether inedible [5]. Traditional grain quality and safety assessments often use microbial experiments (e.g., spore counting, enzyme-linked immunosorbent assays). In spite of their excellent detection accuracy, these methods are time-consuming, labor-intensive, and destructive [6]. As an important basis for evaluating the quality of seeds, the phenotypic detection of seeds is a non-destructive testing method. However, due to the influence of subjective factors in manual testing, the test results vary from person to person, and the detection efficiency is also low, which is easy to misjudge [7,8]. Therefore, quality inspectors urgently need a fast and objective method to detect diseases in corn seeds.

With deep learning's ability to extract features efficiently and accurately, it has become widely used in agriculture, reducing the need for manual feature extraction and analysis and making great progress in crop disease detection. Using hyperspectral imaging technology and deep convolutional neural networks (DCNNs), Zhang et al. [9] classified corn seeds with different degrees of freezing damage and reached a higher than 90% classification accuracy. Javanmardi et al. [10] used deep convolutional neural networks to classify varieties of maize seeds with an accuracy of 98%. Wang et al. [11] used hyperspectral imaging to identify aged maize seeds, which involved a full-spectrum classification model using the support vector machine (SVM) algorithm, principal component analysis (PCA), and ANOVA to reduce the data's dimensionality and extract the feature wavelengths; they classified maize seeds harvested in different years with a prediction accuracy of 97.5%. Yang et al. [12] used hyperspectral imaging (HSI) combined with sparse auto-encoders (SAEs) and convolutional neural network (CNN) algorithms to classify the mold grades of maize kernels. SAEs and a CNN were combined with an SVM classifier to construct the SAE-CNN-SVM model, and the results showed 99.47% and 98.94% correct recognition rates on the training and test sets, respectively.

A real-time method based on deep convolutional neural networks for the identification of maize leaf diseases was proposed by Mishra et al. [13], and the model was deployed on a Raspberry Pi 3. It was used to identify maize leaf diseases, and their model achieved an accuracy of 88.46%. Meng et al. [14] developed a spectral disease indices (SDIs) monitoring model based on in situ leaf reflection spectra to detect southern corn rust (SCR)-infected leaves and to classify the severity of SCR damage. The performance of the developed SCR-SDIs was evaluated by employing a support vector machine (SVM), and the model achieved an overall accuracy of 87% and 70% for SCR detection and severity classification, respectively. The authors also found that these spectral features were associated with the leaf pigments and water content. Albarrak et al. [15] created a date fruit dataset containing eight categories and used the MobileNetV2 model for date fruit classification. The results showed that the classification accuracy was 99%. Padilla et al. [16] used convolutional neural networks and OpenMP to detect leaf blight, leaf rust, and leaf spot in corn crops, and performed validation experiments on a Raspberry Pi, with measured accuracies of 93%, 89%, and 89%, respectively.

All of the above studies have provided positive results in agricultural product quality assessment and classification. However, hyperspectral imaging data acquired has high physical complexity, and analysis of hyperspectral data requires fast computers, sensitive detectors, and large data storage capacity [17]. In addition, large convolutional neural networks or traditional machine learning models are difficult to deploy in agricultural production environments with limited computing resources.

To make deep learning models flexible for deployment on mobile platforms, scholars have proposed lightweight network structures such as GhostNet [18], MobileNet [19], ShuffleNet [20], etc. Existing domestic and international methods for mobile crop disease identification include application realization on mobile devices such as cell phones, intelligent mobile monitoring robots, aerial monitoring drones, and other ground deployment methods [21,22]. In response to the above research, this paper proposes a lightweight corn seed disease identification method with an improved ShuffleNetV2. We started by using CycleGAN [23] to solve the problem of the unbalanced corn seed disease dataset. Then, the ShuffleNetV2 and ECA [24] modules are combined to improve network performance; the network structure is simplified to speed up network inference. Finally, an experimental evaluation was conducted on a corn seeds dataset [25], and the results showed that the improved model was lighter and had better recognition accuracy compared with ShuffleNetV2, which shows the potential for crop pest recognition on mobile platforms with low computational power.

2. Dataset Preparation

The dataset in this paper is the public Corn Seeds Dataset [25] provided by the laboratory in Hyderabad, India, which classifies corn seeds into four categories, pure, broken, discolored, and silkcut, for a total of 17,801 maps. The number of healthy seeds accounts for 40.8% of the original dataset, and the number of diseased seeds that are broken, discolored, and silkcut account for 32%, 17.4%, and 9.8% of the total, respectively. The number of corn seeds in each of the four categories is thus extremely unbalanced. Nagar S. et al. [25] used the BigGAN [26] method to generate 5000 pseudograms, but the dataset suffered from classification inaccuracies, making the recognition model network severely over-fitted. Figure 1 shows a preview of the four categories of the Corn Seeds Dataset.

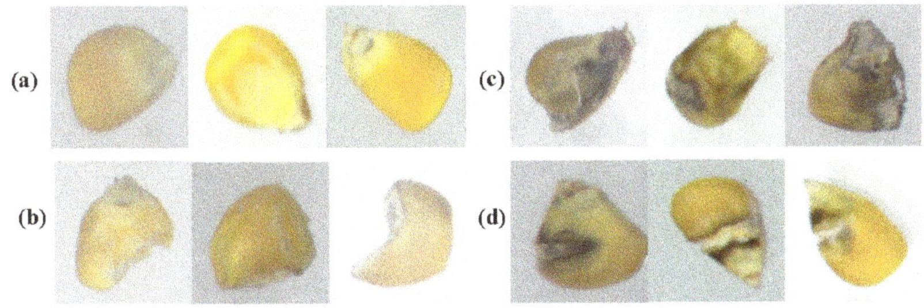

Figure 1. Corn Seeds Dataset: (**a**) pure, (**b**) broken, (**c**) discolored, (**d**) silkcut.

2.1. CycleGAN Data Augmentation

In this paper, each type of seed in the dataset is first manually corrected according to the original criteria:

(1) Pure: the seeds are full in appearance, with no visible breakage, mold, black rot, or cracks;
(2) Broken: the seeds are incomplete in appearance, with visible breaking, accompanied by a few mold infections and discoloration;
(3) Discolored: large areas of mold infection and black rot on the seed surface causing discoloration of the seeds, accompanied by a small amount of breakage;
(4) Silkcut: the basic type of seed is intact, with visible cracks on the surface, with a few accompanying discolorations or breakage at the cracks.

Next, we use the Cycle–Consistent Adversarial Networks (CycleGAN) to generate disease seeds in the dataset to solve the dataset's imbalance problem. CycleGAN is an unsupervised generative adversarial network that does not require a one-to-one mapping relationship between training data for image-to-image translation. Figure 2 illustrates the model structure of CycleGAN, where A and B represent two different styles of image domains; a and b represent the images in the A and B domains, respectively; and G and F represent the generators required for the mutual translation process of image domains A and B. The translation process from A to B can be described as follows: a obtains a forged image $G(a)$ with the style of B through generator G, and the forged image $G(a)$ is input to generator F to obtain the reconstructed image $F(G(a))$. The translation of B to A follows the same process as above. The two discriminators, $Discriminator_A$ and $Discriminator_B$, discriminate the forged image and calculate the probabilities $D_A(a)$ and $D_A(F(b))$ that a and $F(b)$ belong to the A domain and similarly obtain the probabilities $D_B(b)$ and $D_B(G(a))$ that b and $G(a)$ belong to the B domain. The computed probabilities between the discriminators D_A and D_B are used to define the adversarial loss of CycleGAN, which ensures that the generator and discriminator evolve with each other, thus allowing the generator to generate more realistic images; the A and B domains and the reconstructed image domains $F(G(a))$ and $G(F(b)$ are mapped using a cyclic consistent loss function to ensure an efficient mapping from domain A to B.

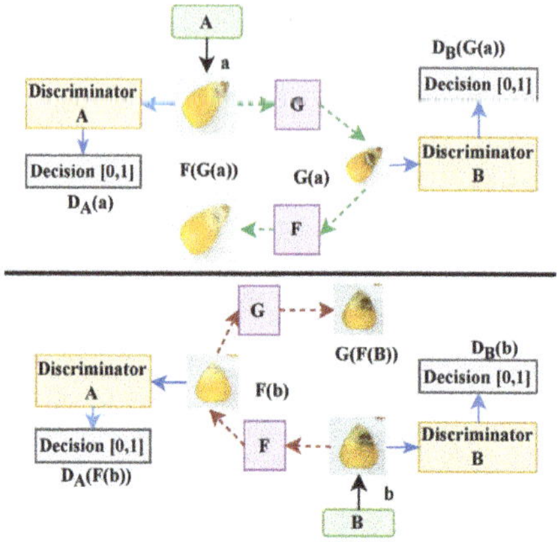

Figure 2. CycleGAN model structure.

2.2. Loss Function

The loss function of CycleGAN consists of the adversarial loss $Loss_{GAN}$ and the cycle consistent loss $Loss_{cycle}$, which ensures an efficient mapping of the two domains. The total adversarial loss is $Loss_{GAN}$ with image domains A and B. G is the mapping generator from image domain A to B with the adversarial loss function denoted as $Loss_{A2B}$; F is the mapping generator from image domain B to A with the adversarial loss function denoted as $Loss_{B2A}$; and the total adversarial loss is $Loss_{GAN}$, as shown in the following Equations (1)–(4):

$$Loss_{GAN} = Loss_{A2B} + Loss_{B2A} \tag{1}$$

$$Loss_{A2B} = \mathbb{E}_{(b \sim pdata(b))}[logD_B(b)] + \mathbb{E}_{(a \sim pdata(a))}[log(1 - D_B(G(a)))] \tag{2}$$

$$Loss_{B2A} = \mathbb{E}_{(a \sim pdata(a))}[logD_A(a)] + \mathbb{E}_{(b \sim pdata(b))}[log(1 - D_A(G(b)))] \tag{3}$$

$$Loss_{Cycle} = \mathbb{E}_{(a \sim pdata(a))}[\| F(G(a)) - a \|_1] + \mathbb{E}_{(b \sim pdata(b))}[\| G(F(b)) - b \|_1] \tag{4}$$

where $pdata(a)$ and $pdata(b)$ are the probability distributions of image domains A and B, respectively.

The cycle consistent loss function $Loss_{cycle}$ is shown in Equation (4). For each image a from domain A, the image translation cycle should be able to bring a back to the original image, i.e., $a \approx F(G(a))$. We call this forward cycle consistency. The cycle consistent loss of the B to A domain is similar: $b \approx G(F(b))$ satisfies the backward cycle consistency, thus avoiding the situation of invalid adversarial loss. From the above Equations (1)–(4), the loss function Loss of CycleGAN can be obtained as in Equation (5):

$$Loss = Loss_{GAN} + \lambda \times Loss_{Cycle} \tag{5}$$

where λ is weight of the cyclic consistency loss, controlling for the relative importance of adversarial and cyclic consistency losses.

2.3. Training Results

Table 1 below shows the distribution of pure, broken, discolored and silkcut seeds in this dataset. This paper uses CycleGAN to supplement the discolored and silkcut classes. The two training processes define image domain A as pure and image domain B as discolored or silkcut.

The CycleGAN image translation models for pure and discolored seeds are first trained. In its model training phase, for the input image domains A (pure) and B (discolored), the corresponding forged and reconstructed images are generated by the generative network; then, the gradient of the generative network is calculated and the weight parameters are updated; next, the gradient of the discriminative network is calculated and the weight parameters are updated; finally, the model parameters are saved. The training process for the pure to silkcut translation is the same as above.

Table 1. Distribution of datasets.

Classification	Original Dataset	After Screening	Participate in Training	CycleGAN Generation	New Dataset
Pure	7265	6972	5473	0	5473
Broken	5670	5489	5489	0	5489
Discolored	3115	2748	2748	2677	5425
Silkcut	1751	1569	1569	4011	5580

The input image is processed by RandomHorizontalFlip, RandomCrop, and Normalize into the generator; the input and output image resolutions are unified at 128 × 128, the number of residual blocks is 6, the Adam optimizer is used, the batch size is 1, and a total of 200 iterations are trained from the beginning. The learning rate remains constant at 0.0002 for the first 100 iterations, and decreases linearly in the direction of 0 for the next 100 iterations. The experimental framework was PaddlePaddle 2.1.2 (Baidu; Beijing, China), Python 3.7 (Centrum Wiskunde & Informatica, Netherlands), and an NVIDIA Tesla V100 graphics card for model training.

The experimental procedure is shown in Figures 3 and 4. The image translation process recordings for each 50 iterations of the two training sessions are shown separately. The first row of the figure are all real training images of image domains *A* and *B*, and the second row shows the corresponding generated images. The figure clearly shows that as the number of training epochs increases, the healthy maize seeds gradually possess the features of the diseased seeds, while the mold and cracks on the diseased seeds gradually disappear, representing that the CycleGAN network has learned the mapping relationship between the two image domains, completing the translation between pure and discolored (or pure and silkcut).

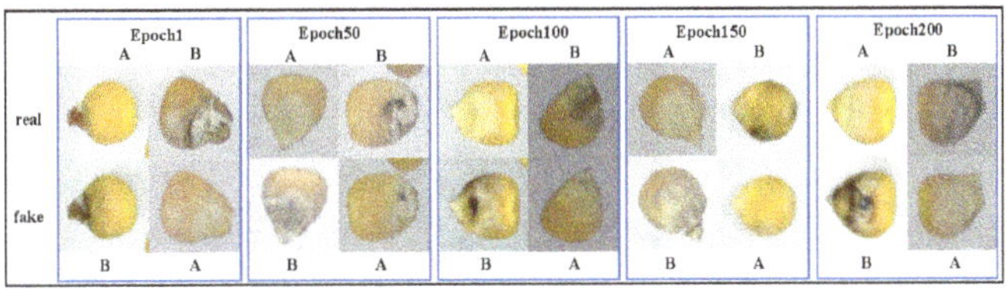

Figure 3. Pure and discolored interconversion training process.

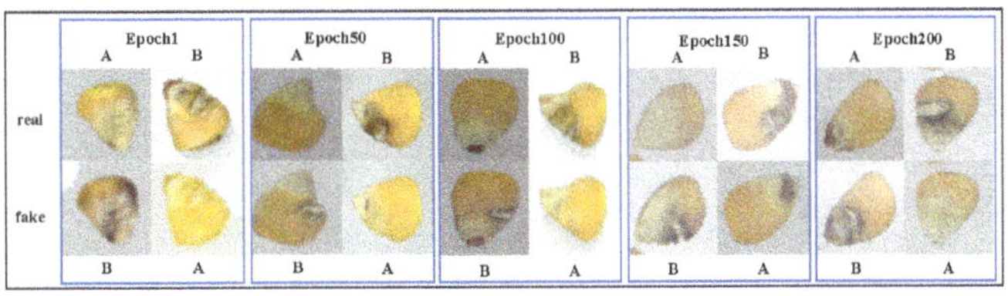

Figure 4. Pure and silkcut interconversion training process.

In the model testing phase, e.g., translating pure seeds to discolored, the latest saved model is first loaded; then, a batch of pure and discolored images are fed into CycleGAN; and after the test is completed, we will obtain a batch of fake images, such as the fake discolored seed images in Figure 5. Finally, we save the resulting fake map and use it to solve the imbalance problem in the dataset.

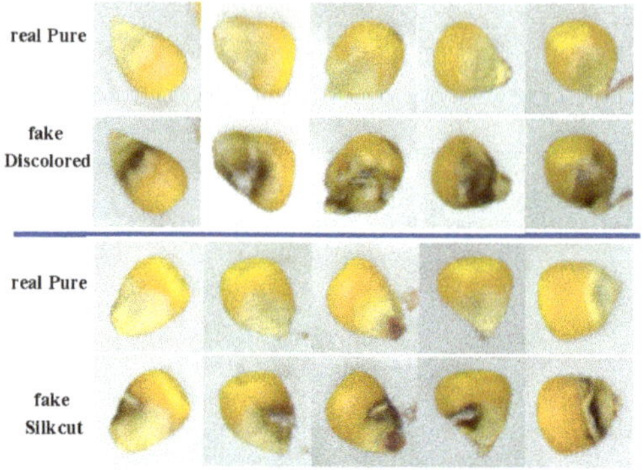

Figure 5. CycleGAN model image translation test results.

Figure 5 shows the results of testing with the trained CycleGAN model, with pure translated into discolored on top and pure translated into silkcut on the bottom. The generated images, noted as fake discolored or fake silkcut, are similar to the discolored and silkcut corn seed features in the real dataset and can be used to balance the dataset.

3. Efficient Channel Attention

Attention mechanisms have been widely used in machine vision in recent years, and attention in neural networks allows the system to pay more attention to focused information and learn better attention weights. The ECA attention mechanism proposed by Wang et al. is an improvement on the SE (Squeeze-and-Excitation) [27] module. The authors found that avoiding dimensionality reduction and a proper local cross-channel interaction strategy helped to improve the performance and efficiency of channel attention.

The ECA structure is shown in Figure 6, with feature χ passing through the Global Average Pooling (GAP) layer to obtain a $1 \times 1 \times C$ feature matrix, and the convolution operation is performed on this feature using a local weight sharing, one-dimensional convolution, followed by a sigmoid activation function to obtain the attention weights; then, the output feature matrix is derived by multiplying this weight with the input feature map. One-dimensional convolution involves an adaptive hyperparameter k (convolution kernel size), which represents the coverage of local cross-channel interactions, as shown in Equation (6):

$$k = \psi(C) = |\frac{log_2(c)}{\gamma} + \frac{b}{\gamma}|_{odd}, (\gamma = 2, b = 1) \quad (6)$$

where $|x|_{odd}$ denotes the oddest number closest to x. ECA attention mechanisms have fewer parameters compared to SE, and appropriate cross-channel information interaction ensures that gains are brought to the network while introducing a small number of parameters.

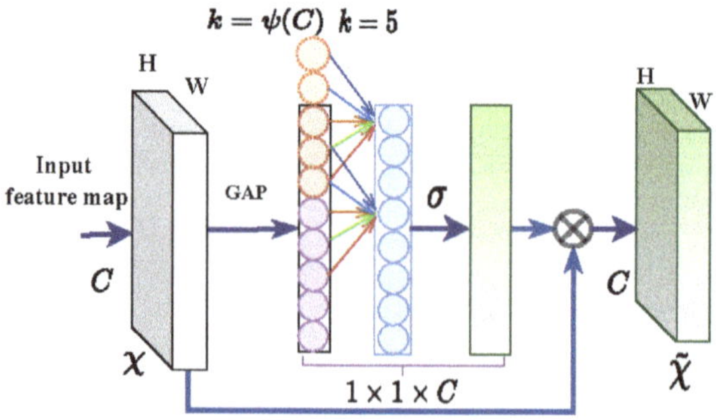

Figure 6. ECA attention mechanism.

4. ECA—ShuffleNetV2 Relevant Theories

The ultimate goal of deep learning development has always been to proceed to practical applications, which makes people more concerned about how to obtain the optimal results with limited resources. In this paper, we compare the Top-1 Accuracy of GhostNet, MobileNet V2, MobileNet V3, and ShuffleNetV2 series networks under CIFAR-10 [28]. The following metrics were counted using the torchstate tool provided by PyTorch: FLOPs, Params, and Memory cost.

To meet the needs of mobile deployments, only lightweight versions of these networks are tested in this article. The detailed model versions are shown in Table 2. We had to make a trade-off between accuracy and speed, and the table shows that MobileNet V2 1× achieved the highest Top-1 Accuracy on CIFAR-10, with ShuffleNetV2 1× following closely behind. However, the FLOPs, Params, and Memory costs of ShuffleNetV2 1× are much lower than those of MobileNet V2 1×. Without a doubt, ShuffleNet V2 1× won this race, and this was further confirmed in subsequent experiments. The subsequent experiments also focused on ShuffleNetV2 and MobileNet V2. To further refine the experiments, the rest of the networks are tested in the final inference speed comparison.

Table 2. Comparison of various parameters of the lightweight model.

Model	FLOPs (M)	Params (M)	Memory Cost (MB)	Top-1 Accuracy (%)
ShuffleNetV2 0.5×	42.63	0.352	10.63	85.7
ShuffleNetV2 1×	149.58	1.257	20.84	88.5
MobileNet V2 0.5×	101.35	0.590	40.12	87.6
MobileNet V2 1×	318.96	2.236	74.25	89.4
MobileNet V3-Small	58.8	1.528	16.20	88.0
GhostNet 0.5×	45.82	1.319	20.03	85.6
GhostNet 1×	149.41	3.914	40.05	87.7

4.1. Depthwise Separable Convolution

Depthwise Separable Convolution (DSC) has been effective in making network structures more lightweight [29]. It consisting of depthwise (DW) and pointwise (PW) convolution and has a relatively low number of parameters and computational cost in extracting features compared to ordinary convolution. One depthwise convolution is computed for only one channel of the input feature map. Pointwise convolution is similar to ordinary convolution with a size of $1 \times 1 \times M$. M is the number of channels in the input feature matrix. Thus, the number of output feature maps is equal to PW's number of convolution kernels. Moreover, PW convolution solves the problem of DW convolution, which is the poor interaction between the feature information of different channels at the same spatial location, shown in Figure 7.

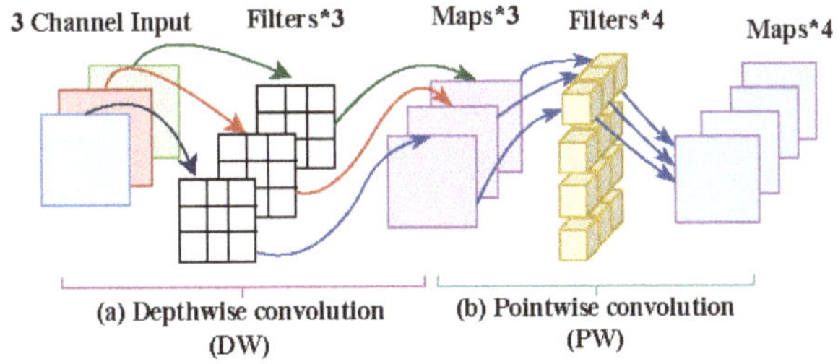

Figure 7. Depthwise separable convolution: (**a**) DW: Depthwise convolution; (**b**) PW: Pointwise convolution.

4.2. Grouped Convolution and Channel Shuffle

Grouped convolution discretizes dense convolutional connections to build a sufficiently deep and wide neural network by replicating the grouped convolution. Compared with standard convolution, grouped convolution has less parameters, lower complexity, and helps to facilitate the parallelism in the model, but the lack of information exchange between different groups weakens the feature extraction ability of the network. ShuffleNet V1 uses the channel shuffle to remedy this deficiency [30]. As shown in Figure 8, Output Features 1 after group convolution are "shuffled" by the channel shuffle operation, which fully integrates the inter-group channel information without increasing the computational effort.

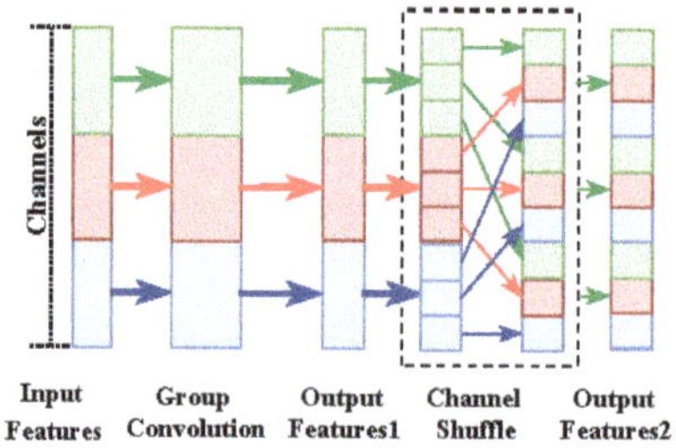

Figure 8. Grouped convolution and channel shuffle.

4.3. *ShuffleNet V2*

ShuffleNetV2 proposes that FLOPs are an indirect metric that cannot be equated with direct metrics such as speed and accuracy and proposes four guidelines for efficient network design:

(1) Maintain a constant number of convolutional input and output channel widths to minimize the memory access cost;
(2) The quantity of groups in group convolution is inversely proportional to the speed of network operation;
(3) Cautious fragmentation operations and reducing the count of network branches can improve operational efficiency;
(4) Reducing element-wise operations which have relatively small FLOPs but high memory access costs.

Starting from the four guidelines mentioned above, the authors present an improved ShuffleNetV2 network structure, illustrated in Figure 9. The authors devised the channel split operation shown in Figure 9a. In front of each unit, the input channel C is divided into two equal branches. To avoid fragmentation, one branch is left unchanged. The other branch follows criterion (1) and consists of three convolutional layers with a constant number of channels and no longer uses 1×1 grouped convolution. Then, the two branches are concatenated in the depth direction, followed by a channel shuffling operation to enhance the information interaction between the channels. Figure 9b shows the module with downsampling, and with channel splits removed, twice the number of channels of the output feature map as the input are obtained.

4.4. *ECA—ShuffleNetV2*

The basic component of a transformer is self-attention, which essentially performs Query-Key-Value operations at a global scale or within a larger window, which is the reason for the superior performance of transformers on downstream tasks [31]. Recently, many scholars have applied large convolution kernels to CNNs. Liu et al. [32] borrowed the large-scale window of the transformer in their paper and changed the size of the convolution kernels in CNNs from 3×3 to 7×7. They experimentally concluded that 7×7 convolution kernels can achieve better detection on the ImageNet dataset with only a small increase in the number of parameters. Similarly Ding et al. [33] state in their article that large convolution kernels are both more accurate and more efficient at this task.

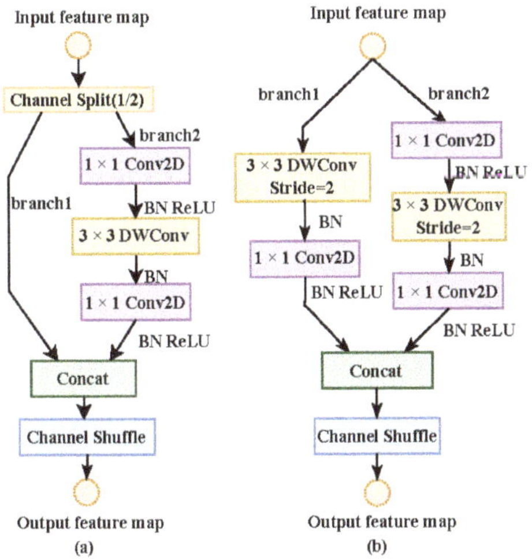

Figure 9. (**a**) ShuffleNetV2 basic unit, ShuffleNetV2 block 1; (**b**) unit for spatial down sampling, ShuffleNetV2 block 2. DWConv: depthwise convolution.

In this paper, the two units of ShuffleNetV2 are improved by moving forward the depthwise convolution in the original branch 2, followed by two pointwise convolutions. The specific structure of branch 2 is shown in Figure 10b, where a depthwise convolution of the size 7 × 7 is used instead of the 3 × 3 depthwise convolution in ShuffleNetV2, depicted in Figure 10c.

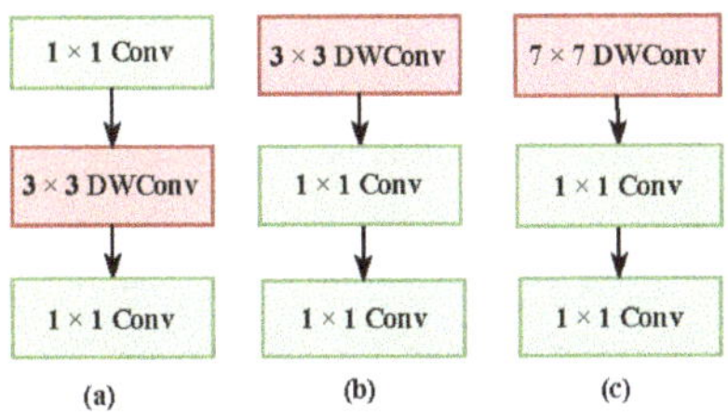

Figure 10. (**a**) Structure for branch 2 in the ShuffleNetV2 unit; (**b**) DW convolutional advance of branch 2's structure; (**c**) structure of branch 2 in the ECA ShuffleNetV2 base unit.

In practical applications, the model needs to be reasonably designed according to the complexity of the task. The deep small kernel network has a large theoretical receptive field, but its effective receptive field is limited. We replace the CNN network with a deep small kernel with a shallow but large kernel and conclude in this paper that the method is effective and feasible.

This is a relatively simple four-classification task, and the model size should be appropriately reduced to improve detection efficiency. Therefore, in this paper, the number of repeats of Block C is reduced. Morevoer, the 3 × 3 DW convolution is connected after the channel shuffling of Block D with the downsampling function to further extract features. The modified unit of the network is shown in Figure 11.

Figure 12 shows a comparison of the ShuffleNetV2 1× and the improved ECA-ShuffleNetV2 1× network structure, with the input image resolution changed to 160 × 160. Since the DW convolution is moved forward in the branch, the number of channels in the output feature layer is changed to [58, 116, 232, 464, 1024] here to guarantee that the number of channels of its output feature matrix is divisible by the number of channels of the input feature matrix. The 3 × 3 convolutional layer and the max pooling at the beginning of the ShuffleNetV2 are replaced by a 4 × 4 convolutional, a stride of 4, and the input image is downsampled 4×. We reduce the size of the network by changing the number of iterations of Block C in each stage from (3, 7, 3) to (1, 1, 1). Conv5 is then followed by the ECA attention module, and the appropriate cross-channel information interaction ensures that a small quantity of parameters are introduced while bringing gains to the network.

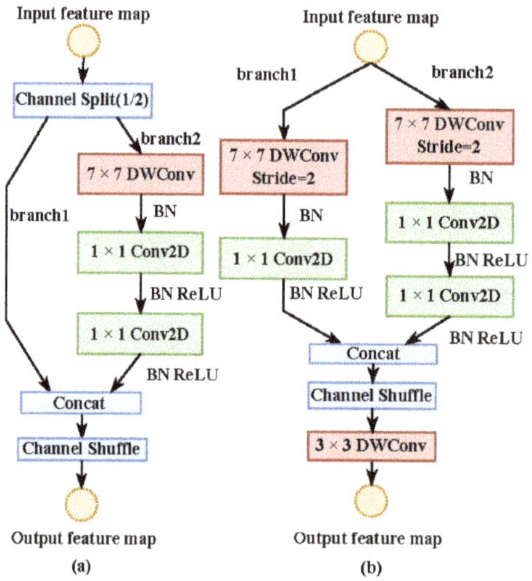

Figure 11. (**a**) ECA-ShuffleNetV2 basic Block C; (**b**) ECA-ShuffleNetV2 Block D for spatial down sampling; DWConv: depthwise convolution.

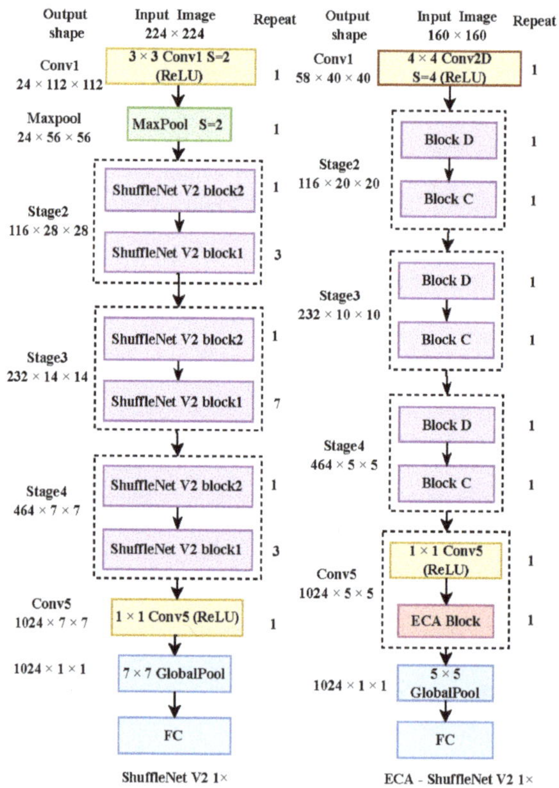

Figure 12. ShuffleNetV2 1× and ECA-ShuffleNetV2 1× network structure.

5. Model Training

5.1. Experimental Details

The dataset in this paper is from the Corn Seeds Dataset provided by the Hyderabad Laboratory in India. The imbalance in the dataset was resolved using CycleGAN, and the distribution of the datasets involved in the training is shown in Table 3, with a total of four categories and 21,967 photos. The balanced dataset was randomly divided into a training set and a validation set according to the 4:1 ratio of each category.

Table 3. Datasets.

No.	Classification	Training Set	Validation Set	Number
1	Pure	4321	1152	5473
2	Broken	4333	1156	5489
3	Discolored	4281	1144	5425
4	Silkcut	4408	1172	5580
Total		17,343	4624	21,967

The experiments were conducted in the Anaconda environment, Python 3.8.1 (Centrum Wiskunde & Informatica, Netherlands), Pytorch 1.10.0 (Facebook Artificial Intelligence Research, Menlo Park, California, United States), CUDA 11.3 (NVIDIA, Santa Clara, California, United States), and trained on an NVIDIA TITAN Xp graphics card. The SGD (stochastic gradient descent) optimizer was used, with parameters set to momentum = 0.9, weight decay = 0.012, and lr = 0.01. The learning rate decay was set to ReduceLROnPlateau with threshold = 0.99, mode = 'min', factor = 0.70, and patience = 3. The input image was resized to 160 × 160 by random horizontal flipping and normalized. The batchsize was set to 200 for a total of 70 epochs of iterative training. Using parameter grouping optimization in training [34], the trainable parameters are divided into two groups based on whether they require L2 regularization processing, where weights in the convolution and fully connected layers are used for L2 regularization. Other parameters, including the biases and γ and β in BN layers, are left unregularized.

5.2. Experimental Results and Analysis

Figure 13 shows the loss and accuracy comparison results of ShuffleNetV2 1×, MobileNet V2, and ECA-ShuffleNetV2 1× models on the train and validation dataset. Obviously, ECA-ShuffleNetV2 1× outperforms the first two in terms of convergence speed and training stability, and has higher accuracy and lower loss value in the training and validation set, which verifies the effectiveness and reliability of the improved method proposed in this paper.

Grade-CAM [35] (Gradient-weighted Class Activation Mapping) is used to visualize the class activation mapping of models as a way of expressing the "visual interpretation" of CNN-based models without the need to modify the model's structure or retrain it. Filling the regions of interest in the CNN model with highlights allows us to analyze whether the network is learning the correct features. In Grade-CAM heat maps, the brighter the color of an activated region indicates its greater relevance to a particular category.

In Figure 14, comparing the grade-CAM heat maps of corn seeds generated by the three models (ECA-ShuffleNetV2 1× (2, 2, 2); ShuffleNetV2 1× (2, 2, 2); ShuffleNetV2 1× (4, 8, 4)), it is not difficult to see that there are differences in the sensitivity of different models to key features. As can be seen in the heat map, the ECA-ShuffleNet V2 model is able to precisely focus on key features that distinguish corn seed disease categories from one another, such as disease spots and cracks. It indicates that this region was helpful for the model to identify the category of the disease and shows that the model extracted the important disease features well. In contrast, the other two models deal with a large amount of redundant information or ignore valid category feature information, which makes the models less effective at classification.

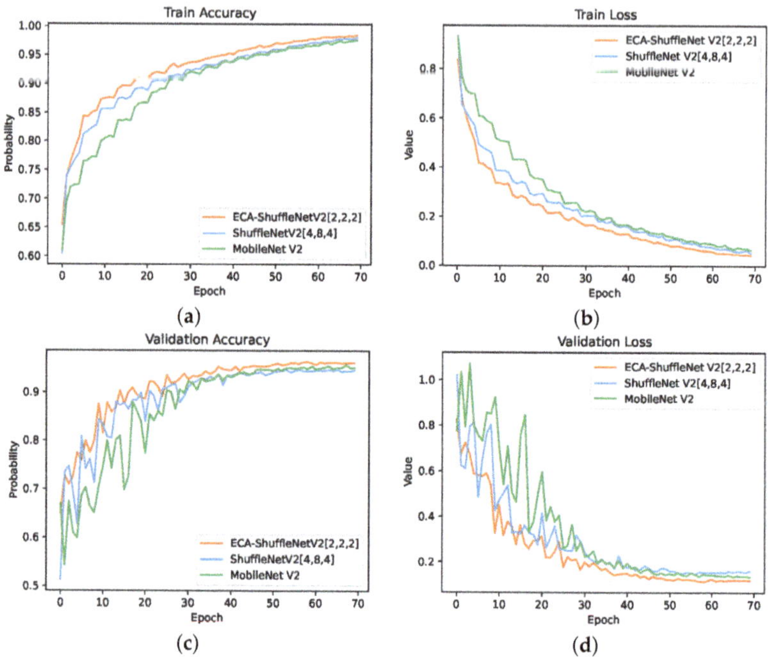

Figure 13. Comparison of accuracy and loss values for ECA-ShuffleNetV2 1×, ShuffleNetV2 1×, and MobileNet V2: (**a**) Training accuracy comparison convolution; (**b**) Training loss comparison; (**c**) Validation accuracy comparison convolution; (**d**) Validation loss comparison.

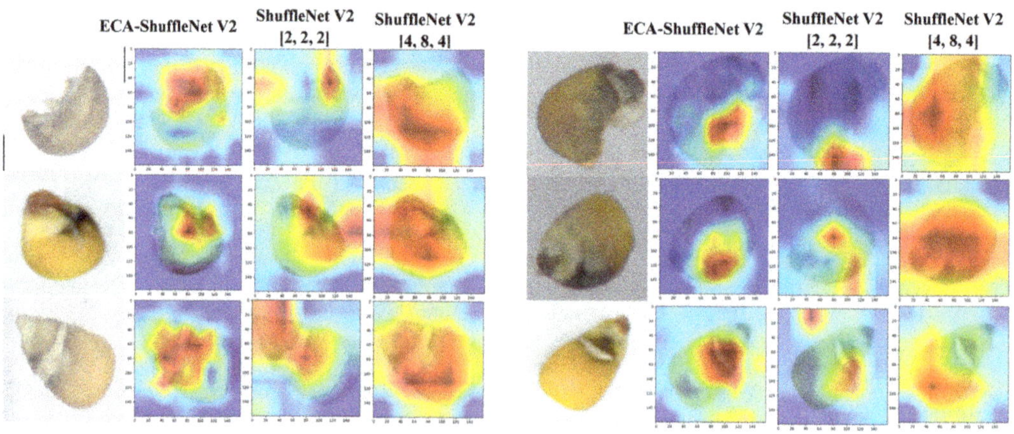

Figure 14. Grad-CAM heat map.

5.3. Ablation Experiments

The results of the ablation experiments of the improved model are shown in Table 4. Model 1 is a ShuffleNetV2 1× network, and Model 2 only reduces the number of iterations of Model block 1 to one, which reduces the depth of the network, reduces the number of parameters and FLOPs by 1/3, and improves accuracy by 0.07%. Model 3 only adds 3 × 3 DW convolution after block 2 of Model 1, which increases accuracy by 0.68% by introducing only a small number of parameters. Model 4 moved the DW convolution in branch 2 forward and modified the output feature layer channels from [24, 116, 232, 464, 1024] to [58, 116, 232, 464, 1024], the accuracy improved to 95.24%. Model 5 replaces the 3 × 3 DW convolution with 7 × 7 DW convolution in the branch, and the accuracy is improved by 0.08. Model 6 adds the ECA attention module after Conv5, with almost no change in the number of parameters and FLOPs and an accuracy increase of 0.87 percentage points, with only a small increase in memory cost. Model 7 replaces the convolution and max pooling layers of Conv1 with an ordinary convolution, a kernel size of 4, and a stride of 4, and the accuracy rose by 0.47%. This method, Model 8 (ECA-ShuffleNetV2), uses the above six improvement strategies simultaneously on the basis of Model 1 (ShuffleNetV2 1×). Relative to Model 1, the accuracy improved to 96.28% with a decrease of nearly 30% in Params and a compression of nearly half in FLOPs, Memory cost, and Madd, resulting in a good improvement. Model 9 (MobileNet V2 1×) achieved 95.69% accuracy on this dataset, but Params, FLOPs, Memory cost, and Madd were all greater than those of ShuffleNetV2.

After training, confusion matrices were created for the three algorithms MobileNet V2 1×, ShuffleNetV2 1×, and ECA-ShuffleNetV2 1×, and the performance evaluation of the model was carried out using the values (TP, TN, FP, FN) on the confusion matrix. The confusion matrix for the validation set of the CNN model is shown in Figure 15. The confusion matrix is used to visualize the performance of the CNN model, with each column representing the true label of the sample and the rows of the matrix representing the predicted class of the classifier. These four metrics typically include true positives (TP), true negatives (TN), false positives (FP), and false negatives (FN) [36]. In this task, TP and TN correspond to the correct identification of diseased maize seed categories, while FP and FN correspond to misclassification. Model performance was evaluated according to the parameters associated with the confusion matrix, i.e., Accuracy, Precision, Recall, Specificity, F1_Score, which are defined as shown in Table 5. Performance was evaluated on the validation set [37].

Table 4. Results of ablation experiments.

Model	Params (M)	FLOPs (M)	Memory Cost (MB)	Madd (M)	Accuracy (%)
1	1.258	76.32	10.63	150.88	94.74
2	0.848	47.24	6.74	93.36	94.81
3	1.267	77.29	12.85	152.67	95.42
4	1.265	74.89	12.04	147.61	95.24
5	1.356	85.75	10.63	169.74	95.61
6	1.258	76.32	10.65	150.88	94.92
7	1.258	73.51	9.17	145.5	95.31
8	0.913	44.75	5.86	88.50	96.28
9	2.229	162.74	37.98	318.95	95.69

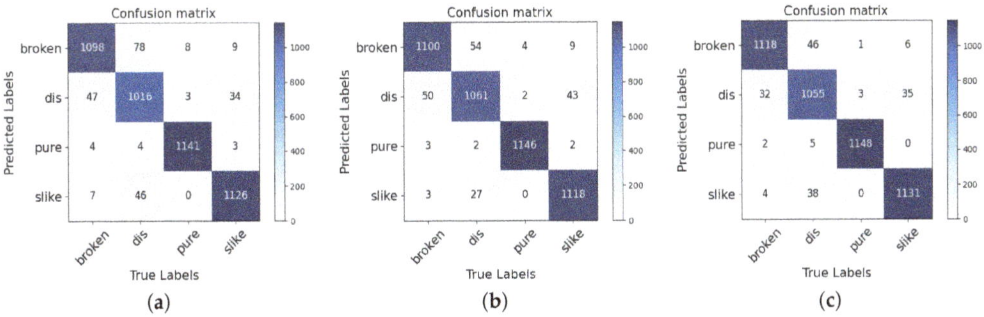

Figure 15. Confusion Matrix for the validation set: (**a**) ShuffleNetV2 1×; (**b**) MobileNet V2; (**c**) ECA-ShuffleNetV2.

A comparison of the confusion matrix's performance metrics for the three models is shown in Tables 4 and 6, which include the Accuracy, Precision, Recall, Specificity, F1_Score of the models. Table 4 demonstrates that the model with the highest accuracy in disease identification is the ECA-ShuffleNetV2 1× (96.28%) proposed in this paper, followed by MobileNet V2 1× (95.69%), and ShuffleNetV2 1× (94.74%). Moreover, it can be concluded that ECA-ShuffleNetV2 1× has excellent detection performance both in terms of speed and accuracy, and this network can extract detailed information about the samples and provide a way of thinking for the processing of similar classification tasks in this area. The experimental results also show that appropriate dataset augmentation has a positive impact on small or unbalanced samples in classifier training. Additionally, CycleGAN effectively contributes to dataset augmentation and increases model accuracy. Meanwhile, the deep learning-based feature extraction method can efficiently and accurately retain the information of corn seed appearance and reduce the information loss caused by manual feature extraction.

Table 5. Performance evaluation metrics.

Metrics	Formula (M)	Evaluation Focus
Accuracy	$\frac{(TP+TN)}{(TP+FP+FN+TN)}$	The ratio of the number of correct predictions made by the classifier to the total number of predictions made by the classifier is measured.
Precision	$\frac{(TP)}{(TP+FP)}$	Denotes the percentage of samples in which the predicted outcome is a positive case and the true case is also a positive case.
Recall	$\frac{(TP)}{(TP+FN)}$	Represents the model's ability to correctly predict the positives out of actual positives.
Specificity	$\frac{(TN)}{(TN+FP)}$	Specificity is the metric that evaluates a model's ability to predict the true negatives of each available category.
F1_Score	$\frac{(2*TP)}{(2*TP+FP+FN)}$	F1_Score gives equal weight to both the Precision and Recall for measuring its performance in terms of accuracy.

Table 6. Classification results of 3 models: SN: ShuffleNetV2 1×; MN V2: MobileNet V2 1×; ECA-SN: ECA-ShuffleNetV2 1×.

Name	Precision (%)			Recall (%)			Specificity (%)			F1_Score (%)		
	SN	MN V2	ECA-SN	SN	MN V2	ECA-SN	SN	MN V2	ECA-SN	SN	MN V2	ECA-SN
Broken	92.04	94.26	95.47	94.98	95.16	96.71	97.26	98.07	98.47	93.49	94.71	96.09
Discolored	92.36	91.78	93.78	88.81	92.74	92.22	97.59	97.27	97.99	90.55	92.26	92.99
Pure	99.05	99.39	99.39	99.05	99.48	99.65	99.68	99.80	99.80	99.05	99.44	99.52
Silkcut	95.50	97.39	96.42	96.08	95.39	96.50	98.46	99.13	98.78	95.79	96.38	96.46

From Table 6, it can be seen that all three classifiers have the excellent recognition of healthy corn seeds, with precision over 99%, and poor recognition of discolored seeds. This is due to the fact that, in most cases, there is more than one form of deterioration in maize seeds during storage; for example, the presence of broken or discolored conditions causes the classifier to become tangled up in identifying specific disease conditions. However, the model identifies healthy corn seeds well, which is sufficient to determine whether corn seeds are up to standard. The ECA-ShuffleNetV2 model proposed in this paper has less than 1 M parameters, and the memory cost is only 5.86 MB; meanwhile, the precision of the model in identifying healthy seeds is 99.39%, and the number of misclassified seeds is small, which can basically meet the needs of mobile applications.

Table 7 shows the single image inference speed of the ShuffleNetV2 1×, MobileNet V2, MobileNet V3-Small, GhostNet 0.5×, and ECA-ShuffleNetV2 1× models on a portable laptop. The device is equipped with an Intel(R) Core(TM) i5-7200U CPU and 4G RAM without GPU acceleration. Input image resolution is 160 × 160. To ensure the accuracy of the test, the single image inference time is the average time it takes for 100 images to be inferred by the model. The table clearly shows that the ECA-ShuffleNetV2 1× achieves an inference time of 9.71 ms per image. This is expected, as ShuffleNetV2 was designed from the outset with inference latency in mind and does not focus solely on FLOPs.

Table 7. The single image inference speed comparison for five models. All results are evaluated with single thread. SN: ShuffleNetV2 1×; MN V2: MobileNet V2 1×; MN V3-S: MobileNet V3-Small; GN: GhostNet 0.5×; ECA-SN: ECA-ShuffleNetV2 1×.

Model	SN	MN V2	MN V3-S	GN	ECA-SN
Time (ms)	16.20	28.11	33.89	28.39	9.71

5.4. Related Work

Previously, Ferentinos [38] developed a deep learning model based on a convolutional neural network architecture specifically adapted to plant leaf detection, trained on a publicly available dataset of 87,848 photos, with experimental results showing that the VGG convolutional neural network achieved 99.53% accuracy (top-1 error of 0.47%) in the classification of the test set. Moreover, the inference time of the model is about 2 ms under a single GPU, which is suitable for the deployment of mobile devices.

Common classifier training methods are transfer learning and training from scratch. For the case of supervised learning, the most time-consuming steps are the production of convincing and relevant sample datasets and the significant time spent on parameter training; the overfitting and convergence states of CNN networks are also of concern. In problems similar to the classification of seed pests and diseases, transfer learning can be used to speed up the training of classifiers [39]. Gulzar et al. [40] built on the VGG16 network and trained the model using transfer learning; the model achieved 99% accuracy on the seed classification dataset. Hamid et al. [41] also performed experiments on this dataset using MobileNetV2 for seed classification, and the results showed an accuracy of 98% and 95% for the training and test sets, respectively. Although the classification accuracy is not as good as that of Gulzar et al.'s method, the MobileNetV2 model has better parameter scale, inference speed, and memory usage than VGG16 and is more suitable for applications in mobile scenarios.

The prediction accuracy of supervised deep learning models depends heavily on the amount and diversity of data available during training. Generally, when dealing with complex tasks, the amount of data available for training models is difficult to obtain. In addition, when there are large differences between research tasks or when the problem of imbalance in sample datasets is severe, ordinary simple dataset augmentation algorithms (e.g., padding, random rotating, darkening or brightening/color modification) may not be sufficient for the experimental needs. It is possible to overcome this challenge by using GAN-based dataset augmentation, such as how CycleGAN was used in this paper [42].

6. Conclusions and Future Work

This paper presents an improved ECA-ShuffleNetV2 network for corn seed disease identification. Production of the dataset requires only a low-cost digital camera. On a CPU device, this model has a single-threaded inference speed of about 9.71 ms and a classification accuracy of 96.28% on the validation set. It has 0.913 M parameters and 44.75 M FLOPs. The model is suitable for deployment on mobile devices, such as smart phones and portable laptops available to growers or quality assessment practitioners, as well as on offline mobile monitoring sites.

Compared with traditional machine learning, deep learning is more effective at pest and disease classification. In addition, the corn seed disease recognition model built in this paper has the advantages of high recognition accuracy, fast recognition speed, and relatively small model parameters. This model can quickly extract the characteristics of diseased seeds, which greatly reduces the workload of manual detection. It is possible for relevant departments to collect and produce different data sets of agricultural samples for quality evaluation model training of other cereals according to the actual situation, which has good application prospects.

Using machine vision, we can only obtain phenotypic information about seeds; however, that information does not describe their internal characteristics. Therefore, in the follow-up work, we will combine our method with hyperspectral technology in order to detect and classify seed diseases and enhance the model's robustness so that it works well on similar tasks.

According to a comparison with related research at this stage, the limited number of seed samples chosen for this study is not representative of all diseases present in Chinese maize seeds today. Particularly, diseased seeds are much less common than healthy seeds, which is a huge test of our work. In the next study, a complete maize seed image acquisition system will be designed to address the problem. To build a comprehensive and accurate data set of maize seed diseases, we will collaborate with relevant departments. We will also ensure its classification quality in order to improve the practical utility of the model.

While the acquisition of datasets is critical, the success of supervised machine learning is not possible without high-quality data annotation. Manual annotation is labor-intensive and time-consuming, so relying only on manual annotation is not a wise decision. Our future research will use unsupervised machine learning to classify and label maize seed diseases. This will further reduce the workload and accelerate the deployment of the model in agricultural quality assessment.

In addition, this experiment did not consider the relationship between external factors, such as different corn growing regions, corn varieties, and climate in China, and corn seed diseases. These external factors are also crucial to the quality assessment of agricultural products and food waste reduction. As a result, follow-up work will include further experiments and investigations related to this topic.

Author Contributions: Conceptualization, L.L. and W.L.; methodology, L.L. and W.L.; validation, L.L. and W.Y.; formal analysis, W.Y. and M.Z.; investigation, L.L., T.J. and W.Y.; resources, W.L.; data curation, L.L., T.J. and W.Y.; writing—original draft preparation, L.L. and T.J.; writing—review and editing, L.L. and W.L.; visualization, L.L. and W.Y.; supervision, W.L.; project administration, L.L. and W.L.; funding acquisition, W.L. All authors have read and agreed to the published version of the manuscript.

Funding: This work was supported by the Yantai Science and Technology Innovation Development Plan Project (Grant No. 2022XDRH015).

Informed Consent Statement: Not applicable.

Data Availability Statement: Not applicable.

Conflicts of Interest: The authors declare no conflict of interest.

References

1. Phupattanasilp, P.; Tong, S.-R. Augmented Reality in the Integrative Internet of Things (AR-IoT): Application for Precision Farming. *Sustainability* **2019**, *11*, 2658.
2. Kong, J.; Wang, H.; Wang, X.; Jin, X.; Fang, X.; Lin, S. Multi-stream hybrid architecture based on cross-level fusion strategy for fine-grained crop species recognition in precision agriculture. *Comput. Electron. Agric.* **2021**, *185*, 106134.
3. Watson, A.; Burgess, L.W.; Summerell, B.; O'Keeffe, K. Fusarium species associated with cob rot of sweet corn and maize in New South Wales. *Australas. Plant Dis. Notes* **2014**, *9*, 142 .
4. Sastry, K.S. *Seed-Borne Plant Virus Diseases*; Springer: Berlin/Heidelberg, Germany, 2013; pp. 85–100.
5. Schmidt, M.; Horstmann, S.; Colli, L.D.; Danaher, M.; Speer, K.; Zannini, E.; Arendt, E.K. Impact of fungal contamination of wheat on grain quality criteria. *J. Cereal Sci.* **2016**, *69*, 95–103.
6. Franco-Duarte, R.; Černáková, L.; Kadam, S.; Kaushik, K.S.; Salehi, B.; Bevilacqua, A.; Corbo, M.R.; Antolak, H.; Dybka-Stępień, K.; Leszczewicz, M.; et al. Advances in Chemical and Biological Methods to Identify Microorganisms—From Past to Present. *Microorganisms* **2019**, *7*, 130.
7. Li, J.; Wu, J.; Lin, J.; Li, C.; Lu, H.; Lin, C. Nondestructive Identification of Litchi Downy Blight at Different Stages Based on Spectroscopy Analysis. *Agriculture* **2022**, *12*, 402.
8. Lu, Z.; Zhao, M.; Luo, J.; Wang, G.; Wang, D. Design of a winter-jujube grading robot based on machine vision. *Comput. Electron. Agric.* **2021**, *186*, 106170.
9. Zhang, J.; Dai, L.; Cheng, F. Classification of Frozen Corn Seeds Using Hyperspectral VIS/NIR Reflectence Imaging. *Molecules* **2019**, *24*, 149.
10. Javanmardi, S.; Ashtiani, S.M.; Verbeek, F.J.; Martynenko, A. Computer-vision classification of corn seed varieties using deep convolutional neural network. *J. Stored Prod. Res.* **2021**, *92*, 101800.

11. Wang, Z.; Huang, W.; Tian, X.; Long, Y.; Li, L.; Fan, S. Rapid and Non-destructive Classification of New and Aged Maize Seeds Using Hyperspectral Image and Chemometric Methods. *Front. Plant Sci.* **2022**, *13*, 849495.
12. Yang, D.; Jiang, J.; Jie, Y.; Li, Q.; Shi, T. Detection of the moldy status of the stored maize kernels using hyperspectral imaging and deep learning algorithms. *Int. J. Food Prop.* **2022**, *25*, 170–186.
13. Mishra, S.; Sachan, R.; Rajpal, D. Deep Convolutional Neural Network based Detection System for Real-time Corn Plant Disease Recognition. *Procedia Comput. Sci.* **2020**, *167*, 2003–2010.
14. Meng, R.; Lv, Z.; Yan, J.; Chen, G.; Zhao, F.; Zeng, L.; Xu, B. Development of Spectral Disease Indices for Southern Corn Rust Detection and Severity Classification. *Remote Sens.* **2020**, *12*, 3233.
15. Albarrak, K.; Gulzar, Y.; Hamid, Y.; Mehmood, A.; Soomro, A.B. A Deep Learning-Based Model for Date Fruit Classification. *Sustainability* **2022**, *14*, 6339.
16. Padilla, D.A.; Pajes, R.A.I.; Guzman, J.T.D. Detection of Corn Leaf Diseases Using Convolutional Neural Network with OpenMP Implementation. In Proceedings of the 2020 IEEE 12th International Conference on Humanoid, Nanotechnology, Information Technology, Communication and Control, Environment, and Management (HNICEM), Manila, Philippines, 3–7 December 2020; pp. 1–6.
17. Caballero, D.; Calvini, R.; Amigo, J.M. Chapter 3.3—hyperspectral imaging in crop fields: Precision agriculture. In *Hyperspectral Imaging*; Amigo, J.M., Ed.; Data handling in science and technology; Elsevier: Amsterdam, The Netherlands, 2019; Volume 32, pp. 453–473.
18. Han, K.; Wang, Y.; Tian, Q.; Guo, J.; Xu, C.; Xu, C. GhostNet: More Features From Cheap Operations. In Proceedings of the 2020 IEEE/CVF Conference on Computer Vision and Pattern Recognition (CVPR), Seattle, WA, USA, 13–19 June 2020; pp. 1577–1586.
19. Sandler, M.; Howard, A.; Zhu, M.; Zhmoginov, A.; Chen, L.-C. MobileNetV2: Inverted Residuals and Linear Bottlenecks. In Proceedings of the 2018 IEEE/CVF Conference on Computer Vision and Pattern Recognition, Salt Lake City, UT, USA, 18–23 June 2018; pp. 4510–4520.
20. Ma, N.; Zhang, X.; Zheng, H.T.; Sun, J. Shufflenet v2: Practical guidelines for efficient cnn architecture design. In Proceedings of the European Conference on Computer Vision (ECCV), Munich, Germany, 8–14 September 2018; pp. 116–131.
21. Che'Ya, N.; Mohidem, N.A.; Roslin, N.; Saberioon, M.; Tarmidi, Z.; Shah, J.; Ilahi, W.; Man, N. Mobile Computing for Pest and Disease Management Using Spectral Signature Analysis: A Review. *Agronomy* **2022**, *12*, 967.
22. Shendryk, Y.; Sofonia, J.; Garrard, R.; Rist, Y.; Skocaj, D.; Thorburn, A.P. Fine-scale prediction of biomass and leaf nitrogen content in sugarcane using UAV LiDAR and multispectral imaging. *Int. J. Appl. Earth Obs. Geoinf.* **2020**, *92*, 102177.
23. Zhu, J.; Park, T.; Isola, P.; Efros, A.A. Unpaired Image-to-Image Translation Using Cycle-Consistent Adversarial Networks. In Proceedings of the 2017 IEEE International Conference on Computer Vision (ICCV), Venice, Italy, 22–29 October 2017; pp. 2242–2251.
24. Wang, Q.; Wu, B.; Zhu, P.; Li, P.; Zuo, W.; Hu, Q. ECA-Net: Efficient Channel Attention for Deep Convolutional Neural Networks. In Proceedings of the 2020 IEEE/CVF Conference on Computer Vision and Pattern Recognition (CVPR), Seattle, WA, USA, 13–19 June 2020; pp. 11531–11539.
25. Nagar, S.; Pani, P.; Nair, R.; Varma, G. Automated Seed Quality Testing System using GAN & Active Learning. *arXiv* **2021**, arXiv:2110.00777.
26. Brock, A.; Donahue, J.; Simonyan, K. Large Scale GAN Training for High Fidelity Natural Image Synthesis. *arXiv* **2018**, arXiv:1809.11096.
27. Hu, J.; Shen, L.; Sun, G. Squeeze-and-Excitation Networks. In Proceedings of the 2018 IEEE/CVF Conference on Computer Vision and Pattern Recognition, Salt Lake City, UT, USA, 18–23 June 2018; pp. 7132–7141.
28. Krizhevsky, A. Learning Multiple Layers of Features from Tiny Images. 2009. Available online: http://www.cs.utoronto.ca/~kriz/learning-features-2009-TR.png (accessed on 13 October 2022).
29. Howard, A.; Zhu, M.; Chen, B.; Kalenichenko, D.; Wang, W.; Weyand, T.; Andreetto, M.; Adam, H. MobileNets: Efficient Convolutional Neural Networks for Mobile Vision Applications. *arXiv* **2017**, arXiv:1704.04861.
30. Zhang, X.; Zhou, X.; Lin, M.; Sun, J. ShuffleNet: An Extremely Efficient Convolutional Neural Network for Mobile Devices. In Proceedings of the 2018 IEEE/CVF Conference on Computer Vision and Pattern Recognition, Salt Lake City, UT, USA, 18–23 June 2018; pp. 6848–6856.
31. Vaswani, A.; Shazeer, N.; Parmar, N.; Uszkoreit, J.; Jones, L.; Gomez, A.N.; Kaiser, L.; Polosukhin, I. Attention is All you Need. In *Advances in Neural Information Processing Systems*; Guyon, I., Luxburg, U.V., Bengio, S., Wallach, H., Fergus, R., Vishwanathan, S., Garnett, R., Eds.; Curran Associates, Inc.: Red Hook, NY, USA, 2017; Volume 30.
32. Liu, Z.; Mao, H.; Wu, C.Y.; Feichtenhofer, C.; Darrell, T.; Xie, S. A convnet for the 2020s. In Proceedings of the IEEE/CVF Conference on Computer Vision and Pattern Recognition, New Orleans, LA, USA, 19–20 June 2022; pp. 11976–11986.
33. Ding, X.; Zhang, X.; Han, J.; Ding, G. Scaling up your kernels to 31 × 31: Revisiting large kernel design in cnns. In Proceedings of the IEEE/CVF Conference on Computer Vision and Pattern Recognition, New Orleans, LA, USA, 19–20 June 2022; pp. 11963–11975.
34. He, T.; Zhang, Z.; Zhang, H.; Zhang, Z.; Xie, J.; Li, M. Bag of Tricks for Image Classification with Convolutional Neural Networks. In Proceedings of the 2019 IEEE/CVF Conference on Computer Vision and Pattern Recognition (CVPR), Long Beach, CA, USA, 15–20 June 2019; pp. 558–567.
35. Selvaraju, R.R.; Cogswell, M.; Das, A.; Vedantam, R.; Parikh, D.; Batra, D. Grad-CAM: Visual Explanations from Deep Networks via Gradient-Based Localization. *Int. J. Comput. Vis.* **2020**, *128*, 336–359.
36. Xu, P.; Yang, R.; Zeng, T.; Zhang, J.; Zhang, Y.; Tan, Q. Varietal classification of maize seeds using computer vision and machine learning techniques. *J. Food Process. Eng.* **2021**, *44*, e13846.

37. Feng, J.; Sun, Y.; Zhang, K.; Zhao, Y.; Ren, Y.; Chen, Y.; Zhuang, H.; Chen, S. Autonomous Detection of Spodoptera frugiperda by Feeding Symptoms Directly from UAV RGB Imagery. *Appl. Sci.* **2022**, *12*, 2592.
38. Ferentinos, K.P. Deep learning models for plant disease detection and diagnosis. *Comput. Electron. Agric.* **2018**, *145*, 311–318.
39. Ullah, N.; Khan, J.; Khan, M.; Khan, W.; Hassan, I.; Obayya, M.; Negm, N.; Salama, A. An Effective Approach to Detect and Identify Brain Tumors Using Transfer Learning. *Appl. Sci.* **2022**, *12*, 5645.
40. Gulzar, Y.; Hamid, Y.; Soomro, A.B.; Alwan, A.A.; Journaux, L. A Convolution Neural Network-Based Seed Classification System. *Symmetry* **2020**, *12*, 2018.
41. Hamid, Y.; Wani, S.; Soomro, A.B.; Alwan, A.A.; Gulzar, Y. Smart Seed Classification System based on MobileNetV2 Architecture. In Proceedings of the 2022 2nd International Conference on Computing and Information Technology (ICCIT), Tabuk, Saudi Arabia, 25–27 January 2022; pp. 217–222.
42. Shorten, C.; Khoshgoftaar, T.M. A survey on Image Data Augmentation for Deep Learning. *J. Big Data* **2019**, *6*, 60.

agriculture

Article

Calibration and Test of Contact Parameters between Chopped Cotton Stalks Using Response Surface Methodology

Bingcheng Zhang [1], Xuegeng Chen [1], Rongqing Liang [1], Xinzhong Wang [2], Hewei Meng [1,*] and Za Kan [1]

[1] College of Mechanical and Electrical Engineering, Shihezi University, Shihezi 832000, China
[2] School of Agricultural Engineering, Jiangsu University, Zhenjiang 212013, China
* Correspondence: mengbai4251982@sina.com; Tel.: +86-13369935035

Abstract: The accuracy of the material parameter settings directly affects the reliability of the results of the discrete element method simulation. It is necessary to calibrate the relevant parameters to obtain accurate discrete element simulation results when separating the cotton stalk particles from the residual film after crushing. The repose angle of the chopped cotton stalk particles was used as the response value to calibrate the contact parameters between particles. Physical tests measured the intrinsic particle and contact parameters between the cotton stalk particles and the contact material, which provided data for the simulation tests. According to the biological structure characteristics of cotton stalk, the discrete element method model of cotton stalk particles was constructed by bonding the elements of nonequal-diameter basic particles. Based on the response surface methodology, the stacking test of particles was simulated. The response model between the contact parameters and repose angle was established, and the effect law of the single-factor terms and interaction terms on the repose angle was analyzed. The optimal combination of contact parameters was obtained through the single-objective and multi-variable optimization methods. Finally, the contact parameter combination was verified by a simulation test of the repose angle. The results showed that the average relative error of the repose angle between the simulation test and the physical test was 1.04%, which verified the accuracy of the calibrated contact parameters and the reliability of the simulation test. These parameters provide a basis for the discrete element simulation study of cotton stalk motion in the separation process of cotton stalks and residual film and the subsequent gas–solid coupling simulation research.

Keywords: discrete element method; cotton stalk particles; contact parameters; calibration; response surface methodology

Citation: Zhang, B.; Chen, X.; Liang, R.; Wang, X.; Meng, H.; Kan, Z. Calibration and Test of Contact Parameters between Chopped Cotton Stalks Using Response Surface Methodology. *Agriculture* **2022**, *12*, 1851. https://doi.org/10.3390/agriculture12111851

Academic Editors: Muhammad Sultan, Redmond R. Shamshiri, Md Shamim Ahamed and Muhammad Farooq

Received: 29 September 2022
Accepted: 1 November 2022
Published: 4 November 2022

1. Introduction

Xinjiang, China's major cotton-producing area, commonly uses film mulching for cotton growing due to its geographical and environmental constraints [1–3]. Over the years, this growing method has increased cotton production and income in Xinjiang at the expense of serious residual plastic mulch pollution in farmlands [4,5]. For this reason, universities and research institutes have conducted research on the mechanized recycling of residual film, helping to alleviate such pollution to some extent. However, the recovered residual film mixed with cotton stalks makes reusing it more difficult [6,7]. Therefore, the separation of the residual film and impurities is necessary.

The cotton stalk is a major component of residual film mixtures. Separation from the residual film through conventional test methods makes it difficult to accurately analyze its force and movement. As computer technology advances, computer simulations are increasingly used in the design and optimization of agricultural machinery, greatly improving the efficiency of machinery development. The simulation technology based on the discrete element method (DEM) is widely used in developing and optimizing agricultural machinery [8,9]. The first step of a DEM-based numerical simulation is to establish a DEM model and define its intrinsic parameters (such as density, Poisson's ratio, and elastic modulus, etc.) and contact parameters (such as the coefficient of restitution, coefficient of static friction, and coefficient of rolling friction, etc.) [10]. The first step of a numerical simulation in EDEM (DEM Solutions, United Kingdom) is to calibrate the parameters of materials; the DEM model's reliability depends on the contact model's accuracy and the parameters chosen for the particle properties. The intrinsic parameters of the materials can be obtained through physical experiments [11]. However, due to factors, such as the test methods, instruments, working conditions and constitutive law for particulate matter (particle shape, size, roughness), the contact parameters of the materials may be inaccurate, resulting in an inconsistency between the simulation results and actual test results [12]. In this case, it is necessary to define and calibrate the contact parameters of cotton stalks first when simulating the separation of the cotton stalks and residual film in EDEM so that the simulation results are consistent with the actual test results of the particle material in a specific area or under specific conditions.

DEM-based research on the physical characteristics and motion law of agricultural materials can provide a theoretical basis for designing agricultural machinery and equipment, which has become a trend in computer simulations in the field of agricultural engineering in recent years. DEM is not only widely used in the modeling, parameter calibration and simulation analysis of agricultural bulk materials [13,14], but it also plays a great role in the simulation analysis of stalk material characteristics. Kattenstroth et al. [15] constructed a stalk model based on the multi-sphere approach and bonded-sphere approach of DEM and then analyzed the effects of relevant parameters on the cutting quality by simulating the stalk-cutting process. Li et al. [16] constructed stalk and grain models in EDEM and then simulated and analyzed the motion and separation of stalks and grain in the airflow field by adopting computational fluid dynamics and the discrete element method (CFD–DEM). Xu et al. [17] constructed cucumber stalk models in EDEM and then simulated the stalk-crushing process. Ramirezgomez et al. [18] constructed the models of corn stalks, rice husks, and forage rape stems in EDEM and then calibrated the contact parameters for the density and models of blocky biomass fuel pressed by such materials. By combining the physical and simulation tests of the repose angle of corn stalk particles, Fang et al. [19] designed a Plackett–Burma test to screen for significant contact parameters and calibrated the static friction coefficient and rolling friction coefficient of particles using the response surface method. In order to obtain more accurate DEM simulation parameters in the compression process of the alfalfa stalk, Ma et al. [20] used the Plackett–Burman test and the Box–Behnken test design to calibrate the contact parameters of the alfalfa stalk based on the repose angle physical test. By combining the linear model of cohesion and the Hertz–Mindlin (no slip) contact model, Feng et al. [21] constructed a stalk model in EDEM, calibrated the simulation parameters of the stalk-adopting orthogonal test and analyzed the motion characteristics of the stalk model in the rotary drum. Liao et al. [22] measured the intrinsic parameters of forage rape stem by adopting physical tests and calibrated the discrete element contact parameters of forage rape stem particles based on DEM, with the repose angle of the stalk particles as the response value. The optimal combination of contact parameters was determined by adopting response surface analysis and optimization. Finally, the accuracy of the calibration parameters was verified by comparative tests. Zhang et al. [23] constructed the corn stalk model by using the Hertz–Mindlin with Bonding contact model in EDEM and calibrated the contact parameters between the corn stalk and its external working parts by combining the physical and simulation tests. Ma et al. [24] constructed the models of rice particles and stalks by using the Hertz–Mindlin contact model and used them to simulate the separation process of rice particles and impurities in the operation of a grain combined harvester. Based on the Hertz–Mindlin with Bonding contact model, Liu et al. [25] constructed a flexible wheat stalk model by DEM and simulated its bending behavior. Wang et al. [26] measured the intrinsic parameters and contact parameters of the citrus stalk by adopting physical tests, constructing the citrus stalk DEM model, and finally calibrating the bonding parameters of the model by adopting the three-point bending test and shearing test. Zeng and Chen [27] constructed a simulation model of wheat straw by DEM without considering the bending and plastic deformations of wheat stalks, assigning the model's intrinsic parameters and contact parameters for use in the simulation experiment of stubble-free tillage. Guo et al. [28] constructed the tomato stalk model with an equal stalk diameter by using the multi-sphere approach and bonded-sphere approach in EDEM, assigning the model's basic contact parameters to simulate the mixing process. The above literature demonstrates the extensive research conducted by researchers on DEM-model construction, simulation parameters and bonding parameters calibration for various types of crop stalks. Among them, research on cotton stalks remains scarce. In addition, the shape and actual structure are quite different, as the stalk DEM model is mainly constructed by bonding multiple equal-diameter particle spheres, resulting in poor consistency between the simulation results and the actual test results.

For that reason, based on the biological structure characteristics of cotton stalks, the single distribution method is adopted to arrange the particle groups in the epidermis and internal tissues (xylem and pith) of the chopped cotton stalks [29], and the method of bonding elements of nonequal-diameter basic particles is adopted to construct a DEM model of cotton stalk particles in this paper. The repose angle of chopped cotton stalks was obtained by adopting the cylinder-lifting method, and the relevant parameters required for the simulation were obtained by physical tests. The contact parameters between cotton stalks were calibrated by adopting the repose angle simulation test, and the second-order response model between the repose angle and the contact parameters was established. The effect of the restitution coefficient, static friction coefficient and rolling friction coefficient between cotton stalks on the repose angle was obtained. On this basis, the optimal combination of contact parameters is obtained by taking the value of the repose angle physical test as the optimization target value, and the simulation verification is carried out. The objectives of this study are as follows: (1) To analyze the effect of the contact parameters between particles on the repose angle of chopped cotton stalks; (2) to obtain the DEM parameters of chopped cotton stalk particles; (3) to verify the accuracy of the calibrated parameters. The research provides a basis for the discrete element simulation study of cotton stalk motion in the separation process of cotton stalks and residual film and the subsequent gas–solid coupling simulation research.

2. Materials and Methods

2.1. Test Materials

The "Xinluzao No. 45" cotton from the cotton planting test field of Shihezi University in Xinjiang was selected as the test material and was sampled on 5 November 2021, with the moisture content of cotton stalks ranging from 22.03% to 36.16% and the diameter ranging from 5 mm to 15 mm. Cotton stalks without pests and bending damage were selected to ensure the experiment effect. In the mechanically-harvested residual film mixtures in the cotton field, the particle sizes of cotton stalks were different after being crushed by hammer-type crushing devices. Therefore, the test samples were prepared by adopting the methods of determining the repose angle of chopped stalks in references Fang et al. [19] and Sun et al. [30]. After the lateral branches were removed, the cotton stalks were chopped according to the length, ranging from 10 mm to 20 mm, and the particles at the stem-bud bulges and lateral branches were removed, as shown in Figure 1a. In terms of biological structure, the cotton stalk was divided into the epidermis, xylem and pith, as shown in Figure 1b.

Figure 1. (**a**) Chopped cotton stalk particles; (**b**) biological structure of cotton stalk.

2.2. *Simulation Contact Model Selection and Model Construction*

2.2.1. Simulation Contact Model Selection

The DEM contact model of the particles in the simulation test had a significant effect on the test results. There is no adhesion between the cotton stalk particles in the simulation, so the effect of surface energy is ignored [31]. Meanwhile, it is assumed that the difference in the small overlaps between particles or between the particles and contact materials determines the changes in parameters, such as the displacement, force and velocity of particles in motion. According to Newton's Second Law of Motion, each particle model produces motion under the action of force and torque, and there is normal motion, tangential motion and rolling between particles. Based on the above assumptions, the Hertz–Mindlin (no slip), which is a default model in EDEM, was selected as the contact model of cotton stalk particles [32–34], as shown in Figure 2.

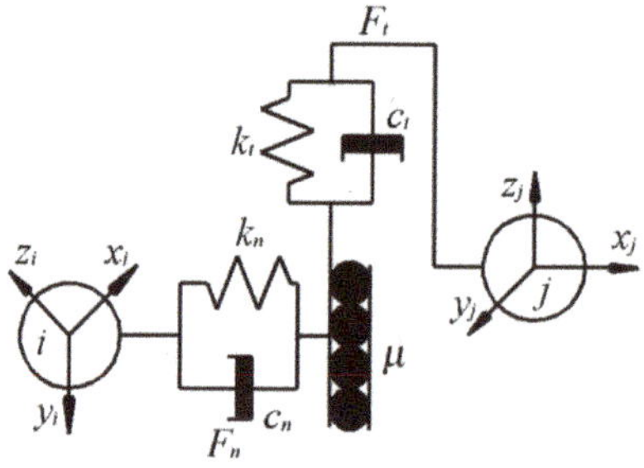

Figure 2. Schematic diagram of Hertz–Mindlin (no slip) contact model.

In this contact model, the relationship equations between the normal force (F_n) and tangential force (F_t) of particles and the overlaps δ_n and δ_t in their respective directions can be expressed as:

$$\begin{cases} F_n = \frac{4}{3}\delta_n^{\frac{3}{2}} E^* \sqrt{r^*} \\ F_t = -S_t\delta_t \end{cases} \tag{1}$$

where E^* is Young's modulus, r^* is the equivalent radius, δ_n is normal overlap, δ_t is tangential overlap, and S_t is tangential stiffness. Where the E^*, r^* and S_t can be given by Equation (2):

$$\begin{cases} E^* = \frac{E_i+E_j}{E_i(1-v_j^2)+E_j(1-v_i^2)} \\ r^* = \frac{r_i r_j}{r_i+r_j} \\ S_t = 8G^*\sqrt{r^*\delta_n} \\ G^* = \frac{E^*}{2(1+\mu)} \end{cases} \tag{2}$$

where E_i and E_j are the Young's modulus of particles i and j, r_i and r_j are the radii of particles i and j, v_i and v_j are the Poisson's ratio of particles i and j, G^* is the equivalent shear modulus, and μ is Poisson's ratio.

Meanwhile, the normal damping force F_n^d and tangential damping force F_t^d of particles can be expressed as:

$$\begin{cases} F_n^d = -v_n\beta\sqrt{\frac{10S_n m^*}{3}} \\ F_t^d = -v_t\beta\sqrt{\frac{10S_t m^*}{3}} \end{cases} \tag{3}$$

where m^* is the equivalent mass, v_n and v_t are the relative normal velocity and relative tangential velocity, and β is the damping ratio. Where m^*, S_n and β can be given by Equation (4):

$$\begin{cases} \beta = \frac{\ln e}{\sqrt{\ln^2 e + \pi^2}} \\ m^* = \left(\frac{1}{m_i} + \frac{1}{m_j}\right)^{-1} \\ S_n = 2E^*\sqrt{r^*\delta_n} \end{cases} \tag{4}$$

where m_i and m_j are the mass of particles i and j, and e is the coefficient of restitution.

The particles are inevitably affected by the rolling friction in motion; thus, the rolling friction can be explained by the torque T_i on the contact surface [35], such as in Equation (5):

$$T_i = -\mu_r F_n R_i \omega_i \tag{5}$$

where μ_r is the coefficient of rolling friction, R_i is the radius of particle i, and ω_i is the angular velocity.

2.2.2. Discrete Element Model Construction of Cotton Stalk Particles

In the simulation, a suitable cotton stalk particle model must be constructed. It should be ensured that the geometric and physical characteristics of cotton stalk particles in the DEM model match those of real ones. The complicated microstructure of cotton stalk particles makes it difficult to construct a DEM model that is exactly the same as the real structure. Therefore, by referring to the method of constructing the stalk DEM model by relevant scholars and thus combining the biological structure characteristics of cotton stalk, a single distribution method was adopted to arrange the particle groups in the epidermis and internal tissues (xylem and pith) of the chopped cotton stalks, and then the DEM model of cotton stalk particles was constructed by bonding the elements of nonequal-diameter basic particles [36,37], as shown in Figure 3. The same particle model was adopted for the xylem and pith of cotton stalks to make the calculation simpler and more efficient in the simulation [38]. The constructed cotton stalk particle model was 20 mm in height and 10 mm in diameter and filled with 612 particles. Among them, the radii of the spherical elements constituting the epidermis and internal tissues of the cotton stalk model were 1 mm and 2 mm, respectively, and their numbers are 494 and 108, respectively. In the simulation test, the diameter and height of the cotton stalk particle model were generated at random, ranging from 0.5 to 1 times the size of the basic cotton stalk model.

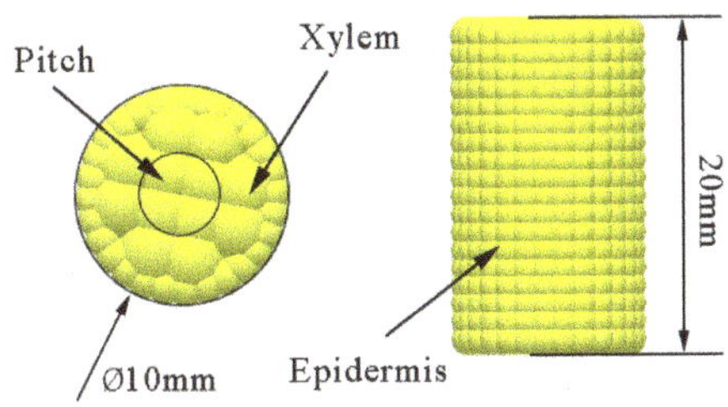

Figure 3. Basic simulation model of cotton stalk particles.

2.3. *Experimental Determination of Some DEM Input Parameters*

The accuracy of the DEM simulation model parameters directly affects the reliability of the simulation results. Because of the irregular shape of particles, the contact parameters between the particles measured by conventional methods are not accurate and have large errors, which cannot be directly used in the simulation test. For this reason, it is necessary to calibrate the contact parameters between the cotton stalk particles. In this paper, the internal parameters of the cotton stalk and its contact parameters with contact materials were measured by physical tests and set as fixed values. The same measurement method was adopted to obtain the range values of the contact parameters between particles so as to provide reference data for calibrating the contact parameters between particles.

2.3.1. Cotton Stalk Density

The ratio of material mass to its volume is defined as density. Referring to ASTM D854-10 [39], the volume of the cotton stalk was measured by the liquid immersion method [19]. The mass of the cotton stalk was measured by a JMB5003 electronic balance (measuring accuracy: 0.001 g; Jiming Weighting and Calibration Equipment Co., Ltd, Yuyao, Zhejiang, China) to determine its density. In order to reduce the test error and avoid the liquid from being absorbed by the cotton stalk, the test was completed in a very short time, and we carried out multiple sets of repeated tests to reduce the test error.

2.3.2. Elastic Modulus and Poisson's Ratio

The compression test was used to calculate the elastic modulus and Poisson's ratio of cotton stalks. Cylindrical cotton stalk particles, with a length of 20 mm and a diameter of 10 mm, were prepared, and uniaxial plate compression tests were conducted by using a universal material testing machine [22]; the compression speed was set to 5 mm/min, and the loading displacement was set to 4 mm. High-speed photography was used to compare and analyze the change rates of the height and diameter before and after compression. The test process was recorded by a FASTEC-TS4 high-speed camera (Fastec Imaging, San Diego, CA, USA, 510 fps), and the distance between the camera and the sample was 500 mm. The test was repeated 20 times, and the average value was calculated. The elastic modulus and Poisson's ratio of cotton stalks were calculated by Equation (6).

$$\begin{cases} E = \frac{F_0 l_0}{S \Delta l} \\ v = \left| \frac{\varepsilon_2}{\varepsilon_1} \right| = \frac{\Delta d l_0}{\Delta l d_0} \end{cases} \tag{6}$$

where E is the elastic modulus (MPa), v is Poisson's ratio, F_0 is the external force on the cotton stalk (N), l_0 is the initial length (mm), S is the initial cross-sectional area (mm), Δl is the length deformation (mm), ε_1 is the longitudinal strain (%); ε_2 is the transverse strain (%), d_0 is the initial diameter (mm), and Δd is the diameter deformation (mm).

2.3.3. Coefficient of Restitution

The coefficient of restitution, which is defined as the ratio of the velocity of an object after a collision to that before the collision, reflects the ability of an object to recover to its original state after a collision. Most studies on the coefficient of restitution of agricultural materials equate irregular materials with spherical particles, thus simplifying the collision model and ignoring the effect of the material's shape on the coefficient of restitution. Therefore, this paper considers the effect of the shape and anisotropy of culm-like materials on the coefficient of restitution. According to the test method in references [40,41], the spatial velocity after a collision between the cotton stalk and the contact material is obtained by combining high-speed photography, and the coefficient of restitution is calculated based on the principle of energy conservation, as shown in Equation (7):

$$e = \sqrt{\frac{\sum E_{k,o}}{\sum E_{k,i}}} = \sqrt{\frac{V_t{}^2\left(57 + \frac{l^2}{r^2}\right)}{48V_0{}^2}} \tag{7}$$

where e is the coefficient of restitution, $\sum E_{k,i}$ is the total kinetic energy before the collision, $\sum E_{k,o}$ is the total kinetic energy after a collision, V_t is the rebound velocity of the material after colliding (m/s), V_0 is the instantaneous approaching velocity before the material collides with the collision material (m/s), r is the radius (mm), and l is the length (mm).

2.3.4. Static Friction Coefficient

The static friction coefficient is the ratio of the maximum static friction force applied to an object to the positive pressure. The oblique plane sliding method is commonly used to measure the static friction coefficient [42,43]. The gravity of an object of mass, m, is decomposed into two forces: a force, F, parallel to the oblique plane and a force, T, perpendicular to the oblique plane. When the inclination angle, γ, of the oblique plane is less than the sliding critical angle, F is less than the static friction force, f, between the object and the oblique plane, and the object remains static. With the increase of γ, F increases. When γ is greater than the sliding critical angle of the object, F is greater than f at this time, and the object will begin to slide along the oblique plane. Among them, the relationship between the static friction coefficient, μ_s, and the inclination angle, γ, is shown in Equation (8):

$$\mu_s = \frac{f}{T} = mg\sin\gamma/(mg\cos\gamma) = \tan\gamma \tag{8}$$

The schematic diagram of the static friction coefficient measuring device is shown in Figure 4. During the test, one end of the lifting plate is connected with the lifting device of the universal testing machine by a rope, where the lifting device is adjusted to be in a horizontal position, and the contact material panel is fixed on the lifting plate at the same time. It should start the universal testing machine to make the lifting plate rise uniformly. The test process was recorded by a FASTEC-TS4 high-speed camera (Fastec Imaging, San Diego, CA, USA). The video was analyzed by ProAnalyst software to obtain the γ value displayed by the electronic goniometer when the cotton stalk sample began to slide, and the static friction coefficient, μs, was calculated by Equation (8). The static friction coefficient measurement tests for each group were repeated 10 times, and the average was calculated.

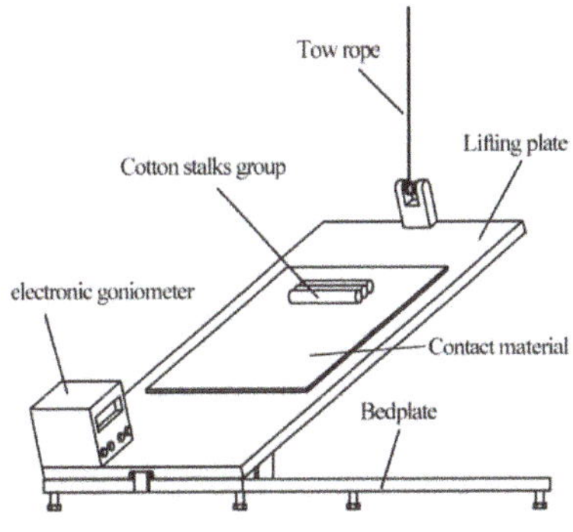

Figure 4. Schematic diagram of the static friction coefficient measuring device.

2.3.5. Rolling Friction Coefficient

According to the theory and experimental method in reference [44], high-speed photography was used to determine the rolling friction coefficient between the cotton stalks and contact materials. Through theoretical analysis, there is a linear relationship between the ratio of energy loss and total energy in the process of particle rolling from the oblique plane and the tangent value of the oblique plane angle, and the slope is the rolling friction coefficient, as shown in Equation (9):

$$C_f = \frac{W_{fr}}{U} = \frac{U - E_k}{U} = \mu_r \cot\theta \tag{9}$$

where C_f is the energy ratio of rolling friction loss (%), U is the initial gravitational potential energy (J), W_{fr} is the effort of friction force (J), E_k is the kinetic energy at the end of particle motion (J), μ_r is the rolling friction coefficient, the radius (mm), and θ is the oblique plane angle (°).

In the test, the cotton stalk was placed on the oblique plane with an oblique angle of θ and a length of L so that it could roll down from a certain height along the oblique plane. The initial gravitational potential energy of the cotton stalk is $U = mgLsin\theta$. The instantaneous speed v_s (m/s), when the cotton stalk moves to the bottom end of the oblique plane, is obtained by high-speed photography technology, and then the instantaneous kinetic energy of the cotton stalk is calculated as $E_k = 0.5\ mv_s{}^2$. In the test, the θ ranges from 15° to 50°, a total of 8 angle values are set, and each angle is repeated 10 times. The linear relationship between C_f and cotθ is fitted out by a linear regression equation, and the linear slope is the rolling friction coefficient.

2.4. *Physical Test of Repose Angle and DEM Simulation Test*

2.4.1. Rolling Friction Coefficient

The cylinder-lifting method was applied for the repose angle test [45,46]. The diameter of the cylinder was set to be 4 to 5 times larger than the maximum length of the cotton stalk particles; thus, the diameter was set to 100 mm and the height to 200 mm. In order to ensure that the repose angle is only affected by the interaction of the contact parameters between cotton stalks, a circular chassis, which was formed by cotton-stalk particle paving, was used as the substrate (Figure 5a), and the universal testing machine was used to lift the cylinder upward at a speed of 50 mm·s^{-1}, making the cotton stalk a particle pile, as shown in Figure 5b.

Figure 5. Physical test of repose angle: (**a**) the circular chassis substrate; (**b**) stacking test results.

A high-definition camera was used in the test to record the boundary information of cotton stalk particle stacking. The MATLAB software was used for preprocessing, gray processing, hole filling and binarization. The boundary information of the repose angle of cotton stalks was extracted. Finally, the least-squares method was used to fit the extracted stacking boundary linearly, so the slope of the fitted line was the tangent of the repose angle. The image processing is shown in Figure 6. The repose angle physical test was conducted 20 times. The average repose angle of the cotton stalk particles is 26.45°, and the standard deviation is 0.57°.

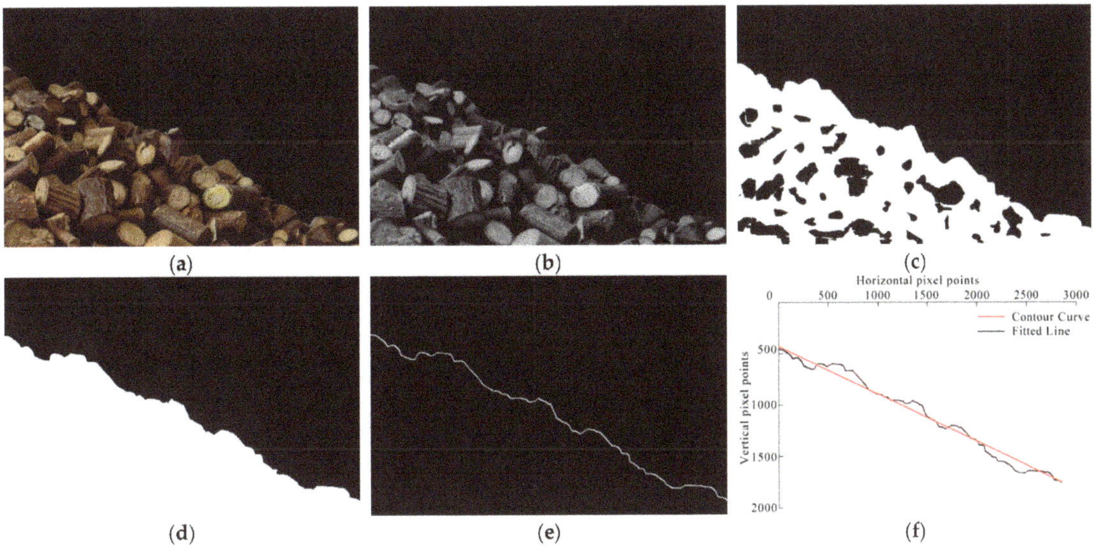

Figure 6. Image processing results: (**a**) preprocessing; (**b**) gray-scale processing; (**c**) hole filling; (**d**) binarization; (**e**) boundary extraction; (**f**) fitting of one-sided repose angle.

2.4.2. DEM Simulation Test

Based on the physical test equipment and discrete element model, the simulation test of the repose angle was conducted in EDEM. The cylinder model was constructed according to the actual size in Solidworks and imported into EDEM, and a particle factory was set at the top of it. In the simulation test, the time step was set to 20% of that of Rayleigh, and the total simulation time was set to 5 s, in which the time of generating the cotton stalk particles model and system stability was 1.8 s. The vertical upward lifting speed of the cylinder was set to 50 $mm \cdot s^{-1}$, and the cotton stalk particles slowly overflowed from the bottom of the cylinder, which finally formed a stable particle pile on the bottom plate, as shown in Figure 7. After the simulation test of the repose angle, the particle pile images under the +X and +Y views were collected. The image processing method in Section 2.4.1 was used to extract and fit the boundary of the repose angle, and the average value was taken as the simulation repose angle.

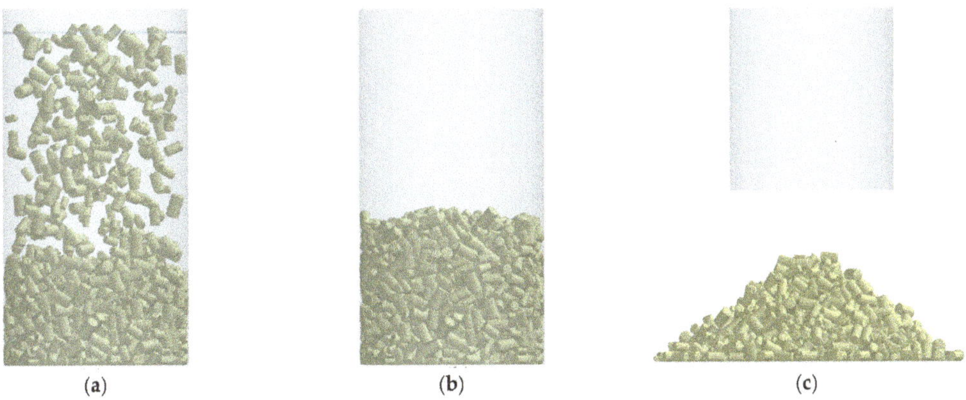

Figure 7. Simulation test process of repose angle: (**a**) particle filling; (**b**) system reaches stability; (**c**) heap completion.

2.5. Determination of DEM Input Parameters and Test Scheme

2.5.1. Determination of DEM Input Parameters

By referring to the literature and determining the input parameters in Section 2.3, the intrinsic parameters of cotton stalks and steel, and the contact parameters between them, were obtained. Because the intrinsic parameters have little effect on the repose angle, and the contact parameters between the cotton stalks and the steel can be determined by tests, those parameters can be set as fixed values. The input parameters under the simulation test are shown in Table 1.

Table 1. The input parameters of the simulation test.

DEM Parameter	Parameter Value	DEM Parameter	Parameter Value
Density of Cotton Stalk (g·cm^{-3})	0.326	Cotton Stalk-Steel coefficient of restitution	0.56
Elastic modulus of Cotton Stalk (Pa)	4.34×10^9	Cotton Stalk-Steel static friction coefficient	0.62
Poisson's ratio of Cotton Stalk	0.35	Cotton Stalk-Steel rolling friction coefficient	0.16
Density of Steel (g·cm^{-3})	7.85	* Cotton Stalk-Cotton stalk coefficient of restitution	0.16~0.48
Elastic modulus of Steel (Pa)	2.06×10^{11}	* Cotton Stalk-Cotton stalk static friction coefficient	0.45~0.65
Poisson's ratio of Steel	0.30	* Cotton Stalk-Cotton stalk rolling friction coefficient	0.10~0.20

Note: * is the parameter to be calibrated.

2.5.2. Response Surface Test Design

The central composite design (CCD), as a design of the experiment commonly used in response surface methodology (RSM), is facilitated to analyze the effects of various factors on the response value and optimize the results. The CCD module in the Design Expert software is used to encode the contact parameters between cotton stalks: the coefficient of restitution (X_1), static friction coefficient (X_2) and rolling friction coefficient (X_3). The coding of the influencing factors and the setting of horizontal values in the simulation test of the repose angle of cotton stalk particles are shown in Table 2.

Table 2. Influencing factors and their level values of simulation test of repose angle.

Level	X_1	X_2	X_3
−1.68	0.05	0.38	0.07
−1	0.16	0.45	0.1
0	0.32	0.55	0.15
1	0.48	0.65	0.20
1.68	0.59	0.71	0.23

According to the influencing factors and their horizontal values set in Table 2, the simulated repose angle (Y) of cotton stalk particles was taken as the response value, and the three-factor five-level central composite test was designed by Design Expert software 8.0.6 (Stat Ease Inc., Minneapolis, MN, USA). The second-order response model between the response values and the contact parameters was constructed to obtain the effect laws of the coefficient of restitution, the static friction coefficient and the rolling friction coefficient on the repose angle. The test scheme is shown in Table 3.

Table 3. The scheme and results of repose angle simulation tests.

Test No.	Coding			Response Value	Test No.	Coding			Response Value
	X_1	X_2	X_3	Y (°)		X_1	X_2	X_3	Y (°)
1	−1	−1	−1	25.52 ± 1.32	11	0	−1.68	0	25.47 ± 1.23
2	1	−1	−1	23.43 ± 2.51	12	0	1.68	0	28.36 ± 1.14
3	−1	1	−1	25.28 ± 1.36	13	0	0	−1.68	23.69 ± 1.46
4	1	1	−1	25.64 ± 1.85	14	0	0	1.68	31.52 ± 2.65
5	−1	−1	1	27.96 ± 1.34	15	0	0	0	25.49 ± 1.28
6	1	−1	1	30.3 ± 2.84	16	0	0	0	25.02 ± 1.25
7	−1	1	1	25.68 ± 3.01	17	0	0	0	25.18 ± 1.64
8	1	1	1	33.76 ± 2.58	18	0	0	0	25.55 ± 0.87
9	−1.68	0	0	25.61 ± 1.47	19	0	0	0	26.51 ± 1.84
10	1.68	0	0	28.26 ± 1.65	20	0	0	0	24.43 ± 1.91

Note: ± is the standard deviation.

2.6. Optimum Condition and Validation

In order to obtain the optimal input parameter combination of contact parameters between cotton stalk particles, by combining with the boundary conditions of the contact parameters in Section 2.5.1, the optimization module in the Design Expert software was utilized to optimize the second-order response model, with the physical test value of the repose angle set as the target values. Taking this parameter combination as the simulation input parameters, the simulation test of the repose angle was conducted. By comparing the relative error of the repose angle in the simulation test and the physical test, the accuracy and reliability of the optimal combination parameters were verified.

3. Results and Discussion

3.1. Analysis of the Simulation Test Result of the Repose Angle

3.1.1. ANOVA and Model Construction

The simulation test was conducted according to the test scheme in Table 3, and the simulated repose angles of the cotton stalk particles under different parameter combinations were obtained, as shown in Table 3. The simulation test results were analyzed by analysis of variance (ANOVA) through the data processing module of the Design Expert software, as shown in Table 4.

Table 4. ANOVA of the test results.

Source	Sum of Squares	Mean Square	*F Value*	*p-Value*
Model	129.06	14.34	29.88	<0.0001 **
X_1	12.66	12.66	26.37	0.0004 **
X_2	4.70	4.70	9.79	0.0107 *
X_3	70.36	70.36	146.59	<0.0001 **
X_1X_2	8.38	8.38	17.47	0.0019 **
X_1X_3	18.45	18.45	38.44	0.0001 **
X_2X_3	0.078	0.078	0.16	0.6953
X_1^2	4.24	4.24	8.83	0.014 *
X_2^2	4.13	4.13	8.60	0.015 *
X_3^2	8.75	8.75	18.22	0.0016 **
Residual	4.80	0.48		
Lack of Fit	2.41	0.48	1.01	0.4959
Pure Error	2.39	0.48		
Cor Total	133.86			
Adeq Precision	22.04			
$R^2 = 0.9641$, $R^2_{adj} = 0.9319$, $R^2_{Pred} = 0.8216$, $C.V = 2.60\%$				

Note: ** = extremely significant factor ($p \leq 0.01$); * = significant factor ($0.01 < p \leq 0.05$); NS = not significant factor ($p > 0.05$).

Through the regression analysis of variance on the test results, if the insignificant influencing factor in the second-order response model was eliminated, the response model between the contact parameters and the repose angle was obtained, as shown in Equation (10):

$$
\begin{aligned} Y = 57.52 - 71.21X_1 - 70.51X_2 - 97.98X_3 + 63.98X_1X_2 + 189.84X_1X_3 \\ -19.75X_2X_3 + 21.18X_1^2 + 53.51X_2^2 + 311.63X_3^2 \end{aligned} \quad (10)
$$

The results of ANOVA are shown in Table 4. The F value (29.88) and p-value ($p < 0.0001$) of the model indicated that the model was extremely significant within the 95% confidence interval. The F value of the Lack of Fit term is 1.01, indicating that the Lack of Fit term is not significant relative to the pure error. The p-value of the Lack of Fit term is 0.4959, which is far greater than 0.05 and extremely insignificant. The coefficient of variation, $C.V$ (2.60%), is low, indicating that the second-order response model between the contact parameters and the repose angle obtained by ANOVA is reliable.

In addition, the determination coefficient (R^2) and the adjusted determination coefficient ($R^2{}_{adj}$) of the second-order response model were 0.9641 and 0.9319, respectively, indicating a high fit of the model to the experimental data. The difference between $R^2{}_{adj}$ (0.9319) and the predicted determination coefficient $R^2{}_{Pred}$ (0.8216) was small (less than 0.2), and the SNR (Adeq precision) was 22.04%, indicating that the test factors have a high degree of interpretation for the response values. The result is further demonstrated by the fact that the scatter of predicted values is very close to the straight line from Figure 8 or the diagram of the normal plot of residuals.

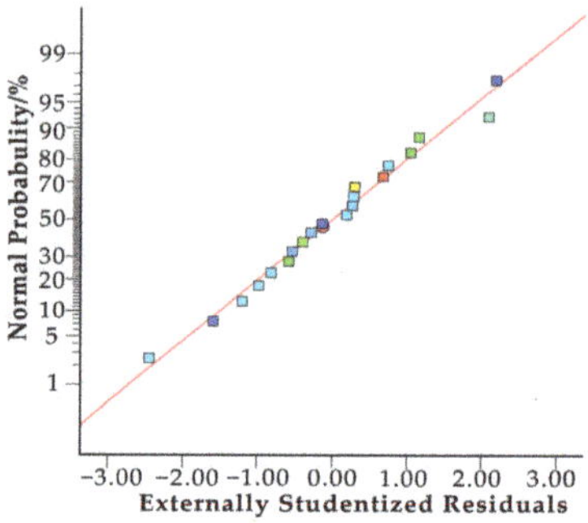

Figure 8. Normal plot of residuals.

3.1.2. Single-Factor Effect Analysis

As shown in Table 4, the single-factor terms X_1, X_3, and secondary term $X_3{}^2$ have an extremely significant ($p < 0.01$) effect on the repose angle. The p-values of X_2, $X_1{}^2$ and $X_2{}^2$ are all from 0.01 to 0.05, so they are the significant effect factors of the repose angle. In order to analyze the effect trend of the test factors X_1, X_2 and X_3, respectively, on the repose angle (Y) more intuitively, any two factors in the second-order response model are set to a central level value of 0, and the effect law of the single-factor on the repose angle is obtained, respectively, as shown in Figure 9.

It can be seen in Figure 9 that when X_1 changes from a low level (−1.68) to a high level (1.68), that is, when the coefficient of restitution gradually increases from 0.05 to 0.59, the repose angle decreases first and then increases. When the level value of X_1 was −0.899, the minimum value of the repose angle was 24.94°. The effect law of the single-factor term X_2 on the repose angle is the same as that of X_1. When the static friction coefficient increases gradually from 0.38 to 0.71, the value of the repose angle also decreases first and then increases. As the level value of X_3 changes from a low level (−1.68) to a high level (1.68), the repose angle gradually increases; that is, when the rolling friction coefficient gradually increases from 0.07 to 0.23, the repose angle gradually increases. In addition, with the level value of X_3 changing from −1.68 to −1, the repose angle increased slowly, and with the level value of X_3 changing from −1 to 1.68, the repose angle increased rapidly.

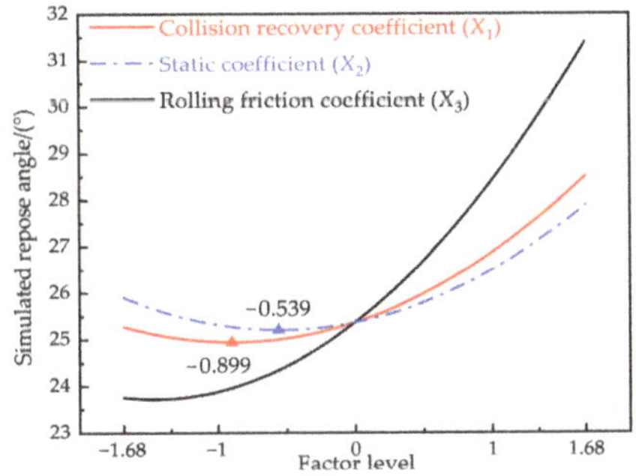

Figure 9. Single-factor response diagram.

3.1.3. Interaction Effect Analysis

As can be seen from the results of ANOVA in Table 4, among the interaction terms, the p-values of X_1X_2 and X_1X_3 were far less than 0.01, so X_1X_2 and X_1X_3 have an extremely significant effect on the repose angle. Because the p-value of X_2X_3 was far greater than 0.05, it showed a non-significant effect on the repose angle. To research the effect of the interaction terms on the repose angle, this paper plotted the response surface of the interaction terms X_1X_2 and X_1X_3 on the interaction of the repose angle and analyzed their effect law on the repose angle. The response surface is shown in Figure 10.

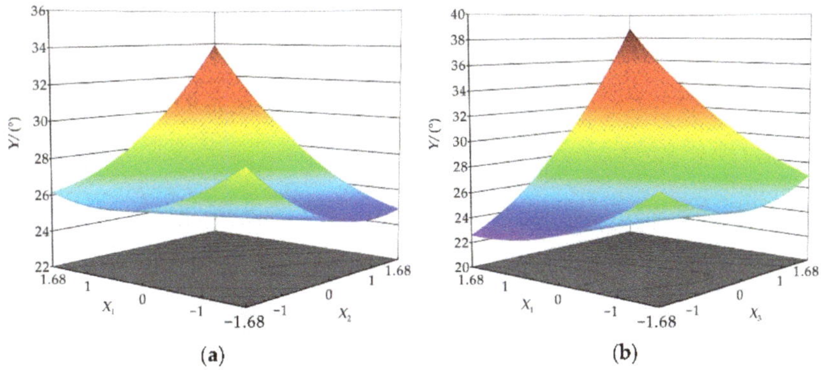

Figure 10. Influence of interaction terms on repose angle: (**a**) interaction term X_1X_2; (**b**) interaction term X_1X_3.

1. Effect of interaction term X_1X_2 on the repose angle

When the level value of single-factor term X_3 is at the center level 0, the effect law of interaction term X_1X_2 on the repose angle is shown in Figure 10a. When the level value of X_1 is at −1.68 and the level value of X_2 increases from −1.68 to 1.68, the repose angle decreases first and then increases. This change is the same as the effect law of the single-factor term X_2 on the repose angle. When the level value of X_1 is at 1.68 and the level value of X_2 increases from −1.68 to 1.68, the repose angle increases sharply. This change differs from the effect law of the single-factor X_2 on the repose angle. Among the two-factor interaction terms X_1X_2, the effect law of single-factor X_2 on the repose angle is the same as that of X_1. However, the slope of the X_1 response surface curve is steeper than the X_2 direction, indicating that the effect of X_1 on the repose angle is more significant than that of X_2.

2. Effect of interaction term X_1X_3 on repose angle

When the level value of single-factor term X_2 is at the center level 0, the effect law of interaction term X_1X_3 on the repose angle is shown in Figure 10b. When the level value of X_1 is at -1.68 and the level value of X_3 increases from -1.68 to 1.68, the repose angle decreases first and then increases. This change differs from the effect law of the single-factor term X_3 on the repose angle. When the level value of X_1 is at 1.68 and the level value of X_3 increases from -1.68 to 1.68, the repose angle increases sharply. This change is the same as the effect law of the single-factor X_3 on the repose angle. Among the two-factor interaction terms X_1X_3, the effect law of the single-factor X_3 on the repose angle is the same as that of X_1. However, the slope of the X_3 response surface curve is steeper than the X_1 direction, indicating that the effect of X_3 on the repose angle is more significant than that of X_1. By comparing Figure 10a,b, it can be concluded that the effect order of the single-factor on the repose angle is $X_3 > X_1 > X_2$, which is consistent with the ANOVA results.

3.2. Determination of Optimal Parameter Combinations and Verification

By taking the physical test values of repose angle (Y) as the optimization target values and X_1, X_2 and X_3 as the optimization values, the optimization module in Design Expert software was used to solve the second-order response model (i.e., Equation (10)) to obtain the optimal parameter combination. The single-objective and multi-variable optimization method was used to establish the objective and constraint functions. The mathematical model of parameter optimization is as follows:

$$\begin{cases} Y(X_1, X_2, X_3) = 26.45^\circ \\ \begin{cases} 0.16 \le X_1 \le 0.48 \\ 0.45 \le X_2 \le 0.65 \\ 0.1 \le X_3 \le 0.2 \end{cases} \end{cases} \tag{11}$$

The optimal combination of contact parameters after calibration is $X_1 = 0.45$, $X_2 = 0.47$ and $X_3 = 0.16$. In order to verify the accuracy of the optimal parameter combination, the simulation test of the repose angle of cotton stalk particle stacking was conducted after entering the obtained contact parameters into EDEM.

The simulation test was repeated five times, and the simulated values of the repose angle were 26.91°, 25.90°, 26.15°, 26.43° and 26.40°, respectively. The relative errors between the simulated value of the repose angle and the physical test value were 1.74%, 2.08%, 1.13%, 0.08%, and 0.19%, respectively. In the simulation test, the average value of the repose angle was 26.36°, and the average relative error was 1.04%. There was no significant difference between the simulated values and the physical test values, indicating that the optimal parameter combination was accurate and reliable, further verifying that the second-order response model between the coefficient of restitution, static friction coefficient, rolling friction coefficient and repose angle is suitable for the calibration of cotton stalk contact parameters.

4. Conclusions

The DEM simulation was used to study the repose angle of chopped cotton stalk particles. The method can be used as an application to analyze the effects of DEM parameters on predicting the repose angle. The following conclusion can be drawn from this study:

(1) The contact parameters between the cotton stalk particles and the contact material (steel) were measured by physical tests. The coefficient of restitution was 0.56, the static friction coefficient was 0.62, and the rolling friction coefficient was 0.16. As for the range of contact parameters between the cotton stalk particles, the coefficient of restitution ranged from 0.16 to 0.48, the static friction coefficient ranged from 0.45 to 0.65, and the rolling friction coefficient ranged from 0.10 to 0.20. The cylinder-lifting method was applied to test the repose angle of chopped cotton stalks, and the average

value (26.45°) and the standard deviation (0.57°) of the repose angle of cotton stalk particles were obtained.

(2) A central composite design test on the response surface methodology was constructed to conduct the simulation test of the repose angle, the second-order response model between contact parameters and repose angle. According to the results of ANOVA, all the figures indicate that the test factors had a high interpretation of the response value.
(3) By analyzing the effects of single-factor and interaction factors on the repose angle, the extremely significant factors affecting the repose angle were the coefficient of restitution and rolling friction coefficient, while the static friction coefficient was the most significant factor. The coefficient of restitution interacted with the static friction coefficient and the rolling friction coefficient on the repose angle, and they had a significant effect on the repose angle.
(4) The optimal combination of contact parameters was determined as follows: the coefficient of restitution was 0.45, the static friction coefficient was 0.47, and the rolling friction coefficient was 0.16. The verification test showed no significant difference between the simulated and physical test values, and the average relative error was 1.04%, indicating that the simulated values agreed well with the physical test values, verifying the authenticity of the simulation test and the reliability of the optimal combination of simulation parameters. The research provides a basis for the discrete element simulation study of cotton stalk motion in the separation process of cotton stalks and residual film and thus could be used for subsequent gas–solid coupling simulation research.

Author Contributions: Conceptualization, methodology, formal analysis, writing—review and editing, writing—original draft preparation, B.Z. and Z.K.; supervision, X.C.; funding acquisition, Z.K.; investigation, H.M.; software, X.W.; data curation, R.L. All authors have read and agreed to the published version of the manuscript.

Funding: This research was funded by the National Natural Science Foundation of China, grant number: 52065058; the Graduate Education Innovation Project of Xinjiang Uygur Autonomous Region, grant number: XJ2022G085; and the Open Fund of Jiangsu Province and Education Ministry Co-sponsored Synergistic Innovation Center of Modern Agricultural Equipment, Grant No. XTCX2006.

Institutional Review Board Statement: Not applicable.

Data Availability Statement: The data presented in this study are available on request from the corresponding author.

Conflicts of Interest: The authors declare no conflict of interest.

References

1. Wang, Z.; He, H.; Zheng, X.; Zhang, J.; Li, W. Effect of cotton stalk returning to fields on residual film distribution in cotton fields under mulched drip irrigation in typical oasis area in Xinjiang. *Trans. Chin. Soc. Agric. Eng.* **2018**, *34*, 120–127.
2. Zhang, D.; Yan, C.; Liu, H.; Wang, H.; Hu, W.; Qin, X.; Ma, X. The status and distribution characteristics of residual mulching film in Xinjiang, China. *J. Integr. Agric.* **2016**, *15*, 2639–2646. [CrossRef]
3. Dai, J.; Dong, H. Intensive cotton farming technologies in China: Achievements, challenges and countermeasures. *Field Crops Res.* **2014**, *155*, 99–110. [CrossRef]
4. Liu, E.; He, W.; Yan, C. 'White revolution' to 'white pollution'—Agricultural plastic film mulch in China. *Environ. Res. Lett.* **2014**, *9*, 091001. [CrossRef]
5. Zhang, D.; Ng, E.; Hu, W.; Wang, H.; Galaviz, P.; Yang, H.; Sun, W.; Li, C.; Ma, X.; Fu, B.; et al. Plastic pollution in croplands threatens long-term food security. *Glob. Chang. Biol.* **2020**, *26*, 3356–3367. [CrossRef]
6. Liang, R.; Zhang, B.; Zhou, P.; Li, Y.; Meng, H.; Kan, Z. Power consumption analysis of the multi-edge toothed device for shredding the residual film and impurity mixture. *Comput. Electron. Agric.* **2022**, *196*, 106898. [CrossRef]
7. Liang, R.; Zhang, B.; Zhou, P.; Li, Y.; Meng, H.; Kan, Z. Cotton length distribution characteristics in the shredded mixture of mechanically recovered residual films and impurities. *Ind. Crop. Prod.* **2022**, *182*, 114917. [CrossRef]
8. Horabik, J.; Molenda, M. Parameters and contact models for DEM simulations of agricultural granular materials: A review. *Biosyst. Eng.* **2016**, *147*, 206–225. [CrossRef]
9. Lenaerts, B.; De Ketelaere, B.; Tijskens, E.; Herman, R.; Baerdemaeker, J.D.; Aertsen, T.; Saeys, W. Simulation of grain-straw separation by discrete element modeling with bendable straw particles. *Comput. Electron. Agric.* **2014**, *101*, 24–33. [CrossRef]

10. Wang, L.; Li, R.; Wu, B.; Wu, Z.; Ding, Z. Determination of the coefficient of rolling friction of an irregularly shaped maize particle group using physical experiment and simulations. *Particuology* **2018**, *38*, 185–195. [CrossRef]
11. Fang, W.; Wang, X.; Han, D.; Chen, X. Review of material parameter calibration method. *Agriculture* **2022**, *12*, 706. [CrossRef]
12. Rackl, M.; Hanley, K. A methodical calibration procedure for discrete element models. *Powder Technol.* **2017**, *307*, 73–83. [CrossRef]
13. Cunha, R.N.; Santos, K.G.; Lima, R.N.; Duarte, C.R.; Barrozo, M. Repose angle of monoparticles and binary mixture: An experimental and simulation study. *Powder Technol.* **2016**, *303*, 203–211. [CrossRef]
14. Radvilaitė, U.; Ramírez-Gómez, Á.; Kačianauskas, R. Determining the shape of agricultural materials using spherical harmonics. *Comput. Electron. Agric.* **2016**, *128*, 160–171. [CrossRef]
15. Kattenstroth, R.; Harms, H.H.; Lang, T. Systematic alignment of straw to optimise the cutting process in a combine's straw chopper. In Proceedings of the 69th International Conference on Agricultural Engineering LAND TECHNIK AgEng, Hannover, Germany, 11–12 November 2011.
16. Li, H.; Li, Y.; Gao, F.; Zhao, Z.; Xu, L. CFD-EDM simulation of material motion in air-and-screen cleaning device. *Comput. Electron. Agric.* **2012**, *88*, 111–119. [CrossRef]
17. Xu, Y.; Zhang, X.; Wu, S.; Chen, C.; Wang, J.M.; Yuan, S.; Chen, B.; Li, P.; Xu, R. Numerical simulation of particle motion at cucumber straw grinding process based on EDEM. *Int. J. Agric. Biol. Eng.* **2020**, *13*, 227–235. [CrossRef]
18. Ramírez-Gómez, Á.; Gallego, E.; Fuentes, J.M.; González-montellano, C.; Ayuga, F. Values for particle-scale properties of biomass briquettes made from agroforestry residues. *Particuology* **2014**, *12*, 100–106. [CrossRef]
19. Fang, M.; Yu, Z.; Zhang, W.; Cao, J.; Liu, W. Friction coefficient calibration of corn stalk particle mixtures using plackett-burman design and response surface methodology. *Powder Technol.* **2022**, *396*, 731–742. [CrossRef]
20. Ma, Y.; Song, C.; Xuan, C.; Wang, H.; Yang, S.; Wu, P. Parameters calibration of discrete element model for alfalfa straw compression simulation. *Trans. Chin. Soc. Agric. Eng.* **2020**, *36*, 22–30.
21. Feng, J.; Lin, J.; Li, S.; Zhou, J.; Zhou, Z. Calibration of discrete element parameters of particle in rotary solid state fermenters. *Trans. Chin. Soc. Agric. Machin.* **2015**, *46*, 208–213.
22. Liao, Y.; Wang, Z.; Liao, Q.; Wan, X.; Zhou, Y.; Liang, F. Calibration of discrete Element Model Parameters of Forage Rape Stalk at Early Pod Stage. *Trans. Chin. Soc. Agric. Machin.* **2020**, *51*, 236–243.
23. Zhang, T.; Liu, F.; Zhao, M.; Ma, Q.; Wang, W.; Fan, Q.; Yan, P. Determination of corn stalk contact parameters and calibration of discrete element method simulation. *J. Chin. Agric. Univer.* **2018**, *23*, 120–127.
24. Ma, Z.; Xu, L.; Li, Y. Discrete-element method simulation of agricultural particles' motion in variable-amplitude screen box. *Comput. Electron. Agric.* **2015**, *118*, 92–99. [CrossRef]
25. Liu, F.; Zhang, J.; Chen, J. Modeling of flexible wheat straw by discrete element method and its parameter calibration. *Int. J. Agric. Biol. Eng.* **2018**, *11*, 42–46. [CrossRef]
26. Wang, Y.; Xu, H.; Zhang, Y.; Yang, Y.; Zhao, H.; Yang, C.; He, Y.; Wang, K.; Liu, D. Discrete element modelling of citrus fruit stalks and its verification. *Biosyst. Eng.* **2020**, *200*, 400–414. [CrossRef]
27. Zeng, Z.; Chen, Y. Simulation of straw movement by discrete element modelling of straw-sweep-soil interaction. *Biosyst. Eng.* **2019**, *180*, 25–35. [CrossRef]
28. Guo, Q.; Zhang, X.; Xu, Y.; Li, P.; Chen, C.; Xie, H. Simulation and experimental study on cutting performance of tomato cane straw based on EDEM. *J. Drain. Irrig. Machin. Eng.* **2018**, *36*, 1017–1022.
29. Quist, J.; Evertsson, C.M. Cone crusher modelling and simulation using DEM. *Minerals Eng.* **2016**, *85*, 92–105. [CrossRef]
30. Sun, X.; Li, B.; Liu, Y.; Gao, X. Parameter measurement of edible sunflower exudates and calibration of discrete element simulation parameters. *Processes* **2022**, *10*, 185. [CrossRef]
31. Paulick, M.; Morgeneyer, M.; Kwade, A. Review on the influence of elastic particle properties on DEM simulation results. *Powder Technol.* **2015**, *283*, 66–76. [CrossRef]
32. Hertz, H. On the contact of elastic solids. *J. Für Die Reine Und Angew. Math. (Crelles J.)* **1880**, *92*, 156–171.
33. Mindlin, R.D. Compliance of elastic bodies in contact. *J. Appl. Mech.* **1949**, *16*, 259–268. [CrossRef]
34. Han, Y.; Jia, F.; Tang, Y.; Liu, Y.; Zhang, Q. Influence of granular coefficient of rolling friction on accumulation characteristics. *Acta Phys. Sin.* **2014**, *63*, 533–538.
35. Liu, F.; Zhang, J.; Chen, J. Construction of visco-elasto-plasticity contact model of vibratory screening and its parameters calibration for wheat. *Trans. Chin. Soc. Agric. Eng.* **2018**, *34*, 37–43.
36. Cabiscol, R.; Finke, J.H.; Kwade, A. Calibration and interpretation of DEM parameters for simulations of cylindrical tablets with multi-sphere approach. *Powder Technol.* **2017**, *327*, 232–245. [CrossRef]
37. Wang, X.; Yu, J.; Lv, F.; Wang, Y.; Fu, W. A multi-sphere based modelling method for maize grain assemblies. *Adv. Powder Technol.* **2017**, *28*, 584–595. [CrossRef]
38. Mcdowell, G.R.; Falagush, O.; Yu, H. A particle refinement method for simulating DEM of cone penetration testing in granular materials. *Geotech. Lett.* **2012**, *2*, 141–147. [CrossRef]
39. *ASTM D854-10*; Standard Test Methods for Specific Gravity of Soil Solids by Water Pycnometer. American Society for Testing and Materials (ASTM): West Conshohocken, PA, USA, 2010.
40. Mueller, P.; Antonyuk, S.; Stasiak, M.; Tomas, J.; Heinrich, S. The normal and oblique impact of three types of wet granules. *Granul. Matter* **2011**, *13*, 455–463. [CrossRef]

41. Zhang, B.; Chen, X.; Liang, R.; Li, J.; Wang, X.; Meng, H.; Kan, Z. Cotton stalk restitution coefficient determination tests based on the binocular high-speed camera technology. *Int. J. Agric. Biol. Eng.* **2022**, *15*, 181–189. [CrossRef]
42. Castiglioni, C.A.; Drei, A.; Carydis, P.; Mouzakis, H. Experimental assessment of static friction between pallet and beams in racking systems. *J. Build. Eng.* **2016**, *6*, 203–214. [CrossRef]
43. Zhang, B.; Liang, R.; Li, J.; Li, Y.; Meng, H.; Kan, Z. Test and Analysis on Friction Characteristics of Major Cotton Stalk Cultivars in Xinjiang. *Agriculture* **2022**, *12*, 906. [CrossRef]
44. Cui, T.; Liu, J.; Yang, L.; Zhang, D.; Zhang, R.; Lan, W. Experiment and simulation of rolling friction characteristic of corn seed based on high-speed photography. *Trans. Chin. Soc. Agric. Eng.* **2013**, *29*, 34–41.
45. Al-Hashemi, H.M.; Bukhary, A.H. Correlation between california bearing ratio (CBR) and angle of repose of granular soil. *Electron. J. Geotech. Eng.* **2016**, *21*, 5655–5660.
46. Derakhshani, S.M.; Schott, D.L.; Lodewijks, G. Micro-macro properties of quartz sand: Experimental investigation and DEM simulation. *Powder Technol.* **2015**, *269*, 127–138. [CrossRef]

agriculture

Article

Numerical Study of Pneumatic Conveying of Rapeseed through a Pipe Bend by DEM-CFD

Yao Xiao [1,2], Zitao Ma [1,2], Mingliang Wu [1,2,3] and Haifeng Luo [1,2,*]

1 College of Mechanical and Electrical, Hunan Agricultural University, Changsha 410128, China
2 Hunan Provincial Engineering Technology Research Center for Modern Agricultural Equipment, Changsha 410128, China
3 Specific Oilseed Crops (Oil-Tea Camellia) Mechanization Research Base, Ministry of Agriculture and Rural Affairs, Changsha 410128, China
* Correspondence: luohaifeng@hunau.edu.cn

Abstract: In the wide-width and high-speed operation of the rapeseed air-feeding planter, the air-feeding seed metering system adopts the top-down seed tube of different structural types in the production process, thus leading to significant differences in the discharge consistency and breakage rate of the respective row in the seeding process. Thus, the corrugated, hole-type, and ordinary round-tube seed tubes were taken as the research objects for this study, and virtual walls were introduced to compare and analyze the movement of seeds after collision with the seed tubes that had different wall structures. The effects of three types of seed tubes on the motion characteristics of seed particles were analyzed using DEM-CFD gas-solid coupling, and the simulation results were verified through bench experiments. The results indicated that when the inlet velocity was 16 m/s, and there was no material; the average error between the simulated value and the examined value of the airflow velocity at the same point of the vertical conveying pipe of the ordinary round-tube seed tube was 6.71%, thus verifying the feasibility of the simulation model built establishment of this study; when the inlet airflow velocity was 16 m/s and the seed particles were generated at the same per second, both the corrugated and hole-type seed tubes had a surge in airflow speed in the elbow part, and the highest airflow of the corrugated and hole-type pipes the velocities were 32.48 and 26.20 m/s, respectively. The corrugated and hole-type structure significantly affected the airflow field characteristics in the seed tube; the corrugated and hole-type seed tubes significantly improved the stable delivery of seeds, and the speed and force of the seeds were similar "sinusoidal fluctuation", and the stagnation time of the seeds in the ordinary round-tube, corrugated, and hole-type seed tubes were 0.3, 0.38, and 0.48 s, respectively, and the seed velocities at the outlet were 5.18, 1.73, and 3.76 m/s, respectively. This study provides a reference for the optimization of the structure of the seed tube of the air-feeding seed metering system.

Keywords: air-feeding seed discharge system; seed tube; gas-solid two-phase flow; computer simulation

Citation: Xiao, Y.; Ma, Z.; Wu, M.; Luo, H. Numerical Study of Pneumatic Conveying of Rapeseed through a Pipe Bend by DEM-CFD. *Agriculture* **2022**, *12*, 1845. https://doi.org/10.3390/agriculture12111845

Academic Editors: Muhammad Sultan, Redmond R. Shamshiri, Md Shamim Ahamed and Muhammad Farooq

Received: 9 September 2022
Accepted: 31 October 2022
Published: 3 November 2022

1. Introduction

Seeding has been confirmed as a vital link in rape planting, and the uniformity of seeding directly affects the yield of rape [1,2]. The air-feeding seed metering system can meet the requirements of the wide-width and high-speed seeding operation of the planter through quantitative seed supply [3,4], air distribution into rows [5,6], and high-speed air-feeding, and significantly improve the seeding efficiency [7]. When the planter is operating at high speeds, the air-feeding seed metering system employs different structural types of seed tubes during seeding due to the difference in the conveying airflow field and the migration trajectory of rapeseed particles, thus affecting the particle distribution characteristics in the distributor. Furthermore, the secondary flow in the seed tube separates the seeds from the airflow under the action of centrifugal force, causing collisions, which even causes the blockage of the seed tube, reduces the consistency and stability of the seeding amount in the respective row, and affects the uniformity of seeding.

To enhance the uniformity of seeding, Lei Xiaolong et al. [8] developed an inclined cone-hole wheel seed feeding device, built the mechanical model of the population in the filling and feeding processes through theoretical analysis, and obtained the major structure of the seed feeding device, as well as operating parameters. Chang Lijin et al. [9–11] designed a precession shaft-moving external sheave seed supply device to meet the demand for large and variable seed supply of the air-supplied seed metering system. In order to realize the function of precise and small seeding, Liu Lijing et al. designed a star-shaped feeder [12,13]. Zhai Gaixia et al. [14] designed a kind of star feeder to realize the matching operation of the seed supply device and the air-feeding seed metering device, and to achieve good adaptability to the seed size. The central external sheave seed feeding device and the sheave parameters corresponding to different seeding rates were produced. Li Yanjun et al. [15–17] used the bionic design method to solve the problem of uniform seed flow distribution during the operation of the air-feeding seed metering system. For crucian carp streamline distributor, the EDEM-Fluent coupling simulation method was used to analyze the internal flow field and the movement trajectory of seeds in the distributor. Senturov [6] designed a horizontal distributor with built-in inclined staggered balls, and studied the seeds in the process of horizontal transportation. Yatskul et al. [18,19] compared and analyzed the effects of operating parameters and structural parameters such as airflow velocity, seeding amount, pipe elbows, etc. on the seeding uniformity of the distributor, and used high-speed photography technology to analyze the movement state of seeds in the seeding process. Dai Yizheng et al. [20–22] used CFD to simulate the velocity flow field distribution in the rice sorter to explore the seeding mechanism of the pneumatic cluster-type rice sorter under the requirement of large seeding quantity and high-speed operation. In summary, at present, in-depth research has been carried out on the structural optimization and performance improvement of mechanical seed feeding devices and distributors (e.g., hole-type seed metering devices, groove wheel type seed metering devices, horizontal distributors, vertical distributors). There has been little research on the effect of the structure of the seed tube on the seeding performance of the air-feeding seed metering system. The adaptability of the seed tube to a larger seeding amount during high-speed and wide-width operation with the elastic collision of the seed particles with the seed tube and the kinematics and dynamics analysis of seed breakage rate and distribution uniformity are less studied, which affects the development and application of air-assisted seed metering systems.

In view of the wide width and high-speed seeding operation of rapeseed air-feeding, the air-feeding seed metering system adopts bottom-up vertical seeding pipes of different structures, and the consistency of the respective row and the breakage rate during seeding are obviously different. In actual production, taking corrugated, hole-type, and ordinary round-tube seed tubes as the research objects, DEM-CFD gas-solid coupling simulation was used to analyze the effects of three types of seed tubes on the movement characteristics of rapeseeds. The bench test was used to test the airflow velocity at the vertical conveying pipe of the ordinary round-tube seed tube without material, and compared with the simulation results to verify the feasibility of the simulation test.

2. Materials and Methods

2.1. Model Description

2.1.1. Pneumatic Seed Metering System

The air-feeding seed metering system is mainly composed of a seed box, a centralized seed supply device, a seed-air flow mixing chamber, a fan, a distribution device, a seed tube, and a seed metering pipe. The seed tube primarily comprises a horizontal conveying pipe and a vertical conveying pipe and elbow. Figure 1 presents the structure of the air-feeding seed metering system and the seed tube. When the air-feeding seed metering system operates, the seed particles enter the centralized seed supply device from the seed box, the high-speed airflow generated by the fan is fully mixed with the seed particles in the gas-solid mixing chamber, and the seed-air flow two-phase flows through the seed tube and enters the distribution device. The air, randomly distributed into rows, ensures the seeds are evenly discharged through the seed tube and fall into the seed ditch, or evenly fall into the terminal precision seed metering device, and the seeds are discharged to the seed ditch by the seed metering device to complete the seeding operation.

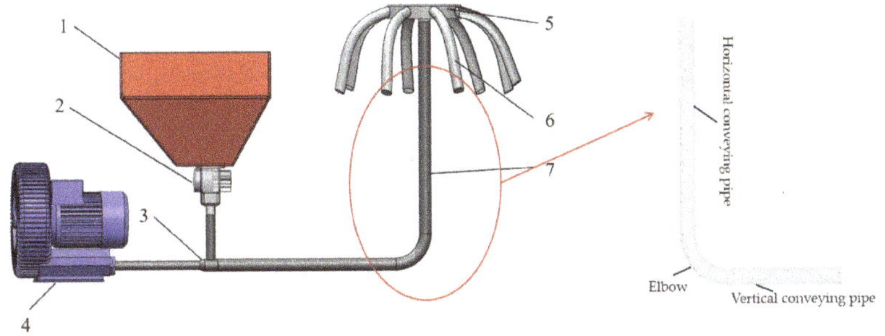

Figure 1. The schematic diagram of the pneumatic seeding system. 1. seed box; 2. centralized seed supply device; 3. seed-air flow mixing chamber; 4. fan; 5. distribution device; 6. seed tube; 7. seed metering pipe.

The seed tube takes on a great significance in the air-feeding seed metering system, and it serves as the major component connecting the gas-solid mixing chamber and the distribution device. Its main function is to enhance the distribution uniformity of the seeds before entering the distribution device, which is considered the basis of uniformity. The seed-air flow two-phase flow in the gas-solid mixing chamber is accelerated through the horizontal conveying pipe of the seed tube and then enters the elbow. When the two-phase flow changes from horizontal to vertical movement, the density of seed particles decreases, and intermittent conveying occurs. After the flow reaches the vertical conveying pipe, the pressurized structure (e.g., corrugations or holes) on the vertical conveying pipe is capable of inducing continuous self-vibration of the air flow and facilitating the continuous and stable entry of the seed group into the distribution device.

2.1.2. Seed Tube Design

The seed tube can improve the mixing uniformity of seed particles and air flow. The concentration of seed particles in the seed tube is correlated with the structure of the seed tube. The relative concentration of seed particles in the seed tube is [23]:

$$m_a = \frac{4Q}{\pi D^2 \rho v} \tag{1}$$

where m_a denotes the relative concentration of seed particles in the seed tube; Q represents the seed delivery volume in the seed tube, g/s; D expresses the inner diameter of the seed tube, m; ρ is the conveying airflow density, 1.225 kg/m^3; v represents the inside of the seed tube airflow velocity, m/s.

From Formula (1), the diameter of the seed tube is expressed as follows:

$$D = \sqrt{\frac{4Q}{m_a \pi \rho v}} \tag{2}$$

According to the years of seeding experience in winter rapeseed areas in the Yangtze River Basin Province, the high-speed operation speed of the seeder is 6~12 km/h, the working width is 2.2 m, the seed amount per mu of rapeseed direct seeding is 300~400 g, the calculated seed amount transported by the seed tube is 1.65~4.41 g/s; the gas-solid two-phase flow in the seed tube is a dilute phase flow, and the concentration of rapeseed particles is 0.8. Considering the overall structure of the air-feeding seed tube, the seed metering system, and the pressure loss of the fan, the inner diameter of the seed tube is determined to be 40 mm. The length of the horizontal conveying pipe of the seed tube is 400 mm, the length of the vertical conveying pipe is 600 mm, and the radius of curvature is 100 mm.

The seed tube is mainly composed of a horizontal conveying pipe, a vertical conveying pipe, and an elbow. According to the wall structure of the vertical conveying pipe, it is divided into corrugated type, hole-type, and ordinary round-tube seed tubes. Figures 2 and 3 are divided into corrugated and hole-type seed tubes, where D is the diameter of the seed tube, H is the depth of the corrugations or holes, P is the spacing of the corrugations or holes, L is the length of the corrugations or holes, φ is the diameter of the holes, and α is the angle between the two holes. The specific parameters of the three pipelines are listed in Table 1.

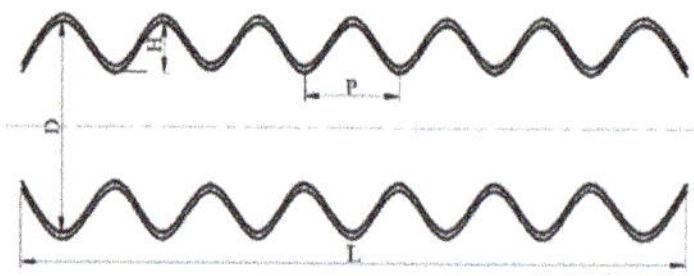

Figure 2. Diagram of corrugated pipe.

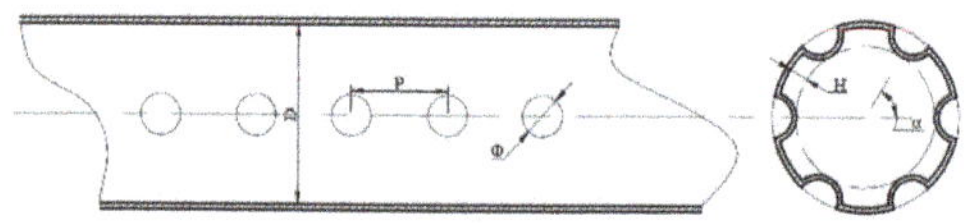

Figure 3. Diagram of hole-type pipe.

Table 1. Dimensional configurations of the tubes.

Parameters	Corrugated Type	Holes-Type	Ordinary Tube Seed Tube
Inside diameter of tube, D (mm)	40	40	40
The depth of the corrugations or holes, H (mm)	5	7	
The spacing of the corrugations or holes, P (mm)	20	20	
The length of the corrugations or holes, L (mm)	167	165	
The diameter of the holes, φ (mm)		6	
The angle between the two holes, α (mm)		72	

2.1.3. Particle–Wall Collision Analysis

It is inevitable that the particles collide with the wall in the pipeline. The collision between the particles and the wall will affect the momentum of rapeseeds, change the direction of movement of the seeds, and have a direct impact on the stable transportation of seeds [24]. The rapeseeds were replaced by spherical particles, and the contact and collision behavior of the rapeseeds was analyzed according to the characteristics of the tube wall. Figure 4a shows the collision process between rapeseeds and the wall of the smooth seed tube. The collision restitution coefficient e and friction coefficient f are used to describe the partial velocity of the seeds after the collision, u_1 and v_1 represent the tangential and normal velocity components of the rapeseed before the collision, respectively, and u_2 and v_2 respectively represent the tangential velocity of the rapeseed after the collision. The velocity component and the normal velocity component, according to the momentum theorem, the velocity component after the seed collision can be obtained:

$$\begin{cases} u_2 = u_1 + f(e+1)v_1 \\ v_2 = -ev_1 \end{cases} \tag{3}$$

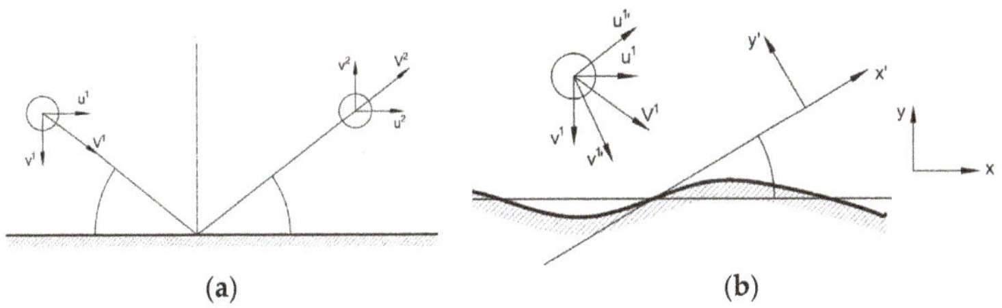

Figure 4. Collision process between rapeseeds and wall surface. (**a**) Collision with smooth wall; (**b**) collision with convex wall.

The collision process between rapeseeds and the concave-convex surface is shown in Figure 4b. Due to the addition of corrugations and hole structures, the wall surface of the seed tube changes from a smooth surface to a concave-convex surface. The random wall roughness angle γ is used to indicate when the seeds randomly collide with the wall surface. The inclination angle of the contact surface, γ, significantly affects the rebound direction and speed of the seeds after the collision. At this time, a virtual wall is introduced, and the coordinates are converted to analyze the speed of the particles after the collision. The tangential velocity component and the normal velocity component after the collision of the eye wall surface, the collision process of rapeseeds based on the virtual wall surface satisfies the following relationship:

$$\begin{cases} u_2' = u_1\left[cos^2\gamma - f(1+e)sin\gamma cos\gamma - esin^2\gamma\right] \\ +v_1(1+e)\left(fcos^2\gamma + sin\gamma cos\gamma\right) \\ v_2' = u_1(1+e)\left(sin\gamma cos\gamma - fsin^2\gamma\right) \\ -v_1\left[ecos^2\gamma - f(1+e)sin\gamma cos\gamma - sin^2\gamma\right] \end{cases} \tag{4}$$

Formulas (3) and (4) suggest that the velocity component after the seed collides is correlated with the collision restitution coefficient, friction coefficient, and roughness angle. The actual collision process between the seed and the wall is difficult to accurately describe and analyze due to the randomness of the roughness angle γ. Therefore, the DEM-CFD gas-solid coupling method was used to predict the movement of particles in the seed tube.

2.2. Numerical Simulation

2.2.1. Model of Particle

The movement of particles in the airflow field follows Newton's second law, and is affected by gravity, buoyancy, fluid resistance, Saffman lift, and Magnus lift, the contact force between particles and the wall, etc. [25]:

$$m_p\frac{dv}{dt} = F_d + F_{GB} + F_{Sa} + F_{Ma} + F_c \tag{5}$$

where m_p represents particle mass, F_d represents the fluid drag force, F_{GB} represents the resultant force of gravity and buoyancy, F_{Sa} represents the Saffman lift force, F_{Ma} represents the Magnus lift force, and F_c represents particle-wall contact force.

The contact force between the particle and the wall is:

$$F_c = F_n{}^d + F_t{}^d \tag{6}$$

$$F_n{}^d = -2\sqrt{\frac{5}{6}}\frac{In\ e}{\sqrt{In^2\ e+\pi^2}}\sqrt{S_n m^*}v_n \tag{7}$$

$$F_t{}^d = -2\sqrt{\frac{5}{6}}\frac{In\ e}{\sqrt{In^2\ e+\pi^2}}\sqrt{S_t m^*}v_t \tag{8}$$

where m^* represents equivalent eccentric mass, e represents restitution coefficient, S_n represents normal stiffness, S_t tangential stiffness, v_n represents normal component of relative velocity, and v_t represents normal component of relative velocity.

The normal stiffness and tangential stiffness in the above formula are calculated as:

$$S_n = 8E^*\sqrt{R^*\delta_n} \tag{9}$$

$$S_t = 8G^*\sqrt{R^*\delta_n} \tag{10}$$

where E^*, G^*, and R^* represent Young's modulus, equivalent shear modulus, and equivalent radius, and δ_n represents normal overlap quantity.

2.2.2. Model of Fluid

The standard k-ε model is selected as the turbulence model. When studying the turbulent motion of the draft tube, the disturbance phenomenon caused by the wall structure to the airflow and the airflow pressure drop in the draft tube need to be paid attention to, such that the Darcy friction coefficient is used to describe it. The transport formula corresponding to the standard k-ε model is as follows:

$$\frac{\partial(\rho_a k)}{\partial t} + \frac{\partial(\rho_a k v_i)}{\partial x_i} = \frac{\partial}{\partial x_j}\left[\left(\mu + \frac{v_i}{\sigma_k}\right)\frac{\partial k}{\partial x_j}\right] + G_k - \rho_a\varepsilon \tag{11}$$

$$\frac{\partial(\rho_a \varepsilon)}{\partial t} + \frac{\partial(\rho_a \varepsilon v_i)}{\partial x_i} = \frac{\partial}{\partial x_j}\left[\left(\mu + \frac{v_i}{\sigma_\varepsilon}\right)\frac{\partial \varepsilon}{\partial x_j}\right] + C_1 G_k - C_2\rho_a\frac{\varepsilon^2}{k} \tag{12}$$

where k represents turbulent kinetic energy, ε represents dissipation rate of turbulence energy, G_k represents Turbulent kinetic energy k produced by mean velocity gradient, σ_k and σ_ε represent Prandt number corresponding to turbulent kinetic energy and turbulent dissipation rate, C_1 and C_2 are empirical constant, x_i and x_j represent motion distance in different directions, v_i represents airflow velocity in different directions.

The Darcy friction coefficient [26] is expressed as follows:

$$\lambda = \frac{2\Delta P}{\rho_a v_i{}^2}\frac{D_h}{l} \tag{13}$$

where ΔP denotes the pressure drop between the two test sections, *Pa*; D_h represents the relative diameter of the pipe diameter D, m; l expresses the distance between the two test sections, m.

2.2.3. DEM-CFD Coupling

The DEM-CFD coupling simulation is to add fluid-particle interaction force on the particles according to Newton's second law, and the drag force is the main fluid-particle interaction force. When only the force of the fluid on the particles is considered, the Freesteam drag model is usually used for calculation, and its calculation formula is written as follows:

$$F_f = \frac{1}{2}C_D\rho_a A\left(\overrightarrow{v_a} - \overrightarrow{v_p}\right)\left|\overrightarrow{v_a} - \overrightarrow{v_p}\right| \tag{14}$$

$$C_D = \frac{8}{Re}\frac{1}{\sqrt{\Phi_\perp}} + \frac{16}{Re}\frac{1}{\sqrt{\phi}} + \frac{8}{\sqrt{Re}}\frac{1}{\Phi^{3/4}} + 0.42\cdot 10^{0.4(-log(\phi))^{0.2}}\frac{1}{\phi_\perp} \tag{15}$$

where v_a is the airflow velocity, m/s; v_p is the particle velocity, m/s; A is the particle cross-sectional area, m^2; C_D is the single particle drag coefficient; *Re* is the Reynolds number; ϕ is the sphericity; $\phi_\perp$ is the cross-section sphericity.

DEM-CFD coupling can be divided into one-way coupling method and two-way coupling method in accordance with the fluid-particle interaction mode. The difference lies in whether the effect of particles on the fluid is considered. When the particles occupy less than 10% of the airflow volume in the seed tube, the particles slightly affect the flow field distribution, and the force of the particles on the fluid can be ignored, such that the one-way coupling method is adopted. The one-way coupling method can save computing resources and shorten the computing time. Since the effect of particles on the fluid is ignored, the fluid motion can be calculated with the continuity formula and the momentum conservation formula with locally averaged variables. The continuity formula for fluid motion is as follows:

$$\frac{\partial}{\partial x_j}\left(\rho_f u_j\right) = 0 \tag{16}$$

The momentum conservation formula is as follows:

$$\frac{\partial}{\partial x_j}\left(\rho_f u_j u_f\right) = -\frac{\partial P}{\partial x_i} + \frac{\partial}{\partial x_j}\left[\mu_{eff}\left(\frac{\partial u_i}{\partial x_j} + \frac{\partial u_j}{\partial x_j}\right)\right] + \rho_f g \tag{17}$$

where ρ_f denotes fluid density; u is the fluid velocity; P is the pressure of fluid; x is the coordinates; μ_{eff} is the effective viscosity.

2.2.4. Numerical Setup

In the EDEM-CFD coupled simulation, the commercial CFD software ANSYS Fluent 2020 was used to solve the fluid phase; with air as the medium carrier, the density of which was 1.225 kg/m^3, the viscosity was 1.789×10^{-5} Pa s, and the airflow velocity was 16 m/s. The particle phase was solved using EDEM 2021. In the case of DEM-CFD coupling, the time step of EDEM should be much lower than that of Fluent, therefore, the time steps of EDEM and Fluent are set to 5×10^{-6} and 1×10^{-3} s, respectively. Set the number of Fluent steps to 2000, that is, the total simulation time is 2 s. In the simulation, the seed tube material is aluminum alloy. The material characteristics of seeds and aluminum alloy and their mechanical properties [7,8] and air flow field parameters are shown in Table 2.

Table 2. Simulation parameters.

Case	Parameters	Values
EDEM software	time step/s	5×10^{-6}
	save interval/s	0.01
	acceleration of gravity/(m/s^2)	9.81
Fluent software	time step/s	2×10^{-3}
	save interval/s	0.05
	current density/(kg/m^3)	1.225
	air viscosity coefficient/(Pa·s)	1.789×10^{-5}
Rapeseed	poisson ratio	0.25
	shear modulus /Pa	1.1×10^7
	density/(kg/m^3)	1060
Seed tube	poisson ratio	0.3
	shear modulus /Pa	2.7
	density/(kg/m^3)	2700×10^{10}
Rapeseed–rapeseed contact parameters	coefficient of restitution	0.6
	coefficient of static friction	0.5
	dynamic friction factor	0.01
Rapeseed–seed tube contact parameters	coefficient of restitution	0.6
	coefficient of static friction	0.3
	dynamic friction factor	0.01

The thousand-grain weight of the rapeseeds in the simulation test was 3.8 g. According to the seeding amount of rapeseed per mu and the forward speed of the machine, the inlet of the seed tube was set to generate 560 rapeseeds per second. The time for generating rapeseed particles at the entrance of the seed tube is 2 s, and the total simulation time was 2 s. Three types of air distribution in the seed tube and the movement characteristics of rapeseed particles were carried out. The velocity and force changes of a single rapeseed, the maximum speed and maximum force of the seed at each moment, and the distribution of rapeseeds in the seed tube were derived from EDEM to analyze the movement characteristics of rapeseed particles.

2.2.5. Mesh Dependence Analysis

Before modeling calculation, the mesh independence verification was performed first, and the optimal number of mesh is determined by considering the calculation workload and calculation accuracy. The seed tube model was divided into 161,088, 237,789, and 376,027 grids to verify the grid independence. The axial position Z = 70 mm was selected for comparison of axial velocity. It can be seen from Figure 5 that the number of mesh had a certain impact on the simulation results. When the number of mesh was 237,789 or 376,027, the calculation results tended to be consistent. Considering the accuracy of simu—lation results and the cost of simulation calculation, the number of mesh of seed tube flow field in this paper was 237,789.

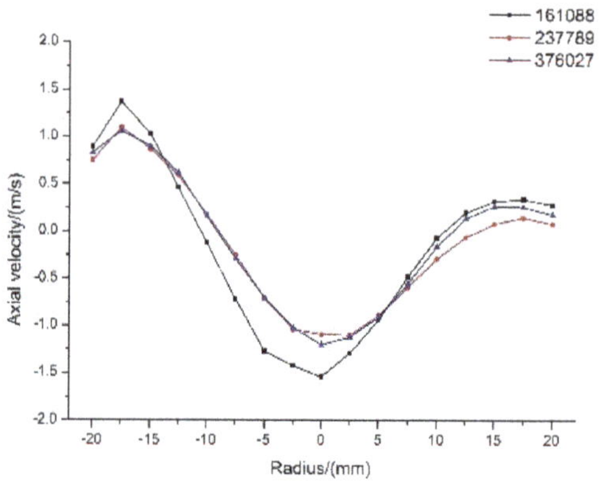

Figure 5. Mesh dependence analysis.

3. Results

3.1. Model Validation

To verify the reliability of the simulation results, an ordinary round-tube seed tube was selected in this experiment, and the airflow velocity at the vertical conveying pipe under the condition of no material when the inlet velocity was 16 m/s was examined. In the test, the L-shaped pitot tube (the pitot tube coefficient: 0.99~1.01, the speed measurement range: 2~70 m/s) and the wind speed, pressure, and air volume meter were used to examine the wind speed. The pressure hole was perpendicular to the airflow direction. Four measurement points were evenly arranged along the outlet direction (including A, B, C, D), and the distance between the measurement points was 100 mm, and the distances from the outlet of the seed tube were 100 mm, 250 mm, 400 mm, and 550 mm, respectively. The distribution of the measurement points is shown in Figure 6. When the inlet wind speed was 16 m/s, the wind speed of different measuring points was examined 10 times, and the examined values are listed in Table 3.

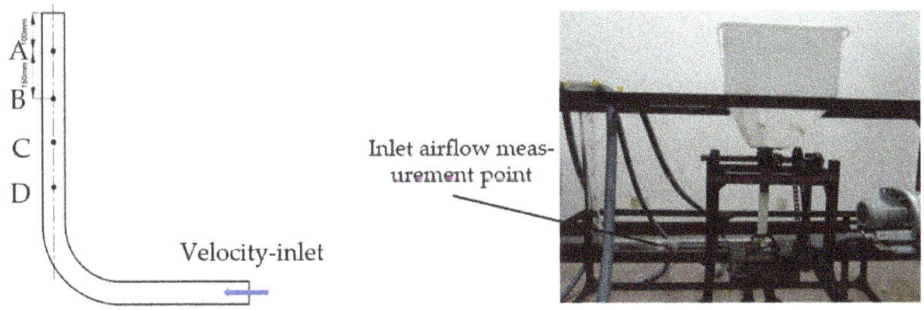

(a) (b)

Figure 6. Measuring point position map. (**a**) Location distribution of measuring points; (**b**) test device diagram.

Table 3. Export speed 10 measurements and their average.

Velocity/(m.s^{-1})	Position/(mm)			
	A	B	C	D
1	14.11	13.09	15.40	15.43
2	13.70	13.97	14.05	15.60
3	14.74	14.65	15.54	15.86
4	14.62	13.03	14.23	16.10
5	12.90	14.11	15.91	15.80
6	14.09	14.56	15.82	15.59
7	13.95	14.64	15.68	15.66
8	14.25	13.86	16.05	15.49
9	14.30	14.29	15.22	16.08
10	13.56	15.33	15.93	15.76
average	14.02	14.15	15.38	15.74
standard deviation	0.51	0.68	0.67	0.22

As depicted in Table 2, the wind speed measurement test produced a certain measurement error, which may be due to the small-scale vibration of the Pitot tube under the action of the wind, resulting in the full pressure hole of the Pitot tube being not parallel to the airflow direction. After many repetitions of the test, its patency may deteriorate, thus causing biased measurement results.

After using the CFD-POST software to extract the data of each test point in the simulation results, the examined and simulated values of the wind speed at each measurement point at the exit were listed in Table 4. Errors in the table are shown as (|Analog value—Examined value|)/Examined value × 100%. It can be seen from the table that the airflow velocity examined by the simulated value and the examined value basically gradually decreased along the outlet direction, indicating that the approximate trend of the simulated result and the actual examined value was the same, and the simulated value at the measurement points B, C, D and the examined value were larger than that of A. The possible reason is that the last three measuring points were close to the curved part of the flow, and the airflow field is relatively disordered, such that there may be a certain deviation between the simulated value and the examined value. After calculation, the average error of the four positions was 6.71%, which is acceptable in agricultural engineering applications with such complex factors. Therefore, it can be confirmed that the air-feeding seed metering system is used in this paper and the simulation results of the internal flow field in the tube are reliable.

Table 4. Comparison of simulated value and examined value of outlet velocity.

Position/(mm)	Gas Velocity/(m.s^{-1})		Errors
	Simulation Value	Examined Value	
A	14.35	14.02	2.35
B	15.48	14.15	9.40
C	16.52	15.38	7.41
D	16.95	15.74	7.69

3.2. Effect of Seed Tube Structure on Airflow Field

Figure 7 presents the airflow velocity and pressure at the respective point extracted along the outline of the seed tube for different types of seed tubes to analyze the airflow velocity close to the wall of the seed tube. As depicted in the figure, the addition of the pressurized structure of the seed tube affected the air pressure and speed. In the elbow part of the seed tube, the airflow speed increased, while the pressure decreased rapidly. The supercharging structure significantly affected the airflow speed of the elbow part. When the corrugated supercharging was used, the airflow speed of the elbow part increased dramatically to 32.48 m/s. When the hole was pressurized, the maximum speed of the elbow part was 26.20 m/s. However, the airflow velocity of the ordinary round-tube seed tube without the pressurized structure slightly fluctuated, whereas the pressurized structure slightly affected the airflow velocity at the horizontal conveying pipe and the vertical conveying pipe. The pressurized structure significantly affected the internal pressure of the seed tube. The pressure value of the seed tube with corrugated pressurization is the highest, the hole-type is the second, and the pressure value of the ordinary roundtube seed tube is the lowest. Sharply reduced, after entering the elbow, the airflow pressure recovers, and the minimum pressure values of the three kinds of seed tubes are basically the same. Among the three kinds of seed tubes, the pressure loss of the corrugated tube was the smallest. It can be seen that the corrugated tube has better flow field characteristics, the pressure loss in the seed tube is small, and the airflow velocity is better.

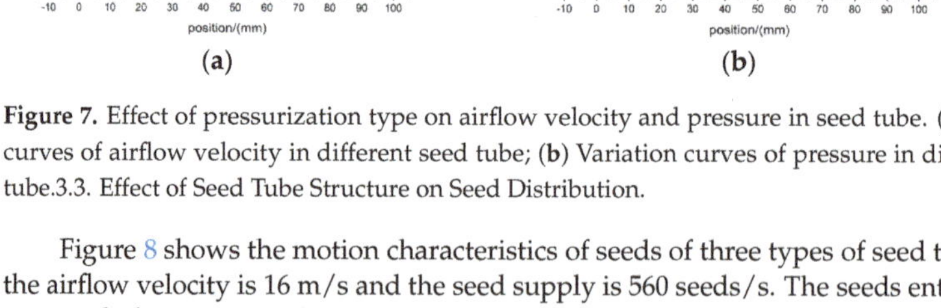

Figure 7. Effect of pressurization type on airflow velocity and pressure in seed tube. (**a**) Variation curves of airflow velocity in different seed tube; (**b**) Variation curves of pressure in different seed tube.3.3. Effect of Seed Tube Structure on Seed Distribution.

Figure 8 shows the motion characteristics of seeds of three types of seed tubes when the airflow velocity is 16 m/s and the seed supply is 560 seeds/s. The seeds enter the pipe in an orderly manner. When passing through the elbow, the seed group moves along the pipe wall. Due to the collision and rebound between the seed group and the pipe wall, the stable seed flow begins to disperse. The density of the conveyed seed particle group is reduced, and the problem of intermittent conveying. The corrugated and hole-type seed tubes induce continuous fluctuations in the air flow due to their intermittent curved surfaces. The airflow fluctuations affect the running of the seeds, such that the dispersed seed flow is re-aggregated and continuously conveyed to the distributor. The seed tube is conducive to the continuous and stable delivery of the seed flow.

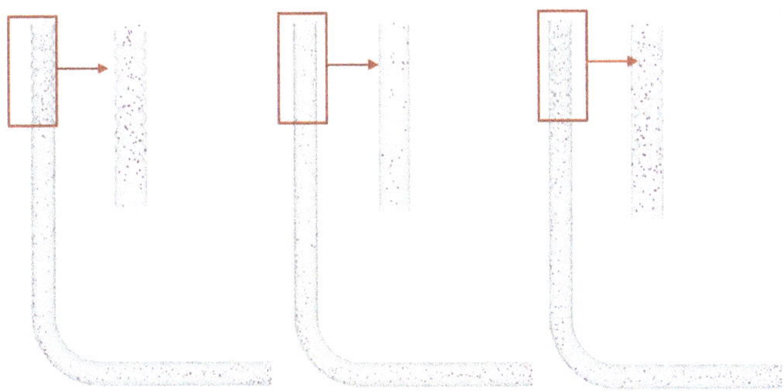

Figure 8. Distribution of rapeseeds in different seed tubes.

3.3. Effect of Seed Tube Structure on Seed Movement Characteristics

Figure 9a presents the test result of tracking the speed of a single seed particle generated in various seed tubes when 560 seed particles are generated per second. As depicted in the figure, the corrugated tube is the same as the ordinary round-tube seed tube when the seed is in the seed tube for 0~0.3 s. The force of the round tube is basically the same, while in the hole-type tube, there is a surge of force due to the collision between the seeds flowing through the elbow and the wall surface. Due to the different stagnation time of the seeds in the seed tube, there is one force change "wave peak" in the corrugated seed tube, and two force change "wave peaks" in the hole-type seed tube, indicating that the seeds are affected by the wall structure. This indicates that the speed of seeds varies alternately due to the influence of the wall structure, which makes the distribution of seeds more uniform in the pressurized tube. It can be seen from Figure 9b that the maximum resultant force on the seed particles in the corrugated, ordinary round-tube seed tube, and hole-type seed tubes is 0.35, 0.14, and 0.43 N, respectively, and the resultant force on the seed particles in the seed tube fluctuates significantly, indicating that the corrugation and the pocket structure increase the probability of the seeds colliding with the wall, resulting in the particle-wall collision force.

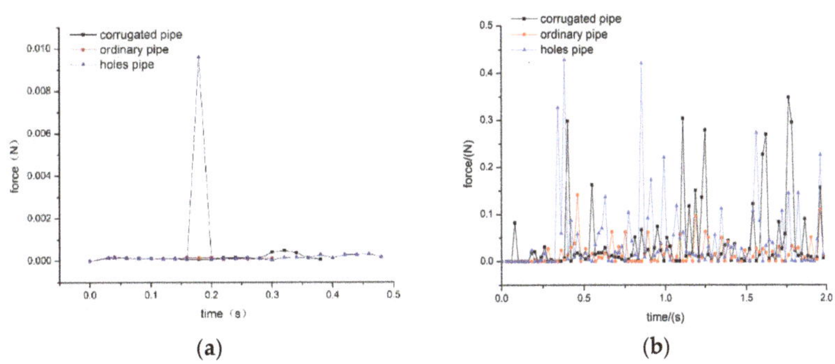

Figure 9. Force variation of single particle and maximum force diagram of particle in seed tube. (**a**) Effect of seed tube on stress of single seed; (**b**) Influence of seed tube on maximum force of seed.

Figure 10a presents the speed test results of tracking the first seed particles generated in various seed tubes when 560 seed particles were generated per second. The maximum speed of seed particles in the tube reached 4.79, 4.35, and 5.18 m/s, respectively, and the stagnation time of seeds in the seed tube was 0.38, 0.48, and 0.30 s, respectively, thus suggesting that the ordinary round-tube seed tube exhibits high transportation efficiency but high speed. The force and speed of the seed particles increase when the seed particles collide with the main body of the distribution device, thus leading to an increase in the seed breaking rate. After the seeds in the ordinary round-tube seed tube pass through the elbow part, the seed speed increases linearly, while the corrugated and nested seeds increase linearly. The speed of the seeds in the eye-type seed tube shows a fluctuating trend, and the speed variation range of the corrugated tube is small, which is conducive to the stable transportation of the seeds. The speed of the seeds at the outlet are 5.18, 1.73, and 3.76 m/s. According to the elastic collision analysis between the fertilizer particles and the distribution device in the literature [27], it can be seen that the crushing rate of the material is correlated with the speed of the material particles and the distributor approaching each other. The breakage rate of the type and hole-type seed tubes is lower than that of the ordinary round-tube seed tube. As depicted in Figure 10b, the maximum velocity distribution in the corrugated and hole-type seed tubes shows a trend similar to "sinusoidal fluctuation", indicating that the airflow velocity fluctuates up and down, which increases the collision between the seeds and the wall of the seed tube, which can promote a more even distribution of rapeseeds before entering the distributor.

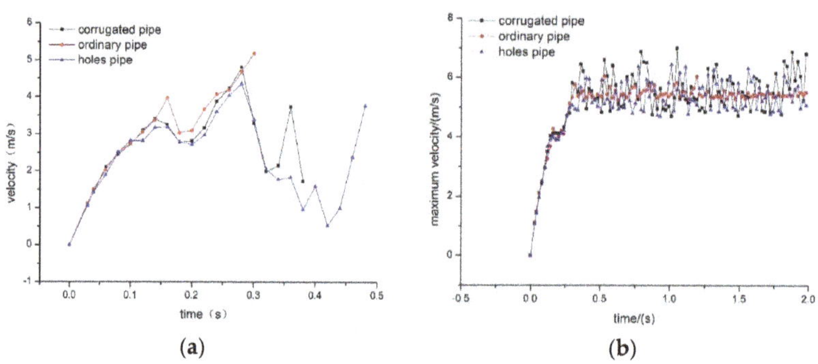

Figure 10. Plot of single particle velocity variation and maximum particle velocity in seed tube. (**a**) Effect of seed tube on stress of single seed; (**b**) Influence of seed tube on maximum force of seed.

4. Conclusions

In this paper, the DEM-CFD coupling method was used to simulate the effect of the structure of different vertical conveying pipes of the air-feeding seed metering system on the seed movement characteristics. It was verified by the bench test that the DEM-CFD coupling simulation can be well described. The movement of the seeds in the seed tube draw the following conclusions:

1. The main structural parameters of the corrugated, ordinary round-tube, and hole-type seed tubes were determined. Based on the Herz contact model, the collision contact between rapeseeds and the wall surface was analyzed when the wall of the seed tube was smooth and uneven. During the process, the virtual wall was used to study the collision of seeds on the concave-convex tube wall, so as to obtain the motion formula after the rapeseeds collided with the tube wall.
2. In this paper, the effect of the vertical conveying pipe structure of the seed guiding pipe on the uniformity of the seed flow was determined. The gas-solid coupling method of DEM-CFD was used to analyze the distribution of the air flow field of the seeds in the corrugated, hole-type, and ordinary round-tube seed tubes, and the movement characteristics of rapeseeds in different seed tubes were analyzed. The results show that the vertical conveying tube of the seed tube plus the corrugated or hole-type shape significantly affected the airflow field in the seed tube. The maximum airflow velocity was 32.48 and 26.20 m/s, respectively; the vertical conveying pipe of the seed tube plus corrugated and hole-shaped structures are beneficial in improving

the damage to the uniform seed flow caused by the near-wall movement of the seed through the elbow part. The speed and force of the seeds in the corrugated and hole-shaped seed tubes were similar to the "sinusoidal fluctuation" trend, thus facilitating the uniform distribution of the seeds. However, the maximum force of the seeds in the hole-type seed tube fluctuated significantly, meaning they were more prone to the phenomenon of a surge in force. For the corrugated type, it will negatively affect the breakage of the seeds.

3. The bench test verifies the feasibility of the DEM-CFD coupled simulation. The airflow velocity at the vertical conveying pipe was examined when the airflow velocity at the entrance of the ordinary round-tube seed tube was 16 m/s and there was no material. Compared with the airflow velocity obtained by simulation, it was found that the average error between the simulated value and the measured value was 6.71%, which is acceptable in the application of such engineering problems with complex factors in agricultural engineering. The reliability of the results of the simulation of the internal flow field in the seed tube of the air-feeding seed metering system.

Using the DEM-CFD coupling method, the movement mechanism of rapeseeds in the airflow field in the seed tube can be better described, and different pressurized structures can be compared. The DEM-CFD coupled simulation method has been extensively used in two-phase flow and gas flow transport to visualize particle motion and investigate the source of differences. Numerical simulation of the movement of seeds in different seed tubes based on the DEM-CFD coupling method can provide a reference for the optimization of the air-feeding seed metering.

Author Contributions: Conceptualization, Y.X. and M.W.; methodology, Y.X.; software, H.L.; validation, Z.M., H.L. and Y.X.; formal analysis, M.W.; investigation, Z.M.; resources, M.W.; data curation, Y.X.; writing—original draft preparation, Y.X.; writing—review and editing, Y.X.; visualization, Y.X.; supervision, M.W.; project administration, H.L.; funding acquisition, H.L. All authors have read and agreed to the published version of the manuscript.

Funding: This research was funded by Postgraduate Scientific Research Innovation Project of Hunan Province. grant number "CX20210660".

Institutional Review Board Statement: Not applicable.

Informed Consent Statement: Not applicable.

Data Availability Statement: All data are presented in this article in the form of figures and tables.

Conflicts of Interest: The authors declare no conflict of interest.

References

1. Yazgi, A.; Degirmencioglu, A. Measurement of seed spacing uniformity performance of a precision metering unit as function of the number of holes on vacuum plate. *Measurement* **2014**, *56*, 128–135. [CrossRef]
2. Lei, X.; Liao, Y.; Wang, L. Simulation of gas-solid two-phase flow and parameter optimization of pressurized tube of air-assisted centralized metering device for rapeseed and wheat. *Trans. Chin. Soc. Agric. Eng.* **2017**, *33*, 67–75.
3. Downs, H.; Taylor, R. Evaluation of Pneumatic Granular Herbicide Applicators for Seeding Small Grains in Oklahoma. *Appl. Eng. Agric.* **1986**, *2*, 58–63. [CrossRef]
4. Guler, I. Effects of flute diameter, fluted roll length, and speed on alfalfa seed flow. *Appl. Eng. Agric.* **2005**, *21*, 5–7. [CrossRef]
5. Shirsath, S.; Padding, T.; Clercx, H. Cross-validation of 3D particle tracking velocimetry for the study of granular flows down rotating chutes. *Chem. Eng. Sci.* **2015**, *134*, 312–323. [CrossRef]
6. Senturov, S. Process Study and Parametres Reasoning of a Centralized Air Eeding. Ph.D. Thesis, Ural State University, Gorki, Russia, 1979.
7. Liao, Q.; Lei, X.; Liao, Y. Research progress of precision seeding for rapeseed. *Trans. Chin. Soc. Agric. Mach.* **2017**, *48*, 1–16.
8. Lei, X.; Liao, Y.; Li, Z. Design and experiment of seed feeding device in air-assisted centralized metering device for rapeseed and wheat. *Trans. Chin. Soc. Agric. Eng.* **2015**, *31*, 10–18.
9. Chang, L.; Zhang, X. Design and test of one-step centralized type pneumatic seeding system. *Trans. Chin. Soc. Agric. Eng.* **2011**, *11*, 136–141.
10. Chang, L. Research and Experimental Analysis of Centralized Pneumatic Seeding System. Master's Thesis, Shandong Agricultural University, Shandong, China, 2007.

11. Chang, L.; Zhang, X.; Chen, Y. Design of the quantitative force feed in the air-stream central-type drill system. *J. Agric. Mech. Research.* **2007**, *6*, 66–67, +89.
12. Liu, L.; Liu, Z.; Yang, X. Design and test on pneumatic No-till wheat planter. *Trans. Chin. Soc. Agric. Mach.* **2011**, *42*, 54–57.
13. Liu, L.; Yang, X.; Li, C. Design of 2BMG-24 No-till wheat planter. *Trans. Chin. Soc. Agric. Mach.* **2009**, *40*, 39–43, +48.
14. Zhai, G.; Bao, D.; Wang, Z. Design for metering device key parts of pneumatic grass seeder. *Trans. Chin. Soc. Agric. Mach.* **2014**, *45*, 47–51.
15. Li, Y.; Liu, Y.; Liu, L. Distribution mechanism of aitflow in seed tube of different lengths in pneumatic seeder. *Trans. Chin. Soc. Agric. Mach.* **2020**, *51*, 55–64.
16. Li, Y.; Liu, R.; Liu, C. Simulation and test of seed velocity coupling in seed tube of pneumatic seed metering device. *Trans. Chin. Soc. Agric. Mach.* **2021**, *52*, 54–61, +133.
17. Li, Y.; Zhao, J.; Zhao, J. Biomimetic design and experiment of the distributor of pneumatic seeding system based on the crucian curve. *Trans. Chin. Soc. Agric. Mach.* **2022**, *52*, 80–87.
18. Yatskul, A.; Lemiere, J.; Cointault, F. Influence of the divider head functioning conditions and geometry on the seed's distribution accuracy of the air seeder. *Biosyst. Eng.* **2017**, *161*, 120–134. [CrossRef]
19. Yatskul, A.; Lemiere, J.; Cointault, F. Establishing the conveying parameters required for the air seeder. *Biosyst. Eng.* **2018**, *166*, 1–12. [CrossRef]
20. Dai, Y.; Luo, X.; Wang, Z. Design and xeperiment of rice pneumatic centralized seed distributor. *Trans. Chin. Soc. Agric. Eng.* **2016**, *32*, 36–42.
21. Dai, Y.; Luo, X.; Zhang, M. Design and experiments of the key components for centralized pneumatic rice dry direct seeding machine. *Trans. Chin. Soc. Agric. Eng.* **2020**, *36*, 1–8.
22. Dai, Y. Design and Experiment of Centralized Pneumatic Racedrilling Machine for Dry Land. Ph.D. Thesis, South China Agricultural University, Guangdong, China, 2018.
23. Yang, L. *Pneumatic Conveying Engineering*; China Machine Press: Beijing, China, 2006.
24. Chen, Q.; Liu, L. Numerical simulation of sphere motion in infinite-length pipe flow. In Proceedings of the 16th National Congress on Hydrodynamics, Jiangsu, China, 30 October 2021.
25. Akhshik, S.; Behzad, M.; Rajabi, M. CFD-DEM approach to investigate the effect of drill pipe rotation on cuttings transport behavior. *J. Pet. Sci. Eng.* **2015**, *127*, 229–244. [CrossRef]
26. Li, M.; Tariq, S.; Ebrahim, A. Geometric optimization for thermal–hydraulic performance of dimpled enhanced tubes for single phase flow. *Appl. Therm. Eng.* **2016**, *103*, 639–650. [CrossRef]
27. Wang, L.; Liao, Q.; Liao, Y. Effects of distributor types on fertilizing performance in an air-assisted applicator. *Trans. Chin. Soc. Agric. Eng.* **2021**, *37*, 24–34+315.

agriculture

Article

The Design of a Mechanized Onion Transplanter for Bangladesh with Functional Testing

Spencer Stubbs and Jonathan Colton *

Georgia Institute of Technology, North Avenue, Atlanta, GA 30332, USA
* Correspondence: jonathan.colton@me.gatech.edu

Abstract: Rising labor costs and the inefficient manual methods of cultivating red onions in Bangladesh contribute to the country's failure to meet is own demand. Mechanizing the process of transplanting red onion seedlings will reduce the manual labor required while increasing crop yields. This paper provides an initial study of a proposed mechanized onion transplanter designed to attach to the back side of a two-wheel tractor with power tiller operated-seeder, commonly used in Bangladesh. Testing of a prototype made from these designs proves that the design is functional but requires further development for commercial/widespread use.

Keywords: mechanization; onion; transplanting

Citation: Stubbs, S.; Colton, J. The Design of a Mechanized Onion Transplanter for Bangladesh with Functional Testing. *Agriculture* **2022**, *12*, 1790. https://doi.org/10.3390/agriculture12111790

Academic Editor: Muhammad Sultan

Received: 11 September 2022
Accepted: 24 October 2022
Published: 28 October 2022

1. Introduction

1.1. Onion Farming in Bangladesh

In recent years, the agricultural sector in Bangladesh has been suffering as many workers are leaving the country's rural areas (i.e., farmlands) in search of more lucrative work in cities. This exodus is leaving many farmers, especially small stakeholder farmers, facing great difficulty in their effort to produce enough food and make a profit [1].

Although shrinking, the agriculture sector remains a powerful driver in the Bangladeshi economy and is vitally important [2]. Unlike more developed countries, smallholder farms are the foundational drivers in Bangladesh's agricultural sector. The average farm size in Bangladesh is estimated to be 0.5 hectares (roughly 1.24 acres), making it one of the smallest average farm sizes in the world. As such, increasing the productivity of smallholder farms is one of the best ways the country can ensure it has an adequate food supply and strong economy [3].

The seeds for red onions, which are the most popular onion crop in Bangladesh, are typically sown in seed beds around November when the Bangladeshi highlands are dry [4]. The conditions for the red onion seeds must neither be too wet nor too dry, leaving only a small window of opportunity for their germination. After 25–45 days, the 8–10 cm seedlings are typically unearthed and transported to markets where they can be purchased by farmers. Then, the seedlings are planted in the lower farmlands where they will remain until they mature and are harvested [5].

The smallholder farmers who buy the onion seedlings generally hire local laborers to transplant the onion seedlings by hand one-by-one. This tedious labor is not only time consuming, but also erodes the profits of the farm owners. Hired labor accounts for 36% of the total cost for onion crops. Although onions produce higher profit margins than most crops in Bangladesh, they also require higher investment than most [6]. By reducing the labor costs associated with onion crops, farmers can increase their profit margins.

Existing transplanters that are capable of transplanting onion seedlings generally have significant drawbacks making them unusable by smallholder farmers in Bangladesh. Several types of machines exist that are capable of transplanting onion seedlings exist, but each of have significant drawbacks rendering them unusable or impractical for the smallholder farmers in Bangladesh.

1.2. Existing Transplanters

Self-propelled machines are generally powered by a 5–6 HP engine and require one or two users for operation. These machines are capable of planting 1–4 rows of seedlings simultaneously, making them size-appropriate for the smallholder farms of Bangladesh.

Fully automatic machines only require a single user during operation. However, they require that seedlings be grown trays that consist of individual pots for each seedling. Although these agricultural trays are common for many types of crops worldwide, they are not used by onion farmers in Bangladesh. Semi-automatic machines require a laborer to separate the seedlings and individually place them in the transplanter while it is moving. Oftentimes, this requires one laborer to operate the machine while another separates and places the seedlings in the transplanter.

Both semi- and fully automatic machines that are self-propelled typically cost 7000 USD or more which makes them impractical for smallholder farms.

Four-Wheel Tractor Attachment Transplanters

Like self-propelled machines, there are also semi- and fully automatic transplanters that exist as attachments for 4WTs. These machines are much larger than self-propelled transplanters and are capable of planting more than 10 rows simultaneously.

Fully automatic transplanters generally require two laborers; one to drive the tractor and another to load trays of seedlings into the transplanter. These machines are large and complex and usually cost a minimum of 40,000 USD.

Semi-automatic attachments typically require 8 or more laborers to separate the seedlings, and another to drive the tractor. Laborers sit atop the transplanter, separate onion seedlings, and drop them into chutes. The seedlings are guided by the chute to the ground where they are planted by using either a plow or a punch-type mechanism. Semi-automatic transplanting attachments typically cost around 4800 USD and above.

1.3. Two-Wheel Tractors in Bangladesh

The small size of average farms in Bangladesh makes 4WTs impractical to use for most farms in the region. Commonly used in more developed countries by commercial and private farmers, 4WTs are more expensive and less maneuverable than two-wheel tractors (2WT). 2WTs, resultingly, have become one of the most common agricultural machines used in Bangladesh.

Two-wheel tractors are single axle tractors commonly driven by a 10–12 HP diesel engine. The handles extend backwards towards the operator who walks, or in some cases—rides, behind the tractor. These machines are very valuable for smallholder farmers because they can pull a variety of different attachments such as trailers, cultivators, plows, seeders, and harvester attachments.

Many onion farmers in Bangladesh have already adopted the use of 2WTs in combination with a power-tiller operated seeder (PTOS) for the use of preparing and seeding fields. PTOS attachments, like the one shown in Figure 1, till the ground while sowing seeds and fertilizer simultaneously.

Figure 1. Two-wheel tractor with a PTOS attachment [7].

1.4. Proposal for an Onion Transplanter for Smallholder Farmers

As part of a large initiative to promote mechanization in Bangladesh's agriculture sector, USAID's Feed the Future Bangladesh Cereal Systems Initiative for South Asia—Mechanization and Extension Activity (CSISA-MEA) is working to develop and promote scale-appropriate agriculture machines that are manufacturable and serviceable in Bangladesh [8]. Stakeholders of the CSISA-MEA project have suggested that a mechanized solution to the labor intensive process of transplanting onion seedlings could be beneficial in promoting onion cultivation amongst smallholder farms in Bangladesh.

This paper presents a preliminary design for a mechanized onion transplanter intended to ease the labor requirements associated with transplanting red onion seedlings in Bangladesh [9]. The transplanter is designed to be attached to, and powered by, the 2WT and PTOS systems commonly used in Bangladesh, making it useful for the smallholder farmers that make up most Bangladeshi farmers. This paper also presents the results of an initial study of the design's capability to perform its intended functions.

The prototype testing results presented in this paper provides evidence that transplanting onion seedlings can be achieved using this design. The functional design presented may serve as a starting point for future researchers interested in developing a more robust design.

2. Materials and Methods

2.1. Technical Requirements

Drawing on requirements from project stakeholders and agricultural standards, the main functional requirements shown in Table 1 are established [9]. Planting spacings and depth were derived from agricultural standards pertaining to best practices for growing onions [10]. Operational movement pace was based on the optimal movement pace for 2WT attachments [8]. Finally, the success rate of transplanting was established by project stakeholders.

Table 1. Target specifications of the onion transplanter.

Onion Planting Spacing (cm)	Row Planting Spacing (cm)	Onion Planting Depth (cm)	Operational Movement Pace (0.3 m/s)	Success Rate of Transplanting
10	20	2.5	0.3	95%

2.2. Design of the Transplanter

The onion transplanter consists of three subsystems, each designed to perform a specific set of functions. The three subsystems that make up the transplanter are the dispensing subsystem, the orienting subsystem, and the planting subsystem. The design of the transplanter is shown in Figure 2, which indicates the subsystems.

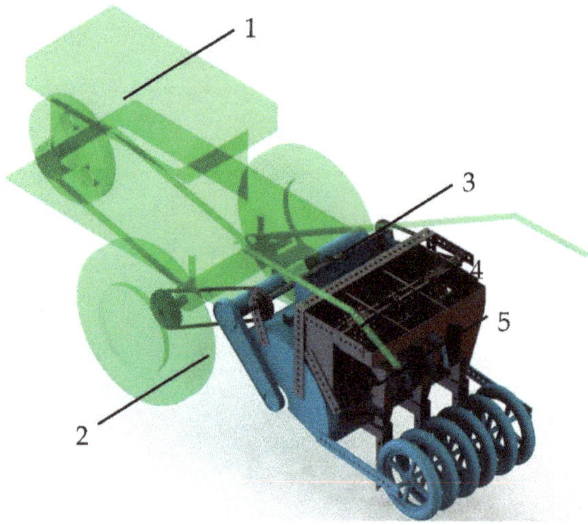

Figure 2. CAD model of the onion transplanter, as attached to a 2WT and PTOS. (**1**) 2WT; (**2**) PTOS; (**3**) Dispensing subsystem; (**4**) Orienting subsystem; (**5**) Planting subsystem. This model was designed using SolidWorks Student Edition 2021.

The transplanter consists of three of each subsystem, placed side by side and attached to the tiller cover of the PTOS system. The side-by-side configuration of the system components allows the user to transplant three rows of onions with each pass of the field. The spacing between system iterations is 20 cm to comply with the technical requirement previously indicated.

Because the transplanter is intended to be manufactured and serviced in Bangladesh, material availability was a primary consideration throughout the design process. Similar machines and equipment, when manufactured in Bangladesh, are composed primarily of mild steel which available and affordable in the region. Therefore, all unique components of the transplanter were designed under the assumption that they would be made of mild steel. Common components such as sprockets, gears, shafts, and screws are also readily available. As such, no material specification is provided for common components.

2.2.1. Design of the Dispensing Subsystem

The dispensing subsystem, shown in Figure 3, mimics the functional method of existing garlic transplanter machines which rely on chain-driven cups to separate and lift garlic cloves out of a hopper [11]. Similarly, onion seedlings are loaded into the onion transplanter's hopper but are uniformly oriented (i.e., roots pointing toward the same wall) as shown in Figure 4. The ramp inside the hopper allows gravity to push the onions downward toward the onion cups, which separate and lift the seedlings as they rotate.

Figure 3a shows the hopper, cups, sprockets, and shafts of the subsystem. Figure 3b shows a cross-sectional view of the subsystem, illustrating the travel path of the onion seedlings as they are separated at the bottom of the hopper and lifted until they are dispensed at the top.

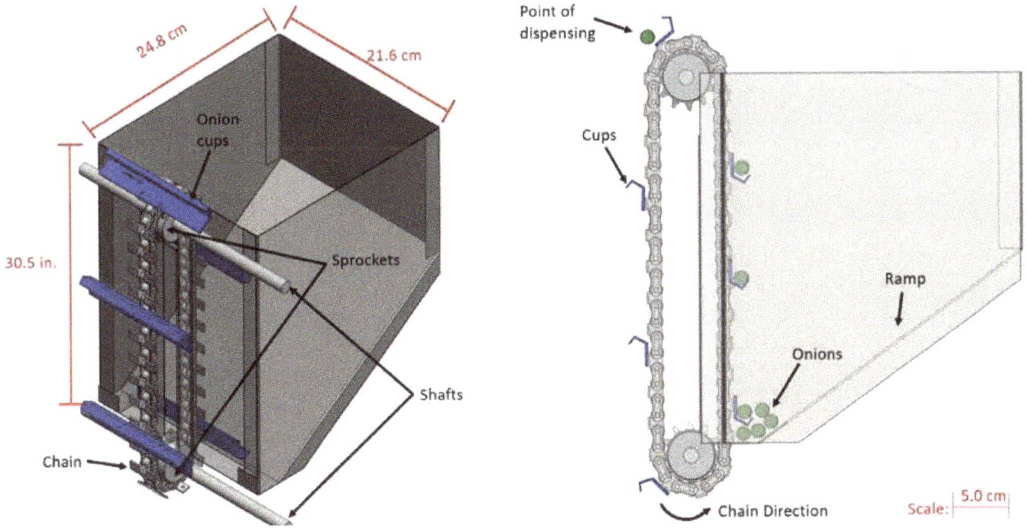

(a) (b)

Figure 3. (**a**) Isometric view of the dispensing subsystem CAD model; (**b**) Side view of the dispensing subsystem showing the various positions of onions during dispensing.

Figure 4. Uniformly oriented onions in the dispensing subsystem.

To ensure that the onions do not fall from the bottom of the hopper, a slotted door is utilized at the bottom, shown in Figure 5. The profile of the slotted door and the onion cups match such that the cups can pass through the door, while the door blocks onions from falling through.

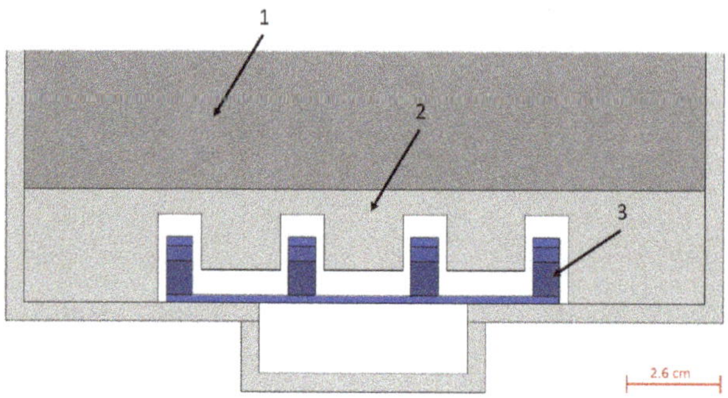

Figure 5. (**1**) Ramp within the hopper; (**2**) Slotted hopper door; (**3**) Matching cups.

2.2.2. Design of the Orienting Subsystem

After the onions in the hopper are all oriented uniformly, they enter the next subsystem, the orienting subsystem. The orienting subsystem consists of a funnel designed to receive the dispensed onions and adjust their orientation such that the roots face downward.

As the onions are dispensed horizontally into the orienting funnel, the stalk end of the onions impacts the interference wall shown in Figure 6, causing the root end of the onion to orient downward (relative to the stalk end), thus orienting the seedling properly for planting.

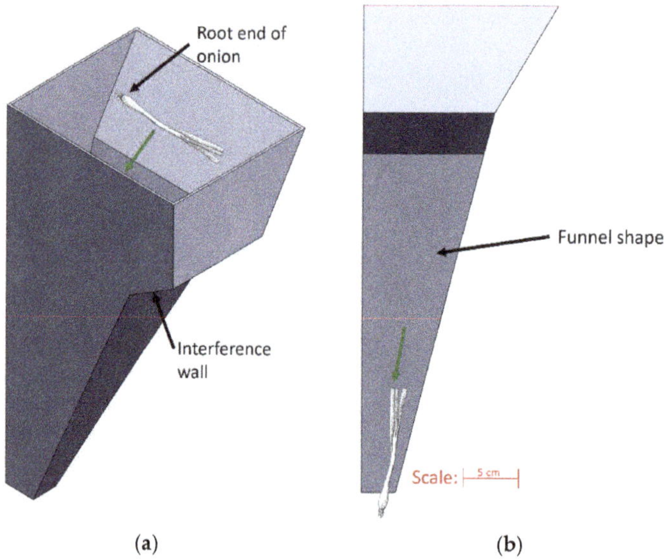

Figure 6. Design of the orienting funnel. (**a**) Isometric view; (**b**) Right-side view.

2.2.3. Design of the Planting Subsystem

Existing onion transplanters typically utilize two methods of planting onion seedlings: punching mechanisms and furrowers. Examples of punching mechanisms and furrowing mechanisms are shown in Figure 7. Whereas punching mechanisms form discrete holes into which the onions are dropped, furrowing mechanisms typically drag a blade or blunt object through the ground which creates a furrow into which the onions are dropped. Given the correct soil conditions, gravity often will cause the furrow to fill itself in and bury the seedlings without additional input from the transplanter [12]. Typically, soils that are somewhat dry and have a fine till provide the ideal soil conditions for furrower transplanters like these. As onions grow best in a sandy loam soil, this method of using a furrower mechanism was implemented in the transplanter.

(a) (b)

Figure 7. (a) Punch mechanism [13]; (b) Furrower [14].

The furrower (Figure 8) is designed to be made of a square hollow tube so that the onions can be dropped through the furrower and into the ground as the furrow is being opened. The wide, blunt shape of the furrower creates a self-filling furrow wide enough for onion seedlings. In favorable soil conditions, as described above, the blunt shape of the furrower creates a furrow that tends to fill itself in, thus burying the seedlings after they enter the furrow.

Figure 8. Hollow furrower of the onion transplanter.

2.2.4. Design of the Powertrain

To ensure that the onion transplanter is active only while the 2WT is in motion, the transplanter is powered by the drive axle of the 2WT. A diagram of the powertrain designed to power the onion transplanter can be seen in Figure 9. Black rectangles represent the system's shafts. Gray and blue boxes represent sprockets/gears used to transmit power between shafts. Gray colored boxes indicate that the components are included in the design scope of the onion transplanter, whereas the blue boxes indicate that the components are existing components associated with the design of the PTOS system [8]. The function of each shaft is described in Table 2.

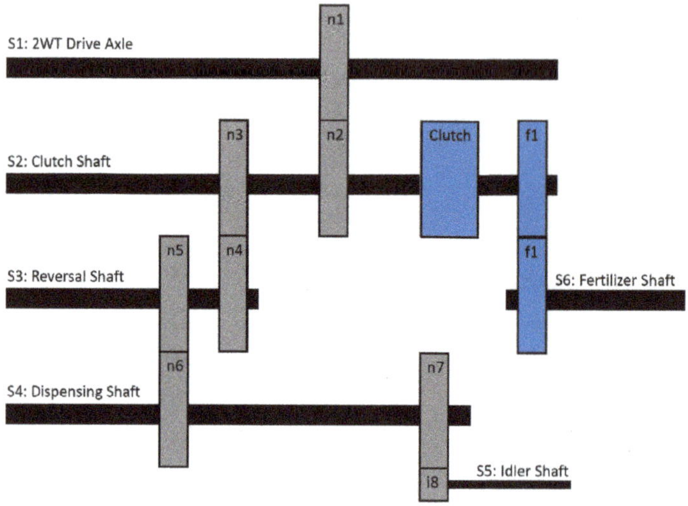

Figure 9. Powertrain schematic.

Table 2. Shaft labels and descriptions.

Shaft Label	Function
S1	Provide power to the transplanter
S2	Allows fertilizer shaft to be engaged
S3	Reverses the rotational motion direction using spur gears
S4	Drives the onion dispensing cups
S5	Holds idler sprocket and maintains tension in chain
S6	Operates the fertilizer mechanism (design by others)

Knowing that the operational movement pace of the 2WT is 1.5 km/h (41.7 cm/s), the angular velocity of shaft *S1*, ω_{S1}, and the dispensing rate of the onions, DR_{onion}, are calculated as follows:

$$\omega_{S1} = \frac{0.417\frac{m}{s}}{0.317\ m} = 1.312\frac{rad}{s} = 12.53\ rpm, \tag{1}$$

$$DR_{onion} = \frac{10\ cm}{41.7\frac{cm}{s}} = 0.25\ \frac{s}{onion}, \tag{2}$$

which can then be used to find the required angular velocity of the dispensing shaft, $\omega_{S4,\ req}$, and the overall gear ratio of the entire system, GR_{req}, knowing that the spacing between cups is 5.08 cm and the shaft diameter of the dispensing shaft is 0.72 cm.

$$\omega_{S4,\ req} = \frac{5.08\ cm}{0.25\frac{s}{onion} * \frac{0.72\ cm}{2}} = 7.37\frac{rad}{s} = 70.4\ rpm. \tag{3}$$

$$GR_{req} = \frac{7.37}{1.32} = 5.58. \tag{4}$$

With the final gear ratio, the number of teeth for sprockets and gears *n1*–*n7* are selected according to Table 3.

Table 3. Number of teeth for each sprocket/gear of the powertrain.

Sprocket/Gear Label	Number of Teeth
n1	32
n2	12
n3	19
n4	9
n5	12
n6	12
n7	12

The actual overall gear ratio is then calculated as:

$$GR_{actual} = \frac{32}{12} \times \frac{19}{9} \times \frac{12}{12} = 5.62, \tag{5}$$

which varies only 0.7% from the gear ratio required to achieve a 10 cm onion spacing.

2.3. Prototyping and Testing Environment

2.3.1. Prototyping

The transplanter design was modified slightly for purposes of testing. Instead of three rows, only one row of onions was planted during testing. It should also be noted that a 2WT was unavailable for testing. Therefore, the transplanter prototype was attached to a garden cart and pulled manually to simulate the motion of the 2WT. Accordingly, the gearing of the prototype was adjusted to accommodate for differences in speed and axle diameter between the 2WT and garden cart. The prototype used in testing is shown in Figure 10.

Figure 10. Prototype of the onion transplanter.

Components of the prototype were made from readily available materials such that it could be made quickly. Although the materials used are likely to differ from the materials available in Bangladesh, they are unlikely to impact the functional performance of the prototype. Some common components such as chains and sprockets were used. The materials used to manufacture the unique components are listed below in Table 4.

Table 4. Material Specifications of unique prototype components.

Component	Material
Hopper (Dispensing Subsystem)	Zinc-Galvanized Low-Carbon Steel
Cups (Dispensing Subsystem)	Polypropylene (PP)
Funnel (Orienting Subsystem)	Zinc-Galvanized Low-Carbon Steel
Furrower (Burying Subsystem)	6061 Aluminum
Brackets/Framing	6061 Aluminum
Shaft Components	1566 Carbon Steel

2.3.2. Testing Environment and Materials

To facilitate controlled soil conditions, a test bed was made that consisted of two different soil types, as shown in Figure 11. Soil on the left-hand side of the test bed was native Georgia soil, which has a high clay content. Soil on the right-hand side of the test bed consisted of a mixture of native Georgia soil with a gravel pavement base, made up of roughly equal parts by volume.

Figure 11. Soil beds created to test the transplanter prototype.

Due to the highly perishable nature of onion seedlings and the short season in which they can be grown, it was not possible to test the transplanter with Bangladeshi onion seedlings. Therefore, two materials available in Georgia were selected as substitutes for testing: wooden dowels and scallions (green onions).

Wooden dowels were chosen for conducting preliminary testing because of their uniform diameter of 0.95 cm (3/8″) (close to the average diameter of scallions) and because they are straight. These attributes made dowels a good representation of the "ideal onion". Hence, they would be easier for the transplanter to handle than a normal onion whose diameter varies and has a curved stalk.

Scallions were a more accurate representation of Bangladeshi onion seedlings. The overall shape of Bangladeshi onion seedlings is very close to scallions, with common features including varying stalk and bulb diameters, divergences in the stalks, and an overall curved profile. Although similar in profile, it is noted that Bangladeshi onion seedlings tend to be smaller than the mature scallions used in testing. Shown in Figure 12, the bulb diameter of measured Bangladeshi onion seedlings varied between 2.5–7 mm.

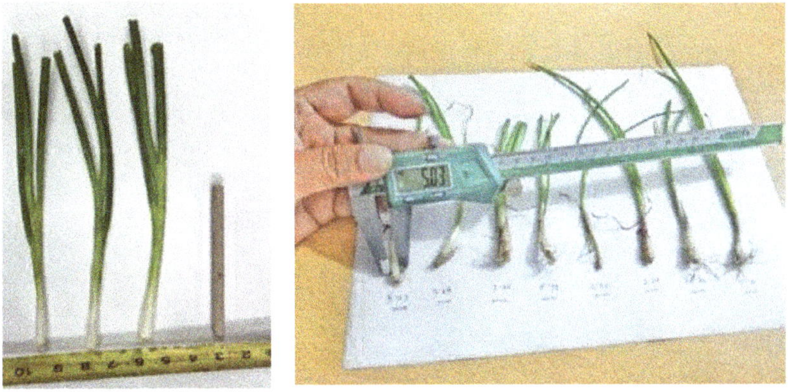

(**a**) (**b**)

Figure 12. (**a**) Scallions next toa 0.95 cm (3/8″) dowel; (**b**) Bangladeshi onion seedlings.

A profile of the scallions was created by taking various length and width measurements, which can be seen in Figure 13.

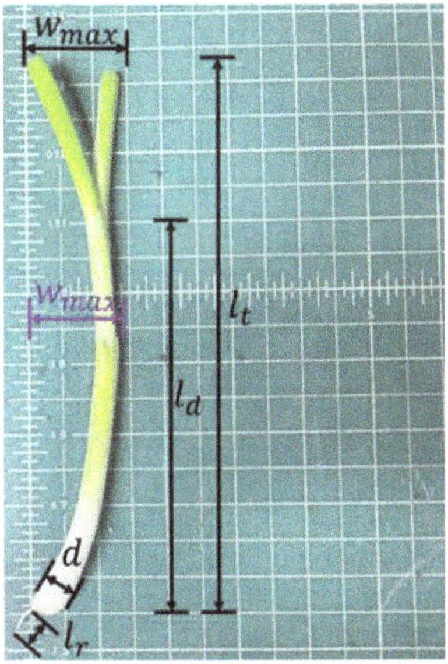

Figure 13. Measurements used to characterize the scallions. The parameters are: w_{max} is the maximum width of the onion which typically occurred in either of the two places shown, l_d is the length from the onion bulb to the point where the stalk diverges, l_r is the length of the roots, d is the bulb diameter, and l_t is the total length of the onion.

Using a sample size of 83 scallions, the average measurements were found and are shown in Table 5. These characteristics were used to adjust the cup size of the transplanter from the original design, for which the cup size was intended for the smaller Bangladeshi seedlings.

Table 5. Average measurements of scallions.

	l_t (cm)	l_d (cm)	l_r (cm)	w_{max} (cm)	d (cm)
Average	19.3	10.4	1.0	4.1	1.0
Standard Deviation	5.6	8.9	1.8	3.9	0.4

2.4. Testing Methods

The prototype functionality was tested using the following test procedure:

1. Record the number of onions placed in the hopper (N).
2. Arrange onions such that all are oriented the same way and place them in the hopper.
3. Pull the prototype the full length of the test bed.
4. Record measurements of:
    a. The number of onions dispensed in the proper orientation (n_p).
    b. The total spacing between each onion in the furrow (s_x).
        i. Include all onions that were dispensed through the chute & furrower, even onions in the wrong orientation.
    c. The burial depth of all onions buried at an angle greater than 45° (d_b).
        i. Burial depth is given a value of 0 if dowel is buried at an angle less than 45°.
        ii. Include onions that are buried upside-down.
    d. The number of failures that occur (F).
        i. Failures include any onion that jams, is dispensed upside down, or falls out of the system by any means other than the funnel.
    e. The number of successes that occur (S).
        i. Include any onion that is planted upright at an angle greater than 45°.
5. Remove any onions from the soil.
6. Without removing any onions remaining in the dispensing cups, return the prototype to the start position.
7. Rake the soil to remove the previous furrow.
8. Repeat steps 2–7 until test is complete.

Previous research has shown that onions have far lower survivability when planted horizontally [15]. Anecdotally, it was observed that onions planted at an angle of 45° or greater, relative to the ground, remained upright after transplanting. Thus, onions planted with angles of 45° or less were not counted as successfully transplanted.

With the recorded data, the metrics in Table 6 are calculated in order to evaluate the overall performance of the transplanter.

Table 6. Metrics and calculations used to evaluate transplanter performance.

Metric	Calculation (Indeed Editorial Team, 2020)
Average spacing ($\overline{s_x}$)	$\overline{s_x} = \frac{\sum s_x}{N} \times 100\%$
Average burial depth ($\overline{d}_b$)	$\overline{d}_b = \frac{\sum d_b}{N} \times 100\%$
Percentage of properly oriented seedlings (O_p)	$O_p = \frac{n_p}{N}$
Overall failure rate (FR)	$FR = \frac{F}{N} \times 100\%$
Overall success rate (SR)	$SR = \frac{S}{N} \times 100\%$

The prototype was tested first with the 0.95 cm (3/8") wooden dowels and then tested with scallions. Each test was performed using a varying number of onions (or dowels). The transplanter was first tested with a single scallion (or dowel) as a proof of concept. In each successive test, the number of onions (or dowels) was increased until the testing was completed. The results of testing are provided in the next section.

3. Results

Testing Results

A total of 50 dowels and 50 scallions were tested over the course of several tests following the procedure previously outlined. Figure 14 shows several onions that were successfully transplanted during the test. Using a normal distribution, 95% and 99% confidence levels were calculated using the normal distribution. Performance metrics and confidence intervals are shown in Table 7.

Figure 14. Onions transplanted during testing.

Table 7. Testing results of the transplanter prototype.

Metrics	Dowels (95% Confidence Interval)	Dowels (99% Confidence Interval)	Scallions (95% Confidence Interval)	Scallions (99% Confidence Interval)
Average spacing ($\overline{s_x}$) (cm)	11.9 ± 2.5	11.9 ± 3.3	34.3 ± 7.1	34.3 ± 9.4
Average burial depth ($\bar{d}_b$) (cm)	4.8 ± 0.5	4.8 ± 0.5	4.8 ± 0.5	4.8 ± 0.5
Percentage of properly oriented seedlings (O_p)	76 ± 12%	76 ± 16%	76 ± 12%	76 ± 15%
Overall failure rate (FR)	26 ± 12%	26 ± 16%	48 ± 14%	48 ± 18%
Overall success rate (SR)	74 ± 12%	74 ± 16%	52 ± 14%	52 ± 18%

Data presented in this section represent testing in which the prototype was actuated under its own power utilizing the drive train made from bicycle sprockets. The dispensing rate was, therefore, dependent on the speed at which the garden cart was pulled. The target speed for the cart was 1 ft/s, but it was not possible to always meet this goal; therefore, deviations are not considered when presenting or discussing these data. The cart speed has the potential to affect any of the metrics of the system but was anecdotally observed to have a direct impact on the burial depth and the system's ability to bury the dowels or onions at an angle greater than 45 degrees. This anecdotal evidence suggested that a lower cart speed would result in a shallower burial depth and in a smaller angle of burial (i.e., less than 45 degrees).

4. Discussion

The performance of the transplanter prototype can be compared to the transplanter's established target specifications using Table 8.

Table 8. Performance compared to transplanter's target specifications.

Metric	Test/Evaluation Results (99% C.I.)	Target Specification
Onion Planting Spacing (cm)	34.3 ± 9.4	10
Row Planting Spacing (cm)	20 (per CAD model)	20
Onion Planting Depth (cm)	4.8 ± 0.5	2.5
Operational Movement Pace (m/s)	0.3 (per powertrain calculations)	0.3
Success Rate of Transplanting (%)	52 ± 18	95

Perhaps the most important metric from Table 8 is the success rate of transplanting, which reflects the transplanter's ability to simultaneously achieve all of its' intended functions. The limits of the confidence interval indicate that the researchers are 99% confident that the transplanter is able to correctly transplant 34–70% of the onions introduced to it. Because the upper limit of the confidence interval is below the target specification of 95%, it is clear that the prototype does not meet the target specification in its current state.

Similarly, the target specifications for onion planting spacing and onion planting depth fall outside of the limits provided by the 99% confidence intervals for their respective test results. This indicates that the prototype was not successful in meeting the target specifications.

It was observed during testing that the dispensing subsystem played the most crucial role in the transplanter's inability to meet the target specifications. On many occasions, the cups would fail to pick up an onion as they travelled upwards through the hopper or onions would fall out of the cups before reaching the top of the hopper. Onions also were observed to become tangled with the cup prongs or with other onions, which could cause them to remain in the cup instead of being properly dispensed into the orienting funnel. Both of these phenomena resulted in irregular dispensing intervals and were a major cause of the high onion planting spacing and low success rate of the transplanter.

Along with the failures previously described, the success rate of transplanting was also reduced by the number of onions buried upside down. Onions that fell out of cups prematurely were the largest cause of improperly oriented transplants as well. Onions that fell out of cups and back into the hopper would often become misoriented as they fell. When they were later picked up and dispensed properly, they would enter the orienting funnel backwards and thus be buried upside down.

Although the prototype was unable to reach the target specification of 2.5 cm for planting depth, this is a shortcoming of the testing materials and environment rather than a failure of the transplanter. Because of the larger size of scallions and the high clay content of the soil, it was found that the onions would not remain standing upright unless buried to a depth of around 4.5 cm or more. This indicates that testing of the prototype will need to be performed with Bangladeshi onion seedlings in Bangladesh to properly assess the transplanter's performance.

Row planting spacing and operational movement pace are validated by analyzing the final dimensions of the transplanter in the CAD model and the powertrain calculations, respectively. Prior to building the final transplanter, these characteristics can be checked and adjusted as needed to ensure fulfillment of the target specifications.

Although unable to meet the established target specifications, the prototype successfully demonstrated that the design is indeed capable of performing the functions required to transplant onion seedlings automatically. The design in this paper provides future researchers with a starting point that may be improved upon to achieve more consistent success and, thus, approach the target specifications.

5. Conclusions

The design of a mechanized onion transplanter for use by smallholder farmers in Bangladesh, is presented in this paper. The onion transplanter is designed to be attached to and powered by the two-wheel tractors that are commonly used throughout the country.

The transplanter is projected to be capable of planting three rows of onion seedlings at the recommended spacing of 20 cm while requiring the labor of only a single individual.

A prototype of the transplanter was created for functional testing. Results of functional testing indicate that onions can be successfully transplanted using the machine, however, the prototype's transplanting success rate of 52% falls short of the target specification. Most failures were attributed to inconsistency in the transplanter's dispensing subsystem. Therefore, it is recommended that further development of the dispensing subsystem be performed to increase the transplanter's overall performance.

Given that the intention of this study wisas to prove functionality of the presented design, some parameters remain unexplored or only partially explored. Future researchers are encouraged to explore ways to improve upon the current component designs in pursuit of a more robust design. Design parameters such as cup shape, funnel shape, furrower size, etc. will provide more opportunities for investigation.

The primary limitations of the study are the materials and testing environment with which the prototype was tested. Future testing should be conducted in Bangladesh with proper Bangladeshi onion seedlings to ensure that the transplanter performs with the materials for which it was designed.

Author Contributions: Conceptualization, S.S. and J.C.; methodology, S.S.; software, S.S.; validation, S.S.; formal analysis, S.S.; investigation, S.S.; resources, S.S. and J.C.; data curation, S.S.; writing—original draft preparation, S.S.; writing—review and editing, J.C.; visualization, S.S.; supervision, J.C.; project administration, J.C.; funding acquisition, J.C. All authors have read and agreed to the published version of the manuscript.

Funding: This research was funded by USAID's Feed the Future Bangladesh Cereal Systems Initiative for South Asia—Mechanization and Extension Activity.

Institutional Review Board Statement: Not applicable.

Data Availability Statement: Data can be found in the thesis associated with this project, located at https://smartech.gatech.edu/handle/1853/67282 (accessed on 1 September 2022).

Conflicts of Interest: The authors declare no conflict of interest. The funders had no role in the design of the study; in the collection, analyses, or interpretation of data; in the writing of the manuscript; or in the decision to publish the results.

References

1. Gregg, E.S.; Colton, J.; Matin, M.A.; Krupnik, T.J. *Efficient and Participatory Design of Scale-Appropriate Agricultural Machinery Workshops in developing countries: A case study in Bangladesh*; Elsevier: Amsterdam, The Netherlands, 2019.
2. Bangladesh Bureau of Statistics. *Report on Agriculture and Rural Statistics*; Bangladesh Bureau of Statistics: Dhaka, Bangladesh, 2019.
3. Gautam, M.; Ahmed, M. *Too Small to be Beautiful? The Farm Size and Productivity Relationship in Bangladesh*; Elsevier: Amsterdam, The Netherlands, 2018.
4. Kajol, R.H. *Farming of Onion Seeds on the Rise in Faridpur*; Dhaka Tribune: Dhaka, Bangladesh, 2021.
5. Farha, A.; Shahanaz, B.; Uddin, J. An Overview of Studies on Summer Onion Production Technology. *Int. J. Bus. Soc. Sci. Res.* **2020**, *8*, 25–27.
6. USAID. *Scaling up of Agricultural Machinery in Bangladesh*; USAID: Washington, DC, USA, 2016.
7. Khan, M.S.; Jahan, S. Cranking, a Thing of the Past. 2019. Retrieved from CIMMYT. Available online: https://www.cimmyt.org/news/cranking-a-thing-of-the-past/ (accessed on 1 September 2022).
8. Krupnik, T.J.; Valle, S.S.; McDonald, A.J.; Justice, S.; Hossain, I.G. *Made in Bangladesh: Scale-Appropriate Machinery for Agricultural Resource Conservation*; Mexico, D.F., Ed.; CIMMYT: Mexico City, Mexico, 2013.
9. Stubbs, S.J. The Design and Testing of a Mechanized Onion Transplanter for Bangladesh. Master's Thesis, Georgia Institute of Technology, Atlanta, GA, USA, 2022.
10. Nikus, O.; Mulugeta, F. *Onion Seed Production Techniques*; Master Printing Press Plc.: Asella, Ethiopia, 2010.
11. Alibaba. Factory Price Garlic Seeds Planting Machine. 2022. Retrieved from Alibaba Website. Available online: https://www.alibaba.com/product-detail/Garlic-Planting-Machine-Factory-Price-Garlic_1700007799756.html?spm=a2700.pccps_detail.normal_offer.d_title.5a3f62577GA8Rn&s=p (accessed on 1 September 2022).
12. Wei. Onion Planting Machine Working Condition. 2017. Retrieved from YouTube. Available online: https://www.youtube.com/watch?v=aSwXCVBpTec (accessed on 1 September 2022).

13. Romiter Food Processing. Automatic 8 Lane Onion Transplanter Machine. 2017. Retrieved from YouTube. Available online: https://www.youtube.com/watch?v=zh9_sMflOHY (accessed on 1 September 2022).
14. Pandirwar, A.; Kumar, A.; Singh, J.; Mani, I.; Bhomik, A.D. Development and Evaluation of a Semi-Automatic Six Row Onion Seedlings Transplanter. *Agric. Mech. Asia Afr. Lat. Am.* **2019**, *50*, 29–35.
15. Lipe, W.N.; Thomas, D. *Effects of Onion Transplant Pruning and Orientation on Plant Survival, Yield, and Size Distribution*; Texas Agricultural Experimnet Station: College Station, TX, USA, 1980; pp. 28–29.

MDPI
St. Alban-Anlage 66
4052 Basel
Switzerland
www.mdpi.com

Agriculture Editorial Office
E-mail: agriculture@mdpi.com
www.mdpi.com/journal/agriculture

www.ingramcontent.com/pod-product-compliance
Lightning Source LLC
Chambersburg PA
CBHW041933260926
51850CB00018B/117